Probability at Saint-Flour

For further volumes:
http://www.springer.com/series/10212

Saint-Flour Probability Summer School

Founded in 1971, the Saint-Flour Probability Summer School is organised every year by the mathematics department of the Université Blaise Pascal at Clermont-Ferrand, France, and held in the pleasant surroundings of an 18th century seminary building in the city of Saint-Flour, located in the French Massif Central, at an altitude of 900 m.

It attracts a mixed audience of up to 70 PhD students, instructors and researchers interested in probability theory, statistics, and their applications, and lasts 2 weeks. Each summer it provides, in three high-level courses presented by international specialists, a comprehensive study of some subfields in probability theory or statistics. The participants thus have the opportunity to interact with these specialists and also to present their own research work in short lectures.

The lecture courses are written up by their authors for publication in the LNM series.

The Saint-Flour Probability Summer School is supported by:

- Université Blaise Pascal
- Centre National de la Recherche Scientifique (C.N.R.S.)
- Ministère délégué à l'Enseignement supérieur et à la Recherche

For more information, see back pages of the book and
http://math.univ-bpclermont.fr/stflour/

Jean Picard
Summer School Chairman
Laboratoire de Mathématiques
Université Blaise Pascal
63177 Aubière Cedex
France

Donald A. Dawson • Edwin Perkins

Superprocesses at Saint-Flour

Springer

Donald A. Dawson
School of Mathematics and Statistics
Carleton University
Ottawa ON, Canada

Edwin Perkins
Department of Mathematics
The University of British Columbia
Vancouver BC, Canada

Reprint of lectures originally published in the Lecture Notes in Mathematics volumes 1541 (1993) and 1781 (2002).

ISBN 978-3-642-25431-4
Springer Heidelberg Dordrecht London New York

Library of Congress Control Number: 2011943170

Mathematics Subject Classification (2010): 60J80; 60-02, 60G57, 60G57; 60K35

Printed on acid-free paper

Springer is part of Springer Science+Business Media (www.springer.com)

Preface

The *École d'Été de Saint-Flour*, founded in 1971 is organised every year by the *Laboratoire de Mathématiques* of the *Université Blaise Pascal* (Clermont-Ferrand II) and the *CNRS*. It is intended for PhD students, teachers and researchers who are interested in probability theory, statistics, and in applications of stochastic techniques. The summer school has been so successful in its 40 years of existence that it has long since become one of the institutions of probability as a field of scholarship.

The school has always had three main simultaneous goals:
1. to provide, in three high-level courses, a comprehensive study of 3 fields of probability theory or statistics;
2. to facilitate exchange and interaction between junior and senior participants;
3. to enable the participants to explain their own work in lectures.

The lecturers and topics of each year are chosen by the Scientific Board of the school. Further information may be found at http://math.univ-bpclermont.fr/stflour/

The published courses of Saint-Flour have, since the school's beginnings, been published in the *Lecture Notes in Mathematics* series, originally and for many years in a single annual volume, collecting 3 courses. More recently, as lecturers chose to write up their courses at greater length, they were published as individual, single-author volumes. See www.springer.com/series/7098. These books have become standard references in many subjects and are cited frequently in the literature.
As probability and statistics evolve over time, and as generations of mathematicians succeed each other, some important subtopics have been revisited more than once at Saint-Flour, at intervals of 10 years or so.

On the occasion of the 40th anniversary of the *École d'Été de Saint-Flour,* a small ad hoc committee was formed to create selections of some courses on related topics from different decades of the school's existence that would seem interesting viewed and read together. As a result Springer is releasing a number of such theme volumes under the collective name "Probability at Saint-Flour".

Jean Bertoin, Erwin Bolthausen and K. David Elworthy

Jean Picard, Pierre Bernard, Paul-Louis Hennequin
(current and past Directors of the *École d'Été de Saint-Flour*)

September 2011

Table of Contents

MEASURE-VALUED MARKOV PROCESSES

Donald A. DAWSON

Originally published in: *Ecole d'Eté de Probabilités de Saint-Flour XXI – 1991*, Lecture Notes in Mathematics, Vol. **1541**, 1–260, DOI: 10.1007/BFb0084190,

2

MEASURE-VALUED MARKOV PROCESSES

Donald Dawson
Department of Mathematics and Statistics
Carleton University
Ottawa, Canada

CONTENTS

3

4

6

1. INTRODUCTION

The central theme of these lectures is the construction and study of measure-valued Markov processes. This subject is in the midst of rapid development and has been stimulated from several different directions including branching processes, population genetics models, interacting particle systems and stochastic partial differential equations. The objective of these notes is to provide an introduction to these different aspects of the subject with some emphasis on their interrelations and also to outline some aspects currently under development. Chapters 1-9 provide an introduction to some of the main ideas and tools in the theory of measure-valued processes. Chapters 10-12 cover topics currently under active development and are primarily intended as an introduction to the growing literature devoted to these aspects of measure-valued processes. Throughout the emphasis is given to outlining the main lines of development rather than attempting a systematic detailed exposition. In Section 1.5 we describe in more detail the structure of these notes (including interrelations between the chapters) as well as mention the principal methods. In the remaining sections of this introduction we outline the roots of the theory of measure-valued Markov processes and the major topics to be discussed in these notes.

By a *measure-valued Markov process* we will always mean a Markov process whose state space is $M(E)$, the space of Radon measures on $(E,\mathcal{E})$, where E is a Polish space and $\mathcal{E} = \mathcal{B}(E)$ is the σ-algebra of Borel subsets of E.

1.1. Particle Systems and their Empirical Measures

Consider a system of N E-valued processes $\{Z_i(t): i=1,\ldots,N,\ t\geq 0\}$. The associated (*normalized*) *empirical measure process* is defined by

$$X(t) = N^{-1}\sum_{i=1}^{N} \delta_{Z_i(t)}. \tag{1.1.1}$$

We will show in the next chapter that empirical measure processes of type (1.1.1) in which the $\{Z_i\}$ form an exchangeable Markov system are in fact measure-valued Markov processes. Exchangeable particle systems arise naturally in many fields including statistical physics, population biology and genetic algorithms.

In Section 2.10 and Chapter 4 we will proceed to study a related class of measure-valued processes on $\mathbb{R}^d$ which arise from spatially distributed population models in which the number of particles, $N(t)$, at time t, is no longer conserved

8

but in which particles undergo birth and death. In this case we consider atomic measures of the form

$$(1.1.2) \qquad X(t) = m(N(0)) \sum_{i=1}^{N(t)} \delta_{Z_i(t)},$$

where $m(N(0))$ is the mass of each particle.

1.2. Limits of Particle Systems and Stochastic Partial Differential Equations

Non-atomic measure-valued processes arise naturally in the limit, $N \to \infty$, of systems of the form (1.1.1) or in the corresponding *high density* limit of (1.1.2). In many such cases a law of large numbers phenomenon (*propagation of chaos*) occurs and the limiting process is deterministic. For example this occurs in the usual *mean-field* or *McKean-Vlasov limit* (cf. Gärtner (1986), Léonard (1986), Sznitman (1989)). The high density limit $N(0) \longrightarrow \infty$, $m(N(0))\cdot N(0) \longrightarrow m$, of systems of the form (1.1.2) can also give rise to (deterministic) linear or nonlinear partial differential equations including reaction diffusion equations, hydrodynamic equations, etc. (cf. e.g. Oelschläger (1989), De Masi and Presutti (1991), Sznitman (1989)).

The main emphasis in these lectures will be to study such limits when the limiting process is itself random. However we do not consider the frequently studied case of fluctuations around a law of large numbers limit in which normalized and centered sequences are studied (cf. Holley and Stroock (1979)), but rather study the non-centered and therefore *non-negative* limits. For example in the context of symmetrically interacting diffusions *random* McKean-Vlasov limits can arise (cf. Sect. 5.8.1). In the next chapter we will consider sequences of finite exchangeable systems which arise in the study of population genetics and genetic algorithms which converge to a random measure-valued limit.

It should be emphasized that in many applications it is the finite particle systems themselves which are of primary interest. However qualitative properties of the limiting process can often provide insight into the collective behavior of the former. In addition these limiting processes possess rich mathematical structures which are of interest in their own right.

If we begin with the empirical measure of a system of particles in $\mathbb{R}^d$ one possibility is that the limiting measure-valued process has the form $X(t,dx) = \tilde{X}(t,x)dx$, where $\tilde{X}(t,x)$ denotes the *density process* and that $\tilde{X}(t,.)$ belongs to an appropriate linear space, V, of non-negative measurable functions. It would also be reasonable to expect that $\tilde{X}(t,x)$ is described as the solution of a *stochastic*

9

partial differential equation. There are a number of different ways to formulate a stochastic partial differential equation (see e.g. Walsh, (1986)). For example the *integral form* of such an equation is given in terms of an $\mathcal{S}'(\mathbb{R}^d)$-valued Wiener process $\{W(t):t\geq 0\}$ with covariance operator Q_0 as follows: for each $\phi\in V^*$, a linear space of test functions in duality with V, with canonical bilinear form $\langle .,.\rangle$ on $V\otimes V^*$,

$$\text{(1.2.1)}\qquad \langle X(t),\phi\rangle-\langle X(0),\phi\rangle = \int_0^t \langle A(s,X(s)),\phi\rangle ds + \int_0^t \langle B(s,X(s))W(ds),\phi\rangle$$

where for each s $A(s,.):V \longrightarrow V$, and for each $v\in V$ $B(s,v)$ is a linear operator such that the Itô integral $\int_0^t \langle B(s,X(s))W(ds),\phi\rangle$ is well-defined yielding a martingale with increasing process $\int_0^t Q(s,X(s))(\phi,\phi)ds$ where $Q(s,v) := B(s,v)Q_0(s,v)B^*(s,v)$.

Under certain natural conditions on $A(s,.)$, $B(s,.)$ and $Q(s,.)$ the solution to equation (1.2.1) is non-negative and hence measure-valued. An important class of stochastic partial differential equations of this form were first developed by Pardoux (1975) and generalized by Krylov and Rozovskii (1981). Equations of this type do occur in some cases including the study of stochastic flows in $\mathbb{R}^d$ (cf. Kunita (1986), Rozovskii (1990)), turbulent flows (cf. Chow (1978)) and random McKean-Vlasov limits. However it turns out that we must also consider more general measure-valued diffusions, that is, *singular* measure-valued as well as *density-valued* processes.

1.3 Some Basic Classes of Measure-valued Processes

To introduce these notions let us consider at a purely formal level the measure-valued analogue, $\{X_t:t\geq 0\}$, of a finite dimensional diffusion process associated with a second order elliptic operator. For functions, F, in an appropriate domain $D(G) \subset bC(M(E))$ (the bounded continuous functions on $M(E)$), a second order infinite dimensional differential operator would have the form

$$\text{(1.3.1)}\qquad G(t)F(\mu) = \int_E A(t,\mu,dx)(\delta F(\mu)/\delta\mu(x)) + 1/2\int_E\int_E (\delta^2 F(\mu)/\delta\mu(x)\delta\mu(y))\; Q(t,\mu;dx,dy)$$

where $\dfrac{\delta F(\mu)}{\delta\mu(x)} := \lim\limits_{\varepsilon\downarrow 0} (F(\mu+\varepsilon\delta_x)-F(\mu))/\varepsilon$, $A(t,.)$ generates a deterministic evolution on $M(E)$, and $Q(t,\mu;dx,dy)$ is a symmetric signed measure on $E\times E$ such that

10

$\iint \phi(x)\phi(y)Q(t,\mu;dx,dy) \geq 0,\ \forall\ \phi \in bC(E)$. Formally,

$$Q(t,\mu;dx,dy) = \frac{d}{ds}\left[E[X_{t+s}(dx)X_{t+s}(dy)\,|\,X_t=\mu]-E[X_{t+s}(dx)\,|\,X_t=\mu]\cdot E[X_{t+s}(dy)\,|\,X_t=\mu]\right]\Bigg|_{s=0}.$$

Some additional structure is required in order to guarantee that X_t is *non-negative*. If $E = \{e_1,\ldots,e_d\}$, then $M(E) = (\mathbb{R}_+)^d$ and

$$G = \frac{1}{2}\sum_{i,j=1}^{d} a_{ij}(x)\frac{\partial}{\partial x_i}\frac{\partial}{\partial x_j} + \sum_{i=1}^{d} b_i(x)\frac{\partial}{\partial x_i}.$$

In this finite dimensional case Stroock and Varadhan (1979) established the existence and uniqueness to the martingale problem in the non-degenerate case $\inf_x \left(\sum_{i,j}\xi_i\xi_j a_{ij}(x)\right) > 0\ \forall\ (\xi_1,\ldots,\xi_d) \neq 0$. The degenerate case is more difficult. Results assuming that there is a Lipschitz square root are found in [EK, p. 374]. In the case $E = \{e_1\}$, then $M(E) = \mathbb{R}_+$ and this reduces to the generator of a one-dimensional diffusion given by

$$Gf(x) = b(x)f'(x) + \frac{1}{2}a(x)f''(x).$$

If $a(.)$ and $b(.)$ are Lipschitz and $b(0)\geq 0$, $a(0)=0$, then the resulting process is non-negative. An analogous condition for the measure-valued case would be that $A(t)$ generates a positivity-preserving evolution and that $Q(t,\mu;A\times A) = 0$ if $\mu(A) = 0$. Hence by their very nature measure-valued processes correspond to the highly degenerate case and this is one reason for the difficulty in developing a general theory. In the subsequent sections we will establish the existence of measure-valued diffusions with generators of form (1.3.1) in a number of cases. For example we will consider the case $A(t,v) = Av + R(t,v)$, $v\in D(A) \subset C(\mathbb{R}^d)$, in which A is the generator of a Feller semigroup on $C_0(\mathbb{R}^d)$ (the *spatial motion* or *mutation semigroup*), the *interaction term* $R(.,.)$ satisfies certain regularity conditions and in which Q has one of the following two basic forms:

(i) sampling-replacement: $Q_S(\mu:dx,dy) = \gamma(\delta_x(dy)\mu(dx) - \mu(dx)\mu(dy))$, $\gamma>0$,

(ii) branching: $Q_B(\mu:dx,dy) = c(x)\delta_x(dy)\mu(dx)$, $c(x)\geq 0$.

We will refer to these two classes of processes (with $R \equiv 0$) as *Fleming-Viot processes* and *continuous superprocesses*, respectively, and will study them in considerable detail in Chapters 2-9. In the case of branching we will also consider more generalized branching mechanisms involving jumps. In both cases special tools are available: infinite divisibility and Laplace functionals in the case of branching,

11

and exchangeability and duality in the case of sampling-replacement. Not only are these special classes of interest in their own right but they are natural building blocks for developing and studying more general classes of measure-valued processes.

A rapidly developing area of research in measure-valued processes is the study of Fleming-Viot and continuous supeprocesses with interaction term, R, given by a nonlinear function. The effect of interactions is to destroy much of the simplicity (e.g. infinite divisibility, closed moment hierarchy, etc.) of these processes and this makes their study much more complicated. In Chapter 10 we give a brief introduction to some approaches and recent literature on this subject.

1.4. Structure and Behavior of Measure-valued Processes

Having introduced some families of measure-valued processes we now consider their local spatial structures and sample path behavior.

By *local spatial structure* is meant the determination as to whether the mass distribution at fixed times is atomic, absolutely continuous or singular and in the last case to investigate the nature of the support. In terms of *sample path behavior* we will consider questions of sample path continuity as well as determine which sets can be "hit" by the process.

Both branching and sampling systems share another important common feature - namely the central role played by *family structures*. In both cases tracing back through the family history of a typical individual turns out to be a powerful tool in the analysis of both the local spatial structure and long-time behavior of the system. This approach is currently an active area of research. In Chapters 11-12 we will give an introduction to the main features of this development.

Although in these notes we will only briefly mention the long-time behavior of measure-valued Markov processes we would like to mention here the importance and scope of this subject. The study of the *long-time behavior* of stochastic systems includes notions of transience, recurrence, invariant measure, ergodicity, large deviations from the invariant measure, and phase transitions. For example in the spatial setting transience is related to local extinction and spatial clumping - two closely related phenomena (cf. Dawson and Perkins (1991), Gorostiza and Wakolbinger (1991)). The long-time behavior of measure-valued processes is also closely related to that for infinite particle systems and many of the same ideas and difficulties arise (cf. Liggett (1985)).

1.5. Structure of these Notes

Now, let us describe in more detail the structure of these notes. First, a sig-

12

nificant part is devoted to the systematic treatment of the two classes of measure-valued processes introduced above, namely *superprocesses* and the *Fleming-Viot processes*. Both classes are of importance in view of various applications as well as due to the central role they play in the development of the whole theory of measure-valued processes. Thus, superprocesses arise as the measure-valued generalization of branching processes (cf. also sections 1.3 - 1.4 above), whereas the Fleming-Viot processes are the measure-valued analogues of various processes arising in population genetics (in particular they describe the genetic structure of a population). These two classes are in fact closely related; the main difference is the fact that the Fleming-Viot processes are *probability measure*-valued processes, and their total mass does not change in time.

Although these two classes of measure-valued processes possess many analogous properties, the methods required to deal with them are often different. The comparison of these classes and attempts to provide a unified approach to them can serve as stimulus for further developments in the theory of measure-valued processes. Processes of both classes can be constructed as weak limits of properly scaled particle systems (i.e. branching particle systems in the case of superprocesses and the Moran particle processes in the case of Fleming-Viot processes). The approach based on particle system approximations is developed in Chapters 2, 4, and 12 on various levels of generality. In order to introduce some basic notions, in Chapter 2 this approach is carried out at a very intuitive level, for the simplest representatives of both classes.

Chapter 3 contains a summary of some important results in the general theory of random measures adapted to the purposes of these notes. We include this chapter just to make the notes to some extent self-contained; the reader familiar with the theory of random measures can omit this chapter in the first reading.

Certain properties of measure-valued processes can be studied by use of the appropriate analytical tools. In particular, path properties of certain superprocesses can be studied in terms of properties of the corresponding log-Laplace (non-linear evolution) equations. The methodology of log-Laplace functionals is developed in Chapter 4. In addition in this chapter we also present in more detail the inter-relations between branching particle systems and superprocesses (cf. sections 4.4 and 4.6).

The analytical methods developed in Chapter 4 are also applied in Chapter 9 to the investigation of the sample path properties of the class of superprocesses known as *super-Brownian motion*. The log-Laplace equation can be used to derive certain

13

path properties of super-Brownian motion which are analogous to those of the classical Brownian motion (e.g. support dimension, carrying dimension, Hausdorff measure, modulus of continuity, range and multiple points, etc.).

A powerful tool systematically used in these notes which provides a unified approach to both classes and which is a basic tool for the development of the theory of measure-valued processes is the martingale problems characterization. Various martingale problems are considered in Chapter 5 (for the Fleming-Viot processes) and Chapter 6 (for superprocesses). In Chapter 5 we also illustrate the usefulness of the martingale problem technique to formulate wider classes of the measure-valued processes including time inhomogeneous versions of Fleming-Viot processes and some random flow models.

The structure of conditional distributions of measure-valued processes is also of interest. In Chapter 8 it is shown that the continuous superprocess conditioned to have constant mass and normalized is a Fleming-Viot process and in general the continuous superprocess conditioned on the total mass processes is a time inhomogeneous Fleming-Viot process. This is carried out using the martingale problem characterizations in Chapters 5 and 6 for both classes of measure-valued processes as well as the *stochastic calculus of measure-valued process* developed in Chapter 7. The latter includes the consideration of martingale measures, stochastic integrals and a Cameron-Martin-Girsanov formula. Applications are also given to continuity of sample paths, the weighted occupation times and a more general class of functionals.

The general theory of random measures, the martingale problem technique, andand the stochastic calculus developed in Chapters 3 - 7 are all used in Chapter 8 to determine conditions under which the main classes of measure-valued processes are either pure atomic or absolutely continuous (cf. sections 8.2 and 8.3).

In the same way as the application of log-Laplace functionals provides a tool for the analysis of superprocesses in Chapters 4 and 7, the *method of duality* provides a tool for the analysis of Fleming-Viot processes. Moreover, this method enables us to reveal the interrelations between the Fleming-Viot process and the Moran particle process in more depth. The method of duality is developed in sections 2.8, 5.5 - 5.6, 5.8 and 10.2 on various levels of generality (which correspond to the consideration of wider and wider classes of Fleming-Viot processes).

In Chapter 10 we build more complex models, namely superprocesses with interactions and Fleming-Viot processes with selection. A number of specific models (e.g. noncritical branching, branching with nonlinear death rates, multitype branching, the Fleming-Viot processes with recombination and selection, stepping stone models

14

etc.) are also discussed in this chapter. The results of Chapter 10 illustrate the power of the combination of the martingale problem technique and either stochastic calculus or duality in the context of measure-valued processes.

In Chapter 11 we collect some powerful techniques for the study of random measures adapted to the purposes of the development of Chapter 12. In particular, in Chapter 11 we obtain *an infinite particle representation of the Fleming-Viot process* (cf. section 11.3) and *a cluster representation for superprocess* (cf. section 11.5).

In Chapter 12 we consider *family* structures of superprocesses and Fleming-Viot processes. Note that in order to distinguish them we refer to the dynamical family structure of superprocesses as *historical processes* and to the dynamical family structure of Fleming-Viot processes as *genealogical processes.* Let us point out that when studying family structures various tools developed in previous chapters have to be adapted to this more complicated historical/genealogical setting (e.g. martingale problem (cf. sections 12.3.3 and 12.5.1), branching particle system approximation (cf. section 12.4), etc.). Chapter 12 is intended as an introduction to the rapidly developing theory of historical (genealogical) processes.

The consideration of family structures provides a setting and level of generality which has already led to the solution of several earlier unsolved problems of the theory of measure-valued processes including some of the results described in Chapter 9. The culmination of Chapter 12 is a complete probabilistic description of the genealogical structure of a population at a fixed time in terms of an infinite random tree for both the continuous superprocess and Fleming-Viot process as well as a proof that the laws of the two resulting infinite random trees are mutually singular.

1.6. Acknowledgements

I would like to thank Professor P.L. Hennequin for the invitation to participate in the 1991 session of the Saint Flour Ecole d'Eté de Calcul des Probabilités and the participants for their encouragement. I would like to thank in particular E.B. Dynkin, E.A. Perkins, and T.G. Kurtz for providing preprints of their recent work on which much of these notes is based. I would also like to thank L.G. Gorostiza, Hao Wang, and V. Vinogradov for their advice concerning the preparation of these notes.

I would also like to thank the Natural Sciences and Engineering Research Council of Canada for financial support which has made possible much of my collaborative work over the years. Above all I thank my family for their continuing moral support.

15

1.7. Index of Some Frequently Used Notation and References

$\mathbb{R}^d$, $\mathbb{Z}^d$ d-dimensional euclidean space, lattice

$\mathbb{N}$ natural numbers

$\binom{n}{k} = n!/(k!(n-k)!)$

E denotes a Polish space (sometimes a compact metric space)

$D_E = D([0,\infty),E)$ denotes the set of càdlàg functions from $[0,\infty)$ into E

Note that $D_{M_F(E)}$ or $D_{M_1(E)}$ is often denoted by **D** for the sake of brevity

$\sigma\{f_1,\ldots,f_n\}$ = σ-algebra generated by functions $f_1,\ldots,f_n$

$\mathcal{F}_1\otimes\mathcal{F}_2 = \sigma\{F_1\times F_2 : F_1\in\mathcal{F}_1, F_2\in\mathcal{F}_2\}$

$\mathcal{F}_1\vee\mathcal{F}_2$ = smallest σ-algebra containing $\mathcal{F}_1$ and $\mathcal{F}_2$

$\mathcal{F}_{t+} = \bigcap_{\varepsilon>0}\mathcal{F}_{t+\varepsilon}$ (if $\mathcal{F}_t$ is an increasing family of σ-algebras)

$\mathcal{F}^{uc}$ = universal completion of σ-algebra $\mathcal{F}$

(the universal completion is assumed where appropriate)

ϕ is $\mathcal{F}_1/\mathcal{F}_2$-measurable $\iff$ $\phi^{-1}(B)\in\mathcal{F}_1\ \forall\ B\in\mathcal{F}_2$

$\mathcal{E} = \mathcal{B}(E)$ = the σ-algebra of Borel subsets of E

$\mathcal{E}$ also denotes the bounded $\mathcal{E}$-measurable functions

$p(\mathcal{C})$ = non-negative functions in class $\mathcal{C}$

$pp(\mathcal{C})$ = functions f in class $\mathcal{C}$ with $\inf f > 0$

$b(\mathcal{C})$ = bounded functions in class $\mathcal{C}$

C(E) continuous functions on E

$bC_b(E)$ bounded continuous functions with bounded supports

$C_c(E)$ continuous functions with compact support

$C_0(\mathbb{R}^d)$, $C_0(E)$ continuous functions tending to zero at infinity (if E is locally compact)

$C^k(\mathbb{R}^d)$ functions whose partial derivatives of order k or less exist and are continuous

$C^\infty(\mathbb{R}^d)$ functions whose partial derivatives of all orders exist and are continuous

$M_1(E)$ probability measures on $\mathcal{E}$

$M_F(E)$ finite measures on E

$M_{1,N}(E)$ probability measures consisting of N atoms of mass 1/N.

$f\circ g$ = composition of functions

f', f" first and second derivatives of the function f

$\langle\mu,f\rangle := \int f\,d\mu$ if $f\in b\mathcal{E}$, $\mu\in M_F(E)$

16

μ^n = n-fold product measure $\mu(dx_1)\ldots\mu(dx_n)$

$\mu_1 * \mu_2$ = convolution of μ_1 and μ_2

supp(μ) = topological support of the measure μ

$\Sigma_{N-1} = \{(s_1,\ldots,s_N): s_i \geq 0,\ \Sigma\, s_i = 1\}$

$\Xi_N(x_1,\ldots,x_N) := N^{-1} \sum_{i=1}^{N} \delta_{x_i}$

B(x,r) denotes the open ball with centre x and radius r in a metric space

$\xrightarrow{bp}$ bounded pointwise convergence

$\Rightarrow$ weak (narrow) convergence of probability measures

f.d.d. in sense of finite dimensional distributions

$\overset{\mathcal{D}}{=}$ equality in distribution

δ_z with $z \in E$ denotes the unit mass at z

1 = constant function 1(x) = 1 $\forall$x

1_A or 1(A) = indicator function of the set A

$a \wedge b$ = min(a,b), $a \vee b$ = max(a,b)

A^c = complement of A

Δ = Laplacian operator

$\Delta_\alpha = -(-\Delta)^{\alpha/2}$

Id = Identity operator

S^* = Adjoint of operator S

E(X) expectation of the random variable X (without reference to law)

P(X) = expectation of X under probability measure P ,P(A) = P(1_A)

P << Q P is absolutely continuous with respect to Q

$\mathcal{L}aw$(X) = probability law of X

$\mathcal{P}ois(\lambda)$ = Poisson distribution with mean λ

$\mathcal{E}xp$(a) = Exponential distribution with mean a

Frequent References:

[D] Dawson (1992)

[DP] Dawson and Perkins (1991)

[EK] Ethier and Kurtz (1985)

[LS] Liptser and Shiryayev (1989)

2. FROM PARTICLE SYSTEMS TO MEASURE-VALUED PROCESSES

In this chapter we introduce two classes of measure-valued processes using only rather elementary methods and in particular the method of moment measures. Our starting point is a class of exchangeable finite particle systems and we study the limits of their normalized empirical measures as the number of initial particles tends to infinity. The resulting limit process is the Fleming-Viot probability-measure-valued process. Before this we introduce the setting of measure-valued Feller processes. The last two sections of this chapter provide a preliminary introduction to weak convergence of processes, martingale problems and branching particle systems, all of which will be developed in some generality in subsequent chapters.

2.1 Measure-Valued Feller Processes.

Let (E,d) be a compact metric space, $C(E)$ the space of continuous functions, $\mathcal{E} = \mathcal{B}(E)$ the σ-algebra of Borel subsets of E, and $M_1(E)$ the space of probability measures on E. We denote by $b\mathcal{E}$ (resp. $pb\mathcal{E}$) the bounded (resp. non-negative bounded) $\mathcal{E}$-measurable functions on E. If $\mu\in M_1(E)$ and $f\in b\mathcal{E}$, we define $\langle\mu,f\rangle := \int_E f d\mu$. Note that $M_1(E)$ furnished with the topology of weak convergence is a compact metric space (recall that $\mu_n \overset{w}{\Longrightarrow} \mu$ if and only if $\langle\mu_n,f\rangle \longrightarrow \langle\mu,f\rangle$ $\forall$ $f\in bC(E)$ cf. [EK, Chap. 3].

Let $\mathbf{D} = D([0,\infty),M_1(E))$ be furnished with the usual Skorohod topology (cf. Section 3.6) and $X_t:\mathbf{D}\longrightarrow M_1(E)$, $X_t(\omega) := \omega(t)$ for $\omega\in\mathbf{D}$. Let $\mathcal{D}^o_t = \sigma(X_s:0\le s\le t)$, $\mathcal{D} = \vee\mathcal{D}^o_t = \mathcal{B}(\mathbf{D})$, $\mathcal{D}_t = \mathcal{D}^o_{t+} := \bigcap_{\varepsilon>0}\mathcal{D}^o_{t+\varepsilon}$. For any $\mathcal{D}_t$-stopping time τ, $\mathcal{D}_\tau := \{A\in\mathcal{D}: A\cap(\tau\le t)\in\mathcal{D}_t\ \forall t\}$. Then $(\mathbf{D},(\mathcal{D}_t)_{t\ge0},\mathcal{D},(X_t)_{t\ge0})$ defines the *canonical probability-measure-valued process.*

Recall that $\mathbf{D}$ and $M_1(\mathbf{D})$ are both Polish spaces. If $P\in M_1(\mathbf{D})$, and $F\in b\mathcal{D}$, we let $P(F) := \int F dP$. (We sometimes also use the notation $E[X]$ to denote the expectation of a random variable X.)

For $t\ge0$, define $\Pi_t:M_1(\mathbf{D})\longrightarrow M_1(M_1(E))$ by $\Pi_t P := P\circ X_t^{-1}$. Then for fixed $P\in M_1(\mathbf{D})$, the mapping $t\longrightarrow \Pi_t P \in D([0,\infty), M_1(M_1(E)))$. (N.B. However the mapping $P \longrightarrow \Pi_\bullet P$ is not continuous from $M_1(\mathbf{D})$ to $D([0,\infty),M_1(M_1(E)))$.)

By an $M_1(E)$-valued *stochastic process* we will mean a family of probability measures $\{P_\mu:\mu\in M_1(E)\}$ on $(\mathbf{D},\mathcal{D},(\mathcal{D}_t)_{t\ge0})$ such that

(i) $P_\mu(X(0)=\mu) = 1$, that is $\Pi_0 P_\mu = \delta_\mu$,

(ii) the mapping $\mu \longrightarrow P_\mu$ from $M_1(E)$ to $M_1(\mathbf{D})$ is measurable.

18

It is said to be *time homogeneous strong Markov* if for every $\{\mathcal{D}_t\}$-stopping time τ, $\mu \in M_1(E)$, with $P_\mu(\tau < \infty) = 1$,

(iii) $P_\mu[F(X(\tau+t))|\mathcal{D}_\tau] = T_t F(X(\tau))$, P_μ-a.s.

for all $F \in b\mathcal{B}(M_1(E))$, $t \geq 0$, where

$$T_t F(\mu) := P_\mu F(X(t)) = \int F(\nu)\mathcal{P}(t,\mu,d\nu).$$

The *transition function* is defined by $\mathcal{P}(t,\mu;.) := \Pi_t P_\mu(.)$. Let $(C(M_1(E)), \|.\|)$ denote the Banach space of continuous functions on $M_1(E)$ with $\|F\| := \sup_\mu |F(\mu)|$. The process is a *Feller process* if in addition $T_t: C(M_1(E)) \longrightarrow C(M_1(E))$ $\forall t>0$ and $\|T_t F - F\| \longrightarrow 0$ as $t \to 0$. Then $\{T_t : t \geq 0\}$ forms a *strongly continuous semigroup of positive contraction operators* on $C(M_1(E))$, that is, $T_t F \geq 0$ if $F \geq 0$ and $\|T_t F\| \leq \|F\|$ for $F \in C(M_1(E))$. Given a Feller semigroup the *strong infinitesimal generator* is defined by

$$\mathfrak{G}F := \lim_{t \downarrow 0} \frac{T_t F - F}{t} \quad \text{(where the limit is taken in the norm topology).}$$

The *domain* $\tilde{\mathfrak{D}}(\mathfrak{G})$ of $\mathfrak{G}$ is the subspace of $C(M_1(E))$ for which this limit exists. Since $\int_0^\infty e^{-\lambda t} T_t F dt \in \tilde{\mathfrak{D}}(\mathfrak{G})$ if $\lambda > 0$ and $F \in C(M_1(E))$, it follows that $\tilde{\mathfrak{D}}(\mathfrak{G})$ is dense in $C(M_1(E))$. A subspace $\mathfrak{D}_0 \subset \tilde{\mathfrak{D}}(\mathfrak{G})$ is a *core* for $\mathfrak{G}$ if the closure of the restriction of $\mathfrak{G}$ to $\mathfrak{D}_0$ is equal to $\mathfrak{G}$.

<u>Lemma 2.1.1.</u> Let $\mathfrak{G}$ be the generator of a strongly continuous contraction semigroup $\{T_t\}$ on $C(M_1(E))$. Let $\mathfrak{D}_0$ be a dense subspace of $C(M_1(E))$ and $\mathfrak{D}_0 \subset \mathfrak{D}(\tilde{\mathfrak{G}})$. If $T_t: \mathfrak{D}_0 \to \mathfrak{D}_0$, then it is a core for $\mathfrak{G}$.

(A similar statement is true for a semigroup $\{S_t\}$ defined on $C(E)$, with generator A and domain $D(A)$.)

<u>Proof.</u> [EK, Ch. 1, Prop. 3.3]

In order to formulate an $M_1(E)$-valued Feller process we must first introduce some appropriate subspaces of $C(M_1(E))$ which can serve as a core for the generator.

The *algebra of polynomials*, $C_p(M_1(E))$, is defined to be the linear span of *monomials* of the form

$$F_{f,n}(\mu) = \int f(x)\mu^n(dx)$$
$$= \int_E \cdots \int_E f(x_1,\ldots,x_n)\mu(dx_1)\ldots\mu(dx_n)$$

where $f \in C(E^n)$.

The function $F \in C(M_1(E))$ is said to be *differentiable* if the limit

19

$$F^{(1)}(\mu;x) := \frac{\delta F(\mu)}{\delta\mu(x)} := \lim_{\varepsilon\downarrow 0}\,(F(\mu+\varepsilon\delta_x)-F(\mu))/\varepsilon \;=\; \frac{\partial}{\partial\varepsilon}F(\mu+\varepsilon\delta_x)\Big|_{\varepsilon=0}$$

exists for each $x\in E$ and belongs to $C(E)$ $\forall\mu\in M_1(E)$. The set of functions for which $F^{(1)}(\mu;x)$ is jointly continuous in μ and x is denoted by $C^{(1)}(M(E))$.

The *second derivative* is defined by

$$F^{(2)}(\mu;x,y) := \frac{\delta^2 F(\mu)}{\delta\mu(x)\delta\mu(y)} = \frac{\partial^2}{\partial\varepsilon_1\partial\varepsilon_2}F(\mu+\varepsilon_1\delta_x+\varepsilon_2\delta_y)\Big|_{\varepsilon_1=\varepsilon_2=0}$$

if it exists for each x and y and belongs to $C(E\times E)$ $\forall$ $\mu\in M_1(E)$.

Let $C^{(k)}(M_1(E))$ denote the set of functions for which $F^{(k)}(\mu;x_1,\ldots,x_k)$ exists and is continuous on $M_1(E)\times E^k$.

Lemma 2.1.2 (a) $C_P(M_1(E))$ is dense in $C(M_1(E))$ and convergence determining in $M_1(M_1(E))$.

(b) Functions in $C_P(M_1(E))$ are infinitely differentiable, and the first and second derivatives are given by

$$\frac{\delta F_{f,n}(\mu)}{\delta\mu(x)} = \sum_{j=1}^{n}\int_E\cdots\int_E f(x_1,..,x_{j-1},x,x_{j+1},\ldots,x_n)\prod_{i\neq j}\mu(dx_i)$$

$$\frac{\delta^2 F_{f,n}(\mu)}{\delta\mu(x)\delta\mu(y)} =$$

$$\sum_{j=1}^{n}\sum_{\substack{k=1\\ j\neq k}}^{n}\int_E\cdots\int_E f(x_1,..,x_{j-1},x,x_{j+1},..,x_{k-1},y,x_{k+1},..,x_n)\prod_{i\neq j,k}\mu(dx_i)$$

Proof. (a) The linear span of the space in question is an algebra of functions on the compact metric space $M_1(E)$. In order to verify that $C_P(M_1(E))$ separates points it suffices to note that $\mu\in M_1(E)$ is uniquely determined by $\{\langle\mu,\phi\rangle: \phi\in C(E)\}$. The first part of the result is then an immediate consequence of the Stone-Weierstrass theorem.

If $\int F(\mu)P_n(d\mu)\to\int F(\mu)P(d\mu)$ as $n\to\infty$ for all F belonging to a dense subset of $C(M_1(E))$, then it is true for all $F\in C(M_1(E))$. This proves that $C_P(M_1(E))$ is convergence determining in $M_1(M_1(E))$.

(b) follows by a simple calculation. □

2.2 Independent Particle Systems: Dynamical Law of Large Numbers

Let $\{S_t: t\geq 0\}$ be a Feller semigroup on the Banach space $(C(E),\|.\|)$ where $\|.\|$ is the supremum norm, with E compact. Then the domain D(A) of the infinitesimal gen-

20

erator A is a dense subspace of $C(E)$. We assume that there exists a separating algebra of functions

$D_0 \subset D(A)$, $S_t D_0 \subset D_0$.

Consequently D_0 is a core for A (cf. Lemma 2.1.1).

Let $P(t,x,dy)$ denote the transition function of $\{S_t\}$, that is,

$$S_t f(x) = \int_E f(y)P(t,x,dy), \quad f \in C(E).$$

It will be convenient to work with a canonical version of the Feller process which will be described in the following result. Let $D_E = D([0,\infty),E)$ denote the space of càdlàg functions from $[0,\infty)$ into E. Then D_E is a Polish space if it is furnished with the Skorohod topology (see Section 3.6 for these definitions).

Proposition 2.2.1. Let $\{S_t\}$ be a Feller semigroup on $C(E)$ with E compact. Then

(a) For each $x \in E$ there exists a probability measure P_x on $\mathcal{B}(D_E)$ satisfying

(2.2.1) $\quad P_x(\omega(0)=x) = 1$, and for $s \le t$,

(2.2.2) $P_x(f(\omega(t))|\sigma(\omega(u):u \le s) = (S_{t-s}f)(\omega(s))$, P_x-a.s. $\forall f \in C(E)$.

(b) There exists a standard probability space $(\Omega,\mathcal{F},Q_A)$ and a measurable mapping ζ: $(E\times\Omega,\mathcal{E}\otimes\mathcal{F}) \longrightarrow (D_E,\mathcal{B}(D_E))$, such that for each $x \in E$

(2.2.3) $\quad Q_A(\{\omega:\zeta(x,\omega)\in B\}) = P_x(B) \quad \forall B \in \mathcal{B}(D_E)$.

Furthermore, $\zeta(.,\omega)$ is continuous at x for Q_A-a.e. ω, for each $x \in E$.

The resulting measurable random function is denoted by $(\Omega,\mathcal{F},Q_A,\{\zeta(x)\}_{x\in E})$.

(A *standard probability space* is one which is isomorphic to $[0,1]$ with Lebesgue measure.)

Proof. (a) Given $x \in E$, the existence of P_x satisfying (2.2.2) is a standard result on the existence of a càdlàg version of the Feller process (e.g. [EK Chapt. 4, Thm 2.7]).

(b) It can also be shown that the mapping $x \longrightarrow P_x$ from E to $M_1(D_E)$ is continuous when the latter is given the weak topology (see Section 3.1). Since the map $x \longrightarrow P_x$ is continuous, the existence of a representation $(\Omega,\mathcal{F},Q_A,\{\zeta(x)\}_{x\in E})$, ξ: $\Omega\times E \longrightarrow D_E$, such that

(i) for each $x \in E$, $\xi(.,x)$ is measurable and has law P_x,

and

(ii) $\xi(\omega,.)$ is continuous at x for Q_A-a.e. ω for each $x \in E$,

follows from the extension of Skorohod's almost sure representation theorem due to Blackwell and Dubins (1983). From this the existence of a jointly measurable version of ζ follows by a standard argument (cf. Neveu (1964, III-4)). □

A system of N independent particles $\{Z(t):t\ge 0\} := \{Z_1(.),\dots,Z_N(.)\}$ each undergoing an A-motion in E and with initial value $Z_i(0)$ having law $\mu \in M_1(E)$ is then

21

realized on $((E\times\Omega)^N,(\mu\otimes Q_A)^N)$ by

$$Z_i(((e_1,\omega_1),\ldots,(e_N,\omega_N)),t) = \zeta(e_i,\omega_i)(t),\ i=1,\ldots,N,$$
$$((e_1,\omega_1),\ldots,(e_N,\omega_N)) \in (E\times\Omega)^N.$$

Then $Z(t)$ is an E^N-valued Markov process with semigroup

$$S_t^N f(x_1,\ldots,x_N) = \int_E\ldots\int_E f(y_1,\ldots,y_N)P(t,x_1,dy_1)\ldots P(t,x_N,dy_N),\ f\in C(E^N).$$

The semigroup $\{S_t^N:t\geq 0\}$ is strongly continuous on the closure of D_0^N (:= algebra generated by $\{f_1(x_1)\ldots f_N(x_N):\ f_i\in D_0,\ i=1,\ldots,N\}$), which is $C(E^N)$, and hence S_t^N is a Feller semigroup associated with a process with values in E^N. The corresponding generator is

$$A^{(N)} := \sum_{i=1}^{N} A_i \quad \text{on} \quad D(A^{(N)}) \subset C(E^N).$$

where A_i denotes the action of A on the ith variable. Furthermore it easily follows that $S_t^N:D_0^N \longrightarrow D_0^N$ and therefore D_0^N is a core for $A^{(N)}$.

The associated *empirical measure process* is given by

$$X^N(t) = \Xi_N(Z_1(t),\ldots,Z_N(t)) := N^{-1}\sum_{i=1}^{N}\delta_{Z_i(t)} \in M_1(E).$$

It will follow from Proposition 2.3.3 that $X^N(\cdot)$ is also a Feller process with state space $M_{1,N}(E)$, the space of measures consisting of atoms whose masses are multiples of 1/N and contained in E. We will denote its generator by $\mathfrak{G}_N^A$.

Let $\mathfrak{D}_0(\mathfrak{G}_N^A) := \{F_{f,n}(\mu) = \langle\mu^n,f\rangle:\ f\in D_0^n,\ n\leq N\}$.

For $F_{f,n}\in\mathfrak{D}_0(\mathfrak{G}_N^A)$ and $\mu_N = \frac{1}{N}\sum_{i=1}^{N}\delta_{z_i}$,

$$F_{f,n}(\mu_N) = N^{-n}\sum_{i_1=1}^{N}\ldots\ldots\sum_{i_n=1}^{N} f(z_{i_1},\ldots,z_{i_n})$$

$$= N^{-n}\sum_{k=1}^{n}\sum_{p\in\wp_k^n}\sum_{j_1=1}^{N}\ldots\sum_{j_k=1}^{N} f(z_{j_{p1}},\ldots,z_{j_{pn}})$$

where for $1\leq k<n$, $\wp_k^n$ denotes the set of partitions $p:\{1,\ldots,n\}\longrightarrow\{1,\ldots,k\}$.

$$\mathfrak{G}_N^A F_{f,n}(\mu_N) = \langle\mu_N^n,A^{(n)}f\rangle$$

22

$$+ N^{-n}\sum_{k=1}^{n-1}\sum_{p\in\rho_k^n}\sum_{j_{p1}=1}^{N}\cdots\sum_{j_{pk}=1}^{N}(A^{(k)}f^{(p)} - A^{(n)}f)(z_{j_{p1}},\dots,z_{j_{pn}})$$

$$= F_{A^{(n)}f,n}(\mu_N) + R(N,n,f)(\mu_N)$$

and for $p\in\rho_k^n$, $f^{(p)}(z_{j_1},\dots,z_{j_n}) = f(z_{j_{p1}},\dots,z_{j_{pn}})$.

Since

$$R(N,n,f)(\mu_N) = N^{-n}\sum_{k=1}^{n-1}\sum_{p\in\rho_k^n}\sum_{j_{p1}=1}^{N}\cdots\sum_{j_{pk}=1}^{N}(A^{(k)}f^{(p)} - A^{(n)}f)(z_{j_{p1}},\dots,z_{j_{pn}})$$

it follows that $|R(N,n,f)(\mu_N)| \le c(n)\|f\|_{A,n}/N$ where for $f\in D_0^n$,

$\|f\|_{A,n} := \|f\| + \max_k \max_{p\in\rho_k^n} \|A^{(k)}f^{(p)}\|$. Note that $\|S_t^N f\|_{A,n} \le \|f\|_{A,n}$ and therefore $S_t^N: D_0^n \to D_0^n$.

By the law of large numbers $X^N(0) \Rightarrow X(0) := \mu$. Using the above expression for the generator we can then show (as a special case of Theorem 2.7.1) that $X^N(t) \Rightarrow X(t)$, which is a deterministic $M_1(E)$-valued process characterized as the unique solution of the *weak equation*

$$\langle X(t),f\rangle = \langle X(0),f\rangle + \int_0^t \langle X(s),Af\rangle ds \quad \forall\ f\in D(A),$$

that is, formally, $\frac{\partial X}{\partial t} = A^*X$ where A^* denotes the adjoint of A.

This implies that

$$\langle X(t),f\rangle = \langle X(0),S_t f\rangle.$$

This is the simplest example of a *dynamical law of large numbers* and is a degenerate case of the McKean-Vlasov limit of exchangeably interacting particle systems. For detailed developments on the McKean Vlasov (or mean-field) limit of interacting particle systems and the related phenomenon of propagation of chaos the reader is referred to Gärtner (1988), Léonard (1986) and Sznitman (1989).

2.3 Exchangeable Particle Systems

Let $\mathcal{Per}(N)$ denote the set of permutations of $\{1,\dots,N\}$. A continuous function $f:E^N\to\mathbb{R}$ is said to be *symmetric*, $f\in C_{sym}(E^N)$, if $f = \tilde{\pi}f \quad \forall\ \pi\in\mathcal{Per}(N)$, where $(\tilde{\pi}f)(z_1,\dots,z_N) := f(z_{\pi 1},\dots,z_{\pi N})$.

Given $z_1,\dots,z_N \in E$ (not necessarily distinct) the associated *empirical measure* is defined by

23

$$\Xi_N(z_1,\ldots,z_N) := N^{-1}\sum_{i=1}^{N}\delta_{z_i} \in M_1(E).$$

The mapping $\Xi_N: E^N \longrightarrow M_1(E)$ is clearly $\sigma(C_{sym}(E^N))$-measurable. On the other hand, given a measure $\mu = \sum_{i=1}^{M} a_i\delta_{z_i} + \nu \in M_1(E)$, with $z_1,\ldots,z_M$ distinct, and ν nonatomic, let $\Sigma(\mu) := \{(z_1;a_1)\ldots,(z_M;a_M)\} \in (E\times\mathbb{R}_+)^M$, $\mathrm{mod}(\mathcal{P}er(M))$ (i.e. unordered M-tuples). The mapping $\mu \longrightarrow \Sigma(\mu)$ is measurable from $(M_1(E),\mathcal{B}(M_1(E)))$ to $\cup_{M=1}^{\infty}(E\times\mathbb{R}_+)^M$ where the latter is furnished with the smallest σ-algebra containing $\sigma(C_{sym}(E\times\mathbb{R}_+)^M)$ for each M (cf. Theorem 3.4.1.1(d)). Consequently, if $\mu\in M_{1,N}(E)$, then the mapping $\mu \longrightarrow ((z'_1,n_1),\ldots,(z'_k,n_k))$ where the $z'_1,\ldots,z'_k$ are the distinct locations of the atoms and the n_k are their multiplicities is $(M(E),\mathcal{B}(M(E)))$-measurable. Then the unordered n-tuple $(z_1,\ldots,z_n)$ is given by listing the distinct $z'_1,\ldots,z'_k$ with the appropriate multiplicities. Thus we obtain the following.

<u>Lemma 2.3.1</u> The sub-σ-algebras $\sigma\{C_{sym}(E^N)\}$ and $\sigma\{\Xi_N\}$ of $\mathcal{B}(E^N)$ coincide. In particular, if $f\in C_{sym}(E^N)$, then $f(z_1,\ldots,z_N)$ is $\sigma(\Xi_N)$-measurable.

<u>Proof.</u> If $\Sigma\left(\Xi_N(z_1,\ldots,z_N)\right) = ((z'_1,n_1),\ldots,(z'_k,n_k))$, and $f\in C_{sym}(E^N)$ then $f(z_1,\ldots,z_N) = f(z'_1,\ldots,z'_1,\ldots,z'_k,\ldots,z'_k)$ (with z'_i repeated n_i times for each $i = 1,\ldots,k$). □

The E-valued random variables $Z_1,\ldots,Z_N$ are *exchangeable* if the joint distributions of $Z_1,\ldots,Z_N$ and $Z_{\pi 1},\ldots,Z_{\pi N}$ are identical for any $\pi\in\mathcal{P}er(N)$. The probability law P on $\mathcal{B}(E^N)$ of the exchangeable random variables $Z_1,\ldots,Z_N$ is uniquely determined by its restriction to the sub-σ-algebra $\sigma(C_{sym}(E^N))$. Let $M_{1,ex}(E^N)$ denote the family of exchangeable probability laws on E^N. Then $C_{sym}(E^N)$ is $M_{1,ex}(E^N)$-determining, that is, if $\mu_1,\mu_2 \in M_{1,ex}(E^N)$ and $\int_{E^N} f(x)\mu_1(dx) = \int_{E^N} f(x)\mu_2(dx)\ \forall\ f\in C_{sym}(E^N)$, then $\mu_1=\mu_2$. Moreover if $\mu\in M_{1,ex}(E^N)$, $g \in pC_{sym}(E^N)$, and $\langle\mu,g\rangle < \infty$, then $\mu_g(A) := \langle\mu,g1_A\rangle/\langle\mu,g\rangle \in M_{1,ex}(E^N)$.

Given a Polish space S let $D_S = D([0,\infty);S)$ denote the space of càdlàg functions from $[0,\infty)$ to S furnished with the Skorohod topology (cf. Ch. 3, Sect. 6). Given $\pi\in\mathcal{P}er(N)$, let $\tilde{\pi}:E^N \longrightarrow E^N$ be defined by $(\tilde{\pi}x)_i = x_{\pi i}$ for $x = (x_1,\ldots,x_N) \in E^N$ and $\tilde{\pi}:D_{E^N} \longrightarrow D_{E^N}$ be defined by $(\tilde{\pi}x)_i(t) = x_{\pi i}(t)$.

24

An *exchangeable system of* N *particles* is defined by an exchangeable probability law P on D_{E^N}, or equivalently,

(i) an exchangeable initial distribution $\Pi_0 P$ on E^N, and

(ii) a family $\{P_y : y \in E^N\}$ of conditional distriubtions on D_{E^N} which satisfies

$P_{\tilde{\pi} y} = P_y \circ \tilde{\pi}^{-1}$ or $P_{\tilde{\pi} y}(\tilde{\pi} A) = P_y(A)$ for every $y \in E^N$, $A \in \mathcal{D}$, and $\pi \in \mathcal{P}er(N)$.

We next give a simple criterion which implies that an E^N-valued Markov process is exchangeable.

Lemma 2.3.2 Let $Z = (Z_1,\dots,Z_N)$ be an E^N-valued càdlàg Markov process with transition function $p(s,x;t,dy)$. Then Z is an exchangeable system provided that the marginal distributions $P(Z(t) \in .)$, $t \in \mathbb{R}_+$ are exchangeable and $p(s,y;t,B) = p(s,\tilde{\pi} y;t,\tilde{\pi} B)$ for every $\pi \in \mathcal{P}er(N)$, $y \in E^N$, $B \in \mathcal{B}(E^N)$, or equivalently,

(2.3.1) $(S_t f(\tilde{\pi}.))(\tilde{\pi}^{-1} x) = (S_t f)(x)$, $f \in C(E^N)$.

Note that in the case of a time homogeneous Feller semigroup $\{S_t\}$ and with generator A and core $D_0(A)$ the above criterion is implied by

(2.3.2) $(Af(\tilde{\pi}.))(\tilde{\pi}^{-1} y) = (Af(.))(y)$, $f \in D_0(A)$.

Proof. Let $m \in \mathbb{Z}_+$, $t_1,\dots,t_m \in \mathbb{R}_+$ and $\pi \in \mathcal{P}er(N)$. Then for $B_i^j \in \mathcal{B}(E)$,

$$P_{y_0}\left(Z(t_i) \in \prod_{j=1}^N B_i^{\pi^{-1} j},\ i=1,\dots,m\right)$$

$$= \int_{\prod_j B_1^{\pi^{-1} j}} \cdots \int_{\prod_j B_m^{\pi^{-1} j}} p(0,y_0;t_1,dy_1) \prod_{i=1}^{m-1} p(t_i,y_i;t_{i+1},dy_{i+1})$$

$$= \int_{\prod_j B_1^{\pi^{-1} j}} \cdots \int_{\prod_j B_m^{\pi^{-1} j}} p(0,\tilde{\pi} y_0;t_1,\tilde{\pi} dy_1) \prod_{i=1}^{m-1} p(t_i,\tilde{\pi} y_i;t_{i+1},\tilde{\pi} dy_{i+1})$$

$$= \int_{\prod_j B_1^j} \cdots \int_{\prod_j B_m^j} p(0,\tilde{\pi} y_0;t_1,dy_1) \prod_{i=1}^{m-1} p(t_i,y_i;t_{i+1},dy_{i+1})$$

$$= P_{\tilde{\pi} y_0}\left(Z(t_i) \in \prod_{j=1}^N B_i^j,\ i=1,\dots,m\right),$$

since by assumption $p(t_i,\tilde{\pi} y_i;t_{i+1},\tilde{\pi} dy_{i+1}) = p(t_i,y_i;t_{i+1},dy_{i+1})$.

Thus the finite dimensional distributions of $P_{\tilde{\pi} y_0}$ and $P_{y_0} \circ \tilde{\pi}^{-1}$ coincide which yields

25

the result. □

Proposition 2.3.3 Let $Z = (Z_1,\dots,Z_N)$ be an E^N-valued càdlàg exchangeable Feller process. Then the *empirical measure process* $X(t) := \Xi_N(Z(t))$ is a càdlàg $M_1(E)$-valued Feller Markov process.

Proof. For each $\phi\in C(E)$, $\int\phi(x)X(t,dx) = N^{-1}\sum_{i=1}^{N}\phi(Z_i(t))$ is càdlàg and hence $X(t)\in D([0,\infty);M_1(E))$, a.s.

Let $\mathcal{F}_t^Z := \sigma\{Z(s):0\le s\le t\}$. Then in order to prove the Markov property for $X(.)$ it suffices to show that

$P(X(t+s)\in.\,|\sigma(X(t))\vee\mathcal{F}_t^Z) = P(X(t+s)\in.\,|\sigma(X(t)),$ a.s.

It follows from the Markov property of Z and the inclusion $\mathcal{F}_t^Z\supset\sigma(X(t))$, that the left hand side equals $P(X(t+s)\in.\,|\sigma(Z(t))$ a.s. Hence it suffices to show that $P(\Xi_N(Z(t+s))\in.\,|\sigma(Z(t)) = P(\Xi_N(Z(t+s))\in.\,|\sigma(X(t))$. Since $(Z(t),Z(t+s))$ forms N $(E\times E)$-valued exchangeable random variables by hypothesis, $P(\Xi_N(Z(t+s))\in.\,|\sigma(Z(t)) = P(\Xi_N(\tilde{\pi}Z(t+s))\in.\,|\sigma(\tilde{\pi}Z(t))) = P(\Xi_N(Z(t+s))\in.\,|\sigma(\tilde{\pi}Z(t)))$ a.s. Thus $P(\Xi_N(Z(t+s))\in.\,|\sigma(Z(t))$ is a symmetric function of $Z(t)$ and by Lemma 2.3.1 we conclude that there is a $\sigma(\Xi_N(Z(t))$-measurable version of $P(\Xi_N(Z(t+s))\in.\,|\sigma(Z(t))$, and this yields the Markov property. Finally, note that the assumption that Z is càdlàg and Feller implies that Ξ_N is also càdlàg. □

2.4 Random Probability Measures, Moment Measures and Exchangeable Sequences.

Let X be a random probability measure on E, E Polish. Then the *nth moment measure* is a probability measure defined on E^n as follows: $M_n(dx_1,\dots,dx_n) = E(X(dx_1)\dots X(dx_n))$. It is the probability law of n-exchangeable E-valued random variables, $\{Z_1,\dots,Z_n\}$. Noting that this is a consistent family and using Kolmogorov's extension theorem we can associate with every random probability measure on E an exchangeable sequence of E-valued random variables, $\{Z_n:n\in\mathbb{N}\}$. The converse result is related to de Finetti's theorem and is formulated in Chapter 11.

Lemma 2.4.1 (a) A random probability measure X on E is uniquely determined by its moment measures of all orders.

(b) The sequence $\{X_n\}$ of random probability measures with moment measures $\{M_{n,m},n,m\in\mathbb{N}\}$ converges weakly to a random probability measure X with moment measures $\{M_m\}$ if and only if $M_{n,m}\Rightarrow M_m$ as $n\to\infty$ for each $m\in\mathbb{N}$.

Proof. This follows from Lemma 2.1.2(a), Corollary 3.2.7 and Lemma 3.2.8. □

2.5 The Moran Particle Process

Let $(A,D(A))$ be the generator of a Feller semigroup $\{S(t)\}$ on $C(E)$ with E compact and consider the N-particle system introduced in Section 2.2.

We now add a pair interaction, called *replacement sampling*, to the independent motions considered above. In this model each particle jumps at rate $(N-1)\gamma/2$, $\gamma>0$, to the location of one of the remaining $(N-1)$ particles, that is, the times between replacements are i.i.d. exponential with mean $2/\gamma N(N-1)$. At the time of a jump a pair (i,j) is chosen at random and the jth particle jumps to the location of the ith particle.

Let $\{\mathbf{Z}^N(t)\} = \{Z_1^N(t),\dots,Z_N^N(t)\} \in E^N$ where $Z_j^N(t)$ denotes the site of the jth individual at time t, where again $\{Z_1^N(0),\dots,Z_N^N(0)\}$ are i.i.d. with law μ. This process is obtained by a *direct construction* on a probability space on which a countable independent collection of exponential random variables and copies of $(\Omega,\mathcal{F},Q_A,\{\zeta(x)\}_{x\in E})$ of Proposition 2.2.1 are defined. To make this precise first consider the probability space

$$\left(\mathbb{R}^{N(N-1)},\mathcal{B}(\mathbb{R}^{N(N-1)}),\nu;\{\eta^{i,j}\}_{\substack{i,j=1,..,N\\ i\neq j}}\right)$$

where $\nu := \prod_{i=1}^{N(N-1)}\left(\frac{\gamma}{2}\cdot e^{-\gamma u_i/2}du_i\right)$ and the $\eta^{i,j}$ are simply the coordinate functions. The basic probability space on which the process is defined is

$$\left\{(E,\mu)^N\otimes(\mathbb{R}^{N(N-1)}\times\Omega,\nu\otimes Q_A)^{\mathbb{N}},\{e_k:k=1,\dots,N\},\{\eta_k,\zeta_k\}_{k\in\mathbb{N}}\right\}.$$

Then $\{\eta_k\}$ is an i.i.d. sequence of copies of $\{\eta^{i,j}\}$ and $\{\zeta_k\}$ is an i.i.d. sequence of copies of the process $(\Omega,\mathcal{F},Q_A,\{\zeta(x)\}_{x\in E})$. The process $\{\mathbf{Z}(t):t\geq 0\}$ = $\{Z_1(.),\dots,Z_N(.)\}$ is defined as follows:

$\tau_0:=0$, $\tau_k := \tau_{k-1} + \min_{i,j} \eta_k^{i,j}$, and without loss of generality we can assume that the pair (i_k,j_k) is uniquely defined by $\min_{i,j}\eta_k^{i,j} = \eta_k^{i_k,j_k}$.

For $0\leq t<\tau_1$ set

$$Z_i(t) := \zeta_i(e_i;t)$$

for $\tau_k\leq t<\tau_{k+1}$, $k\geq 1$, set

$$Z_{j_k}(t) := \zeta_{N+k}(Z_{i_k}(\tau_k);(t-\tau_k)), \quad a(N+k) = i_k,$$

$$Z_i(t) := \zeta_m(Z_{a(m)}(s(m));(t-s(m)) \text{ for } i\neq j_k,$$

$$\text{if } \quad Z_i(t) = \zeta_m(Z_{a(m)}(s(m));(t-s(m)), \text{ for } \tau_{k-1}\leq t<\tau_k$$

27

(where $(Z_{a(m)}, s(m))$ denotes the starting place and time of the mth particle motion; and a(m) denotes the immediate *ancestor* of the mth particle motion).

Remark. This probability space representation also provides information on the *family structure* of the population. For example, consider the ℓth particle at time t. If we define

$$k(\ell,t) := \max\{k: \tau_k \le t,\ \ell \in (i_k, j_k)\},$$

then i_k (or j_k) is the sibling of ℓ. This family structure will be studied in Ch. 12.

Since the sojourn times are independent random variables and the jumps have a Markov chain structure, it is easy to verify that the process $\{Z^N(t)\}$ is a stationary Markov process with state space E^N. We define the semigroup

$$(V_t^N f)(x_1,\dots,x_N) := E[f(Z_1^N(t),\dots,Z_N^N(t)) \mid Z_1^N(0)=x_1,\dots,Z_N^N(0)=x_N],$$

for $f \in C(E^N)$.

Theorem 2.5.1. (a) $\{Z^N(t)\}$ is a Feller process with semigroup $\{V_t^N\}$, where $V_t^N f$, $f \in$ given by the unique solution of the evolution equation

(2.5.1)

$$(V_t^N f)(x_1,\dots,x_N)$$

$$= e^{-N(N-1)\gamma t/2}(S_t^N f)(x_1,\dots,x_N)$$

$$+ \tfrac{1}{2}\gamma \sum_{\substack{i,j\\ i\neq j}} \int_0^t e^{-N(N-1)\gamma s/2}\, S_s^N(\Theta_{ij}(V_{t-s}^N f))(x_1,\dots,x_N)\,ds$$

where $\{S_t^N\}$, D_0, $\|.\|_{A,N}$ are the same as those defined in Sect. 2.2, and $\Theta_{ij}: C(E^N) \to C(E^N)$ is defined by

$$(\Theta_{ij} f)(x_1,\dots,x_N) := f(x_1,\dots,x_i,\dots,x_{j-1},x_i,x_{j+1},\dots,x_N)$$

that is, $(\Theta_{ij} f)(x_1,\dots,x_N)$ depends only on $\{x_k: k \neq j\}$ and is obtained by evaluating f at $\tilde{x}_1,\dots,\tilde{x}_N$ with

$$\tilde{x}_k = x_k \quad \text{if } k \neq j$$
$$\tilde{x}_j = x_i.$$

(b) $V_t^N: D_0^N \to D_0^N$ and $\|V_t^N f\|_{A,N} \le \|f\|_{A,N}$ and consequently D_0^N is a core for $\{V_t^N\}$.

28

The process Z^N is called the *N-particle Moran process*, and the corresponding *Moran generator* has the form

(2.5.2)
$$L_N f(x_1,\dots,x_N) = A^{(N)} f(x_1,\dots,x_N) + \tfrac{1}{2}\gamma\Big(\sum_{i\neq j}\Theta_{ij}f - f\Big)(x_1,\dots,x_N)$$

where $A^{(N)}f(x_1,\dots,x_N) = \sum_{j=1}^{N} A_j f(x_1,\dots,x_N)$ and A_j denotes the action of A on the jth variable.

(Note also that Equation (2.5.1) is formally equivalent to the differential equation $\frac{\partial}{\partial t}V_t^N f = L_N V_t^N f$, $V_0^N f = f$.)

Remarks 2.5.2

(a) Note that $L_2 f(x_1,x_2) = A_1 f(x_1,x_2) + A_2 f(x_1,x_2)$
$$+ \tfrac{1}{2}\gamma(f(x_1,x_1)-f(x_1,x_2)) + \tfrac{1}{2}\gamma(f(x_2,x_2)-f(x_1,x_2)).$$

(b) We also define a function of (N-1)-variables $\tilde{\Theta}_{ij}f$ as follows:
$$(\tilde{\Theta}_{ij}f)(y_1,\dots,y_{N-1}) := f(x_1,\dots,x_N)$$
with $x_k = y_k$ for $k < i\vee j$, $k\neq i\wedge j$
$x_{i\vee j} = x_{i\wedge j} = y_{i\wedge j}$
$x_k = y_{k-1}$ for $k > i\vee j$.

Note that $\Theta_{ij}f$ can be written in terms of $\tilde{\Theta}_{ij}f$ as follows:
$$(\Theta_{ij}f)(x_1,\dots,x_N) = (\tilde{\Theta}_{ij}f)(x_1,\dots,x_{j-1},x_{j+1},\dots,x_N),$$
and if μ is a probability measure, then $\int(\Theta_{ij}f)d\mu^N = \int(\tilde{\Theta}_{ij}f)d\mu^{N-1}$.

Proof of Theorem 2.5.1. Conditioning on the time and position of the first interaction jump, τ_1, we obtain the following equation for the semigroup V_t^N of this process:

(2.5.3) $(V_t^N f)(x_1,\dots,x_N)$
$$= E[E[f(Z_1^N(t),\dots,Z_N^N(t))\,|\,\mathcal{F}_\tau^Z]\,|\,Z_1^N(0)=x_1,\dots,Z_N^N(0)=x_N]$$
$$= E[E[1(\tau_1>t)f(Z_1^N(t),\dots,Z_N^N(t))]\,|\,Z_1^N(0)=x_1,\dots,Z_N^N(0)=x_N]$$
$$+ \tfrac{1}{2}\gamma\sum_{i\neq j}\int_0^t E\Big[1(\tau_1\in ds)\Theta_{ij}\Big(E[f(Z_1^N(t),\dots,Z_N^N(t))\,|\,Z_1^N(s)=y_1,\dots,Z_N^N(s)=y_N]\Big)$$

29

$$\left. |Z_1^N(0)=x_1,\ldots,Z_N^N(0)=x_N] \right\}$$

where $\Theta_{ij}:C(E^N)\to C(E^N)$ is defined as above. Substituting the definition of $V_{t-s}^N f$ and the exponential distribution of the first jump time τ_1 on the right hand side of (2.5.3) we obtain (2.5.1).

If $f(x_1,..,x_N) = \sum_{i=1}^N g_i(x_i)$, then the solution can be written explicitly. Then noting that $\Theta_{ij} \sum_k \sum_\ell g_{k\ell}(x_k,x_\ell)$ can be written in terms of functions of the form $\sum_k \sum_\ell g_{k\ell}(x_k,x_\ell)$ and $\sum_{i=1}^N g_i(x_i)$, we can then solve this system of equations for functions of the form $f(x_1,\ldots,x_N) = \sum_k \sum_\ell g_{k\ell}(x_k,x_\ell)$. Continuing in this way we can prove that this system of equations has a unique solution for any initial $f \in C(E^N)$. It is easy to verify that the solution $\{V_t^N f\}$ defines a strongly continuous semigroup on $C(E^N)$. Moreover, the facts that $\|S_t^N f\|_{A,N} \le \|f\|_{A,N}$ and $\|\Theta_{ij} f\|_{A,N} \le \|f\|_{A,N}$, together with equation (2.5.1) imply that $\|V_t^N f\|_{A,N} \le \|f\|_{A,N}$. Therefore $V_t^N: D_0^N \to D_0^N$ and D_0^N is a core by Lemma 2.1.1. The required expression for the generator L_N (cf. (2.5.2)) is then obtained by differentiation. □

Proposition 2.5.3. L_N satisfies the exchangeability condition (2.3.2).

Proof. In order to emphasize the main point, for the moment, we assume that the generator of the jth particle motion is $A^{(j)}$ and the interaction involving the ith and jth particles is given by $\Theta^{(i,j)}$.

Then $A^{(j)}$ denotes the action of operator $A^{(j)}$ on the kth variable x_k and $\Theta_{\ell k}^{(i,j)}$ denote the action of the operator $\Theta^{(i,j)}$ on the pair (x_ℓ,x_k). Then

$$L_N f(x_1,\ldots,x_N) = \sum_{j=1}^N (A_j^{(j)}f)(x_1,\ldots,x_N) + \tfrac{1}{2}\gamma\cdot\left(\sum_{i\neq j}\Theta_{ij}^{(i,j)}f-f\right)(x_1,\ldots,x_N).$$

We then have

$$L_N[\tilde{\pi}f](\tilde{\pi}^{-1}x) = \sum_{j=1}^N (A_j^{(\pi^{-1}j)}f)(x) + \tfrac{1}{2}\gamma \sum_{i\neq j}\left(\Theta_{ij}^{(\pi^{-1}i,\pi^{-1}j)}f-f\right)(x)$$

30

$$= (Lf)(x)$$

since $A_j^{(\pi^{-1}j)} f = A_j f$ and $\Theta_{ij}^{(\pi^{-1}i,\pi^{-1}j)} f = \Theta_{ij} f$ in the case under consideration. □

2.6 The Measure-valued Moran Process

Now consider the empirical process $X^N(t)$;

$$X^N(t;B) := \Xi_N(Z(t))(B) = \frac{1}{N}\sum_{i=1}^{N} 1_B(Z_i(t)) \quad B \in \mathcal{E},$$

with state space $M_{1,N}(E)$ (with E compact). Using Propositions 2.3.3, 2.5.3 and Theorem 2.5.1 we conclude that X^N is a càdlàg Feller process with state space $M_{1,N}(E)$.

The corresponding semigroup is defined by

$$T_t^N F(\mu) := EF(X^N(t)|X^N(0)=\mu), \quad F\in C(M_1(E)),\ \mu\in M_{1,N}(E).$$

If $F_{f,n}$ is a monomial of degree $n \le N$ (cf. Sect. 2.1), and $\mu_N = \sum_{i=1}^{N} \delta_{x_i}$, then

$$F_{f,n}(\mu_N) = N^{-n} \sum_{x_{i_1}=1}^{N} \cdots \sum_{x_{i_n}=1}^{N} f(x_{i_1},\dots,x_{i_n}).$$

$$= N^{-n}\sum_{k=1}^{n}\sum_{p\in\wp_k^n}\sum_{j_{p1}=1}^{N}\cdots\sum_{j_{pk}=1}^{N} f(x_{j_{p1}},\dots,x_{j_{pn}})$$

where the $\wp_k^n$ are as defined in section 2.2.

Then

(2.6.1)

$$T_t^N F_{f,n}(\mu_N) = N^{-n}\sum_{k=1}^{n}\sum_{p\in\wp_k^n}\sum_{j_{p1}=1}^{N}\cdots\sum_{j_{pk}=1}^{N} (V_t^N f^{(p)})(x_{j_1},\dots,x_{j_n})$$

where V_t^N is as in (2.5.1).

Let $\mathfrak{G}_N^A$ denote the generator of the measure-valued process coming from the empirical measure of N independent A-processes as in Section 2.2. Then the generator of the measure-valued process X^N is given by

(2.6.2) $$\mathfrak{G}_N F(\mu) = \mathfrak{G}_N^A F(\mu)$$

$$+ \tfrac{1}{2}\gamma N(N-1)\int_E\int_E [F(\mu - \delta_x/N + \delta_y/N) - F(\mu)]\mu(dx)\mu(dy).$$

Note that we can also interpret the transitions which determine the second term in $\mathfrak{G}_N$ as the death of an individual and the birth of a new individual whose site is that of a second individual chosen at random from the remaining population - this is the *sampling replacement* mechanism of population genetics.

Let $\mathfrak{D}_0(\mathfrak{G}) = C_{P,D_0}(M_1(E)) := \{F_{f,n}(\cdot): f \in D_0^n, n \in \mathbb{N}\}$.

where $F_{f,n}(\mu) := \int \ldots \int f(x_1,\ldots,x_n)\mu^n(dx)$.

If $F_{f,n} \in \mathfrak{D}_0(\mathfrak{G})$, then

$$\mathfrak{G}_N^A F_{f,n}(\mu_N)$$

$$= \langle \mu_N^n, A^{(n)} f\rangle + N^{-n} \sum_{k=1}^{n-1} \sum_{p \in \wp_k^n} \sum_{j_{p1}=1}^{N} \ldots \sum_{j_{pk}=1}^{N} (A^{(k)}f^{(p)} - A^{(n)}f)(z_{j_1},\ldots,z_{j_n})$$

$$= F_{A^{(n)}f,n}(\mu_N) + R_1(N,n,f)(\mu_N)$$

and the remainder term satisfies $|R_1(N,n,f)(\mu_N)| \le c_1(n)\|f\|_{A,n}/N$.

Using Taylor's formula again , we obtain that for $F = F_{f,n} \in \mathfrak{D}_0(\mathfrak{G})$,

(2.6.3)

$$\mathfrak{G}_N F(\mu) = \int \{A \frac{\delta F(\mu)}{\delta \mu(x)}\}\mu(dx) + \frac{\gamma}{4} \iint \left[\frac{\delta^2 F(\mu)}{\delta \mu(x)^2} + \frac{\delta^2 F(\mu)}{\delta \mu(y)^2} - 2\frac{\delta^2 F(\mu)}{\delta \mu(x)\delta \mu(y)} \right]\mu(dx)\mu(dy)$$

$$+ R(N,n,f)(\mu)$$

where $|R(N,n,f)(\mu_N)| \le c_2(n)\|f\|_{A,n}/N$.

The limiting generator is called the *Fleming-Viot generator:*

$$\mathfrak{G}F(\mu) = \int \left(A \frac{\delta F(\mu)}{\delta \mu(x)}\right)\mu(dx) + \frac{\gamma}{2} \iint \frac{\delta^2 F(\mu)}{\delta \mu(x)\delta \mu(y)} Q_S(\mu;dx,dy) \tag{2.6.4}$$

where $Q_S(\mu;dx,dy) := \mu(dx)\delta_x(dy) - \mu(dx)\mu(dy)$.

For $F_{f,n} \in \mathfrak{D}_0(\mathfrak{G})$,

$$\mathfrak{G}F_{f,n}(\mu) = \langle \mu^n, A^{(n)} f\rangle + \frac{1}{2}\gamma \sum_{\substack{i,j \\ i \ne j}} \left(\langle \tilde{\Theta}_{ij} f, \mu^{(n-1)}\rangle - \langle f, \mu^{(n)}\rangle \right)$$

(2.6.5)

$$= F_{A^{(n)}f,n}(\mu) + \frac{1}{2}\gamma \sum_{\substack{i,j \\ i \ne j}} \left(F_{\Theta_{ij}f,n}(\mu) - F_{f,n}(\mu) \right)$$

32

$= F_{L_n f}(\mu),$

where $\tilde{\Theta}_{ij}:C(E^n)\to C(E^{n-1})$ is defined as in Remark 2.5.2(b) and L_n is Moran particle generator defined by (2.5.2).

2.7 The Fleming-Viot Limit

In this section we show that as $N\to\infty$, the processes X^N converge (in the sense of weak convergence of finite dimensional distributions (f.d.d.) to a measure-valued process introduced by Fleming and Viot (1979).

Let $\mu_N = N^{-1}\sum_{j=1}^{N}\delta_{Z_j}$, with $\{Z_j\}$ exchangeable E-valued random variables and assume that the random measures $\mu_N \Longrightarrow \mu\in M_1(E)$ (for example, $\{Z_j\}$ are i.i.d. μ).

Theorem 2.7.1. Let $\mathfrak{D}_0(\mathfrak{G}) := \{F_{f,n}(\cdot): f\in D_0^n,\ n\in\mathbb{N}\}$. Assume that $\mu_N\Longrightarrow\mu$ as $N\to\infty$. Then as $N\to\infty$ the processes $\{X_N\}$ with generators $\mathfrak{G}_N$ (defined by (2.6.5)) converge (f.d.d.) to a Feller process taking values in $M_1(E)$, with initial value μ, semigroup $\{T_t\}$ defined on $\mathfrak{D}_0(\mathfrak{G})$ by

$$T_tF_{f,n}(\mu) := F_{V_t^nf}(\mu) = \int\dots\int(V_t^nf)(x)\mu^n(dx),\quad f\in D_0^n,$$

where V_t^n is as in Theorem 2.5.1. Moreover $\mathfrak{D}_0(\mathfrak{G})$ is a core for the generator $\mathfrak{G}$. *This process is called the Fleming-Viot process with mutation operator A and sampling rate γ and will be referred to as the S(A,γ)-Fleming-Viot process.*

Proof. We first verify that $\{T_t\}$ defines a Feller semigroup on $C(M_1(E))$. By the same argument as in Lemma 2.1.2 we can verify that $\mathfrak{D}_0(\mathfrak{G})$ is dense in $C(M_1(E))$. For $F_{f,n}\in\mathfrak{D}_0(\mathfrak{G})$,

$$T_tF_{f,n}(\mu) := F_{V_t^nf}(\mu) = \int\dots\int(V_t^nf)(x)\mu^n(dx)$$

is strongly continuous on $\mathfrak{D}_0(\mathfrak{G})$ by Theorem 2.5.1(a) and $V_t^n:D_0^n\longrightarrow D_0^n$ by Theorem 2.5.1(b). Therefore $\{T_t\}$ is a Feller semigroup on $C(M_1(E))$ and $T_t:\mathfrak{D}_0(\mathfrak{G})\longrightarrow\mathfrak{D}_0(\mathfrak{G})$. Thus $\mathfrak{D}_0(\mathfrak{G})$ is a core for the generator $\mathfrak{G}$ (by Lemma 2.1.1).

In order to complete the proof of the weak convergence of the finite dimensional distributions of X^N to this limiting Feller process note that by (2.6.3)

$$\mathfrak{G}_NF_{f,n}(\mu_N) = \mathfrak{G}F_{f,n}(\mu_N) + R(N,n,f)(\mu_N)$$

where $|R(N,n,f)(\mu)| \le c(n)\|f\|_{A,n}/N$ and $\|V_t^nf\|_{A,n}\le\|f\|_{A,n}$.

Then recalling the definition of $\{T_t^N\}$, from the beginning of Section 2.6 we have

33

$$T^N_t F_{f,n}(\mu_N) = T_t F_{f,n}(\mu_N) + \int_0^t T_{t-s} R(N,n,V^N_s f)(\mu_N) ds$$

and as above $|R(N,n,V^N_s f)(\mu_N)| \le c(n)\|f\|_{A,n}/N$.

Therefore if $\mu_N \to \mu$, $T_t F_{f,n}(\mu_N) = \int (V^n_t f)(x)\mu^n_N(dx) \to \int (V^n_t f)(x)\mu^n(dx)$

and $\int_0^t T_{t-s} R_N(N,n,V^N_s f)(\mu_N) ds \to 0$.

This proves the convergence $T^N_t F_{f,n}(\mu_N) \to T_t F_{f,n}(\mu)$.

Thus we have proved that $E(F_{f,n}(X_N(t)) | X_N(0)=\mu_N) \to T_t F_{f,n}(\mu)$ for all polynomials $F_{f,n}$. Since $M_1(E)$ is compact, and $\{X_N(t): N\in\mathbb{N}\}$ are tight random measures for each t, then this proves (since polynomials are convergence determining) that

$$T_t F_{f,n}(\mu) = \int F_{f,n}(\nu) P_{t,\mu}(d\nu)$$

for some probability measure $P_{t,\mu}$ on $M_1(E)$. Thus we have proved that the law of measure-valued Moran process $X_N(t)$ converges to $P_{t,\mu}(d\nu)$. Since $\{V^n_t f\} \in C(E^n)$, the mappings $\mu \to T_t F_{f,n}(\mu)$ and $\mu \to P_{t,\mu}$ are continuous. Using the Markov property the convergence of the finite dimensional distributions follows by a standard argument. □

2.8. The Moment Measures of Fleming-Viot.

Since $\mathfrak{G} F_{t,n}(\mu) = F_{L_n f}(\mu)$, the limiting moment measures, M_n, of the Fleming-Viot process, X(t), satisfy the weak form of the system of equations dual to (2.5.1), that is,

(2.8.1)

$$\partial M_n(t;dx_1,\dots,dx_n)/\partial t$$

$$= \sum_{i=1}^n (A^i)^* M_n(t;dx_1,\dots,dx_n) - \tfrac{1}{2}\gamma n(n-1) M_n(t;dx_1,\dots,dx_n)$$

$$+ \tfrac{1}{2}\gamma \sum_i \sum_{j\ne i} M_{n-1}(t;dx_1,\dots,dx_{i-1},dx_{i+1},\dots,dx_n)\, \delta_{x_j}(dx_i)$$

$$M_n(0;dx_1,\dots,dx_n) = \mu(dx_1)\dots\mu(dx_n).$$

Proposition 2.8.1 The solution of the moment equations is given as follows:

for $f \in C_{sym}(E^k)$,

$$T_t F_{f,k}(\mu) = \int\dots\int f(x_1,\dots,x_k) M_k(t,\mu;dx_1,\dots,dx_k)$$

$$= \langle \mu^k (S^k_t)^*, f\rangle + \sum_{j=1}^{k-1} \int_0^t ds \langle \mu^j (S^j_s)^*, U^k_j(t-s) f\rangle$$

34

where for $f \in C_{sym}(E^k)$

$$(S_t^k f)(x_1,\ldots,x_k)$$

$$:= \quad e^{-\gamma k(k-1)t/2}\int\ldots\int p(t,x_1,dy_1)\ldots p(t,x_k,dy_k)f(y_1,\ldots,y_k)$$

for $\nu \in M_1(E^k)$

$$\nu(S_t^k)^*(dy_1,\ldots,dy_k)$$

$$:= \quad e^{-\gamma k(k-1)t/2}\int \nu(dx_1,\ldots,dx_k)p(t,x_1;dy_1)\ldots p(t,x_k;dy_k),$$

where $p(t,x;dy)$ denotes the transition probability function of the A-process, $*$ denotes the adjoint of an operator and $U_j^k(t)$ is inductively defined by

$$U_{k-1}^k(t) := \tfrac{1}{2}\gamma\Psi_k^{k-1}S_t^k, \quad t\geq 0,$$

$$U_j^k(t) := \int_0^t U_j^{k-1}(s)U_{k-1}^k(t-s)ds \ , \text{ for } k-j > 1,$$

$$(\Psi_k^{k-1}f)(x_1,\ldots,x_{k-1}) := \sum_{\substack{i,j\\ i\neq j}} (\tilde{\Theta}_{ij}f)(x_1,\ldots,x_{k-1}).$$

Proof. This obtained by induction from the above equations (2.8.1). (cf. Dynkin (1989c), Ethier and Griffiths (1992)). □

Corollary 2.8.2 For $f \in C(E^n)$, $t\geq 0$,

$$\int\ldots\int f(y_1,\ldots,y_n)M_n(t,\mu;dy_1,\ldots,dy_n) = \int\ldots\int V_t^n f(y_1,\ldots,y_n)\mu(dy_1)\ldots\mu(dy_n)$$

when $\{V_t^n\}$ is the n-particle Moran semigroup (Theorem 2.5.1) i.e. the moment measure $M_n(t,\mu;.)$ is the probability law at time t of the n particle Moran process with initial distribution i.i.d. μ.

Proof. This follows directly from the proof of Theorem 2.7.1. □

Remarks 2.8.3.

(1) For the moment let us begin with the system of moment equations (2.8.1) whose solution is given by Proposition 2.8.1. For $f \in C(E^n)$ let us define

$$T_t F_{f,n}(\mu) := \int\ldots\int f(x_1,\ldots,x_n)M_n(t,\mu;dx_1,\ldots,dx_n).$$

We can then verify directly that

(i) $\{M_n(t,\mu;.)\in M_1(E^n):n=1,2,\ldots\}$ forms a projective system of probability measures which define a sequence of E-valued exchangeable random variables and consequently by de Finetti's theorem (cf. Section 11.2) define a random probability measure on E,

(ii) $\{T_t\}$ defines a Feller semigroup on $C_{\mathcal{P}}(M_1(E))$.

35

This gives an alternative approach to the construction of the Fleming-Viot process which does not require the consideration any approximating particle system.

(2) In this chapter we have gone full circle - starting with the Moran process we constructed the Fleming-Viot process as a weak limit and then considering the moment measures of the Fleming-Viot we have recovered the Moran process (Corollary 2.8.2). In Chapter 11 we will see that there is a deeper connection here and in fact a consistent family of n-particle Moran processes, $n\in\mathbb{N}$, is embedded in the Fleming-Viot process.

(3) Duality between the Moran and Fleming-Viot processes.

The relation between the Moran processes and the Fleming-Viot process expressed in Corollary 2.8.2 also illustrates a simple example of the type of *duality relation* which will be developed in Chapter 5. To describe this in a more general setting let us be given the following:

(i) $\{T_1(t,s):0\le s\le t\}$ an evolution family on E_1 with generator A^1_s, and domain $D(A^1)$, that is,

Domain of A^1_s contains $D(A^1)$ for each s,

$T_1(t,s):D(A_1)\to D(A_1)$, $T_1(t,t) = Id$,

$$\frac{\partial T_1(t,s)}{\partial s} = -T_1(t,s)A^1_s, \text{ and}$$

$$\frac{\partial T_1(t,s)}{\partial t} = A^1_t T_1(t,s), \text{ for } 0 < s < t.$$

Similarly, let $T_2(t,s)$ be an evolution family on E_2 with generator A^2_s.

Proposition 2.8.4. Assume that $f\in C(E_1\times E_2)$ and $g:[0,\infty)\to C(E_1\times E_2)$ be such that $f(.,y)\in D(A^1)$ for each $y\in E_2$, and $f(x,.)\in D(A^2)$ for each $x\in E_1$, and

$$g(s,x,y) = A^1_s f(x,y) = A^2_s f(x,y)$$

where A^1_s acts on the first variable and A^2_s acts on the second variable. Then

$$T_1(t,0)f(x,y) = T_2(t,0)f(x,y).$$

Proof. From the assumptions we obtain that

$$\frac{d}{ds} T_2(t,s)T_1(s,0)f$$

$$= -T_2(t,s)A^2_s T_1(s,0)f + T_2(t,s)A^1_s T_1(s,0)f \equiv 0$$

and therefore $T_1(t,0)f(x,y) = T_2(t,0)f(x,y)$.

Remark 2.8.5. Let $E_1 = M_1(E)$ and $E_2 = \bigcup_{n=1}^{\infty} C(E^n)\cap D^n_0$, T^1_t be the Fleming-Viot

36

semigroup on $C_P(M_1(E))$, and T_t^2 be the Moran semigroup on $C\left(\bigcup_{n=1}^{\infty} C(E^n)\right)$ (i.e. $T_t^2F(f) := F(V_t^n f)$ if $f \in C(E^n)$. Let $F(\mu,f) := \int f d\mu^n$. Then Proposition 2.8.4 can be applied to show that

$T_t F_{f,n}(\mu) = T_t^1 F(\mu,f) = T_t^2 F(\mu,f) = F(\mu, V_t^n f) = F_{V_t^n f}(\mu)$ if $\mu \in M_1(E)$, $f \in C(E^n)$.

This notion of duality will be systematically developed in Section 5.5.

2.9 Weak Convergence of the Approximating Processes and the Martingale Problem

In subsequent sections we will systematically develop the notions of weak convergence of measure-valued processes and measure-valued martingale problems. In this section we briefly introduce this approach by applying it to the Fleming-Viot process.

In particular we will show that in addition to weak convergence of finite dimensional distributions the laws of the measure-valued Moran processes X_N, their distributions $P^N_{\mu_N}$ are tight in the space of probability measures on $D([0,\infty),M_1(E))$ and consequently weak convergence of processes follows. This implies that the Fleming-Viot process can be realized as a càdlàg process (we will later show that it is actually continuous).

We will now show (by using some results to be proved in the next chapter) that the Fleming-Viot process can also be characterized as the unique solution of the martingale problem for $(\mathfrak{G},\mathfrak{D}_0(\mathfrak{G}))$ defined in Sect. 2.7.

Since $\{X_N(t)\}$ is a Feller process with generator $\mathfrak{G}_N$ and core $\mathfrak{D}_0(\mathfrak{G})$ (cf. Sect. 2.6), it follows that

$$M_N(t) := F(X_N(t)) - \int_0^t \mathfrak{G}_N F(X_N(s))ds, \quad F \in \mathfrak{D}_0(\mathfrak{G}),$$

is a bounded martingale under $P^N_{\mu_N}$.

Therefore for $t \in [0,T]$,

$$(2.9.1) \quad F_{f,n}(X_N(t)) - \int_0^t \mathfrak{G} F_{f,n}(X_N(s))ds = M_N(t) + \int_0^t R(N,F,s)(X_N(s))ds$$

and $\sup_{0 \le s \le T} |R(N,F,s)| \le c(F)/N$.

In order to prove the tightness of the $P^N_{\mu_N}$ on $\mathbf{D} := D([0,\infty),M_1(E))$ we will use Theorem 3.6.4. Noting that $M_1(E)$ is compact (so that condition (i) of Theorem 3.6.4 is automatic) it suffices to show that for $\phi \in D_0(A)$, $\langle X_N(t),\phi\rangle$ are tight in $D([0,\infty),\mathbb{R})$. Applying (2.9.1) to $F_1(\mu) = \langle \mu,\phi\rangle$ and $F_2(\mu) = \langle \mu,\phi\rangle^2$, and then using Corollary 3.6.3 it follows that the laws of $\langle X_N(t),\phi\rangle$ are tight in $D([0,\infty),\mathbb{R})$. Since

37

we already proved the convergence of the finite dimensional distributions in Theorem 2.7.1, this yields the weak convergence of the probability measures $P^N_{\mu_N}$ on **D**.

Now let $F\in \mathcal{D}_0(\mathfrak{G})$. In this case the real-valued functional

$$F(x(t))-F(x(s)) - \int_s^t \mathfrak{G}F(x(u))du$$

is continuous on $D([0,\infty),M_1(E))$. Now let $H\in b\mathcal{D}_s\cap C(\mathbf{D})$. Therefore letting X denote the canonical process we get that

$$\begin{aligned}
&P_\mu\left[\left(F(X(t))-F(X(s)) - \int_s^t \mathfrak{G}F(X(u))du\right)H(X)\right]\\
&= \lim_{N\to\infty} P^N_{\mu_N}\left[\left(F(X(t))-F(X(s)) - \int_s^t \mathfrak{G}F(X(u))du\right)H(X)\right]\\
&= \lim_{N\to\infty} P^N_{\mu_N}\left[\left(F(X(t))-F(X(s)) - \int_s^t \mathfrak{G}_N F(X(u))du + \int_s^t R(N,F,u)du\right)H(X)\right]\\
&= \lim_{N\to\infty} P^N_{\mu_N}\left[\left(M_N(t) - M_N(s) + \int_s^t R(N,F,u)du\right)H(X)\right]\\
&= \lim_{N\to\infty} P^N_{\mu_N}\left[\left(\int_s^t R(N,F,u)du\right)H(X)\right] = 0.
\end{aligned}$$

This implies that

$$M_F(t) := \quad F(X(t))-F(X(s)) - \int_s^t \mathfrak{G}F(X(u))du$$

is also a martingale for each $F\in\mathcal{D}_0(\mathfrak{G})$ under P_μ. Therefore $\{P_\mu:\mu\in M_1(E)\}$ is *a solution to the martingale problem for* $(\mathfrak{G},\mathcal{D}_0(\mathfrak{G}))$. In fact the family $\{P_\mu:\mu\in M_1(E)\}$ is uniquely characterized in this way since any solution to the martingale problem must have the <u>same</u> moment measures as the Fleming-Viot process. This can be verified by applying the martingale problem to polynomials which yield the moment equations solved above. The details of this argument will be given in greater generality below.

2.10 Branching Particle Systems

Let us for the moment continue in the same spirit and consider a simple branching particle system on a compact metric space E. The main difference from the Moran model is that the total number of particles is no longer constant in time. For this reason the basic state space is now $M(E)$ the space of finite Borel measures on E. We will again follow the elementary approach based on moment measures to characterize the transition function for the limiting measure-valued process.

We consider a system of particles in the space E which move, die and produce offspring. We begin by assuming that during its lifetime each particle performs an

38

A-motion independently of the other particles.

In the case of critical branching when particles die they produce k particles with probability p_k, $k=0,1,2,\ldots$, $\sum_k kp_k = 1$. We will also assume in this section that

$$m_2 := \sum k^2 p_k, \text{ and } \sum k^3 p_k < \infty.$$

After branching the resulting set of particles evolve in the same way and independently of each other starting off from the parent particle's branching site. Let N(t) denote the total number of particles at time t. We will denote their locations by $\{x_i(t): 1 \le i \le N(t)\}$.

In order to obtain a measure-valued process by use of an appropriate scaling we assume that particles have mass ε and branch at rate c/ε.

For $B \in \mathcal{E}$, define

$$X_\varepsilon(t,B) := \varepsilon\left(\sum_{i=1}^{N(t)} 1_B(x_i(t))\right).$$

Let C_F denote the class of functions on M(E) of the form $F_f(\mu) = f(\langle\mu,\phi\rangle)$ with $f \in C_b(\mathbb{R})$, $\phi \in C(E)$. Let $\mathfrak{D}(\mathfrak{G}) := \{F_f(\mu): F_f(\mu) = f(\langle\mu,\phi\rangle);\ f \in C_b^\infty(\mathbb{R}),\ \phi \in D_0(A)\}$ where $D_0(A)$ is as in Sect. 2.2. Then $X_\varepsilon(.)$ is an M(E)-valued Feller process. The generator of $X_\varepsilon(\cdot)$ is defined on $\mathfrak{D}(\mathfrak{G})$, by

$$\mathfrak{G}_\varepsilon F(\mu) := G^A F_f(\mu) + c\varepsilon^{-2}\int\left\{\sum_k p_k[f(\langle\mu,\phi\rangle+\varepsilon(k-1)\phi(x)) - f(\langle\mu,\phi\rangle)]\right\}\mu(dx)$$

where G^A denotes the generator of the empirical process associated to particles performing independent A-motions in E.

Then for $F \in \mathfrak{D}(\mathfrak{G})$,

$$\mathfrak{G}_\varepsilon F(\mu) = \mathfrak{G}F(\mu) + O(\varepsilon)$$

where $\mathfrak{G}F(\mu) = f'(\langle\mu,\phi\rangle)\langle\mu,A\phi\rangle + \frac{1}{2}c(m_2-1)f''(\langle\mu,\phi\rangle)\langle\mu,\phi^2\rangle$.

Letting $\varepsilon \to 0$ we obtain a measure-valued process with generator defined on $\mathfrak{D}(\mathfrak{G})$ by

$$\mathfrak{G}F(\mu) := f'(\langle\mu,\phi\rangle)\langle\mu,A\phi\rangle + \tfrac{1}{2}c(m_2-1)f''(\langle\mu,\phi\rangle)\langle\mu,\phi^2\rangle$$

$$= \int A(\delta F(\mu)/\delta\mu(x))\mu(dx) + \tfrac{1}{2}c(m_2-1)\iint(\delta^2 F(\mu)/\delta\mu(x)\delta\mu(y))\delta_x(dy)\mu(dx).$$

To carry out the proof two changes are required to the arguments of section 2.7. First, since the total mass of this process is not uniformly bounded, a martingale inequality is required to verify that for each $\eta, T > 0$ there exists a compact subset K_η of M(E) such that $P(X(t) \in K_\eta\ \forall t \in [0,T]) > 1-\eta$. (This calculation is a special case of the proof of tightness which is given in Ch. 4.) Secondly, to characterize the limiting finite dimensional distributions we can again use moment measures. To

39

do this we must check that the moment measures uniquely determine the random measure $X(t)$ conditioned on $X(s)$ with $s<t$. The condition of Corollary 3.2.10 can be verified for the limiting process $\{X(t)\}$ with generator $\mathfrak{G}$ using the fact that $E(e^{\theta X(t,E)}) < \infty$ for small $\theta > 0$ (see Ch. 4).

In the next chapter we will develop the technical tools necessary to carry out the proofs of limit theorems of this type in a more general setting and then we will return to the study of a general class of measure-valued branching processes in Chapter 4.

40

3. RANDOM MEASURES AND CANONICAL MEASURE-VALUED PROCESSES

The objective of this chapter is to collect some basic material on measures on Polish spaces, random measures, canonical measure-valued processes, homogeneous and inhomogeneous Markov processes and weak convergence of random measures and measure-valued processes which will be needed in subsequent chapters.

3.1. State Spaces for Measure-valued Processes

Let E be a Polish space with metric d, and $\mathcal{E}$ be the Borel σ-algebra.

Let $C(E)$ ($pC(E)$, $bC(E)$, $pbC(E)$, $bC_b(E)$, $C_c(E)$, respectively) denote the spaces of all continuous functions on E (non-negative, bounded, non-negative bounded, with bounded support, with compact support, continuous functions respectively).
Let $\mathcal{E}$ ($p\mathcal{E}$, $b\mathcal{E}$, $bp\mathcal{E}$, $b\mathcal{E}_b$, $\mathcal{E}_c$, etc.) denote the corresponding spaces of Borel measurable functions on E.

Let $M(E)$ denote the space of Radon measures on $(E,\mathcal{E})$, that is the Borel measures, μ, such that $\mu(K) < \infty$ for all compact sets $K \subset E$.

For $\mu \in M(E)$ and $f \in b\mathcal{E}_c$ let $\langle\mu,f\rangle := \int f d\mu$.

Let $M_F(E)$ and $M_1(E)$ denote the finite and probability measures on E, respectively.

3.1.1. Finite measures on a Polish space

Let $(M_F(E),\tau_w)$ denote the finite measures with the *weak topology* defined by

$$\mu_n \Rightarrow \mu \text{ iff } \langle\mu_n,f\rangle \longrightarrow \langle\mu,f\rangle \ \forall f \in bC(E).$$

Then $(M_F(E),\tau_w)$ is a Polish space.

3.1.2. Measures on Compact Metric Spaces

If E is compact then $M_F(E)$ may be compactified $\bar{M}_F(E) = M_F(E) \cup \{\infty_w\}$, where $\{\infty_w\}$ is a compactifying point and the topology τ is defined by

$$\mu_n \Rightarrow \mu \in M_F(E) \text{ iff } \langle\mu_n,f\rangle \longrightarrow \langle\mu,f\rangle \ \forall f \in C(E),$$

$$\mu_n \Rightarrow \infty_w \text{ iff } \langle\mu_n,1\rangle \longrightarrow \infty.$$

Then $(\bar{M}_F(E),\tau)$ is also a compact metrizable space called the *Watanabe compactification* of $M_F(E)$.

3.1.3. Radon measures on a locally compact metric space

Let E be a locally compact metric space. Then the *vague topology* is defined by

$$\mu_n \xrightarrow{\tau_v} \mu \quad \text{iff } \langle\mu_n,f\rangle \longrightarrow \langle\mu,f\rangle \quad \forall f \in C_c(E).)$$

Then $(M(E),\tau_v)$ is a Polish space.

The locally compact space E can be compactified: set $\bar{E}= E\cup\{\infty_a\}$ such that $x_n \to \infty_p$ if and only if $x_n \in K$ for at most finitely many n for each compact set $K\subset E$. Then $\{\mu\in M_F(\bar{E}):\mu(\{\infty_a\})=0\}$ can be identified with the finite Radon measures on E.

3.1.4. Locally finite measures on a Polish space

Let $\mathcal{E}_b$ be the ring of bounded sets in $\mathcal{E}$. Let $(M_{LF}(E),\tau_v)$ denote the collection of *locally finite* Borel measures, μ, that is, measures such that $\mu(A) < \infty$ $\forall A\in \mathcal{E}_b$. We define the *vague topology*

$$\mu_n \xrightarrow{\tau_v} \mu \text{ iff } \langle\mu_n,f\rangle \longrightarrow \langle\mu,f\rangle \quad \forall f\in bC_b(E).$$

Then $(M_{LF}(E),\tau_v)$ is a Polish space.

3.1.5. p-Tempered Measures on $\mathbb{R}^d$

Let $p>0$ and $\phi_p(x) := (1+|x|^2)^{-p}$; $M_p(\mathbb{R}^d) = \{\mu:\langle\mu,\phi_p\rangle < \infty\}$ with the topology τ_p defined by

$$\mu_n \Longrightarrow \mu \quad \text{iff } \langle\mu_n,f\rangle \longrightarrow \langle\mu,f\rangle \ \forall\ f\in K_p(\mathbb{R}^d),$$

where $K_p(\mathbb{R}^d) = \{f:f = g + \alpha\cdot\phi_p, \alpha\in\mathbb{R},\ g\in C_c(\mathbb{R}^d)\}$. $(M_p(\mathbb{R}^d),\tau_p)$ is also a Polish space. Let $\{f_n:n\geq 1\}$ be a dense set in $C_c(\mathbb{R}^d)$ and $f_0 = \phi_p$. We define a metric on $M_p(\mathbb{R}^d)$ by

$$(3.1.1)\qquad d_p(\mu,\nu) := \sum_{m=0}^{\infty} 2^{-m}\left(1 \wedge \left|\int f_m d\mu - \int f_m d\nu\right|\right).$$

Let $\dot{\mathbb{R}}^d := \mathbb{R}^d\cup\{\infty_p\}$, where ∞_p is an isolated point and extend ϕ_p to $\dot{\mathbb{R}}^d$ by defining $\dot{\phi}_p(\infty_p) = 1$. Let $M_p(\dot{\mathbb{R}}^d) = \{\mu:\langle\mu,\dot{\phi}_p\rangle < \infty\}$ with the p-vague topology $\mu_n \Longrightarrow \mu$ iff $\langle\mu_n,f\rangle \longrightarrow \langle\mu,f\rangle$ $\forall$ $f\in K_p(\dot{\mathbb{R}}^d)$ (defined as above). In this case $M_p(\dot{\mathbb{R}}^d)$ is locally compact and sets of the form $\{\mu:\langle\mu,\dot{\phi}_p\rangle \leq k\}$ are compact (cf. Iscoe (1986a)).

3.2 Random Measures and Laplace Functionals.

Let E be Polish, $(M_F(E),\tau_w)$ be as above and $\mathcal{M} = \mathcal{B}(M_F(E))$ denote the Borel subsets of $M_F(E)$. A *finite random measure on* E is given by a probability measure P on $(M_F(E),\mathcal{M})$.

A sequence $\{f_n\}$ in $pb\mathcal{E}$ is said to *converge bp* to f if $f_n(x) \to f(x)$ $\forall x$, and $\exists$ $M < \infty$ such that $\sup_{n,x} f_n(x) \leq M$. Given $H \subset bp\mathcal{E}$, the bounded pointwise closure is the smallest collection of functions containing H which is closed under bp convergence.

Lemma 3.2.1. There exists a countable set $V = \{f_n\} \subseteq bpC(E)$ such that $1\in V$, and V is closed with respect to addition operation, and the bp closure of V is an element

42

of $bp\mathcal{E}$. Then the set V is also convergence determining in $M_F(E)$.

Proof. [EK pages 111-112].

We can then define a metric on $M_F(E)$ in terms of $V = \{f_n\}$ (as in Lemma 3.2.1) as follows

$$(3.2.1) \qquad d(\mu,\nu) := \sum_{m=0}^{\infty} 2^{-m}\left(1 \wedge |\langle\mu,f_m\rangle - \langle\nu,f_m\rangle|\right).$$

Since V is convergence determining it follows that μ_n converges to μ in the topology τ_w if and only if $d(\mu_n,\mu) \longrightarrow 0$ as $n\to\infty$.

Remark 3.2.2 If E is compact then $\langle\mu_n,1\rangle \longrightarrow m < \infty$ implies that $\{\mu_n\}$ is relatively compact. From this we can verify the fact that d is a complete metric if E is compact. To obtain a complete metric in the general case we begin by recalling that every Polish space E is Borel isomorphic to a compact metric space (cf. Cohn (1980) Theorem 8.3.6). We can then put a metric on E such that it becomes a compact metric space and such that the Borel sets for this metric coincide with the Borel subsets of E. We then choose a countable subset $V' \subset bp\mathcal{E}$ which is a dense subset of the space of functions which are *continuous in the new metric*. We can and will assume that $1\in V'$ and that the functions in V' are strictly positive. Then a metric defined as above but using V' in place of V is complete.

Lemma 3.2.3. (i) If $\mathcal{A}$ is a class of Borel sets in E closed under finite intersections and containing a basis for the topology on E, then

$\mathcal{M} = \sigma\{f_A : A\in\mathcal{A}\}$ where $f_A(\mu) = \mu(A)$.

(ii) Let V be as in Lemma 3.2.1. Then $\mathcal{M} = \sigma\{\langle .,f\rangle : f\in V\}$.

Proof. (i) For $f\in bC(E)$, the mapping $\mu \longrightarrow \langle\mu,f\rangle$ is continuous and hence $\mathcal{M}$-measurable. It is easy to check that $\{f : \mu \longrightarrow \langle\mu,f\rangle$ is $\mathcal{M}$-measurable$\}$ is closed under bp limits. Since $bp\mathcal{E}$ is the bp closure of $bC(E)$, this implies that $\mu \longrightarrow \mu(A)$ is $\mathcal{M}$-measurable for any $A\in\mathcal{E}$ and hence $\mathcal{M} \supset \sigma\{f_A : A\in\mathcal{A}\}$. On the other hand for each f of the form $f = \sum_{i=1}^{n} a_i\chi_{A_i}$ with $A_i\in\mathcal{A}$, $a_i \in \mathbb{R}$, the mapping $\mu \longrightarrow \langle\mu,f\rangle$ is $\sigma\{f_A : A\in\mathcal{A}\}$-measurable. By Dynkin's class theorem for functions ([EK p. 497]) the bp-closure of this class of functions contains $bC(E)$. Hence for $f\in bC(E)$, $\mu \longrightarrow \langle\mu,f\rangle$ is $\sigma\{f_A : A\in\mathcal{A}\}$-measurable which implies that $\mathcal{M} \subset \sigma\{f_A : A\in\mathcal{A}\}$.

(ii) The proof is similar to that of (i). □

Lemma 3.2.4. Let $\mathcal{A}$, V be as in Lemma 3.2.3. Then a probability measure, X, on $(M_F,\mathcal{M})$ is uniquely determined by the "finite dimensional distributions" of

43

$\{X(A_1),\ldots,X(A_n)\colon\ A_i\in\mathcal{A}\}$ denoted by $\{P_{A_1\ldots A_n}:n\in\mathbb{N}\}$ or equivalently by $\{\langle X,f_1\rangle,\ldots,\langle X,f_n\rangle : f_i\in V,\ i=1,\ldots,n,\ n\in\mathbb{N}\}$ denoted by $\{P_{f_1,\ldots,f_n}:n\in\mathbb{N}\}$.

Proof. This follows immediately from Lemma 3.2.3. □

Let P be a probability measure on $(M_F(E),\mathcal{M})$. The associated *Laplace functional* is defined by

$$L(f) := \int_{M_F(E)} e^{-\langle\mu,f\rangle} P(d\mu), \quad f\in bp\mathcal{E}.$$

It is easy to check that $L(.)$ is continuous under bp-limits.

Let $\mathcal{V}\subset\mathcal{B}(M_F(E))$ be the linear span of $\{F_f(.):f\in V\}$ where V is as in Lemma 3.2.1 and $F_f(\mu) := e^{-\langle\mu,f\rangle}$. Note that $\mathcal{V}$ is an algebra of functions since V is closed under addition.

Lemma 3.2.5.

(i) The space of $\mathcal{M}$-measurable functions is the bp closure of $\mathcal{V}$ and $\mathcal{M} = \sigma\{F_f:f\in V\}$.

(ii) P is uniquely determined by $L(f)$, $f\in V$.

Proof. (i) Sets of the form $\{\mu:\langle\mu,f_i\rangle\in O_i,\ i=1,\ldots,n\}$ where O_i are open in $\mathbb{R}$, $f_i\in V$ form a fundamental set of neighbourhoods in $M_F(E)$. Note that the bounded pointwise closure of finite linear combinations of the form $\sum a_j e^{-jx}$, a_j rational, contains sets of the form 1_O, O open. Hence the indicator, 1_{O_i}, of each fundamental open set O_i belongs to the bp-closure of $\mathcal{V}$. The result follows by Dynkin's class theorem for functions (cf. [EK, p. 497]).

(ii) then follows from (i). □

Theorem 3.2.6 (i) $\{F_f:f\in V\}$ is convergence determining in $M_1(M_F(E))$, that is, if $\int F_f(\mu)P_n(d\mu) \longrightarrow \int F_f(\mu)P(d\mu)$ as $n\to\infty$ $\forall$ $f\in V$, with $P_n,P\in M_1(M_F(E))$, then P_n converges weakly to P.

(ii) There exists a countable set of strictly positive $\mathcal{E}$-measurable functions V' such that if $\int F_f(\mu)P_n(d\mu) \longrightarrow L(f)$ as $n\to\infty$ $\forall$ $f\in V'$, for $\{P_n\}$ in $M_1(M_F(E))$, then there exists $P\in M_{\le 1}(M_F(E))$ (the subprobability measures) such that

$$L(f) = \int_{M_F(E)} F_f(\mu)\ P(d\mu)\ \forall\ f\in V'.$$

Proof. (i) The linear span of $\{F_f:f\in V\}$ is an algebra. We will show that it is *strongly separating*, that is, for every $\delta>0$ and μ there exists $\{F_{f_1},\ldots,F_{f_N}\}$ such that

44

$$\inf_{\nu\in(B(\mu,\delta))^c} \max_{1\le i\le N} |F_{f_i}(\mu)-F_{f_i}(\nu)| > 0$$

where $B(\mu,\delta)$ is a ball with center μ and radius δ (with respect to the metric d). Let $\mu\in M_F(E)$ and $\delta>0$. Let $2^{-m} < \delta/4$. If $d(\nu,\mu) > \delta$, then

$\sum_{i=1}^m |\langle\nu,f_i\rangle-\langle\mu,f_i\rangle| \ge \delta/2$ and hence $\max_{1\le i\le m} |\langle\nu,f_i\rangle-\langle\mu,f_i\rangle| \ge \delta/2m$. But then $\max_{1\le i\le m} |F_{f_i}(\nu)-F_{f_i}(\mu)| \ge \text{const}\cdot\delta$, where the constant can be chosen independent of ν. Hence the $\{F_f\}$ are strongly separating. The result then follows since a strongly separating algebra of functions is convergence determining (cf. [EK p. 113]).

(ii) We can put a new metric on E making it a compact metric space and for which the Borel sets coincide with $\mathcal{E}$. We then choose (as in Remark 3.2.2) a countable set V' of strictly positive functions on E which are dense in the space of nonnegative functions continuous for the new metric and consequently convergence determining for $M_F(E)$. Recall that a sequence $\{P_n\} \in M_1(M_F(E))$, E compact, can be viewed as elements of $M_1(\overline{M_F(E)})$ where $\overline{M_F(E)}$ denotes the Watanabe compactification of $M_F(E)$. The result (ii) follows by noting that $M_1(\overline{M_F(E)})$ is compact, every element of $M_1(\overline{M_F(E)})$ gives a subprobability on $M_F(E)$, with the subsequent application of the continuity theorem for Laplace transforms and (i). The details are left to the reader.□

<u>Corollary 3.2.7.</u> Let $P_n\in M_1(M_F(E))$. To prove that $P_n \xrightarrow{w} P$ as $n\to\infty$ it suffices to show that

(i) $\{P_n\}$ are tight, that is, for every $\varepsilon>0$ there exists a compact subset $K_\varepsilon \subset M_F(E)$, such that $\sup_n P_n(K_\varepsilon^c) < \varepsilon$,

(ii) $L_n(f) = \int F_f(\mu)P_n(d\mu)$ converges as $n\to\infty$ for each $f\in V$.

Given $P\in M_1(M_F(E))$, we define the moment measures, when they exist, as follows: $M_n \in M_F(E^n)$,

$$M_n(A_1\times\ldots\times A_n) := \int_{M_F(E)} \mu(A_1)\ldots\mu(A_n)P(d\mu), \quad A_1,\ldots,A_n \in \mathcal{E}.$$

Subset A of $M_F(E)$ is *tight* if

(i) $\sup_{\mu\in A} \mu(E) < \infty$, and

(ii) for every $\varepsilon>0$ there exists a compact subset $K_\varepsilon \subset E$ such that $\sup_{\mu\in A} \mu(K_\varepsilon^c) < \varepsilon$.

<u>Lemma 3.2.8.</u> Let $\{P_n\} \subset M_1(M_F(E))$. Consider the first moment measures $M_{1,n}(B) =$

45

$\int_{M_F(E)} \mu(B)P_n(d\mu),\ B\in \mathcal{E}.$

If the $\{M_{1,n}:n=1,2,\ldots\}$ exist and are tight in $M_F(E)$, then the $\{P_n\}$ are tight in $M_1(M_F(E))$.

Proof. Let $\ell := \sup_n M_{1,n}(E) < \infty$ and for $m\in \mathbb{N}$, K_m be a compact such that

$\sup_n M_{1,n}(K_m^c) < 1/2^m$. Let

$$C_m := \{\mu:\mu(E) \le 2^{m+1}\ell,\ \forall k\ge m+2,\ \mu(K_{2k}^c) \le 2^{-k}\}.$$

Then C_m is compact by Prohorov's criterion and

$$P_n(C_m^c) \le \frac{\ell}{2^{m+1}\ell} + \sum_{k=m+2}^{\infty} P(\{\mu:\mu(K_{2k}^c) > 2^{-k}\})$$

$$\le 2^{-(m+1)} + \sum_{k=m+2}^{\infty} 2^k/2^{2k} = 2^{-m} \quad \text{for all } n.$$

Hence the $\{P_n\}$ are tight by Prohorov's criterion.□

Theorem 3.2.9 (a) Let $P \in M_1(M_F(E))$. Then in order that P is uniquely characterized by its moment measures $\{M_n:n=1,2,\ldots\}$ it suffices to show that

$$(3.2.2)\qquad \sum_{n=1}^{\infty} [M_n(E^n)]^{-1/2n} = +\infty.$$

(b) Let $\{P_m:m\in\mathbb{N}\} \subset M_1(M_F(E))$ and all moment measures exist for each m. If for each k, the kth moment measures $M_{k,m}$ of P_m converge weakly to a finite measure M_k on E^k, and $\sum_{k=1}^{\infty} [M_k(E^k)]^{-1/2k} = +\infty$, then there exists a unique $P\in M_1(M_F(E))$ such that P_m converges weakly to P and P has moment measures $\{M_k\}$.

Proof. (a) This is an extension of the classical result of Carleman to the case of random measures (cf. Zessin (1983)).

(b) follows as in Zessin (1983, Theorem 2.2). □

Corollary 3.2.10. A sufficient condition is that

$$\limsup_{n\to\infty} \left(\frac{[M_n(E^n)]^{1/n}}{n}\right) < \infty.$$

Proof. See for example Breiman (1968, Prop. 8.49). □

3.3 Poisson Cluster Random Measures and the Canonical Representation of Infinitely Divisible Random Measures

The *Poisson random measure* (with intensity measure $\Lambda(.) \in M_F(E)$) is given by the Laplace functional

$$(3.3.1)\qquad L_\Lambda(\phi) = \exp\left(-\int_E (1 - e^{-\phi(x)})\Lambda(dx)\right),\quad \phi \in bp\mathcal{E}.$$

46

Let E_1, E_2 be Polish spaces. The *Poisson cluster random measure* with finite intensity measure Λ on E_1 and cluster law $\{P_x\}_{x\in E_1}$ on $M_F(E_2)$ such that $\int_{E_1} P_x \Lambda(dx) \in M_F(E_2)$, is the random measure on E_2 defined by

$$(3.3.2) \qquad L_{\Lambda,\{P_x\}}(\phi) := \exp\left(-\int_{E_1} (1 - P_x e^{-\langle\cdot,\phi\rangle})\Lambda(dx)\right),\ \phi\in bp\mathcal{E}(E_2).$$

A random measure X is said to be *infinitely divisible* if for every positive integer n, $X \overset{D}{=} X_1+\ldots+X_n$ where $X_1,\ldots,X_n$ are independent identically distributed random measures. The Poisson random measure and Poisson cluster random measures are infinitely divisible.

In fact the Poisson cluster random measure is closely related to the canonical representation of infinitely divisible random measures given in the following result.

Theorem 3.3.1. Let X be a random measure with values in $M_{LF}(E)$ with infinitely divisible law P. Then there exists a pair $(M,R) \in M_{LF}(E)\times M(M_{LF}(E))$ such that

(3.3.3a) $R(\{0\}) = 0$, and

$$(3.3.3b) \qquad \int_{M_{LF}(E)} (1-e^{-\mu(A)})R(d\mu) < \infty \quad \forall\ A\in \mathcal{E}_b$$

$$(3.3.3c) \qquad L(\phi) = E(e^{-\langle X,\phi\rangle}) = e^{-u(\phi)} \quad \forall\ \phi\in pb\mathcal{E}_b,$$

and the *log-Laplace functional*, $u(\cdot)$, is represented by

$$(3.3.3d) \qquad u(\phi) := \langle M,\phi\rangle + \int_{M_{LF}(E)} (1-e^{-\langle\nu,\phi\rangle})R(d\nu).$$

Conversely, any functional of this form is the Laplace functional of an infinitely divisible random measure.

Proof. See Matthes, Kerstan and Mecke (1978, Ch. 2), Kallenberg (1983) or Dawson (1991, Theorem 3.4.1).

The representation (3.3.3) is called the *canonical representation* and the measure R is called the *canonical measure* or the *KLM measure.*

To complete this section we introduce another generalization of the notion of Poisson cluster random measures. In this case the intensity Λ is itself a random measure on E_1, i.e. given by a probability measure P_I on $M_{LF}(E_1)$. Then the resulting random measure on E_2 has Laplace functional

$$(3.3.4) \qquad L_{I,\{P_x\}}(\phi) = \int\exp\left(-\int(1-P_x e^{-\langle\cdot,\phi\rangle})\Lambda(dx)\right)P_I(d\Lambda),\ \phi\in bp\mathcal{E}_{2,b}.$$

and is called the *Cox cluster (or doubly stochastic) random measure,* X, on E_2 with random intensity I on E_1 and clustering law $\{P_x: x\in E_1\}$. We assume that $x \to P_x$ is a measurable mapping from E_1 to $M_1(M_F(E_2))$. Moreover I is assumed to be a random measure on E_1 and X be a random measure on $E_1\times E_2$. Assume that the pair (I,X) has

47

joint probability law $P\in M_1(M_F(E_1)\times M_F(E_1\times E_2))$ such that conditioned on I, X is a Poisson cluster random measure with intensity I on E_1 and cluster distribution $\delta_x\times P_x$ on $E_1\times E_2$.

3.4. The Structure of Random Measures

3.4.1 Pure Atomic Random Measure

Every $\mu\in M_F(E)$ can be uniquely decomposed into $\mu = \mu_a+\mu_d$ where μ_a is a *pure atomic measure* and μ_d is a *diffuse* (nonatomic) measure, that is, $\mu_a = \sum a_j\delta_{x_j}$, and $\mu_d(\{x\}) = 0$ for every $x\in E$. We denote $\mu_a = \zeta_a(\mu)$ the mapping $\mu \to \mu_a$ defined on $M_F(E)$ and $M_a(E) := \{\mu\in M_F(E):\mu=\zeta_a(\mu)\}$.

Let $\bar{\mathfrak{A}}_\infty = \{(a_1,a_2,\dots):a_1\ge a_2\ge\dots\ge 0,\ \Sigma\ a_i \le 1\}$, and
$\mathfrak{A}_\infty = \{(a_1,a_2,\dots):a_1\ge a_2\ge\dots\ge 0,\ \Sigma\ a_i = 1\}$.

Define $\vartheta:M_1(E) \to \bar{\mathfrak{A}}_\infty$, by letting $\vartheta(\mu)$ be the vector of *descending order statistics* of the masses of the atoms of $\zeta_a(\mu)$ and let $|\vartheta(\mu)| \le \infty$ denote the number of non-zero atoms. We also define $\tilde{\vartheta}(\mu) := (\vartheta(\mu);x_1,\dots,x_{|\vartheta(\mu)|})$ if $\mu_a = \sum_{i=1}^{|\vartheta(\mu)|} a_i\delta_{x_i}$, $a_1\ge a_2\ge\dots$. Note that $\mu\in M_a(E)\cap M_1(E)$ if and only if $\vartheta(\zeta_a(\mu)) \in \mathfrak{A}_\infty$.

Let E be a separable metric space. A *partition family* for E is given by a collection $\{E^n_j,x^n_j,\rho_n;n,j\in\mathbb{N}\}$ where

(a) each $E^n_j \in \mathcal{E}$, $x^n_j \in E^n_j$, $E^n_j \cap E^n_k = \emptyset$ if $j\ne k$,

(b) $\bigcup_j E^n_j = E$ for each $n\in\mathbb{N}$,

(c) $\{E^{n+1}_j:j\in\mathbb{N}\}$ is a refinement of $\{E^n_j:j\in\mathbb{N}\}$, that is, each $E^{n+1}_j\subset E^n_k$ for some k,

(d) $\rho_n := \sup_j \operatorname{diam}(E^n_j) \to 0$ as $n\to\infty$.

Given $\eta>0$, and the partition family $\{E^n_j,x^n_j\}$ we define a mapping $\xi_n:M_F(E) \to M_F(E)$ as follows:

$$\xi_n(\mu) := \sum_j \mu(E^n_j)\delta_{x^n_j}.$$

Given $\eta > 1$, we define $\zeta_{\eta,n}:M_F(E) \to M_F(E)$ by

$$\zeta_{\eta,n}(\mu) := \sum_j \left(\mu(E^n_j)\right)^\eta \delta_{x^n_j}$$

48

Theorem 3.4.1.1 (a) $\zeta_a(\mu) = \lim_{\eta\downarrow 1}\lim_{n\to\infty} \zeta_{\eta,n}(\mu)$.

(b) The mapping $\mu \longrightarrow \mu_a$ is $\mathcal{B}(M_F(E))/\mathcal{B}(M_F(E))$ measurable.

(c) $M_a(E) := \{\mu:\mu(E) = \zeta_a(\mu)(E)\} \in \mathcal{B}(M_F(E))$.

(d) The mapping $\tilde{\vartheta}:M_1(E) \longrightarrow \bar{\mathcal{A}}_\infty \times \bigcup_{i=0}^{\infty} E^i$ is measurable.

Sketch of Proof. (a) It is easy to verify that if $\mu = \sum a_i\delta_{y_i}$, then

$$\lim_{\eta\downarrow 1}\lim_{n\to\infty} \zeta_{\eta,n}(\mu) = \mu.$$

On the other hand if μ is diffuse then

$$\lim_{\eta\downarrow 1}\lim_{n\to\infty} \zeta_{\eta,n}(\mu) = 0.$$

(b) The mappings $\{\zeta_{\eta,n}\}$ are clearly measurable since the mappings $\mu \longrightarrow \mu(E^n_j)$ are measurable (cf. Lemma 3.2.3). This implies that ζ_a is measurable.

(c) follows immediately from the definition.

(d) For $n\in\mathbb{N}$ let π_n denote the permutation of $\{1,2,3,\ldots\}$ such that

$$\pi_n j < \pi_n i \text{ if } \mu(E^n_j) > \mu(E^n_i)$$

$$\pi_n j < \pi_n i \text{ if } \mu(E^n_j) = \mu(E^n_i) \text{ and } j < i.$$

Let $\tilde{\vartheta}_n(\mu) := (\{\mu(E^n_j)\},\{x^n_j\})$. Clearly $\mu(E^n_{\pi_n^{-1}j}) \longrightarrow a_j$ and x_j is one of the limit points of the sequence of finite sets $\{x^n_{\pi_n^{-1}k} : a_k=a_j\}$. □

We will now give another application of this technique used in the last proof to a class of Cox random measures. Let $x \longrightarrow P_x$ be a measurable mapping from E_1 to $M_1(M_F(E_2))$. Let I be a random measure on E_1 and X be a random measure on $E_1\times E_2$. Assume that the pair (I,X) has joint probability law $P \in M_1(M_F(E_1)\times M_F(E_1\times E_2))$ such that conditioned on I, X is a Poisson cluster random measure with intensity I on E_1 and cluster distribution $\delta_x\times P_x$ on $E_1\times E_2$. Intuitively we can think of the formation of X in two stages, first the choice of a set of points in E_1 and then associating to each of these points a cluster. The following conditioning result allows us to distintegrate the measure in this way.

Let $\mathcal{G}_0 := \sigma\{1_{(0,\infty)}(X(A\times E_2)):A\in\mathcal{E}_1\}$, $\mathcal{G}_1 = \sigma\{I(A):A\in\mathcal{E}_1\}$, and choose a version of the conditional expectation $E(X|\mathcal{G}_0\vee\mathcal{G}_1)$ which is almost surely a random measure on E. Recall that the set $\{I:I \text{ is nonatomic}\} \in \mathcal{G}_1$ (cf. Theorem 3.4.1.1(c)).

Theorem 3.4.1.2. Let X be a Cox cluster random measure on E_2 with random intensity I on E_1 and cluster law $\{P_x:x\in E_1\}$ and let $\{A^n_i,x^n_i,\rho_n:n,i\in\mathbb{N}\}$ be a partition family for E_1 with $\rho_n= 1/n$. Then

(a) P-a.s. on $\{I:I \text{ is nonatomic}\}$

49

$$E(X|\mathcal{G}_0\vee\mathcal{G}_1) = \int\delta_x\times\left(\int\mu P_x(d\mu)\right)\tilde{X}(dx), \quad \text{and}$$

$$E\left(e^{-\langle X,\phi\rangle}|\mathcal{G}_0\vee\mathcal{G}_1\right) = \exp\left(\int\log\left(\int e^{-\int\phi(x,y)\mu(dy)}P_x(d\mu)\right)\tilde{X}(dx)\right), \quad \phi\in pb\mathcal{E}_1\times\mathcal{E}_2,$$

where $\tilde{X}$ is a random measure on E_1 satisfying

$$\tilde{X}(G) = \lim_{n\to\infty}\sum_{i=1}^{\infty}1(A_i^n\subset G)1_{(0,\infty)}(X(A_i^n\times E_2)) \quad \text{for } G \text{ open.}$$

(b) If I is a.s. nonatomic, then conditioned on I, $\tilde{X}$ is distributed as a Poisson random measure on E_1 with intensity I.

Proof. See [DP, Appendix, section 3].

3.4.2 Lebesgue Decomposition and Absolutely Continuous Random Measures.

Let X be a random measure defined on the probability space $(\Omega,\mathcal{F},P)$ and with values in $(M_F(E),\mathcal{B}(M_F(E)))$ where E is a Polish space and $\nu\in M_F(E)$. For each $\omega\in\Omega$ denote the Lebesgue decomposition of $X(\omega)$ with respect to ν by $X = \zeta^\nu_{ac}(X)(\omega) + \zeta^\nu_s(X)(\omega)$ where for each $\omega\in\Omega$, $\zeta^\nu_{ac}(X)(\omega) \ll \nu$ (*absolutely continuous part*) and $\zeta^\nu_s(X)(\omega) \perp \nu$ (*singular part*).

Lemma 3.4.2.1. Then $\omega \longrightarrow \zeta^\nu_{ac}(X)(\omega)$ and $\omega \longrightarrow \zeta^\nu_s(X)(\omega)$ are both measurable maps of $(\Omega,\mathcal{F})$ into $(M_F(E),\mathcal{B}(M_F(E)))$, that is, $X_{ac}(\nu)$ and $X_s(\nu)$ are random measures.

Proof. (See Dai Yonglong (1982) or Cutler (1984)). We will simply sketch the proof that ζ^ν_{ac} is measurable where $\zeta^\nu_{ac}(\mu) + \zeta^\nu_s(\mu)$ denotes the Lebesgue decomposition of $\mu\in M_F(E)$. It suffices to show that ζ^ν_{ac} is $\mathcal{B}(M_F(E))/\mathcal{B}(M_F(E))$-measurable. The first step is to show (cf. Cutler (1984, Theorem 2.1.3)) that $N := \{(\mu_1,\mu_2):\mu_1\ll\nu,\mu_2\perp\nu\}$ is a measurable subset of $M_F(E)\times M_F(E)$. The mapping $\psi:M_F(E)\times M_F(E) \longrightarrow M_F(E)$ defined by $\psi(\mu,\gamma) := (\mu+\gamma)$ is continuous and the usual Lebesgue decomposition theorem implies that $\psi|N$ is one-to-one. Since $M_F(E)\times M_F(E)$ and $M_F(E)$ are Polish spaces we can conclude by Kuratowski's theorem (cf. Parthasarthy (1967, p. 15) that $\psi(B)$ is measurable for any measurable subset of N. For $B\in\mathcal{E}$, let $(\zeta^\nu_{ac})^{-1}(\{\mu:\mu(B)<\alpha\}) = \psi(N\cap\{(\mu_1,\mu_2):\mu_1(B)<\alpha\}) \in \mathcal{B}(M_F(E))$ which yields the result. □

Lemma 3.4.2.2. Let X be a random measure defined on $(\Omega,\mathcal{F},P)$ with values in $(M_F(\mathbb{R}^d),\mathcal{B}(M_F(\mathbb{R}^d)))$, Assume that

(i) there exists a Borel subset $N\subset\mathbb{R}^d$ of Lebesgue measure zero such that for each $z\in\mathbb{R}^d\backslash N$ there is a sequence $\varepsilon_n(z)\longrightarrow 0$ as $n\to\infty$, and as $n\to\infty$ $\dfrac{X(B(z,\varepsilon_n))}{\varepsilon_n^d}$ converges in distribution to a random variable $\eta(z)$ with $E(\eta(z)) < \infty$,

50

(ii)

(3.4.2.1) $E(\langle X,\phi\rangle) := \int_{\mathbb{R}^d\setminus N} E(\eta(z))\phi(z)dz \quad \forall\ \phi \in bC(\mathbb{R}^d).$

Then X is almost surely an absolutely continuous measure on $\mathbb{R}^d$.

Proof. (cf. Dawson, Fleischmann and Roelly (1990)) From Lemma 3.4.2.1 the absolutely continuous component $X_{ac}(\omega) := \zeta_{ac}(X)(\omega)$ (with respect to Lebesgue measure) is measurable and the same is true for the singular component $X_s(\omega) := \zeta_\nu(X)$. Furthermore, for each ω by the Lebesgue density theorem the limit

$$(3.4.2.2)\quad \lim_{n\to\infty} \frac{X(B(z,\varepsilon_n))}{\varepsilon_n^d} = \lim_{n\to\infty} \frac{X_{ac}(B(z,\varepsilon_n))}{\varepsilon_n^d} = \eta_{ac}(\omega,z)$$

exists for all $z\in \mathbb{R}^d\setminus N(\omega)$ where $N(\omega)$ is a Borel subset of $\mathbb{R}^d$ of Lebesgue measure zero and $\eta_{ac}(\omega,z)$ is a version of the Radon-Nikodym derivative of $X_{ac}(\omega)$ with respect to Lebesgue measure. It is easy to verify that $\eta_{ac}:\Omega\times\mathbb{R}^d \to \mathbb{R}_+$ is $(\mathcal{F}\otimes\mathcal{B}(\mathbb{R}^d))$-measurable and (3.4.2.2) holds a.s. with respect to the product measure $P(d\omega)dz$. In particular, for almost all z, the relationships (3.4.2.2) are true with respect to convergence in distribution. Then by assumption (i), we conclude that $\eta_{ac}(.,z)$ coincides in distribution with $\eta(z)$, for almost all $z \in \mathbb{R}^d$. Therefore by (3.4.2.1)

$$\int_{M_F(\mathbb{R}^d)} P(d\omega)\int_{\mathbb{R}^d} \phi(z)X_{ac}(\omega,dz) = \int_{M_F(\mathbb{R}^d)} P(d\omega)\int_{\mathbb{R}^d} \phi(z)\eta_{ac}(\omega,z)dz =$$

$$= \int_{M_F(\mathbb{R}^d)} P(d\omega)\int_{\mathbb{R}^d} \phi(z)X(\omega,dz) \qquad \forall\ \phi\in pbC(\mathbb{R}^d).$$

Therefore since $X_{ac}(\omega) \le X(\omega)\ \forall\ \omega\in\Omega$, this implies that $X = X_{ac}$ P-a.s. and the proof is complete. □

3.5 Markov Transition Kernels and Laplace Functionals

Let E be a compact metric space and let $M_F(E)$ denote the space of finite Borel measures on E furnished with the weak topology. This space is a locally compact separable metric space and can be compactified as in section 3.1. Let ρ be a complete metric on $M_F(E)$, $C_0(M_F(E))$ denote the class of continuous functions that tend to zero at infinity.

We wish to consider possibly time inhomogeneous $M_F(E)$-valued Markov processes with transition probabilities $\mathcal{P}(s,t,\mu;d\nu)$. We can characterize a transition function on $M_F(E)$ in terms of a *Laplace transition functional*

$$L(s,t,\mu;\phi) = \int e^{-\langle\nu,\phi\rangle}\mathcal{P}(s,t,\mu;d\nu), \quad \phi \in bp\mathcal{E},$$

as follows:

(i) for s,t,μ fixed, $\phi \longrightarrow L(s,t,\mu;\phi)$ is the Laplace functional of a random measure on E,

(ii) for each $\phi\in bp\mathcal{E}$, $\mu \longrightarrow L(s,t,\mu;\phi)$ is $\mathcal{M}$-measurable,

(iii) $L(s,t,\mu;\phi) = \int \mathcal{P}(s,u,\mu;d\nu)L(u,t,\nu;\phi)$ for $s < u < t$.

In the time homogeneous case the transition function has the form

$$\mathcal{P}(t,\mu;d\nu) = \mathcal{P}(s,s+t,\mu;d\nu)$$

and we define the corresponding semigroup of contraction operators on $b\mathcal{B}(M_F(E))$ by

$$T_tF(\mu) := \int F(\nu)\mathcal{P}(t,\mu:d\nu).$$

The transition function is said to be *Feller* if

$$T_t:C_0(M_F(E))\longrightarrow C_0(M_F(E))$$

and $\|T_tF-F\| \longrightarrow 0 \;\forall\; F\in C_0(M_F(E))$.

<u>Lemma 3.5.1</u> In order that a time homogeneous transition function on $M_F(E)$ with Laplace transition functional L correspond to a Feller semigroup on $C_0(M_F(E))$ it is necessary and sufficient that

(i) for fixed $t > 0$, the mapping $\mu \longrightarrow \mathcal{P}(t,\mu,.)$ from $M_F(E)$ to $M_1(M(E_F))$ be continuous, or equivalently,

$$\mu \longrightarrow L(t,\mu,\phi) := \int e^{-\langle\nu,\phi\rangle}\mathcal{P}(t,\mu;d\nu)$$

is continuous for each $\phi \in V$ (where V is as in Theorem 3.2.1),

(ii) $\lim_{\mu\to\Delta} \mathcal{P}(t,\mu;.) = \delta_\Delta$, i.e. $\lim_{\mu\to\Delta} L(t,\mu,1) = 0 \quad \forall\; t\geq 0$,

(where Δ is as in the Watanabe compactification given in Sect. 3.1),

(iii) the mapping $t \longrightarrow \mathcal{P}(t,\mu;.)$ must be uniformly stochastically continuous at zero, i.e. for each $\varepsilon > 0$,

$$\lim_{t\downarrow 0} \sup_\mu [1 - \mathcal{P}(t,\mu;N_\varepsilon(\mu))] = 0,$$

where $N_\varepsilon(\mu) := \{\nu:\rho(\mu,\nu)<\varepsilon\}$.

<u>Proof.</u> See [Dynkin (1965 Vol. I, Chapt. 2, Sect. 5)].

<u>Corollary 3.5.2</u> Under the other hypotheses condition (iii) follows from the weaker condition that $\mathcal{P}(t,.,.)$ be stochastically continuous, i.e.

$$\lim_{t\downarrow 0} \mathcal{P}(t,\mu;.) = \delta_\mu, \text{ or equivalently}$$

$$\lim_{t\downarrow 0} L(t,\mu,\phi) = e^{-\langle\mu,\phi\rangle} \text{for each } \mu \in M_F(E) \text{ and } \phi \in V.$$

<u>Proof.</u> [Dynkin (1965, Vol. I, Lemma 2.10)].

Given a Feller transition function there exists a càdlàg *canonical realization*

52

of the measure-valued process (cf. [EK, Ch. 4, Th. 2.7]) which is denoted by $(\mathbf{D},\mathcal{D},(\mathcal{D}_t)_{t\ge0},(X_t)_{t\ge0},(P_\mu)_{\mu\in M_F(E)})$ where $\mathbf{D} = D([0,\infty),M_F(E))$, $X_t:\mathbf{D} \to M_F(E)$, $X_t(\omega) = \omega(t)$ for $\omega\in\mathbf{D}$, $\mathcal{D}_t^o = \sigma(X_s:0\le s\le t)$, $\mathcal{D}_t = \mathcal{D}^o_{t+}$, $\mathcal{D} = \vee\mathcal{D}_t$, P_μ is a probability measure on $\mathcal{D}$, the Borel σ-algebra on $\mathbf{D}$, and the $\{P_\mu:\mu\in M_F(E)\}$ are strong Markov with transition function $\mathcal{P}(t,.;.)$.

We will also require a class of time inhomogeneous Markov process with Polish state space given by the following definition.

<u>Definition</u>. Let $(E,\mathcal{E})$ be a Polish space with its Borel σ-field. Let $\hat{E} \in \mathcal{B}([0,\infty))\times\mathcal{E}$ and set $E^t = \{x:(t,x)\in\hat{E}\}$. ($\hat{E}$ is called the *global state space.*) Let $(\Omega,\mathcal{F})$ be a measurable space and for $0\le s\le t<\infty$ let $\mathcal{F}^o_{[s,t]}$ be a sub-σ-algebra of $\mathcal{F}$ and if $[s,t]\subset[u,v]$ the $\mathcal{F}^o_{[s,t]}\subset\mathcal{F}^o_{[u,v]}$. Let $\mathcal{F}_{s,t} := \bigcap_{\varepsilon>0}\mathcal{F}^o_{[s,t+\varepsilon]}$ and $\mathcal{F}_{s,\infty} := \bigvee_{t\ge s}\mathcal{F}_{s,t}$. If $\tau\ge s$ is a $(\mathcal{F}_{s,t})_{t\ge s}$ stopping time,

$$\mathcal{F}_{s,\tau} := \{A\in\mathcal{F}_{s,\infty}: A\cap\{\tau\le t\}\in\mathcal{F}_{s,t}\ \forall\ t\ge s\}.$$

Then $Z := (\Omega,(\mathcal{F}_{s,\infty}),(\mathcal{F}_{s,t})_{t\ge s},\{Z_t\},(P_{s,z})_{z\in E^s\ge0})$ is called an *inhomogeneous Borel strong Markov process* (IBSMP) with càdlàg paths in $E^t\subset E$ iff:

(i) $\forall\ t\ge0$, $Z_t:(\Omega,\mathcal{F}^o_{[t,t]})\longrightarrow(E,\mathcal{E})$ is measurable and satisfies $\forall\ (s,z)\in\hat{E}$,

(ii) $\forall\ (s,z)\in\hat{E}$, $P_{s,z}$ is a probability measure on $(\Omega,\mathcal{F}_{s,\infty})$ such that

$$P_{s,z}(Z(s)=z,\ Z_t\in E^t\ \forall\ t\ge s \text{ and } Z_\bullet \text{ is càdlàg on } [t,\infty)) = 1,$$

and for all $A\in\mathcal{F}_{u,\infty}$,

$$(s,z)\longrightarrow P_{s,z}(A) \text{ is Borel measurable on } \hat{E}\cap([0,u]\times E),$$

(iii) if $(s,z)\in\hat{E}$, $\psi\in b\mathcal{B}([s,\infty)\times D([s,\infty),E))$ and $\tau\ge s$ is a stopping time with respect to $(\mathcal{F}_{s,t})_{t\ge s}$, then $P_{s,z}$-a.s. on $\{\tau<\infty\}$,

$$P_{s,z}(\psi(\tau,Z(\tau+\bullet))\,|\,\mathcal{F}_{s,\tau})(\omega) = P_{\tau(\omega),Z(\tau)(\omega)}(\psi(\tau(\omega),Z(\tau(\omega)+\bullet)))$$

(strong Markov property).

3.6 Weak Convergence of Processes

We have seen in Chapter 2 that measure-valued processes can arise as limits of particle systems. In order to carry out such limiting procedures it is necessary to develop tools to study the weak convergence of sequences of càdlàg canonical processes. In this section we review some of basic criteria for proving relative compactness for sequences of measure-valued processes. However since these criteria rely on general criteria for processes with values in Polish spaces we first consider the latter case.

Let (E,d) be a Polish space. Let Λ be the set of continuous, strictly increasing Lipschitz continuous functions from $[0,\infty)$ onto $[0,\infty)$ such that

$$\gamma(\lambda) := \sup_{s>t\geq 0} \left| \log \frac{\lambda(s)-\lambda(t)}{s-t} \right| < \infty.$$

For $\omega,\omega' \in D_E = D([0,\infty),E)$,

$$\rho(\omega,\omega') := \inf_{\lambda\in\Lambda} \left(\gamma(\lambda) + \int_0^\infty e^{-u}\left(1 \wedge \sup_{t\geq 0} d(\omega(t\wedge u),\omega'(\lambda(t)\wedge u))\right) du\right)$$

It is easy to verify that ρ defines a metric on D_E. The resulting topology is called the *Skorohod topology* (J_1-topology) on D_E. Let $\mathcal{D} := \mathcal{B}(D([0,\infty);E))$ denote the σ-algebra of Borel subsets with respect to this topology. Let $X_t(\omega) := \omega(t)$ for $\omega \in D$ and $\mathcal{D}_t^o := \sigma(X_s: 0\leq s\leq t)$, $\mathcal{D}_t := \bigcap_{\varepsilon>0} \mathcal{D}_{t+\varepsilon}^o \subset \mathcal{D}$.

Then $(D_E,\mathcal{D},(\mathcal{D}_t)_{t\geq 0},(X_t)_{t\geq 0})$ denotes the *canonical stochastic process* on E. For $f\in C(E)$, let $\tilde{f}: D_E \longrightarrow D([0,\infty),\mathbb{R})$ be defined by $(\tilde{f}x)(t) := f(x(t))$.

<u>Theorem</u> <u>3.6.1</u> Let (E,d) be a Polish space. Then the sequence $\{P_n\}$ of probability measures on D_E is relatively compact if and only if the following two conditions are satisfied:

(a) For every $\varepsilon>0$ and rational $t\geq 0$, there exists a compact set $K_{\varepsilon,t}$ such that

$$\sup_n P_n\{X(t)\in (K_{\varepsilon,t})^c\} < \varepsilon,$$

(b) For every $\varepsilon>0$ and $T>0$ there exists $\delta>0$ such that

$$\sup_n P_n\{w'(X,\delta,T)\geq\varepsilon\} \leq \varepsilon,$$

where for $x \in D_E$,

$$w'(x,\delta,T) := \inf_{\min|t_i-t_{i-1}|\geq\delta} \max_i \sup_{t_i\leq s<t\leq t_{i+1}} d(x(s),x(t))$$

where $0\leq t_i\leq T$.

<u>Proof.</u> See e.g. Billingsley (1968, Th. 15.2), [EK p. 128].

<u>Theorem</u> <u>3.6.2</u> (Kurtz's Tightness Criterion for $D([0,\infty);E)$)

Let (E,d) be a Polish space. Let $\{P_n\}$ be a sequence of probability measures on D_E satisfying condition (a) of Theorem 3.6.1 Assume that for some $\beta>0$, each n and T, and $\delta>0$ $\exists$ a non-negative random variable $\gamma_n^T(\delta)\geq 0$ such that

$$E_n\left([1\wedge d(X(t+u),X(t))]^\beta \,|\, \mathcal{D}_t\right) \leq E_n[\gamma_n^T(\delta)\,|\,\mathcal{D}_t], \quad 0\leq t\leq T,\ 0\leq u\leq\delta,$$

and $\lim\sup_{\delta\to 0} \sup_n E_n[\gamma_n^T(\delta)] = 0.$

Then $\{P_n\}$ is relatively compact.

54

Proof. [EK, p. 138].

Corollary 3.6.3. Let (E,d) be a Polish space and $\{P_\alpha\}$ be a family of probability laws on $(D_E,\mathcal{D},(\mathcal{D}_t)_{t\ge0},(X_t)_{t\ge0})$. Let D_0 be a dense subset of a subalgebra $C_a(E) \subset bC(E)$.

Assume that for each $f\in D_0$ there exists a càdlàg adapted process Z_f such that

$f(X(t)) - \int_0^t Z_f(s)ds$ is a $(\mathcal{D}_t,P_\alpha)$-martingale and that

$$P_\alpha\left(\operatorname{ess\,sup}_{s\le t} |Z_f(s)|\right) < \infty, \text{ or}$$

$$P_\alpha\left(\left(\sup_\alpha E\left(\int_0^t |Z_f(s)|^p ds\right)\right)^{1/p}\right) < \infty \text{ for each } t<\infty \text{ for some } p\in(1,\infty).$$

Then the family $\{P_\alpha\circ(\tilde f)^{-1}\}$ is tight in $D([0,\infty),\mathbb{R})$ for each $f\in C_a$.

Proof. See [EK, Ch. 3, Th. 9.4].

Theorem 3.6.4. (Jakubowski's criterion for tightness for $D([0,\infty),E)$)

Let (E,d) be a Polish space. Let $\mathbb{F}$ be a family of real continuous functions on E that separates points in E and is closed under addition, i.e. $f,g\in\mathbb{F} \Longrightarrow f+g\in\mathbb{F}$. Given $f\in\mathbb{F}$, $\tilde f:D_E \longrightarrow D([0,\infty),\mathbb{R})$ is defined by $(\tilde f x)(t) := f(x(t))$. A sequence $\{P_n\}$ of probability measures on D_E is tight iff the following two conditions hold:

(i) for each T>0 and $\varepsilon>0$ there is a compact $K_{T,\varepsilon}\subset E$ such that

$$P_n(D([0,T],K_{T,\varepsilon})) > 1-\varepsilon,$$

(ii) the family $\{P_n\}$ is $\mathbb{F}$-weakly tight, i.e. for each $f\in\mathbb{F}$ the sequence $\{P_n\circ(\tilde f)^{-1}\}$ of probability measures in $D([0,\infty),\mathbb{R})$ is tight.

Finally, if the sequence $\{P_n\}$ is tight, then it is relatively compact in the weak topology on $M_1(D_E)$.

Proof. Jakubowski (1986).

The criteria of Theorem 3.6.4 reduce the question of the relative compactness of $\{P_n\}$ on D_E to that for the real-valued case. A very useful criterion for the relative compactness in the latter case is given by the following well-known result of Aldous.

Theorem 3.6.5 (Aldous Conditions for tightness in $D([0,\infty),\mathbb{R})$)

Let $\{P_n\}$ be a sequence of probability measures on $D([0,\infty),\mathbb{R})$ and X be the canonical process. Assume that

(i) for each rational $t\ge0$, $P_n\circ X_t^{-1}$ is tight in $\mathbb{R}$,

(ii) given stopping times τ_n bounded by T and $\delta_n\downarrow0$ as $n\to\infty$, then

$$\lim_{n\to\infty} P_n(|X_{\tau_n+\delta_n} - X_{\tau_n}| > \varepsilon) = 0$$

55

or

(ii') $\forall\ \eta>0\ \exists\ \delta, n_0$ such that

$$\sup_{n\geq n_0}\ \sup_{\theta\in[0,\delta]}\ P_n(|X(\tau_n+\theta)-X(\tau_n)| > \varepsilon) \leq \eta.$$

Then the $\{P_n\}$ are tight.

Proof. Aldous (1978).

Based on this result Joffe and Métivier (1986) derived the following criterion for tightness of locally square integrable processes which we will now describe.

A càdlàg adapted process X, defined on $(\Omega,\mathcal{F},\mathcal{F}_t,P)$ with values in $\mathbb{R}$ is called a *D-semimartingale* if there exists an increasing càdlàg function A(t), a linear subspace $D(L)\subset C(\mathbb{R})$, and a mapping $L:(D(L)\times\mathbb{R}\times[0,\infty)\times\Omega) \longrightarrow \mathbb{R}$ with the following properties:

(ai) For every $(x,t,\omega)\in \mathbb{R}\times[0,\infty)\times\Omega$ the mapping $\phi \longrightarrow L(\phi,x,t,\omega)$ is a linear functional on D(L) and $L(\phi,.,t,\omega)\in D(L)$,

(aii) for every $\phi\in D(L)$, $(x,t,\omega) \longrightarrow L(\phi,x,t,\omega)$ is $\mathcal{B}(\mathbb{R})\times\mathcal{P}$-measurable, where $\mathcal{P}$ is the predictable σ-algebra on $[0,\infty)\times\Omega$, ($\mathcal{P}$ is generated by the sets of the form $(s,t]\times F$ where $F\in \mathcal{F}_s$ and s,t are arbitrary)

(bi) for every $\phi\in D(L)$ the process M^ϕ defined by

$$M^\phi(t,\omega) := \phi(X_t(\omega)) - \phi(X_0(\omega)) - \int_0^t L(\phi,X_{s-}(\omega),s,\omega)dA_s$$

is a locally square integrable martingale on $(\Omega,\mathcal{F},\mathcal{F}_t,P)$,

(bii) The functions $\psi(x) := x$ and ψ^2 belong to D(L).

The functions

$$\beta(x,t,\omega) := L(\psi,x,t,\omega)$$

$$\alpha(x,t,\omega) := L((\psi)^2,x,t,\omega) - 2x\beta(x,t,\omega)$$

are called the *local coefficients of first and second order.*

Theorem 3.6.6 (Joffe-Métivier Criterion for tightness of D-semimartingales)
Let $X^m = (\Omega^m,\mathcal{F}^m,\mathcal{F}_t^m,P^m)$ be a sequence of D-semimartingales with common D(L) and associated operators L^m, functions A^m, β_m, and α_m. Then the sequence $\{X^m:m\in\mathbb{N}\}$ is tight in $D([0,\infty),\mathbb{R})$ provided the following conditions hold:

(i) $\sup_m E|X_0^m|^2 < \infty$

(ii) there is a K>0 and a sequence of positive adapted processes $\{C_t^m:t\geq0\}$ (on Ω^m for each m) such that for every $m\in\mathbb{N}$, $x\in\mathbb{R}$, $\omega\in\Omega^m$

(a) $|\beta_m(x,t,\omega)|^2 + \alpha_m(x,t,\omega) \leq K(C_t^m(\omega)+x^2)$

(b) for every T>0

56

$\sup_m \sup_{t\in[0,T]} E[C^m_t] < \infty$ and $\lim_{k\to\infty} \sup_m P^m(\sup_{t\in[0,T]} C^m_t \geq k) = 0$

(iii) there exists a positive function γ on $[0,\infty)$ and a decreasing sequence of numbers $\{\delta_m\}$ such that $\lim_{t\to 0} \gamma(t) = 0$, $\lim_{m\to\infty} \delta_m = 0$, and for all $0<s<t$ and all m,

$$(A^m(t)-A^m(s)) \leq \gamma(t-s) + \delta_m .$$

Further, if we set $M^m_t := X^m_t - X^m_0 - \int_0^t \beta_m(X^m_{s-},s,.)dA^m_s$, then for each T>0 there a constant K_T and m_0 such that for all $m\geq m_0$,

(iv) $E(\sup_{t\in[0,T]} |X^m_t|^2) \leq K_T(1+E|X^m_0|^2)$

and

(v) $E(\sup_{t\in[0,T]} |M^m_t|^2) \leq K_T(1+E|X^m_0|^2)$.

Proof. Joffe and Métivier (1986).

Corollary 3.6.7. Assume that for T>0 there is a constant K_T such that

$$\sup_m \sup_{\substack{t\leq T\\ x\in\mathbb{R}}} (|\alpha_m(t,x)|+|\beta_m(t,x)|) \leq K_T \text{ a.s.},$$

$$\sup_m (A^m(t)-A^m(s)) \leq K_T(t-s) \text{ if } 0\leq s\leq t\leq T,$$

$$\sup_m E|X^m_0|^2 < \infty,$$

and M^m_t is a square integrable martingale with $\sup_m E(|M^m_T|^2) \leq K_T$.

Then the $\{X^m : m\in\mathbb{N}\}$ are tight in $D([0,\infty),\mathbb{R})$.

3.7 Weak Convergence and Continuity of Measure-valued Processes.

Let $M_1[D(\mathbb{R}_+,M_F(E))]$, $M_1[D(\mathbb{R}_+,M_p(\dot{\mathbb{R}}^d)]$ and $M_1[D(\mathbb{R}_+,\mathbb{R})]$ denote the spaces of probability measures on the respective Skorohod spaces with the topology of weak convergence. Note that different versions of the following theorem have been used by several authors (e.g. Roelly-Coppoletta (1986), Vaillancourt (1990)) and Gorostiza and Lopez-Mimbela (1990)).

Theorem 3.7.1 (a) Assume that E is compact and $\mathbb{F}$ is a dense subset of C(E) closed under addition. A sequence $\{P_n\} \subset M_1[D([0,\infty),M_F(E)]$ is tight if and only if $\{P_n\circ\tilde{f}_\phi^{-1}\}$ is tight in $M_1[D([0,\infty),\mathbb{R})]$ for each $\phi\in\mathbb{F}$ where $f_\phi(\mu) = \langle\mu,\phi\rangle$, and $\tilde{f}_\phi : D([0,\infty),M_F(E)) \longrightarrow D([0,\infty),\mathbb{R})$ is defined by $(\tilde{f}_\phi x)(t) := f_\phi(x(t))$.

57

(b) A sequence $\{P_n\}$ in $M_1[D([0,\infty),M_p(\dot{\mathbb{R}}^d))]$ is tight if and only if $\{P_n \circ \tilde{f}_\phi^{-1}\}$ is tight in $M_1[D([0,\infty),\mathbb{R})]$ for each $\phi \in K_p(\dot{\mathbb{R}}^d)$ and $f_\phi(\mu) := \langle\mu,\phi\rangle$ (where $K_p(\dot{\mathbb{R}}^d)$ is defined as in Sect. 3.1).

Proof. The proof of (a) is similar but easier than that of (b) so that we will only prove the latter. The necessity follows from the continuity of the mapping $\tilde{f}_\phi$. By hypothesis $\{P_n \circ \tilde{f}^{-1}_{\dot{\phi}_p}\}$ is tight in $D([0,T],\mathbb{R})$ for each $T > 0$. Hence for each $\varepsilon>0$ there exists a compact $K_\varepsilon \subset D([0,T],\mathbb{R})$ such that $P_n \circ \tilde{f}^{-1}_{\dot{\phi}_p}(K_\varepsilon) \geq 1-\varepsilon$. But then by the characterization of compact sets in $D([0,T],\mathbb{R})$ (cf. Theorem 3.6.4(i)) there exists $k_\varepsilon > 0$ such that $K_\varepsilon \subset D([0,T],[-k_\varepsilon,k_\varepsilon])$. Let $\Gamma_{T,\varepsilon} := \{\mu:\mu \in M_p(\dot{\mathbb{R}}^d): |\langle\mu,\dot{\phi}_p\rangle| \leq k_\varepsilon\}$ which is a compact subset of $M_p(\dot{\mathbb{R}}^d)$. But then $P_n(D([0,T];\Gamma_{T,\varepsilon})) = P_n \circ \tilde{f}^{-1}_{\phi_p}(D([0,T];[-k_\varepsilon,k_\varepsilon]) \geq 1-\varepsilon$ and hence condition (i) of Theorem 3.6.4 is satisfied. Now let $\mathbb{F}$ denote the class of functions on $M_p(\dot{\mathbb{R}}^d)$ of the form $f_\phi(\mu) := \langle\mu,\phi\rangle$ with $\phi \in K_p(\dot{\mathbb{R}}^d)$ and note that $\mathbb{F}$ separates points and is closed under addition. The result then follows by applying Theorem 3.6.4. □

It will also be useful to have a criterion which guarantees that a measure-valued process has continuous sample paths. We first recall the basic criterion for the path continuity which is due to Kolmogorov and Ibragimov.

Lemma 3.7.2. (a) Assume that a real-valued stochastic process $\{Y(t):t\geq 0\}$ satisfies

$$E|Y(t)-Y(s)|^r \leq C|t-s|^{1+\alpha}, \quad 0\leq s,t\leq T \tag{3.7.1}$$

for $r>0$, $\alpha>0$. Then (for fixed T) for any $\gamma \in (2,2+\alpha)$ and $\lambda>0$

$$P\left(\sup_{0\leq s<t\leq T} \frac{|Y(t)-Y(s)|}{|t-s|^\beta} \geq (8\gamma/(\gamma-2))(4\lambda)^{1/r}\right) \leq CA/\lambda \tag{3.7.2}$$

where $\beta = (\gamma-2)/r$ and A is a constant.

In particular Y has a β-Hölder continuous version for any $0 < \beta < \alpha/r$.

(b) Let $\{Z(t):t\in\mathbb{R}^d\}$ be a random field satisfying

$$E|Z(x)-Z(y)|^r \leq C|x-y|^{d+\alpha}$$

for $r>0$, $\alpha>0$. Then Z has a β-Hölder continuous version for any $0 < \beta < \alpha/r$.

Proof. See for example Stroock and Varadhan (1979, p. 49) or Walsh (1986, Chapt. 1, Cor. 1.2). □

Corollary 3.7.3. Let $\{X(t):t\geq 0\}$ be a process with values in $M_F(E)$, E compact, (resp. $M_p(\dot{\mathbb{R}}^d)$). Let $\{f_n\}$ be as in Lemma 3.2.1 (resp. 3.1.5)

58

and assume that

(3.7.3) $P\left(|\langle X(t),f_n\rangle - \langle X(s),f_n\rangle|^r\right) \le C|t-s|^{1+\alpha}$ for $r>1$, $\alpha>0$. Then $X \in C([0,\infty),M_F(E))$ (or $C([0,\infty),M_p(\dot{\mathbb{R}}^d))$, resp.), P-almost surely. In addition, for $\beta < \alpha/r$, there is a version of $\{X(t)t\ge 0\}$ that is β-Hölder continuous in the metric d defined by (3.2.1) (resp. (3.1.1)).

<u>Proof</u>. Let $Z_n := \sup_{0\le s<t\le T} \frac{|\langle X(t),f_n\rangle - \langle X(s),f_n\rangle|}{|t-s|^\beta}$ and $2 > \lambda > 1$. Using (3.7.2) we obtain

$$\sum_{n=0}^{\infty} P\left(Z_n \ge (8\gamma/(\gamma-2))(4)^{1/r}\lambda^{n/r}\right) \le \sum_{n=0}^{\infty} CA/\lambda^n < \infty.$$

Then by the Borel-Cantelli lemma,

$$\sup_{0\le s<t\le T} \frac{d(X(t),X(s))}{|t-s|^\beta} \le \sum Z_n/2^n < \infty, \quad \text{P-a.s.} \ \square$$

59

4. MEASURE-VALUED BRANCHING AND LOG-LAPLACE FUNCTIONALS

4.1 Some Introductory Remarks

Branching processes form one of the basic classes of stochastic processes and arise in a wide variety of applications. In the context of space-time stochastic models branching particle systems have served an important role in providing a rich class of mathematically tractable models. In this chapter we will investigate their measure-valued versions which first appeared in the work of Jirina (1964) and Watanabe (1968). The theory of measure-valued branching processes has developed into a mature subject. The reason for this is two-fold. In the first place there are three complementary mathematical structures which can be exploited - these are the log-Laplace nonlinear evolution equation to which one can apply analytical results, the family structure to which one can apply probabilistic tools and the martingale problem characterization to which one can apply the tools of stochastic calculus. In the second place measure-valued branching processes exhibit interesting dimension dependent local spatial structure and in addition when provided with spatially homogeneous initial conditions interesting long-time scale behavior.

In section 2.10 we introduced a class of branching particle systems and indicated how the corresponding measure-valued limit could be obtained using the methods described in Chapter 2. In this chapter we return to the study of branching systems but for a variety of reasons the setting of section 2.10 is too restrictive for our needs. We will generalize this in a number of directions as follows (together with our reasons for doing so):

(i) we will replace the space and time homogeneous motion and branching mechanism by space and time dependent mechanisms - this is needed for example for the study of branching when the offspring distribution is modulated by a random medium;

(ii) we will replace the "deterministic" clock which determines the branching times by allowing each particle to have its own clock which is given by an additive functional of its trajectory - this is used in the formulation of branching in fractal media,

(iii) we will replace the branching mechanism with finite third moment assumption by a general one in which the offspring distribution need not have even second moments;

(iv) we will replace the assumption that E is a compact metric space with the assumption that E is a Polish space - this is essential to our formulation of the historical process in Chapter 12;

(v) we will replace the space of finite measures by a space of σ-finite measures on E - this is essential to the study of general entrance laws and stationary distributions.

60

Having made all these extensions it is not surprising that the setting is now somewhat more complex and our methods are somewhat different. However even though the technical details are more complex, the simple heuristic ideas of Chapter 2 remain as our guiding principle.

4.2 Markov Transition Kernels and Log-Laplace Semigroups

Let E be a compact metric space and let $M_F(E)$ denote the space of finite Borel measures on E furnished with the weak topology.

An $M_F(E)$-valued Markov process $\{X_t\}$ with Laplace transition functional $L(s,t,\mu;\phi)$ is said to be a *measure-valued branching process* if it can be represented in the form

$$L(s,t,\mu;\phi) = E\left(e^{-\langle X_t,\phi\rangle} \,|X_s=\mu\right) = e^{-\int V_{s,t}\phi(x)\mu(dx)}, \quad \phi\in pb\mathcal{E},\ \mu\in M_F(E)$$

where the family of nonlinear operators $\{V_{s,t}: s\le t\}$ on $pb\mathcal{E}$ satisfies two basic structural properties:

(i) it forms a time inhomogeneous semigroup, that is, for $r\le s\le t$,

(4.2.1) $\quad V_{r,s}(V_{s,t}) = V_{r,t},\ V_{t,t} = I$, and

(ii) for each s,t and $\mu\in M_F(E)$, the mapping $\phi \longrightarrow \int V_{s,t}\phi(x)\mu(dx)$ is the log-Laplace functional of a random measure on E.

A family of operators $\{V_{s,t}\}$ satisfying (i) and (ii) is called a *log-Laplace semigroup* (or *cumulant, Skorohod*, or *Ψ-semigroup*) and determines the finite dimensional distributions of a $M_F(E)$-valued Markov process (whose transition functional is given as above). The theory of Ψ-semigroups was developed in Watanabe (1968) and Silverstein (1969). In this section we establish the existence of measure-valued branching processes that includes most of the currently studied classes. For recent developments on the characterization a general class of measure-valued branching processes see Dynkin, Kuznetsov and Skorohod (1992).

4.3. A General Class of Branching Systems

The main result of this section is to establish the existence of a wide class of log-Laplace semigroups as solutions of a class of nonlinear evolution equations, which we will call "log-Laplace equations". For the moment we consider the homogeneous case and then this is a *mild* (or evolution) equation

$$\text{(4.3.1)}\qquad v(t,x) = S_t\phi(x) + \int_0^t S_{t-s}\Phi(v(s,.))(x)ds, \quad \text{for } \phi\in pb\mathcal{E}$$

where A is the generator of a semigroup $\{S_t\}$ on $pb\mathcal{E}$, and

$$(4.3.2)\qquad \Phi(\lambda) = -\tfrac{1}{2}c\lambda^2 + b\lambda + \int_0^\infty\left(1 - e^{-\lambda u} - \frac{\lambda u}{1+u}\right)n(du)$$

and n(du) is a measure on $(0,\infty)$ satisfying $\int_0^1 u^2 n(du) < \infty$.

Although we do not require the solutions to be differentiable, it is convenient to carry out formal calculations using the following (formally equivalent) differential equation

$$(4.3.3)\qquad \frac{\partial v(t,x)}{\partial t} = Av(t,x) + \Phi(v(t,x)) \quad \textit{(log-Laplace Equation)}$$

$$v(0,x) = \phi(x).$$

(Such formal calculations can then be justified using the mild equation.)

In this chapter we restrict our attention to the *critical case*:

$$S_t 1 = 1,\ b(.) \equiv 0,\ 0 \le c < \infty,$$

$$(4.3.4)\qquad \left(\int_1^\infty u\ n(du) + \int_0^1 u^2 n(du)\right) < \infty,$$

and thus

$$(4.3.5)\qquad \Phi(\lambda) = -\tfrac{1}{2}c\lambda^2 + \int_0^\infty (1 - e^{-\lambda u} - \lambda u)n(du),\ \lambda\ge 0.$$

The restriction $b \equiv 0$ is made for notational simplicity at this point. In Example 10.1.2.2 we will show that non-critical branching can be obtained from critical branching using a Cameron-Martin-Girsanov formula. However if the condition on n is removed new phenomena such as explosions can occur which we do not consider.

Note that for each x, $\Phi(x,.)$ is the log-Laplace function of an infinitely divisible random variable. If in addition we assume that $\int_0^1 un(du) < \infty$, then $e^{-\Phi(\lambda)} = E\left(e^{-\lambda(Z_1+Z_2-E(Z_2))}\right)$ where Z_1 is a normal random variable and Z_2 is a non-negative infinitely divisible random variable. Heuristically, in this case the continuous state branching process which will be constructed below can be interpreted as the corresponding infinitely divisible process run with a clock speed proportional to the current mass.

We next describe the formulation of the general class of critical inhomogeneous superprocesses which we will consider. The basic ingredients are a motion process, branching rate, and branching mechanism.

Motion Process:

Let E be a Polish space, and $D_E = D([0,\infty),E)$ the space of càdlàg functions with the Skorohod topology as defined in Section 3.6. For $\omega\in D_E$ and $t\ge 0$, $W_t(\omega) := \omega(t)$.

62

For each $t\geq 0$, let there be given $E^t \subset E$ and define $\hat{E} := \{(s,x): x\in E^s\}$, and $\hat{D}_E := \{\omega\in D_E, E),\ \omega_t\in E^t\ \forall t\}$, $\mathcal{E}^t := \mathcal{E}\cap E^t$. We assume that $\hat{E} \in \mathcal{B}([0,\infty))\otimes\mathcal{E}$ and $\hat{\mathbf{D}}$ is a Borel subset of D_E. For $0\leq s\leq t$, $\mathcal{D}_{s,t} := \bigcap_{\varepsilon>0} \sigma\{W_u: s\leq u\leq t+\varepsilon\}$, $\mathcal{D} := \mathcal{D}_{0,\infty}$. For $s\geq 0$, let $D_{E,s} := D([s,\infty),E)$.

The process $(\hat{D}_E, \mathcal{D}, \mathcal{D}_{s,t}, \{W_t\}_{t\geq 0}, \{P_{s,x}\}_{s\geq 0, x\in E^s})$ is assumed to be a canonical *time-inhomogeneous Borel strong Markov process*, that is,

(i) $\forall (s,x)\in\hat{E}$, $P_{s,x}$ is a probability measure on $(\hat{D}_E, \mathcal{D}_{s,\infty})$ such that for all $A\in\mathcal{D}_{u,\infty}$, $(s,x) \longrightarrow P_{s,x}(A)$ is Borel measurable on $([0,u]\times E)\cap\hat{E}$ and $P_{s,x}(W_s = x) = 1$,

(ii) *(strong Markov property)* if $(s,x)\in\hat{E}$, $\psi\in b\mathcal{B}([s,\infty)\times D_{E,s})$ and $\tau\geq s$ is a $\{\mathcal{D}_{s,t}: t\geq s\}$-stopping time, then

$$P_{s,x}(\psi(\tau, W_{\tau+\cdot})\,|\,\mathcal{D}_{s,\tau})(\omega) = P_{\tau(\omega), W_{\tau(\omega)}}(\psi(\tau(\omega), W_{\tau(\omega)+\cdot}),$$

for $P_{s,x}$-a.e. ω on $\{\tau < \infty\}$.

The associated *inhomogeneous semigroup* $\{S_{s,t}\}$ is defined by:

$$S_{s,t}\phi(x) = P_{s,x}f(W_t),\quad 0\leq s\leq t,\ x\in E^s,\ \phi\in b\mathcal{E}.$$

Branching Rate:

Let $\kappa(dt)$ be a *positive continuous additive functional* of W, that is, $\kappa: \hat{D}_E \longrightarrow M_{LF}([0,\infty))$, such that $\kappa(r,t)$ is $\mathcal{D}_{r,t}$-measurable. In addition we assume that κ is a.s. absolutely continuous (with respect to Lebesgue measure) and that for $0\leq r\leq t$, $\sup_{x\in E^r} P_{r,x}(e^{\theta\kappa(r,t)}) < \infty\ \ \forall\ \theta > 0$.

Branching Mechanism: For each (t,x), $x\in E^t$, $\Phi(t,x,.)$ is assumed to be of the form

$$\text{(4.3.6)}\qquad \Phi(t,x,\lambda) = -\tfrac{1}{2}c(t,x)\lambda^2 + \int_0^\infty (1 - e^{-\lambda u} - \lambda u)n(t,x,du),\ \lambda\geq 0,$$

where $c \in pb(\mathcal{B}(\mathbb{R}_+)\otimes\mathcal{E})$, and n is a measurable mapping from $(\mathbb{R}_+, \mathcal{B}(\mathbb{R}_+))\times(E,\mathcal{E})$ to $M((0,\infty))$ satisfying

$$\sup_{t,x}\left(\int_1^\infty u\, n(t,x,du) + \int_0^1 u^2 n(t,x,du)\right) < \infty.$$

Note that

$$\text{(4.3.7)}\qquad \Phi(t,x,0) = 0,\ \Phi(t,x,\lambda) \leq 0,$$

$$\Phi'(t,x,\lambda) = -c(t,x)\lambda - \int_0^\infty (1-e^{-\lambda u})u.n(t,x,du) \leq 0,$$

63

$$\Phi''(t,x,\lambda) = - c(t,x) - \int_0^\infty e^{-\lambda u} u^2 n(t,x,du) \le 0,$$

$$(-1)^{k+1}\Phi^{(k)}(t,x,0) \ \ge 0, \ k\ge 2.$$

(Here ', " denote the first and second derivatives with respect to λ and $\Phi^{(k)}$ denotes the kth derivative of Φ with respect to λ.)

The Log-Laplace Equation.

The basic *log-Laplace equation* is given by

$$(4.3.8) \qquad V_{s,t}\phi(x) = P_{s,x}\left[\phi(W_t) + \int_s^t \kappa(dr)\Phi(r,W_r,V_{r,t}\phi(W_r))\right].$$

where $\phi \in bp\mathcal{E}^t$. We will prove that this equation has a unique solution and yields a log-Laplace semigroup.

Strategy of the Proof: We will prove

(i) uniqueness of the solution to (4.3.8) by an analytical argument,

(ii) the existence of a solution to the log-Laplace equation is obtained by a probabilistic method based on a branching particle approximation which yields a family of operators $V^\varepsilon_{s,t}$ which converge to a solution of (4.3.8), that is,

$$V_{s,t}\phi = \lim_{\varepsilon\downarrow 0} V^\varepsilon_{s,t}\phi,$$

(iii) the probabilisitic argument simultaneously shows that $\exp\left(-\int V_{s,t}\phi(x)\mu(dx)\right)$ is the log-Laplace functional of a random measure,

(iv) the semigroup property of $\{V_{s,t}\}$ is proved using the uniqueness of solutions to (4.3.8).

We begin with four technical lemmas.

Lemma 4.3.1 (Dynkin's generalized Gronwall inequality)

Assume that $h:[0,\infty)\times D_E \longrightarrow \mathbb{R}$ is progressively measurable such that

$\sup_{t,x} |h(t,x)| \le M$, and

$$(4.3.9) \qquad h(r,x) \le c_1 + c_2 P_{r,x}\int_r^t h(s,W)\kappa(ds) \quad \text{for} \quad r\in [t_0,t).$$

Then

$$(4.3.10) \quad h(r,x) \le c_1 P_{r,x}\, e^{c_2\kappa(r,t)} \qquad \text{for all} \quad r\in [t_0,t).$$

Proof. Using (4.3.9) and induction it is easy to verify that for each $n \ge 1$

$$h(r,x) \le c_1 \sum_{k=0}^{n} c_2^k\, P_{r,x}\int..\int 1(r<s_1<\ldots<s_k<t)\kappa(ds_1)\ldots\kappa(ds_k)$$

64

$$+ c_2^{n+1} P_{r,x} \int \cdot\cdot \int 1(r<s_1<\ldots<s_{n+1}<t)\kappa(ds_1)\ldots\kappa(ds_{n+1})h(s_{n+1}, W_{s_{n+1}})$$

$$= c_1 \sum_{k=0}^{n} c_2^k P_{r,x}\kappa(r,t)^k/k! + R_{n+1},$$

where $|R_{n+1}| \le M\cdot c_2^{n+1} P_{r,x}\kappa(r,t)^{n+1}/(n+1)! \longrightarrow 0$ and the proof is complete.□

Lemma 4.3.2 (Uniform Lipschitz property)

Under the above assumptions on c and n, given $\lambda_0>0 \ \exists \ K(\lambda_0)>0$ such that

$$|\Phi(t,x,\lambda_1)-\Phi(t,x,\lambda_2)| \le K(\lambda_0)|\lambda_1-\lambda_2| \quad \forall \ \lambda_1,\lambda_2 \in (0,\lambda_0).$$

Proof. If $\lambda_1 < \lambda_2$, then

$$|\Phi(t,x,\lambda_1)-\Phi(t,x,\lambda_2)| \le 2\lambda_0 \sup_{t,x} |\tfrac{1}{2}\, c(t,x)|\,|\lambda_1-\lambda_2|$$

$$+ \int_0^\infty |(e^{-\lambda_1 u}+\lambda_1 u)-(e^{-\lambda_2 u}+\lambda_2 u)|\,n(t,x,du).$$

Writing $|(e^{-\lambda_1 u}+\lambda_1 u)-(e^{-\lambda_2 u}+\lambda_2 u)| = \int_{\lambda_1}^{\lambda_2} u(1-e^{-\lambda u})d\lambda$, we verify that the second term

$$\le \int_{\lambda_1}^{\lambda_2} \left\{ \int_0^1 \lambda u^2 n(t,x,du) + \int_1^\infty u\; n(t,x,du)\right\} d\lambda$$

$$\le \sup_{t,x} \left(\int_0^\infty (u\wedge u^2) n(t,x,du)\right) |\lambda_1-\lambda_2|. \ \square$$

Lemma 4.3.3 Under the assumptions (4.3.6) the log-Laplace equation (4.3.8) has at most one solution.

Proof. Consider two solutions $V^1_{s,t}\phi(x)$ and $V^2_{s,t}\phi(x)$. Note that $\sup_x |V^i_{s,t}\phi(x)| \le \sup_x |\phi(x)| := \lambda_0$. Using Lemma 4.3.2 we obtain that

$$|V^2_{s,t}\phi(x) - V^1_{s,t}\phi(x)|$$

$$\le P_{s,x}\left[\int_s^t \kappa(dr)\left[|\Phi(r,W_r,V^2_{r,t}(W_r)) - \Phi(r,W_r,V^1_{r,t}(W_r))|\right]\right]$$

$$\le K(\lambda_0)P_{s,x}\left\{\int_s^t \left[|V^2_{r,t}(W_r) - V^1_{r,t}(W_r)|\right]\kappa(dr)\right\}.$$

Then Lemma 4.3.1 implies that

$$|V^2_{s,t}\phi(x) - V^1_{s,t}\phi(x)| = 0 \quad \text{for all} \quad s < t. \ \square$$

Lemma 4.3.4 Let $\mathcal{G}$ be a measurable function, $\mathcal{G}:[0,\infty)\times E\times[0,1]:\longrightarrow[0,1]$.

Then the equations

65

$$(4.3.11)\quad w_{r,t}(x) = P_{r,x}\left[e^{-\phi(W_t)}e^{-\kappa(r,t)} + \int_r^t e^{-\kappa(r,s)}\kappa(ds)\mathcal{G}(s,W_s,w_{s,t}(W_s))\right]$$

and

$$(4.3.12)\quad w_{r,t}(x) = P_{r,x}\left[e^{-\phi(W_t)} + \int_r^t \kappa(ds)\,\mathcal{G}(s,W_s,w_{s,t}(W_s)) - \int_r^t \kappa(ds)w_{s,t}(W_s)\right]$$

are equivalent.

Proof. Assume (4.3.12) and now calculate

$$P_{r,x}\left\{\int_r^t e^{-\kappa(r,s)}\kappa(ds)\mathcal{G}(s,W_s,w_{s,t}(W_s))\right\}$$

$$= P_{r,x}\left\{-\int_r^t e^{-\kappa(r,s)}\kappa(ds)\int_s^t \mathcal{G}(u,W_u,w_{u,t}(W_u))\kappa(du)\right.$$

$$\left. + \int_r^t \kappa(ds)\mathcal{G}(s,W_s,w_{s,t}(W_s))\right\} \textit{(integration by parts)}$$

$$= P_{r,x}\left\{-\int_r^t e^{-\kappa(r,s)}\kappa(ds)\left\{w_{s,t}(W_s) - e^{-\phi(W_t)} + \int_s^t \kappa(du)w_{u,t}(W_u)\right\}\right.$$

$$\left. + w_{r,t}(x) - e^{-\phi(W_t)} + \int_r^t \kappa(du)w_{u,t}(W_u)\right\}$$

((4.3.12) and Markov property)

$$= w_{r,t}(x) - P_{r,x}\left\{e^{-\kappa(r,t)}e^{-\phi(W_t)}\right\}$$

by simplification and another integration by parts. □

4.4 Branching Particle Approximations

In this section we will prove the existence of a solution to equation (4.3.8) and establish that it yields a log-Laplace semigroup.

Theorem 4.4.1 Given (W,κ,Φ) there exists a transition function on $M_F(E)$ with transition Laplace functional

$$(4.4.1)\quad P_{s,\mu}\left(e^{-\langle X(t),\phi\rangle}\right) = \exp\left(-\int (V_{s,t}\phi)(x)\mu(dx)\right),\quad \phi\in bp\mathcal{E}^t,\ \mu\in M_F(E),$$

where $V_{s,t}\phi$ is the unique solution of the equation (4.3.8).

The associated $M_F(E)$-valued time inhomogeneous Markov process is called the (W,κ,Φ)-*superprocess*.

66

Proof. We first obtain the existence of a solution to a closely related equation via a probabilistic construction.

Lemma 4.4.2 Consider a branching particle system $\{Z_t\}$ with motion process **W**, branching rate $\kappa(dr)$ and offspring generating function $\mathcal{G}(t,x,.)$, i.e.

$$\mathcal{G}(t,x,z) = \sum_{n=0}^{\infty} p(t,x,n)z^n,$$

$$\sup_{t,x} \sum_{n=0}^{\infty} n.p(t,x,n) \le K,$$

where $p(t,x,n)\kappa(dt)$ is the probability that n offspring are produced by a branch at time t at position x.

(a) Then the branching particle system exists (i.e. no explosion) and the distribution of Z_t starting at time r with one particle at x has Laplace functional

$$w_{r,t}(x) := P^Z_{r,\delta_x} e^{-\langle Z_t,\phi\rangle}$$

which satisfies

$$(4.4.2)\qquad w_{r,t}(x) = P_{r,x}\left[e^{-\phi(W_t)}e^{-\kappa(r,t)} + \int_r^t e^{-\kappa(r,s)}\kappa(ds)\mathcal{G}(s,W_s,w_{s,t}(W_s))\right].$$

(b) $P^Z_{r,x}(\langle Z_t,1\rangle) \le P_{r,x}(e^{(K-1)\kappa(r,t)})$.

Proof. (a) The non-explosion property follows from Harris (1963, Ch. 5, Th. 9.1). To obtain equation (4.4.2) we condition on the existence and the time of the first branch in the interval [s,t] (cf. Dawson and Ivanoff (1978)).□

(b) This follows by taking $\phi \equiv \theta$, in (4.3.12) differentiating with respect to θ, evaluating at θ=0 and then using Lemma 4.3.1.

Proof of Theorem 4.4.1. The proof will be dividied into four steps.

Step 1: The branching particle approximation. Given Φ satisfying (4.3.6) it is easy to check that for any $0 < \varepsilon < 1$, $0\le v\le 1$,

$$(4.4.3)\qquad \mathcal{G}^{\varepsilon}(t,x,v) := a(t,\varepsilon,x)^{-1}[a(t,\varepsilon,x)v - \varepsilon\Phi(t,x,(1-v)/\varepsilon)]$$

with $0 < a(t,\varepsilon,x) := -\Phi'(t,x,\varepsilon^{-1}) \le c_1/\varepsilon + c_2$

is the probability generating function of a non-negative integer-valued random variable with mean one. In fact using (4.3.6) and (4.3.7) it is easy to verify that $\mathcal{G}^{\varepsilon}(t,x,0) = -\varepsilon\Phi(t,x,1/\varepsilon) \ge 0$, $\mathcal{G}^{\varepsilon}(t,x,1) = 1$, $D\mathcal{G}^{\varepsilon}(t,x,v)|_{v=0} = 0$, $D\mathcal{G}^{\varepsilon}(t,x,v)|_{v=1} = 1$, and $D^k\mathcal{G}^{\varepsilon}(t,x,v)|_{v=0} \ge 0$ for all $k \ge 2$ (where D denotes differentiation with respect to v).

We now construct for each $\varepsilon > 0$ an approximating branching particle system $\{Z^{\varepsilon}_t\}$ to the (W,κ,Φ)-superprocess with initial measure μ at time r. We begin with an initial

67

Poisson random measure Z^ε_r with intensity μ/ε. Consider the resulting branching particle field, Z^ε_t with offspring generating function $\mathcal{G}^\varepsilon$, branching rate $\kappa^\varepsilon(ds) = a(s,\varepsilon,W_s)\kappa(ds)$, initial random measure Z^ε_r. and particle mass ε. Combining Lemma 4.4.2 with the Poisson cluster formula (3.3.2) we obtain the Laplace functional of the random measure $\varepsilon Z^\varepsilon_t$ as follows:

$$P_{r,\mathcal{P}ois(\mu/\varepsilon)}\, e^{-\langle Z^\varepsilon_t,\varepsilon\phi\rangle} = \exp\left(- \left\langle \mu, \frac{1-w^\varepsilon_{r,t}}{\varepsilon} \right\rangle\right)$$

where $w^\varepsilon_{r,t}$ is as in Lemma 4.4.2 with $\mathcal{G}$ replaced by $\mathcal{G}^\varepsilon$, and $w^\varepsilon_{r,t}$

$$w^\varepsilon_{r,t}(x) = P_{r,x}\left[e^{-\varepsilon\phi(W_t)}\, e^{-\kappa^\varepsilon(r,t)} + \int_r^t e^{-\kappa^\varepsilon(r,s)}\,\kappa^\varepsilon(ds)\,\mathcal{G}^\varepsilon(s,W_s,w^\varepsilon_{s,t}(W_s))\right].$$

Then by Lemma 4.3.4

$$w^\varepsilon_{r,t}(x) = P_{r,x}\left\{e^{-\varepsilon\phi(W_t)} + \int_r^t \kappa^\varepsilon(ds)\,\mathcal{G}^\varepsilon(s,W_s,w^\varepsilon_{s,t}(W_s)) - \int_r^t \kappa^\varepsilon(ds)\,w^\varepsilon_{s,t}(W_s)\right\}.$$

Putting $v^\varepsilon_{r,t}(x) = \dfrac{1-w^\varepsilon_{r,t}(x)}{\varepsilon}$ and substituting the expression for $\mathcal{G}^\varepsilon$ we get

$$(4.4.4)\quad v^\varepsilon_{r,t}(x) = P_{r,x}\left[\left(\frac{1-e^{-\varepsilon\phi(W_t)}}{\varepsilon}\right) + \int_r^t \kappa(ds)\,\Phi(s,W_s,v^\varepsilon_{s,t}(W_s))\right].$$

Then

$$(4.4.5)\quad 0 \le v^\varepsilon_{r,t}(x) \le P_{r,x}\left(\frac{1-e^{-\varepsilon\phi(W_t)}}{\varepsilon}\right) \le P_{r,x}\,\phi(W_t)$$

since $\Phi \le 0$.

<u>Step 2. The proof of convergence of log-Laplace functionals.</u>

<u>Lemma 4.4.3</u> $v^\varepsilon_{r,t}(x) \xrightarrow{\varepsilon\downarrow 0} v_{r,t}(x)$ exists (uniformly in x) and $v_{r,t}(x)$ is the unique solution to Equation (4.3.8).

<u>Proof.</u> The uniqueness follows from Lemma 4.3.3. It follows from (4.4.5) that $0 \le v^\varepsilon_{r,t}(.) \le \lambda_0$, provided that $0\le\phi\le\lambda_0$. Then using the Lipschitz property of Φ we obtain

68

$$|v^{\varepsilon_1}_{r,t} - v^{\varepsilon_2}_{r,t}|(x) \le P_{r,x} \left\| \frac{1-e^{-\varepsilon_1\phi(W_t)}}{\varepsilon_1} - \frac{1-e^{-\varepsilon_2\phi(W_t)}}{\varepsilon_2} \right\|_\infty$$

$$+ M(\lambda_0)P_{r,x}\int_r^t \kappa(ds)\|v^{\varepsilon_1}_{s,t}(W_s) - v^{\varepsilon_2}_{s,t}(W_s)\|_\infty.$$

Then by Lemma 4.3.1,

$$\|v^{\varepsilon_1}_{r,t} - v^{\varepsilon_2}_{r,t}\|_\infty \le \left\| \frac{1-e^{-\varepsilon_1\phi(.)}}{\varepsilon_1} - \frac{1-e^{-\varepsilon_2\phi(.)}}{\varepsilon_2} \right\|_\infty \sup_x P_{r,x} e^{M(\lambda_0)\kappa(r,t)}.$$

This yields the existence of the limit $v_{r,t}$. The fact that $v_{r,t}$ satisfies the equation (4.3.8) then follows from (4.4.3) by a bounded convergence argument. □

Step 3. Proof of the Semigroup Property of V.

We have $V_{r,t}\phi(y) = P_{r,y}[\phi(W_t) + \int_r^t \kappa(du)\Phi(u,W_u,V_{u,t}\phi(W_u))]$, $y\in E^r$,

$$V_{s,r}[V_{r,t}\phi](x)$$

$$= P_{s,x}[(V_{r,t}\phi)(W_r)+\int_s^r \kappa(du)\Phi(u,W_u,(V_{u,r}V_{r,t}\phi)(W_u))]$$

$$=P_{s,x}[P_{r,W_r}\phi(W_t) + P_{r,W_r}\int_r^t \kappa(du)\Phi(u,W_u,V_{u,t}\phi(W_u))]$$

$$+ P_{s,x}[\int_s^r \kappa(du)\Phi(u,W_u,(V_{u,r}V_{r,t})\phi(W_u))]$$

$$= P_{s,x}[\phi(W_t) + \int_s^r \kappa(du)\Phi(u,W_u,(V_{u,r}V_{r,t})\phi(W_u))$$

$$+ \int_r^t \kappa(du)\Phi(u,W_u,V_{u,t}\phi(W_u))] \quad \textit{(by Markov property of W)}.$$

Hence $U_{u,t}\phi := V_{u,t}\phi$ for $u\ge r$

$:= V_{u,r}V_{r,t}\phi$ for $u<r$

satisfies the same log-Laplace equation as does $V_{u,t}\phi$, $\forall u$. Therefore applying uniqueness yields $V_{s,t} = V_{s,r}V_{r,t}$ for $s<r<t$.□

Step 4. Completion of the proof. We now complete the proof of Theorem 4.4.1. First note that the first moment measures, $P_{r,\mathcal{Pois}(\mu/\varepsilon)}(\varepsilon Z^\varepsilon_t(dx)) = S_{r,t}\mu \in M_F(E)$ due to the criticality of the branching and hence are tight by Lemma 3.2.8. Therefore the random measures $\{\varepsilon Z^\varepsilon_t:\varepsilon > 0\}$ are also tight. Together with Lemma 4.4.3 and Theorem 3.2.6 this implies that $X^\varepsilon_t := \varepsilon\cdot Z^\varepsilon_t$ converges weakly to a random measure with Laplace functional (4.4.1). Note that as $\varepsilon \downarrow 0$, $\varepsilon Z^\varepsilon_r$ converges weakly to μ and that $\mu \longrightarrow \int(V_{s,t}\phi)(x)\mu(dx)$ is a $\mathcal{B}(M_F(E))$-measurable function. This completes the proof that (4.4.1) defines a Laplace transition functional. □

Corollary 4.4.6. As $\varepsilon \to 0$, the processes $\{X^{\varepsilon}_t\}$, $X^{\varepsilon}_t := \varepsilon Z^{\varepsilon}_t$ converge in the sense of finite dimensional distributions to the process $\{X(t)\}$ whose finite dimensional distributions are determined by the joint Laplace functional:

for $s \le t_1 \le \ldots \le t_n$, $\phi_1,\ldots,\phi_n \in pb\mathcal{E}$,

$$(4.4.6)\quad P_{s,\mu}\left(\exp\left(-[\langle X(t_1),\phi_1\rangle+\ldots+\langle X(t_n),\phi_n\rangle]\right)\right) = \exp\left(-\langle \mu, V_{t_1,\ldots,t_n}(\phi_1,\ldots,\phi_n)\rangle\right)$$

and V is recursively defined by $V_{t_1}(\phi_1) = V_{s,t_1}(\phi_1)$ with V_{s,t_1} given by the solution of (4.3.8), and

$$V_{t_1,\ldots,t_n}(\phi_1,\ldots,\phi_n) = V_{t_1,\ldots,t_{n-1}}(\phi_{n-1}+V_{t_{n-1},t_n}\phi_n).$$

Proof. (cf. Gorostiza and López-Mimbela (1992)) The proof is by induction on n. The proof in the case n=1 follows from the argument in the proof of Theorem 4.4.1. Letting $X^{\varepsilon} := \varepsilon Z^{\varepsilon}$, we have applying the Markov property of X^{ε}, and (4.4.3), that

$$P_{s,\mathcal{P}ois(\mu.\varepsilon)}\left(\exp\left(-[\langle X^{\varepsilon}(t_1),\phi_1\rangle+\ldots+\langle X^{\varepsilon}(t_n),\phi_n\rangle]\right)\right)$$

$$= P_{s,\mathcal{P}ois(\mu.\varepsilon)}\Big(\exp\Big(-[\langle X^{\varepsilon}(t_1),\phi_1\rangle+\ldots+\langle X^{\varepsilon}(t_{n-1}),\phi_{n-1}+V_{t_{n-1},t_n}\phi_n\rangle$$

$$+\langle X^{\varepsilon}(t_{n-1}),\varepsilon^{-1}\log(1-V_{t_{n-1},t_n}(1-e^{-\varepsilon\phi_n}))-V_{t_{n-1},t_n}\phi_n\rangle]\Big)\Big).$$

But by Lemma 4.4.3, $\|\varepsilon^{-1}\log(1-V_{t_{n-1},t_n}(1-e^{-\varepsilon\phi_n}))-V_{t_{n-1},t_n}\phi_n\| \to 0$ as $\varepsilon\to 0$. Together with the induction hypothesis we conclude that

$$\lim_{\varepsilon\to\infty} P_{s,\mathcal{P}ois(\mu.\varepsilon)}\left(\exp\left(-[\langle X^{\varepsilon}(t_1),\phi_1\rangle+\ldots+\langle X^{\varepsilon}(t_n),\phi_n\rangle]\right)\right)$$

$$= \exp\left(-\langle \mu, V_{t_1,\ldots,t_n}(\phi_1,\ldots,\phi_n)\rangle\right)$$

and the proof is complete. □

This completes the construction of the finite dimensional distributions of the (W,κ,Φ)-superprocess. We can then construct a Markov process with these finite dimensional distributions on a probability space using Kolmogorov's extension theorem in the usual way. In the time homogeneous case in which $\kappa(ds) = ds$ and W is a right process Fitzsimmons (1988) proved the existence of a $M_F(E)$-valued right process with these finite dimensional distributions. An extension of this to the time inhomogeneous setting is given in [DP]. These results will not be described in detail in these notes. However the existence, in special cases, of versions with regular sam-

70

ple paths and the strong Markov property is discussed in the remainder of this chapter as well as in Chapter 6.

4.5 The (α,d,β)-Superprocess

In this special case the motion process W is a symmetric α-stable process in $E = \mathbb{R}^d$ with semigroup $\{S_t^\alpha\}$, generator $\Delta_\alpha = -(-\Delta)^{\alpha/2}$, $0<\alpha\le 2$, and with paths in $D_{\mathbb{R}^d} = D([0,\infty);\mathbb{R}^d)$. The branching mechanism in this special case is given by

$$\Phi(x,\lambda) = -\gamma\lambda^{1+\beta}, \quad 0 < \beta \le 1 \text{ (spatially homogeneous)}$$

$$\kappa(ds) = ds$$

$$n(du) = \frac{\beta(1+\beta)\gamma}{\Gamma(1-\beta)}\frac{1}{u^{\beta+2}}\,du, \quad c(.) \equiv 0, \text{ if } \beta < 1,$$

$$n(\cdot) \equiv 0, \quad c(x) \equiv 1, \text{ if } \beta = 1.$$

The approximating branching particle systems have offspring generating functions

$$g^\varepsilon(v) = v + (1+\beta)^{-1}(1-v)^{1+\beta}$$

which in this special case are independent of ε. In the case $\beta = 1$, this reduces to $g^\varepsilon(v) = \frac{1}{2} + \frac{1}{2}v^2$ *(binary branching)*. Also $\kappa^\varepsilon(ds) = \gamma(1+\beta)\varepsilon^{-\beta}ds$.

The resulting $M_F(\mathbb{R}^d)$-valued process is called the (α,d,β)-*superprocess* and is characterized by its log-Laplace functional $V_t\phi$ which is given by the unique solution of

$$(4.5.1)\quad V_t\phi(x) = S_t^\alpha\phi(x) - \gamma\int_0^t S_u^\alpha[(V_{t-u}\phi)^{1+\beta}]du$$

that is, $u(t,x) = (V_t\phi)(x)$ satisfies

$$\frac{\partial u}{\partial t} = \Delta_{d,\alpha}u - \gamma u^{1+\beta}, \quad u(0,x) = \phi(x) \in bp\mathcal{E}.$$

If $\phi(\cdot) \equiv \theta$, then

$$(4.5.2)\qquad u(\theta,t) = \frac{\theta}{(1+t\beta\gamma\theta^\beta)^{1/\beta}}.$$

Then

$$\lim_{\theta\uparrow\infty} u(\theta,t) = 1/(\beta\gamma t)^{1/\beta},$$

and consequently the *non-extinction* probability

$$P_\mu(\langle X_t,1\rangle = 0) = 1 - \exp\left(-\frac{\langle\mu,1\rangle}{(\beta\gamma t)^{1/\beta}}\right) \sim \mathit{const}\cdot t^{-1/\beta} \quad \text{as } t \to \infty.$$

Note that this result can be derived from the results of Zolotorev (1957).

The following scaling property of the solution of the nonlinear evolution equation (4.5.1) plays an important role in the study of (α,d,β)-superprocesses.

71

Lemma 4.5.1 For $R > 0$,

(4.5.3) $\quad V_t[R^{-\alpha/\beta}\phi(./R)](y) = R^{-\alpha/\beta}(V_{t/R^\alpha}\phi)(y/R).$

Proof. Using scaling relation $(S^\alpha_{t/R^\alpha}\phi)(y/R) = (S^\alpha_t\phi(./R))(y)$, we obtain

$$(R^{-\alpha/\beta}(V_{t/R^\alpha}\phi)(y/R))$$

$$= R^{-\alpha/\beta}S^\alpha_{t/R^\alpha}\phi(y/R) - \gamma\int_0^{t/R^\alpha} R^{-\alpha/\beta}\, S^\alpha_{(t/R^\alpha)-s}(V_s\phi)^{1+\beta}ds$$

$$= R^{-\alpha/\beta}\, S^\alpha_{t/R^\alpha}\phi(y/R)$$

$$- \gamma\int_0^t S^\alpha_{(t/R^\alpha)-(s'/R^\alpha)}[V_{(s'/R^\alpha)}\phi(./R)]^{1+\beta}\frac{R^{-\alpha/\beta}}{R^\alpha}ds'$$

$$= R^{-\alpha/\beta}(S^\alpha_{t/R^\alpha}\phi)(y/R) - \gamma\int_0^t S^\alpha_{t-s'}[[R^{-\alpha/\beta}V_{s'/R^\alpha}\phi]^{1+\beta}(./R)](y).$$

Hence it satisfies the same equation as $V_t(R^{-\alpha/\beta}\phi(./R))(y)$ and the result follows by uniqueness. □

4.6 Weak Convergence of Branching Particle Systems

In this section we will show that the (α,d,β)-superprocesses can be extended to the space $M_p(\dot{\mathbb{R}}^d)$ and realized as càdlàg strong Markov processes.

Let $p > d/2$ and if $\alpha < 2$, we make the additional assumption $p < (d+\alpha)/2$. Let $M_p(\dot{\mathbb{R}}^d)$ be defined as in Section 3.1.5 and let $C_p(\dot{\mathbb{R}}^d) := \{f \in C(\mathbb{R}^d \cup \{\infty\}): \lim_{|x|\to\infty} f(x)/\phi_p(x) \text{ exists}\}$ with norm $\|f\|_p := \sup_x |f(x)|/\phi_p(x)$ (recall that the added point $\{\infty\}$ is isolated). $M_p(\mathbb{R}^d)$ can be identified with $\{\mu: \mu \in M_p(\dot{\mathbb{R}}^d),\ \mu(\{\infty\}) = 0\}$ and will be furnished with the $C_p(\mathbb{R}^d)$ weak topology. The semigroup $\{S^\alpha_t\}$ can be extended to $C_p(\dot{\mathbb{R}}^d)$ (cf. Iscoe (1986a) in such a way that $S^\alpha_t f(\infty) = f(\infty)$, that is, $\{\infty\}$ is an absorbing point. Also $|\Delta_\alpha\phi_p(x)| \le const\ \phi_p(x)$ (cf. [D, Lemma 5.5.1]).

In order to construct the process we proceed as above but in addition we find μ_ε $(\in M_F(\mathbb{R}^d)) \uparrow \mu \in M_p(\dot{\mathbb{R}}^d)$. We consider the approximating branching particle systems as above in which the initial measure is $\mathcal{P}ois(\mu_\varepsilon/\varepsilon)$. In this section we will show that the laws of the approximating branching particle systems are relatively compact in $M_1(D([0,\infty),M_p(\dot{\mathbb{R}}^d)))$.

72

Let $X^{\varepsilon}(t,du) = \varepsilon\, Z^{\varepsilon}(t,du)$ where $Z^{\varepsilon}(t,du)$ denotes the approximating branching particle system with $Z^{\varepsilon}(0)$ given by a Poisson random measure with intensity measure μ_{ε}, α-symmetric stable motions, $\kappa^{\varepsilon}(w,ds) = 1(w(s)\neq\infty)\gamma(1+\beta)\varepsilon^{-\beta}ds$, and offspring probability generating function

$$\mathcal{G}^{\varepsilon}(v) = v + (1+\beta)^{-1}(1-v)^{1+\beta}.$$

Let $p \in (d/2,(d+\alpha)/2)$ or $(d/2,\infty)$ if $\alpha = 2$.

We first need to state the appropriate maximal inequality.

Lemma 4.6.1 For $0 < \theta < \beta$, and $T > 0$

$$\text{(4.6.1)}\qquad E\sup_{t\le T}\left\{ \langle X^{\varepsilon}(t),\phi_p\rangle^{1+\theta}\right\} \le K\cdot(1+I(X_0^{\varepsilon})) \le K' \quad \forall\ \varepsilon\in(0,1),$$

where $I(X_0^{\varepsilon}) := \sup_{t\le T} E\left(\langle X^{\varepsilon}(0),S_t^{\alpha}\phi_p\rangle^{1+\beta} + \langle X^{\varepsilon}(0),S_t^{\alpha}\phi_p\rangle\right)$ and K' is independent of ε. Further if $I(X_0^{\varepsilon}) < 1$, then

$$\text{(4.6.2)}\qquad E\sup_{t\le T}\left\{ \langle X^{\varepsilon}(t),\phi_p\rangle^{1+\theta}\right\} \le K''\cdot(I(X_0^{\varepsilon}))^{(1+\theta)/(1+\beta)}.$$

Proof. The proof is based on elementary inequalities involving some properties of the Laplace functional and Doob's maximal inequality. For a detailed proof see [D, Lemma 5.5.3].

Theorem 4.6.2. Let $X^{\varepsilon}(t,du) = \varepsilon\, Z^{\varepsilon}(t,du)$ where $Z^{\varepsilon}(t,du)$ denotes the approximating branching particle system with $Z^{\varepsilon}(0)$ given by a Poisson random measure with intensity measure μ_{ε}, α-symmetric stable motions, $\kappa^{\varepsilon}(w,ds) = 1(w(s)\neq\infty)\gamma(1+\beta)\varepsilon^{-\beta}ds$, and offspring probability generating function

$$\mathcal{G}^{\varepsilon}(v) = v + (1+\beta)^{-1}(1-v)^{1+\beta}.$$

Let $p \in (d/2,(d+\alpha)/2)$ if $0<\alpha<2$, or $(d/2,\infty)$ if $\alpha = 2$.

(a) Then the processes $\{X^{\varepsilon}(t,.)\}$ under $P^{Z}_{r,\mathcal{P}ois(\mu_{\varepsilon}/\varepsilon)}$ with $\mu_{\varepsilon} \to \mu \in M_p(\mathbb{R}^d)$ converge weakly to a càdlàg $M_p(\dot{\mathbb{R}}^d)$-valued process whose Laplace functional is given by

$$P_{\mu}\left(e^{-\langle X(t),\phi\rangle}\right) = \exp\left(-\int v_t(x)\mu(dx)\right), \quad \mu\in M_p(\mathbb{R}^d),$$

where

$$\text{(4.6.3)}\qquad \partial v_t(x)/\partial t = \Delta_{\alpha} v_t - \gamma(v_t)^{1+\beta},$$
$$v_0(x) = \phi(x) \in C_p(\mathbb{R}^d).$$

73

(b) If $\mu(\{\infty\}) = 0$, then $\sup_t X(t,\{\infty\}) = 0$, P_μ-a.s.

(c) If $\mu \in M_p(\mathbb{R}^d)$, then X is a.s. càdlàg (for the $C_p(\mathbb{R}^d)$ topology).

<u>Proof.</u> The convergence of the Laplace functional of the finite dimensional distributions of X_t^ε to (4.4.6) when $\mu \in M_F(\mathbb{R}^d)$ was proved in Corollary 4.4.6. The convergence of the finite dimensional distributions when $\mu_\varepsilon \uparrow \mu \in M_p(\mathbb{R}^d)$ follows by verifying the tightness of the first moment measures - this is a simple consequence of the fact that S_t^α maps $C_p(\dot{\mathbb{R}}^d)$ into itself. It thus remains to prove the tightness of the processes X^ε in $D([0,\infty),M_p(\dot{\mathbb{R}}^d))$.

<u>Proof of Tightness.</u> (a) Let $\varepsilon_n \to 0$ as $n \to \infty$. By Theorem 3.7.1, in order to prove tightness of the processes $\{X^{\varepsilon_n}\}$ it suffices to show that for $\phi \in K_p(\mathbb{R}^d)$, the family $\{Z_n(t) := \langle X^{\varepsilon_n}(t),\phi\rangle;\ n \in \mathbb{Z}_+\}$ is tight in $D([0,\infty),\mathbb{R})$. Using the Aldous condition (Theorem 3.6.5) it then suffices to show that

$$Z_n(\tau_n+\delta_n)-Z_n(\tau_n) \longrightarrow 0 \quad \text{in distribution as } n \to \infty.$$

Here δ_n are positive constants converging to zero as $n \to \infty$ and τ_n is any stopping time of the process Z_n with respect to the canonical filtration, satisfying $\tau_n \le T$.

By the strong Markov property applied to the process X^{ε_n} we obtain that for $r,s \ge 0$,

$$L_n(\delta_n;s,r) = E\Big(\exp\{-sZ_n(\tau_n+\delta_n)-rZ_n(\tau_n)\}\Big)$$
$$= E\Big(\exp\{\langle -X^{\varepsilon_n}(\tau_n), v_{\delta_n}(s\phi)+v_0(r\phi)\rangle\}\Big)$$

where $\{v_t(s\phi):t\ge 0\}$ satisfies equation (4.6.3) (with ϕ replaced by $s\phi$).

Therefore

$$|L_n(0;s,r) - L_n(\delta_n;s,r)|$$
$$\le \|v_n(s\phi;\delta_n)-v_n(s\phi;0)\|_p \ E\{ \sup_{t\le T} \langle X^{\varepsilon_n}(t),\phi_p\rangle\}$$
$$\le \ \text{const}\ \|v_n(s\phi;\delta_n)-v_n(s\phi;0)\|_p \qquad \textit{(by Lemma 4.6.1).}$$

A modification of the proof of Lemma 4.4.3 shows that $\|v_n(s\phi;\delta_n)-v_n(s\phi;0)\|_p \longrightarrow 0$ as $n \to \infty$ and therefore $|L_n(0)-L_n(\delta_n)| \longrightarrow 0$.

By Lemma 4.6.1 the sequence $\{Z_n(\tau_n+\delta_n),Z_n(\tau_n)\}$ is tight. Consider a subsequence $\{n_k\}$ such that $(Z_{n_k}(\tau_{n_k}+\delta_{n_k}),Z_{n_k}(\tau_{n_k}))$ converges in distribution. Since $|L_n(0;s,r) - L_n(\delta_n;s,r)| \longrightarrow 0$ we conclude that the limiting distribution has the

74

form (Z_∞, Z_∞). This implies that $Z_n(\tau_n+\delta_n)-Z_n(\tau_n) \longrightarrow 0$ in distribution as $n \to \infty$ and the proof of tightness is complete.

(b) In order to prove that $\mu(\{\infty\}) = 0$ implies $\sup_t X(t,\{\infty\}) = 0$, P_μ-a.s. it suffices to show that for all $\delta > 0$, $T>0$

$$\lim_{R\to\infty} \sup_n E\left(\sup_{0\le t\le T}\left(\int_{|x|>R} \phi_p(x)X^{\varepsilon_n}(t,dx)\right)^{1+\theta}\right) < \delta \text{ for some } 0<\theta<\beta.$$

It follows from the properties of the approximating particle system that

$$P_{\mathcal{Pois}(\mu_\varepsilon/\varepsilon)} = P_{\mathcal{Pois}(\mu_{1,\varepsilon}/\varepsilon)} * P_{\mathcal{Pois}(\mu_{2,\varepsilon}/\varepsilon)} \quad \text{(* denotes convolution)}$$

where $\mu_{1,\varepsilon} = \mu_\varepsilon(dx)1(|x|<K)$, $\mu_{2,\varepsilon} = \mu_\varepsilon(dx)1(|x|\ge K)$ (let $X^{1,\varepsilon}$, $X^{2,\varepsilon}$ denote the corresponding branching particle systems with initial conditions $\mathcal{Pois}(\mu_{1,\varepsilon}/\varepsilon)$, $\mathcal{Pois}(\mu_{2,\varepsilon}/\varepsilon)$, respectively).

Note that

$$\sup_{0<\varepsilon<1} E\left\{\sup_{t\le T} \langle X^{2,\varepsilon}(t),\phi_p 1(|x|>R)\rangle^{1+\theta}\right\}$$

$$\le \sup_{0<\varepsilon<1} E\left\{\sup_{t\le T} \langle X^{2,\varepsilon}(t),\phi_p\rangle^{1+\theta}\right\}$$

$$\le const \sup_{0<\varepsilon<1}\left\{\sup_{0\le t\le T} E\left(\langle X^{2,\varepsilon}(0),S_t^\alpha\phi_p\rangle^{1+\beta} + \langle X^{2,\varepsilon}(0),S_t^\alpha\phi_p\rangle\right)\right\}^{(1+\theta)/(1+\beta)}$$

by (4.6.1)

for $0<\theta<\beta$.

Now, since $\mu_\varepsilon \longrightarrow \mu$ and $\mu(\{\infty_p\}) = 0$, given $\delta > 0$ we can choose K in the definition of $\mu_{2,\varepsilon}$ sufficiently large so as to make this smaller than $\delta/2$ (uniformly for $0 < \varepsilon < 1$).

But for $0 < p' < p$,

$$\sup_{0<\varepsilon<1} E \sup_{t\le T}\left\{\langle X^{1,\varepsilon}(t),\phi_p 1(|x|>R)\rangle^{1+\theta}\right\}$$

$$\le (1+R^2)^{-(p-p')(1+\theta)} \cdot \sup_{0<\varepsilon<1} E \sup_{t\le T}\left\{\langle X^{1,\varepsilon}(t),\phi_{p'}\rangle^{1+\theta}\right\}$$

$$\le const\ (1+R^2)^{-(p-p')(1+\theta)} \cdot \left\{\sup_{0\le t\le T} E\left(\langle X^{1,\varepsilon}(0),S_t^\alpha\phi_{p'}\rangle^{1+\beta} + \langle X^{1,\varepsilon}(0),S_t^\alpha\phi_{p'}\rangle\right)\right\}^{(1+\theta)/(1+\beta)}.$$

We can then choose R to make this smaller than $\delta/2$ (uniformly in $0 < \varepsilon < 1$).

75

(c) The arguments in (b) prove the necessary Prohorov tightness condition for the $C_p(\mathbb{R}^d)$-topology. This completes the proof. □

Remark 4.6.3. Consider the (A,Φ)-superprocess when A is a Feller process on a compact metric space E, $\Phi(x,\lambda) = -\gamma\lambda^{1+\beta}$, $0 < \beta \le 1$ and $\mu \in M_F(E)$. A slightly simplified version of the proof of Theorem 4.6.2 yields tightness on $D([0,\infty),M_F(E))$ for the corresponding family of branching particle systems. In the case of Polish E, a new argument is required to verify the first condition in Jakubowski's criterion. The verification of the second condition can be carried out as in Theorem 4.6.2 provided that the domain of A contains a convergence determining family.

Let $D_0(\Delta_\alpha,p) \subset pC_p(\mathbb{R}^d)$ be closed under positive linear combinations and form a core for the semigroup $\{S_t^\alpha\}$ on $C_p(\mathbb{R}^d)$. Let us extend functions in $D_0(\Delta_\alpha,p)$ to $C_p(\dot{\mathbb{R}}^d)$ by defining

$$\dot{f}(\{\infty_p\}) := \lim_{|x|\to\infty} f(x)/\phi_p(x).$$

Also note that $\dot{S}_t^\alpha \dot{f}(\infty_p) = \dot{f}(\infty_p)$ $\forall t$, $\forall f \in C_p(\dot{\mathbb{R}}^d)$. Let $\{T_t : t\ge 0\}$, the semigroup of operators on $C_0(M_p(\dot{\mathbb{R}}^d))$ associated with the (α,d,β)-superprocess, be defined as follows. Let $\mathcal{D}_0(G)$ denote the collection of functions on $M_p(\dot{\mathbb{R}}^d)$ of the form

$$F(\mu) = \exp\left\{-[\langle\mu,\dot{\phi}\rangle + \theta\langle\mu,\dot{\phi}_p\rangle)]\right\}$$

with $\phi \in D_0(\Delta_\alpha,p)$, $\theta>0$. It is easy to verify that $\mathcal{D}_0(G)$ is a dense subset of $C_0(M_p(\dot{\mathbb{R}}^d))$. For functions $F \in \mathcal{D}_0(G)$,

$$\begin{aligned} T_tF(\mu) &:= EF(X(t)\,|\,X(0)=\mu) \\ &= \exp\left\{-[\langle\mu 1_{\mathbb{R}^d}, V_t(\phi+\theta\phi_p)\rangle + \theta\mu(\{\infty_p\})]\right\}. \end{aligned}$$

Note that this implies that if $F(\mu) = f(\{\mu(\infty_p)\})$ then $T_tF(\mu) = F(\mu)$ for all t (i.e. no branching occurs at $\{\infty_p\}$ - probabilistically this is a law of large numbers phenomenon).

Proposition 4.6.4. (a) The semigroup $\{T_t : t\ge 0\}$ is Feller on $C_0(M_p(\dot{\mathbb{R}}^d))$.

(b) The process $\{X_t : t\ge 0\}$ is strong Markov.

Proof.(a) (cf. Iscoe (1986a, Theorem 1.1)). It follows from Theorem 4.6.2 that $\{T_t\}$ is a contraction semigroup on $b\mathcal{B}(M_p(\dot{\mathbb{R}}^d))$. It remains to show that $T_t : C_0(M_p(\dot{\mathbb{R}}^d)) \to C_0(M_p(\dot{\mathbb{R}}^d))$. To do this it suffices to show that for functions in $\mathcal{D}_0(G)$, $T_tF(\mu)$ is continuous on $M_p(\dot{\mathbb{R}}^d)$.

Thus it suffices to show that for such F, if $\mu_n \to \mu \ne \infty$ in $M_p(\dot{\mathbb{R}}^d)$, then

$$\lim_{n\to\infty} T_tF(\mu_n) = \lim_{n\to\infty} \exp\left\{-[<\mu_n 1_{\mathbb{R}^d}, V_t(\phi+\theta\phi_p)> + \theta\mu_n(\{\infty_p\})]\right\}$$

$$= \exp\left\{-[<\mu 1_{\mathbb{R}^d}, V_t(\phi+\theta\phi_p)> + \theta\mu(\{\infty_p\})]\right\}$$

and

$$\lim_{n\to\infty} T_tF(\mu_n) = 0 \text{ if } \phi=\phi_p \text{ and } \mu_n \longrightarrow \{\infty\}.$$

But this follows from the analytical result that $V_t:C_p(\mathbb{R}^d)\longrightarrow C_p(\mathbb{R}^d)$, $V_t\phi_p \geq \mathit{const}\cdot\phi_p$ and the fact that

$$\lim_{|x|\to\infty} V_t\phi_p(x)/\phi_p(x) = \lim_{|x|\to\infty} S_t^{\alpha}\phi_p(x)/\phi_p(x) = 1$$

(the second equality follows from an elementary argument and the first one follows from the Feynman-Kac formula).

(b) The strong Markov property follows directly from the Feller property and will also be a consequence of the results of Theorem 6.1.3. □

<u>Remarks.</u> 1. The fact that there is no branching at $\{\infty\}$ is a law of large numbers phenomenon and can be derived analytically from the fact that $\lim_{|x|\to\infty} V_t\phi_p(x)/\phi_p(x) =$ 1. In order to get the former note that $\mu_n = a_n\delta_{x_n} \longrightarrow \delta_{\{\infty\}}$ in $M_p(\mathbb{R}^d)$ if and only if $|x_n|\to\infty$ and $a_n\phi_p(x_n)\longrightarrow 1$. But then X_t can be written as the normalized sum of $[\phi_p(x_n)^{-1}]$ i.i.d. random variables (plus a negligible remainder) and the result then follows from the weak law of large numbers). It also arises from the limit of the branching particle systems since we have assumed that $\kappa(w,ds) = 0$ if $w(s) = \infty$.

4.7 <u>The Continuous B(A,c)-Superprocess</u>

In this section we consider in greater detail the class of B(A,c)-superprocesses (which includes the (α,d,1)-superprocesses of Sect. 4.5). First we give some notation. Let E be locally compact with one point compactification $\bar{E}$ and (A,D(A)) denote the generator of the A-Feller process on $\bar{E}$ and let $\beta = 1$. Then the B(A,c)-superprocess is characterized by the log-Laplace equation

$$(4.7.1)\qquad V_t\phi = S_t\phi - \tfrac{1}{2}c\int_0^t S_{t-s}(V_s\phi)^2 ds.$$

<u>Lemma 4.7.1</u> Let $\phi\in b\mathcal{E}$. Then

(a) $P_\mu(<X_t,\phi>) = <\mu,S_t\phi>,$

77

(b) $P_\mu(\langle X_t,\phi\rangle^2) = \int_0^t \langle \mu S_{t-s},(S_s\phi)^2\rangle ds + \langle \mu,S_t\phi\rangle^2$,

(c) if $\phi\in bp\mathcal{E}$, then

$$P_\mu(\langle X_t,\phi\rangle^n) = \sum_{k=0}^{n-1} \binom{n-1}{k} \langle \mu,v_t^{(n-k)}\rangle P_\mu(\langle X_t,\phi\rangle^k)$$

where $v_t^{(1)} = S_t\phi$, and $v_t^{(n)} = \sum_{k=1}^{n-1}\binom{n-1}{k}\int_0^t S_{t-s}(v_s^{(k)}\cdot v_s^{(n-k)})ds$, $n\ge 2$.

Proof. Let $u^{(n)}(t,\theta) := \frac{\partial^n}{\partial\theta^n} V_t(\theta\phi)$ for $\phi\in bp\mathcal{E}$. Then

(4.7.2) $\langle \mu,u^{(1)}(t,\theta)\rangle P_\mu(\exp(-\theta\langle X_t,\phi\rangle)) = P_\mu(\langle X_t,\phi\rangle\exp(-\theta\langle X_t,\phi\rangle))$.

Iteratively differentiating (4.7.2) with respect to θ and evaluating at $\theta=0$ yields (c). (a) and (b) follow as special cases. □

Let $\{f_n:n\in\mathbb{Z}_+\}$ be a convergence determining sequence in $pD(A)$ with $(\|Af_n\|+\|f_n\|) \le 1$ and define

$$d(\mu,\nu) := \sum 2^{-n}\left(1 \wedge \left|\int f_n d\mu - \int f_n d\nu\right|\right).$$

Theorem 4.7.2. Let $\{X_t:t\ge 0\}$ be a $B(A,c)$-superprocess on the compact space $\bar{E}$, where $\bar{E}$ is the one-point compactification of E. Then for each $\mu\in M(\bar{E})$, $\{X_t:t\ge 0\}$ is P_μ-almost surely lies in $C([0,\infty),M(\bar{E}))$. In addition for $\beta < 1/4$ there exists a β-Hölder continuous version of $\{X_t:t\ge 0\}$ in the metric d.

Proof. Using Lemma 4.7.1 it can be verified that for $T<\infty$, $n\in\mathbb{Z}_+$,

(4.7.3) $P_\mu([\langle X_t,f_n\rangle-\langle X_s,f_n\rangle]^4) \le C_T(t-s)^2$, $0\le s<t\le T$.

This implies that for each $n\in\mathbb{Z}_+$, $\langle X_t,f_n\rangle$ is P_μ-almost surely continuous. The β-Hölder continuity follows from (4.7.3) and Corollary 3.7.3. □

Remark. A stronger version of this result (β-Hölder continuity for $\beta < 1/2$) will be obtained in Prop. 7.3.1.

78

5. PROBABILITY MEASURE-VALUED PROCESSES AND MARTINGALE PROBLEMS

5.1. Martingale Problems and Markov processes

There is a number of possible approaches to the construction and characterization of measure-valued Markov processes. For example, in Chapters 2 and 4 we adopted semigroup methods to construct the Fleming-Viot and measure-valued branching processes. In this chapter we develop the martingale problem approach which serves as a unified setting in which to compare these two classes of processes and to use them as building blocks to construct more complex processes.

In this section we give a brief introduction to the martingale problem for E-valued processes with E a Polish space. (However note that we will later apply it with E replaced by $M_1(E)$ or $M_F(E)$.) We frequently refer to Ethier and Kurtz (1986, Chap.4) which should be consulted for a more complete exposition of the martingale problem method.

Let E be a Polish space and $D_{E,s} := D([s,\infty),E)$ with coordinate process $\{X_t\}$ and filtration $(\mathcal{D}_{s,t})_{t\geq s}$ where $\mathcal{D}_{s,t} := \bigcap_{\varepsilon>0} \mathcal{D}^o_{s,t+\varepsilon}$ where $\mathcal{D}^o_{s,t} := \sigma(\{X_u\}: s\leq u\leq t)$. Also let $\mathcal{D}_{s,\infty} := \bigvee_{n\in\mathbb{N}} \mathcal{D}_{s,s+n}$.

A *time inhomogeneous martingale problem* is given by a pair $(\mathfrak{D}(\mathfrak{G}),\mathfrak{G})$ satisfying:

(5.1.1) $\mathfrak{G}:[0,\infty)\times\mathfrak{D}(\mathfrak{G}) \longrightarrow b\mathcal{E}$, $\mathfrak{D}(\mathfrak{G}) \subseteq bC(E)$. and for each $F\in\mathfrak{D}(\mathfrak{G})$,
$(s,x) \longrightarrow (\mathfrak{G}F)(s,x) = (\mathfrak{G}(s)F)(x) \in b(\mathcal{B}(\mathbb{R}_+)\otimes\mathcal{E})$, and
for each $s\in[0,\infty)$, $\mathfrak{G}(s)$ is a linear mapping on $\mathfrak{D}(\mathfrak{G})$.

A probability measure P on $(D_{E,s},\mathcal{D}_{s,\infty},(\mathcal{D}_{s,t})_{t\geq s})$ is said to be a *solution to the s-initial value martingale problem for* $(\mathfrak{G},\mathfrak{D}(\mathfrak{G}))$ if for every $F\in\mathfrak{D}(\mathfrak{G})$,

(5.1.2) $$M_F(t,X) := F(X(t)) - F(X(s)) - \int_s^t (\mathfrak{G}F)(u,X(u))du$$

is a $((\mathcal{D}_{s,t})_{t\geq s},P)$-martingale.

Equivalently, P satisfies:

(5.1.3) $H(X)$ is P-integrable and $$\int_{D_{E,s}} H(X)P(dX) = 0 \quad \forall \quad H\in \mathfrak{H}_s$$

where $\mathfrak{H}_s := \{H:\ H(u,t,F,G_u;X): s\leq u\leq t<\infty,\ F\in\mathfrak{D}(\mathfrak{G}),\ G_u\in b\mathcal{D}_{s,u}\}$, and

$$H(u,t,F,G_u;X) := [M_F(t,X)-M_F(u,X)]G_u(X).$$

The set of solutions to the initial value martingale problem at time 0 is a convex subset of $M_1(D_{E,0})$, denoted by $\mathfrak{P}(\mathfrak{G})$. Let $\mathfrak{P}_e(\mathfrak{G})$ denote the class of extreme points of $\mathfrak{P}(\mathfrak{G})$. For $P\in\mathfrak{P}(\mathfrak{G})$, let

79

$$\mathcal{H}^1(P) := \{M: M \text{ is a P-local martingale, } \|\sup_t |M_t|\|_1 < \infty\}$$

where $\|.\|_1$ is the $L^1(P)$-norm. Let $\mathcal{H}^1_{pr}(P)$ denote the smallest $\mathcal{H}^1(P)$-closed linear space containing $\{\int HdM_F \in \mathcal{H}^1(P)$: H is predictable, $F\in \mathfrak{D}(\mathfrak{G})\}$ and the constant process, where $\int HdM_F$ denotes the Itô stochastic integral.

Theorem 5.1.1 *(Jacod's Predictable Representation Theorem)*

If $P \in \mathfrak{P}(\mathfrak{G})$, then the following conditions are equivalent:

(i) $P\in \mathfrak{P}_e(\mathfrak{G})$,

(ii) $\mathcal{H}^1(P) = \mathcal{H}^1_{pr}(P)$ and $\mathcal{D}_0$ is P-trivial,

(iii) every bounded P-martingale, M, with $M_0=0$, which is orthogonal to M_F, $\forall F\in\mathfrak{D}(\mathfrak{G})$ is zero and $\mathcal{D}_0$ is P-trivial.

Proof. See Jacod (1979, Theorem 11.2).

Proposition 2.3.3 gives an example of a function of a Markov process that is Markov. In order to establish this type of result in the context of martingale problems we include the notion of *restricted martingale problem* which is formulated as follows.

Let $M_0(E) \subset M_1(E)$. Then $\mathcal{A} \subset pbC(E)$ is said to be $M_0(E)$-*determining* if

(i) $pb\sigma(\mathcal{A})$ is the bp closure of $\mathcal{A}$,

(ii) if $\mu_1,\mu_2 \in M_0(E)$ and $\mu_1 = \mu_2$ on $\sigma(\mathcal{A})$, then $\mu_1=\mu_2$, and

(iii) for each $\mu\in M_0(E)$ and $g\in pp\mathcal{A}$, $A \longrightarrow \mu_g(A) := \langle\mu,g1_A\rangle/\langle\mu,g\rangle \in M_0(E)$

(where $pp\mathcal{A} := \{g\in\mathcal{A}: \inf g > 0\}$).

Let $M_0(E) \subset M_1(E)$ and $\mathfrak{D}(\mathfrak{G}) \subset pbC(E)$ be $M_0(E)$-determining. Then a family of probability measures $\{P_{s,\mu}:\mu\in M_0(E)\}$, $M_0(E)\subset M_1(E)$ on $(D_{E,s},\mathcal{D}_{s,\infty},(\mathcal{D}_{s,t})_{t\ge s})$ is said to be a *solution to the restricted s-initial value martingale problem for* $(\mathfrak{G},\mathfrak{D}(\mathfrak{G}),M_0(E))$ if for every $\mu\in M_0(E)$, $\Pi_s P_{s,\mu}=\mu$, $\Pi_t P_{s,\mu} \in M_0(E)$ $\forall t\ge s$, and $P_{s,\mu}$ is a solution to the s-initial value martingale problem for $(\mathfrak{G},\mathfrak{D}(\mathfrak{G}))$. The initial value martingale problem is said to be *well-posed* if for every $s\ge 0$ and $\mu\in M_0(E)$, there is a *unique* solution to the martingale problem satisfying $\Pi_s P_{s,\mu}=\mu$.

The following result is a minor extension of a basic result of Stroock and Varadhan (1979, section 6.2) to include the case of restricted and time inhomogeneous martingale problems. (The extension to restricted martingale problems will be used to characterize exchangeable Markov systems and the time inhomogeneous martingale problem will be used in sections 5.5, 5.7 and Chapter 12.)

Theorem 5.1.2. Assume that

(i) $M_0(E)\subset M_1(E)$, $\mathfrak{D}(\mathfrak{G}) \subset ppbC(E)$ is $M_0(E)$-determining, $\mathfrak{G}:[0,\infty)\times\mathfrak{D}(\mathfrak{G})\longrightarrow$

80

$b(\sigma(\mathfrak{D}(\mathfrak{G})))$, and $(s,x) \longrightarrow (\mathfrak{G}F)(s,x) \in b(\mathcal{B}(\mathbb{R}_+)\otimes\sigma(\mathfrak{D}(\mathfrak{G})))\ \forall\ F\in \mathfrak{D}(\mathfrak{G})$,

(ii) for any $s\geq 0$ and any two solutions to the restricted s-initial value martingale problem for $(\mathfrak{G},\mathfrak{D}(\mathfrak{G}),M_0(E))$, denoted by $\{P^1_{s,\mu}:\mu\in M_0(E)\}$ and $\{P^2_{s,\mu}:\mu\in M_0(E)\}$, the marginal distributions $\Pi_t P^1_{s,\mu}$ and $\Pi_t P^2_{s,\mu}$ for $t\geq s$ agree on $\sigma(\mathfrak{D}(\mathfrak{G}))$ for all $t>s$ and $\mu\in M_0(E)$.

Then

(iii) $P_{s,\mu} := P^1_{s,\mu} = P^2_{s,\mu}$ on $(D_{E,s},\mathcal{D}^{\mathfrak{G}}_{s,\infty},(\mathcal{D}^{\mathfrak{G}}_{s,t})_{t\geq s})\ \forall\ \mu\in M_0(E)$ and $s\geq 0$, where $\mathcal{D}^{\mathfrak{G}}_{s,t} := \bigcap_{\varepsilon>0}\sigma\{F(X_u):F\in\mathfrak{D}(\mathfrak{G}),s\leq u\leq t+\varepsilon\}$ and $\mathcal{D}^{\mathfrak{G}}_{s,\infty} = \bigvee_{t\geq s}\mathcal{D}^{\mathfrak{G}}_{s,t}$, and

(iv) the process $(D_{E,s},\mathcal{D}^{\mathfrak{G}}_{s,t},(\mathcal{D}^{\mathfrak{G}}_{s,t})_{t\geq s},\{X_t\}_{t\geq s},\{P_{s,\mu}:\mu\in M_0(E)\})_{s\geq 0}$ is Markov, that is, if $B\in \mathcal{D}^{\mathfrak{G}}_{t_1,\infty}$, then

$$P_{s,\mu}(B|\mathcal{D}^{\mathfrak{G}}_{s,t_1}) = P_{s,\mu}(B|\sigma(F(X_{t_1}):F\in\mathfrak{D}(\mathfrak{G}))),\ P_{s,\mu}\text{-a.s.}$$

<u>Sketch of Proof.</u> To prove (iii) it suffices to show that the finite dimensional distributions agree. First consider $s\leq t_1<t_2$. In order to prove that the two distributions of $(X(t_1),X(t_2))$ agree on $\sigma(\mathfrak{D}(\mathfrak{G}))\otimes\sigma(\mathfrak{D}(\mathfrak{G}))$ it suffices to show that

$$\text{(5.1.4)}\quad P^1_{s,\mu}(g_1(X(t_1))g_2(X(t_2))) = P^2_{s,\mu}(g_1(X(t_1))g_2(X(t_2)))$$

for all $g_1,g_2\in\mathfrak{D}(\mathfrak{G})$ (because of assumption (i)). Since $g_1 \in ppC(E)$, $P^1_{s,\mu}(g_1(X(t_1)))$ $= P^2_{s,\mu}(g_1(X(t_1))) > 0$. Define probability measures on $\mathcal{D}^{\mathfrak{G}}_{t_1,\infty}$ by

$$Q^i_{t_1}(B) = P^i_{s,\mu}(g_1(X(t_1))1_B)/P^i_{s,\mu}(g_1(X(t_1))),\ i=1,2,\ B\in\mathcal{D}^{\mathfrak{G}}_{t_1,\infty}.$$

Using the assumption (i) it can be verified that $\{M_F(t):t\geq t_1\}$ is a $((\mathcal{D}^{\mathfrak{G}}_{t_1,t})_{t\geq t_1},Q^i_{t_1})$-martingale for each $F\in \mathfrak{D}(\mathfrak{G})$ and $i=1,2$. Thus $Q^1_{t_1}$ and $Q^2_{t_1}$ are solutions to the t_1-initial value martingale problem satisfying $\nu := \Pi_{t_1} Q^1_{t_1} = \Pi_{t_1} Q^2_{t_1}$ which by assumption (ii) belongs to $M_0(E)$. Then using assumption (ii), $Q^1_{t_1}(g_2(X(t_2))) = Q^2_{t_1}(g_2(X(t_2))) = P_{t_1,\nu}(g_2(X(t_2))$ which yields (5.1.4). Continuing in this way we can show that all the finite dimensional distributions of $P^1_{s,\mu}$ and $P^2_{s,\mu}$ agree and satisfy the Markov property (iv). □

<u>Remark.</u> Restricted and time inhomogeneous martingale problems will be needed at several places in these notes. However in order to simplify the exposition a little we will confine our attention to the time homogeneous ($\mathfrak{G}_s\equiv\mathfrak{G}\ \forall\ s\geq 0$) and unrestricted martingale problem in the remainder of this section. The appropriate modification for the time inhomogeneous and restricted case is then relatively straightforward.

81

Theorem 5.1.3 Assume that there is a countable subset $\mathfrak{D}_0(\mathfrak{G})$ of $\mathfrak{D}(\mathfrak{G})$ such the assumptions of Theorem 5.1.2 (time homogeneous version) are satisfied with $\mathfrak{D}(\mathfrak{G})$ replaced by $\mathfrak{D}_0(\mathfrak{G})$ and with $M_0(E) = M_1(E)$. Then

(a) the mapping $\mu \longrightarrow P_\mu$ is measurable, $\mathfrak{P}_e = \{P_{\delta_x} : x \in E\}$ and $P_\mu = \int P_{\delta_x}\,\mu(dx)$,

(b) the process $(D_E, \mathcal{D}, (\mathcal{D}_t)_{t\geq 0}, \{P_\mu : \mu \in M_1(E)\})$ is strong Markov with transition measures $P_x := P_{\delta_x}$.

Outline of Proof. (See [EK, Ch. 4, Thms. 4.2 and 4.6] for details.)

(a) The set of solutions can be written in the form $\{P: \int H(X)P(dX)=0\ \forall H\in\mathfrak{H}_0\}$ where $\mathfrak{H}_0 := \{H(s,t,F,G_s;X);\ \ s,t$ are rational, $F\in\mathfrak{D}_0(\mathfrak{G}),\ G_s \in (b\mathcal{D}_s)_0\}$ and $(b\mathcal{D}_s)_0$ is a countable set whose bounded pointwise closure coincides with $b\mathcal{D}_s$.

Since $\mathfrak{H}_0$ is countable, the set of solutions to the $(\mathfrak{G},\mathfrak{D}_0(\mathfrak{G}))$-martingale problem is a Borel subset, S, of $M_1(D_E)$. Note that the mapping $\Pi_0: S \longrightarrow M_1(E)$ is continuous, one-to-one and onto since the martingale problem is well-posed. Thus by Kuratowski's theorem there is a measurable inverse, P_μ from $M_1(E)$ to S. Since the mapping $x \rightarrow \delta_x$ is measurable this also yields the measurability of $P_x := P_{\delta_x}$. Given $\mu \in M_1(E)$, $P_\mu(H(X)) = \int P_{\delta_x}(H(X))\mu(dx) = 0\ \forall\ H \in \mathfrak{H}$ and hence yields a solution P_μ to the martingale problem satisfying $\Pi_0 P_\mu = \mu$. The second statement in (a) follows by uniqueness.

(b) Let $\mu \in M_1(E)$, τ be a P_μ-a.s. finite stopping time and $g_1 \in pb\mathcal{D}_\tau$, such that $P_\mu(g_1) > 0$. Define probability measures on D_E by

$$P_1(g_2(X(\tau+.))) = P_\mu(g_1 P_\mu(g_2(X(\tau+.))\,|\,\mathcal{D}_\tau))/P_\mu(g_1), \text{ and}$$
$$P_2(g_2(X(\tau+.))) = P_\mu(g_1 P_{X(\tau)}(g_2(X(\tau+.))))/P_\mu(g_1),\ g_2 \in b\mathcal{D}.$$

In order to prove the strong Markov property it suffices to show that $P_1 = P_2$. Since $M_F(t)$ is a càdlàg martingale, the optional sampling theorem implies that $M_F(t+\tau) - M_F(\tau)$ is again a martingale. Hence both P_1 and P_2 are solutions to the martingale problem with the same initial condition. Thus they are almost surely equal by uniqueness. This yields the result. □

Remark. If the conditions of the above theorem are satisfied then $(\mathfrak{D}(\mathfrak{G}),\mathfrak{G})$ characterizes the process and will be called an *MP-generator* of the process.

In the remainder of this section we will assume that E is a compact metric space. We will then work with a martingale problem in which the basic state space is $M_1(E)$ (in place of E). In order to formulate the $M_1(E)$-valued martingale problem we will first introduce some appropriate subspaces of $C(M_1(E))$ which will serve as the domain of the generator.

82

Recall that the *algebra of polynomials*, $C_P(M_1(E))$, (resp. $B_P(M_1(E))$) is defined to be the linear span of *monomials* of the form

$$F_{f,n}(\mu) = \int f(x)\mu^n(dx)$$
$$= \int \dots \int f(x_1,\dots,x_n)\mu(dx_1)\dots\mu(dx_n)$$

where $f \in \mathcal{B}(E^n)$, (resp. $C(E^n)$). We also consider the subspaces $\mathcal{B}_{sym}(M_1(E))$, (resp. $C_{sym}(M_1(E))$), and $\mathcal{B}_{pro}(M_1(E))$, (resp. $C_{pro}(M_1(E))$) in which f is symmetric and of the form $\Pi\phi_i(x_i)$, respectively.

For a positive integer m, $B_{F,m}(M_1(E))$, (resp. $C_{F,m}(M_1(E))$) denotes the algebra of all functions of the form

$$F_{f;\phi_1,..,\phi_m}(\mu) = f(\langle\mu,\phi_1\rangle,\dots,\langle\mu,\phi_m\rangle),\ f \in C^\infty(\mathbb{R}^m),$$

and $\phi_1,\dots,\phi_m \in \mathcal{B}(E)$, (resp. $C(E)$).

The families $C_P(M_1(E))$, $C_{sym}(M_1(E))$, $C_{pro}(M_1(E))$ and $C_{F,m}(M_1(E))$ are all dense in $C(M_1(E))$ if E is compact (cf. Lemma 2.1.2 for the case $C_P(M_1(E))$). The first and second derivatives of functions in $C_P(M_1(E))$ were given in Lemma 2.1.2. For functions in $C_{F,m}(M_1(E))$ we have

$$\delta F_{f;\phi_1,..,\phi_m}(\mu)/\delta\mu(x) = \sum_{j=1}^{m} F_{f_j;\phi_1,..,\phi_m}(\mu)\phi_j(x)$$

$$\delta^2 F_{f;\phi_1,\dots,\phi_m}(\mu)/\delta\mu(x)\delta\mu(y) = \sum_{j=1}^{m}\sum_{k=1}^{m} F_{f_{jk},\phi_1,\dots,\phi_m}(\mu)\phi_j(x)\phi_k(y)$$

and $f_j(z_1,\dots,z_m) = \partial f/\partial z_j$, $f_{jk}(z_1,\dots,z_k) = \partial^2 f/\partial z_j\partial z_k$.

5.2. A Finite Type Version of the Moran Model

We will now revisit the Moran model (cf. section 2.5) from the viewpoint of martingale problems. However in this section we begin with a finite space of types.

Let $E_K = \{e_1,\dots,e_K\}$ be a finite subset of E having K elements. E_K will serve as the *set of types.* We consider a population of N individuals in this space. We assume that the following two types of transition can occur:

(i) an individual changes from type x to type y with rate q^K_{xy} *(mutation)*;

(ii) an individual changes type at rate $\frac{1}{2}(N-1)\gamma$ to a new type equal to that of a second individual chosen at random from the remaining population *(sampling-visitation)* or *(sampling-replacement)*, consequently the total jump rate due to this

83

mechanism is $\frac{1}{2}\gamma N(N-1)$.

We can again consider this process from two points of view. First we can consider the exchangeable particle system $\mathbf{Z}(t) = \{Z_1(t),\ldots,Z_N(t)\} \in (E_K)^N$ where $Z_j(t)$ denotes the type of the jth individual at time t. This process has generator

$$L_N^K f(x_1,\ldots,x_N) = \sum_{i=1}^{N} A_i^K f(x_1,\ldots,x_N) + \frac{1}{2}\gamma \sum_{\substack{i,j \\ i\neq j}} [\Theta_{ji} f(x_1,\ldots,x_N) - f(x_1,\ldots,x_N)]$$

where $A^K g(x) := \sum_{y\in E_K} q_{x,y}^K [g(y)-g(x)]$ and A_i^K denotes the application of A^K to the variable x_i. Also let Θ_{ij} be as in Chapter 2.

Theorem 5.2.1. (a) The martingale problem with state space $E_K^N := (E_K)^N$ and with MP-generator $(L_N^K, C(E^N), M_1(E_K^N))$ is well-posed.

(b) The $(L_N^K, C_{sym}(E^N), M_{1,ex}(E_K^N))$-martingale problem is also well-posed and yields an exchangeable Markov system. (Here $M_{1,ex}(E_K^N)$ denotes the family of exchangeable laws on E_K^N.)

Proof. (a) The uniqueness of the one dimensional distributions follows since the martingale problem determines a finite system of N linear (Kolmogorov) equations for $P_\mu(\mathbf{Z}(t)=\mathbf{z})$, $\mathbf{z}\in(E_K)^N$, $\mu\in M_1((E_K)^N)$, which can be explicitly solved in terms of the matrix exponential.

(b) The solution of (a) is Markov by Theorem 5.1.2. Since $L_N^K(f(\tilde{\pi}\cdot))(\underline{x}) = (L_N^K f)(\tilde{\pi}\underline{x})$ it yields an exchangeable Markov jump process with state space $(E_K)^N$ (cf. Lemma 2.3.2) provided that the initial law $\mu\in M_{1,ex}(E_K^N)$. This proves the existence of a solution to the $(L_N^K, C_{sym}(E^N), M_{1,ex}(E_K^N))$-martingale problem. The uniqueness of the marginal distributions of the $(L_N^K, C_{sym}(E^N), M_{1,ex}(E_K^N))$-martingale problem follows as in (a). Since $C_{sym}(E^N)$ is $M_{1,ex}(E^N)$-determining (cf. Lemma 2.3.1) Theorem 5.1.2 then implies that the restricted martingale problem is well-posed. □

From the second viewpoint we consider the empirical process

$$X^N(t;B) := \Xi_N(\mathbf{Z}(t))(B) = \frac{\sum_{j=1}^N 1_B(Z_j(t))}{N}, \quad B \in \mathcal{E},$$

84

with state space $M_{1,N}(E_K)$.

Assume that $F \in C^3(M_1(E_K))$. Then the generator of the measure-valued process X^N is given by

$$\mathfrak{G}^K_N F(\mu) = N\int_{E_K} \sum_{y\in E_K} q^K_{x,y}[F(\mu - \delta_x/N + \delta_y/N)-F(\mu)]\mu(dx)$$

$$+ \frac{1}{2}\gamma N(N-1)\int_{E_K}\int_{E_K} [F(\mu - \delta_x/N + \delta_y/N)-F(\mu)]\mu(dx)\mu(dy)$$

$$= \int \{Q^K \frac{\delta F(\mu)}{\delta\mu(.)}\}(x)\mu(dx)$$

$$+ \frac{\gamma}{4}\iint \left[\frac{\delta^2 F(\mu)}{\delta\mu(x)^2} + \frac{\delta^2 F(\mu)}{\delta\mu(y)^2} - 2\frac{\delta^2 F(\mu)}{\delta\mu(x)\delta\mu(y)}\right]\mu(dx)\mu(dy)$$

$$+ R_N(F)/N.$$

Here we use Taylor's remainder formula, where $R_N(F)$ is bounded and $Q^K f(x) := \sum_{y\in E_K} q^K_{x,y} f(y)$. For functions of the form

$$F(\mu) = f(\langle\mu,\phi_1\rangle,\ldots,\langle\mu,\phi_K\rangle)$$

where $\phi_j(.) = 1_{\{e_j\}}(.)$ with $e_j \in E_K$, $f \in C^3(\mathbb{R}^K)$, and $\mu(\{e_i\}) = z_i$ this becomes $\mathfrak{G}^K_N F(\mu) = G^K_N f(\mu(\{e_1\}),\ldots,\mu(\{e_K\}))$, where

$$G^K_N f(z_1,\ldots,z_K) = \frac{\gamma}{2}\sum_{i,j=1}^{K} z_i(\delta_{ij} - z_j)\frac{\partial^2}{\partial x_i \partial x_j} f(z_1,\ldots,z_K)$$

$$+ \sum_{i=1}^{K}\left(\sum_{j\neq i} q^K_{ji} z_j - q^K_{ii} z_i\right)\frac{\partial}{\partial x_i} f(z_1,\ldots,z_K) + R_N(f)/N$$

where $R_N(f)$ is bounded in N.

<u>Theorem 5.2.2.</u> The $(\mathfrak{G}^K_N, C_{pro}(M_1(E_K)), M_1(E_K))$-martingale problem is well-posed.

<u>Proof.</u> This follows from Lemma 2.3.1 and Theorem 5.2.1. □

5.3. The Finite Type Diffusion Limit

We now consider the limit of X^N as $N \to \infty$. We get

85

$\mathfrak{G}^K_\infty F(\mu) = G^K_\infty f(\mu(\{e_1\}),\dots,\mu(\{e_K\}))$, where

$$G^K_\infty f(z_1,\dots,z_K) = \frac{\gamma}{2}\sum_{i,j=1}^{K} z_i(\delta_{ij}-z_j)\frac{\partial^2}{\partial x_i\partial x_j}f(z_1,\dots,z_K)$$

$$+\sum_{i=1}^{K}\left(\sum_{j\neq i} q^K_{ji}z_j - q^K_{ii}z_i\right)\frac{\partial}{\partial x_i}f(z_1,\dots,z_K)$$

<u>Theorem 5.3.1.</u> The $(\mathfrak{G}^K_\infty, C_{pro}(M_1(E_K)), M_1(E_K))$-martingale problem has a unique solution and this solution $\{P^K_\mu:\mu\in M_1(E_K)\}$ can be realized on $C([0,\infty),M_1(E_K))$.

<u>Proof.</u> (cf. Ethier (1976), Shiga (1981)). The existence is proved by a weak convergence argument. For each $F\in C_{F,m}(M_1(E_K))$ we have that $M_N(X^N(t))$ are uniformly bounded martingales where

$$M_N(\omega,t) := F(\omega(t)) - \int_0^t \mathfrak{G}^K_N F(\omega(s))ds,\ \ t\geq 0,\ \omega\in D([0,\infty),M_1(E_K)).$$

Hence $F(X^N(t))$ is a sequence of D-semimartingales which satisfies the tightness criteria of Joffe and Métivier (cf. Theorem 3.6.6 and Corollary 3.6.7). Thus the laws P^K_N of X^N are tight in $M_1(D([0,\infty),M_1(E_K)))$. Let P^K_∞ be a limit point. We next verify that P^K_∞ is a solution to the martingale problem for $\mathfrak{G}^K_\infty$. Note that $M_\infty:D([0,\infty),M_1(E_K)) \longrightarrow D([0,\infty),\mathbb{R})$ is continuous and $\|M_N(.)(t)-M_\infty(.)(t)\| \leq N^{-1}t|R_N(F)|$. Then following the same argument as in Section 2.9 we can show that $M_\infty(\omega,t) := F(\omega(t)) - \int_0^t \mathfrak{G}^K_\infty F(\omega(s))ds$ is a martingale (with respect to P^K_∞).

We will defer the uniqueness argument since it will be a special case of the uniqueness argument in the next section. Similarly the proof that the process is a.s. continuous is a special case of Theorem 7.3.1. □

<u>Remarks.</u>

1. The limiting process can also be identified with a diffusion on the simplex $\{(x_1,\dots,x_K):x_i\geq 0,\ i=1,\dots,K;\ \Sigma x_i=1\}$. This is called the *Wright-Fisher diffusion* (with K alleles).
2. Note however that we cannot apply the basic uniqueness result of Stroock and Varadhan (1979) for finite dimensional diffusions since the coefficients are degenerate.

5.4 <u>The Measure-Valued Diffusion Limit</u>

Let $A:D(A)\ (\subset C(E)) \longrightarrow C(E)$ generate a Feller semigroup $\{S_t\}$ on $C(E)$ given

86

by a transition function $P(t,x,\Gamma)$, that is,

$$S_t f(x) = \int_E f(y)P(t,x,dy)$$

The operator A is called the *mutation operator* and $\{S_t\}$ is called the *mutation semigroup* for the process. As in Ch. 2 we extend $\{S_t\}$ to $\bigcup_m \mathcal{B}(E^m)$ by defining

$$S_t^m f(x_1,\dots,x_m) := \int_E \dots \int_E f(y_1,\dots,y_m)P(t,x_1,dy_1)\dots P(t,x_m,dy_m).$$

As in Section 2.2 let D_0 be a separating algebra of functions in $D(A)$ satisfying $S_t:D_0 \longrightarrow D_0$, $A^{(m)}$ denote the generator of $\{S_t^m\}$ and D_0^m also be defined as in Section 2.2.

We assume that there exists $\{E_K:K\in\mathbb{N}\}$ where $E_K = \{e_1,\dots,e_K\} \subset E$ and $\{A_K:K\in\mathbb{N}\}$ which generate Feller semigroups on $C(E_K)$ such that $A_K \longrightarrow A$ in the following sense:

$$Af(x)|_{E_K} = A_K f(x) + o(K)R_K(f), \quad x\in E_K, \ f\in D_0,$$

and $|R_K(f)|$ is bounded for each $f\in D_0$.

Now consider the martingale problem associated with Fleming-Viot generator

$$\text{(5.4.1)} \quad \mathfrak{G}F(\mu) = \int_E \left(A \frac{\delta F(\mu)}{\delta\mu(x)}\right)\mu(dx) + \frac{\gamma}{2}\int_E\int_E \frac{\delta^2 F(\mu)}{\delta\mu(x)\delta\mu(y)}\, Q_S(\mu;dx,dy), \ \mu\in M_1(E)$$

where $Q_S(\mu;dx,dy) := \mu(dx)\delta_x(dy) - \mu(dx)\mu(dy)$.

Let $\mathfrak{D}_0(\mathfrak{G})$ denote the class of functions $F_{f,n}(\mu)$ with $f\in D_0^n$, (cf. section 2.5 for the notation). Then

$$\text{(5.4.2)} \quad \mathfrak{G}F_{f,n}(\mu) = \langle\mu^n, A^{(n)}f\rangle + \frac{\gamma}{2} \sum_{\substack{i,j\\ i\neq j}} \left(\langle\mu^{n-1},\tilde{\Theta}_{ij}f\rangle - \langle\mu^n,f\rangle\right)$$

where $\tilde{\Theta}_{ij}:C(E^n)\longrightarrow C(E^{n-1})$ is defined as in Remark 2.5.2(b). The operator $\mathfrak{G}_K$ will be defined as in (5.4.1) but with E replaced by E_K, A replaced by A_K and $M_1(E)$ replaced by $M_1(E_K)$.

<u>Theorem 5.4.1.</u> The processes X_K with laws $\{P_\mu^K\}$ (given in Theorem 5.3.1) converge weakly to a process with values in $M_1(E)$ which is the unique solution of the martingale problem for $(\mathfrak{G},\mathfrak{D}_0(\mathfrak{G}))$. (The limiting process coincides with the $S(A,\gamma)$-Fleming-Viot process on E.)

<u>Proof.</u> The proof of the weak convergence and the verification that each limit point is a solution for the martingale problem for $(\mathfrak{G},\mathfrak{D}(\mathfrak{G}))$ are essentially the same as that in Theorem 5.3.1. Let P_μ be a solution to the martingaie problem. Then the uniqueness of the marginal distribution $\Pi_t P_\mu$ will be proved by establishing that the moment measures are uniquely determined (and in turn this implies that the mar-

87

tingale problem is well-posed by Theorem 5.1.2).

From the martingale problem applied to functions of class $\mathcal{B}_{pro}(M_1(E))$ we deduce the following equations for the moment measures $M_n(t;dx_1,\ldots,dx_n) := P_\mu(X_t(dx_1)\ldots X_t(dx_n))$:

$$\int\ldots\int \prod_{i=1}^{n} \phi_i(x_i)\, M_n(t;dx_1,\ldots,dx_n)$$

$$= e^{-\gamma n(n-1)t/2}\int\ldots\int\left(\prod_{i=1}^{n} S_t\phi_i(x_i)\right)\mu(dx_1)\ldots\mu(dx_n)$$

$$+ \frac{1}{2}\gamma \sum_{\substack{i,j\\ i\neq j}} \int_0^t ds\Big\{ e^{-\gamma n(n-1)(t-s)/2}\int\ldots\int (S_{t-s}\phi_i S_{t-s}\phi_j)(x_j) \prod_{\ell\neq i,j}\left(S_{t-s}\phi_\ell(x_\ell)\right)$$

$$M_{n-1}(s;dx_1,..,dx_j,..,dx_{i-1},dx_{i+1},..,dx_{n-1})\Big\}$$

But this system of equations is just the evolution form of the system (2.8.1) thus identifying the finite dimensional distributions to be those given in Proposition 2.8.1. □

5.5. The Method of Duality

The method of duality has been widely applied in the study of infinite particle systems (cf. Liggett (1985)). In particular it was used to determine the ergodic behavior of the voter model by Holley and Liggett (1975). It was first applied to the diffusions arising in population genetics by Shiga (1980, 1981). Function-valued duals for measure-valued processes were introduced in Dawson and Hochberg (1982), Dawson and Kurtz (1982) and further developed in Ethier and Kurtz (1987, 1990).

In section 2.8 we introduced the idea of the duality between two evolution families and illustrated it by the relation between the Moran semigroup and the Fleming-Viot semigroup. In this section we give a more systematic development of this idea. In Chapter 10 we will apply it to construct the Fleming-Viot process with recombination and selection - a situation in which the moment measure equations are not closed and therefore the method of Ch. 2 cannot be used.

In this section we assume that E_1 is a Polish space, E_2 is a separable metric space and we consider time inhomogeneous martingale problems for E_1 and E_2-valued processes associated with $(\mathcal{D}(\mathfrak{G}_1),\mathfrak{G}_1)$ and $(\mathcal{D}(\mathfrak{G}_2),\mathfrak{G}_2)$, respectively.

Let $f\in b\mathcal{B}(E_1\times E_2)$. Let $\mathcal{D}(\mathfrak{G}_1)\subset b\mathcal{E}_1$, $\mathfrak{G}_1:[0,\infty)\times\mathcal{D}(\mathfrak{G}_1)\to b\mathcal{E}_1$, $\mathcal{D}(\mathfrak{G}_2)\subset b\mathcal{E}_2$,

88

$\mathfrak{G}_2:[0,\infty)\times\mathfrak{D}(\mathfrak{G}_2)\longrightarrow \mathcal{E}_2$. Assume that $f(.,y) \in \mathfrak{D}(\mathfrak{G}_1)\ \forall\ y\in E_2$, $f(x,.) \in \mathfrak{D}(\mathfrak{G}_2)\ \forall\ x\in E_1$. For $s\geq 0$, let $g(s;x,y) := (\mathfrak{G}_1 f)(s;x,y)$ and $h(s;x,y) := (\mathfrak{G}_2 f)(s;x,y)$.

The t_1-initial value martingale problems for $(\mathfrak{G}_1,\{f(.,y),\ y\in E_2\})$ and $(\mathfrak{G}_2,\{f(x,.),x\in E_1\})$ are said to be *dual* for (μ_1,μ_2) with $\mu_i\in M_1(E_i)$, $i=1,2$ if for each solution $P_{\mu_1} \in M_1(D_{E_1,t_1})$ of the martingale problem for $\mathfrak{G}_1$ and each solution $Q_{\mu_2} \in M_1(D_{E_2,t_1})$ for $\mathfrak{G}_2$, and $t_2\geq t_1$,

$$\int_{E_2} P_{\mu_1}\Big[f(X(t_2),y)\Big]\mu_2(dy) = \int_{E_1} Q_{\mu_2}\Big[f(x,Y(t_2))\Big]\mu_1(dx)$$

where X and Y denote the respective coordinate processes in $D([0,\infty),E_1)$ and $D([0,\infty),E_2)$.

The martingale problem for $\mathfrak{G}_2$ is said to provide a *FK-dual representation* (FK is an abbreviation for Feynman-Kac) for the t_1-initial value martingale problem for $\mathfrak{G}_1$ for the pair (μ_1,μ_2) if there exists a function $\beta \in \mathcal{B}(\mathbb{R}_+)\otimes\mathcal{E}_2$ and a solution Q_{t_1,μ_2} for the t_1-initial value martingale problem for $\mathfrak{G}_2$ satisfying

$$\int_{E_1} Q_{t_1,\mu_2}\Big[|f(x,Y(t_2))|\exp\Big\{\int_{t_1}^{t_2}\beta(s,Y(s))ds\Big\}\Big]\mu_1(dx) < \infty, \quad t_2\geq t_1,$$

such that for <u>each</u> solution P_{t_1,μ_1} of the t_1-initial value martingale problem for $\mathfrak{G}_1$

$$\int_{E_2} P_{t_1,\mu_1}\Big[f(X(t_2),y)\Big]\mu_2(dy) = \int Q_{t_1,\mu_2}\Big[f(x,Y(t_2))\exp\Big\{\int_{t_1}^{t_2}\beta(s,Y(s))ds\Big\}\Big]\mu_1(dx).$$

The following result transforms the uniqueness problem for $\mathfrak{G}_1$ into the existence problem for $\mathfrak{G}_2$.

<u>Lemma 5.5.1.</u> Let $M_0(E_1) \subset M_1(E_1)$. Assume that the family $\{f(.,y):y\in E_2\}$ is $M_0(E_1)$-determining and $\forall y\in E_2$ $g(.,y)$ is $\sigma(\{f(.,y):y\in E_2\})$-measurable. Suppose that $\mathfrak{G}_2$ provides a FK-dual representation for the restricted t_1-initial value martingale problem for $(\{f(.,y):y\in E_2\},\mathfrak{G}_1,M_0(E))$ for every pair (μ,δ_y) with $\mu\in M_0(E_1)$, $y\in E_2$. Then for each $\mu\in M_0(E_1)$ there is a unique solution $P_{t_1,\mu}$ to the t_1-initial value martingale problem for $\mathfrak{G}_1$ satisfying $\Pi_{t_1}P_{t_1,\mu}=\mu$.

<u>Proof.</u> Let Q_{t_1,δ_y} be a solution of the t_1-initial value martingale problem for $(\mathfrak{G}_2,\delta_y)$. If $\mu\in M_0(E_1)$ and $P_{t_1,\mu}$ and $\tilde{P}_{t_1,\mu}$ are solutions of the retricted t_1-initial

89

value martingale problem for $(\mathfrak{G}_1,\mu)$, then by assumption

$$P_{t_1,\mu}[f(X(t_2),y)] = \int_{E_1} Q_{t_1,\delta_y}\left[f(x,Y(t_2))\exp\left\{\int_{t_1}^{t_2}\beta(s,Y(s))ds\right\}\right]\mu(dx)$$

$$= \tilde{P}_{t_1,\mu}[f(X(t_2),y)].$$

Since $\{f(.,y):y\in E_2\}$ is $M_0(E_1)$-determining, then $P_{t_1,\mu}(X(t_2)\in B)=\tilde{P}_{t_1,\mu}(X(t_2)\in B)$ $\forall B\in\mathcal{B}(E_1)$. The result now follows from Theorem 5.1.2. □

The following result provides a criterion for verifying the condition of Lemma 5.5.1.

<u>Theorem 5.5.2.</u> Let $\mu\in M_1(E_1)$, $P_{t_1,\mu} \in M_1((D([t_1,\infty),E_1),(\mathcal{D}^1_{t_1,t})_{t\ge t_1})$ and $\Pi_{t_1}P_{t_1,\mu}=\mu$. Let $\{Q_{t_1,\delta_y}:y\in E_2\} \subset M_1(D([t_1,\infty),E_2),(\mathcal{D}^2_{t_1,t})_{t\ge t_1})$ (with $\Pi_{t_1}Q_{t_1,\delta_y}=\delta_y$).

Let $\sigma\ge t_1$ be a $\{\mathcal{D}^2_{t_1,t}\}$-stopping time. Let $f\in \mathcal{E}_1\otimes\mathcal{E}_2$, $g,h\in \mathcal{B}(\mathbb{R}_+)\otimes\mathcal{E}_1\otimes\mathcal{E}_2$, $\beta\in\mathcal{B}(\mathbb{R}_+)\otimes\mathcal{E}_2$. Suppose that for each $T>t_1$ there exist a random variable Γ_T which is $P_{t_1,\mu}\otimes Q_{t_1,\delta_y}$-integrable for all $y\in E_2$, and a constant C_T such that

$$\sup_{\substack{t_1\le s\le T\\ t_1\le r\le\sigma\wedge T}} (\max|f(X(s),Y(r))|,|g(s;X(s),Y(r)|) \le \Gamma_T,\quad P_{t_1,\mu}\otimes Q_{t_1,\delta_y}\text{-a.s.}\forall y \in E_2,$$

$$\sup_{t_1\le u\le\sigma\wedge T} |\beta(u;Y(u))| \le C_T,\ Q_{t_1,\delta_y}\text{-a.s.},\ \forall y\in E_2.$$

Assume that

(i) $f(X(t),y)-\int_{t_1}^{t} g(s;X(s),y)ds$ is a $(P_{t_1,\mu},\{\mathcal{D}^1_{t_1,t}\}_{t\ge t_1})$-martingale for each $y\in E_2$, and

(ii) $\forall y\in E_2$, $f(x,Y(t\wedge\sigma)) - \int_{t_1}^{t\wedge\sigma} h(s;x,Y(s))ds$ is a $(Q_{t_1,\delta_y},\{\mathcal{D}^2_{t_1,t}\}_{t\ge t_1})$-martingale for each $x\in E_1$, and

(iii) $g(s;x,y) = h(t-s;x,y)+\beta(t-s;y)f(x,y)\quad \forall\ t_1\le s\le t,\ x\in E_1,\ y\in E_2$.

Then for all $t_1\le t\le T$, $\forall y\in E_2$,

$$P_{t_1,\mu}[f(X(t),y)] - \int_{E_1}\mu(dx)\left(Q_{t_1,\delta_y}\left[f(x,Y(t\wedge\sigma))\exp\left\{\int_{t_1}^{t\wedge\sigma}\beta(u;Y(u))du\right\}\right]\right)$$

$$= P_{t_1,\mu}\otimes Q_{t_1,\delta_y}\left[\int_{t_1}^{(t_1+t-\sigma)^+} g(s;X(s),Y(\sigma))\exp\left\{\int_{t_1}^{\sigma}\beta(u;Y(u))du\right\} ds\right].$$

90

<u>Proof.</u> If for $s,t \geq t_1$, $F(s,t) := P_{t_1,\mu} \otimes Q_{t_1,\delta_y}\left\{ f(X(s),Y(t\wedge\sigma)) \exp\left(\int_{t_1}^{t\wedge\sigma} \beta(v;Y(v))dv \right) \right\}$,

then from the hypotheses (i) and (ii), it can be shown that

$$F(s,t) = F(t_1,t) + P_{t_1,\mu} \otimes Q_{t_1,\delta_y}\left\{ \int_{t_1}^{s} \left[g(u;X(u),Y(t\wedge\sigma)) \exp\left(\int_{t_1}^{t\wedge\sigma} \beta(v;Y(v))dv \right) \right] du \right\}.$$

$$F(s,t) = F(s,t_1) + P_{t_1,\mu} \otimes Q_{t_1,\delta_y}\left\{ \int_{t_1}^{t} \left[(h(u;X(s),Y(u)) + \beta(u;Y(u)))1(u\leq\sigma) \exp\left(\int_{t_1}^{u} \beta(v;Y(v))dv \right) \right] du \right\}.$$

Under the above hypotheses, it can verified that F is absolutely continuous in s for each t and absolutely continuous in t for each s. (Refer to Dawson and Kurtz (1982) or [EK p. 193] for the details.)

Let $F_1(s,t) := \partial F(s,t)/\partial s$, $F_2(s,t) := \partial F(s,t)/\partial t$. Then

$$F_1(s,t) = P_{t_1,\mu} \otimes Q_{t_1,\delta_y}\left[g(s;X(s),Y(t\wedge\sigma)) \exp\left(\int_{t_1}^{t\wedge\sigma} \beta(v;Y(v))dv \right) \right], \text{ and}$$

$$F_2(s,t) = P_{t_1,\mu} \otimes Q_{t_1,\delta_y}\left[[h(t;X(s),Y(t\wedge\sigma)) + \beta(t;Y(t))]1(t\leq\sigma) \exp\left(\int_{t_1}^{t} \beta(v;Y(v))dv \right) \right].$$

Then it also follows from the above hypotheses that $\int_{t_1}^{T}\int_{t_1}^{T} |F_i(s,t)| dsdt < \infty$.

Let us next verify that

$$F(t,t_1) - F(t_1,t) = \int_{t_1}^{t} [F_1(s,t_1+t-s) - F_2(s,t_1+t-s)]ds.$$

Note that $\displaystyle\int_{t_1}^{\tau} \left(\int_{t_1}^{t} [F_1(s,t_1+t-s) - F_2(s,t_1+t-s)]ds \right) dt$

$$= \int_{t_1}^{\tau}\int_{s}^{\tau} F_1(t_1+t-s,s)dtds - \int_{t_1}^{\tau}\int_{s}^{\tau} F_2(s,t_1+t-s)dtds$$

$$= \int_{t_1}^{\tau} [F(t_1+\tau-s,s) - F(t_1,s)]ds - \int_{t_1}^{\tau} [F(s,t_1+\tau-s) - F(s,t_1)]ds$$

$$= \int_{t_1}^{\tau} [F(s,t_1) - F(t_1,s)]ds, \quad t_1 \leq \tau \leq T.$$

Differentiating with respect to τ we get

91

$$F(\tau,t_1)-F(t_1,\tau) = \int_{t_1}^{\tau} [F_1(s,t_1+\tau-s)-F_2(s,t_1+\tau-s)]ds$$

for a.e. τ in $[t_1,T]$. The result for all $\tau\in [t_1,T]$ then follows by the continuity of $F(\tau,t_1)$ and $F(t_1,\tau)$.

Therefore

$$F(t,t_1)-F(t_1,t)$$

$$= P_{t_1,\mu}\left\{f(X(t),y)\right\} - \int_{E_1} \mu(dx)\left(Q_{t_1,\delta_y}\left\{f(x,Y(t\wedge\sigma))\exp\left(\int_{t_1}^{t\wedge\sigma} \beta(v;Y(v))dv\right)\right\}\right)$$

$$= \int_{t_1}^{t} [F_1(s,t_1+t-s)-F_2(s,t_1+t-s)]ds$$

$$= \int_{t_1}^{t}\left\{P_{t_1,\mu}\otimes Q_{t_1,\delta_y}\left[g(s;X(s),Y((t_1+t-s)\wedge\sigma))\exp\left(\int_{t_1}^{(t_1+t-s)\wedge\sigma} \beta(v;Y(v))dv\right)\right]\right.$$

$$- P_{t_1,\mu}\otimes Q_{t_1,\delta_y}\left[[h(t_1+t-s;X(s),Y(t_1+t-s))+\beta(t_1+t-s;Y(t_1+t-s))]1(t_1+t-s\le\sigma)\right.$$

$$\left.\left.\cdot\exp\left(\int_{t_1}^{t_1+t-s} \beta(v;Y(v))dv\right)\right]\right\}ds$$

$$= P_{t_1,\mu}\otimes Q_{t_1,\delta_y}\left[\int_{t_1}^{(t_1+t-\sigma)^+} g(s;X(s),Y(\sigma))\exp\left(\int_{t_1}^{\sigma} \beta(v;Y(v))dv\right)ds\right] \quad \text{(by Fubini).} \ \square$$

Corollary 5.5.3. Consider the time-homogeneous case $g(s;x,y) \equiv g(x,y)$, $h(s;x,y) \equiv h(x,y)$, $\beta(s;y) \equiv \beta(y)$ $\forall s$. In addition to the hypotheses of Theorem 5.5.2 assume that

(i) $f \in bC(E_1\times E_2)$, and $\{f(.,y):y\in E_2\}$ is measure-determining on E_1,

(ii) there exist stopping times $\sigma_K\uparrow t$ such that

$$\left\{ (1 + \sup_x |g(x,Y(\sigma_K))|)\cdot\exp\left(\int_0^{\sigma_K} |\beta(Y(u))|du\right)\right\}_K$$

are Q_{δ_y}-uniformly integrable $\forall$ $y\in E_2$, and

(iii) $Q_{\delta_y}(Y(s-)\ne Y(s)) = 0$ for each $s\ge 0$.

Then the $(\mathfrak{G}_1,\{f(.,y):y\in E_2\})$ initial value martingale problem is well-posed and

$$P_\mu[f(X(t),y)] = \int_{E_1} \mu(dx)\left(Q_{\delta_y}\left[f(x,Y(t))\exp\left(\int_0^t \beta(Y(u))du\right)\right]\right).$$

92

<u>Proof.</u> By Theorem 5.5.2

$$P_\mu[f(X(t),y)] - \int_{E_1}\mu(dx)\left(Q_{\delta_y}\left[f(x,Y(t\wedge\sigma_K))\exp\left\{\int_0^{t\wedge\sigma_K}\beta(Y(u))du\right\}\right]\right)$$

$$= P_\mu\otimes Q_{\delta_y}\left[\int_0^{(t-\sigma_K)^+} g(X(s),Y(\sigma_K))\exp\left\{\int_0^{\sigma_K}\beta(Y(u))du\right\} ds\right].$$

By (ii)

$$\lim_{K\to\infty} P_\mu\otimes Q_{\delta_y}\left[\int_0^{(t-\sigma_K)^+}|g(X(s),Y(\sigma_K))|\exp\left\{\int_0^{\sigma_K}\beta(Y(u))du\right\} ds\right]$$

$$\le \mathit{const}\cdot \lim_{K\to\infty} Q_{\delta_y}\left[|t-\sigma_K| \sup_x |g(x,Y(\sigma_K))| \exp\left\{\int_0^{\sigma_K}|\beta(Y(u))|du\right\}\right]$$

$$= 0.$$

Moreover, by (iii)

$$f(x,Y(t\wedge\sigma_K))\exp\left\{\int_0^{t\wedge\sigma_K}\beta(Y(u))du\right\} \longrightarrow f(x,Y(t))\exp\left\{\int_0^{t}\beta(Y(u))du\right\} \text{ as } K\to\infty,$$

Q_{δ_y}-a.s. and by (i),

$$\left|f(x,Y(t\wedge\sigma_K))\exp\left\{\int_0^{t\wedge\sigma_K}\beta(Y(u))du\right\}\right| \le \mathit{const}\ \exp\left\{\int_0^{t}|\beta(Y(u))|du\right\}.$$

The result follows by dominated convergence. □

5.6. A Function-valued Dual to the Fleming-Viot Process.

Let E be a compact metric space. In this section we introduce a *function-valued dual* to the Fleming-Viot $M_1(E)$-valued process. The dual process will take values in the algebra of functions

$$C_{dir}(E^{\mathbb{N}}) = \sum_n C(E^n) \quad \text{(direct sum).}$$

Functions in $C_{dir}(E^{\mathbb{N}})$ are denoted by $f = (f_1,f_2,\ldots)$. A function $f \in C_{dir}(E^{\mathbb{N}})$ is said to be *simple* and $\#(f) = n$ if $f_n \ne 0$, and $f_m = 0$ for all $m\ne n$.

The set of simple functions is denoted by $C_{sim}(E^{\mathbb{N}})$.

Define $F:C_{dir}(E^{\mathbb{N}}) \longrightarrow C(M_1(E))$ by

$$F(f,\mu) := \int_{E^{\mathbb{N}}} F(x)\mu^\infty(dx),\quad \mu^\infty = \prod_n\mu_n \text{ with } \mu_n=\mu\ \forall n$$

and note that $C_P(M_1(E))$ is the range of F.

93

Remark. We can also introduce appropriate normed spaces containing $C_{dir}(E^{\mathbb{N}})$, for example, we can consider the norm $\|f\| := \sum \|f_n\|_n < \infty$ where $\|f_n\|_n$ is the sup norm in $C(E^n)$. In addition certain Banach subspaces of $C(M_1(E))$ are required for the study of some processes related to the Fleming-Viot process (see Dawson and March (1992)).

Let $\{T_t\}$ be a Feller semigroup $\{T_t\}$ on $C(M_1(E))$ when $C(M_1(E))$ is given the supremum norm. If $T_tF(f,\mu)$ can be represented by $F(\hat{T}_tf,\mu)$, where $\{\hat{T}_t\}$ is a Markov semigroup on $C_{dir}(E^{\mathbb{N}})$, then the resulting $C_{dir}(E^{\mathbb{N}})$-valued process is called a *function-valued dual*. At least at the formal level a Markov semigroup on $C(E^{\mathbb{N}})$ can be identified with an infinite particle system in E. In Chapter 10, this idea will be developed for the Fleming-Viot process.

Now consider the Fleming-Viot process with generator $(C_P^A(M_1(E)),\mathfrak{G})$ where $C_P^A(M_1(E)) := \{F(f,\mu) \in C_P(M_1(E)),\ f\in D_0^n\}$ (cf. Sect. 5.4). If f is simple and $\#(f)=n$, we rewrite $F(f,\mu)$ as $F_{f,n}(\mu)$. For functions of this class the Fleming-Viot generator has the form:

$$\mathfrak{G}F_{f,n}(\mu) = F_{A^{(n)}f}(\mu) + \frac{\gamma}{2}\sum_{j=1}^{N}\sum_{k\neq j}[F_{\tilde{\Theta}_{jk}f,n-1}(\mu)-F_{f,n}(\mu)].$$

Now consider

$$KF(f) = A^{(\#(f))}f\cdot F'(f) + \frac{\gamma}{2}\sum_{\substack{j,k=1\\ j\neq k}}^{\#(f)}[F(\tilde{\Theta}_{jk}f)-F(f)]$$

where $\tilde{\Theta}_{jk}:D_0^n \to D_0^{n-1}$ is defined as in Remark 2.5.2b. K is the generator of a càdlàg Markov process $\{Y(t)\}$ with values in $C_{sim}(E^{\mathbb{N}})$ and law $\{Q_f:f \in C_{sim}(E^{\mathbb{N}})\}$ which evolves as follows:

(i) Y(t) jumps from $C(E^n)$ to $C(E^{n-1})$ at rate $\frac{1}{2}\gamma n(n-1)$

(ii) at the time of a jump, f is replaced by $\tilde{\Theta}_{jk}f$

(iii) between jumps, Y(t) is deterministic on $C(E^n)$ and evolves according to the semigroup $\{S_t^n\}$ with generator $A^{(n)}$.

Theorem 5.6.1 (a) Let $(\{X(t)\}_{t\geq 0},\{\tilde{P}_\nu:\nu\in M_1(M_1(E))\})$ be a solution to the Fleming-Viot martingale problem and the process $(\{Y(t)\}_{y\geq 0},\{Q_f:f\in C_{sim}(E^{\mathbb{N}})\})$ be defined as above with $Q_f = Q_{\delta_f}$.

Then

94

(a) these processes are dual, that is,

$$\tilde{P}_\nu(F(f,X(t))) = \int_{M_1(M_1(E))} Q_f\{F(Y(t),\mu)\}\nu(d\mu) \ \forall \nu \in M_1(M_1(E)),\ f \in C_{sim}(E^{\mathbb{N}}),$$

and

(b) the Fleming-Viot martingale problem is well posed and $\{X(t)\}_{t\geq 0}$ is a strong Markov process with transition measures $P_\mu = \tilde{P}_{\delta_\mu}$, $\mu \in M_1(E)$.

<u>Proof.</u> In this case the function β is identically zero and for $f \in C_{sim}(E^{\mathbb{N}})$, $\#(f)=n$,

$$\mathfrak{G}F(f,\mu) = F(A^{(n)}f,\mu) + \frac{\gamma}{2}\sum_{j=1}^{n}\sum_{k\neq j}[F(\tilde{\Theta}_{jk}f,\mu)-F(f,\mu)]$$

$$= KF(f,\mu)$$

(a) then follows from Theorem 5.5.2. (b) then follows from Theorems 5.1.2 and 5.1.3. To apply Theorem 5.1.3 we note that we can choose a countable subset of $\tilde{C} \subset C_{sim}(E^{\mathbb{N}})$ such that $\{F(f,.):f\in\tilde{C}\}$ is $M_1(M_1(E))$-determining. □

<u>Remarks.</u> This argument can be modified to cover the case in which we do not have an algebra D_0 as assumed in Sect. 5.4 - see Ethier and Kurtz (1987).

The dual representation gives an alternative proof that the Fleming-Viot martingale problem is well-posed. However we have already done this using the fact that the moment equations are closed and relatively simple to solve. The real power of the dual process is exhibited when the moment equations are either much more complex or they are not closed. We will illustrate this in the next section and in Chap. 10.

In addition to proving that the martingale problem is well-posed the dual process has other useful applications. To illustrate one of these we will now use it to prove an ergodic theorem for the Fleming-Viot process with ergodic mutation semigroup.

<u>Theorem 5.6.2</u> Let $\{X_t\}$ be a Fleming-Viot process with mutation semigroup $\{S_t\}$ on $C(E)$ and $Y(t)$ be the dual process defined above. Suppose there exists $\pi \in M_1(E)$ such that

$$\lim_{t\to\infty} S_t f(x) = \langle\pi,f\rangle \ \forall x \text{ and } f \in C(E).$$

Then for $f \in C_{sim}(E^{\mathbb{N}})$, $X(0)=\mu$,

$$\lim_{t\to\infty} P_\mu[\langle X(t)^\infty,f\rangle] = Q_f[\langle\pi,Y(\tau_1)\rangle]$$

where $\tau_1 := \inf\{t:\#(Y(t)=1\}$.

<u>Proof.</u> If $\#(Y(0)) = n > 1$, then after $(n-1)$ jumps $\#(Y(t))$ absorbs at $Y(t)=1$.

95

Therefore

$$\lim_{t\to\infty} P_\mu[\langle X(t)^\infty,f\rangle] = \lim_{t\to\infty} E_f[\langle\mu,S(t-\tau_1)Y(\tau_1)\rangle]$$

$$= Q_f\langle\mu,\langle\pi,Y(\tau_1)\rangle\rangle = Q_f\langle\pi,Y(\tau_1)\rangle.\square$$

5.7. Examples and Generalizations of the Fleming-Viot Process

A number of variations of the basic Fleming-Viot model arise in the study of measure-valued processes and certain models in population genetics. In this section we will briefly review some of these.

5.7.1 Time Inhomogeneous Fleming-Viot.

Dynkin (1989c) introduced a Fleming-Viot process with time inhomogeneous mutation semigroup and Perkins (1991b) introduced a Fleming-Viot process in which the sampling rate varies in time and may be unbounded. We will briefly indicate the resulting Fleming-Viot process incorporating both types of time inhomogeneity.

Let $C_+ = \{\gamma:[0,\infty) \longrightarrow [0,\infty): \gamma$ cont., $\exists\ \tau_\gamma \in (0,\infty]$ such that $\gamma(t) > 0$ if $t\in [0,\tau_\gamma)$ and $\gamma(t) = 0$ if $t \ge \tau_\gamma\}$. C_+ is given the compact open topology. If $B \subset C_+$, let $B|_{T-} = \{\gamma|_{[0,T)}:\gamma\in B\}$ and $B|_T = \{\gamma(s)|_{[0,T]}:\gamma\in B\}$. Let $\{S_{r,t}\}$ be a strongly continuous evolution family on C(E) with generator A_t and domain D(A) (as in Prop. 2.8.4). (For example in Chapter 8 we will consider the case $A(t) = \sigma(t)A$ where $\sigma \in pC(\mathbb{R}_+)$.)

Theorem 5.7.1 Assume that $1/\gamma \in C_+$.

(a) There exists a unique solution $\{P_\mu^{S(A,\gamma)}:\mu\in M_1(E)\}$, to the initial value martingale problem: for each $\phi \in D(A)$,

$$M_t^{S,\gamma}(\phi) = \langle X_t,\phi\rangle - \langle\mu,\phi\rangle - \int_0^t \langle X_s,A_s\phi\rangle ds, \quad t < \tau_\gamma,$$

is an $(\mathcal{D}_t)$-martingale starting at 0 such that

$$\langle M^{S,\gamma}(\phi)\rangle_t = \int_0^t (\langle X_s,\phi^2\rangle-\langle X_s,\phi\rangle^2)\gamma(s)ds \quad \forall\ t < \tau_\gamma,$$

$$M_t^{S,\gamma}(\phi) = M_{\tau_\gamma-}^{S,\gamma}(\phi) \text{ for } t\ge\tau_\gamma,$$

and $X_0 = \mu$ and $X_t \equiv X_{\tau_\gamma}$ for all $t\ge\tau_\gamma$.

(b) Let $A_t=A$ be the generator of a Feller semigroup and $(\mu_n,\gamma_n|_{[0,T)}) \longrightarrow (\mu,\gamma|_{[0,T)})$ as $n\to\infty$ in $M_1(E)\times C_+|_{T-}$ $\forall\ T \le \tau_\gamma$. Then

$$P_{\mu_n}^{S(A,\gamma_n)}\Big|_{C([0,T],M_1(E))} \Longrightarrow P_\mu^{S(A,\gamma)}\Big|_{C([0,T],M_1(E))} \quad \text{as } n \to \infty\ \forall\ T < \tau_\gamma.$$

96

(c) Let $\mathcal{P}_\gamma$ denote the transition function associated with the solution of the martingale problem in (a). Then

$$\int \mathcal{P}_\gamma(r,\mu;t,d\nu)\langle\nu^k,h\rangle = \langle\mu^k S_k(r,t),h\rangle + \sum_{j=1}^{k-1}\int_r^t ds\langle\mu^j S_j(r,s),U_j^k(s,t)h\rangle$$

where for $h\in\mathcal{B}_{sym}(E^k)$, $t<t_\gamma$

$$(S_k(r,t)h)(x_1,\dots,x_k)$$
$$:= \int e_k(r,t)p(r,x_1;t,dy_1)\dots p(r,x_k;t,dy_k)h(y_1,\dots,y_k)$$

and for $\nu\in M_1(E^k)$

$$(\nu S_k(r,t))(dy_1,\dots,dy_k) =$$
$$\int e_k(r,t)\nu(dx_1,\dots,dx_k)p(r,x_1;t,dy_1)\dots p(r,x_k;t,dy_k),$$

$$e_n(r,t) := \exp\left(-\tfrac{1}{2}\,n(n-1)\int_r^t\gamma(s)ds\right)$$

Ψ_k^{k-1} is as in Prop. 2.8.1

and $U_{k-1}^k(r,t) := \frac{1}{2}k\gamma(r)\Psi_k^{k-1}S_k(r,t)$ for $r<t$

$:= 0$ for $r\ge t$

$$U_j^k(r,t) = \int U_j^{k-1}(r,s)U_{k-1}^k(s,t)ds\ ,\ \text{for}\ \ k-j>1.$$

<u>Outline of Proof.</u> (a) (cf. Perkins (1991b)) The proof of existence and uniqueness on the interval $[0,T]$ with $T<\tau_\gamma$ is similar to that for the time homogeneous case treated above. The main difference appears in the proof of uniqueness of the law of X_t for $0\le t\le T$. To accomplish this in this case we must use a time inhomogeneous function-valued dual process with generator $\{K_s:0\le s\le t\}$ given by:

$$K_sF(f) = A_{t-s}^{(\#(f))}f\cdot F'(f) + \frac{\gamma(t-s)}{2}\sum_{\substack{j,k=1\\ j\ne k}}^{\#(f)}[F(\tilde{\Theta}_{jk}f)-F(f)].$$

(b) The tightness of $\left\{P_{\mu_n}^{S(A,\gamma_n)}\Big|_{C([0,T],M_1(E))} :n\in\mathbb{N}\right\}$ for $T<\tau_\gamma$ is proved by arguments similar to those used earlier and based on Theorem 3.6.6.

The weak convergence of the corresponding inhomogeneous function-valued dual processes can be proved using their direct contruction. The weak convergence of the finite dimensional distributions of $P^{S(A,\gamma)}_{\mu_n}$ is then obtained by using the Markov property and the dual representation.

(c) This follows by the same arguments as in the proof of Prop. 2.8.1. □

Remark. We will refer to the process considered in (a) as the $S(A,\gamma(.))$- Fleming-Viot process. This process will be used in Chapter 8.

5.7.2. The infinitely many neutral alleles model.

Let $E = [0,1]$ and $\quad (Qf)(x) := \theta\int_0^1 (f(y)-f(x))dy, \; x\in E, \; \theta>0.$

This model satisfies the partition property, namely, given any finite partition $E = \bigcup_{j=1}^{K} E_j$, then $\{X(t,E_j): j=1,\dots,K\}$ is a finite dimensional Markov diffusion process as in Section 5.3. For a detailed discussion see Ethier and Kurtz (1990b, 1992b).

5.7.3. The Size Ordered Atom Process

Let $\mathfrak{A}_K := \{z=(z_1,\dots,z_K): z_1\geq 0,\dots,z_K\geq 0, \sum_{i=1}^{K} z_i \leq 1\}$,

$\sigma_i: C([0,\infty),\mathfrak{A}_K)\longrightarrow [0,\infty], \quad \sigma_i(z) := \inf\{t\geq 0: z_i(t)=0\}$.

Define G_K on $\mathfrak{D}(G_K) = C^2(\mathfrak{A}_K)$ by

$$G_K := \frac{\gamma}{2}\sum_{i,j=1}^{K} z_i(\delta_{ij}-z_j)\frac{\partial^2}{\partial z_i\partial z_j} + \frac{1}{2}\frac{K\theta}{(K-1)}\sum_{i=1}^{K}\left\{\frac{1}{K} - z_i\right\}\frac{\partial}{\partial z_i}, \; \theta>0,$$

$D_K = \{z\in C([0,\infty),\mathfrak{A}_K): \sigma_1(z)\geq\sigma_2(z)\geq\dots\geq\sigma_K(z)\}$.

For $K = 2$ the pair $\{z_1(t),z_2(t)\}$ can be identified with $(\max(z(t),1-z(t)), \min(z(t),1-z(t)))$ where $z(t)$ is the unique strong solution of the s.d.e.

$$dz(t) = \theta(1/2 - z(t))dt + \sqrt{\gamma z(t)(1-z(t))}\, dw(t).$$

This is called the *size-ordered atom process* for the *K-neutral-alleles diffusion*. In the limit of *infinitely many neutral alles* this becomes

$$G_\infty = \frac{1}{2}\sum_{i,j=1}^{\infty} z_i(\delta_{ij}-z_j)\frac{\partial^2}{\partial z_i\partial z_j} - \frac{1}{2}\theta\sum_{i=1}^{\infty} z_i\frac{\partial}{\partial z_i}.$$

(For a detailed discussion of this process see Ethier and Kurtz (1981).)

98

5.7.4. Continuous allele stepwise mutation of Ohta and Kimura

$E = \bar{\mathbb{R}}^d$, the one point compactification of $\mathbb{R}^d$ and

$$Af(x) = \frac{1}{2}\Delta f(x), \quad x \in \mathbb{R}^d$$
$$= 0, \quad x = \infty.$$

This is the model originally studied by Fleming and Viot (1979).

5.7.5. Shimizu's gene conversion model

Shimizu (1990) has studied the following *n-loci* model:

$$E = (E_0)^n,$$

$$Af(x_1,\ldots,x_n) = \sum_{j_1 \neq j_2} \{\beta_{j_1 j_2} f(x_1,\ldots,x_n) - f(x_1,\ldots,x_n)\}$$

where $\beta_{j_1 j_2}$ denotes the replacement of the variable x_{j_1} by x_{j_2}.

5.7.6. Infinitely many neutral alleles with ages

Let $E = [0,1]\times[0,\infty]$. The first component denotes the allele and the second the age of the allele, that is the time from the last mutation.

$$(Af)(x,a) = \frac{\partial}{\partial a} f(x,a) + \frac{1}{2}\,\theta\int(f(\xi,0)-f(x,a))P(x,d\xi)$$

$D(A) = C^{0,1}([0,1]\times[0,\infty])$, the space of functions continuous in the first variable and continuously differentiable in the second.

$P(x,d\xi)$ denotes the distribution of the allele which results from a mutation from an allele of type x. In the case $P(x,d\xi) = d\xi$ we obtain the infinitely many neutral alleles with ages model. For a detailed discussion see Ethier (1988).

5.7.7. Infinitely many sites model

For this model which was introduced by Ethier and Griffiths (1987) $E = [0,1]^{\mathbb{Z}_+}$ and the mutation operator is

$$Af(x_1,x_2,\ldots) := \frac{1}{2}\theta\int(f(y,x_1,x_2,\ldots)-f(x_1,x_2,\ldots))\nu_0(dy), \quad \theta>0.$$

Here the type of a gene is given by $x = (x_0,x_1,\ldots) \in E$ such that $x_0, x_1,\ldots$ is the sequence of sites at which mutations have occurred in the line of descent of the gene in question. (x_0 is the most recent mutation).

5.7.8. Weighted Sampling

99

The basic random sampling mechanism of the Fleming-Viot model can be modified to allow for *weighted sampling*. Here $\theta(x,y) \in C_{sym}(E^2)$ (and $\theta(x,x) = 1$), denotes the rate at which an individual of type x is replaced by one of type y. In the setting of the K-type N-particle approximation of section 5.2 with mutation operator A, we obtain

$$\mathfrak{G}_N^K F(\mu) = N\int_{E_K} \sum_{y\in E_K} q^K_{x,y}[F(\mu - \delta_x/N + \delta_y/N)-F(\mu)]\mu(dx)$$

$$+ \gamma N(N-1)\int_{E_K}\int_{E_K} [F(\mu - \delta_x/N + \delta_y/N)-F(\mu)]\theta(x,y)\mu(dx)\mu(dy)$$

$$= \int\{A^K \frac{\delta F(\mu)}{\delta\mu(.)}\}(x)\mu(dx)$$

$$+ \frac{\gamma}{4}\iint\left[\frac{\delta^2 F(\mu)}{\delta\mu(x)^2} + \frac{\delta^2 F(\mu)}{\delta\mu(y)^2} - 2\frac{\delta^2 F(\mu)}{\delta\mu(x)\delta\mu(y)}\right]\theta(x,y)\mu(dx)\mu(dy)$$

$$+ R_N(F)/N$$

Letting $N \to \infty$ and then $K \to \infty$, we obtain

$$\mathfrak{G}F(\mu) = \int\left(A\frac{\delta F(\mu)}{\delta\mu(.)}\right)(x)\mu(dx)$$

$$+ \frac{\gamma}{4}\iint\left[\frac{\delta^2 F(\mu)}{\delta\mu(x)^2} + \frac{\delta^2 F(\mu)}{\delta\mu(y)^2} - 2\frac{\delta^2 F(\mu)}{\delta\mu(x)\delta\mu(y)}\right]\theta(x,y)\mu(dx)\mu(dy)$$

$$= \int\left(A\frac{\delta F(\mu)}{\delta\mu(.)}\right)(x)\mu(dx)$$

$$+ \frac{\gamma}{2}\iint\frac{\delta^2 F(\mu)}{\delta\mu(x)\delta\mu(y)}[\theta(x,z)\delta_x(dy)\mu(dx)\mu(dz) - \theta(x,y)\mu(dx)\mu(dy)]$$

For functions in $C_P(M_1(E))$, this becomes

$$\mathfrak{G}F_{f,N}(\mu)$$

$$= F_{A^{(N)}f,N}(\mu) + \frac{\gamma}{2}\sum_{\substack{k,\, j=1\\ k\neq j}}^{N} [F_{\Psi_{jk}f,N}(\mu) - F_{f,N}(\mu)] + \frac{\gamma}{2}N(N-1)F_{f,N}(\mu)$$

where $\Psi_{jk}:C(E^N) \longrightarrow C(E^N)$ is defined by

$$(\Psi_{jk}f)(x_1,\ldots,x_N) = \theta(x_j,x_k)(\Theta_{jk}f)(x_1,\ldots,x_N) - \theta(x_j,x_k)f(x_1,\ldots,x_N)$$

and Θ_{jk} is defined as in Section 2.5.

The dual process, $\{Y(t)\}$ then has jumps $f \longrightarrow \Psi_{jk}f$ at rate $\gamma\cdot\#(f)(\#(f)-1)/2$ and satisfies $\#(Y(t)) = \#(Y(0))\ \forall t\geq 0$. Between jumps it evolves deterministically accord-

100

ing to the semigroup $f \longrightarrow S_t^{(\#(f))}f$.
The uniqueness to the martingale problem follows from the FK-dual representation with $\beta(Y(t)) := \frac{\gamma}{2}\#(Y(t))(\#(Y(t)-1)$. The case $\theta(x,y) \equiv 1$ reduces to the usual Fleming-Viot model.

5.7.9. Multilevel Sampling.

In example 6.2.4 we discuss multilevel branching systems. The analogous multilevel sampling systems were introduced in Dawson (1986b).

5.8. Some Other Classes of Measure-valued Processes with Function-Valued Duals.

In order to further clarify both the usefulness and limitations of the duality method in this section we very briefly review some other examples of measure-valued processes. We will give additional examples in which duality can be used to establish uniqueness for martingale problems. For contrast, some closely related examples in which the method of duality is not applicable are also considered.

5.8.1. Exchangeable Diffusions in a Brownian Medium.

In order to motivate this model we first consider a system of particles in a Brownian medium in $\mathbb{R}$. The simplest such system has the form

$$dx_i(t) = \int R(y-x_i(t))W(dy,dt)$$

where W is a cylindrical Brownian motion (cf. Ex. 7.1.2) and R is a symmetric smooth function on $\mathbb{R}$.

Then $\langle x_i, x_j\rangle_t = \int_0^t \rho(x_i(s)-x_j(s))ds$ where $\rho(z) := \int_{\mathbb{R}} R(z-y)R(y)dy$.

Vaillancourt (1988) introduced a model of interacting diffusions which combines features of the standard mean-field model of interacting diffusions (cf. Gärtner (1988)) and this model of particles in a Brownian medium. It leads naturally to the notion of a family of exchangeable diffusions which are characterized in the following result.

Theorem 5.8.1.1. Consider a differential operator on $C^\infty((\mathbb{R}^d)^N)$ of the form

$$L := \tfrac{1}{2}(a\nabla)^T\nabla + b^T\nabla = \tfrac{1}{2}\sum_{i,j=1}^{N}(a_{ij}\nabla_i)^T\nabla_j + \sum_{i=1}^{N} b_i^T\nabla_i$$

where ∇ denotes the gradient on $\mathbb{R}^d$, and T denotes transpose.

101

Let a and b be continuous and assume that the martingale problem for L is well-posed. Now consider a system of N particles in $\mathbb{R}^d$ diffusing according to L. Then the solutions $\{P_y : y \in (\mathbb{R}^d)^N\}$ form an exchangeable family, (i.e. a system of N-exchangeable d-dimensional processes) if and only if

$b_i(\pi y) = b_{\pi i}(y)\ \forall i=1,\ldots,m,\ y \in E^N,\ \pi \in \mathcal{P}er(N)$, and $a_{ij}(\pi y) = a_{\pi i,\pi j}(y)$.

Proof. See Vaillancourt (1988).

Remark: If the conclusion of Theorem 5.8.1.1 is satisfied then we can rewrite the above system in the form

(5.8.1.1)

$b(y_i,\mu_N) := b_i(y)$, $\quad \sigma(y_i,\mu_N) := a_{ii}(y)$ and $\rho(y_i,y_j,\mu_N) := a_{ij}(y)$, $\quad 1 \le i \ne j \le N$

where $\mu_N := N^{-1} \sum_{i=1}^{N} \delta_{y_i} \in M_1(E)$.

The empirical measure process is defined by

$$X^N(t) := \Xi_N(y_1(t),\ldots,y_N(t))$$

where Ξ_N is defined as in Section 2.2. The generator of X^N is denoted by $\mathfrak{G}_N$.

If $\quad F_{f,k}(.) := \int_{\mathbb{R}^d} \ldots \int_{\mathbb{R}^d} f(x_1,\ldots,x_k)\mu(dx_1)\ldots\mu(dx_N)$

with $f \in C_c^\infty((\mathbb{R}^d)^k)$, then

$$\mathfrak{G}_N F_{f,k}(\mu) = F_{A_k(\mu)f,k}(\mu) + \frac{1}{2N} F_{B_k(\mu)f,k}(\mu)$$

where

$$A_k(\mu)f(y) := \sum_{\alpha=1}^{k} b^T(y_\alpha,\mu)\nabla_\alpha f(y) + \frac{1}{2}\sum_{\alpha=1}^{k} (\sigma(y_\alpha,\mu)\nabla_\alpha)^T \nabla_\alpha f(y)$$

$$+ \frac{1}{2} \sum_{\substack{\alpha,\beta=1 \\ \alpha \ne \beta}}^{k} (\rho(y_\alpha,y_\beta,\mu)\nabla_\beta)^T \nabla_\alpha f(y),$$

$$B_k(\mu)f(y) = \sum_{\substack{\alpha,\beta=1 \\ \alpha \ne \beta}}^{k} ((\sigma(y_\alpha,\mu)-\rho(y_\alpha,y_\alpha,\mu)\nabla_\beta)^T \nabla_\alpha f(y_{\alpha\beta})$$

and $y_{\alpha\beta} = (y_1,\ldots,y_{\beta-1},y_\alpha,y_{\beta+1},\ldots,y_k)$.

102

The limiting generator (as $N \longrightarrow \infty$) is given by

$$\mathfrak{G}_{\infty} F_f(\mu) = F_{A_k(\mu)f}(\mu).$$

Note that it is a second order differential operator of the form

$$\mathfrak{G}_{\infty} F(\mu) = \int_{\mathbb{R}^d} \left[b^T(z,\mu)\nabla + \tfrac{1}{2}(\sigma(z,\mu)\nabla)^T\nabla \right] \frac{\delta F(\mu)}{\delta\mu(z)}\,\mu(dz)$$
$$+ \frac{1}{2} \int_{\mathbb{R}^d}\int_{\mathbb{R}^d} \left[\nabla^T_{z_1}\nabla_{z_2} \frac{\delta^2 F(\mu)}{\delta\mu(z_1)\delta\mu(z_2)}\right]\rho(z_1,z_2,\mu)\mu(dz_1)\mu(dz_2).$$

Theorem 5.8.1.2. Assume that b, σ and ρ are continuous and satisfy

(i) $|b(y,\mu)|^2+|\sigma(y,\mu)|^2 \le K(1+|y|^2)$

where $|.|$ denotes the Euclidean norm,

(ii) for each $y \in (\mathbb{R}^d)^N$ the matrix $\mathbf{a}(y)$ defined in (5.8.1.1) is positive definite and symmetric,

(iii) at time zero $(y_1(0),\ldots,y_N(0))$ are i.i.d. (μ) and $\int |y|^4\mu(dy) < \infty$.

Then the laws of the empirical measure-valued processes $\{X^N\}$ are tight and the limit points are solutions to the martingale problem for $\mathfrak{G}_{\infty}$.

Proof. See Vaillancourt (1988).

Remarks.

1. When the matrix a is diagonal, that is, $\rho \equiv 0$, the solution to the martingale problem for G_{∞} is a deterministic measure-valued function which satisfies the so-called *McKean-Vlasov equation*. Under some mild conditions it is known that this martingale problem is well-posed (for example see Gärtner (1988), Léonard (1986)). However when $\rho \neq 0$, the resulting limiting process is a measure-valued stochastic process which formally solves a *random McKean-Vlasov stochastic partial differential equation*.

2. This leaves open the question of the uniqueness of the martingale problem when $\rho \neq 0$. If the $A_k(\mu)$ do not depend on μ, then the solution to the martingale problem of $\mathfrak{G}_{\infty}$ can be proved to be well-posed using duality (see Dawson and Kurtz (1982) and Vaillancourt (1988)). However for the more general case the duality method is not applicable and other methods are required.

The case in which duality does work, namely $A_k(\mu) \equiv A_k$, is of considerable interest and is in fact a generalization of a model of Chow which we discuss in the next example.

103

5.8.2. A model of molecular diffusion with turbulent transport

Chow (1976) introduced a model of molecular diffusion with turbulent transport in $\mathbb{R}^3$. This process also arises in the study of partial differential equations with random coefficients (see H. Watanabe (1989)).

However in order to illustrate the power of the duality method we will also add Fleming-Viot sampling to this mechanism (cf. Dawson and Kurtz (1982)). The resulting generator for the process involving molecular diffusion, turbulent transport and Fleming-Viot sampling is given (in dual form) by

$$\mathcal{G}F_{f,N}(\mu) = F_{G^{(N)}f,N}(\mu) + \frac{\gamma}{2}\sum_{\substack{i,j=1\\ j\neq i}}^{N} (F_{\tilde{\Theta}_{ij}f}(\mu)-F_f(\mu)).$$

For $\alpha,\beta = 1,2,3$ let $\rho_{\alpha\beta}(x)$ be smooth functions on $\mathbb{R}^3$ with

$$\sum_{j=1}^{k}\sum_{i=1}^{k} \rho_{\alpha\beta}(x_j-x_i)\xi_{j\alpha}\xi_{k\beta} \geq 0 \quad \forall \text{ real } \{\xi_{\alpha\beta}\}.$$

If $\mathbb{R}^3 \ni x_j = (x_{j,1},x_{j,2},x_{j,3})$, then

$$G^{(N)}f(x_1,\ldots,x_N) = \sum_{j=1}^{N}\sum_{i=1}^{N}\sum_{\alpha,\beta=1}^{3} \rho_{\alpha\beta}(x_j-x_i)\frac{\partial^2}{\partial x_{j,\alpha}\partial x_{i,\beta}} f(x_1,\ldots,x_N)$$

$$+\,\kappa\sum_{j=1}^{N}\sum_{\alpha=1}^{3}\frac{\partial^2}{\partial x_{j,\alpha}^2}f(x_1,\ldots,x_N).$$

For sufficiently large κ the operators $G^{(N)}$ are uniformly elliptic and generate strongly continuous semigroups on $C_0((\mathbb{R}^3)^N)$. Then a $C_{sim}(E^{\mathbb{N}})$-valued dual process is constructed exactly as for the Fleming-Viot process except that the semigroup $S_t^{(N)}$ on $C((\mathbb{R}^3)^N)$ now has generator $G^{(N)}$ and then the proof that the martingale problem is well-posed follows from Lemma 5.5.1.

In the case $\gamma=0$ (the original model of Chow), the martingale problem is that associated with the bilinear stochastic partial differential equation

$$dX(t) = LX(t)dt + (\nabla X(t))^T\cdot dW(t)$$

where L is an elliptic operator and W is a Brownian motion with values in $(\mathcal{S}'(\mathbb{R}^3))^3$ and covariance structure given by ρ (see Chow (1978) for the details). A systematic development of stochastic partial differential equations of this form can

104

be found in Pardoux (1975), Krylov and Rozovskii (1981) and Rozovskii (1990).

5.8.3. A Bilinear Stochastic Evolution Equation

Let $\{W(t):t\geq 0\}$ denote a Brownian motion such that for each t, W(t) is a generalized random field on $\mathbb{R}^d$. More precisely W is a Gaussian process with values in $\mathcal{S}'(\mathbb{R}^d)$ and covariance structure

$$E(\langle W(t),\phi\rangle\langle W(s),\psi\rangle) = \min(s,t)\int_{\mathbb{R}^d}\int_{\mathbb{R}^d}\phi(x)\psi(y)Q(x-y)dxdy$$

where Q is assumed to be continuous and non-negative definite. Now consider the *bilinear stochastic partial differential equation*

$$dX(t,x) = \tfrac{1}{2}\Delta X(t,x)dt + \sigma X(t,x)W(dt,x),$$

$$X(0,x) \in pbC(\mathbb{R}^d),$$

or in evolution form

$$X(t,x) = \int p(t,x,y)X(0,y)dy + \sigma\int\int_0^t p(t-s,x,y)X(s,y)W(ds,dy)$$

where $p(t,.,.)$ is the transition function of standard Brownian motion in $\mathbb{R}^d$. The moment equations for this model are closed and hence there is a function-valued dual. However in general the moments of solutions of bilinear stochastic partial differential equations may not satisfy the Carleman condition and so other methods are required to characterize the process. The existence and uniqueness of a strong evolution solution to this equation can be established (cf. Dawson and Salehi (1980), Noble (1992)).

5.8.4. Remark. Refer to Sections 10.2-10.4 for further examples of the application of the method of duality as well as other examples of stochastic evolution equations.

105

6. A MARTINGALE PROBLEM FORMULATION OF MEASURE-VALUED BRANCHING

6.1. The (A,Φ)-Superprocess Martingale Problem.

We have constructed a general class of measure-valued branching processes in Ch. 4 exploiting the method of Laplace functionals. In this section we verify that the resulting processes are in fact solutions of well-posed martingale problems. This fact will make possible the development of a stochastic calculus for superprocesses in the next chapter.

For technical reasons we will carry this out for a restricted class of measure-valued branching processes. Using the notation of sections 4.3, 4.4, we assume that

(i) $\kappa(dt) = dt$, and

(ii) E is a locally compact metric space and the motion process is a Markov process with Feller semigroup $\{S_t:t\ge 0\}$ on $C_0(E)$ with generator $(D(A),A)$,

(iii) the critical branching mechanism Φ is time homogeneous,

$$\Phi(x,\lambda) = -\tfrac{1}{2}c(x)\lambda^2 + \int_0^\infty (1 - e^{-\lambda u} - \lambda u)n(x,du)$$

where the measure $n(x,dy)$ satisfies $\sup_x [\int_0^1 u^2 n(x,du) + \int_1^\infty u\, n(x,du)] < \infty$, and $0 \le c(x) \le c_0 < \infty$ $\forall x$.

(iv) $\{X(t):t\ge 0\}$ is the canonical process defined on $(\mathbf{D},\mathcal{D},(\mathcal{D}_t)_{t\ge 0})$ where $\mathbf{D} := D([0,\infty),M_F(E))$, and there exists a family of Markov probability laws $\{P_\mu:\mu\in M_F(E)\}$ on $\mathbf{D}$ with transition function given as in (4.4.1). This means that for $t>s$, $\mu\in M_F(E)$,

$$\text{(6.1.1)}\qquad P_\mu\left(e^{-\langle X_t,\phi\rangle} \mid \mathcal{D}_s\right) = e^{-\langle X_s,V_{t-s}\phi\rangle}, \quad \phi\in pb\mathcal{E}.$$

where $\{V_t:t\ge 0\}$ is the (A,Φ)-nonlinear semigroup. In this case $v_t = V_t\phi$ solves

$$\text{(6.1.2)}\qquad v_t = S_t\phi + \int_0^t S_{t-s}\Phi(v_s)ds,$$

where $\Phi(\phi)(x) := \Phi(x,\phi(x))$. We will refer to this as the (A,Φ)-*superprocess*.

An immediate consequence of (6.1.1) and the semigroup property is the fact that

$$E\left(\exp(-\langle X_t,V_{T-t}\phi\rangle)\mid\mathcal{D}_s\right) = \exp(-\langle X_s,V_{t-s}V_{T-t}\phi\rangle) = \exp(-\langle X_s,V_{T-s}\phi\rangle),$$

that is,

(6.1.3) for $\phi\in bp\mathcal{E}$, $\exp(-\langle X_t,V_{T-t}\phi\rangle)$ is a martingale on $[0,T]$.

106

We next give an expression for an MP-generator for the (A,Φ)-superprocess. The domain $\mathcal{D}(\mathfrak{G})$ will consist of functions of the form $F(\mu) = f(\langle\mu,\phi\rangle)$, $\phi\in D(A)$, $f\in C_b^\infty(\mathbb{R})$, and

(6.1.4)
$$\mathfrak{G}F(\mu) = \iint\mu(dx)c(x)\delta_x(dy)F''(\mu;x,y) + \int\mu(dx)\int_0^\infty n(x,du)[F(\mu+u\delta_x)-F(\mu) - uF'(\mu;x)] + \int\mu(dx)AF'(\mu;x)$$

where $F'(\mu;x) := \delta F(\mu)/\delta\mu(x)$, $F''(\mu;x,y) := \delta^2F(\mu)/\delta\mu(x)\delta\mu(y)$,
or equivalently,

$$\mathfrak{G}F\mu) = f'(\langle\mu,\phi\rangle)\langle\mu,A\phi\rangle + \tfrac{1}{2}f''(\langle\mu,\phi\rangle)\langle\mu,c\phi^2\rangle + \int\mu(dx)\int_0^\infty n(x,du)[f(\langle\mu,\phi\rangle+u\phi(x))-f(\langle\mu,\phi\rangle)-f'(\langle\mu,\phi\rangle)u\phi(x)]$$

We next briefly review some basic facts from stochastic calculus (standard references include Dellacherie and Meyer (1975, 1980), [LS], and Métivier (1982)).

An adapted càdlàg process, X, defined on a filtered probability space $(\Omega,\mathcal{F},(\mathcal{F}_t)_{t\ge0},P)$, is a *semimartingale* if it has a representation

(6.1.5) $$X_t = X_0 + M_t + A_t$$

where M is a local martingale with $M_0 = 0$, and A is a process of locally bounded variation. M can be taken to be a locally square integrable martingale. It is called a *special semimartingale* if A has locally integrable variation.

<u>Lemma 6.1.1.</u> A special semimartingale has a <u>unique</u> representation of the form (6.1.5) under the additional requirement that A be predictable.

<u>Proof.</u> See [LS, p. 85].

Let N be an adapted random point measure on $\mathbb{R}_+\times S$ where (S,d) is a Polish space, i.e. $N(t,B)$ denote the number of points in $[0,t]\times B$, $B\in\mathcal{B}(S)$. Assume that N satisfies the following:

(a) $E(N(t,B)) < \infty$ $\forall$ $t > 0$, $B\in \mathcal{B}(S)_b$ (i.e. bounded Borel subsets of S),

(b) there exists $\{\hat{N}(t,B):t\ge0,\ B\in\mathcal{B}(S)\}$ such that

(i) for $B\in\mathcal{B}(S)_b$, $t \to \hat{N}(t,B)$ is a continuous $(\mathcal{F}_t)$-adapted increasing process,

(ii) for each t and a.e. $\omega\in\Omega$, $B \to \hat{N}(t,B)$ is a locally finite measure on $(S,\mathcal{B}(S))$,

(iii) for $B\in \mathcal{B}(S)_b$, $t\to \tilde{N}(t,B) := N(t,B) - \hat{N}(t,B)$ is an $(\mathcal{F}_t)$-local martingale,

that is, $\tilde{N}$ is a *martingale measure* (refer to section 7.1 for the basic definitions of martingale measures and stochastic integrals with respect to martingale meas-

107

ures). In fact $\tilde{N}$ is an L^2 orthogonal martingale measure (cf. Ikeda and Watanabe (1981, Ch. 2, Theorem 3.1)).

The predictable increasing process, $\hat{N}$, is called the *compensator* (cf. [LS, p. 172])).

Consider the semimartingale

$$Z(t) = Z(0) + A(t) + M_c(t) + \int_0^{t+}\int_S g_1(s,z)\tilde{N}(ds,dz) + \int_0^{t+}\int_S g_2(s,z)N(ds,dz)$$

where A is continuous, adapted, locally of bounded variation and $A_c(0)=0$, M_c is a continuous local martingale with increasing process $\langle M_c\rangle_t$ and $M_c(0)=0$, g_1 is predictable and satisfies the local version of $E[\int_0^t |g_1(s,z)|^2\hat{N}(ds,dx)] < \infty$, g_2 is predictable and $\int_0^{t+}\int_S |g_2(s,z)|N(ds,dz) < \infty$ a.s. for each t and $g_1g_2 = 0$.

Lemma 6.1.2. (Itô's Lemma) For $f \in C^{1,2}(\mathbb{R}\times\mathbb{R})$, $f(t,Z(t))$ is also a semimartingale and

(6.1.6)

$$\begin{aligned}
&f(t,Z(t))-f(0,Z(0))\\
&= \int_0^t f_1(s,Z(s))ds\\
&+ \int_0^t f_2(s,Z(s))dA(s) + \tfrac{1}{2}\int_0^t f_{22}(s,Z(s))d\langle M_c\rangle_s\\
&+ \int_0^{t+}\int_S [f(s,Z(s-)+g_2(s,z))-f(s,Z(s-))]N(ds,dz)\\
&+ \int_0^{t+}\int_S [f(s,Z(s)+g_1(s,z))-f(Z(s))-g_1(s,z)f'(s,Z(s))]\hat{N}(ds,dz)\\
&+ \text{a local martingale,}
\end{aligned}$$

where $f_1(s,.) := \partial f(s,.)/\partial s$, $f_2(.,x) := \partial f(.,x)/\partial x$, $f_{22}(.,x) := \partial^2 f(.,x)/\partial x^2$.

Proof. See Ikeda and Watanabe (1981, p. 66), or [LS, p. 118].

The following result due to El Karoui and Roelly (1991) shows that the martingale problem for $\mathfrak{G}$ is well-posed.

Theorem 6.1.3 Assume (i), (ii), (iii), (iv) above. Let $D_0(A) := \{\phi: \phi=\phi_0+\varepsilon, \phi_0 \in pD(A), \varepsilon > 0\}$. Then

(a) for $F(\mu) = f(\langle\mu,\phi\rangle)$, $f \in C_b^\infty(\mathbb{R})$, $\phi \in D_0(A)$,

$M_F(t) := F(X(t)) - \int_0^t \mathfrak{G}F(X(s))ds$ is a martingale

108

where $\mathfrak{G}$ is given by (6.1.4).

(b) Any solution of the $(\mathfrak{G},\mathcal{D}(\mathfrak{G}))$-martingale problem satisfies (6.1.3) and therefore the martingale problem is well-posed.

<u>Proof.</u> (a) Let $Z_t(\phi) := e^{-\langle X_t,\phi\rangle}$. We first show that $Z_t(\phi)$ is a special semimartingale, that is, a locally square integrable martingale plus a process of locally integrable variation. The main idea is then to use the uniqueness of the representation of a special semimartingale to identify the terms in two semimartingale representations of $Z_t(\phi)$.

Step 1. We first prove that for $\phi\in D_0(A)$, $\phi \ge \kappa > 0$,

(6.1.7) $\quad H_t(\phi) = \exp\left(-\langle X_t,\phi\rangle) + \int_0^t \langle X_s, A\phi+\Phi(\phi)\rangle ds\right)$ is a $P_{X(0)}$-local martingale.

To prove this we will show that for $B\in\mathcal{D}_s$, $s\le t$,

$$\frac{d}{dt} E(1_B\exp(-\langle X_t,\phi\rangle)) = -E(1_B\langle X_t,(A\phi+\Phi(\phi))\rangle\exp(-\langle X_t,\phi\rangle)).$$

But from (6.1.1)

$$\varepsilon^{-1}E\left(1_B\exp(-\langle X_{t+\varepsilon},\phi\rangle) - \exp(-\langle X_t,\phi\rangle)\right)$$

$$= \varepsilon^{-1}E\left(1_B(\exp(-\langle X_t, V_\varepsilon\phi\rangle) - \exp(-\langle X_t,\phi\rangle))\right).$$

To take the limit as $\varepsilon\downarrow 0$ inside the expectation we use dominated convergence and equation (6.1.2). To justify this it suffices to note that

$$D^\varepsilon(\mu) := \varepsilon^{-1}|\exp(-\langle\mu,V_\varepsilon\phi\rangle)-\exp(-\langle\mu,\phi\rangle)|$$

$$\le \exp(-\kappa\langle\mu,1\rangle)\ \varepsilon^{-1}(\exp(K_T(\phi)\varepsilon\langle\mu,1\rangle) - 1)$$

and this is bounded function of $\langle\mu,1\rangle$ uniformly in $0<\varepsilon\le 1$. To obtain the last inequality note that $v_t = S_t\phi + \int_0^t S_{t-s}\Phi(v_s)ds$ and $S_t\phi-\phi = \int_0^t S_sA\phi ds$ which yields

$$\|V_t\phi-\phi\|/t \le (\|A\phi\| + \sup_{r<T}\|\Phi(V_r\phi)\|) \le K_T(\phi).$$

(A similar argument works for $\varepsilon\uparrow 0$.) Taking the limit $\varepsilon\downarrow 0$ we obtain the derivative

$$\lim_{\varepsilon\downarrow 0} \varepsilon^{-1}E\left(1_B(\exp(-\langle X_{t+\varepsilon},\phi\rangle) - \exp(-\langle X_t,\phi\rangle))\right)$$

$$= -E\left(1_B\langle X_t,A\phi+\Phi(\phi)\rangle\exp(-\langle X_t,\phi\rangle)\right).$$

This shows that $\exp(-\langle X_t,\phi\rangle) - \int_0^t \langle X_r,A\phi+\Phi(\phi)\rangle\exp(-\langle X_r,\phi\rangle)dr$ is a local martingale.

It then follows that $H_t(\phi)$ is a local martingale (cf. [EK, Ch. 2, Cor 3.3]) and

109

this completes the proof that $H_t(\phi)$ is a martingale.

Step 2. The semimartingale representations.

Let $Y_t(\phi) = \exp\left(-\int_0^t \langle X_s, A\phi+\Phi(\phi)\rangle ds\right)$ and $Z_t(\phi) = \exp(-\langle X_t,\phi\rangle)$.

Note that $Y_t(\phi)$ is a continuous semimartingale with finite variation and $H_t(\phi)$ is a martingale. Therefore $Z_t(\phi) = H_t(\phi)Y_t(\phi)$ is a semimartingale and the (Itô) integration by parts formula yields

(6.1.8) (*Representation 1)*

$$dZ_t(\phi) = Y_t(\phi)dH_t(\phi) + H_{t-}(\phi)dY_t(\phi)$$
$$= Y_t(\phi)dH_t(\phi) - \langle X_t, A\phi+\Phi(\phi)\rangle Z_{t-}(\phi)dt.$$

Note that the variation of $H_s(\phi)dY_s$ in $[0,t]$ is dominated by

$$\int_0^t \exp\left(\int_0^s |\langle X_u, A\phi+\Phi(\phi)\rangle| du\right) |\langle X_s, A\phi+\Phi(\phi)\rangle| ds$$

which is locally integrable under assumption (iii). Hence $Z_t(\phi)$ is a *special semimartingale.*

On the other hand, for $\phi\in D_0(A)$, using Itô's formula we conclude that $\langle X_t,\phi\rangle = -\log Z_t(\phi)$ is also a semimartingale. Let $N(ds,d\mu)$ be the adapted random point measure on $\mathbb{R}_+\times M_F(E)$ given by $\sum \delta_{(s,\Delta X_s)}$ (where $\Delta X_s := X_s - X_{s-}$), $\hat{N}(ds,d\mu)$ denote its compensator and $\tilde{N}(ds,d\mu)$ the corresponding martingale measure (in order that N satisfy the assumptions stated before Lemma 6.1.2 we equip $M_F(E)$ with a metric, d, under which all bounded subsets, B, have the property that $\{\langle\mu,1\rangle:\mu\in B\}$ is a compact subset of $(0,\infty)$).

Then for $\phi\in D_0(A)$, $\langle X_t,\phi\rangle$ has the canonical representation (cf. [LS. p. 191]) of the form

$$\text{(6.1.9)} \qquad \langle X_t,\phi\rangle = \langle X_0,\phi\rangle + U'_t(\phi) + M^c_t(\phi) + \tilde{N}_t(\phi) + N_t(\phi)$$

where $U'_t(\phi)$ is continuous and locally of bounded variation, $M^c_t(\phi)$ is a continuous local martingale with increasing process $C_t(\phi)$ and where

$$\tilde{N}_t(\phi) = \int_0^t\int_{M_F^{\pm}(E)} 1(|\langle\mu,\phi\rangle|\le\|\phi\|)\langle\mu,\phi\rangle\ \tilde{N}(ds,d\mu),$$

$$N_t(\phi) = \int_0^t\int_{M_F^{\pm}(E)} 1(|\langle\mu,\phi\rangle|>\|\phi\|)\langle\mu,\phi\rangle\ N(ds,d\mu),$$

where $M_F^{\pm}(E)$ denotes the space of signed measures on E of finite variation.

110

Applying Itô's lemma (Lemma 6.1.2) to $\exp(-\langle X_t,\phi\rangle)$ with $\langle X_t,\phi\rangle$ given by (6.1.9) we get

$$dZ_t(\phi) = Z_{t-}(\phi)\Big\{-dU'_t(\phi) + \tfrac{1}{2}dC_t(\phi)$$

$$+ \int_{M_F^{\pm}(E)} \Big((e^{-\langle\mu,\phi\rangle} - 1 + \langle\mu,\phi\rangle)\mathbf{1}(|\langle\mu,\phi\rangle| \le \|\phi\|)\Big)\hat{N}(dt,d\mu)$$

$$+ \int_{M_F^{\pm}(E)} \Big((e^{-\langle\mu,\phi\rangle} - 1)\mathbf{1}(|\langle\mu,\phi\rangle| > \|\phi\|)\Big)N(dt,d\mu)\Big\} + d(\text{loc. mart.})$$

It is easy to verify that under assumption (iii)

$$Z_{t-}(\phi)\Big(\exp(-\langle\Delta X_t,\phi\rangle) - 1 + \langle\Delta X_t,\phi\rangle\Big) \ge 0$$

is integrable and we obtain

(6.1.10) *(Representation 2)*

$$dZ_t(\phi) = Z_{t-}(\phi)\cdot\Big(-dU_t(\phi) + \tfrac{1}{2}dC_t(\phi) + \int_{M_F^{\pm}(E)} (e^{-\langle\mu,\phi\rangle} - 1 + \langle\mu,\phi\rangle)\hat{N}(dt,d\mu)\Big)$$

$$+ d(\text{loc. mart.})$$

Step 3. Identification of the Two Representations.

Since $Z_t(\phi)$ is a special semimartingale we can now identify the predictable components of locally integrable variation in the two decompositions (6.1.8) and (6.1.10) (e.g. [LS] 85) to obtain that

$$-Z_{t-}(\phi)\langle X_t, A\phi+\Phi(\phi)\rangle dt$$

$$= Z_{t-}(\phi)\Big(-dU_t(\phi) + \tfrac{1}{2}dC_t(\phi) + \int_{M_F^{\pm}(E)} (e^{-\langle\mu,\phi\rangle} - 1 + \langle\mu,\phi\rangle)\hat{N}(dt,d\mu)\Big)$$

where $U_t(\phi) := U'_t(\phi) + \int_{M_F^{\pm}(E)} \langle\mu,\phi\rangle\mathbf{1}(|\langle\mu,\phi\rangle| > 1)\hat{N}(dt,d\mu)$

is locally of bounded variation and $U_t(\theta\phi) = \theta U_t(\phi)$ for $\theta \in \mathbb{R}_+$; $C_t(\phi)$ is a continuous increasing process and $C_t(\theta\phi) = \theta^2 C_t(\phi)$. Then the process $\langle X_s, A\phi+\Phi(\phi)\rangle$ is integrable on $[0,t]$ and

$$\int_0^t \langle X_s, A\phi+\Phi(\phi)\rangle ds = U_t(\phi) - \tfrac{1}{2}C_t(\phi) - \int_0^t\int_{M_F^{\pm}(E)} (e^{-\langle\mu,\phi\rangle} - 1 + \langle\mu,\phi\rangle)\hat{N}(ds,d\mu).$$

111

Replacing ϕ by $\theta\phi$, $(\theta>0)$ in $\Phi(\phi)$, we obtain

$$\int_0^t \langle X_s, A\theta\phi+\Phi(\theta\phi)\rangle ds$$

$$= \theta\int_0^t \langle X_s, A\phi\rangle ds - \frac{\theta^2}{2}\int_0^t \langle X_s, c\phi^2\rangle ds$$

$$- \int_0^t ds\int X_s(dx)\int_0^\infty n(x,du)\left(e^{-u\theta\phi(x)}-1+u\theta\phi(x)\right)$$

$$= \theta U_t(\phi) - \frac{\theta^2}{2}C_t(\phi) - \int_0^t\int_{M_F^\pm(E)} (e^{-\langle\mu,\theta\phi\rangle}-1+\langle\mu,\theta\phi\rangle)\hat{N}(ds,d\mu).$$

This allows us to conclude:

(i) in the semimartingale representation of $\langle X_t,\theta\phi\rangle$ (6.1.10)

$$C_t(\phi) = \theta^2\int_0^t \langle X_s, c\phi^2\rangle ds,\quad U_t(\phi) = \theta\int_0^t \langle X_s, A\phi\rangle ds,$$

$$\int_0^t\int_{M_F^\pm(E)} (e^{-\langle\mu,\theta\phi\rangle}-1+\langle\mu,\theta\phi\rangle)\hat{N}(ds,d\mu)$$

$$= \int_0^t ds\int X_s(dx)\int_0^\infty n(x,du)\left(\exp(-u\theta\phi(x))-1+u\theta\phi(x)\right)$$

$$= \int_0^t ds\int X_s(dx)\int_0^\infty n(x,du)\left(\exp(-u\langle\delta_x,\theta\phi\rangle) - 1 + u\langle\delta_x,\theta\phi\rangle\right)\quad \forall\theta > 0,\ \phi\in D_0(A),$$

that is, the jump measure of the process X has compensator

$$\hat{N}(ds,d\mu) = dsX_s(dx)n(x,du)\cdot\delta_{u\delta_x}(d\mu),\quad \mu\in M_F(\mathbb{R}^d).$$

In particular this implies that the jumps of X are almost surely in $M_F(E)$, i.e. positive measures.

Step 4. Proof that the semimartingale representation of $\{X_t\}$ implies that it is a solution of the $\mathfrak{G}$-martingale problem.

<u>Proof.</u> Let $f \in C^\infty(\mathbb{R})$. Then by Ito's formula (Lemma 6.1.2)

$$f(\langle X_t,\phi\rangle) = f(\langle X_0,\phi\rangle)+\int_0^t f'(\langle X_{s-},\phi\rangle)dU_s(\phi) \quad + \tfrac{1}{2}\int_0^t f''(\langle X_s,\phi\rangle)dC_s(\phi)$$

$$+ \int_0^t \int_{M_F^\pm(E)} \left(f(\langle X_s+\mu,\phi\rangle)-f(\langle X_s,\phi\rangle)-f'(\langle X_s,\phi\rangle)\langle\mu,\phi\rangle\right)\hat{N}(ds,d\mu)$$

$$+ M_F(t) \text{ (local martingale).}$$

Hence

112

$$M_F(t) = f(<X_t,\phi>) - \int_0^t \left(f'(<X_s,\phi>)<X_s,A\phi>+ \tfrac{1}{2}f''(<X_s,\phi>)<X_s,c\phi^2>\right) ds$$
$$- \int_0^t <X_s,\int_0^\infty n(.,d\lambda)\left(f(<X_s,\phi>+\lambda\phi(.))-f(<X_s,\phi>)-f'(<X_s,\phi>)\lambda\phi(.)\right)>ds$$

is a local martingale. If f,f' and f'' are bounded this is a martingale. This completes the proof that the process X_t is a solution to the martingale problem for $\mathfrak{G}$.

(b) Assume that $\{X(t)\}$ is a solution of the $(\mathfrak{G},\mathfrak{D}(\mathfrak{G}))$-martingale problem. It remains to prove that this implies (6.1.3). For this, assume that $\psi(.,x) \in C_b^1(\mathbb{R}_+)$, $\psi(t,.) \in D_0(A)$, $t \to A\psi(t,x)$ is continuous, $\psi>0$, and $\frac{\partial\psi}{\partial t}(t,.) \in C_b(E)$. If $f \in C_b^\infty(\mathbb{R})$, then by Lemma 6.1.2 we obtain that

(6.1.11)
$$f(<X_t,\psi(t)>) - \int_0^t \left(f'(<X_s,\psi(s)>)<X_s,A\psi(s)+ \frac{\partial\psi}{\partial s}>+ \tfrac{1}{2}f''(<X_s,\psi(s)>)<X_s,c\psi^2(s)>\right) ds$$
$$- \int_0^t <X_s,\int_0^\infty n(.,d\lambda)$$
$$\cdot\left(f(<X_s,\psi(s)>+\lambda\psi(s,.))-f(<X_s,\psi(s)>)-f'(<X_s,\psi(s)>)\lambda\psi(s,.)\right)>ds$$

is a local martingale.

If $\phi\in D_0(A)$, and and we assume that $\psi(t,x) := V_{T-t}\phi(x)$ satisfies the above conditions, then V_t satisfies the differential form of (6.1.2)

$$(6.1.12)\quad \left(\frac{\partial}{\partial t} + A\right)(V_{T-t}\phi) + \Phi(V_{T-t}\phi) = 0$$

(for example, this is satisfied for the (α,d,β)-superprocess).

If we now apply (6.1.11) to $\exp(-<X_t,V_{T-t}\phi>)$ we can verify that it is a martingale. The result follows for general ϕ by taking limits. Therefore the marginal distribution of X_t is determined and by Theorem 5.1.2 we conclude that the martingale problem is well-posed. The argument in the case in which V_t satisfies (6.1.2) but not necessarily (6.1.12) is a little more involved. One first shows (using 6.1.2, the semigroup property of V_t, and $E(<X_t,\phi>|\mathcal{D}_s) = <X_s,S_{t-s}\phi>$, $t\geq s$) that $<X_t,V_0\phi> -<X_0,V_t\phi>+\int_0^t <X_u,\Phi(V_{t-u}\phi)>du$ is a martingale and then applies Itô's lemma (see Fitzsimmons (1992, Corollary 2.23) for the details). □

<u>Corollary</u> <u>6.1.4.</u> Under the assumptions of Theorem 6.1.3, the (A,Φ)-superprocess is $\{X(t):t\geq 0\}$ is a strong Markov process.

<u>Proof.</u> The arguments above remain valid if $D_0(A)$ is replaced by an appropriate countable subset. The strong Markov property then follows by Theorem 5.1.3. □

113

Remark: The analogue of Theorem 6.1.3 was obtained for (W,Φ)-superprocesses when W is a time homogeneous right processes by Fitzsimmons (1991).

6.2. Examples of Measure-valued Branching Processes

In order to give an indication of the richness of the class of branching systems we give some examples and variations of the general class dealt with in this chapter.

6.2.1. The B(A,c)-Superprocess.

Consider the special case in which E is a compact metric space, A is the generator of a Feller process on E and $\Phi(\lambda) \equiv -\frac{1}{2}c\lambda^2$. This process will be called the B(A,c)-superprocess. In Theorem 4.7.2 we proved that this process has almost surely continuous sample paths. The corresponding martingale problem will be studied in more detail in Chapters 7 and 8.

In the special case $E = \{e\}$, $A = 0$, the B(0,c)-superprocess reduces to the *Feller continuous critical branching process*, that is, a non-negative martingale, Z_t, with increasing process $c\int_0^t Z_s ds$.

6.2.2 The (α,d,β)-superprocess, $\beta \le 1$

Let $D_0(\Delta_\alpha,p) \subset C_p(\dot{\mathbb{R}}^d)$ denote a core of the semigroup $\{S_t^\alpha\}$ on $C_p(\dot{\mathbb{R}}^d)$ (refer to Section 4.5 for notation).

Let $\mathfrak{D}_0(\mathfrak{G})$ consist of functions of the form

$$F(\mu) := e^{-\langle\mu,\phi\rangle}, \quad \mu \in M_p(\dot{\mathbb{R}}^d), \quad \phi \in D_0(\Delta_\alpha,p)$$

$$:= 0 \quad \text{if} \quad \langle\mu,\phi_p\rangle = \infty.$$

Then for $F \in \mathfrak{D}_0(\mathfrak{G})$,

if $0 < \beta < 1$

$$\mathfrak{G}F(\mu) = \frac{\beta(1+\beta)\gamma}{\Gamma(1-\beta)}\int\mu(dx)\int_0^\infty \frac{1}{u^{\beta+2}}[F(\mu+u\delta_x)-F(\mu) - uF'(\mu;x)]d\mu$$

$$+ \int\mu(dx)\Delta_\alpha F'(\mu;x),$$

and if $\beta=1$,

$$\mathfrak{G}F(\mu) = \tfrac{1}{2}c\int\mu(dx)\delta_x(dy)F''(\mu;x,y)+\int\mu(dx)\Delta_\alpha F'(\mu;x).$$

We will show in Prop. 7.3.1 that the $(\alpha,d,1)$ superprocess has continuous sample paths. On the other hand the arguments in step 3 of the proof of Theorem 6.1.3 show that for $\beta<1$, the (α,d,β)-superprocess has jumps corresponding to the creation of mass atoms (note that all jumps have the form $\mu \longrightarrow \mu + u\xi_x$ with $u > 0$, $x \in E$). The

114

set of jump times is dense in $[0,\tau)$ where $\tau := \inf\{t: X_t(\mathbb{R}^d)=0\}$ (because $n((0,1)) = \infty$).

6.2.3. Spatially Inhomogeneous Super-Brownian Motion

Consider a system of critical binary branching Brownian motions in $\mathbb{R}$ but in which particles only branch at the origin. In particular we assume that particles branch only when they are located in the interval $(-\varepsilon/2,\varepsilon/2)$, $\varepsilon>0$, and then at rate ε^{-1}. In the measure-valued limit we obtain a superprocess of the above type in which the additive functional $\kappa(ds)$ is given by the local time at 0. It turns out that the log-Laplace function can also be obtained as the unique solution of the following singular nonlinear partial differential equation:

$$\frac{\partial v}{\partial t} = 2\Delta v - \delta_0 v^2 \tag{6.2.3.1}$$

where δ_0 denotes the delta function at the origin. For details refer to Dawson and Fleischmann (1992).

6.2.4. Superprocesses with Immigration and Superprocesses conditioned on Non-Extinction.

Consider an (A,Φ)-superprocess. In addition to fixed initial conditions one can incorporate the immigration of new mass. For example if $\{\nu(t):t\geq 0\}$ is a continuous $M_F(E)$-valued function, then the superprocess with immigration rate ν is given by the Laplace functional

$$P_{X_0}\left(e^{-\langle X_t,\phi\rangle}\right) = e^{-\langle X_0,V_t\rangle}\exp\left(-\int_0^t \langle \nu(s),V_{t-s}\rangle ds\right).$$

The Laplace functional can also be obtained when the immigration involves an infinitely divisible random measure (cf. Kawazu and Watanabe (1971)).

Let P^* be defined as the law of the process conditioned on non-extinction in the remote future, that is, for $B\in \mathcal{D}_t$,

$$P^*_{X_0}(B) := \lim_{s\to\infty} P_{X_0}(B\,|\,\langle X_{t+s},1\rangle > 0).$$

Roelly-Coppoletta and Rouault (1989) showed that

$$P^*_{X_0}\big|_{\mathcal{D}_t} = \frac{\langle X_t,1\rangle}{\langle X_0,1\rangle}\, e^{-\Phi'(0)t} P_{X_0}\big|_{\mathcal{D}_t}.$$

Furthermore they have shown that the conditioned process can be represented as a process with state dependent immigration.

115

6.2.5. Multitype Superprocesses.

At the particle level we consider a system consisting of particles which can be one of a finite number, K, of types. The process can be viewed as a $M(\cup_{j=1}^{K} E_j)$-valued process where $E_1,\dots,E_K$ are copies of E. We consider a branching particle system in which the motion process consists of an A_j-motion on each E_j together with jumps between the different E_j. Taking the high density limit as above leads to the multitype superprocess. See Gorostiza and Lopez-Mimbela (1990), Gorostiza and Roelly (1990), Gorostiza, Roelly and Wakolbinger (1992), Gorostiza and Wakolbinger (1992), and Li (1992).

6.2.6. Multilevel Branching

The state space for the class of two-level branching process is $M_F(M_F(E_1))$ where E_1 is a Polish space. We can obtain the two level branching process as the limit of a system of branching superparticles which move in $M_F(E_1)$ according to a one level measure-valued branching process.

For example, the $(A_1;\beta_1,\beta_2)$-two-level superprocess is obtained from the construction of Ch. 4 with

$$E_2 := M_F(E_1), \quad \Phi_2(\mu,\lambda) = -\gamma_2\lambda^{1+\beta_2},$$

and the motion process on $M_F(E_1)$ is the (A_1,Φ_1)-superprocess (with $\Phi_1(x,\lambda) = -\gamma_1\lambda^{1+\beta_1}$) with linear semigroup $\{S_t^2 : t\geq 0\}$ and generator A_2 on $bC(M_F(E_1))$ given by:

$$A_2F(\mu) := c(\beta_1)\int\mu(dx)\int_0^\infty \frac{1}{u^{\beta_2+2}}[F(\mu+u\delta_x)-F(\mu) - uF'(\mu;x)]du$$
$$+\int\mu(dx)A_1F'(\mu;x), \text{ if } 0<\beta_2<1$$

or

$$A_2F(\mu) := \gamma_2\int_{E_2}\int_{E_2}\mu(dx)\delta_x(dy)F''(\mu;x,y)+\int_{E_2}\mu(dx)A_1F'(\mu;x) \quad \text{if } \beta_2 = 1.$$

In this case the log-Laplace equation is given by

$$V_t^2\phi(\mu) = S_t^2\phi(\mu) + \gamma_2\int_0^t S_{t-s}^2[V_s^2\phi]^{1+\beta_2}ds, \quad \phi\in bp\mathcal{B}(M_F(E_1)).$$

The higher level processes can then be constructed by iterating this procedure. See Dawson, Hochberg and Wu (1990), Dawson and Hochberg (1991) and Wu (1991, 1992) for further information.

7. STOCHASTIC CALCULUS OF MEASURE-VALUED PROCESSES

7.1. Martingale Measures

Let $(\Omega,\mathcal{F},(\mathcal{F}_t)_{t\geq 0},P)$ be a probability space with a right continuous filtration. The σ-sub-algebra of $\mathcal{B}(\mathbb{R}_+)\otimes\mathcal{F}$ generated by the real-valued continuous adapted processes is called the $\mathcal{F}_t$-*predictable* σ-*algebra* and denoted by $\mathcal{P}$. $\mathcal{P}$ is generated by predictable rectangles of the form $(s,t]\times F$, $F\in\mathcal{F}_s$ and $\{0\}\times F$, $F\in\mathcal{F}_0$.

Let (E,d) be a Polish space and $\mathcal{E} = \mathcal{B}(E)$. Consider a random set function $U(A,\omega)$ defined on $\mathcal{A}\times\Omega$ where $\mathcal{A}$ is a subring of $\mathcal{E}$ which satisfies

$$\|U(A)\|_2^2 = E[U(A)]^2 < \infty \ \forall\ A\in\mathcal{A}$$

$$A\cap B = \emptyset \Longrightarrow U(A)+U(B) = U(A\cup B) \text{ a.s. } \forall A \text{ and } B \text{ in } \mathcal{A}.$$

The set function U is said to be σ-*finite* if there exists an increasing sequence (E_n) of E such that

(a) $\bigcup_n E_n = E$;

(b) $\forall n\ \mathcal{E}_n = \mathcal{E}|_{E_n} \subset \mathcal{A}$;

(c) $\forall\ n$, $\sup\ \{\|U(A)\|_2,\ A\in\mathcal{E}_n\} < \infty$.

Remark. In our applications $\mathcal{A}$ is usually assumed to be the ring of d-bounded Borel subsets of E. In this case the random set function is called *locally finite.*

The set function U is said to be *countably additive* if for each sequence (A_j) decreasing to $\emptyset$, $\|U(A_j)\|_2$ tends to zero. Then it is easy to extend U by $U(A) := \lim_n U(A\cap E_n)$ on every set A of $\mathcal{E}$ such that the limit exists in $L^2(\Omega,\mathcal{F},P)$. A set function which satisfies all these properties is called a σ-*finite* L^2-*valued measure.*

$\{M_t(A),t\geq 0,A\in\mathcal{A}\}$ is a L^2-*martingale measure* with respect to $\mathcal{F}_t$ if and only if:

(a) $M_0(A) = 0$, $\forall A\in\mathcal{A}$

(b) $\{M_t(A),t\geq 0\}$ is a $\mathcal{F}_t$-martingale $\forall A\in\mathcal{A}$

(c) $\forall t>0$, $M_t(.)$ is a L^2-valued σ-finite measure

The *covariance functional* $Q(t;.,.)$ of M is defined by

$$Q_M(t;A,B) := \langle M(A),M(B)\rangle_t$$

(the compensator of the process $M_t(A)M_t(B)$. It defines a covariance measure Q_M on $E\times E\times\mathbb{R}_+$.

The martingale measure M is *worthy* if there exists a random σ-finite measure $K(.,.,.;\omega)$, $\Lambda\in\mathcal{E}\times\mathcal{E}\times\mathcal{B}(\mathbb{R}_+)$, $\omega\in\Omega$, such that

(a) K is symmetric and positive definite, that is, for any $f\in b\mathcal{E}\times\mathcal{B}(\mathbb{R}_+)$,

117

$\iiint f(x,s)f(y,s)K(dx,dy,ds) \geq 0.$

(b) for fixed A,B, $\{K(A\times B\times(0,t]),\ t\geq 0\}$ is $\mathcal{F}_t$-predictable,

(iii) $\exists\ E_n \uparrow E$ such that $E(K(E_n\times E_n\times[0,T])) < \infty\ \forall\ n$,

(iv) $|Q_M(t;A,A)| \leq K(A\times A\times[0,t])$.

K is called the *dominating measure* for M..

M is an *orthogonal martingale measure* if it is a martingale measure and if $A\cap B = \emptyset \Longrightarrow M_t(A)$ and $M_t(B)$ are orthogonal martingales, that is, $M_t(A)M_t(B)$ is a martingale, $\forall A,B \in \mathcal{A}$.

If M is a martingale measure and if for all $A\in\mathcal{A}$, the map $t\longrightarrow M_t(A)$ is continuous we say that M is *continuous*.

Theorem 7.1.1 (a) A worthy martingale measure is orthogonal if and only if $Q_M(t)$ is supported by $\{(x,y):x=y\in E\}$ for all t.

(b) An orthogonal martingale measure, M, is worthy, and there exists a random σ-finite measure $\nu(ds,dx)$ on $\mathbb{R}_+\times E$, such that for each $A \in \mathcal{A}$ the process $(\nu(0,t]\times A)_t$ is predictable and satisfies:

$\forall A\in \mathcal{A},\ \forall t>0,\quad \nu((0,t]\times A) = \langle M(A)\rangle_t,\ $ P.a.s.

The measure ν is called the *intensity* of the orthogonal martingale measure M.

(c) If the orthogonal measure M is continuous, then $\nu(\{t\}\times E_n)=0\ \forall t>0,\ \forall n\in\mathbb{N}$.

Proof. Walsh (1986, Chapt. 2).

Example 7.1.2 Gaussian White Noise (Cylindrical Brownian Motion)

Let $(E\times\mathbb{R}_+,\mathcal{E}\otimes\mathcal{B}(\mathbb{R}_+),\nu(dx)dt)$ be a σ-finite measure space. A *Gaussian white noise* based on $\nu(dx)dt$ is a random set function, W, on $\mathcal{E}\otimes\mathcal{B}(\mathbb{R}_+)$ such that

(a) $W(B\times[a,b])$ is a Gaussian random variable with mean zero and variance $\nu(B)\cdot|b-a|$,

(b) if $(A\times[a_1,a_2])\cap(B\times[b_1,b_2]) = \emptyset$, then $W(A\times[a_1,a_2]\cup(B\times[b_1,b_2])) = W(A\times[a_1,a_2]) + W((B\times[b_1,b_2])$ and $W(A\times[a_1,a_2])$ and $W(B\times[b_1,b_2])$ are independent.

Then for $B\in\mathcal{E}$, $W_t(B) := W(B\times[0,t])$ is an orthogonal martingale measure intensity ν. Walsh (1986) proved that a continuous orthogonal martingale measure is a white noise if and only if its covariance measure is deterministic.

Let $E = \mathbb{R}^d$, ν be Lebesgue measure and $\mathcal{S}(\mathbb{R}^d)$ denote the Schwartz space of C^∞-functions on $\mathbb{R}^d$ which are rapidly decreasing at infinity. Consider the martingale problem: $\forall\phi\in\mathcal{S}(\mathbb{R}^d)$,

$W_t(\phi)$ is a continuous square integrable martingale

with increasing process $\langle W(\phi)\rangle_t = t\int\phi^2(x)dx$.

This martingale problem has a unique solution on $C([0,\infty),\mathcal{S}'(\mathbb{R}^d))$ which is a Gaussian process with $\mathrm{Cov}(W_t(\phi),W_s(\psi)) = (t\wedge s)\int\phi(x)\psi(x)dx$. Then $W_t(\phi)$ can be

118

extended by L^2-continuity to $L^2(\mathbb{R}^d)$ and then $\{W_t(B): t\geq 0,\ B \text{ bounded measurable}\}$ is a white noise on $\mathbb{R}_+\times\mathbb{R}^d$ with intensity measure dtdx. The $\mathscr{S}'(\mathbb{R}^d)$-valued process $\{W_t: t\geq 0\}$ is called *cylindrical Brownian motion.*

Example 7.1.3 The B(A,c)-Superprocess.

Consider the B(A,c)-superprocess as described in section 6.2.1. Let $\phi\in pD(A)$, and $\theta\in\mathbb{R}$. Then by (6.1.6),

$$\exp\left(\theta\left[\langle X_t,\phi\rangle-\int_0^t\langle X_s,A\phi\rangle ds\right] - \tfrac{1}{2}\theta^2 c\cdot\left[\left[\int_0^t\langle X_s,\phi^2\rangle ds\right]\right]\right)$$

is a continuous local martingale. Therefore for $\phi\in D(A)$,

$$M_t^B(\phi) := \langle X_t,\phi\rangle-\langle X_0,\phi\rangle-\int_0^t\langle X_s,A\phi\rangle ds$$

is a $\{\mathcal{D}_t\}$-martingale with increasing process

$$\langle M^B(\phi)\rangle_t = c\int_0^t\langle X_s,\phi^2\rangle ds$$

and

$$\langle M^B(\phi),M^B(\psi)\rangle_t = c\int_0^t\langle X_s,\phi\psi\rangle ds, \quad \phi,\psi \in D(A).$$

Since D(A) is dense in C(E) and therefore bp dense in $b\mathcal{E}$, by taking limits we can extend this to all of $b\mathcal{E}$. This yields a continuous orthogonal martingale measure $M^B(ds,dx)$ with intensity $\nu((0,t]\times A) = c\int_0^t X_s(A)ds$.

Example 7.1.4 The (A,Φ)-Superprocess.

Consider the (A,Φ)-superprocess $\{X_t\}$ as formulated in section 6.1. Recall that for any $\phi\in D(A)$,

$$M_t(\phi) := \langle X_t,\phi\rangle - \langle X_0,\phi\rangle - \int_0^t\langle X_s,A\phi\rangle ds$$

is a martingale with decomposition

$$M_t(\phi) = M_t^c(\phi) + M_t^d(\phi)$$

where $M_t^c(\phi)$ is a continuous martingale with increasing process $\int_0^t\langle X_s,c\phi^2\rangle ds$ and $M_t^d(\phi)$ is a purely discontinuous martingale with predictable compensator $\int_0^t ds\int X_s(dx)\int_0^\infty n(x,du)\left(\exp(-u\phi(x))-1+ u\phi(x)\right)$. Note that in general this does not yield an L^2-martingale measure since $M_t^d(\phi)$ need not belong to $L^2(P_\mu)$.

119

Example 7.1.5. The $S(A,\gamma)$ Fleming-Viot Process.

Consider the $S(A,\gamma)$-Fleming-Viot process $\{Y_t:t\geq 0\}$ introduced in Section 5.7.1 with $\gamma\in C_+$. Then for $\phi\in D(A)$,

$$M_t^{S,\gamma}(\phi) := \langle Y_t,\phi\rangle-\langle\mu,\phi\rangle-\int_0^t\langle Y_s,A\phi\rangle ds$$

is a $\{\mathcal{G}_t\}$-martingale (cf. Section 8.1 for notation) with increasing process

$$\langle M^{S,\gamma}(\phi)\rangle_t = \int_0^t[\langle Y_s,\phi^2\rangle-\langle Y_s,\phi\rangle^2]\gamma(s)ds.$$

Again this can be extended to all $\phi\in b\mathcal{E}$ and yields a continuous martingale measure. Note that for $A,B\in\mathcal{E}$,

$$Q(t;A,B) := \langle M^{S,\gamma}(A),M^{S,\gamma}(B)\rangle_t = \int_0^t[Y_s(A\cap B)-Y_s(A)Y_s(B)]\gamma(s)ds$$

and therefore this martingale measure is not orthogonal. However

$$0\leq Q(t;A,A) = \int_0^t[Y_s(A)-Y_s(A)^2]\gamma(s)ds \leq \int_0^t Y_s(A)\gamma(s)ds.$$

Therefore, $K(dx\times dy\times[0,t]) := \int_0^t\delta_x(dy)Y_s(dx)ds$ is a dominating measure and the martingale measure, $M^{S,\gamma}(ds,dx)$, is worthy.

We next define stochastic integrals with respect to martingale measures. Let M be a worthy measure with dominating measure K. Define

$$(f,g)_K := \iiint_{E\times E\times\mathbb{R}_+} f(x,s)f(y,s)K(dx,dy,ds),\quad f,g\in b\mathcal{E}\times\mathcal{B}(\mathbb{R}_+),$$

and for predictable f, $\|f\|_M := E(|f|,|f|)_K$. Let

$$\mathcal{P}_M := \{f(\omega,s,x) \text{ is } \mathcal{P}\otimes\mathcal{E}\text{-measurable} : \|f\|_M<\infty\}.$$

If M is an orthogonal martingale measure with intensity ν, then

$\mathcal{P}_M = L^2_\nu$ where

$$L^2_\nu := \left\{f(\omega,s,x)\ \mathcal{P}\otimes\mathcal{E}\text{ measurable},\ E\left(\int_{\mathbb{R}_+\times E}f^2(\omega,s,x)\nu(\omega,ds,dx)\right)<\infty\right\}.$$

The collection of *simple predictable functions* is given by

$\mathcal{S} := \{h(\omega,s,x) = \sum_{i=1}^n h_i(\omega)1_{(u_i,v_i]}(s)1_{B_i}(x),\ B_i\in\mathcal{A},\ h_i\in b\mathcal{F}_{u_i}\}$. Then $\mathcal{S}$ is dense in $\mathcal{P}_M$ (Walsh (1986, Prop. 2.3)).

If h is a function of $\mathcal{S}$, we can define a new martingale measure

120

(7.1.1) $h\bullet M_t(A) = \left(\int_0^t hdM\right)(A) := \sum_{i=1}^{n} h_i(M_{v_i\wedge t}(A\cap B_i) - M_{u_i\wedge t}(A\cap B_i)) \quad \forall A\in \mathcal{A}.$

and we can show that $E((h\bullet M_t(A))^2) \le \|h\|^2_M$.
Since $\mathcal{S}$ is dense in $\mathcal{P}_M$ the linear mapping $h \longrightarrow \{h.M_t(A), t\ge 0, A\in\mathcal{A}\}$ can be extended to $h\in\mathcal{P}_M$ by taking a sequence $\{h_m\}$ in $\mathcal{S}$, with $\|h_m - h\| \longrightarrow 0$ and defining $h\bullet M$ as the limit in L^2 of $h_m\bullet M$. (It is easy to verify that the limit is independent of the sequence $\{h_m\}$.) If $h\in \mathcal{P}_M$, $h\bullet M$ is called the *stochastic integral martingale measure* of h with respect to M.

The *usual stochastic integral of the predictable process* h *with respect to the martingale measure* M is defined to be $h\bullet M_t(E)$ and is denoted by

$$\int_0^t\int_E h(s,x)M(ds,dx).$$

Theorem 7.1.6. Let M be a worthy martingale measure and $f\in \mathcal{P}_M$. Then
(a) $f\bullet M$ is a worthy martingale measure. If M is continuous, then $f.M$ is continuous.
(b) If f and g belong to $\mathcal{P}_M$, A and B $\in \mathcal{A}$, then

(7.1.2) $\langle f\bullet M(A), g\bullet M(B)\rangle_t = \int_{(0,t]}\int_{A\times B} f(s,x)g(s,y)Q_M(ds;dx,dy).$

if M is orthogonal,

(7.1.3) $\langle f\bullet M(A), g\bullet M(B)\rangle_t = \int_{(0,t]}\int_{A\cap B} f(s,x)g(s,x)\nu(ds,dx).$

Proof. See Walsh (1986, Chapt.2).

Theorem 7.1.7
(a) Let A_t be a continuous increasing adapted process. Then the following two statements are equivalent:

(7.1.3) M_t is a continuous local martingale with increasing process A_t,

(7.1.4) $Z_t(\theta) := \exp\left[\theta M_t - \frac{1}{2}\theta^2 A_t\right]$ is a continuous local martingale for each $\theta\in\mathbb{R}$.
Furthermore, $Z_t(\theta)$ is a supermartingale, and it is a martingale if and only if $E[Z_t(\theta)] = 1$ for every $t\ge 0$.
(b) If M_t is a local martingale and if either $\exp(\frac{1}{2}\theta M_t)$ is a submartingale or $E[\exp(\frac{1}{2}\theta^2 A_t)] < \infty$ for every t, then $Z_t(\theta)$ is a martingale.
(c) Let M be a continuous worthy martingale measure on E, with covariance functional $Q(t;.,.)$ and dominating measure K and $f\in \mathcal{P}_M$. Then
(7.1.5)

121

$$Z_t(\theta) := E\left(\exp\left\{\int_0^t\int_E \theta f(s,x)M(ds,dx) - 1/2\int_0^t\int_E\int_E \theta^2 f(s,x)f(s,y)Q(ds;dx,dy)\right\}\right)$$

is a continuous local martingale (and also a supermartingale) $\forall\theta\in\mathbb{R}$ and $\forall f\in \mathcal{P}_M$. If $E[Z_t(\theta)] = 1\ \ \forall\ t\geq 0$, then it is a martingale.

Proof. (a) cf. Priouret (1974, Ch. 1, Prop. 17 and Ikeda and Watanabe (1981, p. 142)).

(b) cf. Revuz and Yor (1990, p. 309).

(c) follows from (a), (b) and Theorem 7.1.6. □

7.2 Cameron-Martin-Girsanov Formula for Measure-valued Diffusions

In this section we will consider the Cameron-Martin-Girsanov formula in the setting of continuous measure-valued martingale problems. We begin by recalling the equivalent formulations of the martingale problem for a continuous $M_F(E)$-valued process with E a Polish space.

Assume that $D(A)$ is a dense subspace of $bC(E)$ and $A:D(A)\rightarrow bC(E)$, $Q:M_F(E)\rightarrow \mathcal{Q}_F(E\times E)$ where $\mathcal{Q}_F(E)$ denotes the collection of signed measures, ν, on $E\times E$ satisfying $0\leq \nu(A\times A) < \infty\ \ \forall\ A\in\mathcal{E}$.

Let $\mathfrak{D}(\mathfrak{G}) := \{F:F(\mu):= f(\langle\mu,\phi\rangle),\ \phi\in D(A),\ f\in C_b^\infty(\mathbb{R})\}$ and for $F\in \mathfrak{D}(\mathfrak{G})$,

$$\mathfrak{G}F(\mu) := f'(\langle\mu,\phi\rangle)\langle\mu,A\phi\rangle + \frac{1}{2}\iint f''(\langle\mu,\phi\rangle)\phi(x)\phi(y)Q(\mu;dx,dy).$$

Lemma 7.2.1 Assume that $Q(\mu;A,A) \leq K\mu(A)\ \ \forall\ A\in\mathcal{E}$ where K is a constant. Let P be a probability measure on $(D_E,\mathcal{D},(\mathcal{D}_t)_{t\geq 0})$ with $D_E = D([0,\infty),E)$. Let $\{X_t\}$ be a continuous $M_1(E)$-valued process. Then the following are equivalent:

(i) $\forall F\in \mathfrak{D}(\mathfrak{G})$, $\bar{M}_t(F) := F(X_t)-F(X_0) - \int_0^t \mathfrak{G}F(X_s)ds$ is a P-local-martingale,

(ii) $\forall\ \phi\in D(A)$, $M_t(\phi) := \langle X_t,\phi\rangle-\langle X_0,\phi\rangle-\int_0^t\langle X_s,A\phi\rangle ds$ is a P-local martingale with increasing process

$$\int_0^t\iint\phi(x)\phi(y)Q(X_s;dx,dy)ds,$$

(iii) $\forall\ \phi\in D(A)$,

$$Z_t(\phi) := \exp\left\{\langle X_t,\phi\rangle-\langle X_0,\phi\rangle-\int_0^t\langle X_s,A\phi\rangle ds - \frac{1}{2}\int_0^t\iint\phi(x)\phi(y)Q(X_s;dx,dy)ds\right\}$$

is a P-local martingale.

Proof. The equivalence of (i) and (ii) follows from Itô's lemma (Lemma 6.1.2).

122

The equivalence of (ii) and (iii) follows from Theorem 7.1.7. □

Note that under the conditions of Theorem 7.2.1, $M_t(.)$ extends to a worthy martingale measure.

We next introduce the Cameron-Martin-Girsanov formula for measure-valued processes (cf. Dawson (1978a)). In order to keep the exposition as simple as possible we will state it here for probability measure-valued processes on compact metric spaces. The extension to $M_F(E)$-valued processes is discussed in Section 10.1.

Let $R:M_1(E) \longrightarrow M_1(E)$ be a measurable mapping and let

$$\mathfrak{G}_R F(\mu) := f'(\langle\mu,\phi\rangle)\langle\mu,A\phi\rangle + f'(\langle\mu,\phi\rangle)\langle R(\mu),\phi\rangle + \frac{1}{2}\iint f''(\langle\mu,\phi\rangle)\phi(x)\phi(y)Q(\mu;dx,dy).$$

Note that the statements (i)-(iv) of Theorem 7.2.1 are also valid with $\mathfrak{G}_R$ in place of $\mathfrak{G}$.

<u>Theorem 7.2.2</u> (<u>Cameron-Martin-Girsanov</u>)

Assume that $R(\mu)$ can be written in the integral form

$$R(\mu,dx) = \int r(\mu,y)Q(\mu;dx,dy)$$

where $r:M_1(E) \longrightarrow C(E)$ is bounded and measurable. Then the initial value martingale problem associated with $(\mathfrak{D}(\mathfrak{G}),\mathfrak{G}_R)$ is well-posed if and only if the initial value martingale problem for $(\mathfrak{D}(\mathfrak{G}),\mathfrak{G})$ is well-posed. If $\{\Pi_{[0,T]}P_\mu:\mu\in M_1(E)\}$ and $\{\Pi_{[0,T]}P^R_\mu:\mu\in M_1(E)\}$ denote the respective solutions on $(D([0,\infty),M_1(E)),(\mathcal{D}_t)_{t\ge0},\mathcal{D})$ restricted to $\mathcal{D}_T$, then for each μ, $\Pi_{[0,T]}P_\mu$ and $\Pi_{[0,T]}P^R_\mu$ are equivalent measures and the Radon-Nikodym derivative

$$R_r(T) := \frac{d\,\Pi_{[0,T]}P^R_\mu}{d\,\Pi_{[0,T]}P_\mu} > 0, \quad \Pi_{[0,T]}P_\mu\text{-a.s.}$$

is given by $R_r(T) = Z_r(T)$

where for $g:M_1(E) \longrightarrow bC(E)$ bounded and continuous,

$$Z_g(t) := \exp\left\{\int_0^t\int_E g(X_s,y)M(ds,dy) - 1/2\int_0^t\int_E\int_E g(X_s;x)g(X_s;y)Q(X_s;dx,dy)ds\right\}$$

where $M(ds,dy)$ is the martingale measure obtained from $M_t(\phi)$.

<u>Proof.</u> First assume that the initial value martingale problem for $(\mathfrak{D}(\mathfrak{G}),\mathfrak{G})$ is well posed with solution $\{P_\mu\}$.

Let $g:M_1(E)\longrightarrow C(E)$ be bounded and measurable. Since $\{X_s\}$ is continuous and g is bounded and measurable, $g(X_s,.) \in \mathcal{P}_M$ and by Theorem 7.1.6 the stochastic integral $\int_0^t\int g(X_s,y)M(ds,dy)$ is a continuous martingale with increasing process

$\int_0^t\int_E\int_E g(X_s;x)g(X_s;y)Q(X_s;dx,dy)ds.$

Therefore by Theorem 7.1.7, $Z_g(t)$ is a continuous P_μ-local martingale. Since under our assumptions $\int_0^t\int_E\int_E g(X_s;x)g(X_s;y)Q(X_s;dx,dy)ds$ is a.s. bounded, we conclude that $Z_g(t)$ is a P_μ-martingale from Theorem 7.1.7(b).

Define $P_\mu^R(B) := Z_r(T)P_\mu(B)$ for $B\in \mathcal{D}_T$.

Then

$$\frac{d\,\Pi_{[0,T]}P_\mu^R}{d\,\Pi_{[0,T]}P_\mu} = Z_r(T)$$

and P_μ^R is a probability measure on $\mathcal{D}_T$ since $Z_r(t)$ is a martingale.

If we define $g(\mu) := \phi + r(\mu)$ with $\phi\in D(A)$, then

$$Z_g(t) = Z_r(t).\exp\Big\{\int_0^t \phi(x)M(ds,dx) - \int_0^t\int\int\phi(x)r(X_s;y)Q(X_s;dx,dy)ds - \tfrac{1}{2}\int_0^t\int\int\phi(x)\phi(y)Q(X_s;dx,dy)ds\Big\}$$

$$= Z_r(t)\cdot\exp\Big\{\int_0^t \phi(x)M(ds,dx) - \int_0^t\int\int\phi(x)R(X_s;dx)ds - \tfrac{1}{2}\int_0^t\int\int\phi(x)\phi(y)Q(X_s;dxdy)ds\Big\}$$

is a P_μ-martingale.

Hence for each $\phi\in D(A)$,

$$\exp\Big\{\int_0^t \phi(x)M(ds,dx) - \int_0^t\int\int\phi(x)R(X_s;dx)ds - \tfrac{1}{2}\int_0^t\int\int\phi(x)\phi(y)Q(X_s;dx,dy)ds\Big\}$$

is a P_μ^R-martingale. To verify this one should note that

$$dP_\mu^R|_{\mathcal{D}_t}|_{\mathcal{D}_s} = dP_\mu^R|_{\mathcal{D}_s},\quad s < t.$$

But since

$$dP_\mu^R|_{\mathcal{D}_t} = Z_r(t)dP_\mu|_{\mathcal{D}_t},$$

it follows that

$$dP_\mu^R|_{\mathcal{D}_t}|_{\mathcal{D}_s} = P_\mu[Z_r(t)|_{\mathcal{D}_s}]dP_\mu|_{\mathcal{D}_s} = dP_\mu^R|_{\mathcal{D}_s}$$

since $Z_r(s)$ is a P_μ-martingale.

Hence by Theorem 7.2.2 P_μ^R is a solution to the initial value martingale problem for $(\mathfrak{D}(\mathfrak{G}),\mathfrak{G}_R)$. It remains to verify the uniqueness. Let P_μ^R denote a solution to the initial value martingale problem for $(\mathfrak{D}(\mathfrak{G}),\mathfrak{G})$. Let

124

$$Z^R_{-r}(t) := \exp\left\{\int_0^t\int_E -r(X_s,y)M_R(ds,dy) - 1/2\int_0^t\int_E\int_E r(X_s;x)r(X_s;y)Q(X_s;dx,dy)ds\right\},$$

here $M_R(.)$ denotes the martingale measure coming from the martingale

$$M_{R,t}(\phi) := \langle X_t,\phi\rangle-\langle X_0,\phi\rangle-\int_0^t(\langle X_s,A\phi\rangle+\langle R(X_s),\phi\rangle)ds$$

corresponding to the martingale problem for $(\mathfrak{D}(\mathfrak{G}),\mathfrak{G}_R)$.

Then the same argument as above yields

$$\Pi_{[0,T]}P_\mu = Z^R_{-r}(T)\Pi_{[0,T]}P^R_\mu \quad \text{and} \quad Z^R_{-r}(T) > 0,\ P^R_\mu\text{-a.s.}$$

which means that

$$\frac{dP_\mu}{dP^R_\mu}\Big|_{\mathcal{D}_T} = Z^R_{-r}(T).$$

Therefore if two distinct solutions P^R_μ and $\tilde{P}^R_\mu$ exist, then this would give two distinct solutions P_μ and $\tilde{P}_\mu$ of the initial value martingale problem for $(\mathfrak{D}(\mathfrak{G}),\mathfrak{G})$ leading to a contradiction. Hence the uniqueness is proved. □

7.3. Application to Sample Path Continuity

Using the martingale structure we can now improve the sample path continuity result which was obtained in Theorem 4.7.2.

Proposition 7.3.1

(a) Let $\{X(t):t\ge 0\}$ be a B(A,c)-superprocess where (A,D(A)), and d are as in Theorem 4.7.2. Then with probability one, $\{X(t):t\ge 0\}$ is β-Hölder continuous in the metric d for any $\beta < 1/2$.

(b) The $(\alpha,d,1)$-superprocess has trajectories in $C([0,\infty),M_p(\dot{\mathbb{R}}^d))$ (with p as in Section 4.6) so that $\phi_p \in D(A)$ P_μ-a.s. for all $\mu \in M_p(\mathbb{R}^d)$.

(c) Assume that D(A) is convergence determining in $M_1(E)$. Then the $S(A,\gamma)$-Fleming-Viot process (cf. ex. 7.1.5) has trajectories in $C([0,\infty),M_1(E))$ a.s.

Proof. (a) Let $\phi\in D(A)$. Then it follows from Theorems 6.1.3 and 7.2.1 that

$$M^B_t(\phi) := \langle X_t,\phi\rangle - \langle X_0,\phi\rangle - \int_0^t \langle X_s,A\phi\rangle ds$$

is a martingale and Theorem 4.7.2 implies that it is continuous. Then M^B extends to a continuous martingale measure $M^B(ds,dx)$ and applying Itô's formula (Lemma 6.1.2) to $\psi(s,x) = S_{t-s}\phi(x)$, $0\le s\le t$, we obtain

$$\langle X_t,\phi\rangle - \langle X_0,S_t\phi\rangle = \int_0^t\int S_{t-s}\phi(x)M^B(ds,dx) = \hat{X}_t(\phi) \quad \text{for} \quad \phi\in D(A).$$

By bounded pointwise convergence this can be extended to $\phi\in b\mathcal{E}$. Further if

125

$\psi:\mathbb{R}_+\times E\to\mathbb{R}$ is bounded and measurable, then $\int_0^t\int\psi(s,x)M^B(ds,dx)$ is a continuous martingale with increasing process $\int_0^t\int\psi^2(s,x)X(s,dx)ds$, and by the Burkholder-Davis-Gundy inequality (cf. Revuz and Yor (1991, p. 151)),

$$\text{(7.3.1)}\qquad E\left|\int_0^t\int\psi(s,x)M^B(ds,dx)\right|^q \le C_q\, E\left(\left(\int_0^t\int\psi^2(s,x)X(s,dx)ds\right)^{q/2}\right),\quad q>0.$$

Therefore for $q\ge 2$, $0\le t<u\le T$, $\phi\in D(A)$,

$$E(|\hat{X}_u(\phi)-\hat{X}_t(\phi)|^q)^{1/q}$$

$$\le \left[E\left(\left|\int_t^u\int_{\mathbb{R}^d} S_{u-s}\phi(x)M^B(ds,dx)\right|^q\right)^{1/q} + E\left(\left|\int_0^t\int_{\mathbb{R}^d}(S_{u-s}\phi(x)-S_{t-s}\phi(x))M^B(ds,dx)\right|^q\right)^{1/q}\right]$$

$$\le C_q\left[E\left(\left(\int_t^u\langle X_s,(S_{u-s}\phi)^2\rangle ds\right)^{q/2}\right)^{1/q} + E\left(\left(\int_0^t\langle X_s,S_{t-s}[S_{u-t}\phi-\phi]^2\rangle ds\right)^{q/2}\right)^{1/q}\right]$$

(by (7.3.1))

$$\le C_q\left[\left(\sup_{s\le u}E(\langle X_s,1\rangle^{q/2})\|\phi\|^q(u-t)^{q/2}\right)^{1/q} + \left((u-t)^{q/2}t^{q/2}\sup_{s\le u}E(\langle X_s,1\rangle^{q/2})\;\|A\phi\|^q\right)^{1/q}\right]$$

$$\le \text{const}\;(\|\phi\|+\|A\phi\|)(u-t)^{1/2}.$$

The remaining argument is similar to that of Theorem 4.7.2.

(b) (Sketch of proof). From (a) we know that there exists a continuous $M_F(\mathbb{R}^d)$-valued process. To extend the process to $M_p(\dot{\mathbb{R}}^d)$ we first define $X_n(t)\in M_F(\mathbb{R}^d)$ using the initial condition $X_n(0) := X(0)1(0\le|x|<n)$. It is then easy to check that if $\phi\in D(A)\cap C_p$ and $A\phi\in C_p$, then using Doob's maximal inequality

$$\sup_{t\le t_0}|\langle X(t),\phi\rangle-\langle X_n(t),\phi\rangle| \to 0 \text{ a.s. as } n\to\infty.$$

Recalling that $\phi_p\in D(A)$ we conclude that $\langle X(t),\phi\rangle$ is continuous a.s. for ϕ belonging to a $M_p(\dot{\mathbb{R}}^d)$-convergence determining class. Thus $X(.)\in C([0,\infty),M_p(\dot{\mathbb{R}}^d))$, $P_{X(0)}$-a.s.

(c) The proof of (c) is similar to the proof of (a). □

<u>Remarks</u> <u>7.3.2</u> (a) Reimers (1989) and Perkins (1991) proved that under additional conditions on the semigroup S_t (e.g. analyticity) $\langle X(t),\phi\rangle$ is a.s. continuous on $(0,\infty)$ for any bounded measurable function ϕ and hence $X(t)$ is continuous in the τ-topology (i.e. the $\sigma(M_F(E),b\mathcal{E})$-topology). The idea is that under analyticity, the

126

above inequality in (a) may be replaced by one involving only $\|\phi\|$.

(b) Fitzsimmons (1988) established path regularity (right process, Hunt process) for the (W,Φ)-superprocesses when W is a right (Hunt) process by using the Ray compactification and Rost's theorem on Skorohod embedding. He also obtained a.s. continuity when $n\equiv 0$ in his general setting using the results of Bakry and Emery.

(c) It is reasonable to ask whether $\langle X(t),\phi\rangle$ is a semimartingale for $\phi\in\mathcal{E}$. The fact that this is not necessarily so was established by Tribe in the following theorem.

Theorem 7.3.3 Let $\mu\in M_F(\mathbb{R}^d)$, $d\geq 1$, and X_t be the $(\alpha,d,1)$ superprocess. Let H be the indicator of a halfspace. Define

$\Gamma(\alpha) := 2\alpha/(\alpha+1)$ if $\alpha>1$.

Then for any $T > 0$ we have the following decomposition:

$X_t(H) = \mu(H) + V_t + M_t$ for $0\leq t\leq T$

where M_t is a continuous L^2 martingale satisfying $\langle M\rangle_t = \int_0^t X_s(H)ds$ and V_t is continuous on $(0,T]$.

If $0 < \alpha < 1$ and μ has a bounded density then V_t has integrable variation on $[0,T]$.

If $1 < \alpha \leq 2$ and μ has a bounded density, u, then V_t has integrable $\Gamma(\alpha)$ variation on $[0,T]$. If in addition the density is uniformly Hölder continuous and satisfies $u(0,x) > 0$ for some x on the boundary of the halfspace then with probability one V_t has strictly positive $\Gamma(\alpha)$ variation on $[0,T]$ and hence X_t fails to be a semimartingale.

Proof. Tribe (1989, Theorem 3.2).

7.4. The Weighted Occupation Time Process

Let $\{X_t\}$ denote the (A,Φ)-superprocess on E (cf. Sect. 6.1). Following Iscoe (1986) we define the associated *weighted occupation time process*, $\mathcal{O}_{s,t}$ as follows:

$$(7.4.1)\qquad \mathcal{O}_{s,t}(A) := \int_s^t X_u(A)du,\quad A\in\mathcal{E}.$$

Theorem 7.4.1. Let $\psi,\phi\in C_0(E)$, $\mu\in M_F(E)$. Then

$$(7.4.2)\qquad E_\mu\left(\exp\{-[\langle X_t,\phi\rangle+\langle\mathcal{O}_{0,t},\psi\rangle]\}\right) = \exp\left(-\int u(t,x)\mu(dx)\right)$$

where u is the unique solution of the initial value problem

$$(7.4.3)\qquad \frac{\partial u}{\partial t} = (Au + \Phi(u)) + \psi,\quad u(0,x) = \phi(x)$$

where Φ is as in Section 6.1.

Proof. Assume that $\psi(.,x)\in C_b^1(\mathbb{R}_+)$, $\psi(t,.)\in D_0(A)$, $t\longrightarrow A\psi(t,x)$ is differentiable, $\psi>0$, and $\frac{\partial\psi}{\partial t}(t,.)\in D_0(A)$. If $f\in C_b^\infty(\mathbb{R})$, then by the time inhomogeneous form of

127

Itô's lemma we obtain

(7.4.4)

$$f(\langle X_t,\psi(t)\rangle) - \int_0^t \Big(f'(\langle X_s,\psi(s)\rangle)\langle X_s,A\psi(s)+\frac{\partial\psi}{\partial s}\rangle + \frac{1}{2}f''(\langle X_s,\psi(s)\rangle)\langle X_s,c\psi^2(s)\rangle\Big)ds$$

$$- \int_0^t \langle X_s,\int_0^\infty n(.,d\lambda)\Big(f(\langle X_s,\psi(s)\rangle+\lambda\psi(s,.))-f(\langle X_s,\psi(s)\rangle)-f'(\langle X_s,\psi(s)\rangle)\lambda\psi(s,.)\Big)\rangle ds$$

is a martingale.

For $\phi\in D(A)$, let $\xi(t,x)$ be the unique evolution solution of

(7.4.5) $$\frac{\partial}{\partial t}\xi(t,x) = A\xi(t,x) + \Phi(\xi(t,x)) + \psi(x), \qquad \xi(0,x) = \phi(x).$$

Now we apply (7.4.4) to the function $f(u) = e^{-u}$ and $\psi(s,x) := \xi(t-s,x)$ where ξ is given by (7.4.5) to verify that

$$\exp\{-[\langle X_t,\psi(t)\rangle+\langle \mathcal{O}_{0,t},\psi\rangle\} \quad \text{is a } P_\mu\text{-local martingale.}$$

But then it follows that it is a martingale by bounded convergence. Therefore

$$P_\mu\Big(\exp\{-[\langle X_t,\phi\rangle+\langle \mathcal{O}_{0,t},\psi\rangle]\}\Big) = \exp\{-\langle\mu,\xi(t)\rangle\} \quad \forall\ \phi\in D(A),\psi\in C(E).$$

Then we obtain it for general ϕ by taking limits. □

Remark 7.4.2. It is also possible to obtain the weighted occupation time process by an appropriate rescaling of the analogous process for the branching particle systems (cf. Gorostiza and López-Mimbela (1992)). Since particles in a branching particle system with branching rate n have lifetimes of order 1/n, it turns out that $\mathcal{O}_{st}(A)$ can also be interpreted as the (renormalized) limiting mass of particles which die in the time interval [s,t) and whose location at time of death is in the set A.

Remark 7.4.3. The weighted occupation time of a set is a special case of the class of *J-functionals* introduced by Dynkin (1990). To explain this notion consider a (W,Φ)-superprocess $\{X_t\}$ and let A be an *additive functional* of W (i.e. a random measure A on $\mathcal{B}(\mathbb{R}_+)$ such that $A[r,t)\in\sigma\{W_s:r\le s<t\}$). Let $\mathcal{A}^o_W$ denote the set of all finite additive functionals with compact support. Convergence of additive functionals is defined by $A_n \Rightarrow A$ if $P_{r,x}(A_n[r,\infty)\to A[r,\infty)) = 1\ \forall\ (r,x)$ and there exists b such that $A_n[b,\infty) = 0\ \forall n$ and $\sup_{\omega,n} A_n[0,b] < \infty$.

For $u\in\mathbb{R}_+$, and $f\in\mathcal{E}^u$, $A_t = 1_{u<t}f(W_u)$ defines an element of $\mathcal{A}^o_W$. Denote by $\mathcal{A}^\ell_W$ the minimal subcone of $\mathcal{A}^o_W$ which contains all these functionals. Let $A\in\mathcal{A}^1_W$ if there exists $A_n\in\mathcal{A}^\ell_W$ such that $A_n(B)\uparrow A(B)\ \forall\ B\in\mathcal{B}(\mathbb{R}_+)$.

If $A\in\mathcal{A}^1_W$ has the form $\sum_{t\in\Lambda} f^t(W_t)\delta_t$ where Λ is finite then we define $I(A) =$

128

$\sum_{t\in\Lambda} \langle f^t, X_t\rangle$. For general $A \in \mathcal{A}^1_W$, with $A_n \Longrightarrow A$, (A_n as above), set $I(A) := \text{med} \lim_{n\to\infty} I(A_n)$ (for the definition of the medial limit, cf. Dellacherie and Meyer). I(A) is called the *J-functional* of the superprocess $\{X_t\}$ associated with the additive functional A.

For example, if W is continuous and B is an open set in E, let

$A_n(\{\frac{k}{n}\}) := \frac{1}{n}1_B(W_{k/n})1(k\le nt)$, $k,n \in \mathbb{N}$.

Then $A_n \Longrightarrow A_{s\wedge t}$ where $A_s = \int_0^s 1_B(W_u)du$ (occupation time in B up to s ≤t).

Then $I(A_n) = \sum_{k\le nt} X_{k/n}(B)$ converges in probability to $\mathcal{O}_{0,t}(B)$ and hence the weighted occupation time of the set B at time t, $\mathcal{O}_{0,t}(B)$, is the J-functional associated with the additive functional $s \longrightarrow \int_0^{s\wedge t} 1_B(W_u)du$.

129

8. STRUCTURAL PROPERTIES OF BRANCHING AND SAMPLING MARTINGALE PROBLEMS

8.1. Relations between Branching and Sampling Measure-valued Processes

Let E be a compact metric space and (A,D(A)) be the generator of a Feller process on E. In this chapter we explore the relations between the continuous A-superprocess and the Fleming-Viot process.

Let us first restrict our attention to the class, B(A,c), c > 0, of continuous critical branching superprocesses, that is, (W,Φ)-superprocesses in which W is a Feller process with generator (D(A),A), $\Phi(\lambda) = -\frac{1}{2}c\lambda^2$. By Theorem 4.7.2 the resulting M(E)-valued process can be realized as a continuous process

$$\left\{C([0,\infty),M(E)),\{\mathcal{F}_t\}_{t\in\mathbb{R}_+},\{X_t\}_{t\in\mathbb{R}_+},\{P_\mu^{B(A,c)}:\mu\in M(E)\}\right\}$$

where $\mathcal{F}_t = \bigcap_{s>t}\mathcal{F}_s^o$, $\mathcal{F}_s^o = \sigma\{X_u : u\le s\}$. $\{P_\mu^{B(A,c)}:\mu\in M(E)\}$ is characterized as the unique solution of the martingale problem: for $\phi\in D(A)$,

$$M_t^B(\phi) = \langle X_t,\phi\rangle - \langle\mu,\phi\rangle - \int_0^t \langle X_s,A\phi\rangle ds$$

is a $P_\mu^{B(A,c)}$, $\{\mathcal{F}_t\}$-martingale with increasing process $\langle M^B(\phi)\rangle_t = c\int_0^t \langle X_s,\phi^2\rangle ds$.

The second class of processes we will consider are the time inhomogeneous Fleming-Viot sampling processes, $S(\sigma A,\gamma)$, $\gamma^{-1}\in C_+$, $\sigma\in pC(\mathbb{R}_+)$, constructed in Theorem 5.7.1. They are denoted by

$$\left\{C([0,\infty),M_1(E)),\{\mathcal{G}_t\}_{t\in\mathbb{R}_+},\{Y_t\}_{t\in\mathbb{R}_+},\{P_\mu^{S(\sigma A,\gamma)}:\mu\in M_1(E)\}\right\}$$

where $\mathcal{G}_t = \bigcap_{s>t}\mathcal{G}_s^o$, $\mathcal{G}_s^o = \sigma\{Y_u : u\le s\}$, and $\mathcal{G} = \bigvee_t \mathcal{G}_t$.

$\{P_\mu^{S(\sigma A,\gamma)}:\mu\in M_1(E)\}$ is characterized as the unique solution of the martingale problem: for $\phi\in D(A)$,

$$M_t^{S,\gamma}(\phi) = \langle Y_t,\phi\rangle - \langle\mu,\phi\rangle - \int_0^t \sigma(s)\langle Y_s,A\phi\rangle ds,\quad t<\tau_\gamma$$

$$= M_{\tau_\gamma -}^{S,\gamma}(\phi)\quad\text{for}\quad t\ge\tau_\gamma,\ \text{where}\ \tau_\gamma := \inf\{t:\gamma(t)=0\}$$

is a $\{P_\mu^{S(\sigma A,\gamma)}\}$, $\{\mathcal{G}_t\}$-martingale with increasing process

$$\langle M^{S,\gamma}(\phi)\rangle_t = \int_0^t [\langle Y_s,\phi^2\rangle - \langle Y_s,\phi\rangle^2]\gamma(s)ds,\quad t<\tau_\gamma$$

and $Y_t = Y_{\tau_\gamma}$, $t\ge\tau_\gamma$.

The main difference between the two processes is that the total mass of the Fleming-Viot process, $\langle Y_t,1\rangle$, is constant in time but the total mass process for the critical branching process, $\langle X_t,1\rangle$, is a martingale with increasing process t →

130

$c\int_0^t \langle X_s,1\rangle ds$. In particular the B(A,c) process suffers extinction, that is, $\tau_\infty < \infty$ a.s. where $\tau_\infty := \inf\{t:\langle X_t,1\rangle = 0\}$ (extinction of critical branching). Konno and Shiga (1988) and Shiga (1990a) discovered a relationship between the normalized process associated with a *time changed version*, $\tilde{X}(t)$, of the B(A,c)-superprocess and the $S(\sigma(.)A,\gamma)$-process (with $\sigma(t) = \langle\tilde{X}_t,1\rangle$). Etheridge and March (1990) discovered that the S(A,c) process arises by conditioning the normalized B(A,c)-superprocess to have constant total mass. Perkins (1991) established that the conditional law of the B(A,c)-superprocess given the total mass process $\langle X_t,1\rangle$ is given by a $S(A,\gamma(.))$-process with $\gamma(t) = \langle X_t,1\rangle^{-1}$.

To explain these results let us begin with some formal calculations based on the results of Shiga (1990). Consider the B(A,c)-process X_t and define

$$C_t := \int_0^t \langle X_s,1\rangle^{-1} ds.$$

Let D_t denote the inverse of C_t on $[0,C_{\tau_\infty -})$. Then $\tilde{Z}_t := \langle X_{D_t},1\rangle$ is a martingale with increasing process $c\int_0^t (\tilde{Z}_s)^2 ds$. Hence $\tilde{Z}_t = \tilde{Z}_0 \exp\left(c^{1/2}B_t - \frac{1}{2}ct\right)$ for some Brownian motion B_t and ($\tilde{Z}_t$ is the only martingale whose increasing process is $c\cdot\int_0^t \tilde{Z}_s^2 ds$). Then $\langle X_t,1\rangle = \tilde{Z}_0\exp\left(c^{1/2}B_{C_t} - \frac{1}{2}cC_t\right)$. Therefore $C_t \uparrow \infty$ a.s. since otherwise $\lim_{t\uparrow\tau_\infty} \langle X_t,1\rangle > 0$, which yields a contradiction.

Hence C_t is a homeomorphism between $[0,\tau_\infty)$ and $[0,\infty)$. Let $D_t:[0,\infty) \longrightarrow [0,\tau_\infty)$ be the continuous strictly increasing inverse to C_t.

Now define $\tilde{X}_t := X_{D_t}$, $\tilde{Y}_t = \tilde{X}_{nor}(t) := \tilde{X}_t/\langle\tilde{X}_t,1\rangle$ and $\mathcal{G}_t := \mathcal{F}_{D_t}$. Then $\{\tilde{Y}_t\}$ is a probability measure-valued process. We now derive the martingale problem for $\tilde{Y}_t$. For $\phi\in \mathcal{D}(A)$,

$$\langle\tilde{X}_t,\phi\rangle = \langle\mu,\phi\rangle + \int_0^{D_t} \langle X_s,A\phi\rangle ds + M^B_{D_t}(\phi)$$

$$= \langle\mu,\phi\rangle + \int_0^t \langle\tilde{X}_s,A\phi\rangle\langle\tilde{X}_s,1\rangle ds + \tilde{N}_t(\phi)$$

where, since D_t is a continuous time change, $\tilde{N}_t(\phi)$ is a continuous $\mathcal{G}_t$-local martingale with increasing process

$$\langle\tilde{N}(\phi)\rangle_t = c\int_0^{D_t} \langle X_s,\phi^2\rangle ds$$

$$= c\int_0^t \langle\tilde{X}_s,\phi^2\rangle\langle\tilde{X}_s,1\rangle ds.$$

Then

131

$$<\tilde{X}_t,1> = <\mu,1> + \tilde{N}_t(1)$$

$$<\tilde{N}(1)>_t = \int_0^t c<\tilde{X}_s,1>^2 ds$$

$$<\tilde{N}(\phi),\tilde{N}(1)>_t = \int_0^t c<\tilde{X}_s,\phi><\tilde{X}_s,1>ds.$$

Applying Ito's formula and noting that $<\tilde{X}_t,1> > 0$ for all $t > 0$ we have

$$<\tilde{Y}_t,\phi> = <\mu_{nor},\phi> + \int_0^t <\tilde{X}_s,1><\tilde{Y}_s,A\phi>ds + N_t^{S,1}(\phi)$$

where $\mu_{nor} := \mu/\mu(E)$, and $N_t^{S,1}(\phi)$ is a continuous $\mathcal{F}_t$-local martingale satisfying

$$<N(\phi)>_t = c\int_0^t [<\tilde{Y}_s,\phi^2>-<\tilde{Y}_s,\phi>^2]ds.$$

Hence $\{\tilde{Y}_t\}$ is a $S(\sigma A,\gamma)$-process with "random evolution operator", that is, with γ=c, $\sigma(s) = <\tilde{X}_s,1>$.

We can also reverse the process and obtain a "skew product" representation for $\tilde{X}_t$ with initial measure μ. To do this we begin with $\tilde{Z}_t = \tilde{Z}_0\exp(c^{1/2}B_t - \frac{1}{2}ct)$, with $\tilde{Z}_0 = <\mu,1> > 0$. Note that $\tilde{Z}_s$ is a martingale with increasing process

$$<\tilde{Z}>_t = \int_0^t c(\tilde{Z}_s)^2 ds.$$

Conditioned on $\tilde{Z}_t$, let $\tilde{Y}_t$ be a $S(\sigma A,\gamma)$ process with initial measure μ_{nor}, γ=c, $\sigma(s) = \tilde{Z}_s$. We can then verify that $\tilde{Z}_t\tilde{Y}_t \overset{\mathcal{D}}{=} \tilde{X}(t)$.

We can then do the reverse time change as follows. Let $D_t := \int_0^t \tilde{Z}_s ds$ and note that $D_t \uparrow \tau_\infty < \infty$. Now let C_t denote the inverse of D_t on $[0,\tau_\infty-)$. Then $\tilde{Z}_{C_t}\tilde{Y}_{C_t} \overset{\mathcal{D}}{=} X_t$, thus yielding a skew product representation of the B(A,c)-superprocess. Moreover note that $Z_t := \tilde{Z}_{C_t}$ is a Feller critical continuous branching process (cf. Sect. 6.2.1) and $Y_t = \tilde{Y}_{C_t}$ is a $S(A,\gamma)$-process with $\gamma(s) = c/Z_s$. These formal calculations suggest the following result which will now be proved following Perkins (1991).

<u>Theorem</u> <u>8.1.1.</u> Let $Q_y := P_\mu^{B(A,c)}\circ(<X_\cdot,1>)^{-1}$ with $y:=<\mu,1>$.

If $\mu\in M_F(E)-\{0\}$, and $\mu_{nor}(.) := \mu(.)/\mu(E)$, then $\forall$ $B\in \mathcal{G}$

$$P_\mu^{B(A,c)}(X_{nor} \in B|<X_\cdot,1>=\gamma^{-1}) = P_{\mu_{nor}}^{S(A,\gamma)}(B), \text{ for } Q_{<\mu,1>}\text{-a.a. } \gamma^{-1}.$$

Hence $P_{\mu_{nor}}^{S(A,\gamma)}$ is a regular conditional distribution for $P_\mu^{B(A,c)}\circ(X_{nor})^{-1}$ given $\gamma(.) = <X_\cdot,1>^{-1}$.

132

Proof. Step 1. Representation of the Normalized Process.

Let $\tau_n := \inf\{t:\langle X_t,1\rangle \le 1/n\}$ and $\tau_\infty := \inf\{t:\langle X_t,1\rangle=0\}$. Let $\phi\in D(A)$ be fixed. Then applying Itô's lemma to $\langle X_t,\phi\rangle/\langle X_t,1\rangle$ we obtain

(8.1.1) $$\langle X_{nor}(t\wedge\tau_n),\phi\rangle = \langle\mu_{nor},\phi\rangle + \int_0^t 1(s\le\tau_n)\langle X_{nor}(s),A\phi\rangle ds + \bar{M}_t^n(\phi).$$

where

(8.1.2)

$$\bar{M}_t^n(\phi) := \int_0^t 1(s\le\tau_n)\langle X_s,1\rangle^{-1} dM_s^B(\phi) - \int_0^t 1(s\le\tau_n)\langle X_s,\phi\rangle\langle X_s,1\rangle^{-2} dM_s^B(1).$$

and

$$\langle\bar{M}^n\rangle_t = \int_0^t 1(s<\tau_n)[\langle X_{nor}(s),\phi^2\rangle-\langle X_{nor}(s),\phi\rangle^2]\langle X_s,1\rangle^{-1} ds.$$

From (8.1.1)

(8.1.3) $$\sup_{t\le T,\, n\in\mathbb{N}} |\bar{M}_t^n(\phi)| \le 2\|\phi\|_\infty + T\|A\phi\|_\infty .$$

Using Doob's maximal inequality we obtain

(8.1.4) $$E\left[\sup_{t\le T} (\bar{M}_t^n(\phi)-\bar{M}_t^m(\phi))^2\right] \le 4E(\bar{M}_T^n(\phi)-\bar{M}_T^m(\phi))^2.$$

For each ϕ and T, $\{\bar{M}_T^n(\phi):n\in\mathbb{N}\}$ is a bounded martingale (in n) (cf. (8.1.2) and (8.1.3)). Therefore by the martingale convergence theorem $\bar{M}_T^n(\phi)$ converges a.s. and in L^2 as $n\to\infty$. Using (8.1.4) and the usual Borel-Cantelli argument we can choose a subsequence $\{n_k\}$ such that almost surely $\{\bar{M}_t^{n_k}(\phi)\}$ is a Cauchy sequence in $C([0,T])$ with continuous limit $\bar{M}_t(\phi)$. In addition $\{\bar{M}_t(\phi):t\ge0\}$ is a continuous martingale that satisfies

(8.1.5) $$\bar{M}_t^n(\phi) = \bar{M}_{t\wedge\tau_n}(\phi) \text{ and } \bar{M}_t(\phi) = \bar{M}_{t\wedge\tau_\infty}(\phi), \quad \forall t\ge0, \text{ a.s.}$$

(8.1.6) $$\sup_{t\le T} |\bar{M}_t(\phi)| \le 2\|\phi\|_\infty + T\|A\phi\|_\infty, \text{ a.s.}$$

Letting $n\longrightarrow\infty$ in (8.1.2) we obtain

(8.1.7)

$$\langle X_{nor}(t),\phi\rangle = \langle\mu_{nor},\phi\rangle + \int_0^t 1(s<\tau_\infty)\langle X_{nor}(s),A\phi\rangle ds + \bar{M}_t(\phi) , \quad t\ge0, \text{ a.s.}$$

Step 2. Martingale Problem in Enlarged Filtration.

Let $\mathcal{H}_t = \mathcal{F}_t\vee\sigma\{\langle X_s,1\rangle:s\ge0\}$. We will now prove that $\bar{M}_t(\phi)$ is an $\mathcal{H}_t$-martingale.

133

Let s<t and let F be a bounded $\sigma(\langle X_s,1\rangle:s\ge 0)$-measurable random variable. Recall that $\langle X_t,1\rangle$ is the unique solution to the martingale problem:

$\langle X_t,1\rangle$ is a continuous martingale with increasing process $c\int_0^t \langle X_s,1\rangle ds$.

Then the predictable representation theorem (cf. Theorem 5.1.1(ii))) applied to the martingale $E[F|\sigma(\langle X_s,1\rangle:0\le s\le t)]$ provides the representation

$$(8.1.8)\qquad F = P_\mu(F) + \int_0^\infty f(s)d\langle X_s,1\rangle$$

for some $\sigma(\langle X_s,1\rangle:s\le t)$-predictable f. Therefore

$$P_\mu^{B(A,c)}((\bar{M}_{t\wedge\tau_n}(\phi)-\bar{M}_{s\wedge\tau_n}(\phi))F|\mathcal{F}_s)$$

$$= P_\mu^{B(A,c)}((\bar{M}_t^n(\phi)-\bar{M}_s^n(\phi))\int_0^\infty f(u)dM_u^B(1)|\mathcal{F}_s) \text{ by (8.1.5) and (8.1.8)}$$

$$= P_\mu^{B(A,c)}\Bigg(\Bigg[\int_s^t 1(u\le\tau_n)\langle X_u,1\rangle^{-1}dM_u^B(\phi)$$

$$- \int_s^t 1(u\le\tau_n)\langle X_u,\phi\rangle\langle X_u,1\rangle^{-2}dM_u^B(1)\Bigg]\int_s^t f(u)dM_u^B(1)|\mathcal{F}_s\Bigg)$$

(by definition of $\bar{M}_t^n(\phi)$*)*

$$= P_\mu^{B(A,c)}\Bigg(\int_s^t 1(u\le\tau_n)(\langle X_u,\phi\rangle\langle X_u,1\rangle^{-1}-\langle X_u,\phi\rangle\langle X_u,1\rangle^{-1})f(u)du|\mathcal{F}_s\Bigg)$$

(by definition of $M_u^B(\phi)$, $M_u^B(1)$*)*

$= 0.$

Let $n\to\infty$ in the above formula, then an application of (8.1.5) and (8.1.6) yields$P_\mu((\bar{M}_t(\phi)-\bar{M}_s(\phi))F|\mathcal{F}_s) = 0$. This completes the proof that $\bar{M}_t(\phi)$ is an $\mathcal{H}_t$-martingale. Moreover it follows from (8.1.2) and (8.1.5) that

(8.1.9)

$$\langle\bar{M}(\phi)\rangle_t = \int_0^t 1(s<\tau_\infty)(\langle X_{nor}(s),\phi^2\rangle-\langle X_{nor}(s),\phi\rangle^2)\langle X_s,1\rangle^{-1}ds,\ P_\mu^{B(A,c)}\text{-a.s.}$$

<u>Step 3.</u> The Conditional Martingale Problem. Let $\{P(B|\gamma):B\in\mathcal{G},\gamma^{-1}\in C_+\}$ be a regular conditional probability for X_{nor} given $\{\langle X.,1\rangle = \gamma^{-1}\}$ (under $P_\mu^{B(A,c)}$). We will now verify that $P(B|\gamma)$ satisfies the $S(A,\gamma)$-martingale problem for $Q_{\langle\mu,1\rangle}$-a.e. γ^{-1}.

If $\gamma^{-1}\in C_+$ and $\phi\in D(A)$, define

$$M_t^{S,\gamma}(\phi) := \langle Y_t,\phi\rangle - \langle\mu,\phi\rangle - \int_0^t \langle Y_s,A\phi\rangle ds,\ t<\tau_\gamma,$$

$$M_t^{S,\gamma}(\phi) := M_{\tau_\gamma-}^{S,\gamma}(\phi)\quad\text{for}\quad t\ge\tau_\gamma,$$

134

on $\left\{C([0,\infty),M_1(E)),(\mathcal{G}_t)_{t\in\mathbb{R}_+},(Y_t)_{t\in\mathbb{R}_+},(P_\mu^{S(A,\gamma)}:\mu\in M_1(E))\right\}$.

By (8.1.5) and (8.1.7) we conclude that

$$\bar{M}_t(\phi) = M_t^{S,\langle X.,1\rangle^{-1}}(\phi)(X_{nor})\ \forall\ t\ge 0,\ P_\mu^{B(A,c)}\text{-a.s.} \tag{8.1.10}$$

If $G\in b\mathcal{G}_s^o$, and s<t, then the $\mathcal{H}_t$-martingale property of $\bar{M}_t(\phi)$ shows that

$$P_\mu^{B(A,c)}((\bar{M}_t(\phi)-\bar{M}_s(\phi))G(X_{nor})\,|\,\langle X.,1\rangle) = 0, \qquad P_\mu^{B(A,c)}\text{-a.s.}$$

and hence by (8.1.10),

$$P((M_t^{S,\gamma}(\phi)-M_s^{S,\gamma}(\phi))G\,|\,\gamma) = 0, \quad \text{for } Q_{\langle\mu,1\rangle}\text{-a.e. } \gamma^{-1}. \tag{8.1.11}$$

We can choose Λ, with $Q_{\langle\mu,1\rangle}(\Lambda) = 0$ such that (8.1.11) holds on Λ^c for all rational s<t and all G in a countable set in $b\mathcal{G}_s^o$ with pointwise bounded closure $b\mathcal{G}_s^o$. Taking limits in s and G we obtain that

$$\{M_t^{S,\gamma}(u):t\ge 0\} \text{ is an } (\mathcal{G}_t)\text{-martingale under } P(.\,|\gamma) \text{ for } \gamma^{-1}\in\Lambda^c. \tag{8.1.12}$$

If $\tau_n^\gamma = \inf\{u:\gamma(u)\ge n\}$, then (8.1.9) implies that

(8.1.13)

$$M_{t\wedge\tau_n^\gamma}^{S,\gamma}(\phi)^2-\int_0^t 1(s<\tau_n^\gamma)[\langle Y_s,\phi^2\rangle-\langle Y_s,\phi\rangle^2]\gamma(s)ds \quad \text{is an } \mathcal{G}_t\text{-martingale}$$

under $P(.\,|\gamma)$ $\forall n\in\mathbb{N}$, $Q_{\langle\mu,1\rangle}$-a.e. γ^{-1}. Now consider a countable core, D_0, for A and fix γ^{-1} outside a $Q_{\langle\mu,1\rangle}$-null set so that (8.1.12) and (8.1.13) hold for all $\phi\in D_0$. Take uniform limits in $(\phi,A\phi)$ to see that (8.1.12) and (8.1.13) hold for all ϕ in D(A) and for $Q_{\langle\mu,1\rangle}$-a.e. γ^{-1}. Therefore $P(.\,|\gamma)$ solves the $S(A,\gamma)$-martingale problem for $Q_{\langle\mu,1\rangle}$-a.e. γ^{-1} and hence $P(.\,|\gamma) = P_{\mu_{nor}}^{S(A,\gamma)}$ for $Q_{\langle\mu,1\rangle}$-a.e. γ^{-1} by Theorem 5.7.1. □

<u>Theorem</u> <u>8.1.2</u> (Etheridge and March (1990))

For $1>\varepsilon>0$ let $\tau(\varepsilon) := \inf\{t>0:|\langle X_t,1\rangle-1|\ge\varepsilon\}$ and fix $0<T<T'<\infty$.

Let $\varepsilon_n\to 0$, and $\mu_n \longrightarrow \mu$ in $M_F(E)$ with $\mu\in M_1(E)$. Let $B_n := \{\tau(\varepsilon_n) > T'\}$ and note that $P_{\mu_n}^{B(A,c)}(B_n) > 0$.

Then the conditional law $\{P_{\mu_n}^{B(A,c)}(X_{nor}\in\cdot\,|B_n)\}|_{\mathcal{F}_T}$ is well-defined and

$$\{P_{\mu_n}^{B(A,c)}(X_{nor}\in\cdot\,|B_n)\}|_{\mathcal{F}_T} \Longrightarrow P_\mu^{S(A,\gamma)}|_{\mathcal{G}_T} \quad \text{as } n\to\infty, \text{ with } \gamma(.)\equiv c$$

where $P_\mu^{S(A,\gamma)}|_{\mathcal{G}_T}$ denotes the restriction of $P_\mu^{S(A,\gamma)}(B)$ to $\mathcal{G}_T$.

<u>Proof.</u> Let $\phi:C([0,T],M_1(E)) \longrightarrow \mathbb{R}$ be bounded and continuous. Then by Theorem

8.1.1,

$$|P_{\mu_n}^{B(A,c)}(\phi(X_{nor})|B_n) - P_\mu^{S(A,c)}(\phi)| = |\int_{B_n}\left(P_{\mu_{n,nor}}^{S(A,\gamma')}(\phi) - P_\mu^{S(A,c)}(\phi)\right)Q^{(n)}(d\gamma')|$$

$$\text{where } Q^{(n)}(d\gamma') := Q_{<\mu_n,1>}(d\gamma')/Q_{<\mu_n,1>}(B_n).$$

Note that $B_n = \{<X_t,1> \in \tilde{B}_n\}$ with $\tilde{B}_n := \{f\in C_+, \sup_{0\le t\le T'} |f(t)-1| < \varepsilon_n\}$.

Hence

$$|\int_{B_n}\left(P_{\mu_{n,nor}}^{S(A,\gamma')}(\phi) - P_\mu^{S(A,c)}(\phi)\right)Q^{(n)}(d\gamma')| \le \sup_{f\in\tilde{B}_n} |P_{\mu_{n,nor}}^{S(A,\gamma')}(\phi) - P_\mu^{S(A,c)}(\phi)|.$$

But $\sup_{f\in\tilde{B}_n} |P_{\mu_{n,nor}}^{S(A,\gamma')}(\phi) - P_\mu^{S(A,c)}(\phi)| \longrightarrow 0$ as $n \to \infty$ by Theorem 5.7.1(b). This completes the proof. □

Remark. We can use the above conditional process together with a more general total mass process which is a one dimensional diffusion to build a larger class of measure-valued diffusions.

8.2. Pure Atomic Measure-valued Processes

In the remainder of this chapter we will use the tools of stochastic calculus developed in the last chapter to identify classes of measure-valued processes which belong to one of two extreme cases, namely, pure atomic processes and absolutely continuous processes.

In this section we consider the $S(A,\gamma)$-Fleming-Viot process on a compact metric space E with $\gamma(.) \equiv \gamma > 0$ and show that the process is pure atomic if the mutation generator is bounded.

Refer to Section 3.4.1 for the definitions of $M_a(E)$, ζ_a, $\mathfrak{A}_\infty$, $\bar{\mathfrak{A}}_\infty$, ϑ and the partition family $\{E_j^n, x_j^n; n,m,j\in\mathbb{N}\}$.

Theorem 8.2.1 Assume that $\gamma(.) \equiv \gamma > 0$.

(a) If the mutation operator A is a bounded linear operator on $\mathcal{E}$ of the form

$$Af(x) = \theta(x)\int(f(y)-f(x))P(x,dy)$$

where $\theta\in pb\mathcal{E}$ and $P(x,.)$ is a transition function on E, then

$$P_\mu^{S(A,\gamma)}\{Y_t\in M_a(E) \text{ for all } t > 0\} = 1.$$

(b) Assume in addition that θ is constant and for each $x \in E$, $P(x,.)$ has no atoms. Let $\vartheta(t) := \vartheta(Y_t)$ for $t\ge0$. Then $\vartheta(.)$ is a solution of the $C([0,\infty);\bar{\mathfrak{A}}_\infty)$-martingale problem for the infinite dimensional differential operator

136

$$(8.2.1)\qquad G_\infty = \frac{\gamma}{2}\sum_{i=1}^{\infty}\sum_{j=1}^{\infty} a_i(\delta_{ij}-a_j)\frac{\partial^2}{\partial a_i\partial a_j} - \theta\sum_{i=1}^{\infty} a_i\frac{\partial}{\partial a_i}$$

with $\mathcal{D}(G_\infty)$ taken to be the algebra generated by $\{1,f^2,f^3,\dots\}$ where for $\eta > 0$,

$$(8.2.2)\qquad f^\eta(\mathbf{a}) := \sum a_i^\eta.$$

Then Y is the size ordered atom process for the infinitely-many-neutral-alleles diffusion (cf. Example 5.7.3).

Proof. (cf. Ethier and Kurtz (1981, and 1987 p. 441))

We first reduce (a) to (b). Let $\tilde{E} = E\times[0,1]$, $\bar{\theta} = \sup\theta(x)$,

$$\tilde{P}(x,u;dydv) := \frac{\theta(x)}{\bar{\theta}}P(x,dy)dv + \frac{\bar{\theta}-\theta(x)}{\bar{\theta}}\delta_x(dy)dv, \text{ and set}$$

$$\tilde{A}\phi(x,u) := \bar{\theta}\left[\int_0^1\int(\phi(y,v)-\phi(x,u))\tilde{P}(x,u;dy)dv\right].$$

Then the $\tilde{A}$ martingale problem is of type of (b). Let $\tilde{Y}_t$ be the solution to the $\tilde{A}$-martingale problem. Then $Y(t,A) = \tilde{Y}(t,A\times[0,1])$ and the atomicity result for $Y(.)$ clearly follows from the corresponding result for $\tilde{Y}(.)$. Hence it suffices to prove (b).

(b) Let $\{Y_t\}$ be a solution of the martingale problem for $\mathfrak{G}$ (defined by (5.4.1)). Since A is a bounded operator we have that for each $\phi\in b\mathcal{E}$ $\langle X_t,\phi\rangle$ is a semimartingale and

$$(8.2.3)\qquad \tilde{M}_t(\phi) := \langle Y_t,\phi\rangle - \int_0^t [\langle Y_s,\theta(.)\int(\phi(y)-\phi(.))P(.,dy)\rangle]ds$$

is a continuous martingale with increasing process

$$\langle\tilde{M}(\phi)\rangle_t = \gamma\int[\langle Y_s,\phi^2\rangle - \langle Y_s,\phi\rangle^2]ds.$$

Let $\{E_j^n : n, j\in\mathbb{N}\}$, $n\in\mathbb{N}$, be the sets of a partition family as defined in Section 3.4.1. For $\mu\in M_1(E)$ let

$$\xi_n(\mu) := \sum_j \mu(E_j^n)\delta_{x_j^n}.$$

If $\vartheta^n(t) := \vartheta(\xi_n(Y(t))$, that is, the descending order statistics of $\{Y(t,E_1^n), Y(t,E_2^n),\dots\}$, then $\vartheta(t) := \vartheta(Y_t) = \lim_{n\to\infty}\vartheta^n(t)$.

For $\eta>1$, let

$$f^\eta(\mathbf{a}) := \sum_j a_i^\eta \quad\text{for } \mathbf{a}\in\bar{\mathfrak{A}}_\infty, \text{ and}$$

$$F_n^\eta(\mu) := \sum_j \mu(E_j^n)^\eta = f^\eta(\xi_n(\mu)).$$

137

For $\eta \geq 2$, by Itô's formula

$$(8.2.4)\quad M_{\eta,n}(t) = F^{\eta}_{n}(Y_t) - \int_0^t \mathfrak{G}F^{\eta}_{n}(Y_s)ds$$

is a bounded (uniformly in n) continuous square integrable martingale where

$$\mathfrak{G}F^{\eta}_{n}(\mu) = \sum_{i=1}^{\infty}\left[\eta\theta(\langle P(.,E_i^n),\mu\rangle - \mu(E_i^n))\mu(E_i^n)^{\eta-1} + \frac{\gamma}{2}\eta(\eta-1)(\mu(E_i^n)-\mu(E_i^n)^2)\mu(E_i^n)^{\eta-2}\right].$$

Since $P(x,.)$ has no atoms, then

$$(8.2.5)\qquad \mathfrak{G}F^{\eta}_{n}(\mu) = G_{\infty}f^{\eta}(\xi_n(\mu)) + R_n(\mu) \quad \text{where} \quad |R_n(\mu)| \xrightarrow{bp} 0 \text{ as } n\to\infty.$$

Therefore as $n \to \infty$, $M_{\eta,n}(t)$ converges to

$$(8.2.6)\quad M_{\eta}(t) = f^{\eta}(\vartheta(t)) - \int_0^t G_{\infty}f^{\eta}(\vartheta(s))ds$$

which is therefore a $\{\mathcal{F}^{\vartheta}_t\}$-martingale where $\mathcal{F}^{\vartheta}_t := \sigma(\vartheta(Y_s):0\leq s\leq t)$. Similar calculations show that the analogues of (8.2.5), (8.2.6) are valid for all elements of $\mathfrak{D}(G_{\infty})$ and therefore $\vartheta(.)$ is a solution of the martingale problem for G_{∞}. The result (b) will then follow if we verify that

$$(8.2.7)\qquad P^{S(A,\gamma)}_{\mu}(\vartheta(t) \in \mathfrak{A}_{\infty} \text{ for all } t > 0) = 1.$$

Using Itô's lemma we can verify that for $\eta \geq 2$,

$$(8.2.8)\quad Z_{\eta}(t) := f^{\eta}(\vartheta(t))-f^{\eta}(\vartheta(0)) - \int_0^t\left\{\tfrac{1}{2}\gamma\eta(\eta-1)(f^{\eta-1}(\vartheta(s)) - f^{\eta}(\vartheta(s))) + \tfrac{1}{2}\eta\theta(f^{\eta-1}(\vartheta(s))\right\}ds$$

is a square integrable continuous martingale with increasing process

$$(8.2.9)\qquad I_{\eta}(t) = \eta^2\gamma^2\int_0^t (f^{2\eta-1}(\vartheta(s))-f^{\eta}(\vartheta(s))^2)ds.$$

In fact (8.2.8) and (8.2.9) remain valid for $\eta \in (1,2)$. To verify this consider the C^2-function

$$f^{\eta}_{\varepsilon}(x) := \sum h_{\varepsilon}(x_i) \quad \text{with} \quad h_{\varepsilon}(u) := (u+\varepsilon)^{\eta}-\varepsilon^{\eta}-\eta\varepsilon^{\eta-1}u,$$

apply Itô's lemma to $f^{\eta}_{\varepsilon}(\vartheta(s))$ for $\varepsilon>0$ and then take the limit $\varepsilon \to 0$.

(The details can be found in Ethier and Kurtz (1981).)

Now let $f^{1+}(\mu) := \text{bp-}\lim_{\eta\downarrow 1} f^{\eta}(\mu) = \sum \mu(\{x\})$, and note that $f^1(\mu) := 1$.

Then

138

(8.2.10) $$0 = \lim_{\eta\to 2+} E(Z_\eta(t)-Z_2(t)) = E\left[\int_0^t \gamma(1-f^{1+}(\vartheta(s)))ds\right] \text{ for } t\geq 0.$$

Then (8.2.10) yields $P_\mu^{S(A,c)}(\vartheta(t) \in \mathfrak{A}_\infty \text{ a.e. } t > 0) = 1$. To show that $\vartheta(t) \in \mathfrak{A}_\infty$ for all t, a.s. note that

(8.2.11) $$\lim_{\eta\downarrow 1} E[Z_\eta(t_0)]^2 = \lim_{\eta\downarrow 1} E[I_\eta(t_0)]$$

$$= \gamma^2 E\left[\int_0^{t_0} f^{1+}(\vartheta(s))[1-f^{1+}(\vartheta(s))]ds\right] = 0.$$

Therefore by Doob's maximal inequality

(8.2.12) $$\sup_{0\leq t\leq t_0} |Z_\eta(t)| \longrightarrow 0 \quad \text{in probability as } \eta \longrightarrow 1+,$$

and hence there exists a sequence $\eta_n \to 1+$ such that

(8.2.13) $$\sup_{0\leq t\leq t_0} |Z_{\eta_n}(t)| \longrightarrow 0 \text{ a.s.}$$

Letting $\rho(t) := \limsup_{n\to\infty} \frac{1}{2}\gamma\eta_n(1-\eta_n)\int_0^t f^{\eta_n-1}(\vartheta(s))ds$

we obtain from (8.2.8)

(8.2.14) $$f^{1+}(\vartheta(t))-f^{1+}(\vartheta(0)) - \rho(t) - \tfrac{1}{2}\theta t = 0 \quad \text{for } 0\leq t\leq t_0.$$

Since $\rho(t)$ is nondecreasing in t, (8.2.13) implies that $P(\vartheta(t) \in \mathfrak{A}_\infty \ \forall\ t>0) = 1$. □

<u>Remarks.</u> (1) In the previous result we can replace the constant function γ by any function γ such that $\gamma^{-1} \in C_+$. This together with Theorem 8.1.1 allows us to conclude that the B(A,c)-superprocess (c > 0), X_t, t>0, is a.s. pure atomic if A is a bounded operator satisfying the hypotheses of Theorem 8.2.1.

(2) Ethier and Kurtz (1992) introduced a stronger topology on $M_1(E)$, called the *weak atomic topology* such that convergence in this topology implies weak convergence as well as the convergence of the sizes and locations of the atoms. They also showed under the conditions of Theorem 8.2.1 that the Fleming-Viot process has continuous sample paths in this topology. Under the same conditions Shiga (1990) proved continuity in the total variation norm (except at t=0).

(3) Consider the special case of Theorem 8.2.1 in which E = [0,1] and

$$Af(x) = \theta\int_{[0,1]} (f(y)-f(x))dy.$$

In this case there exits a unique ergodic reversible stationary distribution (cf. Ethier and Kurtz (1992)). It is then natural to pose the question as to *how many* non-zero atoms there are. Schmuland (1991) proved using Dirichlet form techniques

139

that the stationary process $\vartheta(t)$ hits the the (k-1) dimensional simplex $\{a: \sum_{i=1}^{k} a_i = 1\}$ if and only if $\gamma > 4\theta$.

8.3 Absolutely Continuous Superprocesses and Stochastic Evolution Equations

In this section we consider the canonical (α,d,β)-superprocess $(D,\mathcal{D},(\mathcal{D}_t)_{t\ge0},(X_t)_{t\ge0},(P_\mu)_{\mu\in M_F(\mathbb{R}^d)})$. The objective is to determine when the random measures X_t on $\mathbb{R}^d$ are absolutely continuous and in this case to discuss the formulation in terms of stochastic evolution equations.

Theorem 8.3.1 For $t > 0$, $X(t,dx)$ is absolutely continuous with continuous density $X(t,x)$, P_μ-a.s. $\forall\ \mu \in M_F(\mathbb{R}^d)$ if and only if $d < \alpha/\beta$.

Proof. Some special cases of this result were proved in Dawson and Hochberg (1979), and Roelly-Coppoletta (1986). We will now sketch the main steps of the proof following Fleischmann (1988).

Step 1. Log-Laplace Equation with Measure-valued Initial Condition.

Let $\{S_t\}$ denote the α-symmetric stable semigroup and consider the log-Laplace equation

$$(8.3.1)\qquad v_t = S_t\varphi - \int_0^t S_{t-s}v_s^{1+\beta}ds.$$

Given $\varphi\in pC_c(\mathbb{R}^d)$ the mapping $\varphi \longrightarrow v$ is denoted by $V[\varphi]$.

For $\rho\ge1$, let $(L^{\rho,T},\|.\|_{\rho,T})$ denote the Banach space of measurable real-valued functions, f, on $[0,T]\times\mathbb{R}^d$ with

$$\|f\|_{\rho,T} := \int_0^T\int_{\mathbb{R}^d} |f(s,x)|^\rho dsdx \ < \infty.$$

If $\varphi\in pC_c(\mathbb{R}^d)$, then it is easy to check that $V[\varphi]|_{[0,T]} \in L^{1,T}\cap L^{1+\beta,T}$. The mapping $V:pC_c(\mathbb{R}^d) \longrightarrow L^{1,T}\cap L^{1+\beta,T}$ extended by continuity from $pC_c(\mathbb{R}^d)$ to $M_F(\mathbb{R}^d)$ (equipped with the weak topology) such that

$$(8.3.2)\qquad \forall\varphi\in M_F(\mathbb{R}^d),\ V[\varphi] \text{ is a solution to (8.3.1)},$$

$$\varphi \longrightarrow V[\varphi] \text{ is continuous, and}$$

$$\lim_{t\downarrow0} V[\varphi]_t = \varphi \quad (\text{in } M_F(\mathbb{R}^d)).$$

Moreover the function $\theta \longrightarrow V[\theta\delta_0]_t$ for $\theta\ge0$, $t>0$ is differentiable in θ at $\theta=0$ and

$$(8.3.3)\qquad -\frac{\partial V[\theta\delta_0]_t}{\partial\theta}\Big|_{\theta=0} = S_t\delta_0.$$

For the proofs of (8.3.2) and (8.3.3) refer to Fleischmann (1988) or [D, Lemma

140

7.1.2].

<u>Step 2. Proof of Absolute Continuity</u>

Using Lemma 3.4.2.2 the proof of absolute continuity in the case $d < \alpha/\beta$ reduces to verifying certain analytical properties for the log-Laplace equation (8.3.1). In particular condition (i) of Lemma 3.4.2.2 follows from (8.3.2) which implies that

$$V[\theta\varphi_n] \longrightarrow L_z(t,\theta,\cdot) := V[\theta\delta_z],$$

provided that $\varphi_n \longrightarrow \delta_z$ as $n \to \infty$,

and condition (ii) of Lemma 3.4.2.2 follows from (8.3.3) which implies that

$$-\partial L_z(t,\theta,x)/\partial\theta|_{\theta=0} = p_\alpha(t,z-x).$$

<u>Step 3. Proof of Singularity.</u>

The a.s. singularity of X_t for $t > 0$, $d > \alpha/\beta$ will be proved in the case $\alpha = 2$ in Cor. 9.3.3.5. (The proof for the case $\alpha<2$ can be found in Dawson (1992, Theorem 7.3.4)). For the case $d = \alpha/\beta$ see Dawson and Hochberg (1979) and Fleischmann (1988). □

We next consider in more detail the $B(\Delta_\alpha,c)$-superprocess on $\mathbb{R}^d$, $\{X_t:t\geq 0\}$. By Theorem 6.1.3 this process is characterized as the solution of the martingale problem: for $\psi\in C^{1,2}([0,\infty),\mathbb{R}^d)$,

$$\langle X_t,\psi_t\rangle-\langle X_0,\psi_0\rangle-\int_0^t \left\langle X_s,\left(\frac{\partial\psi}{\partial s}+\Delta_\alpha\psi\right)\right\rangle ds = M_t(\psi)$$

where $M_t(\psi)$ is a continuous martingale with increasing process

$$\langle M(\psi)\rangle_t = \int_0^t \psi^2(s,x)X_s(dx).$$

We denote by $M_t(dx)$ the corresponding orthogonal martingale measure and $\int_0^t\int_{\mathbb{R}^d} f(s,x)M(ds,dx)$ the corresponding stochastic integral constructed in Section 7.1. The covariance functional of M_t is given by $Q(t;dx,dy) = \int_0^t \delta_x(dy)X_s(dx)ds$.

The following theorem was proved independently by Konno and Shiga (1986), and Reimers (1986). It establishes the fact that in one dimension the $(\alpha,1,1)$-superprocess has a density which in an appropriate sense satisfies the stochastic partial differential (or pseudo differential if $\alpha < 2$) equation:

(8.3.4) $$d\tilde{X}_t(x) = \Delta_\alpha\tilde{X}_t(x)dt + \sqrt{\tilde{X}_t(x)}\,W(dt,dx)$$

where $W(.,.)$ is space-time white noise.

141

Theorem 8.3.2. Let $d = 1$, $\beta = 1$, $A = \Delta$ or Δ_α, $0<\alpha\le 2$ and $\{S_t\}$ denote the corresponding semigroup. Let $\mu\in M_F(\mathbb{R}^d)$ and P_μ denote the law of the $B(\Delta_\alpha,1)$-superprocess. Assume that $\mu = X_0(dx) = \tilde{X}_0(x)dx$ where $\tilde{X}_0(x)$ is bounded and continuous. Then there exist a jointly continuous density process $\tilde{X}_t(x)$ and a Gaussian white noise, $W(ds,dx)$ (or cylindrical Brownian motion (cf. Example 7.1.2)) defined on an extension of the canonical probability space $(D,\mathcal{D},(\mathcal{D}_+)_{t\ge 0},P_\mu)$ such that $\forall\ \phi \in C_c^\infty(\mathbb{R}^1)$ and $t > 0$,

(8.3.5a)

$$\int_{\mathbb{R}} X_t(x)\phi(x)dx = \int_{\mathbb{R}} X_0(x)\phi(x)dx + \int_0^t\int_{\mathbb{R}} X_s(x)(A\phi)(x)dxds + \int_0^t\int_{\mathbb{R}} \sqrt{\tilde{X}_s(x)}\ \phi(x)W(ds,dx),$$

P_μ-a.s., that is,. $\tilde{X}_t(x)$ is a *weak* D'(ℝ)-valued *solution* of (8.3.4).

In addition, for fixed $t\ge 0$, and $x \in \mathbb{R}$,

$$\text{(8.3.5b)}\quad \tilde{X}_t(x) = \int p_t(y-x)X_0(y)dy + \int_0^t\int_{\mathbb{R}} p_{t-s}(y-x)\sqrt{\tilde{X}_s(y)}\ W(ds,dy),\ P_\mu\text{-a.s.}$$

where $p_t(.)$ is the appropriate α-stable (or Brownian) transition density, that is, $\tilde{X}_t(x)$ is also a *weak evolution (mild) solution* of (8.3.4).

In addition, P_μ-a.s. $\tilde{X}_t(x)$ is Hölder continuous in t with any modulus less than 1/4 and Hölder continuous in x with any modulus less than 1/2.

Sketch of Proof. We will not give a complete proof but will sketch the main ideas. We begin with the the canonical probability space $(D,\mathcal{D},(\mathcal{D}_t)_{t\ge 0},P_\mu)$.

Step 1. Construction of the Gaussian White Noise.

Lemma 8.3.3 There exist a Gaussian white noise, W, and an $(\mathcal{F}_t)$-predictable function $\tilde{X}_t(x,\omega):[0,\infty)\times\mathbb{R}\times\Omega \longrightarrow [0,\infty)$ defined on an extension of the canonical probability space, denoted by $(\Omega,\mathcal{F},\mathcal{F}_t,\bar{P}_\mu)$, such that

$$\text{(8.3.6)}\quad \int_{\mathbb{R}} \tilde{X}_t(x)\phi(x)dx - \int_{\mathbb{R}} \tilde{X}_0(x)\phi(x)dx = \int_0^t\int_{\mathbb{R}} \sqrt{\tilde{X}_s(x)}\ \phi(x)W(ds,dx) + \int_0^t\int_{\mathbb{R}} \tilde{X}_s(x)(A\phi)(x)dxds$$

holds for every $\phi\in C_c^\infty(\mathbb{R})$, P_μ-a.s.

Proof. By Theorem 8.3.1 $X_t(dx)$ is almost surely absolutely continuous. Therefore $X_t(dx) = \tilde{X}_t(x)dx$, P_μ-a.s., where $\tilde{X}_t(z) := \lim_{\varepsilon\downarrow 0} \frac{X_t(B(z,\varepsilon_n))}{\varepsilon_n}$. Since X_t is a continuous process $\tilde{X}_t(x)$ is $\mathcal{F}_t$-predictable (cf. Section 7.1).

142

We first construct a cylindrical Brownian motion $\bar{W}_t$ and associated white noise $\bar{W}(ds,dx)$ on a probability space $(\bar{\Omega},\bar{\mathcal{F}},\bar{\mathcal{F}}_t,\bar{P}^W)$ independent of $(D,\mathcal{D},(\mathcal{D}_t)_{t\geq 0},P_\mu)$ and define the extended probability space $(\Omega,\mathcal{F},\mathcal{F}_t,\bar{P}_\mu)$ to be the product space $(\Omega,\mathcal{F},\mathcal{F}_t,P^W)\otimes(D,\mathcal{D},(\mathcal{D}_t)_{t\geq 0},P_\mu)$. Then for $\phi\in C_c^\infty(\mathbb{R})$ define

$$(8.3.7)\quad W_t(\phi) := \int_0^t\int_{\mathbb{R}} \frac{1}{\sqrt{\tilde{X}_s(x)}}1(\tilde{X}_s(x)\neq 0)\phi(x)M(ds,dx) + \int_0^t\int_{\mathbb{R}} 1(\tilde{X}_s(x)=0)\phi(x)\bar{W}(ds,dx)$$

where $M(ds,dx)$ is the martingale measure associated with the Fleming-Viot process. Then W_t is a cylindrical Brownian motion which extends to a Gaussian white noise $W(ds,dx)$ defined on $(\Omega,\mathcal{F},\mathcal{F}_t,P_\mu)$, and

$$(8.3.8)\qquad M_t(\phi) = \int_0^t\int_{\mathbb{R}} \sqrt{\tilde{X}_s(x)}\ \phi(x)W(ds,dx)\quad \forall\ \phi\in C_c^\infty(\mathbb{R}).$$

Thus

$$(8.3.9)\quad \int_{\mathbb{R}}\tilde{X}_t(x)\phi(x)dx - \int_{\mathbb{R}}\tilde{X}_0(x)\phi(x)dx = \int_0^t\int_{\mathbb{R}}\sqrt{\tilde{X}_s(x)}\ \phi(x)W(ds,dx) + \int_0^t\int_{\mathbb{R}}\tilde{X}_s(x)(A\phi)(x)dxds$$

P_μ-a.s. $\forall\ t\geq 0$. □

<u>Step 2.</u> $\tilde{X}$ is an evolution solution and a.s. jointly continuous.

<u>Proof.</u> For $\phi,\psi\in C_c^\infty(\mathbb{R})$, we have applying (8.3.6) that

$$P_\mu\left(\left(\int_{\mathbb{R}}\tilde{X}_t(x)\phi(x)dx\right)\cdot\int_0^t\int_{\mathbb{R}}\sqrt{\tilde{X}_s(x)}\ \psi(x)W(ds,dx)\right)$$

$$= P_\mu\left(\int_{\mathbb{R}}\tilde{X}_t(x)\phi(x)dx\ \left(\int_{\mathbb{R}}\tilde{X}_t(x)\psi(x)dx-\int_{\mathbb{R}}\tilde{X}_0(x)\psi(x)dx - \int_0^t\int_{\mathbb{R}}\tilde{X}_s(x)(A\psi)(x)dxds\right)\right).$$

But then by Lemma 4.7.1,

$$(8.3.10)\qquad P_\mu(\langle X_t,\phi\rangle) = \langle\mu,S_t\phi\rangle,$$

$$P_\mu(\langle X_t,\phi\rangle\langle X_t,\psi\rangle) = \int_0^t ds\langle\mu S_{t-s},(S_s\phi)(S_s\psi)\rangle + \langle\mu,S_t\phi\rangle\langle\mu,S_t\psi\rangle$$

and therefore

$$(8.3.11)\quad P_\mu\left(\left(\int_{\mathbb{R}}\tilde{X}_t(x)\phi(x)dx\right)\cdot\int_0^t\int\sqrt{\tilde{X}_s(x)}\ \psi(x)W(ds,dx)\right) = \int_0^t ds\langle\mu S_s,\psi S_t\phi\rangle.$$

Using (8.3.11) and a piecewise constant approximation (cf. Konno and Shiga (1986) for the details) we can obtain

143

(8.3.12) $$P_\mu\left(\left(\int_{\mathbb{R}}\tilde{X}_t(x)\phi(x)dx\right)\cdot\int_0^t\int\sqrt{\tilde{X}_s(x)}\ S_{t-s}\phi(x)W(ds,dx)\right) = \int_0^t ds\langle\mu S_s,(S_{t-s}\phi)^2\rangle.$$

Using (8.3.10), (8.3.11) and (8.3.12) we can then verify by direct calculation that

$$P_\mu\left(\int_{\mathbb{R}}\tilde{X}_t(x)\phi(x)dx - \int_{\mathbb{R}}\tilde{X}_0(x)S_t\phi(x)dx - \int_0^t\int_{\mathbb{R}}\int_{\mathbb{R}}\sqrt{\tilde{X}_s(x)}\ p_{t-s}(x-y)\phi(y)dyW(ds,dx)\right)^2 = 0.$$

This proves that for $\phi\in C_c^\infty(\mathbb{R})$,

$$\int_{\mathbb{R}}\tilde{X}_t(x)\phi(x)dx = \int_{\mathbb{R}}\tilde{X}_0(x)S_t\phi(x)dx + \int_0^t\int_{\mathbb{R}}\sqrt{\tilde{X}_s(x)}\ (S_{t-s}\phi)(x)W(ds,dx),\quad P_\mu\text{-a.s.}$$

This completes the construction of the white noise W and the verification of the fact that $\tilde{X}$ is both a weak $D'(\mathbb{R})$ solution and an evolution solution ofequation (8.3.4).

To prove the a.s. joint continuity of the mapping $(t,x) \longrightarrow \tilde{X}_t(x)$ it suffices to show the a.s. joint continuity of

$$Z_t(x) := \int_{t_0}^t\int_{\mathbb{R}} p_{t-s}(y-x)\sqrt{\tilde{X}_s(y)}\ W(ds,dy).$$

This is proved by verifying that (cf. argument in proof of Proposition 7.3.1) for some $\delta,\delta'>0$,

$$P_\mu\left([Z_t(x)-Z_s(y)]^{2n}\right) \le \text{const}\cdot(|t-s|^\delta+|x-y|^{\delta'})^{n-1}$$

and then using Lemma 3.7.2(b). The Hölder continuity is obtained by a refinement of this argument (see Reimers (1986)). □

Corollary 8.3.4. X_t is absolutely continuous for all t>0, with probability one.

Proof. This follows immediately from the a.s. continuity as a function of s of both the measure-valued process $X_s(dx)$ and the function-valued density process $\tilde{X}_s(.)$. □

8.4. Some Remarks on the Stochastic Evolution Equation

Remarks. 1. Konno and Shiga (1986) also established that a similar stochastic evolution equation can be obtained for the $S(A,\gamma)$-Fleming-Viot process $\{Y_t:t\ge0\}$ in $\mathbb{R}^1$. In this case $Y_t(dx) = \tilde{Y}_t(x)dx$ and $\tilde{Y}_t(x)$ satisfies the stochastic integral equation (evolution solution)

$$\tilde{Y}_t(x) = \int_{\mathbb{R}} p_t(x-y)\tilde{Y}_0(y)dy + \gamma\int_0^t\int_{\mathbb{R}} p_{t-s}(y-x)\sqrt{\tilde{Y}_s(y)}\ W(ds,dy)$$
$$- \gamma\int_0^t\int_{\mathbb{R}} p_{t-s}(x-y)\tilde{Y}_s(y)dy\int_{\mathbb{R}}\sqrt{\tilde{Y}_s(z)}\ W(ds,dz).$$

In particular this implies that the Fleming-Viot process is also a.s. absolutely continuous in $\mathbb{R}$.

144

2. An important method for establishing that certain finite dimensional martingale problems are well-posed involves the notion of pathwise uniqueness. We will explain this in the present context. Starting with any solution to the martingale problem we have obtained a weak solution, X, of a stochastic evolution equation driven by Gaussian white noise, W(ds,dx).

Given any two such weak solutions (X^1,W) and (X^2,W) of the stochastic evolution equation defined on the same probability space $(\Omega,\mathcal{F},P)$ with the *same* Wiener process W, assume that we can show that

$$P(\{\omega: X_t^1(\omega)=X_t^2(\omega),\ 0\le t\le T\}) = 1, \text{ for each } T < \infty.$$

Then we say that the stochastic evolution equation has the *pathwise uniqueness property*.

Theorem 8.4.1 The pathwise uniqueness property implies that the initial value martingale problem is well-posed.

Proof. See Yamada and Watanabe (1971).

Pathwise uniqueness has been established for some classes of stochastic partial differential equations. Of particular importance is the case in which the coefficients are Lipschitz (see e.g. Walsh (1986)). Another important class of stochastic partial differential equations for which pathwise uniqueness can be established is that satisfying the coercivity and monotonicity conditions formulated in Pardoux (1975) and Krylov and Rozovskii (1981).

The question of pathwise uniqueness for the stochastic evolution equation (8.3.4) associated with $(\alpha,1,1)$-superprocess is open. However in the case in which E is finite (and A is a matrix) Watanabe (1969) proved pathwise uniqueness for the resulting system of stochastic differential equations.

Remark. Theorem 8.3.2 suggests (in the one dimensional case, d=1) the study of non-negative function-valued (that is, absolutely continuous measure-valued) solutions of the broader class of stochastic partial differential equations:

$$(8.4.1)\quad X_t(x) = \int_{\mathbb{R}} p_t(x-y)X_0(y)dy + \int_0^t\int_{\mathbb{R}} p_{t-s}(y-x)q(X(s,y))W(ds,dy)$$

where $q(0)=0$, q is monotone increasing and continuous and W(ds,dy) is the white noise.

In the cases in which q is Lipschitz continuous or local Lipschitz existence and pathwise uniqueness of solutions to equation (8.4.1) was studied by many authors (for example, Dawson (1972), Walsh (1986), Iwata (1987), Kotelenez (1989), Gyöngy and Pardoux (1991)).

145

In the special case $q(u) = u^{\gamma}$ with $\gamma>0$ equation (8.4.1) was studied in detail by Mueller (1991a,b) using comparison methods, scaling ideas, etc. Another special case of interest is that in which q is Lipschitz and has compact support - see for example Theorem 10.4.1.

8.5 Absolute Continuity of the Occupation Time Measure.

Let Y be the occupation time process associated to the $B(\Delta,c)$-superprocess in $\mathbb{R}^d$ constructed in Section 7.4.

Theorem 8.5.1 Let $d \le 3$. Then the occupation time $\mathcal{O}_{0,t}(dx)$ has a density $\mathcal{O}(t,x)$ (the *local time*) which is jointly continuous in (t,x) a.s.

Proof. Iscoe (1986, Theorem 4) and Fleischmann (1988) proved the existence of a density and Sugitani (1987) proved that it is a.s. jointly continuous using Kolmogorov's criterion. □

Adler and Lewin (1990) also obtained a Tanaka-type formula for the local time of both super-Brownian motion and super-stable processes.

146

9. SAMPLE PATH PROPERTIES OF SUPER-BROWNIAN MOTION

9.1 Some Introductory Remarks

In this chapter we will survey results on the sample path behavior of super-Brownian motion. Throughout the chapter we consider the $(2,d,\beta)$-superprocess $\{X_t: t\geq 0\}$ with probability law $(\{P^{\beta}_{\mu}:\mu\in M_F(\mathbb{R}^d)\})$, that is, the measure-valued branching process $(\mathbf{D},\mathcal{D},(\mathcal{D}_t)_{t\geq 0},(X_t)_{t\geq 0},\{P_{\mu}:\mu\in M_F(E)\})$, associated with the log-Laplace equation

(9.1.1) $\dfrac{\partial v}{\partial t} = \frac{1}{2}\Delta v - \dfrac{1}{1+\beta}v^{1+\beta}$, Δ = d-dimensional Laplacian,

where $0<\beta\leq 1$. The case $\beta=1$ will be referred to as *continuous super-Brownian motion.*

Outlines of proofs are included in some cases as illustrations of the ideas and techniques involved but many results are described without proof. However we hope that this incomplete survey will convince the reader that super-Brownian motion is a rich subject for research and will provide an introduction to the growing literature.

9.2. Probability Estimates from the log-Laplace Equation

The study of Brownian motion is intimately linked with the heat equation and Laplace equation. It is therefore not surprising that the study of $(2,d,\beta)$-superprocesses is intimately related to the study of the nonlinear parabolic equation (9.1.1) and the associated non-linear elliptic equation

(9.2.1) $\Delta v = \dfrac{2}{1+\beta}\, v^{1+\beta}$.

For example, equation 9.1.1 was used to derive the exact form of the log-Laplace function of the total mass (4.5.2) and the scaling relation (4.5.3).

Another important role of the log-Laplace functional is the determination of the probability of the events $\{X_t(A) = 0\}$ and $\{X_s(A) = 0\ \forall\ 0\leq s<t\}$. This method was developed by Iscoe (1988) (in the case $\beta=1$) and has as its starting point the following.

Lemma 9.2.1. Let $B := B(x_0,r) := \{y: d(y,x_0) < r\}$ and for $\varepsilon < r$ let $\phi_\varepsilon \in C(E)$ be defined by

$$\begin{aligned}\phi_\varepsilon(x) &= 1 \text{ if } d(x_0,x) \leq r-\varepsilon\\ &= (r-d(x_0,x))/\varepsilon \text{ if } r-\varepsilon < d(x_0,x) < r\\ &= 0 \text{ if } d(x_0,x) > r.\end{aligned}$$

Then

147

(a) For fixed $t > 0$,

(9.2.2) $$P_\mu^\beta(X_t(B)=0) = \lim_{\theta\to\infty}\lim_{\varepsilon\to 0} P_\mu^\beta\left(e^{-\langle\mu,V_t(\theta\phi_\varepsilon)\rangle}\right),$$

where $V_t\phi$ is the unique solution of the initial value problem

(9.2.3) $$\frac{\partial v}{\partial t} = \frac{1}{2}\Delta v - \frac{1}{1+\beta}v^{1+\beta}, \quad v(0,x) \equiv \phi.$$

(b) For $t > 0$,

(9.2.4) $$P_\mu^\beta(X_s(B)=0\ \forall\ 0\le s<t) = \lim_{\theta\to\infty}\lim_{\varepsilon\to 0} P_\mu^\beta\left(e^{-\langle\mu,U_t(\theta\phi_\varepsilon)\rangle}\right).$$

where $U_t\phi$ is the unique solution of the inhomogeneous equation

$$\frac{\partial v}{\partial t} = \frac{1}{2}\Delta v - \frac{1}{1+\beta}v^{1+\beta} + \phi, \quad v(0,x) \equiv 0.$$

Proof. (a) We obtain from the properties of the Laplace functional and the monotone convergence theorem that

$$P_\mu^\beta(X_t(B)=0) = \lim_{\theta\to\infty} P_\mu^\beta\left(e^{-\theta X_t(B)}\right) = \lim_{\theta\to\infty}\lim_{\varepsilon\to 0} P_\mu^\beta\left(e^{-\theta\langle X_t,\phi_\varepsilon\rangle}\right)$$

$$= \lim_{\theta\to\infty}\lim_{\varepsilon\to 0} P_\mu^\beta\left(e^{-\langle\mu,V_t(\theta\phi_\varepsilon)\rangle}\right).$$

(b) Since $\{X_s:s\ge 0\}$ is right continuous in $M_F(E)$ and B is open in E, then $\liminf_{r\downarrow s} X_r(B) \ge X_s(B)$ and hence $X_s(B) > 0$ for some $0\le s<t$ implies that $\int_0^t X_s(B)ds > 0$. Therefore

$$P_\mu^\beta(X_s(B)=0\ \forall\ 0\le s<t) = P_\mu^\beta\left(\int_0^t X_s(B)ds = 0\right)$$

$$= \lim_{\theta\to\infty} P_\mu^\beta\left(\exp\left(-\theta\int_0^t X_s(B)ds\right)\right)$$

$$= \lim_{\theta\to\infty}\lim_{\varepsilon\to 0} P_\mu^\beta\left(\exp\left(-\theta\int_0^t \langle X_s,\phi_\varepsilon\rangle ds\right)\right). \quad \square$$

A similar result can be obtained for the complement of a closed ball. As an illustration we have the following result.

Theorem 9.2.2. Consider the $(2,d,\beta)$-superprocess $\{X_t:t\ge 0\}$ and let $\mu(\mathbb{R}^d)<\infty$ with supp

148

$\mu \subset B(0,R_0)$ and let $R_0 < R$. Then

$$P^{\beta}_{\mu}(X_{\cdot} \text{ ever charges } \overline{B(0,R)}^c) = 1 - \exp(-R^{-2/\beta}\langle u(R^{-1}.),\mu\rangle)$$

where u is the unique positive solution of the singular elliptic boundary value problem

$$\Delta u(x) = \frac{2}{1+\beta} u^{1+\beta}(x), \quad x \in B(0,1)$$

$$u(x) \longrightarrow \infty \text{ as } x \to \partial B(0,1).$$

In particular if $\mu = \delta_0$, then

$$P^{\beta}_{\delta_0}(X_{\cdot} \text{ ever charges } \overline{B(0,R)}^c) = 1 - \exp(-u(0)R^{-2/\beta}).$$

Proof. Iscoe (1988, Theorem 1) for $\beta=1$. The proof for $0<\beta<1$ is essentially the same.□

Using the same technique we now derive an estimate of the probability that a propagating initial point mass charges the exterior of a small ball in a short interval. But first we state a preliminary result on the solution of the related non-linear elliptic problem.

Lemma 9.2.3 Let $u(.,R)$ denote the solution of the equation

(9.2.1) $$\Delta u = \frac{2}{1+\beta} u^{1+\beta}, \quad u(x) \uparrow \infty \text{ as } |x| \uparrow R. \text{ Then}$$

(9.2.2) $$u(x,R) = R^{-2/\beta}u(x/R,1),$$

and

(9.2.3) $$\lim_{|x|\uparrow 1} \frac{u(x,1)}{1/(1-|x|^{2/\beta})} = \left(\frac{(2+\beta)(1+\beta)}{\beta^2}\right)^{1/\beta}.$$

Proof. This is obtained by an o.d.e. argument - [Dawson, Iscoe and Perkins, (1989, Lemma 3.6), Dawson and Vinogradov (1992a, Appendix 1, Lemma 4.4)]. □

Theorem 9.2.4. Let $a>0$, and $R > (2/\beta)t^{1/2}$. Then

(9.2.4) $$P^{\beta}_{a\delta_0}(X_s(\overline{B(0;R)}^c) > 0 \text{ for some } s\le t)$$

$$\le c(\beta,d)\, aR^{-2/\beta}\left[\frac{R}{t^{1/2}}\right]^{d+(4/\beta)-2} \exp\left\{-\frac{R^2}{2t}\right\}.$$

Outline of Proof. (See Dawson, Iscoe and Perkins (1989, Theorem 3.3(b)) or Dawson and Vinogradov (1992a, Proposition 1.1) for details.) Set

149

$$P := P_{a\delta_0}(X_s(\overline{B(0,R)^c}) > 0 \text{ for some } s\le t)$$

$$= \lim_{\theta\to\infty} 1 - e^{-u^\theta(t,0)a} \le \lim_{\theta\to\infty} a\, u^\theta(t,0,R)$$

where u^θ satisfies

$$\frac{\partial u^\theta}{\partial t} = \frac{1}{2}\Delta u^\theta - \frac{1}{1+\beta}(u^\theta)^{1+\beta} + \theta\psi$$

$$u^\theta(0)=0, \quad \psi \approx 1(|x|>R).$$

In order to apply a Feynman-Kac type representation of the solution of this equation let $\{B_t;t\ge 0\}$ denote a Brownian motion in $\mathbb{R}^d$ and set $\tau := \inf\{t:B_t\notin B(0;R_1)\}$, $R_1<R$. Then using the strong Markov property of Brownian motion,

$$u^\theta(t,0) = E_0\left\{1(\tau<t)\exp\left[-\int_0^\tau (1+\beta)^{-1}(u^\theta(t-v,B_v))^\beta dv\right]\right.$$

$$\left.\cdot E_{B(\tau)}\left\{\int_0^{t-\tau} \theta\psi(B_s)\exp\left[-\int_0^s (1+\beta)^{-1}(u^\theta(t-\tau-v,B_v))^\beta dv\right]ds\right\}\right\}$$

$$\le E_0\left[1(\tau<t)\; u^\theta(t-\tau,R_1)\right], \quad \text{since } |B(\tau)|=R_1.$$

But $u^\theta(t,x,R) \le u(x,R)$ where $u(x,R)$ is the solution of (9.2.1).
By Lemma 9.2.1 $u(x,R) = R^{-2/\beta}u(x/R,1)$ and

$$\lim_{|x|\uparrow R} \frac{R^{-2/\beta}u(x/R,1)}{1/(1-|x/R|^{2/\beta})} = R^{-2/\beta}\left(\frac{(2+\beta)(1+\beta)}{\beta^2}\right)^{1/\beta}. \tag{9.2.5}$$

Hence $\quad P \le \; a\,P(\tau_{R_1}\le t)\cdot u(R_1,R).$

Then optimizing on R_1 we obtain

$$P \le c(\beta,d)\, aR^{-2/\beta}\left[\frac{R}{t^{1/2}}\right]^{d+(4/\beta)-2} e^{-R^2/2t}. \;\square$$

9.3. The support process of super-Brownian motion

9.3.1. The Support Map, Support Dimension and Carrying Dimension.

Let (E,d) be a separable metric space. Without loss of generality we may assume that d is bounded (e.g. $d(x,y)\le 1\ \forall\ x,y$). Let $\mathfrak{F}(E)$ (resp. $\mathfrak{K}(E)$) denote the space of closed (resp. compact) subsets of E.
The induced *Hausdorff metric* ρ on $\mathfrak{F}(E)$ is defined by

$\rho(A,B) := \max(\rho_1(A,B),\rho_1(B,A))$, if A and B are non-empty, where

150

$$\rho_1(A,B) := \sup_{x\in A} d(x,B), \qquad d(x,B) := \inf_{y\in B} d(x,y).$$

$$\rho(A,\emptyset) := 1 \text{ if } A\neq\emptyset.$$

Note that $\rho_1(A,B) \le r$ implies $A \subset B^r$ where $B^r := \{x:d(x,B)\le r\}$.

N.B. ρ_1 is not symmetric and hence not a metric. However it satisfies the triangle inequality

$$\rho_1(A_1,A_3) \le \rho_1(A_1,A_2) + \rho_1(A_2,A_3).$$

If $\mu\in M_F(E)$, $\mu\neq 0$, the *support of* μ, denoted by $S(\mu)$, is defined to be the smallest closed set $C\subset E$ satisfying $\mu(E\backslash C)=0$. Equivalently,

$S(\mu) := \{x\in E:\mu(B(x)) > 0$ for every open ball $B(x)$ centered at $x\}$, if $\mu\neq 0$,

$S(\mu) := \emptyset$ if $\mu = 0$.

Now consider $\mathfrak{F}(E)$, (resp. $\mathfrak{K}(E)$) where $E = \mathbb{R}^d$ (or any locally compact separable metric space). The *compact topology* on $\mathfrak{F}(E)$ is defined by the subbasis consisting of the sets $\mathfrak{F}_G$, $\mathfrak{F}^K$, where G runs over all the open sets of E and K runs over all the compact sets of E. Here

$\mathfrak{F}_A := \{F\in\mathfrak{F}:F\cap A\neq\emptyset\}$, A is a subset of E,

$\mathfrak{F}^A := \{F\in\mathfrak{F}:F\cap A=\emptyset\}$, A is a subset of E.

Since $\mathfrak{F}^{K_1}\cap\mathfrak{F}^{K_2} = \mathfrak{F}^{K_1\cup K_2}$ a basis element in the compact topology is of the form:

$\mathfrak{F}^K_{G_1\ldots G_n} = \{F\in\mathfrak{F}:F\cap K=\emptyset,\ F\cap G_i\neq\emptyset,\ i=1,\ldots,n\}$ where

$n\ge 0$, K is compact and G_i is open.

<u>Theorem</u> <u>9.3.1.1</u> (a) $\mathfrak{F}(E)$ is second-countable, compact and Hausdorff in the compact topology.

(b) The compact topology on $\mathfrak{F}(E)$ is coarser than the Hausdorff topology.

<u>Proof.</u> Cutler (1984, p. 74, 80), Matheron (1975).

<u>Theorem</u> <u>9.3.1.2.</u> (a) Suppose E is a separable metric space. Then the mapping $S:M_F(E)\longrightarrow(\mathfrak{F}(E),\rho)$ is measurable.

(b) If E is also locally compact, then $S:M_F(E)\longrightarrow(\mathfrak{F}(E),\text{compact})$ is also *lower semi-continuous* (in the sense of Matheron), i.e. $\psi^{-1}(\mathfrak{F}_G)$ is open for every open set G.

(c) Let $\mu,\mu_n\in M_F(E)$ and suppose $\mu_n \xrightarrow{w} \mu$. Then

151

$S(\mu) \le \underline{\lim}_{n\to\infty} S(\mu_n)$.

Proof. (a) Cutler (1984, Theorem 4.4.1, Cor. 4.4.1.2, Theorem 4.4.2.)

Let φ be a continuous, strictly increasing function on $[0,\infty)$ with $\varphi(0)=0$. Let $A \subset E$.

The *Hausdorff φ-measure* of A is defined as

$$\varphi\text{-m}(A) := \lim_{\delta\downarrow 0} \varphi_\delta\text{-m}(A)$$

where $\varphi_\delta\text{-m}(A) := \inf\left\{ \sum_k \varphi(r_k) : \bigcup_k B(x_k,r_k) \supset A,\ r_k < \delta/2 \right\}$,

(where $\{B(x_k,r_k)\}$ are open balls with centres x_k and radii r_k.

In particular set $\varphi_\alpha(r) := r^\alpha$, for $0 < \alpha < \infty$, and

$$\mathcal{H}^\alpha(A) = \varphi_\alpha\text{-m}(A) := \lim_{\delta\downarrow 0} \mathcal{H}^\alpha_\delta(A)$$

where $\mathcal{H}^\alpha_\delta = \inf\left\{ \sum_k (r_k)^\alpha : \bigcup_k B(x_k,r_k) \supset A,\ r_k < \delta/2 \right\}$.

The *Hausdorff-Besicovitch* dimension of a set A, dim(A), is defined by means of

$$\mathcal{H}^\alpha(A) := +\infty \quad \text{if} \quad \alpha < \dim(A)$$

$$:= 0 \quad \text{if} \quad \alpha > \dim(A),$$

i.e. $\dim(A) := \inf\{\alpha\ge 0 : \mathcal{H}^\alpha(A)=0\}$.

Note that if $\mathcal{H}^\alpha(A) < \infty$, then $\dim(A) \le \alpha$.

Definition. A measure $\mu \in M_F(E)$ is said to have *carrying dimension* $\text{cardim}(\mu) = d$, if there exists a d-dimensional Borel set A such that $\mu(A^c) = 0$ and $\dim(A) = d$, and this fails for $d' < d$. It is said to have *support dimension* $\text{supdim}(\mu) = d$ if $\dim(S(\mu)) = d$.

Remarks. 1. Clearly $\text{supdim}(\mu) \ge \text{cardim}(\mu)$.

2. If $E = \mathbb{R}^d$ equipped with the Euclidean metric, then $\dim(A) \le d\ \forall\ A\subset E$.

Theorem 9.3.1.3. (a) The dimension map $\dim:\mathfrak{F}(E)\to \mathbb{R}_+$ is measurable (with respect to the Borel σ-algebra generated by the compact (and also the Hausdorff) topology.

(b) The map $\mu \to \dim(S(\mu))$ from $M_F(E) \to \mathbb{R}_+$ is measurable.

Proof. (a) Cutler, (1984, Theorem 4.5.1).

(b) follows from (a) and Theorem 9.3.1.2. □

Remark 9.3.1.4. Consider a measure $\mu \in M_F(\mathbb{R}^d)$. If $\text{cardim}(\mu) < d$, then μ is singular with respect to Lebesgue measure.

152

9.3.2. A Modulus of Continuity for the Support Process

Let $\{X_t : t \geq 0\}$ and $S(X_t)$ be a $(2,d,\beta)$-superprocess and the closed support of X_t, respectively.

Recall Lévy's classical result on the modulus of continuity for standard Brownian motion:

$$\lim_{\delta \downarrow 0} \frac{\sup_{0 \leq s \leq 1-\delta} \sup_{0 < u \leq \delta} |W(s+u) - W(s)|}{\sqrt{2\delta \log 1/\delta}} = 1 \quad \text{a.s.} \tag{9.3.2.1}$$

In this section we will derive a partial analogue to this result for the support process.

Lemma 9.3.2.1. Let $X_0 = \mu \in M_F(\mathbb{R}^d)$, and $R > (2/\beta)s^{1/2}$. Then

$$P^\beta_{X_0}\left(\{\rho_1(S(X_{t+u}), S(X_t)) > R\} \cap \{\langle X_{t+u}, 1\rangle \neq 0\} \text{ for some } u \leq s \,|\, \mathcal{D}_t\right)$$

$$= P^\beta_{X_t}\left(\{\rho_1(S(X_u), S(\mu)) > R\} \cap \{\langle X_{t+u}, 1\rangle \neq 0\} \text{ for some } u \leq s\right)$$

$$\leq c(\beta,d)\, R^{-2/\beta}\left[\frac{R}{s^{1/2}}\right]^{d+(4/\beta)-2} \exp\left\{-\frac{R^2}{2s}\right\} \cdot X_t(\mathbb{R}^d), \quad P^\beta_{X(0)}\text{-a.s.}$$

Proof. We first approximate $X_t(dx)$ by atomic measures of the form

$$\mu_n = \sum_{i=1}^{N(n)} a_i^n \cdot \delta_{x_i^n}, \quad \text{with } x_i^n \in S(X_t) \text{ and } \sum a_i^n = X_t(\mathbb{R}^d),$$

$$\mu_n \Rightarrow X_t(dx) \text{ as } n \to \infty.$$

Then $P^\beta_{\mu_n} \Rightarrow P^\beta_{X_t}$ (by Feller property of $(2,d,\beta)$-superprocesses).

Since $S(\mu)$ is closed, then for $\eta > 0$,

$$\{\{\rho_1(S(X_u), S(\mu)) > R\} \cap \{\langle X_u, 1\rangle \neq 0\} \text{ for some } 0 \leq u < s+\eta\}$$

$$= \{\nu(.) \in D([0,\infty), M_F(\mathbb{R}^d)) : \nu(u, S(\mu)^c)) > 0 \text{ for some } 0 \leq u < s+\eta\}$$

is an open set in the Skorohod topology.

Therefore by the Markov property

$$P^\beta_{X_0}\left(\rho_1(S(X_{t+u}), S(X_t)) > R \text{ for some } 0 \leq u \leq s \,|\, \mathcal{D}_t\right)$$

153

$$= P^{\beta}_{\mu}\Big(\rho_1(S(X_u),S(\mu)) > R \text{ for some } 0\le u\le s\Big) \quad \text{with } \mu = X_t$$

$$\le \liminf_{n\to\infty} P^{\beta}_{\mu_n}\Big(\rho_1(S(X_u),S(\mu)) > R \text{ for some } 0\le u<s+\eta\Big).$$

Since $P^{\beta}_{\mu_n} = \mathop{*}\limits_{i=1}^{N(n)} P^{\beta}_{a_i^n\delta_{x_i^n}}$, then $X_t \overset{\mathcal{D}}{=} X_t^1+\ldots X_t^{N(n)}$ where $\{X_u^i: i=1,\ldots,N(n)\}$ are independent and X_t^i is a version of the $(2,d,\beta)$-superprocess with initial measure $a_i^n\cdot\delta_{x_i^n}$. (Here $*$ denotes the operation of convolution.) Moreover

$$\{\rho_1(S(X(u),S(\mu)) > R \text{ for some } 0\le u<s+\eta\}$$
$$\subset \Big\{\bigcup_{i=1}^{N(n)} \rho_1(S(X_u^i),\{x_i\})) > R \text{ for some } 0\le u<s+\eta\Big\}.$$

Therefore, by Theorem 9.2.4,

$$\liminf_{n\to\infty} P^{\beta}_{\mu_n}\Big(\rho_1(S(X_u),S(\mu)) > R \text{ for some } 0\le u<s+\eta\Big)$$

$$\le \liminf_{n\to\infty} \sum_{i=1}^{N(n)} P^{\beta}_{a_i^n\cdot\delta_{x_i^n}}\Big(\rho_1(S(X_u^i),\{x_i\}) > R \text{ for some } 0\le u<s+\eta\Big)$$

$$\le c(\beta,d)\, R^{-2/\beta}\left[\frac{R}{(s+\eta)^{1/2}}\right]^{d+(4/\beta)-2} \exp\Big\{-\frac{R^2}{2(s+\eta)}\Big\}\cdot X_t(\mathbb{R}^d).$$

The result follows since $\eta>0$ is arbitrary. □

Let

(9.3.2.2) $$h_\kappa(t) := \left[\frac{2}{\beta}(1+\beta)\min\Big(1,\ t\Big(\log\frac{1}{t} + \kappa\log\log\frac{1}{t}\Big)\Big)\right]^{1/2}.$$

The following result is an analogue of Lévy's global modulus of continuity for the path behavior of the $(2,d,\beta)$-superprocess on the interval $[0,T]$.

<u>Theorem 9.3.2.2</u> Let $\mu\in M_F(\mathbb{R}^d)$, $\kappa > \frac{\beta d+2}{2(1+\beta)}$ and $B_K := \{X_T\neq 0\}\cap\{\sup_{0\le t\le T} X_t(\mathbb{R}^d) \le K\}$. Then for P^{β}_{μ}-a.s. $\omega\in B_K$, there is a $\delta = \delta(\omega,\kappa,K,T)>0$ such that if $0\le s,t\le T$ satisfy $0 < t-s < \delta$, then

(9.3.2.3) $$S(X_t) \subset S(X_s)^{h_\kappa(t-s)}, \text{ i.e. } \rho_1(S(X_s),S(X_t)) < h_\kappa(t-s).$$

<u>Remarks:</u> Note the difference in the constant $\sqrt{2(1+\beta)/\beta}$ in $h_\kappa(t)$ and $\sqrt{2}$ in the

154

classical Lévy modulus of continuity result for Brownian motion (in the latter case this is sharp). Theorem 9.3.2.2 can be stated in a more precise form in the setting of the historical process (which will be introduced in Ch. 12). In the case of $\beta=1$, the constant $\sqrt{4}$ (=2) is known to be sharp (cf. [DP, Theorem 8.7]).

<u>Proof of Theorem 9.3.2.2.</u> For simplicity we take T=1. Since by Theorem 4.6.2(c) $\{X_t:t\ge 0\}\in D([0,\infty),M_F(\mathbb{R}^d))$, P^β_μ-a.s., then $\sup_{0\le t\le 1} X_t(\mathbb{R}^d)<\infty$. It therefore suffices to show that for any $K>0$,

$$P^\beta_\mu\left(\left\{S(X_t)\subset S(X_s)^{h_\kappa(t-s)}\right\}^c\cap B_K\right)=0.$$

Given $\kappa>\frac{\beta d+2}{2(1+\beta)}$ we choose any $0<\varepsilon(\beta,\kappa)<(\kappa\cdot\frac{1+\beta}{\beta}-d/2-1/\beta)/2$.
We first derive an almost-sure result for the grid:

(9.3.2.4)

$$P^\beta_\mu\left\{B_K\cap\max_{\substack{0<k=j-i\le(\log N)^\varepsilon\\ 0\le i<j\le N}}\frac{\rho_1(S(X_{j/N}),S(X_{i/N}))}{h_\kappa(k/N)}\le 1\ \text{for all } N=2^n \text{ large enough}\right\}$$

$$=P^\beta_\mu(B_K).$$

To get (9.3.2.4), we first obtain upper estimates for the probabilities

$$(9.3.2.5)\qquad p_N:=P^\beta_\mu\left\{B_K\cap\max_{\substack{0<k=j-i\le(\log N)^\varepsilon\\ 0\le i<j\le N}}\frac{\rho_1(S(X_{j/N}),S(X_{i/N}))}{h_\kappa(k/N)}>1\right\}.$$

It can be shown by use of Lemma 9.3.2.1 that

$$p_N\le C_1(\beta,d,K)\cdot\sum_{0<k\le(\log N)^\varepsilon}N\cdot\left(\frac{k}{N}\cdot\log\frac{N}{k}\right)^{-1/\beta}\cdot\left(\log\frac{N}{k}\right)^{d/2-1+2/\beta}$$

$$\cdot\exp\left\{-\frac{1+\beta}{\beta}\left(\log\frac{N}{k}+\kappa\log\log\frac{N}{k}\right)\right\}$$

$$\le C_2(\beta,d,K)\cdot\sum_{0<k\le(\log N)^\varepsilon}k\cdot(\log N)^{-1-(\kappa\cdot\frac{1+\beta}{\beta}-\frac{d}{2}-\frac{1}{\beta})}$$

$$\le C_3(\beta,d,K)\cdot(\log N)^{-1+2\varepsilon-(\kappa\cdot\frac{1+\beta}{\beta}-\frac{d}{2}-\frac{1}{\beta})}.$$

Recall that $0<\varepsilon<(\kappa\cdot\frac{1+\beta}{\beta}-d/2-1/\beta)/2$, i.e. the power of the logarithm in the latter expression is less than -1. This implies that (taking $N = 2^n$)

$$p_{2^n} \le C_4(\beta,d,\kappa,\delta,K)\cdot n^{-1+2\varepsilon-(\kappa\cdot\frac{1+\beta}{\beta} - \frac{d}{2} - \frac{1}{\beta})}$$

i.e. it is the general term of a convergent series. Then the Borel-Cantelli Lemma implies that for the subsequence $N = 2^n$ at most a finite number of events

$$\left\{B_K \cap \max_{\substack{0<k=j-i\le(\log N)^\varepsilon \\ 0\le i<j\le N}} \frac{\rho_1(S(X_{j/N}),S(X_{i/N}))}{h_\kappa(k/N)} > 1\right\}$$

occurs. This yields (9.3.2.4).

The remainder of the proof is carried out by standard argument (cf. McKean (1969) p.16) and is based on estimate (9.3.2.4) for maximum over the grid, and the triangle inequality $\rho_1(A_1,A_3) \le \rho_1(A_1,A_2) + \rho_1(A_2,A_3)$.

Given $0 < \varepsilon(\beta,\kappa) < (\kappa\cdot\frac{1+\beta}{\beta} - d/2 - 1/\beta)/2$ choose $n_0(\varepsilon)$ so that

(9.3.2.6) $$\max_{\substack{0<k=j-i\le(\log N)^\varepsilon \\ 0\le i<j\le N}} \frac{\rho_1(S(X_{j/N}),S(X_{i/N}))}{h_\kappa(k/N)} \le 1$$

$\forall\ N = 2^n$ with $n \ge n_0$. Now let

$$\delta(\beta,\kappa) \le (\log 2)^\varepsilon\cdot n_0^\varepsilon / 2^{n_0}.$$

(We will pick $\delta(\beta,\kappa)$ later.)

Now consider a pair (s,t): $0 \le s < t \le 1$ such that $u := t-s < \delta(\beta,\kappa)$.

Pick $n = n(u) \ge n_0$ so that $(\log 2)^\varepsilon(n+1)^\varepsilon/2^{n+1} \le u < (\log 2)^\varepsilon n^\varepsilon/2^n$ and note that $n(u) \longrightarrow \infty$ as $u \to 0$.

We can then choose sequences $n < p_1 < p_2 <\ldots$, and $n < q_1 < q_2 <\ldots$ such that

(9.3.2.7) $$s_k := i\cdot 2^{-n} - 2^{-p_1} - 2^{-p_2} - \ldots - 2^{-p_k},$$

(9.3.2.7') $$t_k := j\cdot 2^{-n} + 2^{-q_1} + 2^{-q_2} + \ldots + 2^{-q_k},$$

$$|s_k-s| \le 2^{-p_k},\quad |t_k-t| \le 2^{-q_k},$$

$$s\le s_k \le i\cdot 2^{-n} < j\cdot 2^{-n} \le t_k \le t,\ \text{ and } 0 < k = j - i \le u\cdot 2^n < (\log 2)^\varepsilon\cdot n^\varepsilon.$$

The triangle inequality then yields

156

(9.3.2.8)

$$\rho_1(S(X_t),S(X_s)) \le \rho_1(S(X_{i/_{2^n}}),S(X_s)) + \rho_1(S(X_{j/_{2^n}}),S(X_{i/_{2^n}}))$$

$$+ \rho_1(S(X_t),S(X_{j/_{2^n}})).$$

To estimate the middle term on the right-hand side of (9.3.2.8) we apply monotonicity of function $h_\kappa(\cdot)$ and (9.3.2.4):

(9.3.2.9) $\rho_1(S(X_{j/_{2^n}}),S(X_{i/_{2^n}})) \le h_\kappa(k/2^n) \le h_\kappa(t-s) \quad \text{for} \quad n \ge n_0.$

Also note that

(9.3.2.10') $$\rho_1(S(X_{i/_{2^n}}),S(X_s)) \le \sum_{k=1}^{\infty} \rho_1(S(X_{s_k}),S(X_{s_{k+1}}))$$

(9.3.2.10') $$\rho_1(S(X_t),S(X_{j/_{2^n}})) \le \sum_{k=1}^{\infty} \rho_1(S(X_{t_{k+1}}),S(X_{t_k})).$$

By (9.3.2.4) and the monotonicity of $h_\kappa(.)$ it is easily seen that the each of these two expressions does not exceed

(9.3.2.11) $$\sum_{\ell=n+1}^{\infty} h_\kappa(1/2^\ell).$$

Therefore

$$\rho_1(S(X_{i/_{2^n}}),S(X_s)) \le \sum_{\ell=n+1}^{\infty} h_\kappa(1/2^\ell)$$

(9.3.2.12)

$$\le C_1(\beta,\kappa) \sum_{\ell=n+1}^{\infty} \frac{\sqrt{\ell}}{2^{\ell/2}}$$

$$\le C_1(\beta,\kappa) \int_n^{\infty} x^{1/2}\cdot 2^{-x/2}\cdot dx.$$

The latter integral is not difficult to estimate by use of Laplace's method since it is equivalent as $n \to \infty$ (up to some positive constant) to $n^{1/2}\cdot 2^{-n/2}$. The latter expression in its turn is equivalent as $n \to \infty$ (up to some positive constant) to $h_\kappa(1/2^n)$. These arguments yield the upper estimate for the first term on the right-hand side of (9.3.2.8):

157

$$(9.3.2.13)\quad \rho_1(S(X_{i/2^n}),S(X_s)) \le C_2(\beta,\kappa)\cdot h_\kappa(1/2^n)$$

and a similar estimate can be obtained for the third term on the right hand side of (9.3.2.8).

Note that due to our choice of n, $1/2^n \le \dfrac{2}{(\log 2)^\varepsilon(n+1)^\varepsilon}\cdot(t-s)$. Then by the monotonicity of $h_\kappa(\cdot)$ we obtain $h_\kappa(1/2^n) \le h_\kappa(2\cdot(t-s)/((\log 2)^\varepsilon(n+1)^\varepsilon))$. Combining (9.3.2.8, 9.3.2.9, 9.3.2.13) we obtain

$$(9.3.2.14)\quad \rho_1(S(X_t),S(X_s)) \le h_\kappa(t-s) + C_2(\beta,\kappa)\cdot h_\kappa(C(\varepsilon)(t-s)/(n+1)^\varepsilon)$$

for $n \ge n_0$, where $C(\varepsilon) := 2/(\log 2)^\varepsilon$ (< 3).

Given any $\eta > 0$, the expression $C_2(\beta,\kappa)\cdot h_\kappa(C(\varepsilon)(t-s)/(n+1)^\varepsilon)$ can be made less than $\eta\cdot h_\kappa(t-s)$, by choosing $\delta(\beta,\kappa) > 0$ sufficiently small. Then (9.3.2.14) yields

$$P^\beta_\mu\left\{ \exists\, \delta > 0:\ \rho_1(S(X_t),S(X_s)) \le h_\kappa(t-s)\ \forall\ 0<t-s<\delta \right\} = 1. \quad \square$$

<u>Remark</u> Note that this result yields the *compact support property*, that is, if X_0 has compact support, so does X_t for $t>0$. It also implies that $P^\beta_{\delta_0}(\lim_{t\downarrow 0} \operatorname{diam}(S(X_t)) = 0) = 1$. This can be sharpened as follows:

<u>Theorem 9.3.2.3</u> Let $r(t) := \inf\{R:\ S_t \subset B(0,R)\}$. Then

$$P^\beta_{\delta_0}\left\{ \lim_{t\downarrow 0} \frac{\sup_{0\le u\le t} r(u)}{\sqrt{\frac{2}{\beta}\, t \log \frac{1}{t}}} = 1 \right\} = 1.$$

<u>Proof.</u> Tribe (1989, Theorem 2.1) for the case $\beta=1$, Dawson and Vinogradov (1992a, Formula (1.16")).

<u>Remarks.</u> 1. In fact an exact almost-sure rate of convergence of $\dfrac{\sup_{0\le u\le t} r(u)}{\sqrt{\frac{2}{\beta}\, t \log \frac{1}{t}}}$ to 1 as $t\to 0$ can be obtained (see Dawson and Vinogradov (1992a) Theorem 1.4); this result is analogous to the Chung, Erdös and Sirao (1959) result for standard Brownian motion (cf. Itô and McKean (1965), Section 1.9).

2. A local version of Theorem 9.3.2.2 is also proved in Dawson and Vinogradov

158

(1992a, Theorem 1.3) with constant $\sqrt{2/\beta}$ instead of $\sqrt{2(1+\beta)/\beta}$. The fact that this result is sharp follows from Theorem 9.3.2.3.

Theorem 9.3.2.4. If $\mu \in M_F(\mathbb{R}^d)$ then for P_μ^β-a.e. ω

(a) $\{S(X_t):t\geq 0\}$ is a $\mathfrak{K}(\mathbb{R}^d)$-valued process having right continuous paths with left limits.

If in addition $\beta=1$, then

(b) $\lim_{s\uparrow t} S(X_s) \supset S(X_t)$ $\forall t>0$

(c) If $\beta = 1$, then $S_{t-}-S(X_t)$ is empty or a singleton for all $t>0$ where $S_{t-} := \lim_{s\uparrow t} S(X_s)$.

Proof. Theorem 9.3.2.1 shows that if $t\downarrow s$, then $\rho_1(S_t,S_s) \longrightarrow 0$. On the other hand the compactness of S_s and right continuity of X_t imply that

$$\rho_1(S_s,S_t) \longrightarrow 0 \text{ if } t \to s.$$

This proves the right continuity.

The existence of left limits is a consequence of Theorem 9.3.2.1 and the following criterion of Perkins (1990, Lemma 4.1):

if $f:(0,\infty) \longrightarrow \mathfrak{K}(\mathbb{R}^d)$ is such that

$$\forall\varepsilon>0 \ \exists\ \delta>0 \text{ such that } 0\leq u-t<\delta \text{ implies } f(u)\subset f(t)^\varepsilon,$$

then f possesses left and right limits at all t in $(0,\infty)$.

(b) If $\beta=1$, then the process X_t is continuous and therefore if $X_t(B(x,\varepsilon)) > 0$, then $X_u(B(x,\varepsilon)) > 0$ for u near t by the continuity of X. Then (b) follows immediately.

(c) is proved in Perkins (1990, Prop. 4.6).□

Remarks: (1) In the case $\beta=1$, the countable set of discontinuities of $\{S_t:t>0\}$ occurs when an isolated colony becomes extinct (cf. Perkins (1990, Theorem 4.7). Note that Perkins (1990, Theorem 4.8) also proved that the space-time set of these extinction points form a dense subset of the graph of S if $d\geq 3$.

Tribe (1989) obtained a detailed description of the behavior of continuous super-Brownian motion (and super-stable-processes) near extinction using the Shiga's result on the relation between Fleming-Viot process and the superprocess (cf. Section 8.1).

(2) In the case of (α,d,β)-superprocess with $\alpha<2$ (i.e. super-stable processes) the support behavior is completely different. In fact Perkins has proved the following *instantaneous propagation of support* property.

Theorem 9.3.2.5. Consider the $(\alpha,d,1)$ process with $\alpha < 2$. Then $S_t \neq \emptyset$ implies that $S_t = \mathbb{R}^d$, Q_μ-a.s. for all $t>0$ and $\mu \in M_F(\mathbb{R}^d)$.
Proof. Perkins (1990, Cor. 1.6).

9.3.3. Application to the carrying and support dimensions.

Theorem 9.3.3.1. Let $\{X(t):t\geq 0\}$ be an (α,d,β)-superprocess. Then for each $t>0$,

$$P^\beta_\mu(\text{cardim}(X_t) = \alpha/\beta \,|\, X_t(\mathbb{R}^d)>0) = 1.$$

Proof. See [D, Section 7.3].

For continuous super-Brownian motion the following sharp result was obtained by Perkins (as well as a weaker result in d=2).

Theorem 9.3.3.2. If $d\geq 3$, then there exist constants $0 < c(d)\leq C(d)<\infty$ such that for any $\mu\in M_F(\mathbb{R}^d)$, and P^1_μ-a.e. ω

$$c(d)\varphi_{2,2}\text{-}m(A\cap S(X_t))\leq X_t(A)\leq C(d)\varphi_{2,2}\text{-}m(A\cap S(X_t)) \;\; \forall\; A\in\mathcal{B}(\mathbb{R}^d) \text{ and all } t>0$$

where $\varphi_{2,2}(r) = r^2\log\log 1/r$.
Proof. Perkins (1989, Theorem 1).

For fixed times this can be even further refined as follows.

Theorem 9.3.3.3 If X_t is the $(2,d,1)$-superprocess, with $d\geq 3$, $X_0=\mu\in M_F(\mathbb{R}^d)$, and $t > 0$, then P^1_μ-a.s. there is a $c(d)\in(0,\infty)$ such that

$$X_t(A) = c(d)\;\varphi_{2,2}\text{-}m(A\cap S(X_t)) \quad \forall\; A\in \mathcal{B}(\mathbb{R}^d).$$

Proof. The proof (see [DP, Theorem 5.2]) is based on a zero-one law for the of the historical process (cf. Ch. 12 for an introduction to the historical process).□

Remark 9.3.3.4. By Theorem 8.1.1, that is, the identification of the $(\frac{1}{2}\Delta,f)$-Fleming-Viot process as the conditional law of the branching process conditioned to have $\langle X_t,1\rangle = f(t)$, and Theorem 9.3.3.1 we deduce: for $d\geq 3$, for $P_\mu\circ\langle .,1\rangle^{-1}$-a.a. f,

$$Y_t(A) = f(t)^{-1}c_d\phi\text{-}m(A\cap S(Y_t)) \quad \forall\; A\in \mathcal{B}(\mathbb{R}^d), \quad P^{FV}_{\mu,f}\text{-a.s.}$$

Perkins (1991b Formula (15)) conjectures that this is also true for $f(t)\equiv 1$ but with another unknown constant c'_d. However we will remark at the end of Ch. 12 that one cannot obtain an immediate proof of this from the corresponding result for the continuous superprocess.

The proofs of Theorems 9.3.3.2-9.3.3.3 are too long to be included here. An introduction to the main tools and in particular the lower bound method is given in [D, Section 7.2]. In this section we will derive an uniform in time upper bound on

160

the support dimension for the (2,d,β)-superprocess using the upper estimate of Theorem 9.2.4.

Theorem 9.3.3.5. Let $\{X(t):t\ge 0\}$ be a (2,d,β)-superprocess and $\{S_t\}$ the corresponding support process.

(a) If $\varphi_\beta(x) := x^{2/\beta}\left(\log(1/x)\right)^{-1/\beta}$, then

$$P^\beta_\mu(\varphi_\beta\text{-m}(S_t)<\infty\ \forall t>0) = 1.$$

(b) $P^\beta_\mu(\dim(S_t) \le 2/\beta\ \forall t>0) = 1.$

(c) If $d > 2/\beta$ and $t > 0$, then

$$P^\beta_\mu(X_t \text{ is singular} \mid X_t(\mathbb{R}^d) > 0) = 1$$

(where singularity is understood to be with respect to d-dimensional Lebesgue measure).

Proof. (b) follows immediately from (a) and the definition of dimension.
(c) follows immediately from (b) and Remark 9.3.1.4.

The proof of (a) is based on the representation of $X_{t+\varepsilon}$ as a Cox cluster random measure with intensity $\frac{X(t)}{\varepsilon^{1/\beta}}$ (cf. Corollary 11.5.3 and Theorem 12.4.3). Although the formal development of this approach is deferred to Sections 11 and 12, we will outline the intuitive ideas here and then proceed to the relevant computations.

Let N(s,t) denote the number of clusters starting at time s whose lifetimes are greater than t. Intuitively each cluster represents a subpopulation alive at time (s+t) and having a single common ancestor located at a point x at time s. We will denote by $P^{\varepsilon *}_{s,x}$ the law of this cluster. We will obtain a covering of S_t in the time interval $(i2^{-n},(i+1)2^{-n})$ by open balls by decomposing X_t into a finite number, $N_n(i)$, of clusters starting at time $(i-1)2^{-n}$ at points $\{x_\ell\}_{\ell=1,\dots,N_n(i)}$ of radius r_n. We can then use Theorem 9.2.4 to obtain an estimate of the probability that the ball $B(x_\ell,r_n)$ covers the evolving cluster during the time interval $(i2^{-n},(i+1)2^{-n})$. To carry this out we first formulate and prove two lemmas.

Lemma 9.3.3.6. Let $t > 0$ and $\tilde X_s$ denote a cluster with lifetime greater than ε with cluster law $P^{\varepsilon *}_{s,x}$ and $R > (2/\beta)t^{1/2}$. Then for $0 < \varepsilon < t$, $P^{\varepsilon *}_{s,x}(\tilde X_u(\overline{B(x;R)}^c) > 0$ for some $s\le u\le s+t)$

$$\le c(\beta,d)(\varepsilon/t)^{1/\beta}\left[\frac{R}{t^{1/2}}\right]^{d+(2/\beta)-2}\exp\left\{-\frac{R^2}{2t}\right\}.$$

Proof. We obtain this by letting $a \downarrow 0$ in (9.2.4) and conditioning

161

on non-extinction by time s+ε. □

We denote the number of clusters at instant $i/2^n$ of age 2^{-n} by $N_n(i) := N((i-1)2^{-n}, 2^{-n})$. Note that conditioned on the total mass $\langle X((i-1)2^{-n}),1\rangle \le M$ at time $t=(i-1)2^{-n}$, $N_n(i)$ is Poisson with mean

$$\lambda = \frac{\langle X((i-1)2^{-n}),1\rangle}{c_\beta 2^{-n/\beta}} \le 2^{n/\beta} M/c_\beta$$

where $c_\beta = (\beta/(\beta+1))^{1/\beta}$.

Lemma 9.3.3.7. Let $\tau_M := \min(\inf\{t:\langle X_t,1\rangle \ge M\},1)$.

(i) If $u>2M$, then we have the following estimate for the number of clusters of age 2^{-n} at instants $j/2^n < \tau_M$:

$$\text{(9.3.3.1)}\qquad P^\beta_\mu(N_n(j) > u2^{n/\beta},\ j/2^n < \tau_M,\ j\in\mathbb{N}) \le 2^n \exp\left\{-2^{n/\beta}\left(u \log\frac{c_\beta u}{M} - u + \frac{M}{c_\beta}\right)\right\}.$$

(ii) There exists $n_0(w)$ such that for $n> n_0(\omega)$,

$$\text{(9.3.3.2)}\qquad \max_{i \le 2^n\tau_M} N_n(i) \le M\cdot\left(1+\frac{1}{\beta}\right)^{1/\beta} 2^{n/\beta}, \quad P^\beta_\mu\text{-a.s.}$$

Proof. (i) Obviously

$$P^\beta_\mu(N_n(j) > u2^{n/\beta},\ j/2^n < \tau_M,\ j\in\mathbb{N}) \le \sum_{j=1}^{2^n[\tau_M]} P\{N_n(j) > u2^{n/\beta}\}.$$

To estimate the tail probabilities of the Poisson distributions we note that the cumulant generating function for the Poisson distribution with parameter λ and its Legendre transform are equal to $\lambda(e^s-1)$ and $H_\lambda(v) = v\log\frac{v}{\lambda} - v + \lambda$ respectively. Applying the exponential Chebyshev inequality we obtain

$$P\{N_n(j) > u2^{n/\beta}\} \le \exp\left\{-H_{\lambda_{n,j}}(u2^{n/\beta})\right\}$$

where $\lambda_{n,j} = E(N_n(j)) \le 2^{n/\beta}M/c_\beta$. Differentiating with respect to λ shows that $-H_\lambda(v)$ is increasing in λ if $\lambda \le v$. Hence

$$P\{N_n(j) > u2^{n/\beta}\} \le \exp\left\{-2^{n/\beta}\left(u\log\frac{c_\beta u}{M} - u + \frac{M}{c_\beta}\right)\right\}.$$

(ii) Due to (9.3.3.1) the probability of the event

$\left\{\max_{i\le 2^n\tau_M} N_n(i) > M\cdot\left(1+\frac{1}{\beta}\right)^{1/\beta} 2^{n/\beta}\right\}$ is the general term of a convergent series. Apply-

162

ing the Borel-Cantelli lemma we obtain (9.3.3.2). □

Completion of the Proof Theorem 9.3.3.5(a)

We will cover each of the $N_n(i-1)$ cluster birth points at time $(i-1)2^{-n}$ that have descendents at time $i2^{-n}$ by balls of radius $r_n = C_1 \cdot 2^{-n/2} n^{1/2}$.

Recall that the number of such cluster birth points does not exceed $C_2(M)\, 2^{n/\beta}$, at each $j/2^n < M$. Set

$$\Xi_n(i) := \bigcup_{j=1}^{N_n(i)} B(x_{i,j}, r_n) .$$

Applying Lemma 9.3.3.6 to each of these clusters we obtain

$$P\left(X_u(\overline{\Xi_n(i)}^c > 0 \text{ for some } 0 \le u \le t \ \forall\ t \in (i2^{-n},(i+1)2^{-n}) \forall\ i/2^n \le \tau_M \right\}$$

$$\le C_3 \cdot 2^{n(1+1/\beta)} \cdot n^{((d-2)/2+1/\beta)} \cdot \exp\left(-C_1 2^n \cdot 2^{-n} n\right)$$

$$\le C_3 2^{n(1+1/\beta)} \cdot n^{[(d-2)/2+1/\beta-2\kappa/\beta]} . \exp\left(-C_1 n\right)$$

Hence it is the general term of a convergent series for sufficiently large C_1. Therefore

$$P^\beta_\mu\left\{S(X_t) \subset \Xi_n(i)\ \forall t \in (i2^{-n},(i+1)2^{-n})\ \forall\ i/2^n \le \tau_M,\ i \in \mathbb{N}\ \forall \text{ suff. large } n\right\} = 1.$$

But then for all sufficiently large n,

$$\varphi_\beta\text{-}m(S(X_t)) \le \lim_{n\to\infty} \varphi_{\beta,2r_n}\text{-}m(S_t) \le \lim_{n\to\infty} \varphi_{\beta,2r_n}\text{-}m(\Xi_n(i))$$

$$\le \lim_{n\to\infty} c_4(M) \cdot 2^{n/\beta} \varphi_\beta(2r_n)$$

$$\le C_4(M) \cdot \lim_{n\to\infty} \frac{2^{n/\beta} 2^{-n/\beta} n^{1/\beta}}{(\log 2^n/n)^{1/\beta}} < \infty. \quad □$$

9.4 Charging and Hitting Sets.

9.4.1. Some Basic Definitions and Probability Estimates.

Consider the $(2,d,\beta)$-superprocess with log-Laplace equation (9.1.1). The canonical process X is said to *charge* the Borel subset B of $\mathbb{R}^d$ if $X_t(B) > 0$ for some $t>0$.

163

It turns out that the notion of charging a set is not the natural analogue of the notion of hitting a set for Brownian motion. To introduce the appropriate definition let $\bar{\mathcal{R}}_+(0,t) := \bigcup_{t>s>0} \bar{\mathcal{R}}(s,t)$ for $0<t\le\infty$,

where

$$\bar{\mathcal{R}}(s,t) = \text{closure of } \mathcal{R}(s,t) := \Big[\bigcup_{s\le u\le t} S(X_u)\Big].$$

Equivalently, $\bar{\mathcal{R}}(s,t) := S(\mathcal{O}_{s,t})$ where $\mathcal{O}_{s,t}$ denotes the occupation time measure as defined in Section 7.4. The *range* of super-Brownian motion is defined by $\bar{\mathcal{R}} := \bar{\mathcal{R}}_+(0,\infty)$. Theorem 9.2.2 implies that the range is bounded P^β_μ-a.s. provided that μ has compact support.

The canonical process X is said to *hit* the set $A \subset \mathbb{R}^d$ if $A\cap\bar{\mathcal{R}} \neq \phi$. If A is an analytic subset of $\mathbb{R}^d$, then the event $\{\omega:\bar{\mathcal{R}}(\omega)\cap B\neq\emptyset\}$ belongs to the universal completion $\mathcal{D}^{uc}$ of the σ-algebra $\mathcal{D}$.

Theorem 9.4.1.1. Let $\varepsilon > 0$ and $x_0\in\mathbb{R}^d$.

(a) Let $d>4$. Then there exists a constant $c_2(d)$ such that

$$(9.4.1.1) \qquad P^1_{\delta_x}[X_t(B(x_0;\varepsilon)) > 0 \text{ for some } t\ge 0\,] \le c_2(d)\varepsilon^{d-4}|x-x_0|^{2-d}$$

provided that $x\neq x_0$.

$$\text{(b)}\quad P^1_{\delta_0}(X \text{ charges } B(x;\varepsilon)) \sim \begin{cases} 6/x^2 & d=1 \\ 4/x^2 & d=2 \\ 2/x^2 & d=3 \\ 2/[x^2\log|x|] & d=4. \end{cases}$$

Proof. (a) We will outline the main ideas of the proof and refer the reader to Dawson, Iscoe and Perkins (1989, Section 3) for the details.

First note that

$$P^1_{\delta_x}\exp\Big\{-\int_0^t\!\!\int \theta 1_{B(0;\varepsilon)}(x)X(s,dx)ds\Big\} = \exp\{-u(t,x,\theta,\varepsilon)\}$$

where us is the solution of the initial value problem

$$\partial u/\partial t = 1/2(\Delta u-u^2) + \theta 1_{B(0;\varepsilon)}(.),\quad u(0) = 0.$$

Note that

$$P^1_{\delta_x}\Big\{\int_0^\infty X(s,B(0;\varepsilon))ds = 0\Big\} = \lim_{\theta\uparrow\infty}\lim_{T\uparrow\infty}\exp\{-u(T,x,\theta,\varepsilon)\} = \exp\{-u_\infty(x,\varepsilon)\}$$

where $\Delta u_\infty(x,\varepsilon) = u_\infty(x,\varepsilon)^2$ for $x\notin B(0;\varepsilon)$,

$$u_\infty(x,\varepsilon) \longrightarrow \infty \text{ as } x \to \partial B(0;\varepsilon).$$

By a scaling argument $u_\infty(x,\varepsilon) = \varepsilon^{-2}u_\infty(x/\varepsilon,1)$.

164

Therefore

$$P^1_{\delta_x}(X \text{ charges } B(0;\varepsilon)) = 1 - \exp\{-u_\infty(x,\varepsilon)\} = 1 - \exp\{-\varepsilon^{-2}u_\infty(x/\varepsilon,1)\}.$$

But in d>4 an analytic argument yields

$$u_\infty(x,1) \sim \frac{1}{|x|^{d-2}} \qquad \text{as} \quad |x| \to \infty.$$

Therefore

$$P^1_{\delta_x}(X \text{ charges } B(0;\varepsilon)) \sim 1 - \exp\left\{-\frac{\varepsilon^{-2}\varepsilon^{d-2}}{|x|^{d-2}}\right\}.$$

(b) See Iscoe (1988, Theorem 2). □

If $d \le 3$, and $\beta=1$, then by Theorem 9.4.1.1(b),

$$P^1_{\delta_x}(X \text{ charges } B(0;\varepsilon)) \sim \frac{\text{const}}{|x|^2} \quad \text{(independent of } \varepsilon > 0).$$

This, as well as the the existence of a jointly continuous local time (cf. Theorem 8.5.1), implies that X hits points. This is made precise as follows.

Theorem 9.4.1.2. Let $d \le 3$, $\beta=1$. Then

$$P^\beta_\mu(X \text{ hits } \{x\}) = 1 - \exp\left\{-2(4-d)\int|x-y|^{-2}\mu(dy)\right\} > \text{ if } \mu\ne 0.$$

Proof. The proof of this result is given in [Dawson, Iscoe and Perkins (1989) Theorem 1.3].

On the other hand if $d > 4$ and A is sufficently small then it will not be hit. For example the following result is an easy consequence of (9.4.1.1).

Proposition 9.4.1.3. Let d>4 and $A \subset \mathbb{R}^d$. Then if $x^{(d-4)}\text{-}m(A) = 0$, then $P^1_\mu(X \text{ hits } A) = 0$.

Proof. Let $\{B_j\}$ be a covering of A. Then

$$\begin{aligned} P^1_\mu(X \text{ hits } A) &\le P^1_\mu(X \text{ hits } \textstyle\bigcup_j B_j) \\ &\le \sum_j P^1_\mu(X \text{ hits } B_j) \\ &\le c_2(d) \sum_j (\text{diam}(B_j))^{d-4} \quad \text{(by Theorem 9.4.1.1(a)).} \end{aligned}$$

But since the $x^{d-4}\text{-}m(A) = 0$, this can be made arbitrarily small. □

We should emphasize the difference between the hitting a set and *charging* a set, B. By Remark 7.3.2(a) $X_t(B) = 0 \ \forall \ t>0$, a.s. if $|B|=0$ and hence X does not

165

charge any Lebesgue null sets.

9.4.2. $\mathcal{R}$-Polar Sets

For d-dimensional Brownian motion there is a classical necessary and sufficient condition for a set to be not hit in terms of capacity. In this section we will formulate an analogous necessary and sufficient condition for super-Brownian motion which is due to Dynkin. The main step is to reduce the hitting problem to an analytical problem for the nonlinear elliptic equation (9.2.1).

Definition. The analytic set A is said to be R_β-*polar* if $A^c \neq \emptyset$ and $P^\beta_\mu(A \cap \bar{\mathcal{R}} = \emptyset) = 1$, $\forall \mu \in M_F(\mathbb{R}^d)$. (It is actually sufficient to check that $P^\beta_{\delta_x}(A \cap \bar{\mathcal{R}} = \emptyset) = 1$ $x \notin A$, (cf. Dynkin (1991c, Lemma 2.4)). An arbitrary set is called R_β-polar if it is a subset of a Borel R_β-polar set B.

Let $p_t(.,.)$ denote the standard Brownian transition density in $\mathbb{R}^d$ and

$$G_d(x,y) = \int_0^\infty e^{-t/2} p_t(x,y)dt$$

$$\sim |x-y|^{2-d} \text{ for small } |x-y| \text{ if } d>2.$$

$$\sim \log^+|x-y|^{-1} \text{ if } d=2.$$

For $\beta \geq 0$ the $(1+\beta)$-*capacity* of a subset of $\mathbb{R}^d$ is defined as follows:

$$Cap_{d-2,1+\beta}(A) := \sup\ \nu(A)$$

where the supremum is taken over $\nu \in M(\mathbb{R}^d)$ with $\nu(A^c)=0$ and such that

$$\int_{\mathbb{R}^d} \left[\int_{\mathbb{R}^d} \nu(dx) G_d(x,y)\right]^{1+\beta} dy \leq 1$$

If $\beta=0$, $Cap_{d-2,1}$ coincides with the *classical Newtonian capacity*.

Theorem 9.4.2.1 (Kakutani) Let $d \geq 2$ and $\{B_t : t \geq 0\}$ denote the standard d-dimensional Brownian motion. Then the following assertions are equivalent

(i) $P_x(\tau_A = +\infty) = 1$ $\forall x \notin A$, where $\tau_A := \inf\{t>0 : B_t \in A\}$,

(ii) $Cap_{d-2,1}(A) = 0$.

Proof. See Itô and McKean (1965, Section 7.8).

Theorem 9.4.2.2 (Dynkin(1991c)). A set $A \subset \mathbb{R}^d$ is R_β-polar if and only if $C_{d-2,1+\beta}(A) = 0$.

166

Corollary 9.4.2.3. A single point $\{x\}$ is R-polar if and only if $d \geq \frac{2(1+\beta)}{\beta}$.

Remarks on the Proof. This is proved in Dynkin (1991c). The proof is based on probabilistic arguments which reduce the problem to an analytical problem for semilinear elliptic partial differential equations and then by an application of analytical results of Baras and Pierre (1984), and Brezis and Veron (1980).

We will give an indication of the probabilistic step which involves an interesting aspect of superprocesses which we have not yet considered. To explain this in heuristic terms let us go back to the construction of the superprocess in terms of the branching particle approximations.

Let B be a bounded regular domain in $\mathbb{R}^d$ and for $w \in C([0,\infty),\mathbb{R}^d)$, let $\tau_B := \inf\{t: w_t \notin B\}$. Let us introduce the additive functional $\kappa_B(dt) = 1_{[0,\tau_B]}(t)$ and let W_B denote Brownian motion in $\mathbb{R}^d$ stopped at time τ_B. Now consider the $(W_B,\kappa_B,\Phi_\beta)$-superprocess $\{X^B_t: t\geq 0\}$ with $\Phi_\beta(\lambda) = -\lambda^{1+\beta}$.

We define $X_{\tau_B} := \lim_{t\to\infty} X^B_t$. Then X_{τ_B} is a random measure and for $\phi \in bpC(\partial B)$

$$v(x) := -\log\left(P_{\delta_x} \exp(-\langle X_{\tau_B},\phi\rangle)\right)$$

is the unique solution of

$$(9.4.2.1) \qquad \Delta v = v^{1+\beta} \quad \text{in } B$$

$$v(x) \longrightarrow \phi(a) \text{ as } x\to a\in\partial B,\ x\in B.$$

With a little more work one can construct an enriched version giving the joint law of X_{τ_B} and the full superprocess $\{X_t: t\geq 0\}$. In fact in Dynkin's proof of the result below, an increasing sequence of open sets $B_n \uparrow B$ and a sequence of random measures $\{X_{\tau_{B_n}}: n=1,2,\ldots\}$ defined on a common probability space (cf. Section 12.3.5) is involved.

The main step in the reduction of the proof of Theorem 9.4.2.2 to an analytical problem is summarized in the following.

Theorem 9.4.2.4.

(a) Let B be a bounded regular domain in $\mathbb{R}^d$. Then

$$v(x) = -\log P^{\beta}_{\delta_x}\{X_{\tau_B} = 0\}$$

is the minimal positive solution to the boundary problem

$$\Delta v(x) = v(x)^{1+\beta} \quad \text{in } B$$

$$v(x) \longrightarrow \infty \text{ as } x\to a\in\partial B,\ x\in B.$$

(b) For an arbitrary open set B

167

$$v(x) = -\log P^{\beta}_{\delta_x} \{\bar{\mathcal{R}} \subset B\}$$

determines the maximal positive solution of

(9.4.2.2) $\Delta v(x) = v(x)^{1+\beta}$ in B.

(c) A necessary and sufficient condition for a closed set F to be R_β-polar is that one of the following equivalent statements holds

(i) if $v \geq 0$ satisfies (9.4.2.1) in $B = F^c$, then $v=0$,

(ii) the maximal solution of (9.4.2.1) in $B = F^c$ is bounded.

Proof. See Dynkin (1991c, Theorems 1.2, 1.3, 1.4). □

Remark: For a probabilistic approach to these questions in the case of continuous branching, see Perkins (1990). Also Le Gall (1991b) has recently developed an alternative approach to the study of these questions for continuous branching using his path-valued Markov process. For a survey of the relations between superprocesses and partial differential equations, see Dynkin (1992b).

9.5 Multiple Points and Intersections

In this section we consider *multiple points*, that is, points visited at two or more distinct times by a given (2,d,1)-superprocess and *intersections* (or *collisions*) of independent (2,d,1)-superprocesses, that is, points visited by two or more superprocesses at the same time. One of the main ideas involved is to regard the range of a superprocess over a given time interval as a *target* and to use the criteria of section 9.4.2 to determine if this is a R-polar set. For the sake of comparison we begin with the corresponding results for ordinary Brownian motion.

Theorem 9.5.1 Let $\{B_t : t \geq 0\}$ be a d-dimensional Brownian motion andset

$$\phi_0(x) := x^2 \log \frac{1}{x} \,\mathrm{logloglog}\, \frac{1}{x} \quad \text{and} \quad \phi(x) := x^2 \,\mathrm{loglog}\, \frac{1}{x}.$$

Then

(a) In the case $d \geq 3$, $\phi\text{-}m(B[0,t)) = c_d t \;\; \forall t$ a.s.

(b) In the case $d=2$, $\phi_0\text{-}m(B[0,t)) = c_2 t \;\; \forall t>0$ a.s.

Proof. (a). (Ciesielski and Taylor (1962)

(b) Taylor (1964). □

Theorem 9.5.2 (a) Two independent d-dimensional Brownian motions intersect if and only if $d \leq 3$.

(b) Let Γ_k := set of k-multiple points of a d-dimensional Brownian motion. Then

(i) $\dim \Gamma_k(B) = d-k(d-2)$ for $d>2$.

(ii) $\dim \Gamma_k(B) = 2$ for all k if $d=2$.

168

(iii) $\dim \Gamma_2(B) = 1$ if $d=3$

(iv) $\dim \Gamma_3(B) = \emptyset$ if $d=3$.

Proof. (a) follows from Kakutani's criterion (Theorem 9.4.2.1).

(b) See Taylor (1966), and Le Gall (1987). □

The analogues of these results for super-Brownian motion are contained in the following three theorems.

Theorem 9.5.3. Let $\beta=1$ and $\bar{\mathcal{R}}_+(0,s)$ denote the range of the continuous super-Brownian motion in $\mathbb{R}^d$ with $d > 4$. Let $\phi(x) := x^4\log^+\log^+(1/x)$.

Then there are positive constants $c_1(d)$ and $c_2(d)$ such that

$$c_1\phi\text{-}m(\bar{R}_+(0,s)\cap A) \le Y_{0,s}(A) \le c_2\phi\text{-}m(\bar{R}_+(0,s)\cap A)$$

$\forall A\in \mathcal{B}(\mathbb{R}^d)$ w.p.1.

Proof. (cf. [Dawson, Iscoe and Perkins (1989), Theorem 1.4]). □

Definition. The set of *k-multiple points* of X is denoted by

$$\bar{R}_k := \bigcup\left\{ \bigcap_{j=1}^{k} \bar{R}(I_j): I_1,\dots,I_k \text{ disjoint, compact} \subset (0,\infty)\right\}$$

Theorem 9.5.4. Let $d\ge 4$. Then

(a) If $t>0$, then $\dim(\bar{R}_+(0,t)) = 4$ w.p.1.

(b) $\bar{R}_k = \phi$ if $k \ge \dfrac{d}{d-4}$, and

$\dim \bar{R}_k \le d - k(d-4)$ a.s.

Proof. (b) [Dawson, Iscoe and Perkins (1989), Theorem 1.5]

Remark: This means that there are no double points if $d\ge 8$, triple points if $d\ge 6$ or quintuple points if $d\ge 5$!

The *closed graph* of the (2,d,1)-superprocess is defined to be the closure of the random space-time set

$$\bar{G}(X) := \{(t,x):t>0,\ x\in S(X_t)\} \in \mathcal{B}((0,\infty)\times\mathbb{R}^d).$$

An analytic subset of $\mathbb{R}_+\times\mathbb{R}^d$ is called *G-polar* if $P^1_\mu(\bar{G}(X)\cap A=\emptyset)=1\ \forall\ \mu\in M_F(\mathbb{R}^d)$. Dynkin (1992a) established a criterion for G-polarity following a program similar to that described in section 9.4 for R-polarity. An important application of the notion of G-polarity is the following result of Barlow, Evans and Perkins.

Theorem 9.5.5 Let $X^{(1)}$ and $X^{(2)}$ denote two independent (2,d,1)-superprocesses with $X^{(1)}(0) = \mu_1$ and $X^{(2)}(0) = \mu_2$. If $d\ge 6$, then $\bar{G}(X^{(1)})\cap\bar{G}(X^{(2)}) = \emptyset$ for all $t > 0$, $P^{(1)}_{\mu_1}\times P^{(2)}_{\mu_2}$-a.s.

Proof. Barlow, Evans and Perkins (1991, Theorem 3.6).

169

<u>Remark.</u> An immediate consequence of Theorem 9.5.5 is the *non-intersection* property $S(X_t^{(1)}) \cap S(X_t^{(2)}) = \emptyset \ \forall \ t>0$. Barlow, Evans and Perkins (1991) also establish the existence of a *collision local time* in dimensions $d \le 5$.

9.6 Some Further Comments

The objective of this section has been to give an introduction to the sample path properties of $(2,d,\beta)$-superprocesses. As we have mentioned at several places the known results for (2,d,1)-superprocesses, that is, continuous super-Brownian motion are considerably more complete. References for some recent developments in this area are Adler and Lewin (1992), Barlow, Evans and Perkins (1991), Dawson, Iscoe and Perkins (1989), Dawson and Perkins (1991), Dynkin (1991a,b,c,d, 19921,b), Evans and Perkins (1991), Iscoe (1988), Le Gall (1991a,b), Perkins (1988, 1989, 1990, 1991a,b), and Tribe (1989, 1991, 1992).

There still remain some open questions concerning the structure of $S(X_t)$ for the (2,d,1)-superprocess even at fixed times. For example is $S(X_t)$ a totally disconnected set? In this direction, Tribe (1989) showed that at fixed times $t > 0$ for X_t-a.e. x, the connected component containing x is simply $\{x\}$.

Theorem 9.3.3.3 implies that in dimensions $d \ge 3$, the finite dimensional distributions of $\{X_t : t>0\}$ can be obtained from those of the support process $\{S(X_t):t>0\}$. This implies that $\{S(X_t):t>0\}$ is a $\mathfrak{K}(\mathbb{R}^d)$-valued Markov process. However the question as to whether $\{S(X_t)\}$ is strong Markov is open and is in fact closely linked to the question as to whether or not the statement of Theorem 9.3.3.3 is valid for all $t>0$, P^1_μ-a.s. which is also open. To explain the possibility that $\{S(X_t)\}$ might not be strong Markov consider the same question for the measure-valued process in $\mathbb{R}$, $Z_t = \delta_{B_1(t)} + 2\delta_{B_2(t)} + 4\delta_{B_3(t)}$ where $B_1(.)$, $B_2(.)$ and $B_3(t)$ are independent Brownian motions. In this case $\{Z_t\}$ is strong Markov but the continuous $\mathfrak{K}(\mathbb{R})$-valued process, $\{S(Z_t)\}$, is Markov but not strong Markov.

170

10. BUILDING MEASURE-VALUED PROCESSES WITH INTERACTIONS. VARIOUS EXAMPLES.

To this point we have explored measure-valued branching processes and Fleming-Viot processes and have discovered that they have a rich mathematical structure. However from the point of view of population modelling, they are rather oversimplified. In order to have models exhibiting some of the more interesting features of real populations it is necessary to introduce interactions and additional spatial structure. For example in population models phenomena such as competition for resources, predation, genetic selection and ecological niches are of considerable interest. In this chapter we consider the basic measure-valued branching process and Fleming-Viot process as building blocks from which we can build more complex models. In particular we will consider certain natural types of interaction that can be studied via the Cameron-Martin-Girsanov formula and the FK-dual representation.

10.1. Application of the Cameron-Martin-Girsanov Theorem.

10.1.1 Fleming-Viot with Selection

The class of Fleming-Viot processes arise from the *neutral theory of evolution* (e.g. Kimura (1983)). However *selection (competitive interaction)* plays the central role in the Darwinian theory of evolution and in this theory variability arises from mutation and recombination. From this viewpoint finite population sampling provides a random perturbation which can lead to *tunnelling* between locally optimal points in the selective landscape (cf. Wright (1949)). In order to study such questions in a mathematical setting it is necessary to consider a modification of the Fleming-Viot process to include these effects.

We begin by considering the Fleming-Viot model with *selection.*

Let $\mathfrak{D}(\mathfrak{G}) := \{F:F(\mu) = f(\langle\mu,\phi\rangle),\ \phi\in D(A),\ f\in C_b^{\infty}(\mathbb{R})\}$ and for $F\in\mathfrak{D}(\mathfrak{G})$,

$$\mathfrak{G}_R F(\mu) := f'(\langle\mu,\phi\rangle)\langle\mu,A\phi\rangle + f'(\langle\mu,\phi\rangle)\langle R(\mu),\phi\rangle + \frac{1}{2}\iint f''(\langle\mu,\phi\rangle)\phi(x)\phi(y)Q(\mu;dx,dy)$$

with

$$Q(\mu;dx,dy) = \gamma[\delta_x(dy)\mu(dx)-\mu(dx)\mu(dy)],$$

and

$$R(\mu,dx) = \mu(dx)\left[\int_E V(x,y)\mu(dy)\right] - \mu(dx)\left[\int_E\int_E V(y,z)\mu(dy)\mu(dz)\right]$$

$$= \gamma^{-1}\int_E\left[\int_E V(y,z)\mu(dz)\right]Q(\mu;dx,dy).$$

171

The function V is called the *fitness function* and it is assumed to belong to $b\mathcal{E}\times\mathcal{E}$.

By applying Theorem 7.2.2 (the Cameron-Martin-Girsanov theorem) it follows that the corresponding martingale problem for $(\mathfrak{D}(\mathfrak{G}),\mathfrak{G}_R)$ is well-posed. The resulting Markov process is called the Fleming-Viot process with mutation operator A, sampling intensity γ, and fitness function V.

10.1.2. Branching with Interaction

Let us now return to the B(A,c)-superprocess, X_t, described in section 4.7. Recall that X_t is $M_F(E)$-valued and

$$M_t^B(\phi) = \langle X_t,\phi\rangle-\langle X_0,\phi\rangle-\int_0^t \langle X_s,A\phi\rangle ds, \quad \phi\in D(A),$$

is a square integrable martingale with increasing process

$$\langle M^B(\phi)\rangle_t = c\int_0^t \langle X_s,\phi^2\rangle ds$$

that is, $Q(\mu;dx,dy) = c\delta_x(dy)\mu(dx)$ with $c > 0$.

Let

$$\begin{aligned} R(\mu,dx) &= c\int_E r(\mu,y)\delta_x(dy)\mu(dx) \\ &= \int_E r(\mu,y)Q(\mu;dx,dy) = \; cr(\mu,x)\mu(dx) \end{aligned}$$

and assume that $r \in \mathcal{B}(M_F(E)\times E)$ satisfies

(10.1.2.1) $c\cdot\sup_{\mu,x} r(\mu,x) := K < \infty$, and

$\sup\{|r(\mu,x)|:x\in E, \langle\mu,1\rangle \le n\} < \infty$ for each $n \in \mathbb{N}$.

Let $\mathfrak{D}(\mathfrak{G}) = \{F(\mu) := f(\langle\mu,\phi\rangle)$ with $\phi\in D(A)$, $f\in C_b^\infty(\mathbb{R})\}$ and for $F\in \mathfrak{D}(\mathfrak{G})$, let

$$\begin{aligned} \mathfrak{G}_R F(\mu) \; &= \; f'(\langle\mu,\phi\rangle)\langle\mu,A\phi\rangle \;\; +f'(\langle\mu,\phi\rangle)\langle R(\mu),\phi\rangle \;\; + \\ &\quad \frac{1}{2}\iint f''(\langle\mu,\phi\rangle)\phi(x)\phi(y)Q(\mu;dx,dy). \end{aligned}$$

We would like to apply the same argument as in Section 10.1.1 to establish that the martingale problem is well-posed. To do this we must extend the result of Theorem 7.2.2 to cover processes with values in $M_F(E)$ in which case $\langle X(t),1\rangle$ is not necessarily bounded. This means that

$$Z_r(t) := \exp\left\{\int_0^t\int_E r(X(s),y)M(ds,dy) - \frac{c}{2}\int_0^t\int_E [r(X(s),x)]^2X(s,dx)ds\right\}$$

is a non-negative P_μ-local martingale (and thus a supermartingale) but an additional argument is necessary to prove that it is actually a martingale.

172

Recall that $Z_r(.)$ is a martingale if and only if $P_\mu[Z_r(t)] = 1\ \forall t\ge 0$ (cf. Theorem 7.1.7) and this will now be established.

Lemma 10.1.2.1 Assume that $R(\mu,dx) = cr(\mu,x)\mu(dx)$ satisfies (10.1.2.1). Then $Z_r(t)$ is a P_μ-martingale and the martingale problem for $(\mathcal{D}(\mathfrak{G}),\mathfrak{G})$ is well-posed.

Proof. Let $\tau_n := \inf\{u:\langle X_u,1\rangle + \int_0^u \langle X_s,1\rangle ds \ge n\}$. By the assumption (10.1.2.1), $\tau_n \uparrow \infty$, P_μ-a.s. Fix t and let $B_n := \{\tau_n \le t\}$. Then $Z_r(u\wedge\tau_n)$ is a continuous local P_μ-martingale and

$$P_\mu\left[\exp\left(\frac{1}{2}\int_0^{t\wedge\tau_n}\int_E\int_E r(X(s);x)r(X(s);y)Q(X(s);dx,dy)ds\right)\right] < \infty.$$

Using the criterion of Theorem 7.1.7(b) we conclude that $Z_r(u\wedge\tau_n)$ is a P_μ-martingale and therefore

$$P_\mu[Z_r(t\wedge\tau_n)] = P_\mu[1_{B_n} Z_r(\tau_n)] + P_\mu[1_{B_n^c} Z_r(t)] = 1.$$

Note that $P_\mu[1_{B_n^c} Z_r(t)] \uparrow P_\mu[Z_r(t)]$ as $n\to\infty$ since $\tau_n \uparrow \infty$, P_μ-a.s.

Consider the solution, $P_\mu^r = Z_r(t)P_\mu$, to the local martingale problem for $\mathfrak{G}_R$. If

$$M_u(\phi) := \langle X_u,\phi\rangle - \langle X_0,\phi\rangle - \int_0^u \langle X_s,A\phi\rangle ds - \int_0^u \langle R(X_s),\phi\rangle ds,\quad \phi\in D(A),$$

then $M_{u\wedge\tau_n}(\phi)$ is a P_μ^r-martingale with $\langle M(\phi)\rangle_{u\wedge\tau_n} = c\int_0^{u\wedge\tau_n}\langle X_s,\phi^2\rangle ds$.

If $\phi = 1$ we get

$$P_\mu^r(\langle X_{u\wedge\tau_n},1\rangle) = \langle\mu,1\rangle + P_\mu^r\left(\int_0^{u\wedge\tau_n}\langle R(X_{s\wedge\tau_n}),1\rangle ds\right)$$

$$\le \langle\mu,1\rangle + K\cdot\left(\int_0^u P_\mu^r(\langle X_{s\wedge\tau_n},1\rangle)ds\right).$$

Hence $P_\mu^r(\langle X_{t\wedge\tau_n},1\rangle) \le \langle\mu,1\rangle e^{Kt}$ by Gronwall's inequality. Therefore we also have

$$P_\mu^r\left(\int_0^t \langle X_{u\wedge\tau_n},1\rangle du\right) \le K^{-1}\langle\mu,1\rangle(e^{Kt}-1)\ .$$

Next, note that

$$P_\mu^r(B_n) := P_\mu[1_{B_n} Z_r(\tau_n)] \le P_\mu^r\left\{\int_0^t \langle X_{s\wedge\tau_n},1\rangle ds \ge n/2\right\} + P_\mu^r\left\{\sup_{u\le t\wedge\tau_n}\langle X_u,1\rangle \ge n/2\right\},$$

$$P_\mu^r\left\{\int_0^t \langle X_{s\wedge\tau_n},1\rangle ds \ge n/2\right\} \le 2n^{-1}P_\mu^r\left\{\int_0^t\langle X_{s\wedge\tau_n},1\rangle ds\right\} \le 2n^{-1}K^{-1}\langle\mu,1\rangle(e^{Kt}-1).$$

173

Since $P^r_\mu\left\{\left(\sup_{u\le t\wedge\tau_n} M_u(1)\right)^2\right\} \le const\cdot K^{-1}\langle\mu,1\rangle(e^{Kt}-1)$ by the Burkholder-Davis-Gundy inequality, therefore $P^r_\mu\left\{\sup_{u\le t\wedge\tau_n}\langle X_u,1\rangle \ge n/2\right\} \longrightarrow 0$ as $n\to\infty$. Hence $P_\mu[1_{B_n} Z_r(\tau_n)] \longrightarrow 0$ as $n\to\infty$. Therefore $P_\mu[Z_r(t)] = 1\ \forall\ t$ and therefore $Z_r(t)$ is a martingale. □

Example 10.1.2.2. Noncritical Branching

It was convenient in the earlier sections to restrict our attention to *critical branching*, i.e. $P_\mu[\langle X_t,1\rangle] = \langle\mu,1\rangle\quad \forall t\ge 0$. However the above result applied to $R(\mu) \equiv c\mu$, $c\in\mathbb{R}$, shows that the corresponding martingale problems for subcritical and supercritical branching are also well-posed. It is then an easy exercise to show that the corresponding log-Laplace equation is also satisfied. Moreover the resulting law $\Pi_{[0,T]}P^c_\mu \ll \Pi_{[0,T]}P_\mu$ where the latter is the law of the critical B(A,c)-superprocess. This shows that all a.s. sample path properties in [0,T] derived for the critical B(A,c)-superprocess are also true for the subcritical and supercritical cases.

Example 10.1.2.3. Branching with Nonlinear Death Rates

Let $h\in pbC(E\times E)$, $\mu\in M_F(E)$ and let $P_{0,\mu}$ denote the law of the B(A,c)-superprocess. Consider the following B(A,c;h)-martingale problem:

(10.1.2.2a)

$$M_t(\phi) = \langle X_t,\phi\rangle - \langle X_0,\phi\rangle - \int_0^t \langle X_s, A\phi\rangle ds + \int_0^t\int_E\left(\int_E h(x,y)X_s(dy)\right)\phi(x)X_s(dx)$$

is a continuous martingale with increasing process

(10.1.2.2b)

$$\langle M(\phi)\rangle_t = c\int_0^t \langle X_s,\phi^2\rangle ds.$$

We can then obtain from Theorem 7.2.2 and Lemma 10.1.2.1 the following result.

Theorem 10.1.2.4. The martingale problem (10.1.2.2) is well-posed and the unique solution $\{P_h\}$ is characterized by

$$\frac{\Pi_{[0,T]}\, dP_{h,\mu}}{\Pi_{[0,T]}\, dP_{0,\mu}} = \exp\left(M^h_T - \tfrac{1}{2}\langle M^h\rangle_T\right),$$

where $\quad M^h_t := -\int_0^t\int_E\left(\int_E h(x,y)X_s(dy)\right)M(ds,dx)$

and $M(ds,dx)$ is the orthogonal martingale measure extending M_t.

<u>Example 10.1.2.5. Multitype Branching with Nonlinear Death Rates</u>

Consider the following martingale problem M_{h_1,h_2}:

(10.1.2.3a)

$$M_t^i(\phi) = \langle X_t^i,\phi\rangle - \langle X_0^i,\phi\rangle - \int_0^t \langle X_s^i,A\phi\rangle ds + A_t^i(\phi), \quad i=1,2,$$

where

(10.1.2.3b) $\langle M^i(\phi),M^j(\psi)\rangle_t = \delta_{ij}c\int_0^t \langle X_s^i,\phi\psi\rangle ds$

and $A_t^i(\phi) = \int_0^t\int_E\int_E \phi(x)\int h_i(x-y)X_s^1(dy)X_s^2(dx)ds, \quad h_i \in pbC(E).$

<u>Theorem 10.1.2.6.</u> The martingale problem (10.1.2.3) is well-posed and the unique solution is characterized by

$$\frac{\Pi_{[0,T]}\, dP_{h_1h_2}}{\Pi_{[0,T]}\, dP_{0,0}} = \exp((M_1^h+M_2^h)_T - \tfrac{1}{2}(\langle M_1^h\rangle_T+\langle M_2^h\rangle_T),$$

where $M_1^h(t) := -\int_0^t\iint h_1(x,y)X_s^2(dy)M^1(ds,dx)$

$$M_2^h(t) := -\int_0^t\iint h_2(x,y)X_s^1(dy)M^2(ds,dx)$$

<u>Proof.</u> See Evans and Perkins (1992). □

Now consider the case $h_i(x-y) = p_\varepsilon(x-y)$, $\varepsilon > 0$, and note that $\langle M_1^h,M_1^h\rangle_t = \int_0^t\int\left[\int p_\varepsilon(x-y)X_s^2(dy)\right]^2 X_s^1(dx)ds \longrightarrow \infty$ if $\varepsilon\to 0$.

This suggests that the limit, if it exists, will be singular if $d\geq 2$. In fact,in dimensions $d = 2,3$, Evans and Perkins (1992) have established the existence of a non-trivial limit (at least in the case $h_2 \equiv 0$). This involves the notion of *collision local time* (cf. Barlow, Evans and Perkins (1991)). On the other hand in dimension $d\geq 6$ Theorem 9.5.5 implies that two independent superprocesses never intersect (collide) and consequently the limit is degenerate (i.e. the interaction disappears and we are left with two independent super-Brownian motions). In fact asimilar phenomenon occurs in dimensions $d = 4$ and 5 but for more subtle reasons (related to the lack of a collision local time between an ordinary Brownian particle and a super-Brownian motion). Thus the study of *local interactions* of this type involves a strong interplay with the sample path properties of the underlying super-processes.

175

10.2 An FK-Dual Representation for Fleming-Viot with Recombination and Selection.

In this section we build new processes from the basic Fleming-Viot process studied in previous chapters by introducing interactions corresponding to the mechanisms of selection and recombination in population genetics. The existence of the corresponding process can be obtained by a weak convergence argument similar to that in Chap. 2. However in this case the moment equations are *not closed* and uniqueness cannot be obtained by the arguments used there. The uniqueness is established by the use of a dual process which we will now introduce along with the description of the generator for the measure-valued process. (*See Section 5.6 for the appropriate notation - in particular* Θ_{ij} is as defined in Ch. 2 and $\{S_t\}$ is as in Section 5.6.)

Let $\rho \ge 0$ and let $\eta(x_1,x_2,\Gamma)$ be a transition function from $E\times E$ to E. For $i=1,\dots,m$ define $R_{im}:\mathcal{B}(E^m)\to\mathcal{B}(E^{m+1})$ by

$$R_{im}f(x_1,\dots,x_{m+1}) = \int f(x_1,\dots,x_{i-1},z,x_{i+1},\dots,x_m)\eta(x_i,x_{m+1},dz)$$

and assume that $R_{im}:C_b(E^m)\to C_b(E^{m+1})$. The R_{im} are called the *recombination operators* for the process and ρ is called the *recombination rate*.

For $V\in\mathcal{B}_{sym}(E\times E)$, set $\bar{V} = \sup_{x,y,z}|V(x,y)-V(y,z)|$, and for $i=1,\dots,m$ define the *selection operators* $V_{im}:\mathcal{B}(E^m)\to\mathcal{B}(E^{m+2})$ by

$$V_{im}f(x_1,\dots,x_{m+2}) = \frac{V(x_i,x_{m+1})-V(x_{m+1},x_{m+2})}{\bar{V}}\, f(x_1,\dots,x_m).$$

For $f\in\mathcal{D}(A^{(n)})\cap\mathcal{B}(E^n)$, define $F(f,\mu) := \langle\mu^n,f\rangle$ and

$$\mathfrak{G}F(f,\mu) := F(A^{(n)}f,\mu) + \gamma\sum_{1\le i<j\le n}\Big(F(\Theta_{ij}f,\mu)-F(f,\mu)\Big)$$

$$+\ \rho\sum_{i=1}^{n}\Big(F(R_{in}f,\mu)-F(f,\mu)\Big) + \bar{V}\sum_{i=1}^{n}F(V_{in}f,\mu).$$

For $f\in C_{slm}(E^{\mathbb{N}})$, with $\#(f)=n$, and $f\in\mathcal{D}(A^{(n)})\cap\mathcal{B}(E^n)$ let

$$Kf := \sum_{i=1}^{n}A_if + \gamma\sum_{j=1}^{n}\sum_{k\ne j}[\Theta_{jk}f-f] + \rho\sum_{i=1}^{n}[R_{in}f-f] + \bar{V}\sum_{i=1}^{n}[V_{in}f-f]$$

If $\beta(f) := \bar{V}\#(f)$, then

$$\mathfrak{G}F(f,\mu) = F(Kf,\mu) + \bar{V}(\#(f))F(f,\mu)$$

and $\sup_{\mu\in M_1(E)} |F(Kf,\mu)| \le const\cdot\#(f)$.

Remark. There is considerable current interest in *genetic algorithms* (cf. Goldberg (1989)). These are a class of random optimization algorithms. They are discrete time analogues of Fleming-Viot processes with a selection operator describing *pure search* and the latter is defined by

$$V_{im}f(x_1,\dots,x_{m+2}) := \frac{V(x_i)-V(x_{m+1})}{\bar{V}} f(x_1,\dots,x_m).$$

In addition recombination also plays a central role in the study of this class of algorithms.

Theorem 10.2.1. Let $\mathfrak{G}$ satisfy the above conditions and suppose that the mutation process with generator A has a version with sample paths in $D_E[0,\infty)$. Then for each initial distribution in $M_1(M_1(E))$ there exists a unique solution of the martingale problem for $\mathfrak{G}$.

Proof. The proof (due to Ethier and Kurtz (1987)) is based on duality. We will first construct the appropriate function-valued process.

Let N be a jump Markov process taking non-negative integer values with transition intensities $q_{m,m-1} = \gamma m(m-1)$, $q_{m,m+2} = \bar{V}m$, $q_{m,m+1} = \rho m$, and $q_{i,j} = 0$ otherwise. Let $\{\tau_k\}$ be the jump times of N ($\tau_0=0$) and let $\{\Gamma_k\}$ be a sequence of random operators which are conditionally independent given M and satisfy

$$P(\Gamma_k = \Theta_{ij}|N) = [2/(N(\tau_k-)N(\tau_k))]1_{\{N(\tau_k-)-N(\tau_k)=1\}},\ 1\le i<j\le N(\tau_k-),$$

$$P(\Gamma_k=R_{im}|N) = m^{-1}1_{\{N(\tau_k-)=m,\ N(\tau_k)=m+1\}}$$

$$P(\Gamma_k=V_{im}|N) = m^{-1}1_{\{N(\tau_k-)=m,\ N(\tau_k)=m+2\}}$$

for $1\le i\le m$.

For $f\in C_{sym}(E^{\mathbb{N}})$, define the $C_{sym}(E^{\mathbb{N}})$-valued process Y with $Y(0)=f$ by

$$Y(t) = S_{t-\tau_k}\Gamma_k S_{\tau_k-\tau_{k-1}}\Gamma_{k-1}\dots\Gamma_1 S_{\tau_1}f,\quad \tau_k\le t<\tau_{k+1}.$$

Then for any solution P_μ of the martingale problem for $\mathfrak{G}$ and $f\in C_{sim}(E^{\mathbb{N}})$ we have the FK-dual representation

$$P_\mu[F(f,X(t))] = Q_f\left[F(Y(t),\mu)\exp\left(\bar{V}\int_0^t \#(Y(u))du\right)\right]$$

which establishes that the martingale problem for $\mathfrak{G}$ is well-posed. Since the

177

function $\beta(f) = \tilde{V}\#(f)$ is not bounded we must verify the condition (ii) of Corollary 5.5.3. This follows from the following lemma due to Ethier and Kurtz (1990). □

<u>Lemma 10.2.2.</u> Let $N(t) = \#(Y(t))$ be as above, $\tau_K := \inf\{t: N(t) \geq K\}$ and $\theta > 0$. Then there exists a function $F(n) \geq const \cdot n^2$ and a constant $L > 0$ such that

$$E\left[F(N(t\wedge\tau_K))\exp\left\{\theta\int_0^{t\wedge\tau_K} N(s)ds\right\} | N(0)=n\right] \leq F(n)e^{Lt} \quad \forall\, K \geq 1,$$

and given $N(0)=n$, $\left\{N(t\wedge\tau_K)\exp\left\{\overline{V}\int_0^{t\wedge\tau_K} N(s)ds\right\}: K\geq 1\right\}$ are uniformly integrable.

<u>Proof.</u> Take $F(m) := (m!)^\beta$, with $\beta < 1/2$. Then

$$QF(m) + \theta mF(m) = \gamma m(m-1)(F(m-1)-F(m)) + \rho m(F(m+1)-F(m))$$
$$+ \theta m(F(m+2)-F(m)) + \theta mF(m)$$
$$= \gamma m(m-1)(((m-1)!)^\beta - (m!)^\beta) + \rho m(((m+1)!)^\beta - (m!)^\beta) + \theta m((m+2)!)^\beta + \theta m$$
$$\leq m(m!)^\beta[\rho((m+1)^\beta - 1) + \theta\sigma(m+1)^\beta(m+2)^\beta + \theta(m!)^{-\beta} - \gamma(m-1)(1-m^{-\beta})]$$
$$\leq ((m-1)!)^\beta\{m^\beta[\rho(m+1)^\beta - 1] + \theta m[m(m+1)(m+2)]^\beta + \gamma m(m-1)[1-m^\beta]\}.$$

Since the last negative term dominates for large m if $\gamma>0$, we can choose $L > 0$ such that

$$QF(m) + \theta mF(m) \leq L.$$

The optional sampling theorem (cf. [EK, p 61]) implies that for $\tau_K = \inf\{t: N(t) \geq K\}$ and $N(0)=m$

$$E\left[\exp\left\{\theta\int_0^{t\wedge\tau_K} N(s)ds\right\}\Big| N(0)=m\right]$$
$$\leq E\left[F(N(t\wedge\tau_K))\exp\left\{\theta\int_0^{t\wedge\tau_K} N(s)ds\right\}\Big| N(0)=m\right]$$
$$= F(m) + E\left[\int_0^{t\wedge\tau_K} \exp\left\{\theta\int_0^u N(s)ds\right\}(QF(N(u))+\theta N(u)F(N(u)))du\Big| N(0)=m\right]$$
$$\leq F(m) + L\,E\left[\int_0^{t\wedge\tau_K} \exp\left\{\theta\int_0^u N(s)ds\right\}du\Big| N(0)=m\right]$$

and the lemma follows by Gronwall's inequality.□

<u>10.3 Stepping Stone Models in Population Genetics</u>

In this section we give a brief introduction to the class of stepping stone models of spatially distributed populations and indicate the role of duality in their study.

178

Stepping stone models describe multitype populations which are divided into geographically separated colonies in which the total population size of each colony does not change in time. The set of colonies will be indexed by a countable set S. Each colony consists of a population described by an element of $M_1(E)$ (the distribution of types within the colony). These models can also be considered as mathematical idealizations of a spatially distributed population with locally finite carrying capacity (such as would arise if we considered a supercritical branching model in which the individual death rate is proportional to the population within the colony). The idealization that the population in each colony is fixed simplifies the analysis but it is reasonable to expect that it will have the same long-time qualitative behavior as the more complex branching-interaction model.

10.3.1. The Stepping Stone Model with Two Types.

We first consider the case in which $E = (E_1,E_2)$. In this case it suffices to prescribe $\{x_k(.):k\in S\}$ where

$x_k(t)$:= proportion of type 1 individuals in colony.

The $\{x_k(.)\}$ are assumed to satisfy the following system of Itô stochastic differential equations:

$$(10.3.1)\quad dx_k(t) = \sqrt{\gamma x_k(t)(1-x_k(t))}\, dw_k(t) + b_k(x(t))dt, \quad k\in S,$$

where the $\{w_k(.):k\in S\}$ are independent standard Brownian motions and

$$b_k(x) = a_2(1-x_k) - a_1x_k + Vx_k(1-x_k) + \rho \sum_{j\in S} q_{jk}(x_j-x_k).$$

Here $a_i \geq 0$ denotes the mutation rate of type i, V denotes the selective advantage of type 1 and $\rho q_{jk} \geq 0$ denotes the migration rate from colony j to colony k. The existence of a solution to this system of equations was obtained in Notohara and Shiga (1980).

The corresponding generator is given by

$$\mathfrak{G} = \frac{\gamma}{2} \sum_{i\in S} x_i(1-x_i)\frac{\partial^2}{\partial x_i^2} + \sum_{i\in S} b_i(x)\frac{\partial}{\partial x_i}$$

where $b_i(x) = \sum_{j\neq i} \rho q_{ji}(x_j-x_i) + a_2 - (a_1+a_2)x_i + Vx_i(1-x_i).$

The dual process. The state space for the dual process is

$$\mathbb{Z}_+^S := \{n= (n_i)_{i\in S}: n_i=0,1,2,\ldots,\ |n|=\sum n_i < \infty\}.$$

179

Let $(e_i)_j := 1$ if $i=j$

$:= 0$ if $i \neq j$.

The dual process is a finite particle system on S with births, deaths and particle motion having the following rates:

$n \to n + e_i$ rate $(-V)n_i$ (assume $-V \geq 0$) *(birth at site i)*

$n \to n - e_i$ rate $a_2 n_i + \frac{1}{2}\gamma n_i(n_i-1)$ *(death at site i)*

$n \to n - e_i + e_j$ rate $\rho q_{ij} n_i$ *(particle moves from i to j)*

(N.B. If $V > 0$ we reverse the roles of x_i and $(1-x_i)$.)

Let L denote the generator of this finite particle system.

Define $F \in \mathcal{B}(\mathbb{N}^S \times [0,1]^S)$ by

$$F(n,x) := \prod_{i \in S} (x_i)^{n_i} \text{ and } F(0,x) \equiv 1.$$

Then

$$\begin{aligned}\mathfrak{G}F(n,x) &= \sum_i (\tfrac{\gamma}{2} n_i(1-n_i)+a_2)[F(n-e_i,x)-F(n,x)] \\ &\quad + \sum_i n_i \Big\{ \sum_j \rho q_{ji}[F(n+e_j-e_i,x)-F(n,x)] + (-V)[F(n+e_i,x)-F(n,x)] - a_1 F(n,x) \Big\} \\ &= LF(n,x) - a_1 F(n,x).\end{aligned}$$

The following result of Shiga (1982) establishes the uniqueness for the martingale problem for $x(t)$ as well as a useful tool in studying the ergodic behavior of measure-valued processes.

<u>Theorem</u> <u>10.3.1</u> Let $\{x(t)\}$ be a solution of the martingale problem for this system and $\{n(t)\}$ the $\mathbb{N}^S$-valued process constructed above. Then for each $n \in \mathbb{N}^S$

$$E_x(F(n,x(t))) = E_n\left[F(n(t),x) \exp\left(\int_0^t -a_1 |n(s)| ds\right)\right].$$

<u>Proof.</u> This again follows from Theorem 5.1.2. □

<u>Remark:</u> If V=0 (neutral case) the dual process can be considered to be a system of (slowly) coalescing random walks. The voter model (cf. Holley and Liggett (1975)) can be viewed as a limiting case of this model - the dual process has infinite *coalescing rate* γ, that is, it is a coalescing random walk in which particles coalesce as soon as they meet.

<u>10.3.2. The general stepping stone model.</u>

180

In this section we extend the stepping stone model to the case in which the space of types is a general compact metric space E. This was done for finite systems of colonies by Vaillancourt (1987, 1990) and for countable systems of colonies by Handa (1990). In this section we give an outline of this development.

The state space is now $M_1(E)^S$ where again S is a countable set. Let Π^m denote the set of all mappings from {1,2,...,m} to S and $\mathcal{A} :=$ $\bigcup_{m\geq 1}(D(A^{(m)})\cap C(E^m)\times\Pi^m)$, and for each $(f,\pi) \in \mathcal{A}$,

$$\text{(10.3.2.1)}\quad F((f,\pi),\underline{\mu}) := \int_{E^m} f(x_1,\dots,x_m) \prod_{i=1}^{m} \mu_{\pi(i)}(dx_i), \qquad \underline{\mu} \in (M_1(E))^S.$$

Let $\mathfrak{D}(\mathfrak{G})$ denote the class of functions on $(M_1(E))^S$ of the form (10.3.2.1).

For $F\in\mathfrak{D}(\mathfrak{G})$, we define the operator

$$\mathfrak{G}F(\underline{\mu}) := \sum_{j\in S}\left\{\int_E A\left(\frac{\delta F(\underline{\mu})}{\delta\mu_j(x)}\right)\mu_j(dx) + \gamma\int_E\int_E\left(\frac{\delta^2 F(\underline{\mu})}{\delta\mu_j(x)\delta\mu_j(y)}\right)Q_S(\mu_j;dx,dy)\right.$$
$$\left. + \rho\sum_{k=1}^{m} q_{jk}\int_E\left(\frac{\delta F(\underline{\mu})}{\delta\mu_j(x)}\right)(\mu_k(dx)-\mu_j(dx))\right\}$$

where $Q_S(\mu;dx,dy) = \mu(dx)\delta_x(dy) - \mu(dx)\mu(dy)$. Here $\{q_{jk}\}$ denotes the rate of migration from the jth to the kth colony. Here A is the mutation operator, γ is the sampling rate and ρ is the migration rate.

The existence of an $(M_1(E))^S$-valued solution, $\{\Xi(t)\}$ to the initial value martingale problem for $\mathfrak{G}$ can again be obtained using approximating particle systems (cf. Vaillancourt (1987, 1990) when $|S|$ is finite). The uniqueness is again proved by duality. We next describe an $\mathcal{A}$-valued process, denoted by $(f,\pi)_t = (f_t,\pi_t)$, which serves as the dual to the stepping stone process. Intuitively, jumps in π_t correspond to migration. If the cardinality of the range of π_{t_0} is k, then up to the next jump of π_t, f_t is the dual process associated with k independent $M_1(E)$-valued Fleming-Viot processes.

We first recursively build the process π_t, started at $\pi_0 \in \Pi^m$, consisting of independent coordinates $\pi_t(j)$, j=1,...,m. Each of these is a continuous time Markov chain on S with transition matrix Q and successive $\mathcal{E}xp(1/\rho)$ holding times $\tau_{j,k+1}$, for k=0,1,2,... and with $\tau_{j,0}=0$.

We now turn to the construction of the process f_t, obtained via the construction of an operator-valued process as follows.

Each $\pi\in\Pi^m$ generates a partition $\{\pi^{-1}(k):k\in S\}$ of {1,2,...,m}. Considering every $p\in S$ such that $\pi^{-1}(p)$ has cardinality not less than 2, and for every pair (i,j) of distinct elements of $\pi^{-1}(p)$, we construct a family $\{\sigma^p_{(i,j)}:p\in S;i,j\in$

181

$\pi^{-1}(p), i\neq j\}$ of independent $\mathcal{E}xp(1/\gamma)$-random variables, and order it as follows.

Define a finite set of stopping times $\mathcal{U} := \{v^p_\ell : p=1,2,..,r; \ell=1,..,u(p)-1\}$ where $u(p) = \mathrm{card}(\pi^{-1}(p))\vee 1$ denotes the number of elements of $\pi^{-1}(p)$, with $u(p) = 1$ in the case $\pi^{-1}(p) = \emptyset$, by

$$v^p_1 := \min\{\sigma^p_{(i,j)} : i,j \in \pi^{-1}(p),\ i\neq j\} := \sigma_{(i^p_1, j^p_1)}$$

$$v^p_2 := \min\{\sigma^p_{(i,j)} : i,j \in \pi^{-1}(p)\setminus\{j^p_1\},\ i\neq j\} := \sigma_{(i^p_2, j^p_2)}$$

...

$$v^p_{\ell+1} := \min\{\sigma^p_{(i,j)} : i,j \in \pi^{-1}(p)\setminus\{j^p_1,..,j^p_\ell\},\ i\neq j\} := \sigma_{(i^p_{\ell+1}, j^p_{\ell+1})}$$

for $\ell = 0,1,\ldots,u(p)-2$, where $i^p_1, i^p_2, \ldots$ and $j^p_1, j^p_2, \ldots$ are well-defined random indices.

Let $0 < \zeta_1 < \ldots < \zeta_u < \zeta_{u+1} = \infty$ (with $u = \sum_{p\in S} (u(p)-1)$) be a reordering of the set $\mathcal{U}$, and define $\Phi_\ell f := \Theta_{ij} f$, where $(i,j) = (i(\ell), j(\ell))$ is the pair of random indices associated with jump ζ_ℓ, $\ell=1,2,\ldots,u$.

The operator-valued process R^π_t is then defined recursively on $C(E^m)$ by

$$R^\pi_t := \begin{cases} H^\pi_t & \text{for } t\in[0,\zeta_1) \\ H^\pi_{t-\zeta_t} \circ \Phi_\ell \circ R^\pi_{\zeta_\ell -} & \text{for } t\in[\zeta_\ell, \zeta_{\ell+1}), \ell=1,2,\ldots,u, \end{cases}$$

where $H^\pi_t : C(E^m) \longrightarrow C^\infty(E^m)$ is a semigroup defined by

$$H^\pi_t g(x_1,\ldots,x_m) := (S^{(m)}_t g)(x_{\pi 1},\ldots,x_{\pi m}).$$

Then R^π is a well-defined stochastic bounded linear operator since $\|R^\pi_t g\| \le \|g\|$ for all $t\ge 0$ and H^π_t is strongly continuous in t.

Let $0=\tau_0<\tau_1<\tau_2<\ldots$ denote the successive jump times of process π. Given a countable collection $\{{}^kR^\pi : k=0,1,2,..; \pi\in\Pi^m\}$ of independent copies of each process R^π, define recursively f_t, started at $f_0\in C(E^m)$ by

$$f_t := {}^kR^{\pi_{\tau_k}}_{t-\tau_k}(f_{\tau_k -}) \quad \text{for } t\in[\tau_k, \tau_{k+1}),\ k=0,1,2,\ldots$$

with $f_{0-} = f_0$.

182

Theorem 10.3.2.1. The $(\mathfrak{D}(\mathfrak{G}),\mathfrak{G},(M_1(E))^S)$-martingale problem is well-posed.

Proof. If K is the generator of the $\mathcal{A}$-valued process $(f,\pi)_t$ defined above, then it can be verified that

$$\mathfrak{G}F((f,\pi),\underline{\mu}) = KF((f,\pi),\underline{\mu}).$$

Then applying Theorem 5.5.2 we obtain the duality relation

$$Q_{(f,\pi)_0}F((f,\pi)_t,\Xi(0),) = P_{\{\mu_j:j\in S\}}F((f,\pi)_0,\Xi(t)).$$

(For details consult Vaillancourt (1987, 1990)). □

Remark: It is also possible to incorporate selection into the general stepping stone model combining the ideas of sections 10.2 and 10.3. The appropriate uniqueness argument in this case is proved in Handa (1990, Theorem 3.5).

10.4 A Two-type Model in a Continuous Environment and a Stochastic Evolution Equation

Finally we will consider the situation in which there are two types and in which the geographical space is S (e.g. $\mathbb{R}^d$ or a domain in $\mathbb{R}^d$). In this case $X(t,x)$ denotes the proportion of the population of type 1 at $x\in S$. The proposed state space for the process is $C(S;[0,1])$. A study of this model with $S = \mathbb{R}^d$ will provide some additional insight into one of the sources of dimension-dependent qualitative behavior. To describe it we will introduce a family of nonlinear stochastic evolution equations.

Let $W(t,dx)$ be a cylindrical Brownian motion, that is, W has independent increments and $W(t+s)-W(s)$ is Gaussian with covariance structure given by

$$E[\langle W(t),\phi\rangle\langle W(t),\psi\rangle] = t\int_{\mathbb{R}^d}\phi(x)\psi(x)dx \quad \text{for} \quad \phi,\psi \in L^2(\mathbb{R}^d).$$

We denote the resulting Gaussian white noise martingale measure by $W(ds,dx)$.

Suppose $a(y):[0,1] \to \mathbb{R}$ is continuous and $a(0)=a(1)=0$, and $b(y):[0,1]\to \mathbb{R}$ is continuous and $b(0)\geq 0$, $b(1) \leq 0$. We assume that $(A,\mathcal{D}(A))$ is the generator of a strongly continuous Markov semigroup on $C(\mathbb{R}^d)$ and that a subset of $C_c(\mathbb{R}^d)$ is a core. We also assume that this process has a transition density $p(t,x;y)$ which is jointly continuous in $(0,\infty)\times\mathbb{R}^d\times\mathbb{R}^d$.

Consider the formal stochastic integral equation

$$(10.4.1)\quad X(t,x) = X(0,x) + \int_0^t a(X(s,x))W(ds,dx) + \int_0^t b(X(s,x))ds + \int_0^t AX(s,x)ds.$$

Since we are looking for solutions with general values in $C(\mathbb{R}^d;[0,1])$ and not necessarily belonging to $\mathcal{D}(A)$ the last term may be undefined. For this reason we formulate the following *evolution (mild) form* of this equation:

183

$$(10.4.2)\quad X(t,x) = \int p(t,x,y)X(0,y)dy + \int_0^t\int p(t-s,x,y)a(X(s,y))W(ds,dy)$$

$$+ \int_0^t\int p(t-s,x,y)b(X(s,y))dsdy.$$

We first consider the case in which the coefficients are Lipschitz and the process with generator A satisfies a certain condition that guarantees the existence of a local time at the diagonal for the two particle motion of independent A-processes.

Theorem 10.4.1. *(Stochastic evolution equation)*

Suppose that a(.) and b(.) are Lipschitz and A satisfies one of

$$(10.4.3)\quad S = \mathbb{R} \text{ and } A = -(-\Delta)^{\alpha/2},\ 0 < \alpha \le 2,$$

or

(10.4.4)

(i) for some $0<\alpha<1$, $\sup_{0\le y\le T} \sup_x p_t p_t^*(x,x)/t^\alpha < \infty$.

(ii) for every compact K, every $h_0 > 0$ and every $t > 0$ there exist constants $C > 0$ and $\beta > 0$ such that

$$\int_0^t\int (p(s+h,x;y)-p(s,x';y))^2 dsdy \le C(h^\beta + |x-x'|^\beta)$$

for every $0 < h < h_0$, $x\in K$, $x'\in K$. The the stochastic evolution equation (10.4.2) has a unique solution. Furthermore, the associated martingale problem is well-posed.

Proof. The existence and pathwise uniqueness under condition (10.4.3) (as well as under weaker conditions) has been intensively studied (see e.g. Walsh (1986), Gyöngy and Pardoux (1991). The fact that the associated martingale problem is well-posed follows by an extension of an argument due to Yamada and Watanabe (1971) (cf. Theorem 8.4.1). The proof under condition (10.4.4) is given in Shiga (1988) and under a closely related condition in Kotelenez (1989). □

Following Shiga (1988) we next consider an important special case in which the method of duality applies.

Theorem 10.4.2. Consider Equation (10.4.2) under the assumption (10.4.3) or (10.4.4) and with the following (stepping stone with mutation and selection) coefficients:

$$(10.4.5)\quad a(y) = \sqrt{y(1-y)} \text{ and } b(y) = c_0(1-y)-c_1y+c_2y(1-y),$$

$c_0, c_1, c_2 \ge 0$. Then the associated martingale problem has a unique solution.

Proof. The key step is again a duality argument which we only sketch. The state space for the dual process is the collection of (unordered) finite subsets

184

$x=(x_1,\dots,x_n)$ of $\mathbb{R}^d$. The function F is defined by

$$F(\mathbf{x},X) := \prod_{i=1}^{n} (1-X(x_i)), \quad F(\emptyset,X) = 1.$$

The formal generator of the dual process is

$$Lf(\mathbf{x}) = \sum_{i=1}^{n} A_i f(\mathbf{x}) + c_1 \sum_{i=1}^{n} [f(\Phi_i \mathbf{x})-f(\mathbf{x})] + c_2 \sum_{i=1}^{n} [f(\Psi_i \mathbf{x})-f(\mathbf{x})]$$
$$+ \frac{1}{2} \sum_{i \neq j}\sum \delta(x_i,x_j)[f(\Phi_i \mathbf{x})-f(\mathbf{x})]$$

where Φ_i denotes the deletion of x_i, Ψ_i denotes the addition of a second copy of x_i, and $\delta(x,y)$ is a delta function at the diagonal set $\{(x,y)\in\mathbb{R}^d\times\mathbb{R}^d, x=y\}$. The duality relation

$$E_{X(0)}(F(\mathbf{x},X(t,.)) = E_{\mathbf{x}}\left(F(\mathbf{x}(t),X(0))\cdot\exp\left(-c_0\int_0^t |\mathbf{x}(s)|\,ds\right)\right)$$

is then obtained using Corollary 5.5.3. □

<u>Remarks:</u> The condition (10.4.3) guarantees the existence of the *local time at the diagonal for the two particle motion* of independent A-processes and it is this fact that leads to a nontrivial behavior. For example if $A = \Delta_d$, the d-dimensional Laplacian, then condition (10.4.4) is satisfied if and only if d=1. It then makes sense to ask what happens in higher dimensions if we start with the corresponding particle systems. In fact it turns out that the resulting particle systems are tight and converge but that the limiting process is deterministic and given by the solution of the partial differential equation obtained by deleting the cylindrical Brownian motion term in (10.4.2) (cf. Reimers (1986)). This can be interpreted as a law of large numbers phenomenon which in part results from the independence of particles at different locations.

There are at least two ways in which non-trivial stochastic behavior can occurs in higher dimensions. The first is the removal of the assumption that densities are bounded and this then leads to the singular measure-valued processes such as measure-valued branching and Fleming-Viot processes. The second is the removal of the assumtion that particles at different locations are stochastically independent. If this assumption is removed stochastic evolution equations of the above type can arise in higher dimensions - however the cylindrical Brownian motion is then replaced by one with (for example) nuclear covariance. The existence of a unique strong solution to a class of equations of this type was established by Viot (1976).

185

10.5. Some Other Interaction Models

It is clear that there are many further possible types of interaction which are not covered by the above examples. This is an active area of research having relations with the theory of interacting particle systems.

In particular there have been a number of recent developments concerning superprocesses with interaction. A partial list of these is the following:

(1) Shiga (1990) introduced a strong equation for an atomic measure-valued process incorporating branching and interaction in a Poisson setting.

(2) Perkins (1992) introduced a strong equation with respect to the historical process for a superprocess in which the particle motions are influenced by the mass distribution. This is based on the development of a new stochastic calculus for historical processes.

(3) Adler (1991) introduced and simulated the so-called "goat" superprocess in which the particles move towards or against the gradient of the occupation time measure.

(4) Dawson and March (1992) studied the Fleming-Viot process with non-constant coefficients.

186

11. DE FINETTI AND POISSON CLUSTER REPRESENTATIONS, PALM DISTRIBUTIONS

11.1 Campbell Measures and Palm Distributions.

Let E be a Polish space and $M_{LF}(E)$ denote the set of locally finite measures on E. Now let X be a random measure with law $P \in M_1(M_{LF}(E))$ and locally finite intensity (mean) measure $I(B) := \int \mu(B)\, P(d\mu),\ B \in \mathcal{E}_b$.
The associated *Campbell measure*, $\bar{P}$ is a measure on $\mathcal{B}(M_{LF}(E)) \times \mathcal{E}_b$ defined by

$$\text{(11.1.1)} \qquad \bar{P}(B\times A) := \int_B \mu(A)\, P(d\mu), \quad B \in \mathcal{B}(M_{LF}(E)), \quad A \in \mathcal{E}_b.$$

The associated *Palm distributions* $\{(P)_x : x \in E\}$ are a version of the Radon-Nikodym derivatives

$$(P)_x = \frac{\bar{P}(d\mu\times dx)}{I(dx)}, \quad \text{I-a.s.}$$

and can be assumed to form a family of regular conditional probabilities (for the random measure X given $x \in E$) for the Campbell measure $\bar{P}$, that is, $(P)_x \in M_1(M_{LF}(E))$ for every x. Then for any bounded $\mathcal{E}\times\mathcal{B}(M_{LF}(E))$-measurable function, g, satisfying $\{x : \sup_\mu g(x,\mu) \neq 0\} \in \mathcal{E}_b$,

$$\text{(11.1.2)} \qquad \int_E I(dx)\left[\int_{M_{LF}(E)} g(x,\mu)(P)_x(d\mu)\right] = \int_{M_{LF}(E)} \int_E g(x,\mu)\mu(dx)P(d\mu)$$

and this system of identities uniquely characterizes the $(P)_x$, I-a.s.

In order to get some intuitive feeling for Palm measures we first consider the case of a random probability measure. Consider a two stage experiment in which at the first stage a $X \in M_1(M_1(E))$ is chosen with law P and in the second stage a point Z in E is chosen according to the law X. The Palm distribution $(P)_z$ is then simply the conditional (Bayesian posterior) distribution of X given Z=z. More generally, if X is a finite random measure, then there is an additional weighting on the posterior measures proportional to their total mass. This is indicated by the following consequence of (11.1.2)

$$\int (P)_x(\langle\mu,1\rangle \geq a)I(dx) = \int_a^\infty r\cdot P(\langle\mu,1\rangle \in dr).$$

11.2 De Finetti's Representation of a Random Probability Measure.

Let E be a Polish space and $M_1(E)$ the space of probability measures on E. A sequence of E-valued random variables $\{Z_n : n \in \mathbb{N}\}$ defined on a probability space $(\Omega, \mathcal{F}, P_0)$ is said to be *exchangeable* if $(Z_1,\dots,Z_n,Z_{n+1},Z_{n+2},\dots)$ has the same joint distribution as $(Z_{\pi(1)},\dots,Z_{\pi(n)},Z_{n+1},Z_{n+2},\dots)$ for every $\pi \in \mathcal{P}er(n)$ and $n \in \mathbb{N}$. Given $\{Z_n\}$, let $\mathcal{G}_n = \sigma\{f_n(Z_1,\dots,Z_n) : f_n \in C_{sym}(E^n)\}$, $\mathcal{H}_n := \sigma\{\mathcal{G}_n, Z_{n+1}, Z_{n+2},\dots\}$ and $\mathcal{H}_\infty :=$

187

$\bigcap_n \mathcal{H}_n$. Note that $\mathcal{H}_\infty$ is the σ-algebra of *exchangeable events.*

Theorem 11.2.1 (de Finetti's Theorem)

(a) Let $P \in M_1(M_1(E))$. Then there exists a sequence $\{Z_n\}$ of E-valued exchangeable random variables defined on a probability space $(E^{\mathbb{N}}, \mathcal{E}^{\mathbb{N}}, P_{dF})$ such that $(Z_1,\ldots,Z_n)$ has joint distribution

$$P^{(n)}(dx_1,\ldots,dx_n) := \int_{M_1(E)} \mu(dx_1)\ldots\mu(dx_n)P(d\mu), \quad n \in \mathbb{N},$$

(b) Consider the sequence $\{Z_n\}$ of E-valued exchangeable random variables. Let $X_n(\omega) := \Xi_n(Z_1,\ldots,Z_n)$. Then

$$X(\omega) = \lim_{n\to\infty} X_n(\omega) \in M_1(E) \text{ exists for a.e. } \omega,$$

where the limit is taken in the weak topology on $M_1(E)$ and X has probability law P.

(c) X is $\mathcal{H}_\infty$-measurable and conditioned on $\mathcal{H}_\infty$, $\{Z_n\}$ is a sequence of i.i.d. random variables with marginal distribution X.

Proof. (See e.g. Aldous (1985), Dawson and Hochberg (1979))

(a) Given $P \in M_1(M_1(E))$, we can construct a consistent family $\{P^{(n)} \in M_1(E^n)\}$ of probability measures $P^{(n)}(dx_1,\ldots,dx_n) := \int_{M_1(E)} X(dx_1)\ldots X(dx_n)P(d\mu)$, $n \in \mathbb{N}$.

By Kolmogorov's extension theorem we can then construct a canonical process $(E^{\mathbb{N}}, \mathcal{E}^{\mathbb{N}}, \{Z_n\}_{n\in\mathbb{N}}, P_{dF})$ such that $\{Z_n\}$ form a sequence of exchangeable E-valued random variables and the joint distribution of $(Z_1,\ldots,Z_n)$ is given by $P^{(n)}$.

(b) By exchangeability we have $(Z_i,Y) \overset{\mathcal{D}}{=} (Z_1,Y)$, $1 \le i \le n$, when Y is of the form $(f_n(Z_1,\ldots,Z_n),Z_{n+1},Z_{n+2},..)$ and f_n is symmetric. Therefore conditioned on $\mathcal{H}_n$ the $Z_1,\ldots,Z_n$ are a.s. identically distributed. Hence for $\phi \in b\mathcal{E}$

(11.2.1) $$P_{dF}(\phi(Z_1)|\mathcal{H}_n) = P_{dF}\left(n^{-1}\sum_{i=1}^{n}\phi(Z_i)|\mathcal{H}_n\right)$$

$$= \langle X_n,\phi\rangle \text{ since this is } \mathcal{H}_n\text{-measurable.}$$

Since $\mathcal{H}_n \supset \mathcal{H}_{n+1}$, and $P_{dF}(\phi(Z_1)|\mathcal{H}_n)$ is a reverse martingale, we can apply the reverse martingale convergence theorem to obtain

(11.2.2) $$\langle X_n,\phi\rangle = n^{-1}\sum_{i=1}^{n}\phi(Z_i) \longrightarrow P_{dF}(\phi(Z_1)|\mathcal{H}_\infty) \text{ a.s.}$$

Thus in particular for each compact set, $K \subset E$, $X_n(\omega,K) \longrightarrow X(\omega,K)$ for a.e. ω and $P_{dF}(X(K)) = P_{dF}(Z_1 \in K)$. Then $P_{dF}(X(K^c) > 2^{-n}) \le 2^n P_{dF}(Z_1 \in K^c)$ and we can find a sequence of compact sets K_n such that $X(K_n^c) < 2^{-n}$ for all sufficiently large n, a.s., that is, given $\varepsilon > 0$, and ω there exists let $n(\omega)$ such that $X(\omega,K_n^c) < 2^{-n}$ for all

188

$n \geq n(\omega)$. Given $\varepsilon > 0$ we choose $n_1 \geq n(\omega)$ so that $2^{-n_1} < \varepsilon/2$, and then we choose $n_2 \geq n_1$ such that $X_n(\omega, K^c_{n_1}) \leq \varepsilon$ for all $n \geq n_2$. This gives the almost sure tightness of the $\{X_n\}$. Then applying (11.2.2) to a countable subset of $b\mathcal{E}$ which is convergence determining for $M_1(E)$ establishes the a.s. convergence of X_n in the weak topology on $M_1(E)$. Denoting the limit by X, we conclude that the random probability measure, $X(\cdot)$, defined in this way is a version of the conditional distribution $P_{dF}(Z_1 \in \cdot \,|\, \mathcal{H}_\infty)$.

(c) Fix $k \in \mathbb{N}$ and let $\phi \in bC(E^k)$. Then an argument similar to the one at the beginning of the proof of (b) shows that for $n>k$,

$$P_{dF}(\phi(Z_1,\dots,Z_k)\,|\,\mathcal{H}_n)$$
$$= |\Lambda_{n,k}|^{-1} \sum \dots \sum_{(j_1,\dots,j_k)\in\Lambda_{n,k}} \phi(Z_{j_1},\dots,Z_{j_k})$$

where $\Lambda_{n,k} := \{(j_1,\dots,j_k): 1 \leq j_r \leq n;\ j_m \neq j_r \text{ if } m \neq r\}$, $|\Lambda_{n,k}| = [n(n-1)\dots(n-k+1)]$. Again using the reverse martingale convergence theorem and the fact that $|\Lambda_{n,k}|/n^k \to 1$ as $n \to \infty$, we obtain

$$\text{(11.2.3)}\qquad n^{-k} \sum_{j_1=1}^{n} \dots \sum_{j_k=1}^{n} \phi(Z_{j_1},\dots,Z_{j_k}) \to P_{df}(\phi(Z_1,\dots,Z_k)\,|\,\mathcal{H}_\infty) \text{ a.s.}$$

Taking $\phi(x_1,\dots,x_k) := \prod_{j=1}^{k} \phi_j(x_j)$, and using (11.2.2) and (11.2.3) we conclude

$$P_{dF}\Big(\prod_{j=1}^{k} \phi_j(Z_j)\,|\,\mathcal{H}_\infty\Big) = \prod_{j=1}^{k} P_{dF}(\phi_j(Z_1)\,|\,\mathcal{H}_\infty) \quad \text{a.s.}$$

This implies that the $\{Z_i\}$ conditioned on $\mathcal{H}_\infty$ are i.i.d. □

<u>Corollary 11.2.2</u> (a) Let $\{Z_n(.): n \in \mathbb{N}\}$ be an exchangeable sequence of $D([0,\infty),E)$-valued random variables and $X_n := \Xi_n(Z_1(.), Z_2(.),\dots,Z_n(.))$. Then

(i) $P_{dF}\left\{m^{-1} \sum_{n=1}^{m} \delta_{Z_n(.)} \Rightarrow X\right\} = 1$ where X is a random probability measure on $D([0,\infty),E)$, and

(ii) for $k \in \mathbb{N}$, $0 \leq t_1 < \dots < t_k$,

$$P_{dF}\left\{m^{-1} \sum_{n=1}^{m} \delta_{Z_n(t_1),\dots,Z_n(t_k)} \Rightarrow X_{\{t_1,\dots,t_k\}}\right\} = 1$$

where $X_{\{t_1,\dots,t_k\}}$ is a random probability measure on E^k.

(b) If for each $t \geq 0$, $\{Z_n(t): n \in \mathbb{N}\}$ is exchangeable, then

$$\left\{m^{-1}\sum_{n=1}^{m}\delta_{Z_n(t_1)},\ldots,m^{-1}\sum_{n=1}^{m}\delta_{Z_n(t_k)}\right\}\Longrightarrow\{(X_{t_1},\ldots,X_{t_k})\},$$

where the laws of $\{(X_{t_1},\ldots,X_{t_k})\}$ form a set of finite dimensional distributions of a $M_1(E)$-valued stochastic process.

Proof. (a) and (b) both follow from the almost sure convergence in Theorem 11.2.1. □

Corollary 11.2.3. Let $P\in M_1(M_1(E))$. Then the Palm distributions $\{(P)_z: z\in E\}$ are given by

$$(P)_z(B) = P_{dF}(X\in B\,|\,Z_1=z),\quad B\in\mathcal{E}.$$

Proof. First note that the mean measure $I(B) = P_{dF}(Z_1\in B)$. Then in view of the characterization (11.1.2) it suffices to prove that for $\phi\in b\mathcal{E}$, $f\in b\mathcal{E}^k$, $k\in\mathbb{N}$,

$$\int_E\phi(x_1)\int_{M_1(E)}\left(\int_E\ldots\int_E f(x_2,\ldots,x_{k+1})\mu(dx_2)\ldots\mu(dx_{k+1})\right)P_{dF}(X\in d\mu\,|\,Z_1=x_1)P_{dF}(Z_1\in dx_1)$$

$$=\int_{M_1(E)}\left(\int_E\ldots\int_E\phi(x_1)f(x_2,\ldots,x_{k+1})\mu(dx_1)\ldots\mu(dx_{k+1})\right)P_{dF}(d\mu).$$

But this is true since both sides are equal to $P_{dF}[\phi(Z_1)f(Z_2,\ldots,Z_{k+1})]$. □

11.3 An Infinite Particle Representation of Fleming-Viot Processes

In view of the results of the last section it is reasonable to look for an exchangeable infinite particle system representation for the Fleming-Viot process. This idea was used in Dawson and Hochberg (1982) to study the support dimension of the Fleming-Viot process at fixed times. Donnelly and Kurtz (1992) introduced an infinite particle system which is not exchangeable but which does yield a representation of the full Fleming-Viot process. We will now outline the main ideas of their construction.

For simplicity we return to the setting of Ch. 2. Let (E,d) be a compact metric space, $P(t,x,\Gamma)$ be the transition function, and $(A,D(A))$ the generator of a Feller process with sample paths in $D_E = D([0,\infty),E)$. We begin with a description of an n-particle process, $\{Y^n(t):t\geq 0\}$ called by Donnelly and Kurtz (1992) the n-particle *look-down process*. This is a Markov jump process with state space E^n and generator:

(11.3.1)

$$K_nf(x_1,\ldots,x_n) = \sum_{i=1}^{n}A_if(x_1,\ldots,x_n) + \gamma\sum_{i<j}[\Theta_{ij}f(x_1,\ldots,x_n)-f(x_1,\ldots,x_n)].$$

190

Theorem 11.3.1. Let $\{Z^n(t):t\geq 0\}$ be the n-particle Moran process with generator L_n defined by (2.5.2) and $\{Y^n(t):t\geq 0\}$ the n-particle look-down process with generator K_n defined by (11.3.1). Then the $M_1(E)$-valued Markov processes $\{\Xi_n(Y^n(t)):t\geq 0\}$ and $\{\Xi_n(Z^n(t)):t\geq 0\}$ have the same distribution provided that $Y^n(0) \stackrel{\mathcal{D}}{=} Z^n(0)$ is exchangeable.

Proof. (cf. Donnelly and Kurtz (1992) for different proof.) Let $f \in C_{sym}(E^n)\cap D(A^n)$. Then $f(x_1,\dots,x_n) = (n!)^{-1} \sum_{\pi\in\mathcal{P}er(n)} f(x_{\pi 1},\dots,x_{\pi n})$. Substituting this expression into the generator L_n we obtain

$$K_n f(x_1,\dots,x_n) = (n!)^{-1} \sum_{\pi} K_n f(x_{\pi 1},\dots,x_{\pi n})$$

$$= \sum_{i=1}^{n} A_i f(x_1,\dots,x_n) + \tfrac{1}{2}\gamma \sum_{i\neq j} \Theta_{ij}[f(x_1,\dots,x_n)-f(x_1,\dots,x_n)].$$

$$= L_n f(x_1,\dots,x_n) \in C_{sym}(E^n).$$

Hence the restricted initial value martingale problem for $(K_n,C_{sym}(E^n),M_{1,ex}(E^n))$ martingale problem is identical to the restricted initial value martingale problem for $(L_n,C_{sym}(E^n),M_{1,ex}(E^n))$ where $M_{1,ex}(E^n)$ denotes the set of exchangeable probability laws on E^n. The uniqueness of the marginal distributions follows by solving the moment equations as in Theorem 2.5.1 and Section 2.6 (or using a function-valued dual). The identity in law of the processes then follows from Theorem 5.1.2. □

Remark. Let $\tilde{\mu} \in M_{1,ex}(E^{\mathbb{N}})$. The n-particle look-down process with initial law given by the appropriate marginal of $\tilde{\mu}$ can be realized as $(D([0,\infty),E^n),\mathcal{D},P_{n,\tilde{\mu}})$.

Corollary 11.3.2. Let $\tilde{\mu} \in M_{1,ex}(E^{\mathbb{N}})$. Let $\mathbf{D}_\infty := D([0,\infty),E^{\mathbb{N}})$, $\mathcal{D}_\infty$ denote the Borel σ-algebra on $\mathbf{D}_\infty$ and for $n\in\mathbb{N}$ let $Y_n(.)$ denote the nth coordinate of the canonical process $\mathbf{Y}^\infty$. Then there exists a probability measure $P^{DK}_{\tilde{\mu}}$ on $(\mathbf{D}_\infty,\mathcal{D}_\infty)$ such that the distribution of $(Y_1(.),\dots,Y_n(.))$ coincides with that of the process $\mathbf{Y}^n(.)$ and $\mathbf{Y}^\infty(0)$ has law $\tilde{\mu}$. (This infinite particle system is called the *infinite look-down process* of Donnelly-Kurtz.)

Proof. Observe that if we consider functions f depending only on $x_1,\dots,x_{n-1}$, then $K_n f(x_1,\dots,x_{n-1}) = K_{n-1} f(x_1,\dots,x_{n-1})$. This implies that the laws of $\{Y_1(.),\dots,Y_n(.)\}$, $n\in\mathbb{N}$, form a consistent family of probability measures. Then the infinite look-down process with initial law $\tilde{\mu}$, can be realized by taking the projective limit $(D([0,\infty),E^n),P_{n,\tilde{\mu}})$ which will be denoted by $(\mathbf{D}_\infty,\mathcal{D}_\infty,P^{DK}_{\tilde{\mu}})$. □

Remarks. (1) Let $\mathbf{Y}(t) := (Y_1(t),Y_2(t),\dots:t\geq 0)$ denote the infinite particle look-

191

down process of Donnelly and Kurtz and assume that it has exchangeable initial condition $Y(0)$. It is then easy to verify that $Y(t)$ is an exchangeable system for each $t\geq 0$.

(3) Donnelly and Kurtz (1992) observed that if the mutation process has a stationary distribution then there is a stationary version of the infinite particle system $\{Y_n(t): -\infty<t<\infty,\ n\in\mathbb{N}\}$ and if the mutation process has stationary independent increments, then there is a stationary version of $\{0, Y_2-Y_1, Y_3-Y_1, \ldots\}$.

Given $\{Y_n(.)\}$, consider the empirical process

$$X_m(t) := \Xi_m(Y_1(t),\ldots,Y_m(t)) = \frac{1}{m}\sum_{i=1}^{m}\delta_{Y_i(t)},$$

$$\mathcal{H}_{m,t} := \sigma(X_m(s),(Y_{m+1}(s),Y_{m+2}(s),\ldots),s\leq t)$$

and $\mathcal{H}_{\infty,t} := \bigcap_{m=1}^{\infty}\mathcal{H}_{m,t}$.

<u>Theorem 11.3.3</u> (Donnelly and Kurtz (1992))

Let $\{Y_k(0):k\in\mathbb{N}\}$ have law $\tilde{\mu}\in M_{1,ex}(E^{\mathbb{N}})$. Then $X_m(t)$ converges, as $m\to\infty$, uniformly on bounded intervals to a càdlàg $M_1(E)$-valued process $\{X_t:t\geq 0\}$ with probability one. The limit process $\{X_t\}$ is a Fleming-Viot process with generator $\mathfrak{G}$ (given by (5.4.1)).

Moreover, for $B\in\mathcal{E}$, $t\geq 0$, and $k\in\mathbb{N}$,

$$\text{(11.3.2))}\quad P^{DK}_{\tilde{\mu}}\{Y_1(t)\in B_1,\ldots,Y_k(t)\in B_k\,|\,\mathcal{H}_{\infty,t}\} = \prod_{j=1}^{k} X(t,B_j).$$

<u>Proof.</u> By Theorem 11.2.1 , for fixed t, $X_m(t)$ converges almost surely in the weak topology on $M_1(E)$ and for fixed $k\geq 1$

$$\text{(11.3.3)}\quad P^{DK}_{\tilde{\mu}}\{Y_k(t)\in B\,|\,\mathcal{H}^m_t\} = X_m(t,B)\quad\text{for } m\geq k.$$

Then letting $m\to\infty$ and again using the reverse martingale convergence theorem on the left hand side we obtain

$$P^{DK}_{\tilde{\mu}}\{Y_1(t)\in B\,|\,\mathcal{H}_{\infty,t}\} = X(t,B).$$

(11.3.2) then follows from Theorem 11.2.1(c).

The fact that the resulting $M_1(E)$-valued process is Fleming-Viot process follows from Theorems 11.3.1 and 2.7.1.

It remains to prove the uniform convergence on bounded intervals. Let $f\in bC(E)$, $t>0$ and $\varepsilon>0$. By Theorem 11.2.1 $\{Y_n(t):n\in\mathbb{N}\}$ conditioned on $\mathcal{H}_{\infty,t}$ are i.i.d. Therefore by Cramér's Theorem (cf. Varadhan (1984, Section 3)) applied to the resulting independent random variables there exist C and $D>0$ such that

192

$$P^{DK}_{\tilde{\mu}}\left\{\ \left|\int f(x)X_m(t,dx) - \int f(x)X(t,dx)\right| > \varepsilon \,|\, \mathcal{H}_{\infty,t}\right\} \le e^{C-Dm},\ P^{DK}_{\tilde{\mu}}\text{-a.s.}$$

where C,D are $\mathcal{H}_{\infty,t}$-measurable. However since the random variables $\{f(Z_n):n\in\mathbb{N}\}$ are uniformly bounded by $\|f\|$, in fact there exist constants c,δ (depending only on $\|f\|$) such that $C \le c$, $D \ge d$, and such that

$$(11.3.4)\qquad P^{DK}_{\tilde{\mu}}\left\{\ \left|\int f(x)X_m(t,dx) - \int f(x)X(t,dx)\right| > \varepsilon \,|\, \mathcal{H}_{\infty,t}\right\} \le e^{c-\delta m},\ P^{DK}_{\tilde{\mu}}\text{-a.s.}$$

Since Y^m is the solution of the martingale problem for K_n,

$$M_f(t) := m^{-1}\sum_{i=1}^{m}\left(\left[f(Y_i(s))-f(Y_i(t)-\int_s^t Af(Y_i(u))du\right]1(\tau_i>t)\right)$$

where τ_i denotes the time of the first look-down of Y_i after time s, is a martingale. Then Doob's maximal inequality applied to the submartingales $\exp[\pm\lambda M_f(t)]$, with $\lambda>0$, the exponential Chebyshev inequality (cf. [LS, 13.2,13.27]) and an appropriate choice of λ yield an inequality of the form

$$(11.3.5)\qquad P^{DK}_{\tilde{\mu}}\left\{\ \sup_{t\le T}\ \sup_{t\le s\le t+h_m}\ \left|\int f(x)X_m(s,dx)-\int f(x)X_m(t,dx)\right| > \varepsilon\right\}$$

$$\le\ e^{c-\delta m} + P^{DK}_{\tilde{\mu}}\left(m^{-1}\sum_{j=1}^{m} 1(\tau_j\le h_m) > \frac{\varepsilon}{2(\|f\|+\|Af\|)}\right).$$

But then by the exponential Chebyshev inequality we obtain

$$P^{DK}_{\tilde{\mu}}\left(\sum_{j=1}^{m} 1(\tau_j\le h) > m\cdot a\right) \le e^{-ma}\cdot P^{DK}_{\tilde{\mu}}\left(\exp\left(a\cdot\sum_{j=1}^{m} 1(\tau_j\le h)\right)\right)$$

$$= e^{-ma}\prod_{i=1}^{m}\left[e^{a}(1-e^{-(j-1)\gamma h})+e^{-(j-1)\gamma h}\right].$$

Hence
$$\log\left(P^{DK}_{\tilde{\mu}}\left(\sum_{j=1}^{m} 1(\tau_j\le h) > m\cdot a\right)\right) \le \sum_{j=1}^{m}\log\left(1-(1-e^{-a})e^{-(j-1)\gamma h}\right)$$

$$\le -\sum_{i=1}^{m}(1-e^{-a})e^{-(j-1)\gamma h}$$

$$= -\left((1-e^{-a})\frac{(1-e^{-mh})}{(1-e^{-h})}\right).$$

Therefore

193

$$P^{DK}_{\tilde{\mu}}\left(\sum_{j=1}^{m} 1(\tau_j \le h) > m\cdot a \right) \le \exp\left(-(1-e^{-a}) \frac{(1-e^{-mh})}{(1-e^{-h})} \right).$$

If $a > 0$, we can choose c and δ so that

$$(11.3.6)\qquad P^{DK}_{\tilde{\mu}}\left(\sum_{j=1}^{m} 1(\tau_j \le h_m) > m\cdot a \right) \le e^{c-m\delta}$$

for all m provided that $mh_m \to 0$.

The almost sure uniform convergence result is obtained by choosing a grid $\{kh_m,\ k<T/h_m\}$, with $m\cdot h_m \longrightarrow 0$ and $\sum e^{-\delta m}/h_m < \infty$. Using the continuity of $\{X(t):t\ge 0\}$, and (11.3.4)-(11.3.6) it then follows that for $T < \infty$ and $\varepsilon>0$,

$$P^{DK}_{\tilde{\mu}}\left(\lim_{m\to\infty} \sup_{t\le T} \left| \int f(x)X_m(t,dx) - \int f(x)X(t,dx) \right| > \varepsilon \right) = 0$$

(cf. Donnelly and Kurtz (1992) for the details). □

11.4 Campbell Measures and Palm Distributions for Infinitely Divisible Random Measures.

In this section we review the notions of Campbell and Palm measures associated with an infinitely divisible random measure. Standard sources for this material are Kallenberg (1983), Matthes, Kerstan and Mecke (1978), and Daley and Vere-Jones (1988).

Let E be a Polish space and $M_{LF}(E)$ the space of locally finite random measures introduced in section 3.1. We recall from Theorem 3.3.1 that an $M_{LF}(E)$-valued infinitely divisible random measure has the following canonical representation of its log-Laplace functional:

$$(11.4.1)\quad \log(L(\phi)) = \langle M,\phi\rangle + \int_{M_{LF}(E)} (1-e^{-\langle\nu,\phi\rangle})R(d\nu),\quad \phi\in pb\mathcal{E}_b,$$

where $(M,R) \in M_{LF}(E)\times M_{2,LF}(E)$ and $M_{2,LF}(E)$ denotes the set of measures ν on $M_{LF}(E)$ such that

(i) $\nu(\{0\}) = 0$

(ii) $\int_{M_{LF}(E)} (1-e^{-\mu(A)})\nu(d\mu) < \infty\ \forall\ A\in \mathcal{E}_b$.

The notions of Campbell measure and Palm distribution defined in Section 11.1 for probability measures on $M_{LF}(E)$ can be extended to $M_{2,LF}(E)$ and in particular will be defined for the canonical measure R of an infinitely divisible random measure with locally finite intensity.

194

Let $R \in M_{2,LF}(E)$ in (11.4.1) and $I(A) := \int \mu(A)R(d\mu)$ be the corresponding intensity measure. For $A \in \mathcal{E}_b$, $B \in \mathcal{B}(M_{LF}(E))$ the *Campbell measure* is defined as follows

$$\bar{R}(A\times B) := \int \mu(A)1_B(\mu)R(d\mu) \leq I(A) = \int \mu(A)R(d\mu) < \infty.$$

Lemma 11.4.1. Assume that $I \in M_{LF}(E)$. Then

(a) $\bar{R}$ extends to a measure on $\mathcal{E}_b \otimes \mathcal{B}(M_{LF}(E))$.

(b) There exists a mapping $(e,B) \longrightarrow R_e(B)$ from $E\times\mathcal{B}(M_{LF}(E))$ to $[0,1]$ such that for each e, $R_e(.)$ is a probability measure on $\mathcal{B}(M_{LF}(E))$ and for each $B\in \mathcal{B}(M_{LF}(E))$ $R_{\bullet}(B)$ is $\mathcal{E}$-measurable such that the following disintegration applies

$$\bar{R}(A\times B) = \int_A R_e(B)I(de), \quad A\in\mathcal{E}_b.$$

(c) Further for any $\mathcal{E}_b\otimes\mathcal{B}(M_{LF}(E))$-measurable function g,

$$\int_E\int_{M_{LF}(E)} g(e,\mu)\bar{R}(de\times d\mu)=\int_E\left[\int_{M_{LF}(E)} g(e,\mu)R_e(d\mu)\right]I(de).$$

Proof. (a) Since by definition $\bar{R}$ is a bimeasure, this follows the lines of [EK, Appendix, Theorem 8.1].

(b) The existence of a version of the Radon-Nikodym derivatives

$$R_e(B) = \frac{\bar{R}(de\times B)}{I(de)}, \quad I\text{-a.e. } e,$$

which yields "regular conditional probabilities" follows by a standard argument (see [DP] for details).

(c) follows by approximating by simple functions and passing to the limit. □

The family of probability measures $\{R_e: e\in E\}$ whose existence is established in Lemma 11.4.1 are known as the *Palm distributions* associated with R.

Lemma 11.4.2. If X is an infinitely divisible random measure with law $P \in M_1(M_{LF}(E))$, with canonical representation (11.4.1) with $M\equiv 0$ and $R\in M_{2,LF}(E)$, then the relation between the Palm distributions $(P)_x$ of the random measure X, and those associated with the canonical measure, $(R)_x$, is given by

$$(P)_x(A) = \int_{M_{LF}(E)}\int_{M_{LF}(E)} 1_A(\mu_1+\mu_2)\; P(d\mu_1)(R)_x(d\mu_2),$$

$\forall\; A\in \mathcal{B}(M_{LF}(E))$ for I-a.e. x.

Proof. See Kallenberg [1983, Lemma 10.6].

11.5. Cluster Representation for (A,Φ)-Superprocesses.

Let $\{X_t:t\geq 0\}$ denote an (A,Φ)-superprocess with $X_0 = \mu\in M_F(E)$. Then for each t, X_t is an infinitely divisible random measure on E. In this section we will see

195

that for a certain class of Φ, X_t can be represented as a Poisson cluster random measure and we will identify the cluster law. This will be carried out starting from the canonical representation of X_t. To show that we have a Poisson cluster representation we will show that the canonical measure R, which in general is σ-finite, is a finite measure. It will turn out (cf. Chapter 12) that each clusters obtained in this way can be interpreted as a submass having a common ancestor thus involving the family structure of the population.

Recall that

(11.5.1) $$P_{s,\mu}(e^{-\langle X_t,\phi\rangle}) = e^{-\langle \mu, V_{s,t}\phi\rangle}$$

and by the canonical representation theorem (cf. Theorem 3.3.1)

(11.5.2) $$\langle \mu, V_{s,t}\phi\rangle = \langle m(s,t,\mu),\phi\rangle + \int_{M_{LF}(E)} (1-e^{-\langle \nu,\phi\rangle}) R_{s,t}(\mu,d\nu)$$

where

$$m(s,t,\mu) \in M_{LF}(E),\quad R_{s,t}(\mu,.) \in M_{2,LF}(E).$$

Remark By an application of (11.5.1) and (11.5.2), it is easy to verify that

(11.5.3) $$I_t(A) := E_{s,\mu}(X_t(A)) = \int (T_{s,t}1_A)(x)\mu(dx)$$

$$= \int \nu(A) R_{s,t}(\mu,d\nu) + m(s,t,\mu)(A) < \infty,\quad A \in \mathcal{E}, \text{ and}$$

$$V_{s,t}\phi(x) = \langle m(s,t,x),\phi\rangle + \int (1-e^{-\langle \nu,\phi\rangle}) R_{s,t}(x,d\nu)$$

where $m(s,t,x) = m(s,t,\delta_x)$, $R_{s,t}(x,.) = R_{s,t}(\delta_x,.)$.

The measurability follows from the measurability of $(s,t,x) \longrightarrow V_{s,t}\phi(x)$.

Let $\Phi(x,\lambda)$ be defined as in (4.3.2); recall that $\Phi(x,\lambda) \le 0$ and is monotone decreasing in λ. We next obtain a condition on Φ which guarantees that the canonical measure is finite.

Let $\Psi(\lambda) := \inf_x |\Phi(x,\lambda)|$ and assume that Ψ satisfies (4.3.6). Consider the log-Laplace equation

$$\partial u(t,y)/\partial t = Au(t,y) + \Phi(y,u(t,y)),\quad u(0,y) = \theta,$$

and the ordinary differential equation

$$dv(\theta,t)/dt = -\Psi(v(\theta,t)),\qquad v(\theta,0) = \theta.$$

Under rather general conditions on A and Φ the *comparison theorem*: $u(t,y) \le v(\theta,t)$ $\forall$ y can be proved (see, for example Pardoux and Gyöngy (1991)). We will not state explicit hypotheses for the comparison theorem but will assume that it holds.

Lemma 11.5.1. Let $\Psi(\lambda) := \inf_x |\Phi(x,\lambda)|$. Assume that

$$(11.5.4) \quad \int_\theta^\infty \frac{1}{\Psi(\lambda)}\, d\lambda \; < \infty \quad \text{for } \theta>0.$$

(a) If $t > s$ and $\theta>0$, then $\limsup_{\theta\to\infty}\sup_y V_{s,t}(\theta 1)(y) < \infty$.

(b) $m(s,t,y) \equiv 0$ and $R_{s,t}(y,M_F(E)) = R_{s,t}(y,M_{LF}(E)) < \infty$.

(c) $P_{s,\mu}(\langle X_t,1\rangle = 0) > 0 \;\forall\; \mu\in M_F(E)$.

<u>Proof.</u> (a) Let $u(t,y) = V_{s,t}(\theta 1)(y)$, $\theta>0$, and recall that for $t>s$,

$$\partial u(t,y)/\partial t = Au(t,y) + \Phi(y,u(t,y)), \quad u(0,y) = \theta.$$

Consider the ordinary differential equation

$$dv(\theta,t)/dt = - \Psi(v(\theta,t)), \qquad v(\theta,0) = \theta.$$

Then $\tilde{v}(t) := \lim_{\theta\to\infty} v(\theta,t)$ is given by

$$\int_{\tilde{v}(t)}^\infty 1/\Psi(s)ds = t.$$

Using (11.5.4) and noting that $\int_0^\infty \frac{1}{\Psi(\lambda)}\, d\lambda = \infty$ (by (4.3.6)), we conclude that $\tilde{v}(t)$ is well defined $\forall\; t > 0$ and that $0 < \tilde{v}(t) < \infty$. (a) then follows from the comparison theorem.

(b) follows from (a), (11.5.3) and the assumption that $P_{s,\mu}(\langle X_t,1\rangle) < \infty$.

$$\text{(c)} \quad P_{s,\mu}(\langle X_t,1\rangle = 0) = \lim_{\theta\to\infty} P_{s,\mu}(e^{-\theta\langle X_t,1\rangle}) = \lim_{\theta\to\infty} e^{-\langle\mu,V_{s,t}(\theta 1)\rangle} > 0. \quad \square$$

<u>Remark 11.5.2.</u> A sufficient condition for (11.5.4) is that

$$\inf_x \liminf_{\lambda\to\infty} -\Phi(x,\lambda)/\lambda^{1+\delta} > 0 \text{ for some } \delta > 0$$

and this is satisfied if $c(x) \ge c_0 > 0$ or there exists $\beta_0 > 0$, $u_0 > 0$ such that $n(x,du) = n(x,u)du$ with $n(x,u) \ge cu^{-(\beta+2)}du \;\forall\; u\le u_0$ near zero for some $\beta > 0$.

<u>Corollary 11.5.3.</u> *Cox Cluster Representation*

Assume (11.5.4), $X_s = \mu \in M_F(E)$ and $t > s$. Then

(a) X_t can be represented as a Cox cluster random measure (cf. (3.3.4)) with $E_1 = E_2 = E$, intensity $X_s(dx)R_{s,t}(x,M_F(E))$ and cluster probability law

$$P^*_{s,t;y}(e^{-\langle\mu,\phi\rangle}) = \frac{R_{s,t}(x,d\nu)}{R_{s,t}(x,M_F(E))} = 1 - \frac{V_{s,t}\phi(y)}{R_{s,t}(y,M_F(E))}.$$

197

(b)

$$(11.5.5)\quad X_t(A) \overset{\mathcal{D}}{=} \int_E\int_{M_F(E)\setminus\{0\}} \nu(A)\ N(dx,d\nu),\ A\in\mathcal{E},\qquad \textit{(cluster representation)}$$

where N is a Poisson random measure on $E\times(M_F(E)\setminus\{0\})$ with intensity measure $n(dx,d\nu) = X_s(dx)R_{s,t}(x,d\nu)$. Thus X_t is represented as the superposition of a random number, N, of clusters where N is Poisson with mean $\int X_s(dx)R_{s,t}(x,M_F(E)) < \infty$.

11.6. Palm Distributions of (Φ,κ,W)-Superprocesses: Analytical Representation.

To introduce the intuitive content of this section let us choose a cluster, ν, with law $R_{s,t}(x,d\nu)/R_{s,t}(x,M_F(E))$ and then choose a point $y \in E$ with law $\nu/\nu(E)$. The corresponding Palm distribution is the conditional distribution of ν given y. The purpose of this section is to obtain two representations of this conditional distribution. Roughly speaking, this will be obtained by first setting down a path of the A-process conditioned to begin at x at time s and end at y at time t. Then along this path mass is produced which then evolves according to the measure-valued branching mechanism yielding sub-clusters of mass at time t. Intuitively we can think of a subcluster that breaks off at time $s\le u<t$ as the submass having a last common ancestor with the mass at the point y at time u. This intuitive picture will be made mathematically precise in Chapter 12.

We next give a simplified version of the calculations of this section. Consider the log-Laplace functional $V(\phi) = \int(1-e^{-\langle\nu,\phi\rangle})R(d\nu)$ of an infinitely divisible random measure with canonical measure R. Consider

$$\begin{aligned}U(\phi;\psi) &:= \frac{\partial V(\phi+\varepsilon\psi)}{\partial\varepsilon}\Big|_{\varepsilon=0} = \int\langle\nu,\psi\rangle e^{-\langle\nu,\phi\rangle}R(d\nu)\\ &= \iint\psi(x)e^{-\langle\nu,\phi\rangle}\overline{R}(dx,d\nu)\\ &= \iint\psi(x)e^{-\langle\nu,\phi\rangle}R_x(d\nu)I(dx)\end{aligned}$$

and hence formally the Laplace functional of the Palm distribution R_x is given by

$$\int e^{-\langle\nu,\phi\rangle}R_x(d\nu) = U(\phi,\delta_x)\quad\text{for I-a.e. x.}$$

The second observation is that if V_t satisfies an equation of the form (which for simplicity we assume to be time homogeneous)

$$\frac{\partial V_t(\phi)}{\partial t} = AV_t+\Phi(V_t(\phi)),\quad V_0(\phi) = \phi,$$

then U_t satisfies the linear equation

$$\frac{\partial U_t(\phi,\psi)}{\partial t} = AU_t(\phi,\psi)+\Phi'(V_t(\phi))U_t(\phi,\psi),\quad U_0(\phi,\psi) = \psi.$$

198

If A is the generator of a Markov process then this linear equation can be solved by the Feynman-Kac formula. We will now apply these ideas to determine the Laplace functional for the Palm distributions for the general (Φ,κ,W)-superprocess associated with log-Laplace equation 4.3.8 and canonical representation 11.5.2.

Theorem 11.6.1. Assume that the canonical measure of the (W,κ,Φ)-superprocess starting at δ_x at time s is given by $m \equiv 0$ and $R_{s,t,x} = R_{s,t}(x,.)$. Then

$$\int e^{-\langle\mu,\phi\rangle}(R_{s,t,x})_y(d\mu)$$

$$= E_{s,x}\left(\exp\left\{\int_s^t \Phi_2(r,W_r,V_{r,t}\phi(W_r))\kappa(dr)\right\}\middle| W_t=y\right), \quad P_{s,x}(W_t\in .)\text{-a.e. } y$$

where $\Phi_2(r,x,\lambda) = \frac{\partial}{\partial\lambda}\Phi(r,x,\lambda)$.

Proof. It suffices to show that for all $\psi\in b p\mathcal{E}$,

$$\int\psi(y)\int e^{-\langle\mu,\phi\rangle}(R_{s,t,x})_y(d\mu)P_{s,x}(W_t\in dy) = E_{s,x}\left[\exp\left(\int_s^t \Phi_2(r,W_r,V_{r,t}\phi(W_r))\kappa(dr)\right)\psi(W_t)\right].$$

Using the fact that $P_{s,x}(W_t\in dy) = \int\mu(dy)R_{s,t,x}(d\mu)$, Lemma 11.4.1(c) and (11.5.2), it follows that the left hand side is equal to

$$Z_{s,t}(\phi,\psi,z) := \int\langle\mu,\psi\rangle e^{-\langle\mu,\phi\rangle}R_{s,t,x}(d\mu) = \frac{\partial}{\partial\varepsilon}(V_{s,t}(\phi+\varepsilon\psi))|_{\varepsilon=0}.$$

Using equation (4.3.8) we can verify that Z is the (non-negative) solution of the linear equation,

$$Z_{s,t}(\phi,\psi,x) = E_{s,x}\left[\psi(W_t) + \int_s^t\kappa(dr)\Phi_2(r,W_r,V_{r,t}\phi(W_r))Z_{r,t}(\phi,\psi,W_r)\right].$$

We now verify that

$$Z_{s,t}(\phi,\psi,x) = E_{s,x}\left[\exp\left(\int_s^t\Phi_2(r,W_r,V_{r,t}\phi(W_r))\kappa(dr)\right)\psi(W_t)\right]$$

by substitution in the right hand side. Using the Markov property of W_t this yields

$$E_{s,x}\left[\psi(W_t) + \int_s^t\kappa(dr)\Phi_2(r,W_r,V_{r,t}\phi(W_r))Z_{r,t}(\phi,\psi,W_r)\right]$$

$$= E_{s,x}\left[\psi(W_t) + \int_s^t\kappa(dr)\Phi_2(r,W_r,V_{r,t}\phi(W_r))\exp\left(\int_r^t\Phi_2(u,W_u,V_{u,t}\phi(W_u))\kappa(du)\right)\psi(W_t)\right]$$

$$= E_{s,x}\left[\psi(W_t) - \int_s^t\kappa(dr)\,\frac{d}{d\kappa(r)}\exp\left(\int_r^t\Phi_2(u,W_u,V_{u,t}\phi(W_u))\kappa(du)\right)\psi(W_t)\right]$$

$$= E_{s,x}\left[\exp\left(\int_s^t\Phi_2(r,W_r,V_{r,t}\phi(W_r))\kappa(dr)\right)\psi(W_t)\right]. \quad \square$$

Remark 11.6.2. For the historical process, $\{W_u : s\le u\le t\}$ is measurable with respect

199

to $\sigma(W_t)$ and hence we obtain

$$\int e^{-\langle\mu,\phi\rangle}(R_{s,t})_{y}{}_t(d\mu) = \exp\left(\int_s^t \Phi_2(r,y^r,V_{r,t}(y^r))\kappa(dr)\right)$$

for $P_{s,m}$ a.a. y (cf. [DP, Section 4]).

11.7 A Probabilistic Representation of the Palm Distribution

Note that $-\Phi_2(r,x,\lambda) = 2c(r,x)\lambda + \int_0^\infty (1-e^{-\lambda u})\, u\cdot n(r,x,du)$ is the log-Laplace function of an infinitely divisible positive random variable for each pair (r,x). Given $\{W_r: s\le r\le t\}$ with $W_s=x$, $W_t=y$, we construct a random measure on $\mathbb{R}$, $T(W,dr)$ with independent increments and with Laplace functional

$$E\left(e^{-\int f(r)T(W,dr)}\right) = \exp\left(\int_s^t \Phi_2(r,W_r,f(r))\kappa(dr)\right).$$

Then given W and T(W,dr) we define a Poisson measure, N(W,.), on $\mathbb{R}\times(M_F(E)\backslash\{0\})$ with intensity measure

$$n(W,dr,d\nu) = R_{r,t}(W_r,d\nu)T(W,dr).$$

We assume that (W,T,N) are defined on a common probability space with law $P_{s,x}(dW)P(W,.)$.

Theorem 11.7.1. For $A \in \mathcal{B}(M_F(E))$

$$(R_{s,t,x})_y(A) = E_{s,x}\left\{P\left(\int_s^t\int_{M_F(E)\backslash\{0\}} \nu\, N(W,dr,d\nu) \in A\right)\middle| W_t=y\right\}.$$

Proof. By the construction of (T,N) given W we obtain

$$E_{s,x}\left[E\left[\exp\left(-\int_s^t\int_{M_F(E)\backslash\{0\}} \langle\nu,\phi\rangle N(W,dr,d\nu)\right)\middle| W_u: s\le u\le t\right]\middle| W_t=y\right]$$

$$= E_{s,x}\left[E\left(\exp\left(-\int_s^t\int_{M_F(E)\backslash\{0\}} (1-e^{-\langle\nu,\phi\rangle})R_{r,t}(W_r,d\nu)T(W,dr)\right) \middle| W_u: s\le u\le t\right)\middle| W_t=y\right)$$

$$= E_{s,x}\left[E\left[\exp\left(-\int_s^t (V_{r,t}\phi)(W_r)T(W,dr)\right)\middle| W_u: s\le u\le t\right]\middle| W_t=y\right]$$

$$= E_{s,x}\left(\exp\left(\int_s^t \Phi_2(r,W_r,V_{r,t}\phi(W_r))\kappa(dr)\right)\middle| W_t=y\right)$$

$$= \int e^{-\langle\mu,\phi\rangle}(R_{s,t,x})_y(d\mu)$$

by Theorem 11.6.1 and the proof is complete. □

Remarks. The analogue of the representations for the (Φ,κ,W)-superprocess for the approximating branching particle systems are established in Gorostiza and Wakolbinger (1991). For applications of the representation of the Palm distribution to both the local and long-time behavior of the superprocess refer to [DP] and [D].

200

12. FAMILY STRUCTURE AND HISTORICAL PROCESSES

12.1 Some Introductory Remarks

The purpose of this chapter is to introduce a richer structure possessed by both the branching and sampling classes of processes. This structure involves not only the current distribution of the population but also the family relationships and past history of the population and plays an important role in studying both the sample path behavior and long time behavior of superprocesses.

Although the basic ideas of family structure in population models go back to the origins of the subject, in the context of measure-valued processes the subject is undergoing rapid development in several different directions at this time. This chapter is an attempt to give a unified introduction to these developments. We will outline the main ideas and make frequent references to the recent literature on this subject and parallel arguments in earlier chapters of these notes rather than attempt a complete discussion.

Since the precise formulation of the historical process involves a number of technicalities we begin with a brief informal explanation of the main ideas. In the setting of the N-particle Moran process (Chapt. 2), we can consider an enriched version of the empirical process, namely,

$$G_t^N := N^{-1} \sum_{i=1}^{N} \delta_{\hat{Z}_i(t)} \qquad \textit{(empirical measure on particle histories)}$$

where $\hat{Z}_i \in D_E$ is the path in E followed by the ith particle alive at time t and its ancestors *(its history)*. Note that $\hat{Z}_i(t)(s) := \hat{Z}_i(s\wedge t)$ (path stopped at t) and that when a particle jumps it assumes the entire past history of the particle onto which it jumps.

Then $G_t^N \in M_1(D_E)$ and describes the law of the path which would have been followed by one particle chosen at random at time t from the population. Now considertwo particles chosen at random (with replacement) from the probability distribution G_t^N. Even though these particles are chosen independently there is a positive probability that both will follow a common path in [0,s] and then follow separate paths in (s,t]. If the motion process with generator A is such that two independent particles never follow identical paths, then this would be exactly the situation which would occur if and only if they had a common ancestor up to time s but were descendents of two different offspring of that ancestor at time s. Taking larger and larger random samples from $G^N(t)$ reveals more and more information about the *family structure or genealogy* of the particles alive at time t.

201

If we consider the branching particle system instead of the Moran model, conditioned by non-extinction at time t and consider the corresponding normalized measure on D_E, then exactly the same situation would occur.

For both sampling and branching systems, if we consider the measure-valued limits ($N\to\infty$, resp. $\varepsilon\to 0$) as in Chapter 2, we obtain nonatomic measures, G_t, (resp. H_t on D_E). *The key point is that even in these limiting cases,* G_t *restricted to* $[0,s]$ for $s<t$, *is almost surely concentrated on a finite number of paths.* This reflects the fact that, in the $N\to\infty$ limit, the number of individuals alive at time $s<t$ having descendents alive at time t converges in distribution to a finite random variable. This means that we can represent G_t in terms of a branching tree structure and that, as the family tree branches, the mass (representing the total population alive at time t) is divided into smaller and smaller atoms (representing more and more closely related groups of individuals). The main objective of this chapter is to outline the mathematical formulation of this representation.

Much of the recent development of historical process has its origins in the papers of Durrett (1978), Fleischmann and Prehn (1974, 1975) Fleischmann and Sigmund-Schultze (1977, 1978), Gorostiza (1981), Jagers and Nerman (1984), Kallenberg (1977), and Neveu (1986) on branching processes and Cannings (1974), Kingman (1982), Tavaré (1984) and Watterson (1984) on population genetics.

12.2 The Historical Branching Particle System.

In order to keep things as simple as possible in this section we consider a critical binary branching particle system with motion process W given by a Feller process on a compact metric space E and with representation $(\hat{\Omega},\hat{\mathcal{F}},\hat{Q}_A)$ as in section 2.2 (see remark 12.2.2.2). The family structure of a branching particle system is naturally represented by a tree and this structure was used in an essential way by Neveu (1986), Chauvin (1986a,b), Perkins (1988) and Le Gall (1989b, 1991a,b).

12.2.1 The Binary Tree System

Let us consider a critical binary branching particle system in which the particle motions are given by W. In order to keep track of the family structure of the particles, we introduce Neveu's formulation of the *binary tree system.*

Let $K := \bigcup_{n=0}^{\infty} \{1,2\}^n$ where by convention $\{1,2\}^0 = \{\partial\}$. $k\in K$ is represented as $k := k_1k_2\ldots k_n$ where each $k_i\in\{1,2\}$. Set $|k|=n$, $|\partial|=0$, and if $|k|\geq 1$, set $\bar{k} := k_1\ldots k_{n-1}$. For $k=k_1\ldots k_n$, $h=h_1\ldots h_m \in K$ set $hk = h_1\ldots h_m k_1\ldots k_n$ and $\partial k=k\partial=k$. Finally, $h < h'$ if there exists some $k\in K$ such that $h'=hk$.

A (finite binary) tree is a finite subset κ of K that satisfies

202

(i) $\partial\in\kappa$

(ii) $\bar{k}\in\kappa$ whenever $k\in\kappa$ and $|k|\geq 1$

(iii) if $k=k_1\ldots k_n \in K$, then either $k_1\ldots k_n 1 \in \kappa$, and $k_1\ldots k_n 2 \in \kappa$ or $k_1\ldots k_n 1 \notin \kappa$, and $k_1\ldots k_n 2 \notin \kappa$.

Denote by $\mathfrak{T}$ the set of all (finite binary) trees. If $\kappa \in \mathfrak{T}$ and $k\in \kappa$, we set $v_k(\kappa) = 1_{\{k1\in\kappa\}} = 1_{\{k2\in\kappa\}}$.

We next introduce the notion of a marked tree. A *marked tree* is a pair (κ,ϑ) where κ is a tree and ϑ is a map from κ into a *space of marks* E^*.

12.2.2. The A-Path Process.

Assume that the motion process $Y = (D_E,\mathcal{D},\mathcal{D}_t,Y_t,\{P^x\}_{x\in E})$, filtration defined as in Sect. 3.6 and coordinate process Y_t) is an E-valued Feller process with generator $(D(A),A)$ and laws $\{P^x\}_{x\in E}$.

Notation. If $y,w\in D_E = D([0,\infty),E)$, and $s\geq 0$ let $Y_s(y) := y(s)$ and let $(y/s/w)\in D_E$ be defined by

$$(y/s/w) := \begin{cases} y(u) & \text{if } u < s \\ w(u-s) & \text{if } u \geq s. \end{cases}$$

Let $\pi_t:D_E\to E$ be defined by $\pi_t y := y(t)$. Also let $D_E^t := \{y:y=y^t\}$ where $y^t(\cdot) := y(\cdot\wedge t)$ (path stopped at t).

We now define the *path process associated with* Y to be the D_E-valued process defined on $(D_E,\mathcal{D},\mathcal{D}_t,Y_t,P^x)$ as follows:

$$W_t(y) := \{Y_{t\wedge s}(y):\ s\in \mathbb{R}_+\} \in D_E^t.$$

Then $y(t) = \pi_t W_t(y)$.

Of course with appropriate regularity conditions the path process is a Markov process even when the underlying Y process is not Markov. We do not develop this point of view here which goes back to Harris (1963, Sect. 24 and 27) but refer to Wentzell (1985, 1989, 1992) and Bulycheva and Wentzell (1989) for this as well as for a more systematic study of path processes.

Let $\hat{D}_{E,s} := \{(t,y):t\in[s,\infty),y\in D_E^t\}$ and $\hat{D}_E := \hat{D}_{E,0}$. Let $M_F(D_E)^s := \{\mu\in M_F(D_E):\mu((D_E^s)^c)=0\}$.

The corresponding *canonical path process* is defined by

$(D_E,\mathcal{D},\mathcal{D}_{s,t},\{W_t\}_{t\geq 0},(P_{s,y})_{(s,y)\in\hat{D}_E})$ with $\mathcal{D}_{s,t} := \bigcap_{\varepsilon>0}\sigma\{W_u:s\leq u\leq t+\varepsilon\}$, and

$$(12.2.2.1) \qquad P_{s,y}(B) := P^{y(s)}(y/s/W \in B).$$

Similarly,

$$P_{s,\mu}(B) := \int P^{y(s)}(y/s/W\in B)\mu(dy) \quad \text{if} \quad \mu\in M_F(D_E)^s.$$

203

<u>Remark:</u> Note that the path process, W, has the important property that

(12.2.2.2) for t fixed the σ-algebras of subsets of D_E generated by $\sigma\{W_s : s \le t\}$ and $\sigma\{W_t\}$ are identical.

Although the path process W is not Feller, it does possess nice properties from the point of view of the general theory of Markov processes in the following sense (cf. Section 3.5 for the appropriate definitions).

<u>Theorem 12.2.2.1.</u> Assume that Y and W are as above.
Then $(D_E, \mathcal{D}, \mathcal{D}_{s,t}, \{W_t\}_{t\ge 0}, (P_{s,y})_{(s,y)\in \hat{D}_E})$ is a canonical inhomogeneous Borel strong Markov process with càdlàg paths in $D_E^t \subset D_E$ and inhomogeneous semigroup

(12.2.2.3) $\quad S_{s,t} f(y) = P^{y(s)}(f(y/s/Y^{t-s})), \quad (s,y)\in \hat{D}_E, \ t\ge s, \quad f \in b\mathcal{D}.$

<u>Proof.</u> [DP Prop. 2.1.2]. (In fact this is true for a much wider class of underlying processes, Y.)

We now consider a martingale problem which characterizes the path process. Consider the algebra of functions, denoted by $D_0(\tilde{A})$, described as follows: $\phi \in D_0(\tilde{A})$ if for some $n\in\mathbb{N}$, $t_1<t_2<\ldots<t_n$, g_j and $(\frac{\partial}{\partial s} + A)g_j \in bC([0,\infty)\times E)$ for $j=1,\ldots,n$, and

$$\phi(s,y) = \prod_{j=1}^{n} g_j(s,y(s\wedge t_j))$$

Note that $\{\phi(t,.):\phi\in D_0(\tilde{A})\}$ is bp dense in $\mathcal{D}_t$.

For $\phi \in D_0(\tilde{A})$, set

(12.2.2.4) $$(\tilde{A}\phi)(s,y) := \prod_{\ell=1}^{k} g_\ell(s,y(t_\ell))\Big(\frac{\partial}{\partial s} + A\Big)\prod_{\ell=k+1}^{n} g_\ell(s,y(s)) \quad \text{if } t_k\le s<t_{k+1}$$
$$:= 0 \quad \text{if } s > t_n.$$

If $\phi\in D_0(\tilde{A})$, then $(\tilde{A}\phi)(s,y) \in bC([0,\infty)\times D_E)$ and

(12.2.2.5) $$M_t(\phi) := \phi(t,W_t)-\phi(s,W_s) - \int_s^t (\tilde{A}\phi)(r,W_r)dr \quad \text{is a } P_{s,W_s}\text{-martingale.}$$

$\tilde{A}$ is called the *compensating operator* by Wentzell (1985, 1992). If the path process can be characterized as the unique solution of the $(D_0(\tilde{A}),\tilde{A})$-martingale problem then we call $(D_0(\tilde{A}),\tilde{A})$ an *MP-generator*.

<u>Remark 12.2.2.2.</u> It can be verified that the $(D_0(\tilde{A}),\tilde{A})$-martingale problem is well posed if A is a Feller generator (apply [EK, Ch. 4, Th. 4.1] to verify that the finite dimensional distributions of the t-marginal distribution of W coincide with with those of Y stopped at t and then using Theorem 5.1.2). Also by extending the

204

arguments used in the proof of Proposition 2.2.1, in the Feller case we can construct a D_E-valued measurable random function, ζ, on $\hat{D}_E$ (defined on a standard probability space $(\hat{\Omega},\hat{\mathcal{F}},\hat{Q}_A)$ such that $\zeta((s,y))$ has law $P_{s,y}$.

For certain applications it is desirable to identify a larger class of functionals for which an analogue of (12.2.2.5) is valid. Mueller and Perkins (1991) defined a natural extension of $(D_0(\tilde{A}),\tilde{A})$ as follows. Let

$F_{s,\mu} := \{\phi:\hat{D}_{E,s}\to\mathbb{R},\ \phi$ is Borel measurable, $\phi(s,W_s)$ is right continuous, $P_{s,\mu}$-a.s. and $|\phi(s,W_s)|\le K$, $P_{s,\mu}$-a.s. for some K$\}$,

and define

$$D(\tilde{A}_{s,\mu}) := \Big\{\phi\in F_{s,\mu}:\exists\tilde{A}_{s,\mu}\phi\in F_{s,\mu} \text{ such that } M^{\phi}(t,W)\equiv\phi(t,W_t)-\phi(s,W_s)-\int_s^t(\tilde{A}_{s,\mu}\phi)(r,W_r)dr \text{ is a } P_{s,\mu}\text{-martingale on the } P_{s,\mu}\text{-completion of the filtration } (\mathcal{D}_t)_{t\ge s}\Big\}.$$

For $s\ge0$, $\mu\in M_F(D_E)^s$, $\tilde{A}_{s,\mu}$ is called the $P_{s,\mu}$-*weak generator* of W by Mueller and Perkins (1991). A wide class of functionals which belong to the domain of the weak generator was identified by Wentzell (1992).

<u>Definition:</u> The path process $\{W_t\}$ is said to have *Property S* if

(12.2.2.6) $P_{s,y}\times P_{s,y}(W^1_u=W^2_u) = 0\ \forall\ u>s,\ y\in D^s$.

Note that we can always consider an enriched version of the original process for which the path process will satisfy this property. For example, we can take $E' = E\times\mathbb{R}^1$ and generator $A' := A + \Delta$ (i.e. the components are independent, the first component is an A-process and the second component is a standard Wiener process).

<u>12.2.3. Binary branching A-motions.</u>

We will now formulate a binary branching A-motion in terms of a probability measure on a set of marked trees.

As the appropriate space of marks we take $E^* = \mathbb{R}_+\times D_E$. The set of all E^*-marked binary trees is denoted by $\mathfrak{T}^*$, that is, $\mathfrak{T}^* = (E^*)^{\mathfrak{T}}$ and a typical element of $\mathfrak{T}^*$ will be written as

$$\mathfrak{t} = (\kappa,(\tau_k,W_k)_{k\in\kappa})$$

where $\kappa\in\mathfrak{T}$ and for every $k\in\kappa$, $\tau_k\in\mathbb{R}_+$, $W_k\in D_E$. For $\mathfrak{t} = (\kappa,(\tau_k,W_k)_{k\in\kappa})\in\mathfrak{T}^*$ let $s_h(\mathfrak{t}) := \sum_{k<h}\tau_k$, and $t_h(\mathfrak{t}) := s_h(\mathfrak{t}) + \tau_h$.

Finally we need to introduce translation operators T_h. For any $h\in K$, the

205

mapping T_h is defined on the subset $\{h\in\kappa\}$ of $\mathfrak{I}^*$, by

$$T_h((\kappa,(\tau_k,W_k)_{k\in\kappa}) := (\kappa_h,(\tau_{hk},W_{hk})_{k\in\kappa_h})$$

where $\kappa_h = \{k\in K:\ hk\in\kappa\}$.

<u>Proposition</u> <u>12.2.3.1</u> Let $\{N^{\nu}(s):s\geq 0\}$ be a time inhomogeneous Poisson process with intensity measure $\nu(ds)\in M_{LF}([0,T))$, $T\leq\infty$ and corresponding law Q_N and $0\leq p\leq 1$. Then there exists a unique measureable mapping $(t,z)\longrightarrow\Lambda_{t,z}$ from $[0,\infty)\times D^t$ to $M_1(\mathfrak{I}^*)$, such that the following properties hold:

(i) for every $z\in D^t$, the random variables v_∂,τ_∂ are independent and

(12.2.3.1) $\Lambda_{t,z}[v_\partial=1] = p$, $\Lambda_{t,z}[v_\partial=0] = 1-p$,

(that is, the offspring generating function is $G(z) = 1-p+pz^2)$, $0\leq z\leq 1$),

(12.2.3.2) $\Lambda_{t,z}[\tau_\partial>t+s] = P(N^{\nu}(t+s)-N^{\nu}(t) = 0)$,

(ii) the conditional law of W_∂, given $(v_\partial,\tau_\partial)$, is $P_{t,z}(W(t+\tau_\partial)\in.)$, where $P_{t,z}$ is defined by (12.2.2.1),

(iii) the conditional distribution under $\Lambda_{t,z}(dw)$ of the pair (T_1w,T_2w), given $v_\partial=1$ and $(\tau_\partial,W_\partial)$ is $\Lambda_{\tau_\partial,W_\partial(\tau_\partial)}\otimes\Lambda_{\tau_\partial,W_\partial(\tau_\partial)}$.

<u>Proof.</u> (cf. Neveu (1986), Chauvin (1986a), Le Gall (1991a)).

The process can be constructed in the natural way as in Ch. 2 on a probability space $(\Omega,\mathcal{F},P)$ containing three sequences of independent random variables, $\Big\{\{v_k\},\ \{N_k\},\ \{\zeta_k\}:\ k\in K\Big\}$, $P(v_k=0) = 1-p$, $P(v_k=1) = p$, $\{N_k\}$ have distribution Q_N, and $\{\zeta_k\}$ have distribution $\hat{Q}_A$. □

<u>Remark.</u> Note that for $\mathfrak{t} = (\kappa,(\tau_k,W_k)_{k\in\kappa})\in\mathfrak{I}^*$, $W_k\in D_E^{t_k}$ and $W_k(t,\mathfrak{t}) = W_h(t,t\wedge t_k)$ if $k<h$. The random variables $W_h(t)$ for all $h\in\kappa$ such that $s_h\leq t<t_h$ represent the paths followed by the particles alive at time t and before the instants of their births by their ancestors, up to time t. Let $\pi_t:D_E\longrightarrow E$ be defined by $\pi_t(y) := y(t)$. Then $\pi_tW_h(t)$ denotes the locations of the particles at time t.

We then consider the following measure-valued processes

$$H_t^{*\nu}(t) := \sum_{\substack{h\in\kappa\\ s_h\leq t<t_h}}\delta_{W_h(t)}\quad\text{(counting measure on } D_E^t),$$

206

$$X_t^{*\nu}(t) := \sum_{\substack{h\in\kappa \\ s_h\le t<t_h}} \delta_{\pi_t W_h(t)}.$$

In the case in which $\nu(ds) := c ds$ is a homogeneous Poisson process with c>0, and p = 1/2, the $M_F(D_E)$-valued process $\{H_t^{*\nu}:t\ge 0\}$ is called the *historical critical binary branching particle system*. It is an enriched version of the process $\{X_t^{*\nu}\}$ where the latter is the branching particle system introduced in Section 2.10. Note that this is a *trimmed tree* in which all the branches which have become extinct by time t (i.e. $v_h = 0$ for some $s_h<t$) are trimmed off.

Remark: If $\{W_t\}$ satisfies Property S, then P-a.s. we can reconstruct the marked subtree, $(\kappa,(\tau_k,W_k(t))_{k\in\kappa}:k<h$, for some h with $s_h\le t<t_h)$ (up to relabelling of the two offspring as 1 and 2) from $H_t^{*\nu}(t)$. In particular, if p = 1 (i.e. no deaths) we can view this as follows. Let $N_t(u)$ denote the number of atoms in $r_u H_t^{*\nu}(t)$, $0\le u\le t$. Then $\{N_t(u):0\le u\le t\}$ is a monotone increasing integer-valued function. Furthermore under Property S, there is a one-to-one correspondence between the jumps of $N_t(u)$ and the set $\{t_k$: k < h for some h with $s_h\le t<t_h\}$. Then in order to determine which path split at t_h we look for the unique $W|_{[0,t_h)}$ which has two distinct extensions on $[0,t_h+\varepsilon)$ $\forall\ \varepsilon>0$.

12.2.4. The Generating Function Equation.

Let $N_F(D_E^t)$ denote the set of finite integer-valued measures on D_E^t. Then $\{H_t^{*\nu}\}$ is a càdlàg $N_F(D_E^t)$-valued process.

Let $\Omega^* := D(N_F(D_E))$ with its Borel σ-field $\mathcal{G}^*$, $H_t^{*\nu}(\omega) = \omega(t)$, and $\mathcal{G}^*_{[s,r]} := \sigma(H_t^*:s\le t\le r)$, $\mathcal{G}^*_{[s,\infty)} := \bigvee_{r\ge s} \mathcal{G}^*_{[s,r]}$. If $H_s^{*\nu} = m\in N_F(D_E^s)$, then the resulting law of $H_\cdot^{*\nu}$ on $\mathcal{G}^*_{[s,\infty)}$ is denoted by $Q^{*\nu}_{s,m}$.

If $H_s^{*\nu} = \delta_y$, $y\in D^s$, then the probability generating function of $H_t^{*\nu}$ is defined by

$$(12.2.4.1)\qquad G_{s,t}\xi(y) := Q^{*\nu}_{s,\delta_y}\left(e^{\langle H_t^{*\nu},\log\xi\rangle}\right),\quad \xi\in\mathcal{D},\ 0<\xi\le 1,\ t>s.$$

If $H_s^{*\nu} = \sum \delta_{y_i}$, then from each of these initial particles, δ_{y_i}, we construct independent copies of the above process and $H_t^{*\nu}$ is then given by the superposition of these. Therefore the probability generating function of $H_t^{*\nu}$ is given by

207

(12.2.4.2) $\prod_i G_{s,t}\xi(y_i)$.

Theorem 12.2.4.1 (a) $G_{s,t}\xi$, $0\le s\le t$, is the unique solution of the equation

(12.2.4.3) $$G^{\nu}_{s,t}\xi(y) = e^{-\nu((s,t])} S_{s,t}\xi(y) + \int_s^t S_{s,u} e^{-\nu((s,u])}\Big[G(G^{\nu}_{u,t}\xi(y))\Big]\nu(du).$$

If the intensity measure $\nu(ds) = \nu(s)ds$, then this is the evolution solution of the formal equation

(12.2.4.4) $$G^{\nu}_{t,t}\xi(y) = \xi(y),$$
$$-\frac{\partial G^{\nu}_{s,t}}{\partial s} = (A_s - \nu(s))G^{\nu}_{s,t} + \nu(s)G(G^{\nu}_{s,t}).$$

(b) Let $\mathcal{Pois}_m$ denote the law of a Poisson random measure on D^s_E with intensity m. If we assume that $H^{*\nu}_s$ has distribution $\mathcal{Pois}_m$, then $H^{*\nu}_t$ has Laplace functional

(12.2.4.5)
$$\int Q^{*\nu}_{s,\mu}\Big[e^{-\langle H^{*\nu}_t,\phi\rangle}\Big]\mathcal{Pois}_m(d\mu) = \exp\Big\{-\int\Big(1 - (G^{\nu}_{s,t}e^{-\phi})(y)\Big)m(dy)\Big\},$$

that is, $H^{*\nu}_t$ is a Poisson cluster random measure with intensity m and cluster Laplace functional $(G^{\nu}_{s,t}e^{-\phi})(y)$.

Proof. (a) Equation (12.2.4.3) is obtained by conditioning on the time and place of the first branching event. The uniqueness then follows as in Lemma 4.3.3.

(b) This follows from (a) and the Poisson cluster formula (3.3.2). □

Notation. In the case $\nu(ds) := (c/\varepsilon)ds$, with $\varepsilon>0$, we denote the process $H^{*\nu}$ by $H^{*\varepsilon}$.

12.3. The (A,c)-Historical Process

12.3.1 Construction and Characterization.

Consider the collection of historical critical binary branching particle systems, $\{H^{*\varepsilon}_t:\varepsilon>0\}$, and Poisson random field initial condition at time s with intensity μ/ε, $\mu\in M_F(D_E)^s$. Finally we consider the measure-valued process obtained by letting each particle to have mass ε:

$$H^{\varepsilon}_t := \varepsilon H^{*\varepsilon}_t.$$

Note that this is a special case of the general time inhomogeneous set-up of Chapter 4 in which the motion process is the A-path process W, $G(z) = \frac{1}{2}(1+z^2)$, $0\le z\le 1$, and $\kappa(ds) = \varepsilon^{-1}ds$. For the sake of brevity herafter we denote the (A,c)-historical process by H(A,c).

Theorem 12.3.1.1. There exists a transition function on $M_F(D_E)$ with Laplace transi-

208

tion functional

(12.3.1.1) $Q^{H(A,c)}_{s,\mu}\left(\exp(-\langle H_t,\phi\rangle)\right) = \exp(-\langle\mu,V_{s,t}\phi\rangle)$

$\forall\ \phi \in bp\mathcal{D}$, $\mu\in M_F(D)^S$, $s\le t$, where $V_{s,t}\phi(y) := v_{s,t}(y)$ is the unique solution of

(12.3.1.2) $$V_{s,t}\phi(y) = S_{s,t}\phi(y) - \tfrac{1}{2}c\int_s^t S_{s,r}((V_{r,t}\phi)^2)dr$$

which is Borel measurable in $(s,y,t) \in \{(s,y,t),\ s\in[0,\infty),\ y\in D^S,\ t\ge s\}$ and is bounded if $(t-s)$ is bounded.

Proof. Exactly as in the proof of Theorem 4.4.1 (cf. Lemma 4.4.3) we conclude that $\{H^\varepsilon_t\}$ converges, as $\varepsilon\to 0$, to a time inhomogeneous $M_F(D_E)$-valued Markov process $\{H_t\}$ (in the sense of convergence of finite dimensional distributions) and that the transition function is given by (12.3.1.1, 12.3.1.2). □

The resulting process $\{H_t\}$ is called the (A,c)-*historical process* and is denoted by H(A,c). In fact it coincides with the (W,κ,Φ)-superprocess with W given by the A-path process, $\kappa(ds) = ds$, and $\Phi(\lambda) = -\frac{1}{2}c\lambda^2$.

Since D_E is not locally compact and the process is time inhomogeneous we cannot directly apply the tightness argument of Section 4.6 to obtain weak convergence on $D([0,\infty),M_F(D_E))$. The question of the tightness of the sequence of approximating historical branching particle systems was discussed in [DP, Section 7]. The main obstacle is to verify the first condition in Theorem 3.6.4, namely, given T>0 and $\eta>0$ to verify the existence of a compact subset, $K_{T,\eta}$, of $M_F(D_E)$ such that $P(H^\varepsilon_\bullet \in D([0,T],K_{T,\eta})) > 1-\eta\ \forall\ \varepsilon>0$. Since $\sup_\varepsilon P(\sup_{0\le t\le T} H^\varepsilon_t(M_F(D_E)) > k) \to 0$ as $k\to\infty$ (cf. Lemma 4.6.1), it suffices to find a sequence of compact subsets, $\{K_{T,n}\}$, of D_E such that

$$\sup_\varepsilon P(\forall n\in\mathbb{N},\ \sup_{0\le t\le T} H^\varepsilon_t(K^c_{T,n}) \le 1/n\ ) > 1-\eta.$$

Hence it suffices to find for each n a compact set $K_{T,n} \subset D_E$ such that $\sup_\varepsilon P(\sup_{0\le t\le T} H^\varepsilon_t(K^c_{T,n}) > 1/n) < \eta/2^n$. The existence of such a compact subset is established in [DP, Prop. 7.7]. Thus it can be established that the historical branching particle systems converge weakly to the historical process $\{H_t: t\ge 0\}$ and the latter has càdlàg paths in $M_F(D_E)$. The existence of nice version of the historical process is summarized in the following theorem.

Theorem 12.3.1.2. There is an inhomogeneous Borel strong Markov process with càdlàg

209

paths in $M_F(D_E)$, $\{H_t:t\ge s\}$ with laws $\{Q_{s,\mu}\}_{s\ge 0,\mu\in M_F(D_E)^s}$, whose finite dimensional distributions are determined by the transition function of Theorem 12.3.1.1.

Proof. This can be proved using the construction of Fitzsimmons (1988) and is given in [DP, Theorems 2.1.5, 2.2.3]. It can also be proved by the above weak convergence argument together with the uniqueness to the martingale problem (see Theorem 12.3.3.1). □

The *canonical* (A,c)-*historical process* is denoted by $H := (\Omega,\mathcal{G},\mathcal{G}_{s,t},(H_t)_{t\ge 0},(Q^{H(A,c)}_{s,\mu})_{s\ge 0,\mu\in M_F(D_E)^s})$ where $\Omega = D([0,\infty),M_F(D_E))$, $H_t(\omega) := \omega(t)$ for $\omega\in\Omega$, $\mathcal{G}_{s,t} := \bigcap_{u>t}\sigma(H_r:s\le r\le u)$ and $Q^{H(A,c)}_{s,\mu}$ is a probability measure on $\mathcal{G}$.

Then $H_t \in M_F(D_E)^t$ $\forall t\ge s$, $Q^{H(A,c)}_{s,\mu}$-a.s. If $\phi:\hat{D}_E \to \mathbb{R}$ is measurable, then $\langle H_t,\phi(t)\rangle := \int_{D_E}\phi(t,y)H_t(dy)$.

In Theorem 12.3.4.2 below we will verify that $X_t(B) := H_t(\{y:y(t)\in B\})$, $B\in\mathcal{B}(E)$, is a version of the (A,c)-superprocess. We can thus regard H_t as an *enriched* version of X_t which contains information on the genealogy or family structure of the population.

12.3.2. Remarks on Alternative Formulations of the Historical Process

The above construction of historical processes as limits of branching particle systems was introduced in Dynkin (1991a) and their characterization as inhomogeneous Borel strong Markov processes was established in [DP]. There have also been alternative formulations of historical processes.

Perkins (1988) (also see Dawson, Iscoe and Perkins (1989)) developed a nonstandard model of superprocesses. His idea was to construct a binary branching Brownian motion with infinitely many particles having infinitesimal mass and infinite branching rate. The nonstandard approach has the advantage of allowing one to work with the intuitive particle picture and in particular to incorporate the historical information on each particle. Perkins used this formulation to obtain the precise results on Hausdorff measure function which were described in Section 9.3.3.

Another approach is that due to Le Gall (1991a). His approach to the construction of continuous superprocesses is based on relation between continuous branching and standard Brownian motion as reflected for example in the Ray-Knight Theorem. He has constructed a path-valued Markov process as a tool in studying the problem of polar sets for super-Brownian motion. In particular his construction is based on a detailed representation of the tree of excursions of a standard Brownian motion and is related to the following result of Neveu and Pitman (1989). Let w be

210

a Brownian excursion from 0 which reaches level h > 0 and let $(N_x)_{x\geq 0}$ be the number of upcrossings from x to x+h of the process w. Then Neveu and Pitman (1989) showed that N_x is a continuous time Galton-Watson critical binary branching process (with particle lifetime exponentially distributed with mean $\frac{1}{2}h$). An h-minimum exists at time t if there exists s<t<u such that $w_s=w_u=w_t+h$ and $w_v\geq w_t$ for $s\leq v\leq u$. The probability that there exists an h-minimum is $\frac{1}{2}$ and conditioned on the existence of an h-minimum the processes $\{w(\sigma_\alpha+s):0\leq s\leq\rho_\alpha-\sigma_\alpha\}$ and $\{w(\rho_\alpha+s):0\leq s\leq\tau_\alpha-\rho_\alpha\}$ are independent copies of w. Here ρ_α is the time when the lowest h-minimum is attained, $\sigma_\alpha=\sup\{u<\rho_\alpha:w(u)=w(\rho_\alpha)\}$ and $\tau_\alpha = \inf\{t>\rho_\alpha:w(t)=w(\rho_\alpha)\}$. Le Gall (1991a,b) used these objects to yield an explicit construction of the (2,d,1)-superprocess and also the (2,d,1)-historical process. In this setting the tree structure of the historical process at a fixed time is directly related to a set of excursions of the Brownian motion above a level a-ε to hit a (upcrossings from a-ε to a with a fixed and ε>0) (see Le Gall (1991a, Lemma 8.5)).

12.3.3. Martingale Problem

In the spirit of Chapter 6, it is natural to ask if the historical process can be characterized as the unique solution of a martingale problem. In this subsection we will simply give an outline of the main steps in establishing that this is in fact possible.

Let $H = (\Omega,\mathcal{G},\mathcal{G}_{s,t},\{H_t\},\{Q^{H(A,c)}_{s,\mu}\})$ denote the canonical historical process defined above and let $(D_0(\tilde{A}),\tilde{A})$ denote the MP-generator for the A-path process defined in Section 12.2.2. Using arguments similar to those developed in Chapter 6 we can verify the following: for each $\phi\in D_0(\tilde{A})$, and for $t\geq s$,

$$Z_t(\phi) := \langle H_t,\phi(t)\rangle - \langle H_s,\phi(s)\rangle - \int_s^t \langle H_r,(\tilde{A}\phi)(r)\rangle dr \tag{12.3.3.1}$$

is a $Q^{H(A,c)}_{s,H_s}$, $\{\mathcal{G}_{s,t}:t\geq s\}$-martingale with increasing process

$$\langle Z(\phi)\rangle_t = c\int_0^t\int_{D_E} \phi(r,y)^2 H_r(dy)dr. \tag{12.3.3.2}$$

For example, if $\psi\in b\mathcal{D}_u$, then $\{\langle H_t,\psi\rangle:t\geq u\}$ is a continuous martingale with increasing process

$$\langle H(\psi)\rangle_t = \int_u^t\int\psi^2(y)H_s(dy)ds.$$

In other words $\{Q_{s,\mu}\}$ is a solution to the martingale problem (12.3.3.1), (12.3.3.2). Using an extension of the argument of the proof of Theorem 6.1.3(b) it can be verified that any solution $Q_{s,\mu}$ to the martingale problem for $(D_0(\tilde{A}),\tilde{A})$ start-

211

ing from μ at time s satisfies

$$Q^{H(A,c)}_{s,\mu}\left(\exp(-\langle H_t,\phi(t)\rangle)\right) = \exp\{-\langle\mu,V_{s,t}\phi\rangle\} \quad \text{for} \quad \phi \in D_0(\tilde{A})$$

and consequently this holds for all $\phi\in \mathcal{D}_t$.

Thus the one dimensional marginal distributions to this martingale problem are uniquely determined and applying Theorem 5.1.2 we conclude that the martingale problem is well-posed. Since in the above argument $D_0(\tilde{A})$ can clearly be replaced with a countable subset we can apply Theorem 5.1.2 to establish the strong Markov property. Thus we have the following.

Theorem 12.3.3.1. The law $Q^{H(A,c)}_{s,\mu}$ is the unique probability measure on $\mathcal{G}_{s,\infty}$ such that $\forall\ \phi\in D_0(\tilde{A})$,

$$Z_t(\phi) = \langle H_t,\phi(t)\rangle-\langle\mu,\phi(s)\rangle - \int_s^t \langle H_r,(\tilde{A}_{s,\mu}\phi)(r)\rangle dr, \quad t\geq s,$$

is a continuous $\{\mathcal{G}_{s,t}:t\geq s\}$-martingale such that $Z_0(\phi)=0$ and

$$\langle Z(\phi)\rangle_t = c\int_0^t\int_{D_E} \phi(r,y)^2 H_r(dy)dr.$$

Proof. The uniqueness follows as above. □

Mueller and Perkins (1991) established the analogue of (12.3.3.1) for the richer class of functions, $D(\tilde{A}_{s,\mu})$, which was defined in section 12.2.2 (and also proved that this is valid for a much wider class of underlying processes Y).

12.3.4. Transition Function - Finite Dimensional Distributions.

The purpose of this subsection is to give an analytical description of the transition function for the (A,c)-historical process hereafter denoted by $\{H_t\}$. Note that if $f_t(y) := g(y(t))$, $g\in bp\mathcal{E}$, then

(12.3.4.1) $\qquad S_{s,t}f_t(y) = P^{y(s)}g(y(t-s))$, and

(12.3.4.2) $\qquad V_{s,t}f_t(y) = U_{t-s}g(y(s)) \quad \forall(s,y)\in \hat{D}_E,\ s\leq t$.

Notation. $U_t^{(n)}:bp\mathcal{E}^n \longrightarrow bp\mathcal{E}^{n-1}$ is defined by

$$(U_t^{(n)}g)(x_1,\dots,x_{n-1}) = U_t g(x_1,\dots,x_{n-1},.)(x_{n-1}).$$

We define $\bar{U}_t^{(2)}:pb\mathcal{D}_{0,s}\times\mathcal{E} \longrightarrow pb\mathcal{D}_{0,s}$ by

$$\bar{U}_t^{(2)}g(y) = U_t g(y^s,.)(y(s)).$$

Theorem 12.3.4.1. (a) Let $f_{s,t}(y) = g(y,y(t))$, $g\in bp\mathcal{D}_{0,s}\times\mathcal{E}$. Then

(12.3.4.3) $$V_{s,t}f_{s,t}(y) = (\bar{U}^{(2)}_{t-s}g)(y) \in \mathcal{D}_{0,s} \quad \forall (s,y)\in \hat{D}_E,\ s\le t.$$

(b) If $g\in bp\mathcal{E}^n$, $t_1\le t_2\le\ldots\le t_n$, and $f_{t_1,\ldots,t_n}(y) = g(y(t_1),\ldots,y(t_n))$,

then for any $1\le k\le n$,

(12.3.4.4) $$V_{t_k,t_n}f_{t_1,\ldots,t_n}(y) = (U^{(k+1)}_{t_{k+1}-t_k}\cdots U^{(n)}_{t_n-t_{n-1}}g)(y(t_1),y(t_2),\ldots,y(t_k)).$$

Outline of Proof. (a) For $t\ge s$, $V_{s,t}f_{s,t}(y)$ satisfies (12.3.1.2).
We will verify that the right hand side of (12.3.4.3) also satisfies (12.3.1.1).
Let $\tilde{v}_{s,t}(y) := U^{(2)}_{t-s}g(y)$.
Since

$$(U^{(2)}_{t-s}g)(y) = P_{t-s}g(y,.)(y(s)) - \tfrac{1}{2}c\int_0^{t-s} P_u((U^{(2)}_{t-s-u}(g(y,.))^2)(y(s))du$$

it follows that

$$\tilde{v}_{s,t}(y) = P^{y(s)}(f_{s,t}(y/s/Y^{t-s})) - \tfrac{1}{2}c\int_0^{t-s} P^{y(s)}((\tilde{v}_{u+s,t}(y/s/Y^u))^2)du.$$

The result follows by uniqueness to 12.3.1.2.
(b) The proof follows by induction using repeatedly $V_{t_k,t_n} = V_{t_k,t_{n-1}}V_{t_{n-1},t_n}$ and (a). □

We next verify that $\{X_t\}$ can be embedded in $\{H_t\}$. Let $\Pi_t:M_F(D_E)\to M_F(E)$ be defined by $\Pi_t\mu(B) := \mu(\{y\in D_E:y(t)\in B\})$.

Theorem 12.3.4.2. $X_t = \Pi_t(H_t)$ is a version of the (A,c)-superprocess. Moreover if $\mu\in M_F(D_E)^s$, $\tau\ge s$ is a $\{\mathcal{G}_{s,t}:t\ge s\}$-stopping time and $\psi\in b\mathcal{B}(D(M_F(E)))$, then

(12.3.4.5) $$Q^{H(A,c)}_{s,\mu}(\psi(X(\tau+.)|\mathcal{G}_{s,\tau}) = P^{B(A,c)}_{X(\tau)}(\psi),\quad Q^{H(A,c)}_{s,\mu}\text{-a.s.}$$

on $\{\tau<\infty\}$ where $P^{B(A,c)}_\mu$ is defined as in section 8.1.
Remark on Proof. This follows by an argument similar to the proof of Theorem 12.3.4.1. See [DP Theorem 2.2.4] for the details. □

12.3.5. A Class of Functionals of the Historical Process

Since the historical process involves random measures on D_E it is natural to consider $\int F(y)H_t(dy)$ where F is a measurable functional of interest. In this section we will consider a class of functionals which were introduced by Dynkin (1991b).

Let B be a domain in E and define $\tau_B := \inf\{t:Y_t\notin B\}$. In Section 9.4 we described the random measure X_{τ_B} on $(E,\mathcal{E})$ which was used in Dynkin (1991c) to

213

establish a necessary and sufficient condition for R-polarity. In this section we briefly indicate how this random measure can be constructed as a measure functional of the historical process.

To understand this idea we return to the historical particle system. Consider a modified system in which each particle of a branching particle system evolves according to the motion process stopped at τ_B, W_B, and the additive functional is defined by $\kappa_B(dt) := 1_{[0,\tau_B]}(t)dt$. We then consider the (W_B,κ_B,Φ) particle system and superprocess.

The history of the ith particle at time t, $W_i(t)$, consists of its own trajectory pieced together with the trajectories of all its ancestors (i.e. the law of the particle is simply the motion process). The mass distribution at time t is

$$X_t(A) = \varepsilon \sum_i 1_A(W_{B,i}(t)), \quad A\in\mathcal{E},$$

where ε is the particle mass.
Then we define

$$X_{\tau_B}(A) := \varepsilon\cdot \sum_i 1(\tau_B(W_{B,i})<\infty)\cdot 1_A(W_{B,i}(\infty)),$$

$$\mathcal{O}(\tau,A) := \varepsilon \sum_i \int_0^{\tau_B} 1_A(W_{B,i}(s))ds$$

where the sum is over all particles alive at time infinity.
We can then again take the measure-valued limit as $\varepsilon \to 0$, (by analogy with the argument in Chapter 4) yielding $X_{\tau_B}, \mathcal{O}_{\tau_B}$. This leads to the following theorem.

<u>Theorem</u> <u>12.3.5.1</u> The joint Laplace functional of $(X_{\tau_B},\mathcal{O}_{\tau_B})$ is given by:

$$(12.3.5.1)\quad P_\mu \exp\{-\langle\mathcal{O}_{\tau_B},\psi\rangle-\langle X_{\tau_B},\phi\rangle\} = \exp(-\langle\mu,v\rangle),$$

$$(12.3.5.2)\quad v(x) + P_{0,x}\left(\int_0^{\tau_B}\Phi(W_s,v(W_s))ds\right) = P_{0,x}\left[\int_0^{\tau_B}\psi(W_s)ds+\phi(W_{\tau_B})\right].$$

(In fact $X_{\tau_B},\mathcal{O}_{\tau_B}$ can be defined for all coanalytic sets B.)

<u>Proof.</u> Dynkin (1991c).

What is less clear is whether or not the pair $\{X_{\tau_B},\mathcal{O}_{\tau_B}\}$ can be represented as random variables defined in terms of the canonical process $\{X_t:t\geq 0\}$. Note however that if $\{W_t\}$ denotes the Y_t-path process and Δ is an interval in $[0,\infty)$, then formally $1_\Delta(\tau_B)1_A(Y_{\tau_B}) = \int_\Delta 1_A(\pi_t W_t)A_B(W,dt)$ where $A_B(W,[0,t]) = 1(\tau_B(W_t)\leq t)$.

214

Therefore, formally, $X_{\tau_B}(A) = \int_0^\infty\int_{D_E} 1_A(\pi_t W_t)A_B(W,dt)H_t(dW)$. It is not immediately clear whether the right hand side of this expression is well-defined. However in the case in which A_B is replaced by an additive functional A which has finite support $\{t_i\}$ this is well-defined, namely,

(12.3.5.3)

$$\int_\Delta\int_{D_E} 1_A(\pi_t W_t)A(W,dt)H_t(dW) = \sum_{t_i\in\Delta} \int_{D_E} 1_A(\pi_{t_i} W_{t_i})A(W,\{t_i\})H_{t_i}(dW).$$

Dynkin (1991c) showed that this could be extended to the general case by taking limits in an appropriate sense. We briefly state the main tool used to complete this step.

Let $(L,\mathcal{L})$ be a measurable space and $\zeta\in p\mathcal{L}$. *A measure functional* of the path process with values in $M_F(L)$ is given by a measurable mapping η: $(D_E,\mathcal{D},\mathcal{D}_{s,t},W_t,P_{s,y}) \longrightarrow M_F(L)$. We say that $\eta \in \Gamma^o_W(L)$ if it satisfies

(i) $\exists$ $b\in\mathbb{R}_+$ such that $\eta\{\zeta > b\} = 0$ and $\sup_{y\in D_E} \eta\{y,\zeta \le b\} < \infty$,

(ii) $\eta\{B,\zeta<t\}$ is measurable with respect to $\mathcal{D}^{uc}_{s,t}$, $\forall$ $B\in\mathcal{L}$, $t\in\mathbb{R}_+$.

Then we define $\Gamma_W(L) := \{\eta\colon \exists\ \eta_n\in\Gamma^o_W(L),\ \eta_n\uparrow\eta\}$.

<u>Theorem 12.3.5.2</u> Let $\{H_t\}$ be the (A,c)-historical process. Then to every random measure $\eta\in\Gamma_W(L)$ there corresponds a kernel H^η from $(\Omega,\mathcal{G}^{uc})$ to $(L,\mathcal{L})$ such that: if $\eta\in\Gamma^o_W(L)$, then for every $B\in\mathcal{L}$

$$H^\eta(B\times[0,t))) = \text{medial}\lim_{n\to\infty}\sum_{\substack{i\ge1\\ i/2^n\le t}} \int_{D_E} \eta(W,B,(i-1)2^{-n}\le\zeta<i2^{-n})\, H_{i/2^n}(dW)$$

a.s.

<u>Proof.</u> Dynkin (1991c, Thm. 1.3).

<u>Example:</u> The expression (12.3.5.3) above corresponds to the special case in which

$L = E\times\mathbb{R}_+$, $\zeta(e,t)=t$, and

$$\eta(W,A,\zeta<t) = \sum_{t_i<t} 1_A(\pi_{t_i} W_{t_i})A(W,\{t_i\}), \quad A\in \mathcal{E}.$$

<u>Remark:</u> In particular Theorem 12.3.5.2 can be applied to the random measures with values in $M_F(E)$:

$$\eta_1(W,B,\zeta<t) = \int_0^t 1_B(\pi_t W_t)A(W,dt),\ B\in \mathcal{E},\ \ A(W,[0,t]) = 1(\tau(W_t)\le t);$$

215

$$\eta_2(W,B,\zeta<t) = \int_0^t\int_0^s 1_B(\pi_u W_u)duA(W,dt), \quad B\in \mathcal{E}, \quad A(W,[0,t]) = 1(\tau(W_t)\le t)$$

to yield the random variables X_{τ_B}, $\mathcal{O}_{\tau_B}$, respectively on the canonical probability space of the historical process. This was carried out by Dynkin (1991b) and he showed that the joint Laplace functional satisfies (12.3.5.1), and (12.3.5.2).

12.3.6. Application to Sample Path Behavior.

Since the historical process H_t is an enriched version of the superprocess X_t, there are many potential applications of it. As an illustration the following theorem gives a more complete statement (in a certain sense) of the modulus of continuity result of Theorem 9.3.2.2 for the special case $\beta = 1$.

Theorem 12.3.6.1. Let $Q^{H(\Delta/2,1)}_{s,\mu}$ be the law of the (2,d,1)-historical process. If $\lambda > 2$, then $Q^{H(\Delta/2,1)}_{s,\mu}$-a.s. there is a $\delta(\lambda,\omega)$ such that

$S(H_t) \subset K(\delta(\lambda),\lambda h_0) \quad \forall\ t > 0$

where $h_0(u) := [u(\log 1/u)\vee 1]^{1/2}$ and

$K(\delta,h) := \{y\in C([0,\infty),\mathbb{R}^d), |y(t)-y(s)| \le h(t-s)\ \forall\ (s,t),\ |t-s|\le\delta\}$.

Proof. See [DP, Theorem 8.7].

12.4. The Embedding of Branching Particle Systems in the Historical Process

Let H_t denote the H(A,c)-historical process. *We will assume throughout this section that the A-path process satisfies Property S (cf. 12.2.2.6) and that c=2, that is,* $\Phi(\lambda) = -\lambda^2$.

The purpose of this section is to obtain a probabilistic description of H_t for fixed t. We know that H_t consists of a Poisson number of clusters beginning at time zero which are non-extinct at time t. These clusters can be viewed as measure-valued excusions with lifetime greater than t and form a subpopulation with a "common ancestor" at time 0. In any case it suffices to obtain the probabilistic description of the cluster and three complementary descriptions will be derived. The first is an infinite genealogical tree (Theorem 12.4.4), the second is given by an embedded time-inhomogeneous branching particle system which "explodes" at time t, and the third is in terms of an initial atom having a random mass which splits into smaller and smaller atoms (with conservation of mass) as time approaches t (Theorem 12.4.6).

We can obtain a Cox cluster representation of H_t exactly as we did for X_t in Corollary 11.5.3. In particular, conditioned on H_s, H_t can be represented as H_t =

216

$\sum_j H_{s,t,y_j}$, where $\{y_j\} \subset D_E^s$ are the points of a Poisson random measure with intensity $H(ds)/(t-s)$ and $\{H_{s,t,y_j}\} \subset M_F(D_E^t)$ denote the clusters of age (t-s) associated with the points $\{y_j\}$. The cluster H_{s,t,y_j} can be identified with the subpopulation having a common ancestor with path y_j up to time s. It follows from (4.5.2) (with $\beta=1$) that the masses of the clusters $\langle H_{s,t,y_j},1\rangle$ are independent exponential random variables with mean (t-s). The law of $H_{s,t,y}$, that is, the corresponding *cluster law* is denoted by $\{\Lambda^*_{s,t;y}\}_{0\le s<t;y\in D_E^s}$. For each s,t, and y, $\Lambda^*_{s,t;y}$ is a probability measure on $M_F(D_E^t)$ which describes the distribution of paths up to time t of a subpopulation of individuals alive at time t with a common ancestor at time s whose path up to time s is given by y.

The main objective of this section is to give a probabilistic representation of the *cluster conditioned on its mass*, that is,

$$\Lambda^*_{s,t}(m,y^s;d\nu) := \Lambda^*_{s,t;y}(.\,|\langle H_{s,t,y},1\rangle = m)$$

(this will be done in Theorem 12.4.6).

Most of the results of this section are from [DP, Section 3] and some will be stated without proof. The main objective is to establish the existence of a strong embedding of a hierarchy of historical branching particle systems $\{H_t^{*\varepsilon}:\varepsilon>0\}$ (which were defined in Section 12.3.1) in the historical process H_t. We begin with a proof of the first step in this development.

Let $\mathcal{G}_t^{*\varepsilon} = \sigma\{1(0,\infty)(H_t(A)):A\in\mathcal{D}_{t-\varepsilon}\}$ and let

$$\tilde{H}_t^\varepsilon := Q_{s,\mu}^{H(A,2)}(H_t|\mathcal{G}_t^{*\varepsilon})/\varepsilon \tag{12.4.1}$$

(normalized version of conditional expectation).

Given a measure μ on D^t, let $r_{t-\varepsilon}\mu(A) := \mu(\{w:w^{t-\varepsilon}\in A\})$.

Recall that ε is the expected mass of a cluster of age ε. Therefore $r_{t-\varepsilon}\tilde{H}_t^\varepsilon$ yields a random counting measure on $\mathcal{D}_{t-\varepsilon}$.

<u>Theorem</u> <u>12.4.1.</u> Assume that the path process satisfies property S, $H_s = \mu \in M_F(D_E)^s$ and that $H_s^{*\varepsilon}$ is a Poisson random measure with intensity μ/ε. Then for $t \ge s+\varepsilon$, both $H_{t-\varepsilon}^{*\varepsilon}$ and $r_{t-\varepsilon}\tilde{H}_t^\varepsilon$ have a Poisson cluster representation of the form

$$L_{\Lambda,\{P_x\}}(\phi) = \exp\left(-\int_{E_1}(1-P_x e^{-\langle .,\phi\rangle})\Lambda(dx)\right),\quad \phi\in bp\mathcal{E}_2 \tag{12.4.2}$$

where $E_1 = D^s$, $E_2 = D^{t-\varepsilon}$, with the same intensity, μ/ε, and with the same cluster measures given by

217

$$(12.4.3)\qquad P_y(A) = Q^{*\varepsilon}_{s,\delta_y}(H^{*\varepsilon}_{t-\varepsilon} \in A),\quad A\in \mathcal{B}(M_F(D^{t-\varepsilon})),\quad y\in D^s.$$

Proof. By the analogue of Corollary 11.5.3 we have for $\phi \in bp\mathcal{D}_{t-\varepsilon}$,

$$Q_{s,\mu}(e^{-\langle \tilde{H}^{\varepsilon}_t,\phi\rangle}) = Q_{s,\mu}\left(\exp(-\langle H_{h-\varepsilon}/\varepsilon,(1-e^{-\phi})\rangle)\right)$$
$$= \exp\left(-\int V_{s,t-\varepsilon}((1-e^{-\phi})/\varepsilon)(y)\mu(dy)\right).$$

We now carry out a formal calculation using the differential equation form of the log-Laplace equation

$$-\frac{\partial V_{s,t-\varepsilon}}{\partial s} = \tilde{A}_s V_{s,t-\varepsilon} - (V_{s,t-\varepsilon})^2,\quad V_{t-\varepsilon,t-\varepsilon} = (1-e^{-\phi})/\varepsilon.$$

If $v_{s,t-\varepsilon} := 1-\varepsilon V_{s,t-\varepsilon}$, $v_{t-\varepsilon,t-\varepsilon} = e^{-\phi}$, then

$$-\frac{\partial v_{s,t-\varepsilon}}{\partial s} = \tilde{A}_s v_{s,t-\varepsilon} + \varepsilon\,((1-v_{s,t-\varepsilon})/\varepsilon))^2$$
$$= (\tilde{A}_s - 2\varepsilon^{-1})v_{s,t-\varepsilon} + 2\varepsilon^{-1}G(v_{s,t-\varepsilon}).$$

This calculation can be fully justified in its evolution form and yields

$$(12.4.4)\qquad Q^{*}_{s,\mu}(e^{-\langle \tilde{H}^{\varepsilon}_t,\phi\rangle}) = \exp\left\{-(\varepsilon)^{-1}\int (1-v_{s,t-\varepsilon})(y)\mu(dy)\right\}.$$

On the other hand by (12.2.4.5)

$$(12.4.5)\qquad \int Q^{*\varepsilon}_{s,\mu}\left(e^{-\langle H^{*\varepsilon}_{t-\varepsilon},\phi\rangle}\right)\mathcal{P}ois_{\mu/c\varepsilon}(d\mu)$$
$$= \exp\left(-(\varepsilon)^{-1}\int (1-(G_{s,t-\varepsilon}e^{-\phi})(y))\mu(dy)\right)$$

and $G_{u,t-\varepsilon}e^{-\phi}$ satisfies (12.2.4.3) with $\xi = e^{-\phi}$. Hence

$$(G_{u,t-\varepsilon}e^{-\phi})(y) = v_{u,t-\varepsilon}(y),\quad s\le u\le t-\varepsilon, \text{ by uniqueness.}$$

It follows from (12.4.4) and (12.4.5) that both $r_{t-\varepsilon}\tilde{H}^{\varepsilon}_t$ and $H^{*\varepsilon}_{t-\varepsilon}$ are Poisson cluster random measures with intensity μ/ε and cluster measures $Q^{*\varepsilon}_{s,\delta_y}(H^{*\varepsilon}_{t-\varepsilon} \in .)$.

□

Corollary 12.4.2. (a) For $\varepsilon>0$, the random measure $r_{t-\varepsilon}H_t$ is pure atomic $Q_{s,\mu}$-a.s. with $s\le t-\varepsilon$.
(b) Conditioned on $r_{t-\varepsilon}\tilde{H}^{\varepsilon}_t$, H_t is the sum of independent non-zero clusters with law $\Lambda^{*}_{t-\varepsilon,t;y}$ (cf. Theorem 12.4.6 below) one for each atom of $r_{t-\varepsilon}\tilde{H}^{\varepsilon}_t$, that is,

218

$$Q_{s,\mu}^{H(A,2)}(e^{-\langle H_t,\phi\rangle}|r_{t-\varepsilon}\tilde{H}_t^{\varepsilon})$$

$$= \exp\left\{\int \log \Lambda^*_{t-\varepsilon,t;y}(e^{-\langle.,\phi\rangle})r_{t-\varepsilon}\tilde{H}_t^{\varepsilon}(dy)\right\}, \quad \phi\in p\mathcal{D}.$$

Proof. (a) This follows from the fact that $r_{t-\varepsilon}H_t(A) = 0$, a.s. if $r_{t-\varepsilon}\tilde{H}_t^{\varepsilon}(A) = 0$.

(b) We apply Theorem 3.4.1.2 to the Cox cluster random measure (cf. 3.3.4) H_t with $I = H_{t-\varepsilon}/\varepsilon$, $E_1=E_2=D_E$, and $X(A\times B) := H_t(\{w:(w^{t-\varepsilon},w)\in A\times B\})$. In the terminology of Theorem 3.4.1.2 we have $\mathcal{G}_0 = \mathcal{G}_t^{*\varepsilon}$ and $\mathcal{G}_1 = \sigma(H_{t-\varepsilon})$.

Corollary 11.5.3 implies that H_t is a Cox cluster random measure with cluster law $P_y = \Lambda^*_{t-\varepsilon,t;y}$ and intensity $I = H_{t-\varepsilon}/\varepsilon$ is a.s. nonatomic. Therefore we may use Theorem 3.4.1.2 to conclude that

$$Q_{s,m}(r_{t-\varepsilon}H_t(A)|\sigma(H_{t-\varepsilon})\vee\mathcal{G}_t^{*\varepsilon})/\varepsilon$$

$$= \int 1_A(y)\int \mu(D_E)\Lambda^*_{t-\varepsilon,t;y}(d\mu)\tilde{X}(dy)/\varepsilon$$

$$= \tilde{X}(A).$$

Since $\tilde{X}$ is $\mathcal{G}_t^{*\varepsilon}$-measurable (see the expression for $\tilde{X}$ in Theorem 3.4.1.2) this shows that $r_{t-\varepsilon}\tilde{H}_t^{\varepsilon}(A) = \tilde{X}(A)$ a.s. Then Theorem 3.4.1.2 and the Markov property of H give the first part of (b). The expression for the conditional Laplace functional of H_t now follows from Theorem 3.4.1.2 and the equality $r_{t-\varepsilon}\tilde{H}_t^{\varepsilon} = \tilde{X}$ a.s. □

Corollary 12.4.2 would suggest that this analogy is also valid in the sense of processes. This is established in the next theorem whose proof is too long to be included here.

Theorem 12.4.3 Let $H_0 = \mu\in M_F(D_E)^0$.

(a) (*Ancestral process of age* ε.) $r_0\tilde{H}_\varepsilon^{\varepsilon}$ is a Poisson random measure with intensity $\mu/c\varepsilon$ and the process $\{r_t\tilde{H}_{t+\varepsilon}^{\varepsilon}:t\ge 0\}$ is a branching particle system with transition function given by (12.2.4.3), (12.2.4.4).

(b) (*Probabilistic representation of* H_t.)

For $0\le s\le t$, let

$$r_s\tilde{H}_t^{t-s,x} = r_s\Lambda^*_{0,t;x}(H_t|\mathcal{G}_t^{*t-s})/(t-s) \tag{12.4.6}$$

(i.e. a non-zero cluster of $r_s\tilde{H}_t^{t-s}$ starting at $x\in\mathbb{R}^d$).

Then the process $s\to r_s\tilde{H}_t^{t-s,x}$, $0\le s\le t$, has the same law as the time inhomogeneous binary branching historical particle system $\{H_{s,t}^{*\nu}:0\le s\le t\}$ with initial measure $H_{0,t}^{*\nu} = \delta_x$, $\nu(ds) = 1/(t-s)$.

Proof. [DP, Theorem 3.9].

219

We will next formulate the probabilistic description of the historical cluster first (Theorem 12.4.4) in the language of genealogical trees which are associated with the binary branching A-motions via Proposition 12.2.3.1 and then under condition S in the language of the splitting atomic measures (Theorem 12.4.5).

In the former case however we will now consider the infinite binary tree $K := \bigcup_{n=0}^{\infty} \{1,2\}^n$ and $K_\infty := \{1,2\}^{\mathbb{N}\setminus\{0\}}$. $K^* := (E^*)^K$ and $K^*_\infty := (E^*)^{K_\infty}$ where E^* is as in section 12.2.1. For $k_\infty \in K_\infty$ and $n \in \mathbb{N}$, let $[k_\infty]_n := k_1 k_2 \ldots k_n$. For $k \in K$, we say that $k < k_\infty$ if $k = [k_\infty]_n$ for some n. A typical element of K^* is denoted by $(\tau_k, W_k)_{k \in K}$.

Theorem 12.4.4 *The Infinite Genealogical Tree*

There exists a unique measurable collection $\{\Lambda^a_{t,y}: a > 0,\ t \ge 0,\ y \in D^t_E\}$ of probability measures on K^* such that the following properties hold:

(a) (i) Under $\Lambda^a_{t,y}$, τ_∂ is uniformly distributed over $[t,t+a]$ and conditioned on τ_∂, W_∂ is distributed as an A-path process starting from (t,y) stopped at time $t+\tau_\partial$.

(ii) Under $\Lambda^a_{t,y}$, conditioned on $(\tau_\partial, W_\partial)$ the translated trees $(T_1 w, T_2 w)$ are independent and follow the law $\Lambda^{a-\tau_\partial}_{t+\tau_\partial, W_\partial(t+\tau_\partial)}$.

(b) (i) $\lim_{n\to\infty} \pi_{t_{[k_\infty]_n}} W_{k_\infty}$ exists for every $k_\infty \in K_\infty$, $\Lambda^a_{t,y}$-a.s.

(ii) With $\Lambda^a_{t,y}$-probability one there exists a unique measure ϑ on K_∞ such that for every $h \in K$

$$\vartheta^a_t(\{k_\infty : h < k_\infty\}) = \lim_{\varepsilon\to 0} 2\varepsilon \left(\sum_{\substack{k\in K\\ h<k}} 1(s_k \le a-\varepsilon < t_k) \right). \tag{12.4.7}$$

Proof. Theorem 12.4.4 is proved in Le Gall (1991a, Proposition 8.1 and Theorem 8.2) and is essentially equivalent to the following.

Theorem 12.4.5. *The Branching Particle Picture*

Assume that A satisfies Property S; let c=2, and 0<s<t. Then

(a) Under the law $\Lambda^*_{0,t;x}$, $r_s \tilde{H}^{t-s}_t$ consists of $N^B_t(s)$ atoms of masses $M_1(s), \ldots, M_{N^B_t(s)}(s)$.

(i) $\{N^B_t(s): 0 \le s \le t\}$ is a time inhomogeneous Markov process with $N^B_t(0) = 1$, and

$$P(N^B_t(r_2 t)=k \mid N^B_t(r_1 t)=j) = \binom{k-1}{k-j} \left(\frac{1-r_2}{1-r_1}\right)^j \left(\frac{r_2-r_1}{1-r_1}\right)^{k-j}, \tag{12.4.8}$$

$$1 \le j \le k,\ k \in \mathbb{Z}_+,\ 0 \le r_1 < r_2 < 1,$$

220

$P(N_t^B(0)=k) = e^{-<m,1>/t}(<m,1>/t)^k/k!, \quad k\in\mathbb{Z}_+$.

(ii) Conditioned on $N_t^B(s) = k$, $M_1(s),\ldots,M_k(s)$ are i.i.d. exponential random variables with mean $(t-s)$.

(iii) Given $N_t^B(s) = k$, let τ_i $i=1,\ldots,k$ denote the splitting times of the particles. Then $\{\tau_i\}$ are i.i.d. and uniformly distributed on $[s,t]$.

(b) We have $\Lambda^*_{0,t,y}$-a.s.

(12.4.9) $\quad \varepsilon\cdot\tilde{H}_t^{\varepsilon} \longrightarrow H_t$ as $\varepsilon \to 0$ (weak convergence of measures).

<u>Proof.</u> (a) (i) The total mass of the cluster of age t, $M = M_1$, is exponentially distributed with mean t (cf. (4.5.2) with $\beta=1$). From Theorem 12.4.3(b) $N_t^B(s)$, $0\le s\le t$, is equal in law to $\langle H^{*\nu}_{s,t},1\rangle$, $0\le s<t$, and the latter is a time inhomogeneous pure birth process with birth rate at time u equal to $\nu(u)=1/(t-u)$, $s\le u<t$. Then by Theorem 12.2.4.1 the generating function $G(s,t;\theta) = E(\theta^{N_t^B(t-\varepsilon)} \mid N_t^B(s)=1)$, $s < t-\varepsilon$, satisfies the differential equation

$$\frac{-\partial G(s,t-\varepsilon;\theta)}{\partial s} = \frac{G^2-G}{t-s}, \qquad G(t-\varepsilon,t-\varepsilon;\theta)=\theta.$$

which can be solved to obtain

$$(12.4.10) \qquad G(s,t-\varepsilon;\theta) = \frac{\varepsilon\theta/(t-s)}{1-\theta(t-s-\varepsilon)/(t-s)},$$

which is the generating function of a geometric random variable with parameter $\varepsilon/(t-s)$. Letting $\varepsilon=t-rt$, $s=0$, we obtain for $0 < r < 1$

$$P(N_t^B(rt)=k\,|\,N_t^B(0)=1) = r^{k-1}(1-r), \quad k\ge 1.$$

In addition for $1 > r_2 > r_1 > 0$, conditioned on $N_t^B(r_1 t)$, $N_t^B(r_2 t)$ is distributed as the sum of $N_t^B(r_1 t)$ independent geometric random variables with parameter $[1-(r_2-r_1)/(1-r_1)]$, that is, (12.4.8) holds.

The fact that $N_t^B(0)$ is $\mathcal{Pois}(<m,1>/t)$ follows from Theorem 12.4.1.

(ii) $P(\tau_1 > r(t-s)\,|\,N_t^B(s)=1) = P(N_t^B(s+r(t-s))=1\,|\,N_t^B(s)=1) = 1-r$, and similarly $P(\min_{i\le k} \tau_i > r(t-s)\,|\,N_t^B(s)=k) = (1-r)^k$, both by (12.4.8).

(b) is proved in [DP, Theorem 3.10].

<u>Remarks:</u>

(1) Under $\Lambda^*_{0,t,y}$, the time τ_∂ of the first branch is uniformly distributed over

221

$[0,t]$. The life times τ_1 and τ_2 of its immediate descendents are, conditioned on τ_∂, independent and uniformly distributed over $[t-\tau_1,t]$, etc. The atom positions (marks) W_∂ are obtained by running A-path processes along the branches of the tree, starting from (t,y) for the ancestor.

(2) Under Property S we can recover the full tree structure (up to relabelling of the offspring as 1 and 2) at time t from the binary branching particle system $s \longrightarrow r_s\tilde{H}_t^{t-s,x}$, $0 \le s \le t$ as in the remark at the end of section 12.2.3.

On the other hand as a consequence of Theorem 12.4.4(b) the measure H_t can be reconstructed as the limit of the (suitably normalized) counting measures on the branches of the tree at $(t-\varepsilon)$. In fact the cluster law $\Lambda^*_{t,t+a;y}(B)$, $B \in \mathcal{B}(M_F(D_E)^{t+a})$ can be represented by $\Lambda^a_{t,y}\left(\int d\vartheta^a_t(k_\infty)\delta_{W_{k_\infty}} \in B\right)$ where ϑ^a_t is defined as in (12.4.7). Thus the measure H_t and the complete genealogical tree at time t can be viewed as two different representations of the same information. Each has its own advantages. For example, the tree representation is more natural when A is a bounded operator. On the other hand the branching particle representation generalizes without difficulty to the general (A,Φ)-superprocess whereas the appropriate tree may no longer be a binary tree and then its description becomes much more complicated.

(3) The infinite genealogical tree associated with super-Brownian motion is an example of a *continuum random tree* with the *leaf-tight* as defined by Aldous (1991a,c, 1992).

Finally, we come to the main result of this section which is the determination of the probabilistic structure of a cluster of given age conditioned on its mass.

<u>Theorem</u> <u>12.4.6.</u> *The Splitting Atom Process - Conditioned Cluster Law*

Under the assumptions of Theorem 12.4.5, the cluster law $\Lambda^*_{s,t;y}$ on $M_F(D_E)^t$ can be disintegrated as follows

(12.4.11) $$\Lambda^*_{s,t;y}(d\nu) = \iint \frac{1}{t-s}\, e^{-\eta/(t-s)} \Lambda^*_{s,t}(\eta,y^s;d\nu)d\eta$$

where

(12.4.12) $\Lambda^*_{s,t}(m,y^s;d\nu) := \Lambda^*_{s,t;y}(\,.\,|\langle H_t,1\rangle = m)$ *(conditioned cluster).*

$\Lambda^*_{s,t}(m,y^s;d\nu)$ is the law of a *branching atom process* $\sum_{j=1}^{N^B_t(s)} M_j(s)\delta_{y_j(s)}$, $y_j(s) \in D^s_E$, constructed as follows:

starting with one particle of mass m at time s further particles are produced by a process of subdivision. A particle of mass m_i at time $r<t$ divides with Poisson

222

rate $[m_i t/(t-r)^2]dr$ and on splitting produces two particles with masses m_i' and $m_i - m_i'$ and the ratio m_i'/m_i is uniformly distributed on [0,1]. Between divisions the atoms perform independent A-motions.

Proof. Assume that $N_t^B(0) = 1$ and let $M = M_1(0)$. We first prove that conditioned on $\{M=m\}$, $N_t^B(rt)-1$ (which represents the number of particle splits by time rt) is an inhomogeneous Poisson process with intensity $\frac{m}{t(1-r)^2}dr$. From Theorem 12.4.5 M is the sum of $N_t^B(rt)$ (geometric with mean $1/(1-r)$) masses which are independent $\mathcal{E}xp((1-r)t)$. From these facts we can compute

$$P(N_t^B(rt)=k|M=m)$$

$$= \frac{P(N_t^B(rt)=k)\cdot P(M=m|N_t^B(rt)=k)}{P(M=m)} = \left(\frac{rm}{(1-r)t}\right)^{k-1} \frac{e^{-mr/(1-r)t}}{(k-1)!}$$

$= \mathcal{P}ois(rm/(1-r)t)$. Similarly for $r_2 > r_1$,

$$P(N_t^B(r_2 t)=k|N_t^B(r_1 t)=j, M=m)$$

$$= \frac{(r_2-r_1)^{k-j}(m/t)^{k-j}}{[(1-r_2)(1-r_1)]^{k-j}(k-j)!} \exp\left(-\frac{m(r_2-r_1)}{t(1-r_1)(1-r_2)}\right),$$

that is, the conditional distribution is $\mathcal{P}ois\left(\frac{(r_2-r_1)m}{(1-r_1)(1-r_2)t}\right)$.

By Theorem 12.4.5 the joint density of the time of splitting, τ, of a particle and the masses of the two resulting particles, M_1 and M_2, starting at time $r < t$ is

$$\frac{1}{t-r}\ \frac{2\ e^{-(m_1+m_2)/(t-s)}}{(t-s)^2} \qquad r\le s<t,\quad m_i>0.$$

From this we easily obtain that the marginal density of M is $\mathcal{E}xp(t-r)$, and

$$P(M_1 \le \eta M) = \eta,\ \eta\in[0,1].$$

Refer to [DP, Theorem 3.11] for a more detailed proof. □

Corollary 12.4.7. Under the law $\Lambda^*_{s,t}(m,y^s;d\nu)$, that is, conditioned that the cluster has mass m, $\{N_t^B(s): s\le r<t\}$ is an inhomogeneous Poisson process with $N_t^B(s)=1$ and rate $mt/(t-r)^2$.

223

12.5. The Fleming-Viot-Genealogical Process

Genealogical processes have been extensively studied in recent years in the literature on mathematical population genetics (e.g. Kingman (1982), Tavaré (1984), Donnelly and Tavaré (1986, 1987), Donnelly (1991), Donnelly and Joyce 1992)). In this section we will develop the genealogical process by analogy with the historical process and also indicate how the genealogy of the Fleming-Viot process can be represented via an enriched version of the Donnelly-Kurtz infinite particle representation of section 11.3. Since much of the development follows along the same lines as for the historical process we will simply sketch the main ideas.

12.5.1. Martingale Problem Formulation.

Recall that (cf. Sect. 12.1) at the level of the Moran process we can introduce the empirical measures on histories as we did for branching particle systems. Carrying out the same limiting procedure as in section 2.7 we obtain a probability measure-valued process called the (A,γ)-*genealogical process* which is denoted by

$$G = (\tilde{\Omega},\tilde{\mathcal{G}},\tilde{\mathcal{G}}_{s,t},(G_t)_{t\geq 0},\{P^{G(A,\gamma)}_{s,\mu}\}_{s\geq 0,\mu\in M_1(D_E)})$$

where $\tilde{\Omega} := D([0,\infty),M_1(D_E))$, $G_t(\tilde{\omega}) = \tilde{\omega}(t)$, $\tilde{\mathcal{G}}_{s,t} = \bigcap_{u>t}\sigma(G_r:s\leq r\leq u)$.

Furthermore, $\{P^{G(A,\gamma)}_{s,\mu}\}_{s\geq 0,\mu\in M_1(D_E)}$ is characterized as the unique solution to the following martingale problem.

For each $\phi\in D_0(\tilde{A})$, and for $t\geq s$,

$$\text{(12.5.1.1)}\qquad Z_t(\phi) := \langle G_t,\phi(t)\rangle - \langle G_s,\phi(s)\rangle - \int_s^t \langle G_r,(\tilde{A}\phi)(r)\rangle dr$$

is a $Q^{G(A,\gamma)}_{s,G_s}$-martingale with increasing process

$$\text{(12.5.1.2)}\qquad \langle Z(\phi)\rangle_t = \gamma\int_0^t\left[\int_{D_E}\phi(r,y)^2G_r(dy)dr - \left(\int_{D_E}\phi(r,y)G_r(dy)\right)^2\right]dr.$$

The fact that this martingale problem uniquely characterizes the genealogical process can be verified by combining the remarks made in section 12.3.3 and the method of moment measures much as in the proof of Theorem 5.5.1 or by the construction of a time inhomogeneous dual process as in the proof of Theorem 5.7.1.

Remarks: The relationship between the H(A,c)-historical process and $G(A,\gamma)$-genealogical process is parallel to that between $B(A,c)$ and $S(A,\gamma)$, and the results of Theorems 8.1.1 and 8.1.2 can be obtained in an analogous manner. In other

224

words heuristically the process G_t can be obtained as H_t/m conditioned on $\{\langle H_s,1\rangle \approx m,\ 0\le s\le t\}$. In fact the arguments leading to the proof of Theorem 8.1.2 can be adapted to establish this.

12.5.2. The Particle Representation of the Fixed Time Genealogy.

The purpose of this section is to obtain a probabilistic description at a fixed time t of the $G(A,\gamma)$-genealogical process, $\{G_t\}$. *Throughout this section we assume that A satisfies property* S *and* $\gamma=2$.

In fact we will obtain a description of the evolution of the genealogy of the population starting from the time at which the entire population had a common ancestor. Two such descriptions are derived. The first (Theorem 12.5.2.4) is a splitting atom model analogous to that derived for the branching cluster in Theorem 12.4.6. The second (Theorem 12.5.2.5) is in terms of a hierarchy of Polya urns which in turn is closely related to the infinite genealogical tree.

The de Finetti representation is the key to a probabilistic description of G_t for fixed t in much the same way as the cluster representation was the key ingredient in obtaining the probabilistic representation of H_t. In this section we describe the infinite particle representation of G_t following Donnelly and Kurtz (1992).

G_t can be viewed as the Fleming-Viot process in which the mutation process is given by the A-path process. Next consider the associated $D_E^{\mathbb{N}}$-valued look-down process $\{Y_k(t): t\ge 0,\ k\in\mathbb{N}\}$ constructed exactly as in Section 11.3 except that the A-motion in E is replaced by the A-path process in D_E. The N-particle genealogical look-down process has the following MP-generator: for $\phi \in D_0(\tilde{A}^N)$

(12.5.2.1)

$$(\tilde{K}_N\phi)(s,y_1,\dots,y_N) = (\tilde{A}^N\phi)(s,y_1,\dots,y_N) + 2\sum_{i<j}[(\Theta_{ij}\phi)(s,y_1,\dots,y_N)-\phi(s,y_1,\dots,y_N)]$$

where $\tilde{A}^N$ is the N-particle path process generator and for $y_1,\dots,y_N \in D_E$,

$$(\Theta_{ij}\phi)(s,y_1,\dots,y_N) = \phi(s,y_1,\dots,y_i,\dots,y_{j-1},y_i,y_{j+1},\dots,y_N).$$

In other words at the time of a jump the particle assumes the entire past history of the particle onto which it jumps.

We can also give the following alternative description of the genealogical look-down process. At each fixed time t we consider the $(D_E)^{\mathbb{N}}$-valued random variable obtained by assigning to each $j = 1,\dots,N$ a path $W_j(t;s)$ defined by

$W_j(t;s) = W_j(t;t)$ for $s\ge t$,

$W_j(t;s) = W_{n(j,s,t)}(t;s)$ for $0\le s\le t$, where

225

$n(j,s,t) := \min\{i \le j: j \text{ is a descendent of } i \text{ in } (s,t]\}$

where j is recursively defined to be a descendent of i in (s,t] if j jumped down to i in (s,t] or if j jumped down to k in (r,t], $s<r\le t$ and k is a descendent of i in (s,r]. Under condition S, two particle paths, $W_i(t;.)$ and $W_j(t;.)$, share an identical path *only* on the interval $[0,s_{ij}]\subset[0,t]$ where s_{ij} is the last time that i and j shared a *common ancestor*.

The collection $\{\{W_j(t;.):j=1,\ldots,m,\ t\ge 0\}:m\in\mathbb{N}\}$ yields a consistent family of D_E-valued Markov processes. Furthermore under an exchangeable initial distribution $\{W_j(t;0):j\in\mathbb{N}\}$, for fixed t the D_E-valued random variables $\{W_j(t;.):j\in\mathbb{N}\}$ are exchangeable. Thus we can consider the resulting empirical measure on D_E given by

$$G_m(t,.) := \frac{1}{m}\sum_{j=1}^{m}\delta_{W_j(t;.)}$$

and following the argument at the beginning of the proof of Theorem 11.3.3 it can be verified that $\{G_m(t):t\ge 0\} \Longrightarrow \{G_t:t\ge 0\}$ as $m\to\infty$ where $\{G_t\}$ denotes the (A,2)-genealogical process.

We will next discuss the probabilistic structure of the $\{W_j(t,.):j\in\mathbb{N}\}$. Given n distinct particles $\{W_1(t;.),\ldots,W_n(t;.)\}$ at time t, for $0\le s\le t$, let $N^S_{t,n}(s)$ denote the number of distinct individuals (ancestors) at time s that have descendents among $\{W_1(t;.),\ldots,W_n(t;.)\}$ and let $\Gamma_n(s,t)$ denote the collection of indices of these $N^S_{t,n}(s)$ particles. Since $N^S_{t,n}(s)$ is monotone increasing in s, we can associate with $\{W_j(t;.)\}$ a binary branching A-system in a natural way (cf. remark at the end of section 12.2.3). For $u>0$ let $R^n(u)$ denote the equivalence relation on $\{1,\ldots,n\}$ where i and j are in the same equivalence class if and only if they have the same ancestor at time $t-u$. Let $\mathbb{E}^n$ denote the set of equivalence classes on $\{1,\ldots,n\}$. Let $D^S_{t,n}(u) := N^S_{t,n}((t-u)-)$, $0\le u\le t$, the number of equivalence classes in $R^n(u)$ (and for convenience we take right continuous versions of all processes).

<u>Theorem 12.5.2.1.</u>

(a) The $\mathbb{N}$-valued process $\{D^S_{t,n}(u):u\ge 0\}$ is a pure death process with death rates $d_k = k(k-1)$ and $N^S_{t,n}(t-0) = n$.

(b) Let $N^S_t(s) := \lim_{n\to\infty} N^S_{t,n}(s)$ denote the number of distinct ancestors at time s of the infinite set of particles $\{W_1(t;.),W_2(t;.),\ldots\}$. Then for $s<t$, $N^S_t(s) < \infty$, a.s.

(c) Let $D^S_t(u) := N^S_t((t-u)-)$. Then $\{D^S_t(u):u>0\}$ is a Markov death process started from an entrance boundary at ∞ with death rates $d_k = k(k-1)$.

(d) The process $\{R^n(.)\}$ coincides with Kingman's n-coalescent, that is, it is an

226

$\mathfrak{E}^n$-valued continuous time Markov chain with transition rates:

$q_{\xi\eta}$ = 1 if η is obtained by coalescing two of the equivalence classes of ξ

= 0 otherwise.

Proof. (Also see Dawson and Hochberg (1982, Lemma 6.5)), Donnelly (1991), Donnelly and Joyce (1992).)

(a) The times between jumps in the n-particle look-down process are i.i.d. exponential random variables with mean $1/(n(n-1))$. Therefore the time since the last look-down is exponential with mean $1/(n(n-1))$. To obtain the second-to-last look-down time we then consider the resulting (n-1)-particle look-down system and the distribution of its last look-down is exponential with mean $1/((n-1)(n-2))$. Continuing in this way we get that the time between the (k-1)st last and kth last look-down is an exponential random variable with mean $1/((n-k+1)(n-k))$. Since the times between these look-downs are also independent we conclude that $\{D^S_{t,n}(s):s\geq 0\}$ is a pure death process with death rates $d_k = k(k-1)$.

(b) Let $\tau_{n,k} := \inf\{s:D^S_{t,n}(s) = k\}$. Then $\tau_{n,k} = E_1/(n(n-1)) + \tau_{n-1,k}$ where E_1 and $\tau_{n-1,k}$ are independent random variables and E_1 has an exponential distribution with mean 1. From this we obtain the representation $\tau_{n,k} = E_1/(n(n-1)) + \ldots + E_{n-k}/((k+1)k)$ where $\{E_m\}$ are i.i.d. $\mathcal{E}xp(1)$ random variables. Since $\sum_{j=k}^{\infty} 1/((j+1)j) < \infty$, we conclude that $\lim_{n\to\infty} \tau_{n,k} < \infty$ a.s. Consequently $N^S_t(s) < \infty$ a.s. if $s < t$.

(c) This follows from (a), the consistency of the processes $\{N^S_{t,n}(.):n\in\mathbb{N}\}$ and the construction of N^S_t as the projective limit of the $\{N^S_{t,n}(.):n\in\mathbb{N}\}$.

(d) Transitions in $R^n(.)$ correspond to the coalescence of two equivalence classes and the rate is $k(k-1)$ when $D^S_{t,n}(u)=k$. In terms of the original look-down process it corresponds to a jump in a k-particle look-down process (where each of the k-particles corresponds to an equivalence class). Since in this case each pair (i,j) with $k\geq i>j\geq 1$ experiences a jump $i\to j$ (with the resulting coalescence of the ith and jth classes) after an independent $\mathcal{E}xp(1/2)$-distributed time. This implies that $R^n(.)$ is a Markov process with transition rates equal to $\{q_{\xi\eta}\}$. □

Now let $\Gamma(s,t)$ be the collection of indices of particles at time s<t that have a descendent at time t in the infinite look-down process. By Theorem 12.5.2.1(b) $\Gamma(s,t)$ is a.s. finite and is therefore associated with an equivalence relation on $\mathbb{N}$ having a finite number of equivalence classes which we denote by $R(t-s)$. In other words, $R(u)$ is the equivalence relation on $\mathbb{N}$ in which i and j belong to the same

227

equivalence class if and only if they have the same ancestor at time t-u.

Let $\mathfrak{E} \subset 2^{\mathbb{N}\times\mathbb{N}}$ denote the set of equivalence relations on $\mathbb{N}$ with the subspace topology when $2^{\mathbb{N}\times\mathbb{N}}$ is given the product topology. Then $\mathfrak{E}$ is a compact metrizable space (cf. Kingman (1982a)). A probability measure on $\mathfrak{E}$ is called *exchangeable* if it is invariant under transformations induced by permutations on $\mathbb{N}$.

From the limiting argument above we conclude that R(s) is a $\mathfrak{E}$-valued continuous time Markov chain called the *coalescent* which is characterized by the property that its restriction to {1,...,n} is the n-coalescent described above. Having identified R(.) as a coalescent we can then obtain detailed information on the probabilistic structure of G_t by using the following results of Kingman.

Theorem 12.5.2.2. Let {R(u):u>0} denote the coalescent process, D(u) denote the number of equivalence classes in R(u) and $\mathcal{R}_k := R(\tau_k)$ where $\tau_k := \inf\{u:D(u)=k\}$ (and hence $R(s) = \mathcal{R}_{D(s)}$).

Then

(a) $\{\mathcal{R}_k : k\in\mathbb{N}\}$ is a reverse time Markov chain with state space $\mathfrak{E}$ and transition probabilities

$$P(\mathcal{R}_{k-1}=\eta\,|\,\mathcal{R}_k=\xi) = 2/k(k-1)$$

for each of the $\binom{k}{2}$ equivalences η which can be obtained by coalescing two of the equivalence classes of ξ.

(b) The law $\{\mathcal{P}_k\}$ of $\{\mathcal{R}_k\}$ is for each k an exchangeable probability measure on $\mathfrak{E}$, such that

$$\mathcal{P}^k = \int_{\Sigma_{k-1}} P^x \lambda(dx)$$

where λ is the uniform distribution on Σ_{k-1} and for $x = (x_1,\ldots,x_k)$, P^x is the law of the exchangeable equivalence relation on $\mathbb{N}$ induced by i.i.d. random variables with values in {1,...,k} and distribution $(x_1,\ldots,x_k)$.

(c) $\{D(u):u\geq 0\}$ is independent of $\{\mathcal{R}_k : k\in\mathbb{N}\}$.

(d) If C_j is one of the equivalence classes in R(u), then

$$M_{C_j}(u) := \lim_{n\to\infty} \frac{1}{n}\sum_{k=1}^{n} 1_{C_j}(k) > 0, \quad \text{exists a.s.}$$

(e) Let $C_1(u),\ldots,C_{D(u)}$ denote the equivalence classes of R(u). Then for $j_1 \neq j_2 \neq \ldots \neq j_m$,

$$P(j_k \in C_{i_k}(u), k=1,\ldots,m\,|\,D(u), M_{C_1(u)},\ldots,M_{C_{D(u)}}(u)) = \prod_{k=1}^{m} M_{C_{i_k}}(u).$$

Proof. See Kingman (1982a). Note that the proofs of (d) and (e) involve an ana-

228

logue of de Finetti's theorem for exchangeable equivalence classes (cf. Kingman (1982a, Theorem 2)).□

It is also of interest to regard the genealogical development in "forward time". Note that $t-\tau_1$ = sup {s: all particles at time t have a common ancestor at time s} (in the original direction of time). Define $\bar{D}(s) = D((\tau_1-s)-)$, $\bar{R}(s) = R((\tau_1-s)-)$. Then $\bar{D}$ is a pure birth process with $\bar{D}(0) = 2$ and birth rates $k(k-1)$ and a.s. finite explosion time $\hat{\tau}_\infty := \lim_{k\to\infty} \hat{\tau}_k$ where $\hat{\tau}_k := \inf\{s:\bar{D}(s)=k\}$. We denote by $\bar{D}^t$, $\bar{R}^t$ the corresponding processes conditioned on $\{\hat{\tau}_\infty=t\}$.

Starting with the conditioned pure birth process $\bar{D}^t$ there are two alternate routes to the construction of the branching atom representation of G_t. The first of these is based on properties of the coalescent contained in the following result of Kingman.

Corollary 12.5.2.3. (a) At the time of a split, one equivalence class of mass M_C is split into two equivalence classes of masses $M_{C'} = UM_C$ and $M_{C''}=(1-U)M_C$ where U is an independent uniform (0,1) random variable.

(b) For each C, the probability that C splits is given by M_C.

Proof. Refer to Kingman (1982a, Section 5).

The application of Corollary 12.5.2.3 to G_t yields the following result.

Theorem 12.5.2.4. *Splitting Atom Process*

Assume that A satisfies condition S and let r_s be defined as in section 12.4. Then for s>0, $\{r_s G_t : s<t\}$ is a pure atomic random measure of the form

$$(12.5.2.2) \qquad r_s G_t = \sum_{j=1}^{N_t^S(s)} M_j(s)\delta_{y_j(s)}$$

with $y_j(s) \in D_E^s$, and $(M_1(s),\dots,M_{N_t^S(s)}(s)) \in \Sigma_{N_t^S(s)-1}$.

At the jump times of $N_t^S(s)$ one of the atoms splits into two smaller atoms. If a jump occurs at time s, then for $j=1,\dots,N_t^S(s-)$ the probability that it is atom j that splits is given by $M_j(s-)$. The masses of the two atoms resulting from the split of an atom of mass M_j are uniformly distributed on $[0,M_j]$, that is, $M' = UM_j$ and $M'' =(1-U)M_j$ where U is a random variable uniformly distributed on [0,1] and independent of everything else. Between jumps the atoms perform independent A-motions.

229

Sketch of the Proof.
It follows from Theorem 12.5.2.1 and the construction of the process G_t from the look-down process that the number of atoms is non-decreasing. Property S of A guarantees that there exists a one-to-one correspondence between atoms in $r_s G_t$ and equivalence classes as described by the coalescent $\bar{R}^{-t}(s)$. Furthermore by the construction of G_t using de Finetti's theorem (as in Theorem 11.3.3), the mass assigned to an equivalence class C_j, $j=1,\dots,\bar{D}^{-t}(t-s)$ coincides with $M_j(s)$ as defined above. Hence the mechanism of the splitting of the masses is the same as in Corollary 12.5.2.3. Since the mutation process is independent of the sampling process, we can also conclude that the atom locations in E (which correspond to single particle locations in the look-down process) follow A-motions. □

We now verify that the branching atom process can also be obtained from the hierarchy of Polya urns introduced in Dawson and Hochberg (1982) and also use this viewpoint to obtain another characterization of $M_j(s)$.

Corollary 12.5.2.5. *Polya Urn Hierarchy Representation*
(a) The quantities $M_j(.)$ are also given by

$$(12.5.2.3)\qquad M_j(s) = \lim_{u\uparrow t} \frac{N_t^S(s,u,j)}{\sum_k N_t^S(s,u,k)}\ ,\ j=1,\dots,N_t^S(s)$$

where $N_t^S(s,u;j)$:= number of atoms at time u with common ancestor j at time s.
(b) At each time s, The random vector $(M_{C_1}(s),\dots,M_{C_{D(s)}}(s))$ is uniformly distributed over the simplex $\Sigma_{D(s)-1}$.

Proof. (a) For fixed s>0 we consider an urn model involving $N_t^S(s)$ types. At each jump time, u, of $N_t^S(.)$, s<u<t, a particle of type j is added with probability

$$(12.5.2.4)\qquad \hat{M}_j(u) := \frac{N_t^S(s,u;j)}{\sum_k N_t^S(s,u;k)}\ ,\ j=1,\dots,N_t^S(s),$$

that is, it is an $N_t^S(s)$-type Polya urn. Then it follows from the theory of Polya urns (cf. Blackwell and Kendall (1964) or Johnson and Kotz (1977)) (and the fact that we have conditioned on $\{\tau_\infty = t\}$) that

$$(12.5.2.5)\qquad \tilde{M}_j(s) := \lim_{u\to t} \frac{N_t^S(s,u;j)}{\sum_j N_t^S(s,u;j)}\ \text{exists}$$

and that the vector $(\tilde{M}_1(s),\dots,\tilde{M}_{N_t^S(s)})$ is uniformly distributed over the simplex

230

$\Sigma_{N_t^S(s)-1}$ (cf. Blackwell and Kendall (1964)). Moreover the same result implies that at the time of a split, the mass $\tilde{M}_j(u)$ is divided exactly as in Corollary 12.5.2.3(a). Recall that if an m-type Polya urn is started with n_j initial particles of type j, then the joint distribution of the limiting proportions $(x_1,\dots,x_m)$ is an absolutely continuous distribution on Σ_{m-1} with Dirichlet density

$$f(x_1,\dots,x_m) = \frac{\Gamma(\Sigma n_i)}{\Pi\Gamma(n_i)} \Pi\, x_i^{n_i-1} \qquad \text{(cf. Athreya (1969)).}$$

Using this we can show that if a split occurs at time s, then

$$P(\text{jth class splits}\,|\,\tilde{M}_1(\tau_k-),\dots,\tilde{M}_{k-1}(\tau_k-)) = \tilde{M}_j(\tau_k-),\ j=1,\dots,k-1,$$

in other words, this is in agreement with Corollary 12.5.2.3b). Taking advantage of the fact that merging types in a Pólya urn yields a new Pólya urn, it suffices to to prove this for a two-type urn in which case we obtain

$$f_{m,n}(x,1-x) = \frac{\Gamma(m+n)}{\Gamma(m)\Gamma(n)}\, x^{m-1}(1-x)^{n-1},\ x\in(0,1).$$

Then $P(\text{1st class splits}\,|\,\tilde{M}_1=x,\tilde{M}_2=1-x)$

$$= \frac{\frac{m}{m+n} f_{m+1,n}(x,1-x)}{\frac{m}{m+n} f_{m+1,n}(x,1-x) + \frac{n}{m+n} f_{m,n+1}(x,1-x)} = x.$$

Finally it is clear from the above construction that the $\{M_j(\hat{\tau}_k-):\ j=1,\dots,k-1;\ k\in\mathbb{N}\}$ is independent of $\{\bar{D}^t(u):u>0\}$. This completes the proof that the Polya urn scheme and coalescent yield the same probabilistic mechanism, that is,

$$\mathcal{L}aw((\tilde{M}_1(s),\dots,\tilde{M}_{N_t^S(s)}):s<t) = \mathcal{L}aw((M_1(s),\dots,M_{N_t^S(s)}):s<t)$$

Then (12.5.2.3) follows since it is equivalent to (12.5.2.5) which has already been established.

(b) follows either from Theorem 12.5.2.2(b) or from the result of Blackwell and Kendall stated above. □

Note that Theorem 12.5.2.4 is an analogue of Theorem 12.4.5. On the other hand from the Polya urn systems of Corollary 12.5.2.5 together with $N_t^S(s)$ from Theorem 12.5.2.1 we can construct an *infinite Fleming-Viot genealogical tree* analogous to Theorem 12.4.4 (see also Aldous (1992, Section 4.1)).

The whole study of the genealogy of genetic models is a rapidly growing subject which is developing in a number of directions which we cannot describe here in detail. The following remarks will indicate just a few of these.

231

Remarks.

(1) The Polya urn hierarchy and atomic measures with Dirichlet distributions which arise in Corollary 12.5.2.5 are closely related to the family of Dirichlet processes introduced by Ferguson (1973) and studied by Blackwell and MacQueen (1973) and Feigen and Tweedie (1989).

(2) It is possible to extend $\{N_t^S(s):0\le s\le t\}$ to $\{N_t^S(s):s\ge 0\}$ to yield the genealogical stucture for an "infinitely old population" and then to consider for simplicity $\{N_0^S(s):s\ge 0\}$ as well as the corresponding infinite genealogy, $\tilde{G}_0$, which now becomes a random measure on $D((-\infty,0],E)$. If the A-process has a stationary distribution we denote by $W_1(t,.)$ to the corresponding stationary path process. Having done this it is easy to construct a stationary version of the full infinite system $\{W_j(t):t\in\mathbb{R}, j\in\mathbb{N}\}$. If the underlying process W_1 is ergodic, so is the resulting system (cf. Donnelly and Kurtz (1992)).

(3) Note that we can also disintegrate the law of G_t with respect to $W_1(.)$ yielding the Palm distribution of the random probability measure G_t (cf. Corollary 11.2.3). Given the infinite particle representation $\{W_k(.)\}$ the Palm measure of G_t at $y\in D_E^t$ is given by

(12.5.2.6) $\quad (P)_y = P(G_t\in.\,|W_1(t)=y).$

The representation of Corollary 12.5.2.5 and (12.5.2.4) was (implicitly) used in Dawson and Hochberg (1982) to obtain an upper bound on the carrying dimension of the $S(\Delta,\gamma)$-Fleming-Viot process in $\mathbb{R}^d$ and to verify the compact support property.

(4) A more complete development of the genealogical process in the context of Fleming-Viot processes can be found in Dawson and Hochberg (1982) and Donnelly and Kurtz (1992). For applications to population genetics refer to Kingman (1980, 1982a,b,c), Watterson (1984), Donnelly and Tavaré (1987), and Tavaré (1984, 1989).

(5) It turns out that many of the genealogical processes studied in the literature can be naturally embedded into the FV-genealogical process. Of particular relevance to mathematical population genetics is the genealogical process associated to the infinitely many alleles model in which the mutation operator A does not satisfy condition S but is bounded, as in Section 8.2. In this case the object of interest is the *number of lines of descent*, where a line of descent is defined to be the number of ancestral classes without intervening mutation (cf. Griffiths (1980), Tavaré (1984)). This is closely related to the general genetic model (cf. Cannings (1974)) which is given by a discrete time (generation) process in which there is a fixed population size N, fixed mutation probability and in which a mutation always gives rise to a new type. Let $A_N(m)$ denote the number of ancestral classes originating m

232

generations back into the past (where two individuals are said to be in the same such class if they have the same ancestor m generations back with no intervening mutation along their lines of descen)t. Let $M_{N,j}(m)$, $j=1,\dots,A_N(m)$ denote the proportion of the population in each of these classes. Donnelly and Joyce (1992) have proved under suitable technical assumptions that as $N \to \infty$, the processes $(A_N([N\sigma^{-2}.]),M_N([N\sigma^{-2}.])) \Longrightarrow (D(.),M(.))$ as $N\to\infty$ (weak convergence on D_E) where $E := \{0,1,\dots,\infty\}\times\bar{\mathfrak{A}}_\infty$ and σ^2 is the limiting variance of the offspring distribution of an individual. The limiting process $\{D(t),M(t):t\geq 0\}$ can be embedded in the infinite genealogy $\tilde{G}_0$ associated with the infinitely many alleles model. This unifying weak convergence result was the culmination of a series of papers beginning with Kingman (1982a). Here $\{D(t):t>0\}$ is the Markov death process started from an entrance boundary at ∞ with death rates $k(k+\theta-1)/2$ where θ is the mutation rate. If $\theta>0$, then $D(u)$ eventually reaches 0 (cf. Donnelly (1991), and Donnelly and Joyce (1992)). The use of Polya type urns for the study of these questions has been systematically developed in Hoppe (1987). We will now complete our study of the relation between the Fleming-Viot and branching systems with a comparison of the structures of the associated genealogical-historical processes.

12.6. Comparison of the Branching and Sampling Historical Structures.

It was established in Theorems 12.4.6 and 12.5.2.4 that the fixed time distributions of the B(A,c)-historical process $\{H_t\}$ and $S(A,\gamma)$-genealogical procees $\{G_t\}$ can be described in terms of splitting atom processes. We again assume throughout this section that A satisfies property S and $c=\gamma=2$. Then conditioned on $N_t^S(s)$, $N_t^B(s)$, these two splitting atom processes are identical. Thus the difference between the historical and genealogical processes lies in the processes describing the times of the branching, $N_t^S(s)$ and $N_t^B(s)$.

Moreover, in view of the sampling analogue of Theorem 8.1 we expect that the historical process conditioned on the given trajectory of the total mass process coincides with the genealogical process associated to a time inhomogeneous Fleming-Viot process with highly fluctuating sampling rate. However the latter may not have exactly the same genealogical structure as the Fleming-Viot process with *constant* sampling rate.

Let us begin with a closer look at the process $\{N_t^S(.)\}$. By Theorem 12.5.2.1 the sojourn times $\{\max\{s:N_t^S(s)=k\} - \min\{s:N_t^S(s)=k\} : k\in\mathbb{N}\setminus\{1\}\}$ can be represented by $\{E_k/(k(k-1)): k\in\mathbb{N}\setminus\{1\}\}$ where the $\{E_k\}$ are independent mean one exponential random

233

variables and this yields a representation of $\{N_t^S\}$ on $(\mathbb{R}^{\mathbb{N}},\mathbb{P},\{E_k\}_{k\in\mathbb{N}})$ where $\mathbb{P}$ is an infinite product of mean one exponential probability measures. The remaining time after $N_t^S(.)$ first reaches n, $R_S(n)$, is then represented by $\sum_{k=n+1}^{\infty} E_k/(k(k-1))$.

Note that $\int_{k-1}^{k} x^{-2}dx = 1/(k(k-1))$ and therefore

$$\sum_{k=n+1}^{\infty} 1/(k(k-1)) = 1/n.$$

Then $\lim_{n\to\infty} E(nR_S(n)) = 1$, and

$$\lim_{n\to\infty} n\mathrm{Var}(nR_S(n)) = \lim_{n\to\infty} n^3\cdot \sum_{k=n+1}^{\infty} \mathrm{Var}(E_k)/(k\cdot(k-1))^2 = 3.$$

We therefore get $nR_S(n) \longrightarrow 1$ in probability as $n \to \infty$ and if $N_t^S(s) := \min\{n:R_S(n)\geq s\}$, then $(t-s)N_t^S(s) \longrightarrow 1$ in probability as $s\to 0$.

Consider the B(A,c)-branching process under the assumptions that A satisfies property S and c=2. Consider a single cluster of age t conditioned to have mass m, that is, we consider N_t^B under the law $\Lambda^*_{0,t}(m,y;.)$ defined by (12.4.12). By Corollary 12.4.7, $N_t^B(s)-1$ is an inhomogeneous Poisson process with intensity $\frac{mt}{(t-s)^2}$, that is, it has independent increments and $N_t^B(s)-1$ is Poisson with mean $sm/(t(t-s))$. Hence $\lim_{s\to t} (t-s)N_t^B(s) = m$ (in probability).

We will now compare the H(A,2)-historical process H_t with m=1 to the G(A,2)-genealogical process G_t, both at t=1. From the above we have $\lim_{s\to t} (t-s)N_t^B(s) = \lim_{s\to t} (t-s)N_t^S(s) = 1$.

For $0 \leq u < \infty$ define $Z(u) = N_1^B(u/(1+u))$ (and then $N_1^B(s) = Z(s/(1-s))$). Then Z is a standard Poisson process and $\tilde{E}_n := \tilde{T}_B(n+1)-\tilde{T}_B(n)$ are i.i.d. exponential (1) r.v.'s. Then $\tilde{T}_B(n) = \sum_{i=1}^{n} \tilde{E}_i$, $T_B(n) := \inf\{u:N_1^B(u)\geq n\} = \tilde{T}_B(n)/(1+\tilde{T}_B(n))$ and therefore the remaining time is $R_B(n) := 1-T_B(n) = 1/(1+\tilde{T}_B(n))$.

Then $\tilde{T}_B(n) = \frac{1}{R_B(n)} -1$. By the law of large numbers, $\lim_{n\to\infty} \tilde{T}_B(n)/n = nR_B(n) = 1$ a.s.

Now consider $\tilde{T}_S(n) = \frac{1}{R_S(n)} -1$. Then

(12.6.1)

234

$$F_{n+1} := \tilde{T}_S(n+1) - \tilde{T}_S(n) = \frac{1}{R_S(n+1)} - \frac{1}{R_S(n)}$$

$$= \left\{ \left(\sum_{k=n+2}^{\infty} E_k/k(k-1) \right)^{-1} - \left(\sum_{k=n+1}^{\infty} E_k/k(k-1) \right)^{-1} \right\}$$

$$= \left\{ E_{n+1} \cdot \left((n+1) \sum_{k=n+2}^{\infty} E_k/k(k-1) \right)^{-1} \cdot \left(E_{n+1}/(n+1) + n \sum_{k=n+2}^{\infty} E_k/k(k-1) \right)^{-1} \right\}.$$

The main result of this section is obtained by an analysis of the two sequences $\{\tilde{E}_n\}$ and $\{F_n\}$. It concerns the marginal probability laws in $M_1(M_1(\mathbf{D}([0,1],E)))$, given by

$$P_H := \Lambda^*_{0,t}(1,y;.)\circ(H_1^{-1}) \text{ and } P_G := P^{G(A,1)}_{0,\delta_y}\circ(G_1^{-1})\left(\,\cdot \mid N_1^S(0) = 1 \right).$$

<u>Theorem 12.6.1.</u> P_H and P_G are mutually singular.

First we proceed with three auxiliary lemmas.

<u>Lemma 12.6.2.</u> $B_n := n \sum_{k=n+1}^{\infty} E_k/(k(k-1)) < \infty$, P_H-a.s, and

(a) For $0 < \nu < 1/2$

$$P(|B_n - 1| > n^{-\nu}) \le 2e^{-3(n^{1-2\nu})/4} + c/n^{1+\nu};$$

(b) $|B_n - 1| \le \mathit{const}\; n^{-\nu}$ for all $n \ge n(\nu)$ P_H-a.s.

(c) Set $A_k^N := \sum_{n=k}^{N} (B_n - 1)^2 = \sum_{n=k}^{N} n^2 \left(\sum_{\ell=n+1}^{\infty} \frac{(E_\ell - 1)^2}{\ell^2(\ell-1)^2} \right)$ where $2 \le k \le N$.

Then $A_2^N/(\frac{1}{3}\log N) \longrightarrow 1$ as $N \to \infty$, P_H-a.s. which implies that for any fixed integer $k \ge 2$, A_k^N diverges to $+\infty$ as $N \to \infty$, P_H-a.s.

<u>Outline of Proof.</u> First note that $P_H(B_n) = 1$ so that $B_n < \infty$, P_H-a.s.
(a) is then proved using a standard exponential Chebyshev estimates (see Lemma 5.1 in Dawson and Vinogradov (1992b) for more details).
(b) follows from (a) by a Borel-Cantelli argument.
(c) is straightforward and relies on a modification of Chebyshev's inequality, estimates for moments, a Borel-Cantelli argument and the verification of conditions of the two-series theorem. □

235

Lemma 12.6.3. For any integer $n \ge 2$, $\sigma\{F_k : k \ge n\} = \sigma\{E_k : k \ge n\}$.

Outline of Proof. (See Lemma 5.2 in Dawson and Vinogradov (1992b) for the details.) Obviously, $\sigma\{F_k : k \ge n\} \subset \sigma\{E_k : k \ge n\}$. The proof of the reverse inclusion is proved by induction on n (taken in descending order): from the definition of F_n (cf. (12.6.1)) we easily derive that for an arbitrary fixed integer $n \ge 2$

$$F_n = E_n B_n^{-1}\left(E_n/n + \frac{n-1}{n} B_n\right)^{-1}, \text{ and for any integer } 2 \le k \le n-1,$$

$$F_n = E_n\left(k \sum_{\ell=k+1}^{n} E_\ell/(\ell(\ell-1)) + \frac{k}{n} B_n\right)^{-1}$$

$$\times\left(E_k/k + (k-1)\sum_{\ell=k+1}^{n} E_\ell/(\ell(\ell-1)) + \frac{k-1}{n} B_n\right)^{-1}$$

Solving these equations (by induction) with respect to $E_n, E_{n-1}, \ldots, E_2$ we obtain that for $2 \le k \le n$ E_k is measurable with respect to $\sigma(F_2, \ldots, F_n, B_n)$. The rest of the proof involves the zero-one law for $\{E_n : n \ge 1\}$ and some properties of B_n established in Lemma 12.6.2. □

Lemma 12.6.4. Set

$$M_k^N := \sum_{m=k}^{N} (E_m - 1)(B_m - 1) = \sum_{m=k}^{N} (E_m - 1)\left(m \sum_{\ell=m+1}^{\infty} \frac{(E_\ell - 1)}{\ell(\ell-1)}\right), \quad 2 \le k \le N,$$

and $M_{N+1}^N := 0$. Then $\{M_{N+1}^N, M_N^N, \ldots, M_3^N, M_2^N\}$ is a square integrable martingale for any integer $N \ge 2$ with the increasing process A_n^N defined in Lemma 12.6.2.c ($A_{N+1}^N := 0$), and the ratio $M_2^N / A_2^N \to 0$ as $N \to \infty$, P_H-a.s.

Proof. The proof of this lemmas is straightforward and involves Lemma 12.6.2, Borel-Cantelli arguments and the derivation of exponential upper bounds for the probabilities of large deviations for the family of reverse martingales $\{M_n^N\}$ (similar to those obtained in Chapter 4, Section 13 of Liptser and Shiryayev (1989)). The details are given in Lemma 5.3 of Dawson and Vinogradov (1992b). □

Corollary 12.6.5. $\sum_{n=2}^{\infty}\left\{2(1-E_n)(B_n - 1) - (1+E_n)(B_n - 1)^2/2\right\}$ diverges to $-\infty$, P_H-a.s.

Note that this corollary easily follows from Lemmas 12.6.2 and 12.6.4.

Outline of Proof of Theorem 12.6.1. (See Section 6 of Dawson and Vinogradov (1992b) for a detailed proof.)

It clearly suffices to show that the laws of $\{N_1^B(s) : 0 \le s < 1\}$ and $\{N_1^S(s) : 0 \le s < 1\}$ are mutually singular. To demonstrate the latter it suffices to show that the laws of the infinite sequences $\{\tilde{E}_n\}$ and $\{F_n\}$ defined above are mutually singular. Consider the basic probability space $(\Omega,\mathcal{F},\mathbb{P})$ where $\Omega = \mathbb{R}^{\mathbb{N}}$ and $\mathbb{P}$ is the product of exponential(1) distributions and let $\{E_n\}$ be the canonical process on $(\Omega,\mathcal{F},\mathbb{P})$. Let $\mathbb{Q}$ denote the law of the $\{F_n\}$. Let $\mathcal{F}_{n,\infty} := \sigma\{E_k : k\geq n\}$ and $\mathcal{F}_\infty := \bigcap_{n\geq 2} \mathcal{F}_{n,\infty}$. Note that $\mathcal{F}_\infty$ is P_H-trivial. Then Lemma 12.6.3 implies that $\mathcal{F}_{n,\infty} = \sigma\{F_k : k\geq n\}$.

Now in order to prove that the two genealogical processes are mutually singular it suffices to prove that $\mathbb{P}$ and $\mathbb{Q}$ are mutually singular. (Actually since we wish to compare genealogies of clusters of age 1 starting with exactly one common ancestor we should actually consider $\mathbb{Q}(.|B_2>0)$ instead of $\mathbb{Q}$ but since $\mathbb{Q}(B_2>0) > 0$ this would not change in any way the following arguments.) Also since the sequences $\{E_n\}$ and $\{\tilde{E}_n\}$ both have law $\mathbb{P}$ we will supress the "$\tilde{}$" in the sequel.

Consider the function $g(y;b) := \dfrac{ny}{b((n-1)b+y)}$. We then have that

$$F_n = g(E_n, B_n)$$

where $B_n = n\sum_{k=n+1}^{\infty} E_k/(k(k-1))$ and $\mathbb{P}[B_n] = 1$. Furthermore the conditional density of F_n given $\mathcal{F}_{n+1,\infty}$ is given by

$$\text{(12.6.2)}\quad f_{F_n}(y_n|\mathcal{F}_{n+1,\infty}) = \exp\left\{\frac{-(n-1)B_n^2 y_n}{n-B_n y_n}\right\}\cdot\frac{n(n-1)\cdot B_n^2}{(n-B_n y_n)^2}, \quad \text{if } 0\leq y_n < n/B_n$$

$$= 0 \text{ otherwise.}$$

Then the conditional density of the distribution of $(F_2,\dots,F_n)$ with respect to the product measure $\prod_{j=2}^{n} f_{E_j}(y_j)dy_j$ is given by

(12.6.3)

$$R_n(y_2,\dots,y_n|\mathcal{F}_{n+1,\infty}) := f_{F_2,\dots,F_n}(y_2,\dots,y_n|\mathcal{F}_{n+1,\infty}) \Big/ \prod_{j=2}^{n} f_{E_j}(y_j)$$

$$= \prod_{j=2}^{n} \exp\left\{\frac{-j(B_j^2-1)y_j + B_j y_j(B_j-y_j)}{j-B_j y_j}\right\}\cdot\left(\frac{j(j-1)B_j^2}{(j-B_j y_j)^2}\right)\cdot 1(0\leq y_j<j/B_j).$$

If $\mathbb{Q} \ll \mathbb{P}$ with $d\mathbb{Q}/d\mathbb{P} = R$, then using a conditional expectation argument and Lemma 12.6.3 it can be verified that

$$\text{(12.6.4)}\quad R(y_2,y_3,\dots) = R_n(y_2,\dots,y_n|\mathcal{F}_{n+1,\infty})\cdot\mathbb{P}(R|\mathcal{F}_{n+1,\infty}),\ \mathbb{P}\text{-a.e. } (y_2,y_3,\dots).$$

237

Since $\mathbb{P}(R) = 1$ and $\mathcal{F}_\infty$ is $\mathbb{P}$-trivial, the martingale convergence theorem implies that $\mathbb{P}(R\,|\,\mathcal{F}_{n+1,\infty}) \longrightarrow 1$ as $n\to\infty$, $\mathbb{P}$-a.s. In turn this implies the $\mathbb{P}$-almost sure convergence of

$$(12.6.5)\quad \prod_{j=2}^{n} \exp\left\{\frac{-j(B_j^2 - 1)y_j + B_j y_j(B_j - y_j)}{j - B_j y_j}\right\}\cdot\left(\frac{j(j-1)B_j^2}{(j-B_j y_j)^2}\right)\cdot 1(0\le y_j < j/B_j)$$

on $\{R>0\}$ as $n\to\infty$, and the following representation for R:

$$(12.6.6)\quad R = \prod_{j=2}^{\infty} \exp\left\{\frac{-j(B_j^2 - 1)y_j + B_j y_j(B_j - y_j)}{j - B_j y_j}\right\}\cdot\left(\frac{j(j-1)B_j^2}{(j-B_j y_j)^2}\right)\cdot 1(0\le y_j < j/B_j).$$

Since $\mathcal{F}_\infty$ is $\mathbb{P}$-trivial, the usual singularity-equivalence dichotomy holds and $\mathbb{Q} \ll \mathbb{P}$ if and only if $\mathbb{P}(A\cap A_0) = 1$ where

(12.6.7)

$$A := \left\{0 < \prod_{j=2}^{\infty} \exp\left\{\frac{-j(B_j^2 - 1)y_j + B_j y_j(B_j - y_j)}{j - B_j y_j}\right\}\cdot\left(\frac{j(j-1)B_j^2}{(j-B_j y_j)^2}\right)\cdot 1(0\le y_j < j/B_j) < \infty\right\}$$

and $A_0 := \prod_{n=2}^{\infty} 1(0\le y_n < n/B_n) > 0$ (cf. Shiryayev (1984), p. 496 for a similar criterion). Note that for any $k \ge 1$, $\sum_n \mathbb{P}\{E_n > n/k\} < \infty$ so that $\mathbb{P}(A_0) > 0$. Since $A\cap A_0 \in \mathcal{F}_\infty$, then $\mathbb{P}(A\cap A_0) = 0$ or 1.

Taking logs we wish to show that

$$A\cap A_0 = \left\{-\infty < \sum_{n=2}^{\infty}\left(\frac{-n(B_n^2 - 1)y_n + B_n y_n(B_n - y_n)}{n - B_n y_n} + \log\left[\frac{n(n-1)B_n^2}{(n-nB_n y_n)^2}\right]\right)\cdot 1(0\le y_n < n/B_n) < \infty\right\}$$

has $\mathbb{P}$-measure zero.

Therefore in order to prove mutual singularity it suffices to verify the a.s. non-convergence of the series

$$\sum_{n=2}^{\infty} f_n(B_n, E_n)\ 1(0\le E_n < n/B_n)$$

238

$$= \sum_{n=2}^{\infty} \left(\frac{-n(B_n^2-1)y_n + B_n y_n (B_n - y_n)}{n - B_n y_n} + \log\left(\frac{n(n-1)B_n^2}{(n-nB_n y_n)^2}\right)\right) \cdot 1(0 \le y_n < n/B_n)$$

(recall that $B_n = n \sum_{k=n+1}^{\infty} E_k/(k(k-1)) \to 1$ as $n \to \infty$, $\mathbb{P}$-a.s.).

Using a Taylor expansion for f_n as a function in B_n around $B_n=1$, and some standard probabilistic arguments we can show that this is equivalent to verifying that

$$\sum_{n=2}^{N} \left[2\cdot(1-E_n)(B_n-1) - (1+E_n)(B_n-1)^2/2 \right] \tag{12.6.8}$$

does not converge $\mathbb{P}$-a.s. to a finite limit as $N \to \infty$. But this series in fact diverges to $-\infty$, P_H-a.s. (due to Corollary 12.6.5). □

<u>Remark:</u> In [DP, section 4.2] a 0-1 law was obtained for (the Palm measure) of H_t restricted to an appropriately defined germ σ-algebra which contains the information about motions and branching just immediately before time t. This was used already in these notes to obtain a precise description of the (2,d,1)-superprocess in dimensions d≥3, namely Theorem 9.3.3.3. A similar result is expected for the Fleming-Viot process. However Theorem 12.6.1 shows that we cannot obtain such results directly from the corresponding results for the superprocess as would have been the case if equivalence of P_H and P_G were satisfied.

239

REFERENCES

R.J. Adler and M. Lewin (1991) An evolution equation for the intersection local times of superdiffusions, Stochastic Analysis, Cambridge Univ. Press, 1-22.

R.J. Adler and M. Lewin (1992). Local time and Tanaka Formulae for super-Brownian motion and super stable processes, Stoch. Proc. Appl., 41, 45-68.

D.J. Aldous (1985). *Exchangeability and related topics*, Lecture Notes in Math. 1117, 2-198.

D.J. Aldous (1991a). Asymptotic fringe distributions for general families of random trees, Ann. Appl. Prob. 1, 228-266.

D.J. Aldous (1991b). The continuum random tree II: an overview, in *Stochastic Analysis,* ed. M.T. Barlow and N.H. Bingham, 23-70, Cambridge Univ. Press.

D.J. Aldous (1992). The continuum random tree III, preprint.

K.B. Athreya (1969). On a characteristic property of Pólya's urn, Stud. Sci. Math. Hung., 4, 31-35.

K.B. Athreya and P.E. Ney (1977). *Branching Processes,* Springer-Verlag.

P. Baras, M. Pierre (1984). Singularités éliminables pour des équations semi-linéaires, Ann. INst. Fourier 34, 185-206.

M.T. Barlow, S.N. Evans and E.A. Perkins (1991) Collision local times and measure-valued processes, Can. J. Math. 43, 897-938.

C. Berg, J.P.R. Christensen and P. Ressel (1984). *Harmonic Analysis on Semigroups,* Springer-Verlag.

P. Billingsley (1968). *Convergence of Probability Measures,* John Wiley.

D. Blackwell and D.G. Kendall (1964). The Martin boundary for Polya's urn scheme and an application to stochastic population growth, J. Appl. Prob. 1, 284-296.

D. Blackwell and J.B. MacQueen (1973). Ferguson distributions via Pólya urn schemes, Ann. Stat. 1, 353-355.

D. Blackwell and L.E. Dubins (1983). An extension of Skorohod's almost sure representation theorem, Proc. A.M.S. 89, 691-692.

R.M. Blumenthal and R.K. Getoor (1968). *Markov Processes and Potential Theory,* Academic Press, New York.

V.S. Borkar (1984). Evolution of interacting particles in a Brownian medium, Stochastics 14, 33-79.

A. Bose and I. Kaj (1991a) Diffusion approximation for an age-structured population, LRSP Tech. Report 148, Carleton Univ.

A. Bose and I. Kaj (1991b). Measure-valued age-structured processes, LRSP Tech. Report 161, Carleton Univ.

240

L. Breiman (1968). *Probability*, Addison-Wesley.

H. Brezis, L.A. Peletier and D. Terman (1986). A very singular solution of the heat equation with absorption, Arch. Rational Mech. Anal. 95, 185-209.

H. Brezis and L. Veron (1980). Removable singularities of some nonlinear elliptic equations, Arch. Rational Mech. Anal. 75, 1-6.

H. Brezis and A. Friedman (1983) Nonlinear parabolic equations involving measures as initial conditions, J. Math. pures et appl. 62, 73-97.

O.G. Bulycheva and A.D. Vent-tsel' (1989). On the differentiability of expectations of functionals of a Wiener process, Th. Prob. Appl. 34, 509-512.

C. Cannings (1974). The latent roots of certain Markov chains arising in genetics: A new approach 1. Haploid models. Adv. Appl. Probab. 6, 260-290.

B. Chauvin (1986a). Arbres et processus de Bellman-Harris, Ann. Inst. Henri Poincaré 22, 209-232.

B. Chauvin (1986b). Sur la propriéte de branchement, Ann. Inst. Henri Poincaré 22, 233-236.

B. Chauvin, A. Rouault and A. Wakolbinger (1989). Growing conditioned trees, Stoch. Proc. Appl. 39, 117-130.

P.L. Chow (1976). Function space differential equations associated with a stochastic partial differential equation, Indiana Univ. Math. J. 25, 609-627.

P.L. Chow (1978). Stochastic partial differential equations in turbulence related problems. In *Probabilistic Analysis and Related Topics*, Vol. 1, Academic Press.

K.L. Chung, P. Erdos and T. Sirao (1959). On the Lipschitz condition for Brownian motions, J. Math. Soc. Japan 11, 263-274.

Z. Ciesielski and S.J. Taylor (1962). First passage times and sojourn times for Brownian motion and exact Hausdorff measure of the sample path, Trans. Amer. Math. Soc. 103, 434-450.

J.T. Cox and D. Griffeath (1985). Occupation times for critical branching Brownian motions, Ann. Probab. 13, 1108-1132.

J.T. Cox and D. Griffeath (1987). Recent results on the stepping stone model,in Percolation Theory and Ergodic Theory of Infinite Particle Systems, 73-83, IMA Volume 8, ed. H. Kesten, Springer-Verlag.

J.T. Cox and D. Griffeath (1990). Mean-field asymptotics for the planar stepping stone model, Proc. London Math. Soc. 61, 189-208.

J.F. Crow and M. Kimura (1970). *An Introduction to Population Genetics*, Burgess.

C. Cutler (1984a). *Some measure-theoretic and topological results for measure-valued and set-valued stochastic processes*, Ph.D. Thesis, Carleton University.

C. Cutler (1984b). A Lebesgue decomposition theorem for random measures and random measure processes, Tech Report 23, LRSP, Carleton University.

241

Dai Yonglong (1982). On absolute continuity and singularity of random measures (in Chinese), Chinese Annals of Mathematics 3, 241-246.

Yu. Dalecky and S. Fomin (1991). *The measures and differential equations in infinite dimensional spaces*, Kluwer.

D.J. Daley and D. Vere-Jones (1988). *An Introduction to the Theory of Point Processes*, Springer-Verlag.

D.A. Dawson (1975). Stochastic evolution equations and related measure-valued processes, J. Multivariate Analysis 5, 1-52.

D.A. Dawson (1977). The critical measure diffusion, Z. Wahr. verw Geb. 40, 125-145.

D.A. Dawson (1978a). Geostochastic calculus, Canadian Journal of Statistics 6, 143-168.

D.A. Dawson (1978b). Limit theorems for interaction free geostochastic systems, Colloquia Math. Soc. J. Bolyai, 24, 27-47.

D.A. Dawson (1986a). Measure-valued stochastic processes: construction, qualitative behavior and stochastic geometry, Proc. Workshop on Spatial Stochastic Models, Lecture Notes in Mathematics 1212, 69-93, Springer-Verlag.

D.A. Dawson (1986b). Stochastic ensembles and hierarchies, Lecture Notes in Mathematics 1203, 20-37, Springer-Verlag.

D.A. Dawson (1992). Infinitely Divisible Random Measures and Superprocesses, in *Proc. 1990 Workshop on Stochastic Analysis and Related Topics, Silivri, Turkey*.

D.A. Dawson and K. Fleischmann (1991) Critical branching in a highly fluctuating random medium, Probab. Theory Rel. Fields, 90, 241-274.

D.A. Dawson and K. Fleischmann (1992). Diffusion and reaction caused by point catalysts, SIAM J. Appl. Math. 52, 163-180.

D.A. Dawson, K. Fleischmann, R.D. Foley and L.A. Peletier (1986). A critical measure-valued branching process with infinite mean, Stoch. Anal. Appl. 4, 117-129.

D.A. Dawson, K. Fleischmann, and L.G. Gorostiza, (1989). Stable hydrodynamic limit fluctuations of a critical branching particle system, Ann. Probab. 17, 1083-1117.

D.A. Dawson, K. Fleischmann and S. Roelly (1991). Absolute continuity of the measure states in a branching model with catalysts, Seminar on Stochastic processes 1990, Birkhäuser, 117-160.

D.A. Dawson and K.J. Hochberg (1979). The carrying dimmension of a stochastic measure diffusion, Ann. Prob. 7, 693-703.

D.A. Dawson and K.J. Hochberg (1982). Wandering random measures in the Fleming-Viot model, Ann. Prob. 10, 554-580.

D.A. Dawson and K.J. Hochberg (1985). Function-valued duals for measure-valued processes with applications, Contemporary Mathematics 41, 55-69.

242

D.A. Dawson, K.J. Hochberg and Y. Wu (1990). Multilevel branching systems, in Proc. Bielefeld Encounters in Mathematics and Physics 1989, World Scientific, 93-107.

D.A. Dawson and K.J. Hochberg (1991). A multilevel branching model, Adv. Appl. Prob. 23, 701-715.

D.A. Dawson, I. Iscoe and E.A. Perkins (1989). Super-Brownian motion: path properties and hitting probabilities, Probab. Th. Rel. Fields 83, 135-205.

D.A. Dawson and B.G. Ivanoff (1978). Branching diffusions and random measures. In Stochastic Processes, ed. A. Joffe and P. Ney, 61-104, Dekker, New York.

D.A. Dawson and T.G. Kurtz (1982). Applications of duality to measure-valued processes, Lecture Notes in Control and Inform. Sci. 42, 177-191.

D.A. Dawson and P. March (1992). In preparation.

D.A. Dawson and E.A. Perkins (1991). *Historical processes*, Memoirs of the American Mathematical Society 93, no. 454.

D.A. Dawson and H. Salehi (1980). Spatially homogeneous random evolutions, J. Mult. Anal. 10, 141-180.

D.A. Dawson and V. Vinogradov (1992a). Almost sure path properties of (2,d,β) superprocesses, LRSP Tech. Report 195.

D.A. Dawson and V. Vinogradov (1992b). Mutual singularity of genealogical structures of Fleming-Viot and continuous branching processes, LRSP Tech Report 204.

C. Dellacherie and P.A. Meyer (1976). *Probabilités et potentiel*, Hermann, Vol. I 1976, Vol. II 1980, Vol. III 1983, Vol. IV 1987.

A. De Masi and E. Presutti (1991). Mathematical Methods for Hydrodynamic Limits, Lecture Notes in Mathematics 1501, Springer Verlag.

P. Donnelly (1984). The transient behavior of the Moran model in population genetics, Math. Proc. Camb. Phil Soc. 95, 349-358.

P. Donnelly (1985). Dual processes and an invariance result for exchangeable models in population genetics, J. Math. Biol.

P. Donnelly (1986) Partition structures, Polya urns, the Ewens sampling formula and the ages of alleles, Theor. Pop. Biol. 30, 271-288.

P. Donnelly (1991). Weak convergence to a Markov chain with an entrance boundary: ancestral processes in population genetics, Ann. Probab. 19, 1102-1117.

P. Donnelly and P. Joyce (1992). Weak convergence of population genealogical processes to the coalescent with ages, Ann. Prob. 20, 322-341.

P. Donnelly and T.G. Kurtz (1992) The Fleming Viot measure-valued diffusion as an interactive particle system, preprint.

P. Donnelly and S. Tavaré (1986). The ages of alleles and a coalescent, Adv. Appl. Prob. 18, 1-19.

P. Donnelly and S. Tavaré (1987). The population genealogy of the infinitely-many

243

neutral alleles model, J. Math. Biol. 25, 381-391.

J.L. Doob (1984). *Classical Potential Theory and Its Probabilistic Counterpart*, Springer-Verlag.

R. Durrett (1978). The genealogy of critical branching processes, Stoch. Proc. Appl. 8, 101-116.

R. Durrett (1988). *Lecture Notes on Particle Systems and Percolation*, Wadsworth and Brooks/Cole.

E.B. Dynkin (1965). *Markov Processes*, Volumes I and II, Springer-Verlag.

E.B. Dynkin, (1988). Representation for functionals of superprocesses by multiple stochastic integrals, with applications to self intersection local times, Astérisque 157-158, 147-171.

E.B. Dynkin, (1989a). Superprocesses and their linear additive functionals, Trans. Amer. Math. Soc. , 314, 255-282.

E.B. Dynkin, (1989b). Regular transition functions and regular superprocesses, Trans. Amer. Math. Soc., 316, 623-634.

E.B. Dynkin, (1989c). Three classes of infinite dimensional diffusions, J. Funct. Anal. 86, 75-110.

E.B. Dynkin, (1991a). Branching particle systems and superprocesses, Ann. Probab., 19, 1157-1194.

E.B. Dynkin (1991b), Path processes and historical superprocesses, Probab. Th. Rel. Fields 90, 1-36.

E.B. Dynkin (1991c) A probabilistic approach to one class of nonlinear differential equations, Probab. Th. Rel. Fields 89, 89-115.

E.B. Dynkin (1991d) Additive functionals of superdiffusion processes, in *Random Walks, Brownian Motion and Interacting Particle Systems, A Festschrift in Honor of Frank Spitzer*, 269-282, R. Durrett and H. Kesten, eds., Birkhäuser.

E.B. Dynkin (1992a) Superdiffusions and parabolic nonlinear differential equations. Ann. Probab. 20, 942-962.

E.B. Dynkin (1992b). Superprocesses and partial differential equations, (1991 Wald Memorial Lectures).

E.B. Dynkin, S.E. Kuznetsov and A.V. Skorohod (1992). Branching measure-valued processes, preprint.

N. El Karoui (1985). Non-linear evolution equations and functionals of measure-valued branching processes. In *Stochastic Differential Systems*, ed. M. Metivier and E. Pardoux, Lect. Notes Control and Inf. Sci. 69, 25-34., Springer-Verlag.

N. El Karoui and S. Roelly (1991). Proprietes de martingales, explosion et representation de Lévy-Khinchine d'une classe de processus de branchement à valeurs mesures, Stoch. Proc. Appl. 38, 239-266.

N. El Karoui and S. Méléard (1990) Martingale measures and stochastic calculus,

244

Prob. Th. Rel Fields. 84, 83-101.

A. Etheridge and P. March (1991) A note on superprocesses, Probab. Theory Rel. Fields, 89, 141-147.

S.N. Ethier (1976). A class of degenerate diffusion processes occurring in population genetics, Comm. Pure Appl. Math. 29, 483-493.

S.N. Ethier (1979). Limit theorems for absorption times of genetic models, Ann. Prob. 7, 622-638.

S.N. Ethier (1981). A class of infinite-dimensional diffusions occurring in population genetics, Indiana Univ. Math. J. 30,925-935.

S.N. Ethier (1988). The infinitely-many-neutral-alleles diffusion model with ages, Adv. Appl. Prob. 22, 1-24.

S.N. Ethier (1990a) On the stationary distribution of the neutral one-locus diffusion model in population genetics, Ann. Appl. Prob. 2, 24-35.

S.N. Ethier (1990b) The distribution of the frequencies of age-ordered alleles in a diffusion model, Adv. Appl. Prob. 22, 519-532.

S.N. Ethier and R.C. Griffiths (1987). The infinitely many sites model as a measure-valued diffusion, Ann. Prob. 15, 515-545.

S.N. Ethier and R.C. Griffiths (1988). The two locus infinitely many neutral alleles diffusion model, preprint.

S.N. Ethier and R.C. Griffiths (1990) The neutral two locus model as a measure-valued diffusion, Adv. Appl. Prob.

S.N. Ethier and R.C. Griffiths (1992) The transition function of a Fleming-Viot process, preprint.

S.N. Ethier and T.G. Kurtz (1981). The infinitely many neutral alleles diffusion model, Adv. Appl. Prob. 13, 429-452.

S.N. Ethier and T.G. Kurtz (1985). *Markov processes: characterization and convergence*, Wiley.

S.N. Ethier and T.G. Kurtz (1987). The infinitely many alleles model with selection as a measure-valued diffusion, Lecture Notes in Biomathematics 70, 72-86.

S.N. Ethier and T.G. Kurtz (1990a) Coupling and ergodic theorems for Fleming-Viot processes, preprint.

S.N. Ethier and T.G. Kurtz (1990b) Convergence to Fleming-Viot processes in the weak atomic topology, Stochatic Proc. Appl. to appear.

S.N. Ethier and T.G. Kurtz (1992a) On the stationary distribution of the neutral diffusion model in population genetics, Ann. Appl. Prob. 2.

S.N. Ethier and T.G. Kurtz (1992b). Fleming-Viot processes in population genetics, preprint.

S.N. Evans (1990). The entrance space of a measure-valued Markov branching process

245

conditioned on non-extinction. Tech. Rept. 230, Dept. of Stat., Univ. of California at Berkeley.

S.N. Evans (1991) Trapping a measure-valued branching process conditioned on non-extinction, Ann. Inst. Henri Poincaré 27, 215-220.

S.N. Evans (1992) The entrance space of a measure-valued Markov branching proces conditioned on non-extinction, Can. Math. Bull., to appear.

S. Evans and E. Perkins (1990). Measure-valued Markov branching processes conditioned on non-extinction, Israel J. Math., 71, 329-337.

S. Evans and E. Perkins (1991). Absolute continuity results for superprocesses with some applications, Trans. Amer. Math. Soc., 325, 661-681.

S. Evans and E.A. Perkins (1992). Measure-valued branching diffusions with singular interaction, preprint.

W.J. Ewens (1979). *Mathematical Population Genetics*, Springer-Verlag.

K.J. Falconer (1985). *The Geometry of Fractal Sets*, Cambridge Univ. Press.

H. Federer (1969). *Geometric measure theory*, Springer-Verlag.

P.D. Feigen and R.L. Tweedie (1989). Linear functionals and Markov chains associated with Dirichlet processes, Math. Proc. Camb. Phil. Soc. 105, 579-585.

W. Feller (1951). Diffusion processes in genetics, Proc. Second Berkeley Symp., Univ. of Calif. Press, Berkeley, 227-246.

T.S. Ferguson (1973). A Bayesian analysis of some nonparametric problems, Ann. Stat. 1, 209-230.

X. Fernique (199.) Fonctions aléatoires à valeurs dans les espaces lusiniens, Expositiones Math.

R.A. Fisher (1958). *The genetic theory of natural selection*, Dover.

P.J. Fitzsimmons (1988). Construction and regularity of measure-valued branching processes, Israel J. Math. 64, 337-361.

P.J. Fitzsimmons (1991). Correction to Construction and regularity of measure-valued branching processes, Israel J. Math. 73, 127.

P.J. Fitzsimmons (1992). On the martingale problem for measure-valued Markov branching processes, in *Seminar on Stochastic Processes, 1991*, E. Cinlar, K.L. Chung and M.J. Sharpe, eds., Birkhäuser.

K. Fleischmann (1988). Critical behavior of some measure-valued processes, Math. Nachr. 135, 131-147.

K. Fleischmann and J. Gärtner (1986). Occupation time process at a critical point, Math. Nachr. 125, 275-290.

K. Fleischmann and U. Prehn (1974). Ein Grenzwertsatz für subkritische Verzweigungsprozesse mit endlich vielen Typen von Teilchen, Math. Nachr. 64, 357-362.

246

K. Fleischmann and U. Prehn (1975). Subkritische räumlich homogene Verzweigungsprozesse, Math. Nachr. 70, 231-250.

K. Fleischmann and R. Sigmund-Schultze (1977). The structure of reduced critical Galton-Watson processes, Math. Nachr. 74, 233-241.

K. Fleischmann and R. Sigmund-Schultze (1978). An invariance principle for reduced family trees of critically spatially homogeneous branching processes (with discussion), Serdica Bulg. Math. 4, 11-134.

W.H. Fleming and M. Viot (1979). Some measure-valued Markov processes in population genetics theory, Indiana Univ. Math. J. 28, 817-843.

J. Gärtner (1988). On the McKean-Vlasov limit for interacting diffusions, Math. Nachr. 137, 197-248.

R.K. Getoor (1974). *Markov processes: Ray processes and right processes,* Lecture Notes in Math. 440, Springer-Verlag.

R.K. Getoor (1975). On the construction of kernels, Sem. de Prob. IX., Lecture Notes in Mathematics 465, 441-463, Springer-Verlag.

A. Gmira, L. Veron (1984). Large time behavior of the solutions of a semilinear parabolic equation in $\mathbb{R}^N$, J. Diff. Equations 53, 258-276.

D.E. Goldberg (1989). *Genetic Algorithms in Search, Optimization, and Machine Learning,* Addison-Wesley.

L.G. Gorostiza (1981). Limites gaussiennes pour les champs aléatoires ramifiés supercritiques, Colloque CNRS *Aspects statistiques et aspects physiques des processus gaussiens,* 385-398.

L.G. Gorostiza and J.A. López-Mimbela (1990). The multitype measure branching process, Adv. Appl. Prob. 22, 49-67.

L.G. Gorostiza and J. A. López-Mimbela (1992). A convergence criterion for measure-valued processes, and application to continuous superprocesses, Prog. in Probab., Birkhäuser, to appear.

L.G. Gorostiza and S. Roelly-Coppoletta (1990) Some properties of the multitype measure branching process, Stoch. Proc. Appl. 37, 259-274.

L.G. Gorostiza, S. Roelly-Coppoletta and A. Wakolbinger (1990). Sur la persistence du processus de Dawson-Watanabe stable; intervention del la limite en temps et de la renormalization, Sém. Probab. XXIV, Lecture Notes in Math. 1426, 275-281.

L.G. Gorostiza, S. Roelly and A. Wakolbinger (1992) Persistence of critical multitype particle and measure branching processes, Prob. Th. Rel. Fields.

L.G. Gorostiza and A. Wakolbinger (1991). Persistence criteria for a class of critical branching particle systems in continuous time, Ann. Probab. 19, 266-288.

L.G. Gorostiza and A. Wakolbinger (1992). Convergence to equilibrium of critical branching particle systems and superprocesses, and related nonlinear partial differential equations, Acta Appl. Math., to appear.

R.C. Griffiths (1979) A transition density expansion for a multi-allele diffusion

247

model, Adv. Appl. Prob. 11, 310-325.

I. Gyöngy and E. Pardoux (1991). On quasi-linear stochastic partial differential equations, Probab. Th. Rel. Fields.

K. Handa (1990) A measure-valued diffusion process describing the stepping stone model with infinitely many alleles, Stoch. Proc. Appl. 36, 269-296.

T.E. Harris (1963). *The Theory of Branching Processes*, Springer-Verlag.

K.J. Hochberg (1991) Measure-valued processes: techniques and applications. In *Selected Proc. Sheffield Symp. Appl. Probab.* IMS Lecture Notes - Monograph Series 18, 212-235.

K.J. Hochberg (1986). Stochastic population theory: Mathematical evolution of a genetical model, in *New Directions in Applied and Computational Mathematics*, 101-115, Springer.

R.A. Holley and D.W. Stroock (1978). Generalized Ornstein-Uhlenbeck processes and infinite particle branching Brownian motion, Publ. R.I.M.S. Kyoto Univ. 14, 741-788.

R.A. Holley and D.W. Stroock (1979). Central limit phenomena of various interacting systems, Ann. Math. 110, 333-393.

R. Holley and T. Liggett (1975). Ergodic theorems for weakly interacting systems and the voter model, Ann. Prob. 3, 643-663.

F.M. Hoppe (1987). The sampling theory of neutral alleles and an urn model in population genetics, J. Math. Biol. 25, 123-159.

N. Ikeda, M. Nagasawa and S. Watanabe (1968), (1969). Branching Markov processes I,II,III, J. Math. Kyoto Univ. 8, 233-278, 9, 95-160.

N. Ikeda and S. Watanabe (1981). *Stochastic differential equations and diffusion processes*, North Holland.

I. Iscoe (1980). *The man-hour process associated with measure-valued branching random motions in* $\mathbb{R}^d$, Ph.D. thesis, Carleton University.

I. Iscoe (1986a). A weighted occupation time for a class of measure-valued critical branching Brownian motion, Probab. Th. Rel. Fields 71, 85-116.

I. Iscoe (1986b). Ergodic theory and a local occupation time for measure-valued branching processes, Stochastics 18, 197-143.

I. Iscoe (1988). On the supports of measure-valued critical branching Brownian motion, Ann. Prob. 16, 200-221.

S. Itatsu (1981). Equilibrium measures of the stepping stone model in population genetics, Nagoya Math. J. 83, 37-51.

K. Itô and H.P. McKean (1965). *Diffusion processes and their sample paths*, Springer-Verlag.

K. Itô (1984). *Foundations of stochastic differential equations in infinite dimensional space*, SIAM.

248

B.G. Ivanoff (1981). The multitype branching diffusion, J. Mult. Anal. 11, 289-318.

B.G. Ivanoff (1989). The multitype branching random walk: temporal and spatial limit theorems, preprint.

K. Iwata (1987). An infinite dimensional stochastic differential equation with state space C(ℝ), Prob. Th. Rel. Fields 74, 141-159.

J. Jacod (1979). *Calcul Stochastiques et Problèmes de Martingales*, LNM 714, Springer-Verlag.

J. Jacod and A.N. Shiryaev (1987). *Limit theorems for stochastic processes*, Springer-Verlag.

P. Jagers (1974). Aspects of random measures and point processes. *In Advances in Probability*, P. Ney and S. Port, eds., M. Dekker, 179-238.

P. Jagers (1975). *Branching processes with biological applications*, Wiley.

P. Jagers and O. Nerman (1984). The growth and composition of branching processes, Adv. Appl. Prob. 16, 221-259.

A. Jakubowski (1986). On the Skorohod topology, Ann. Inst. H. Poincaré B22, 263-285.

M. Jirina (1958). Stochastic branching processes with continuous state space, Czechoslovak Math. J. 8., 292-313.

M. Jirina (1964). Branching processes with measure-valued states, In. Trans. Third Prague Conf. on Inf. Th., 333-357.

A. Joffe and M. Métivier (1986). Weak convergence of sequences of semimartingales with applications to multitype branching processes, Adv. Appl. Prob. 18, 20-65.

N.L. Johnson and S. Kotz (1977). *Urn Models and Their Applications*, Wiley.

O. Kallenberg (1977). Stability of critical cluster fields, Math. Nachr. 77, 7-43.

O. Kallenberg (1983). *Random measures*, 3rd ed., Akademie Verlag and Academic Press.

N.L. Kaplan, T. Darden and R.R. Hudson (1988) The coalescent process in models with selction, Genetics 120, 819-829.

K. Kawazu and S. Watanabe (1971). Branching processes with immigration and related limit theorems, Th. Prob. Appl. 26, 36-54.

M. Kimura (1983a). *The neutral theory of molecular evolution*, Cambridge Univ. Press.

M. Kimura (1983b). Diffusion model of intergroup selection, with special reference to evolution of an altruistic character, Proc. Nat. Acad. Sci. USA 80, 6317-6321.

J.F.C. Kingman (1975). Random discrete distributions, J.R. Statist. Soc. B37, 1-22.

J.F.C. Kingman (1978). Uses of exchangeability, Ann. Probab. 6, 183-197.

J.F.C. Kingman (1980) *The mathematics of Genetic Diversity*, CBMS Regional Conf. Series in Appl. Math. 34, SIAM.

249

J.F.C. Kingman (1982a). The coalescent, Stoch. Proc. Appl. 13, 235-248.

J.F.C. Kingman (1982b) On the genealogy of large populations, J. Appl. Prob. 19A, 27-43.

J.F.C. Kingman (1982c). Exchangeability and the evolution of large populations, in *Exchangeability in Probability and Statistics*, eds. G. Koch and F. Spizzichino, 97-112, North Holland.

F. Knight (1981). *Essentials of Brownian Motion and Diffusion*, Amer. Math. Soc., Providence.

N. Konno and T. Shiga (1988). Stochastic differential equations for some measure-valued diffusions, Prob. Th. Rel Fields 79, 201-225.

P. Kotelenez (1988). High density limit theorems for nonlinear chemical reactions with diffusion, Probab. Th. Rel. Fields 78, 11-37.

P. Kotelenez (1989). A class of function and density valued stochastic partial differential equations driven by space-time white noise, preprint.

S. Krone (1990) Local times for superdiffusions (Abstract), Stoch. Proc. Appl. 35, 199-200.

N.V. Krylov and B.L. Rozovskii (1981). Stochastic evolution equations, J. Soviet Math. (Itogi Nauki i Techniki 14), 1233-1277.

H. Kunita (1986). Stochastic flows and applications, Tata Institute and Springer-Verlag.

H. Kunita (1990). Stochastic flows and stochastic differential equations, Cambridge Univ. Press.

T.G. Kurtz and D. Ocone (1988). A martingale problem for conditional distributions and uniqueness for the nonlinear filtering equations, Ann. Probab.

T.G. Kurtz (1981). *Approximation of Population Processes*, SIAM.

S.E. Kuznetsov (1984). Nonhomogeneous Markov processes, J. Soviet Math. 25, 1380-1498.

J. Lamperti (1967). Continuous state branching processes, Bull. Amer. Math. Soc. 73, 382-386.

T.-Y. Lee (1990).Some limit theorems for critical branching Bessel processes and related semilinear differential equations, Probab. Th. Rel. Fields 84, 505-520.

J.F. Le Gall (1987). Exact Hausdorff measure of Brownian multiple points, in *Seminar on Stochastic Processes, 1986*, E. Cinlar, K.L. Chung and R.K. Getoor, eds., Birkhäuser.

J.F. Le Gall (1989a). Marches aléatoires, mouvement brownien et processes de branchement, L.N.M. 1372, 258-274.

J.F. Le Gall (1989b). Une construction de certains processus de Markov à valeurs mesures, C.R. Acad. Sci. Paris 308, Série I, 533-538.

250

J.F. Le Gall (1991a). Brownian excursions, trees and measure-valued branching processes, Ann. Probab. 19., 1399-1439.

J.F. Le Gall (1991b). A class of path-valued Markov processes and its applications to superprocesses, preprint.

Y. Le Jan (1989). Limites projectives de processus de branchement markoviens, C.R. Acad. Sci. Paris 309 Série 1, 377-381.

Y. Le Jan (1991). Superprocesses and projective limits of branching Markov processes, Ann. Inst. H. Poincaré 27, 91-106.

C. Léonard (1986). Une loi des grands nombres pour des systèmes de diffusions avec interaction à coefficients non bornés, Ann. Inst. Henri Poincaré 22, 237-262.

Z.-H. Li (1992). A note on the multitype measure branching process, Adv. Appl. Prob. 24, 496-498.

A. Liemant, K. Matthes and A. Wakolbinger (1988). *Equilibrium Distributions of Branching Processes*, Akademie-Verlag, Berlin, and Kluwer Academic Publ., Dordrecht.

T.M. Liggett (1985). *Interacting Particle Systems*, Springer-Verlag.

R. Sh. Liptser and A.N. Shiryayev (1989). *Theory of Martingales*, Kluwer.

R.A. Littler and A.J. Good (1978). Ages, extinction times and first passage probabilities for a multiallele diffusion model with irreversible mutation, Theor. Pop. Biol. 13, 214-225.

L. Liu and C. Mueller (1989). On the extinction of measure valued critical branching Brownian motion, Ann. Probab. 17, 1463-1465.

R. Marcus (1979). Stochastic diffusion on an unbounded domain, Pacific J. Math. 84, 143-153.

G. Matheron (1975). *Random sets and integral geometry*, Wiley.

K. Matthes, J. Kerstan and J. Mecke (1978). *Infinitely Divisible Point Processes*, Wiley.

H.P. McKean (1969). *Stochastic Integrals*, Academic Press.

S. Méléard and S. Roelly-Coppoletta (1990). A generalized equation for a continuous measure branching process, L.N. Math. 1390, 171-186.

S. Méléard and S. Roelly (1991). Discontinuous measure-valued branching processes and generalized stochastic equations, Math. Nachr. 154, 141-156.

M. Métivier (1982). *Semimartingales*, W. de Gruyter.

M. Métivier and J. Pellaumail (1980). *Stochastic integration*, Academic Press.

M. Métivier (1984). Convergence faible et principe d'invariance pour des martingales à valeurs dans des espaces de Sobolev, Ann. Inst. Henri Poincaré 20, 329-348.

M. Métivier (1985). Weak convergence of measure-valued processes using Sobolev-

251

imbedding techniques, L.N. Math. 1236, 172-183.

M. Métivier (1986). Quelques problemes liés aux systèmes infini de particules et leur limites, Springer L.N.M., 426-446.

M. Métivier and M. Viot (1987). On weak solutions of stochastic partial differential equations, Springer L.N.M. 1322, 139-150.

N.G. Meyers (1970). A theory of capacities for potentials of functions in Lebesgue classes, Math. Scand. 26, 255-292.

C. Mueller (1991a). Limit results for two stochastic partial differential equations, Stochastics 37, 175-199.

C. Mueller (1991b) On the supports of solutions to the heat equation with noise, Stochastics, 37, 225-246.

C. Mueller (1991). Long time existence for the heat equation with noise, Probab. Th. Rel. Fields 90, 505-518.

C. Mueller and E.A. Perkins (1991). The compact support property for solutions to the heat equation with noise, preprint.

J. Neveu (1964). *Bases Mathématiques du Calcul des Probabilités*, Masson et. Cie, Paris.

J. Neveu (1975). *Discrete-Parameter Martingales*, North-Holland.

J. Neveu (1986). Arbres et processus de Galton-Watson, Ann. Inst. H. Poincaré 22, 199-207.

J. Neveu and J.W. Pitman (1980). The branching process in a Brownian excursion, LNM 1372, 248-257, Springer-Verlag.

J.M. Noble (1992). Evolution equations with random potential, private communication.

M. Notohara and T. Shiga (1980). Convergence to genetically uniform state in stepping stone models of population genetics, J. Math. Biol. 10, 281-294.

K. Oelschläger (1989). On the derivation of reaction-diffusion equations as limit dynamics of systems of moderately interacting stochastic processes, Probab. Th. Rel. Fields 82, 565-586.

K. Oelschläger (1990) Limit theorems for age-structured populations, Ann. Probab. 18, 290-318.

T. Ohta and M. Kimura (1973). A model of mutation appropriate to estimate the number of electrophoretically detectable alleles in a finite population, Genet. Res. 22, 201-204.

E. Pardoux (1975). Equations aux dérivées partielles stochastiques non lineaires monotone. Etude des solutions forte de type Ito, Thèse, Univ. de Paris Sud, Orsay.

K.R. Parthasarathy (1967). *Probability Measures on Metric Spaces*, Academic Press.

A. Pazy (1983). *Semigroups of linear operators and applications to partial differential equations*, Springer-Verlag.

252

E.A. Perkins (1988). A space-time property of a class of measure-valued branching diffusions, Trans. Amer. Math. Soc., 305, 743-795.

E.A. Perkins (1989). The Hausdorff measure of the closed support of super-Brownian motion, Ann. Inst. Henri Poincaré 25, 205-224.

E.A. Perkins (1990). Polar sets and multiple points for super-Brownian motion, Ann. Probab. 18, 453-491.

E.A. Perkins (1991a) On the continuity of measure-valued processes, Seminar on Stochastic Processes 1990, Birkhauser, 261-268.

E.A. Perkins (1991b) Conditional Dawson-Watanabe processes and Fleming-Viot processes, Seminar in Stochastic Processes, 1991, Birkhauser, 142-155.

E.A. Perkins (1992). Measure-valued branching diffusions with spatial interactions, Probab. Th. Rel. Fields, to appear.

P. Priouret (1974). Processus de diffusion et equations differentielles stochastiques, Lecture Notes in Math. 390, 38-111, Springer-Verlag.

P. Protter (1990). *Stochastic Integration and Differential Equations*, Springer-Verlag.

M. Reimers (1986). *Hyper-finite methods for multi-dimensional stochastic processes*, Ph.D. thesis, U.B.C.

M. Reimers (1987). Hyperfinite methods applied to the critical branching diffusion, Probab. Th. Rel. Fields 81, 11-27.

M. Reimers (1989). One dimenional stochastic partial differential equations and the branching measure diffusion, Probab. Th. Rel. Fields 81, 319-340.

M. Reimers (1992) A new result on the support of the Fleming-Viot process proved by non-standard construction, preprint.

P. Ressel and W. Schmidtechen (1991). A new characterization of Laplace functionals and probability generating functionals, Prob. Th. Rel. Fields 88, 195-213.

D. Revuz and M. Yor (1991). *Continuous Martingales and Brownian Motion*, Springer-Verlag.

S. Roelly-Coppoletta (1986). A criterion of convergence of measure-valued processes: application to measure branching processes, Stochastics 17, 43-65.

S. Roelly and S. Méléard (1990) Interacting branching measure processes, *Proceedings: Stochastic Partial Differential Equations and Applications III*, Trento, Italy, Springer-Verlag.

S. Roelly-Coppoletta and A. Rouault (1989). Processus de Dawson-Watanabe conditioné par le futur lointain, C.R. Acad. Sci. Paris 309, 867-872.

S. Roelly and A. Rouault (1990). Construction et propriétés de martingales des branchements spatiaux interactifs, Int. Stat. Rev. 58, 173-189.

C.A. Rogers (1970). *Hausdorff measures*, Cambridge Univ. Press.

253

L.C.G. Rogers and D. Williams (1987). *Diffusions, Markov processes and Martingales, Vol. 2, Itl Calculus*, Wiley.

J. Rosen (1990). Renormalization and limit theorems for self-intersections of super-processes, preprint.

B.L. Rozovskii (1990) *Stochastic Evolution Equations*, D. Reidel.

S.M. Sagitov (1990). Multi-dimensional critical branching processes generated by large numbers of identical particles, Th. Prob. Appl. 35.

K. Sato (1976a) Diffusion processes and a class of Markov chains related to population genetics, Osaka J. Math. 13, 631-659.

K. Sato (1976b). A class of Markov chains related to selection in population genetics, J. Math. Soc. Japan 28, 621-636.

K. Sato (1978) Convergence to a diffusion of a multi-allelic model in population genetics, Adv. Appl. Prob. 10, 538-562.

K.I. Sato (1983). Limit diffusion of some stepping stone models, J. Appl. Prob. 20, 460-471.

S. Sawyer (1976). Results for the stepping stone model for migration in population genetics, Ann. Prob. 4, 699-728.

S. Sawyer (1979). A limit theorem for patch size in a selectively neutral migration model, J. Appl. Prob. 16, 482-495.

M.J. Sharpe (1988). *General theory of Markov processes*, Academic Press.

B. Schmuland (1991). A result on the infinitely many neutral alleles diffusion model, J. Appl. Prob.

T. Shiga (1980) An interacting system in population genetics, J. Math. Kyoto Univ. 20, 213-242.

T. Shiga (1981) Diffusion processes in population genetics, J. Math. Kyoto Univ. 21, 133-151.

T. Shiga (1982) Wandering phenomena in infinite allelic diffusion models, Adv. Appl. Prob. 14, 457-483.

T. Shiga (1982), Continuous time multi-allelic stepping stone models in population genetics, J. Math. Kyoto Univ. 22, 1-40.

T. Shiga (1985) Mathematical results on the stepping stone model in population genetics, in *Population Genetics and Molecular evolution*, T. Ohta and K. Aoki, eds., Springer-Verlag.

T. Shiga (1987a). Existence and uniqueness of solutions for a class of non-linear diffusion equations, J. Math. Kyoto Univ. 27-2, 195-215.

T. Shiga (1987b). A certain class of infinite dimensional diffusion processes arising in population genetics, J. Math. Soc. Japan 30, 17-25.

254

T. Shiga (1988) Stepping stone models in population genetics and population dynamics, in *S. Albeverio et al (eds.) Stochastic Processes in Physics and Engineering,* 345-355.

T. Shiga (1990a) A stochastic equation based on a Poisson system for a class of measure-valued diffusions, J. Math. Kyoto Univ. 30(1990), 245-279.

T. Shiga (1990b) Two contrastive properties of solutions for one-dimensional stochastic partial differential equations, preprint.

T. Shiga and A. Shimizu (1980) Infinite dimensional stochastic differential equations and their applications, J. Math. Kyoto Univ. 20, 395-416.

T. Shiga and K. Uchiyama (1986). Stationary states and the stability of the stepping stone model involving mutation and selection, Prob. Th. Rel. Fields 73, 87-117.

N. Shimakura (1985). Existence and uniqueness of solutions for a diffusion model of intergroup selection, J. Math. Kyoto Univ. 25, 775-788.

A. Shimizu (1985). Diffusion approximation of an infinite allele model incorporating gene conversion, in *Population genetics and molecular evolution*, eds. T. Ohta and K. Aoki. Japan Sci. Soc. Press and Springer-Verlag.

A. Shimizu (1987). Stationary distribution of a diffusion process taking values in probability distributions on the partitions, Lecture Notes in Biomath. 70, 100-114.

A. Shimizu (1990). A measure valued diffusion process describing an n locus model incorporating gene conversion, Nagoya Math. J. 119, 81-92.

A.N. Shiryayev (1984). *Probability*, Springer-Verlag.

M.L. Silverstein (1969). Continuous state branching semigroups, Z. Wahr. verw. Geb. 14, 96-112.

D.W. Stroock and S.R.S. Varadhan (1979). *Multidimensional diffusion processes,* Springer-Verlag.

S. Sugitani (1987). Some properties for the measure-valued branching diffusion processes, J. Math. Soc. Japan 41, 437-462.

A.S. Sznitman (1991). Topics in Propagation of Chaos, *Ecole d'été de Probabilités* de *Saint Flour*, L.N.M. 1464, 165-251.

S.J. Taylor (1966). Multiple points for the sample paths of the symmetric stable process, Z. Wahr. verw. Geb. 5, 247-258.

S. Tavaré (1984). Line of descent and genealogical processes, and their applications in population genetics models, Theor. Pop. Biol. 26, 119-164.

S. Tavaré (1989). The genealogy of the birth, death and immigration process, in *Mathematical Evolutionary Theory*, ed. M.W. Feldman, 41-56.

R. Tribe (1989). *Path properties of superprocesses*, Ph.D. thesis, U.B.C.

R. Tribe (1991). The connected components of the closed support of super Brownian motion, Probab. Th. Rel. Fields 89, 75-87.

255

R. Tribe (1992). The behavior of superprocesses near extinction, Ann. Probab. 20, 286-311.

J. Vaillancourt (1987). *Interacting Fleming-Viot processes and related measure-valued processes*, Ph.D. thesis, Carleton University.

J. Vaillancourt (1988). On the existence of random McKean-Vlasov limits for triangular arrays of exchangeable diffusions, Stoch. Anal.

J. Vaillancourt (1990a). Interacting Fleming-Viot processes, Stoch. Proc. Appl. 36, 45-57.

J. Vaillancourt (1990b). On the scaling theorem for interacting Fleming-Viot processes, Stoch. Proc. Appl. 36, 263-267.

S.R.S. Varadhan (1984). *Large Deviations and Applications*, CBMS-NSF Regional Conf. 46, SIAM.

A.D. Venttsel' (1985). Infinitesimal characteristics of Markov processes in a function space which describes the past, Th. Prob. Appl. 30, 661-676.

A.D. Vent-tsel (1989). Refinement of the functional central limit theorem for stationary processes, Th. Prob. Appl. 34, 402-415.

L. Véron (1981). Singular solutions of some nonlinear elliptic equations, Nonlinear Anal. Theory, Math. Appl. 5, 225-242.

M. Viot (1976). *Solutions faibles d'equations aux dwrivwes partielles non lineaires*, Thèse, Univ. Pierre et Marie Curie, Paris.

J.B. Walsh (1986). An introduction to stochastic partial differential equations, in P.L. Hennequin (ed.), *Ecole d'été de Probabilités de Saint-Flour XIV- 1984*, L.N.M. 1180, 265-439.

F.S. Wang (1982a). Diffusion approximations of age-and-position dependent branching processes, Stoch. Proc. Appl. 13, 59-74.

F.S. Wang (1982b). Probabilities of extinction of multiplicative measure diffusion processes with absorbing boundary, Indiana Univ. Math J. 31, 97-107.

H. Watanabe (1988). Averaging and fluctuations for parabolic equations with rapidly oscillating random coefficients, Probab. Th. Rel. Fields 77, 359-378.

H. Watanabe (1989). On the convergence of partial differential equations of parabolic type with rapidly oscillating coefficients, Appl. Math. Optim. 20, 81-96.

S. Watanabe (1968). A limit theorem of branching processes and continuous state branching, J. Math. Kyoto Univ. 8, 141-167.

S. Watanabe (1969). On two dimensional Markov processes with branching property, Trans. Amer. Math. Soc. 136, 447-466.

G.A. Watterson (1976a) Reversibility and the age of an allele I. Moran's infinitely many neutral alleles model, Theor. Pop. Biol. 10, 239-253.

G.A. Watterson (1976b). The stationary distribution of the infinitely many neutral alleles model, J. Appl. Prob. 13, 639-651.

256

G.A. Watterson (1984) Lines of descent and the coalescent, Theor. Pop. Biol. 10, 239-253.

A.D. Wentzell (1992). On differentiability of the expectation of functionals of a Markov process, Stochastics and Stochastic Reports 39, 53-65.

S. Wright (1943). Isolation by distance, Genetics 28, 114-138.

S. Wright (1949) Adaptation and selection. In *Genetics, Paleontology and Evolution*, ed. G.L. Jepson et al, 365-389, Princeton Univ. Press.

Y. Wu (1991). Asymptotic behavior of two level branching processes, LRSP Tech. Report 179, Carleton Univ.

Y. Wu (1991). Multilevel birth and death particle system and its continuous diffusion, LRSP Tech. Report 186, Carleton Univ.

Y. Wu (1992). *Dynamic particle systems and multilevel measure branching processes.* Ph.D. thesis, Carleton University.

T. Yamada and S. Watanabe (1971). On the uniqueness of solutions of stochastic differential equations, J. Math. Kyoto Univ. 11, 155-167, 553-563.

M. Yor (1974). Existence et unicité de diffusions à valeurs dans un espace de Hilbert, Ann. Inst. Henri Poincaré 10, 55-88.

U. Zähle (1988a). Self-similar random measures I. Notion,carrying Hausdorff dimension and hyperbolic distribution, Probab. Th. Rel. Fields 80, 79-100.

U. Zähle (1988b). The fractal character of localizable measure-valued processes I - random measures on product spaces, Math. Nachr. 136, 149-155.

U. Zähle (1988c). The fractal character of localizable measure-valued processes II, Localizable processes and backward trees, Math. Nachr. 137, 35-48.

U. Zähle (1988d). The fractal character of localizable measure-valued processes III. Fractal carrying sets of branching diffusions, Math. Nachr. 138, 293-311.

H. Zessin (1983). The method of moments for random measures, Z. Wahr. verw. Geb. 62, 395-409.

V.M. Zolotarev (1957). More exact statements of several theorems in the theory of branching processes, Th. Prob. Appl. 2, 245-253.

257

Subject Index

258

259

260

Part II

Edwin Perkins: Dawson–Watanabe Superprocesses and Measure-valued Diffusions

Originally published in: *Ecole d'Eté de Probabilités de Saint-Flour XXIX – 1999*, Lecture Notes in Mathematics, Vol. **1781**, 125–131, DOI: 10.1007/b93152,

Table of Contents

Outline of Lectures at St. Flour

1. Introduction
2. Particle Sytems and Tightness (II.1-II.4)
3. The Martingale Problem and Non-linear Equation (II.4-II.8)
4. Path Properties of the Support of Super-Brownian Motion (III.1-III.3)
5. Polar Sets (III.5-III.6)
6. Interactive Drifts (IV)
7. Spatial Interactions 1.
   Stochastic Integration on Trees and a Strong Equation (V.1-V.3)
8. Spatial Interactions 2.
   Pathwise Existence & Uniqueness, and the Historical Martingale Problem (V.4-V.5)
9. Interacting Particle Systems 1. The Voter Model
10. Interacting Particle Systems 2. The Contact Process

Note. A working document with Ted Cox and Rick Durrett was distributed to provide background material for lectures 9 and 10.

Glossary of Notation

$\alpha \sim t$	$\lvert\alpha\rvert/N \le t < (\lvert\alpha\rvert+1)/N = \zeta^\alpha$, i.e., α labels a particle alive at time t
$\lvert A\rvert$	Lebesgue measure of A
A^δ	the set of points less than a distance δ from the set A
$A^g\phi$	$A\phi + g\phi$
$\hat{A}$	weak generator of path-valued Brownian motion–see Lemma V.2.1
$\bar{A}_{\tau,m}$	see Proposition V.2.6
$\vec{A}$	generator of space-time process–see prior to Proposition II.5.8
$\xrightarrow{bp}$	bounded pointwise convergence
$b\mathcal{E}$	the set of bounded $\mathcal{E}$-measurable functions
BSMP	a cadlag strong Markov process with $x \mapsto P^x(A)$ Borel measurable for each measurable A in path space
$\mathcal{B}(E)$	the Borel σ-field on E
C	$C(\mathbb{R}^d)$
$C(E)$	continuous E-valued functions on $\mathbb{R}_+$ with the topology of uniform convergence on compacts
$C_b(E)$	bounded continuous E-valued functions on $\mathbb{R}_+$ with the supnorm topology
$C_b^k(\mathbb{R}^d)$	functions in $C_b(\mathbb{R}^d)$ with bounded continuous partials of order k or less
$C_b^\infty(\mathbb{R}^d)$	functions in $C_b(\mathbb{R}^d)$ with bounded continuous partials of any order
$C_K(E)$	continuous functions with compact support on E with the supnorm topology
$C_\ell(E)$	continuous functions on a locally compact E with a finite limit at ∞
$C(g)(A)$	the g-capacity of a set A–see prior to Theorem III.5.2
$\mathcal{C}$	Borel σ-field for $C = C(\mathbb{R}^d)$
$\mathcal{C}_t$	sub-σ-field of $\mathcal{C}$ generated by coordinate maps y_s, $s \le t$
$\stackrel{\mathcal{D}}{=}$	equal in distribution
$D(E)$	the space of cadlag paths from $\mathbb{R}_+$ to E with the Skorokhod J_1 topology
D^s	the set of paths in $D(E)$ which are constant after time s
D_{fd}	smooth functions of finitely many coordinates on $\mathbb{R}_+ \times C$ –see Example V.2.8
$D(n,d)$	space of $\mathbb{R}^{n\times d}$-valued integrands–see after Proposition V.3.1
Δ	cemetary state added to E as a discrete point
$\mathcal{D}(A)$	domain of the weak generator A–see II.2 and Proposition II.2.2
$\mathcal{D}(\vec{A})_T$	domain of weak space-time generator–see prior to Proposition II.5.7
$\mathcal{D}(\hat{A})$	domain of the weak generator for path-valued Brownian motion –see Lemma V.2.1
$\mathcal{D}(\Delta/2)$	domain of the weak generator of Brownian motion–see Example II.2.4
$\mathcal{D}$	the Borel σ-field on the Skorokhod space $D(E)$
$(\mathcal{D}_t)_{t\ge 0}$	the canonical right-continuous filtration on $D(E)$
$e_\phi(W)$	$e^{-W(\phi)}$
$\mathcal{E}$	the Borel σ-field on E
$\mathcal{E}_+$	the non-negative $\mathcal{E}$-measurable functions
$\hat{E}$	$\{(t, y(\cdot \wedge t)) : y \in D(E), t \ge 0\}$
$f_\beta(r)$	r^β if $\beta > 0$, $(\log 1/r)^{-1}$ if $\beta = 0$
$\hat{\mathcal{F}}$	$\mathcal{F} \times \mathcal{B}(C(\mathbb{R}^d))$
$\hat{\mathcal{F}}_t$	$\mathcal{F}_t \times \mathcal{C}_t$
$\hat{\mathcal{F}}_t^*$	the universal completion of $\hat{\mathcal{F}}_t$

$\mathcal{F}_X$	the Borel σ-field on Ω_X
$g_\beta(r)$	$r^{-\beta}$ if $\beta > 0$, $1 + (\log 1/r)^+$, if $\beta = 0$ and 1, if $\beta < 0$
$G_\varepsilon\phi$	see (IV.3.4)
$G(f,t)$	$\int_0^t \sup_x P_s f(x) ds$
$G(X)$	$\cup_{\delta>0}\mathrm{cl}\{(t,x) : t \geq \delta, x \in S(X_t)\}$, the graph of X
$h - m$	the Hausdorff h-measure–see Section III.3
$h(r)$	Lévy's modulus function $(r\log(1/r))^{1/2}$
$h_d(r)$	$r^2 \log^+ \log^+ 1/r$ if $d \geq 3$, $r^2(\log^+ 1/r)(\log^+ \log^+ \log^+ 1/r)$ if $d = 2$
$H_t^{s,y}$	the H_t measure of $\{w : w = y \text{ on } [0,s]\}$, $s \leq t$, $y(\cdot) = y(\cdot \wedge s)$
$\overline{\mathcal{H}}^{bp}$	the bounded pointwise closure of $\mathcal{H}$
$\mathcal{H}_+$	the set of non-negative functions in $\mathcal{H}$
I	$\bigcup_{n=0}^{\infty} \mathbb{N}^{\{0,\ldots,n\}} = \{(\alpha_0,\ldots,\alpha_n) : \alpha_i \in \mathbb{N}, n \in \mathbb{Z}_+\}$
IBSMP	time inhomogeneous Borel strong Markov process–see after Lemma II.8.1
$I(f,t)$	stochastic integral of f on a Brownian tree–see Proposition V.3.2
$\mathcal{K}$	the compact subsets of $\mathbb{R}^d$
Lip_1	Lipschitz continuous functions with Lipschitz constant and supnorm ≤ 1
(LE)	Laplace functional equation–see prior to Theorem II.5.11
$(LMP)_\nu$	local martingale problem for Dawson-Watanabe superprocess with initial law ν–see prior to Theorem II.5.1
$L_t(X)$	the collision local time of $X = (X^1, X^2)$–see prior to Remarks IV.3.1
$\log^+(x)$	$(\log x) \vee e^e$
$L_W(\phi)$	the Laplace functional of the random measure W, i.e., $E(e^{-W(\phi)})$
$\mathcal{L}^2_{\mathrm{loc}}$	see after Lemma II.5.2
$M_1(E)$	space of probabilities on E with the topology of weak convergence
$M_F(E)$	the space of finite measures on E with the topology of weak convergence
$M_F^t(D)$	the set of finite measures on $D(E)$ supported by paths which are constant after time t
$\mathcal{M}_F$	the Borel σ-field on $M_F(E)$
$\mathcal{M}_{\mathrm{loc}}$	the space of continuous $(\mathcal{F}_t)$-local martingales starting at 0
(ME)	mild form of the nonlinear equation–see prior to Theorem II.5.11
$(MP)_{X_0}$	martingale problem for Dawson-Watanabe superprocess with initial state X_0 –see Proposition II.4.2
$\hat{\Omega}$	$\Omega \times C(\mathbb{R}^d)$
$\Omega_H[\tau,\infty)$	$\Big\{H_\cdot \in C\Big([\tau,\infty), M_F(D(E))\Big) : H_t \in M_F^t(D) \quad \forall t \geq \tau\Big\}$
Ω_H	$\Omega_H[0,\infty)$
Ω_X	the space of continuous $M_F(E)$-valued paths
Ω_D	the space of cadlag $M_F(E)$-valued paths
$p_t(x)$	standard Brownian density
$p_t^x(y)$	$p_t(x-y)$
$\mathcal{P}$	the σ-field of $(\mathcal{F}_t)$-predictable subsets of $\mathbb{R}_+ \times \Omega$
$P_t^g\phi(x)$	$E^x\Big(\phi(Y_t)\exp\Big\{\int_0^t g(Y_s)\,ds\Big\}\Big)$
$\mathbb{P}_{X_0}$	the law of the DW superprocess on $(\Omega_X, \mathcal{F}_X)$ with initial state X_0 –see Theorem II.5.1
$\mathbb{P}_\nu$	the law of the DW superprocess with initial law ν

$\hat{\mathbb{P}}_T$	the normalized Campbell measure associated with K_T, i.e., $\hat{\mathbb{P}}_T(A \times B) = \mathbb{P}(1_A K_T(B))/m(1)$
(PC)	$x \mapsto P^x$ is continuous
(QLC)	quasi-left continuity, i.e., Y is a Hunt process–see Section II.2
$\mathbb{Q}_{\tau,m}$	the law of the historical process starting at time τ in state m–see Section II.8
$\mathcal{R}(I)$	$\bigcup_{t\in I} S(X_t)$, the range of X on I
$\overline{\mathcal{R}}(I)$	$\overline{\mathcal{R}(I)}$ is the closed range of X on I
$\mathcal{R}$	$\bigcup_{\delta>0} \overline{\mathcal{R}}([\delta,\infty))$ is the range of X.
$S(\mu)$	the closed support of a measure μ
S_t	$S(X_t)$
$\mathcal{S}$	simple $\mathcal{P} \times \mathcal{E}$-measurable integrands–see after Lemma II.5.2
(SE)	strong form of nonlinear equation–see prior to Theorem II.5.11
$\underline{t}$	$[Nt]/N$
$\mathcal{T}_b$	bounded $(\mathcal{F}_t)_{t\geq\tau}$-stopping times
$\stackrel{ucb}{\to}$	convergence on E which is uniform on compacts and bounded on E
U_λ	the λ resolvent of a Markov process
$\stackrel{w}{\Rightarrow}$	weak convergence of finite (usually probability) measures
W_t	the coordinate maps on $D(\hat{E})$
$y/s/w$	the path equaling y up to s and $w(t-s)$ thereafter
$y^t(\cdot)$	$y(t \wedge \cdot)$
ζ_α	the lifetime of the αth branch–see after Remark II.3.2

I. Introduction

Over the years I have heard a number of complaints about the impenetrable literature on measure-valued branching processes or Dawson-Watanabe superprocesses. These concerns have in part been addressed by some recent publications including Don Dawson's St. Flour notes (Dawson (1993)), Eugene Dynkin's monograph (Dynkin (1994)) and Jean-Francois Le Gall's ETH Lecture Notes (Le Gall (1999)). Nonetheless, one still hears that several topics are only accessible to experts. However, each time I asked a colleague what topics they would like to see treated in these notes, I got a different suggestion. Although there are some other less flattering explanations, I would like to think the lack of a clear consensus is a reflection of the large number of different entry points to the subject. The Fleming-Viot processes, used to model genotype frequencies in population genetics, arise by conditioning the total mass of a superprocess to be one (Etheridge and March (1991)). When densities exist (as for super-Brownian motion in one spatial dimension) they typically are solutions of parabolic stochastic pde's driven by a white noise and methods developed for their study often have application to large classes of stochastic pde's (e.g. Mueller and Perkins (1992), Krylov (1997b), Mytnik (1998) and Section III.4). Dawson-Watanabe superprocesses arise as scaling limits of interacting particle systems (Cox, Durrett and Perkins (1999, 2000)) and of oriented percolation at criticality (recent work of van der Hofstad and Slade (2000)). Rescaled lattice trees above eight dimensions converge to the integral of the super-Brownian cluster conditioned to have mass one (Derbez and Slade (1998)). There are close connections with class of nonlinear pde's and the interaction between these fields has led to results for both (Dynkin and Kuznetsov (1996,1998), Le Gall (1999) and Section III.5). They provide a rich source of exotic path properties and an interesting collection of random fractals which are amenable to detailed study (Perkins (1988), Perkins and Taylor (1998), and Chapter III).

Those looking for an overview of all of these developments will not find them here. If you are looking for "the big picture" you should consult Dawson (1993) or Etheridge (2000). My goal in these notes is two-fold. The first is to give a largely self-contained graduate level course on what John Walsh would call "the worm's-eye view of superprocesses". The second is to present some of the topics and methods used in the study of interactive measure-valued models.

Chapters II and III grew out of a set of notes I used in a one-semester graduate course on Superprocesses. A version of these notes, recorded by John Walsh in a legible and accurate hand, has found its way to parts of the community and in fact been referenced in a number of papers. Although I have updated parts of these notes I have not tried to introduce a good deal of the more modern machinery, notably Le Gall's snake and Donnelly and Kurtz's particle representation. In part this is pedagogical. I felt a direct manipulation of branching particle systems (as in II.3,II.4) allows one to quickly gain a good intuition for superprocesses, historical processes, their martingale problems and canonical measures. All of these topics are described in Chapter II. In the case of Le Gall's snake, Le Gall (1999) gives an excellent and authoritative treatment. Chapter III takes a look at the qualitative properties of Dawson-Watanabe superprocesses. Aside from answering a number of natural questions, this allows us to demonstrate the effectiveness of the various tools used to study branching diffusions including the related nonlinear parabolic pde,

Originally published in: *Ecole d'Eté de Probabilités de Saint-Flour XXIX – 1999*, Lecture Notes in Mathematics, Vol. **1781**, 132–134, DOI: 10.1007/3-540-47944-9_5,

historical processes, cluster representations and the martingale problem. Although many of the results presented here are definitive, a number of open problems and conjectures are stated. Most of the Exercises in these Chapters play a crucial role in the presentation and are highly recommended.

My objective in Chapters II and III is to present the basic theory in a middling degree of generality. The researcher looking for a good reference may be disappointed that we are only considering finite variance branching mechanisms, finite initial conditions and Markov spatial motions with a semigroup acting on the space of bounded continuous functions on a Polish space E. The graduate student learning the subject or the instructor teaching a course, may be thankful for the same restrictions. I have included such appendages as location dependent branching and drifts as they motivate some of the interactions studied in Chapters IV and V. Aside from the survey in Section III.7, every effort has been made to provide complete proofs in Chapters II and III. The reader is assumed to have a good understanding of continuous parameter Markov processes and stochastic calculus–for example, the first five Chapters of Ethier and Kurtz (1986) provide ample background. Some of the general tightness results for stochastic processes are stated with references (notably Lemma II.4.5 and (II.4.10), (II.4.11)) but these are topics best dealt with in another course. Finally, although the Hausdorff measure and polar set results in Sections III.3 and III.5 are first stated in their most general forms, complete proofs are then given for slightly weaker versions. This means that at times when these results are used, the proofs may not be self-contained in the critical dimensions (e.g. in Theorem III.6.3 when $d = 4$).

A topic which was included in the original notes but not here is the Fleming-Viot process (but see Exercise IV.1.2). The interplay between these two runs throughout Don Dawson's St. Flour notes. The reader should really consult the article by Ethier and Kurtz (1993) to complete the course.

The fact that we are able to give such a complete theory and description of Dawson-Watanabe superprocesses stems from the strong independence assumptions underlying the model which in turn produces a rather large tool kit for their study. Chapters IV and V study measure-valued processes which may have state-dependent drifts, spatial motions and branching rates (the latter is discussed only briefly). All of the techniques used to study ordinary superprocesses become invalid or must be substantially altered if such interactions are introduced into the model. This is an ongoing area of active research and the emphasis here is on introducing some approaches which are currently being used. In Chapter IV, a competing species model is used to motivate careful presentations of Dawson's Girsanov theorem for interactive drifts and of the construction of collision local time for a class of measure-valued processes. In Chapter V, a strong equation driven by a historical Brownian motion is used to model state dependent spatial motions. Section IV.4 gives a discussion of the competing species models in higher dimensions and Section V.5 describes what is known about the martingale problems for these spatial interactions. The other sections in these chapters are again self-contained with complete arguments.

There are no new results contained in these notes. Some of the results although stated and well-known are perhaps not actually proved in the literature (e.g. the disconnectedness results in III.6) and some of the proofs presented here are, I hope, cleaner and shorter. I noticed that some of the theorems were originally derived using nonstandard analysis and I have standardized the arguments (often using the historical process) to make them more accessible. This saddens me a bit as I feel the nonstandard view, clumsy as it is at times, is pedagogically superior and allows one to come up with novel insights.

As one can see from the outline of the actual lectures, at St. Flour some time was spent on rescaled limits of the voter model and the contact process, but these topics have not made it into these notes. A copy of some notes prepared with Ted Cox and Rick Durrett on this subject was distributed at St. Flour and is available from me (or them) upon request. We were trying to unify and extend these results. As new applications are still emerging, I decided it would be better to wait until they find a more definitive form than rush and include them here. Those who have seen earlier versions of these notes will know that I also had planned to include a detailed presentation of the particle representations of Donnelly and Kurtz (1999). In this case I have no real excuse for not including them aside from running out of time and a desire to keep the total number of pages under control. They certainly are one of the most important techniques available for treating interactive measure-valued models and hence should have been included in the second part of these notes.

There a number of people to thank. First the organizers and audience of the 1999 St. Flour Summer School in Probability for an enjoyable and at times exhausting $2\frac{1}{2}$ weeks. A number of suggestions and corrections from the participants has improved these notes. The Fields Institute invited me to present a shortened and dry run of these lectures in February and March, and the audience tolerated some experiments which were not entirely successful. Thanks especially to Siva Athreya, Eric Derbez, Min Kang, George Skoulakis, Dean Slonowsky, Vrontos Spyros, Hanno Treial and Xiaowen Zhou. Most of my own contributions to the subject have been joint and a sincere thanks goes to my co-authors who have contributed to the results presented at St. Flour and who have made the subject so enjoyable for me: Martin Barlow, Ted Cox, Don Dawson, Rick Durrett, Steve Evans, Jean-Francois Le Gall and Carl Mueller. Finally a special thanks to Don Dawson and John Walsh who introduced me to the subject and have provided ideas which can be seen throughout these notes.

II. Branching Particle Systems and Dawson-Watanabe Superprocesses
1. Feller's Theorem

Let $\{X_i^k : i \in \mathbb{N}, k \in \mathbb{Z}_+\}$ be i.i.d. $\mathbb{Z}_+$-valued random variables with mean 1 and variance $\gamma > 0$. We think of X_i^k as the number of offspring of the ith individual in the kth generation, so that $Z_{k+1} = \sum_{i=1}^{Z_k} X_i^k$ (set $Z_0 \equiv 1$) is the size of the $k+1^{\text{st}}$ generation of a Galton-Watson branching process with offspring distribution $\mathcal{L}(X_i^k)$, the law of X_i^k.

We write $a_n \sim b_n$ iff $\lim_{n\to\infty} a_n/b_n = 1$ and let $\overset{w}{\Rightarrow}$ denote weak convergence of finite (usually probability) measures.

Theorem II.1.1. (a) (Kolmogorov (1938)) $P(Z_n > 0) \sim 2/n\gamma$ as $n \to \infty$.
(b) (Yaglom (1947)) $P\left(\frac{Z_n}{n} \in \cdot \mid Z_n > 0\right) \overset{w}{\Rightarrow} Z$, where Z is exponential with mean $\gamma/2$.

Proof. (a) This is a calculus exercise in generating functions but it will be used on several occasions and so we provide a proof. Let $f_n(t) = E(t^{Z_n})$ for $t \in [0,1]$, and $f(t) = f_1(t)$ (here $0^0 = 1$ as usual). A simple induction shows that f_n is the n-fold composition of f with itself. Then Dominated Convergence shows that f' and f'' are continuous on $[0,1]$, where the appropriate one-sided derivatives are taken at the endpoints. Moreover $f'(1) = E(X_i^k) = 1$ and $f''(1) = \text{var}(X_i^k) = \gamma$. As f is increasing and strictly positive at 0 (the latter because X_i^k has mean 1 and is not constant), we must have

$$0 < f(0) \le f_n(0) \uparrow L \le 1 \text{ and } f(L) = L.$$

Note that $f'(t) = E(Z_1 t^{Z_1 - 1}) < 1$ for $t < 1$ and so the Mean Value Theorem implies that $f(1) - f(t) < 1 - t$ and therefore, $f(t) > t$, for $t < 1$. This proves that $L = 1$ (as you probably already know from the a.s. extinction of the critical branching process).

Set $x_n = f_n(0)$ and

$$y_n = n(1 - x_n) = nP(Z_n > 0). \tag{II.1.1}$$

A second order Taylor expansion shows that

$$1 - x_{n+1} = f(1) - f(x_n) = 1 - x_n - \frac{f''(z_n)}{2}(1 - x_n)^2 \text{ for some } z_n \in [x_n, 1].$$

Therefore

$$\begin{aligned} y_{n+1} &= (n+1)(1 - x_{n+1}) \\ &= (n+1)[1 - x_n - \frac{f''(z_n)}{2}(1 - x_n)^2] \\ &= (n+1)\left[\frac{y_n}{n} - \frac{y_n^2}{n^2}\frac{f''(z_n)}{2}\right] \\ &= y_n\left[1 + \frac{1}{n}\left(1 - y_n(1 + n^{-1})\frac{f''(z_n)}{2}\right)\right]. \end{aligned}$$

Originally published in: *Ecole d'Eté de Probabilités de Saint-Flour XXIX – 1999*, Lecture Notes in Mathematics, Vol. **1781**, 135–192, DOI: 10.1007/3-540-47944-9_6,

Now let $\gamma_1 < \gamma < \gamma_2$ and $\delta > 0$. Note that $\lim_{n\to\infty} x_n = 1$ and $\lim_{n\to\infty} f''(z_n) = \gamma$, and so we may choose n_0 so that for $n \geq n_0$,

$$(II.1.2) \qquad (1 - x_n)\gamma_2/2 < \delta,$$

$$(II.1.3) \qquad y_n\Big[1 + \frac{1}{n}(1 - y_n\gamma_2/2)\Big] \leq y_{n+1} \leq y_n\Big[1 + \frac{1}{n}(1 - y_n\gamma_1/2)\Big],$$

and therefore,

$$(II.1.4) \qquad y_{n+1} > y_n \text{ if } y_n < 2/\gamma_2, \text{ and } y_{n+1} < y_n \text{ if } y_n > 2/\gamma_1.$$

Claim $y_n > \frac{2}{\gamma_2}(1 - \delta)$ eventually. Note first that if $n_1 \geq n_0$ satisfies $y_{n_1} \leq \frac{2}{\gamma_2}(1 - \delta/2)$ (there is nothing to prove if no such n_1 exists), then the lower bound in (II.1.3) shows that $y_{n_1+1} \geq y_{n_1}\left(1 + \frac{\delta}{2n_1}\right)$. Iterating this bound, we see that there is an $n_2 > n_1$ for which $y_{n_2} > \frac{2}{\gamma_2}(1 - \delta/2)$. Now let n_3 be the first $n > n_2$ for which $y_n \leq \frac{2}{\gamma_2}(1 - \delta)$ (again we are done if no such n exists). Then $y_{n_3-1} > \frac{2}{\gamma_2}(1 - \delta) \geq y_{n_3}$ and so (II.1.4) implies that $y_{n_3-1} \geq \frac{2}{\gamma_2}$. Therefore (II.1.3) shows that

$$\begin{aligned}
y_{n_3} &\geq y_{n_3-1}\Big[1 + \frac{1}{n_3 - 1}(1 - y_{n_3-1}\gamma_2/2)\Big] \\
&\geq \frac{2}{\gamma_2}\Big[1 - \frac{y_{n_3-1}}{n_3 - 1}\frac{\gamma_2}{2}\Big] \\
&= \frac{2}{\gamma_2}\Big[1 - \frac{(1 - x_{n_3-1})\gamma_2}{2}\Big] \\
&> \frac{2}{\gamma_2}(1 - \delta),
\end{aligned}$$

the last by (II.1.2). This contradicts the choice of n_3 and hence proves the required inequality for $n \geq n_2$. A similar argument shows that $y_n \leq \frac{2}{\gamma_1}(1 - \delta)$ eventually. We thus have shown that $\lim_{n\to\infty} y_n = 2/\gamma$ and hence are done by (II.1.1).

(b) will be a simple consequence of Theorem II.7.2 below, the proof of which will use (a). See also Section II.10 of Harris (1963). ∎

These results suggest we consider a sequence of critical Galton-Watson branching processes $\{Z_0^{(n)} : n \in \mathbb{N}\}$ as above but with initial conditions $Z_0^{(n)}$ satisfying $Z_0^{(n)}/n \to x$, and define $X_t^{(n)} = Z_{[nt]}^{(n)}/n$. Indeed it is an easy exercise to see from the above that $X_1^{(n)}$ converges weakly to a Poisson sum of independent exponential masses.

Notation. E denotes a Polish space. Let $D(E) = D(\mathbb{R}_+, E)$ be the Polish space of cadlag paths from $\mathbb{R}_+$ to E with the Skorokhod J_1-topology. Let $C(E) = C(\mathbb{R}_+, E)$ be the Polish space of continuous E-valued paths with the topology of uniform convergence on compacts. Let $Y_t(y) = y(t)$ for $y \in D(E)$.

Theorem II.1.2. (Feller (1939, 1951)) $X^{(n)} \overset{w}{\Rightarrow} X$ in $D(\mathbb{R})$, where X is the unique solution of

$$(FE) \qquad X_t = x + \int_0^t \sqrt{\gamma X_s}\, dB_s,$$

where B is a one-dimensional Brownian motion.

Proof. We will prove much more below in Theorem II.5.1. The uniqueness holds by Yamada-Watanabe (1971). ∎

We call the above process Feller's branching diffusion with parameter γ.

2. Spatial Motion

We now give our branching population a spatial structure. Individuals are "located" at a point in a Polish space E. This structure will also usually allow us to trace back the genealogy of individuals in the population.

Notation. $\mathcal{E}$ = Borel subsets of $E \equiv \mathcal{B}(E)$,

$C_b(E) = \{f : E \to \mathbb{R} : f$ bounded and continuous$\}$ with the supnorm, $\|\ \|$,

$\mathcal{D} = \mathcal{B}(D(E))$, $(\mathcal{D}_t)_{t\geq 0}$ is the canonical right-continuous filtration on $D(E)$,

$\mu(f) = \int f d\mu$ for a measure μ and integrable function f.

Assume

$$(II.2.1) \qquad Y = (D, \mathcal{D}, \mathcal{D}_t, Y_t, P^x) \text{ is a Borel strong Markov process (BSMP)}$$

with semigroup $P_t f(x) = P^x(f(Y_t))$. "Borel" means $x \to P^x(A)$ is $\mathcal{E}$-measurable for all $A \in \mathcal{D}$. The other required properties here are $Y_0 = x$ P^x–a.s. and the strong Markov property. Evidently our BSMP's have cadlag paths. These assumptions are either much too restrictive or far too abstract, depending on your upbringing. At the risk of offending one of these groups we impose an additional condition:

$$(II.2.2) \qquad P_t : C_b(E) \to C_b(E).$$

This is only needed to facilitate our construction of Dawson-Watanabe superprocesses as limits of branching particle systems and keep fine topologies and Ray compactifications at bay.

Standard arguments (see Exercise II.2.1 below or the proof of Theorem I.9.21 in Sharpe (1988)) show that (II.2.2) implies

Y is a Hunt process, i.e., if $\{T_n\}$ are $\{\mathcal{D}_t\}$-stopping times such that

(QLC) $T_n \uparrow T < \infty$ $P^x a.s.$, then $Y(T_n) \to Y(T)$ $P^x - a.s.$

In particular, $Y_t = Y_{t-}$ $P^x - a.s.$ for all $t > 0$.

Definition. $\phi \in \mathcal{D}(A)$ iff $\phi \in C_b(E)$ and for some $\psi \in C_b(E)$,

$$\phi(Y_t) - \phi(Y_0) - \int_0^t \psi(Y_s) ds \text{ is a } P^x\text{-martingale for all } x \text{ in } E.$$

It is easy to see ψ is unique if it exists and so we write $\psi = A\phi$ for $\phi \in \mathcal{D}(A)$.

Notation. $\xrightarrow{bp}$ denotes bounded pointwise convergence.

Proposition II.2.1. $\phi \in \mathcal{D}(A) \Leftrightarrow \phi \in C_b(E)$ and $\frac{P_t\phi - \phi}{t} \xrightarrow{bp} \psi$ as $t \downarrow 0$ for some $\psi \in C_b(E)$. In this case, $\psi = A\phi$ and for any $s \geq 0$, $P_s\phi \in \mathcal{D}(A)$ and

$$AP_s\phi = P_sA\phi = \frac{\partial}{\partial s}P_s\phi. \tag{II.2.3}$$

Proof. ($\Leftarrow$) If $s \geq 0$, our assumption and the semigroup property show that

$$\frac{P_{s+t}\phi - P_s\phi}{t} \xrightarrow{bp} P_s\psi \quad \text{as} \quad t \downarrow 0 \ \forall s \geq 0.$$

(QLC) implies $P_s\psi(x)$ is continuous in s for each x. An easy calculus exercise shows that a continuous function with a continuous right derivative is differentiable, and so from the above we have

$$\frac{\partial}{\partial s}P_s\phi = P_s\psi, \tag{II.2.4}$$

and so

$$P^x\left(\phi(Y_t) - \phi(Y_0) - \int_0^t \psi(Y_s)ds\right) = P_t\phi(x) - \phi(x) - \int_0^t P_s\psi(x)ds = 0.$$

The Markov property now shows the process in the above expectation is a martingale and so $\phi \in \mathcal{D}(A)$ with $A\phi = \psi$.

($\Rightarrow$) Let $\phi \in \mathcal{D}(A)$ and $s \geq 0$.

$$(P_{t+s}\phi(x) - P_s\phi(x))/t = P^x\left(P^{Y_s}\left(\int_0^t A\phi(Y_r)dr/t\right)\right) \xrightarrow{bp} P_s(A\phi)(x) \quad \text{as} \quad t \downarrow 0,$$

where the last limit holds by Dominated Convergence. Taking $s = 0$, one completes the proof of ($\Rightarrow$). For $s > 0$, one may use the above argument and ($\Leftarrow$) to see $P_s\phi \in \mathcal{D}(A)$ and get the first equality in (II.2.3). The second equality follows from (II.2.4) with $\psi = A\phi$. ∎

Let $U_\lambda f(x) = E^x\left(\int_0^\infty e^{-\lambda t} f(Y_t)dt\right)$ denote the λ-resolvent of Y for $\lambda > 0$. Clearly $U_\lambda : C_b(E) \to C_b(E)$ by (II.2.2).

Proposition II.2.2.
(a) $\forall \phi \in \mathcal{D}(A) \qquad U_\lambda(\lambda - A)\phi = \phi$.
(b) $\forall \phi \in C_b(E) \qquad U_\lambda\phi \in \mathcal{D}(A)$ and $(\lambda - A)U_\lambda\phi = \phi$.

Proof. (a)

$$U_\lambda A\phi(x) = \int_0^\infty e^{-\lambda t} P_t A\phi(x)dt$$

$$= \int_0^\infty e^{-\lambda t} \frac{\partial}{\partial t}(P_t\phi(x))dt \quad \text{(by II.2.3)}$$

$$= e^{-\lambda t} P_t\phi(x)\Big|_{t=0}^{t=\infty} + \lambda \int_0^\infty e^{-\lambda t} P_t\phi(x)dt$$

$$= -\phi(x) + \lambda U_\lambda \phi(x).$$

(b)

$$U_\lambda\phi(Y_t) = E^x\Big(\int_t^\infty e^{-\lambda u}\phi(Y_u)du\Big|\mathcal{D}_t\Big)e^{\lambda t}$$

$$(II.2.5) \qquad = e^{\lambda t}\Big[M_t - \int_0^t e^{-\lambda u}\phi(Y_u)du\Big],$$

where M_t denotes the martingale $E^x\big(\int_0^\infty e^{-\lambda u}\phi(Y_u)du \mid \mathcal{D}_t\big)$. Some stochastic calculus shows that ($\overset{m}{=}$ means equal up to martingales)

$$U_\lambda\phi(Y_t) \overset{m}{=} \int_0^t \lambda e^{\lambda s}\left(M_s - \int_0^s e^{-\lambda u}\phi(Y_u)du\right)ds - \int_0^t \phi(Y_s)ds = \int_0^t \lambda U_\lambda\phi(Y_s) - \phi(Y_s)ds,$$

where we have used (II.2.5). This implies $U_\lambda\phi \in \mathcal{D}(A)$ and $AU_\lambda\phi = \lambda U_\lambda\phi - \phi$. ∎

Notation. $b\mathcal{E}$ (respectively, $\mathcal{E}_+$) is the set of bounded (respectively, non-negative) $\mathcal{E}$-measurable functions. If $\mathcal{H} \subset b\mathcal{E}$, $\overline{\mathcal{H}}^{bp}$ is the smallest set containing $\mathcal{H}$ and closed under $\overset{bp}{\to}$, and $\mathcal{H}_+$ is the set of non-negative functions in $\mathcal{H}$.

Corollary II.2.3. $\overline{\mathcal{D}(A)}^{bp} = b\mathcal{E}$, $\overline{(\mathcal{D}(A)_+)}^{bp} = b\mathcal{E}_+$.

Proof. If $\phi \in C_b(E)$, $P_t\phi \overset{bp}{\to} \phi$ as $t \downarrow 0$ and so it follows that $\lambda U_\lambda\phi \overset{bp}{\to} \phi$ as $\lambda \to \infty$, and so $\phi \in \overline{\mathcal{D}(A)}^{bp}$. The result follows trivially. ∎

Exercise II.2.1. Prove that Y satisfies (QLC).

Hint. (Following the proof of Theorem I.9.21 of Sharpe (1988).) Let

$$X = \lim Y(T_n) \in \{Y(T-), Y(T)\}.$$

It suffices to consider T bounded and show $E^x(g(X)h(Y_T)) = E^x(g(X)h(X))$ for all $g, h \in C_b(E)$ (why?). As in the proof of Corollary II.2.3 it suffices to consider $h = U_\lambda f$, where $f \in C_b(E)$ and $\lambda > 0$. Complete the required argument by using the strong Markov property of Y and the continuity of $U_\lambda f$.

Here are some possible choices of Y.

Examples II.2.4. (a) $Y_t \in \mathbb{R}^d$ is d-dimensional Brownian motion.

$$C_b^2(\mathbb{R}^d) = \{\phi : \mathbb{R}^d \to \mathbb{R}, \phi \text{ is } C^2 \text{ with bounded partials of order 2 or less}\} \subset \mathcal{D}(A)$$

and $A = \frac{\Delta\phi}{2}$ for $\phi \in C_b^2(\mathbb{R}^d)$ by Itô's Lemma. In this case we will write $\mathcal{D}(\Delta/2)$ for $\mathcal{D}(A)$.
(b) $Y_t \in \mathbb{R}^d$ is the d-dimensional symmetric stable process of index $\alpha \in (0,2)$ scaled so that $P^x(e^{i\theta\cdot Y_t}) = e^{i\theta\cdot x - t|\theta|^\alpha}$, where $|y|$ is the Euclidean norm of y. If $\nu(dy) = c|y|^{-d-\alpha}dy$ for an appropriate $c > 0$, then for $\phi \in C_b^2(\mathbb{R}^d) \subset \mathcal{D}(A)$

$$A\phi(x) = \int \left[\phi(x+y) - \phi(x) - \vec{\nabla}\phi(x) \cdot \frac{y}{1+|y|^2}\right] \nu(dy),$$

as can be easily seen, e.g., from the stochastic calculus for point processes in Ikeda-Watanabe (1981, p. 65–67) (see also Revuz-Yor (1990, p. 263)).

In both the above examples

$$C_b^\infty(\mathbb{R}^d) = \{\phi \in C_b(\mathbb{R}^d) : \text{all partial derivatives of } \phi \text{ are in } C_b(\mathbb{R}^d)\}$$

is a core for A in that the *bp*-closure of $\{(\phi, A\phi) : \phi \in C_b^\infty(\mathbb{R}^d)\}$ contains $\{(\phi, A\phi) : \phi \in \mathcal{D}(A)\}$. To see this first note that if $\phi \in \mathcal{D}(A)$ has compact support, then

$$P_t\phi(x) = \int p_t(y-x)\phi(y)dy \in C_b^\infty(\mathbb{R}^d) \text{ for } t > 0$$

because Y has a transition density, p_t, all of whose spatial derivatives are bounded and continuous. In the stable case the latter is clear from Fourier inversion because

$$|\theta|^m \int p_t(y-x)e^{i\theta\cdot y}dy = |\theta|^m e^{i\theta\cdot x - t|\theta|^\alpha}$$

is bounded and integrable in θ for all $m \in \mathbb{N}$. Now choose $\{\psi_n\} \subset C_b^\infty$ with compact support so that $\psi_n \uparrow 1$ and

$$\{|x| \le n\} \subset \{\psi_n = 1\} \subset \{\psi_n > 0\} \subset \{|x| < n+1\}.$$

If $\phi \in \mathcal{D}(A)$ and $\phi_n = \phi\psi_n$, then an integration by parts shows that $\phi_n \in \mathcal{D}(A)$ and $A\phi_n = \psi_n A\phi + \phi A\psi_n$. The above shows that $P_{1/n}\phi_n \in C_b^\infty$. Dominated Convergence implies that $P_{1/n}\phi_n \xrightarrow{bp} \phi$, and (II.2.3) and a short calculation shows that

$$AP_{1/n}\phi_n = P_{1/n}A\phi_n = P_{1/n}(\psi_n A\phi + \phi A\psi_n) \xrightarrow{bp} A\phi.$$

This proves the above claim.

Notation. $M_1(E)$ is the space of probabilities on a Polish space E and its Borel σ-field, equipped with the topology of weak convergence. $C_K(E)$ is the space of continuous function on E with compact support, equipped with the topology of uniform convergence.

Exercise II.2.2. Assume $Y_t \in \mathbb{R}^d$ is d-dimensional Brownian motion with $d > 2$ and $U_0 f$ is defined as above but with $\lambda = 0$.

(a) Show that if $f \geq 0$ is Borel measurable on $\mathbb{R}^d$, then as $\lambda \downarrow 0$, $U_\lambda f(x) \uparrow U_0 f(x) = k_d \int |y-x|^{2-d} f(y) ds \leq \infty$, for some $k_d > 0$.

(b) Show that $U_0 : C_K(\mathbb{R}^d) \to \mathcal{D}(A)$, $AU_0\phi = -\phi$ for all $\phi \in C_K(\mathbb{R}^d)$, and $U_0 A\phi = -\phi$ for all $\phi \in C_K(\mathbb{R}^d) \cap \mathcal{D}(A)$.

Hint. One approach to (b) is to show that for $\phi \in C_K(\mathbb{R}^d)$, as $\lambda \downarrow 0$, $U_\lambda \phi \overset{bp}{\to} U_0\phi \in C_b$ and $AU_\lambda \phi \overset{bp}{\to} -\phi$.

Example II.2.4. (c) Suppose, in addition, that our BSMP Y satisfies

$$(PC) \qquad x \to P^x \quad \text{is continuous from } E \text{ to } M_1(D(E)).$$

This clearly implies (II.2.2).

Under (PC) we claim that the previous hypotheses are satisfied by the path-valued process $t \to (t, Y^t) \equiv (t, Y(\cdot \wedge t)) \in \mathbb{R}_+ \times D(E)$. To be more precise, let $\hat{E} = \{(t, y^t) : t \geq 0, y \in D(E)\}$ with the subspace topology it inherits from $\mathbb{R}_+ \times D(E)$, $\hat{\mathcal{E}} = \mathcal{B}(\hat{E})$, and if $y, w \in D(E)$ and $s \geq 0$, let

$$(y/s/w)(t) = \begin{cases} y(t) & t < s \\ w(t-s) & t \geq s \end{cases} \quad \Big(\in D(E)\Big).$$

Note that $\hat{E}$ is Polish as it is a closed subset of the Polish space $\mathbb{R}_+ \times D(E)$.

Definition. Let $W_t : D(\hat{E}) \to \hat{E}$ denote the coordinate maps and for $(s,y) \in \hat{E}$, define $\hat{P}_{s,y}$ on $D(\hat{E})$ with its Borel σ-field, $\hat{\mathcal{D}}$, by

$$\hat{P}_{s,y}(W_\cdot \in A) = P^{y(s)}((s + \cdot, y/s/Y^\cdot) \in A),$$

i.e., under $\hat{P}_{s,y}$ we run y up to time s and then tag on a copy of Y starting at $y(s)$.

Proposition II.2.5. $(W, (\hat{P}_{s,y})_{(s,y)\in\hat{E}})$ is a BSMP with semigroup

$$\hat{P}_t : C_b(\hat{E}) \to C_b(\hat{E}).$$

Proof. This is a routine if somewhat tedious exercise. Let $(\hat{\mathcal{D}}_t)$ be the canonical right-continuous filtration on $D(\hat{E})$. Fix $u \geq 0$ and to check the Markov property at time u, let $(s,y) \in \hat{E}$, $A \in \hat{\mathcal{D}}_u$, ψ be a bounded measurable function on $\hat{E}$ and $T \geq u$. Also set

$$\tilde{A} = \{w \in D(E) : (v \to (s+v, (y/s/w^v)) \in A\} \in \mathcal{D}_u,$$

and

$$\tilde{\psi}(w) = \psi(s+u+t, y/s/(w^{u+t})), \; w \in D(E).$$

Then

$$\int 1_A\psi(W_{u+t})d\hat{P}_{s,y} = \int 1_{\tilde{A}}(Y)\tilde{\psi}(Y^{u+t})dP^{y(s)}$$
$$= \int 1_{\tilde{A}}(Y(\omega))P^{Y_u(\omega)}(\tilde{\psi}(Y^u(\omega)/u/Y^t)dP^{y(s)}(\omega) \quad \text{(by the Markov property for } Y)$$
$$= \int 1_{\tilde{A}}(Y(\omega))P^{Y_u(\omega)}(\psi(s+u+t,(y/s/Y^u(\omega))/s+u/Y^t))\,dP^{y(s)}(\omega)$$
$$= \int 1_A(s+\cdot,y/s/Y^{\cdot}(\omega))\hat{P}_{s+u,y/s/Y^u(\omega)}(\psi(W_t))dP^{y(s)}(\omega)$$
$$= \int 1_A\hat{P}_{W_u(\omega)}(\psi(W_t))d\hat{P}_{s,y}(\omega).$$

This proves the Markov property at time u.

Turning now to the semigroup $\hat{P}_t$, let $f \in C_b(\hat{E})$ and suppose $(s_n, y_n) \to (s_\infty, y_\infty)$ in $\hat{E}$. Note that if $T > \sup_n s_n$ is a continuity point of y_∞, then

$$y_n(s_n) = y_n(T) \to y_\infty(T) = y_\infty(s_\infty).$$

Therefore by (PC) and Skorohod's representation (see Theorem 3.1.8 of Ethier and Kurtz (1986)) we may construct a sequence of processes, $\{Y_n : n \in \mathbb{N} \cup \{\infty\}\}$ such that Y_n has law $P^{y_n(s_n)}$ and $\lim Y_n = Y_\infty$ in $D(\hat{E})$ a.s. Now use the fact that

$$s_n \to s_\infty,\ (y_n, Y_n) \to (y_\infty, Y_\infty) \text{ in } D(E)^2, \text{ and } Y_n(0) = y_n(s_n)$$
$$\text{imply} \quad y_n/s_n/Y_n \to y_\infty/s_\infty/Y_\infty \text{ as } n \to \infty \text{ in } D(E).$$

This is a standard exercise in the Skorokhod topology on $D(E)$ which is best left to the reader. Note that the only issue is the convergence near s_∞ and here the condition $Y_n(0) = y_n(s_n)$ avoids the possibility of having distinct jump times approach s_∞ in the limit. The above implies that $\lim_{n\to\infty} f(y_n/s_n/Y_n^t) = f(y_\infty/s_\infty/Y_\infty^t)$ a.s. and therefore

$$\lim_{n\to\infty} \hat{P}_t f(s_n, y_n) = \lim_{n\to\infty} E(f(y_n/s_n/Y_n^t)) = E(f(y_\infty/s_\infty/Y_\infty^t)) = \hat{P}_t f(s_\infty, y_\infty).$$

This shows that $\hat{P}_t : C_b(\hat{E}) \to C_b(\hat{E})$. The strong Markov property of W now follows from this and the ordinary Markov property by a standard approximation of a stopping time by a sequence of countably-valued stopping times. Also the Borel measurability of $(s, y) \to \hat{P}_{s,y}(\psi(W))$ is now clear for ψ a bounded continuous function of finitely many coordinates, in which case this function is continuous by the above, and hence for all bounded and measurable ψ by the usual bootstrapping argument. Finally $\hat{P}_{s,y}(W(0) = (s, y)) = 1$ is clear from the definitions. ∎

Exercise II.2.3. Let Y_t be a Feller process with a strongly continuous semigroup $P_t : C_\ell(E) \to C_\ell(E)$, where $C_\ell(E)$ is the space of continuous functions on a locally compact metric space (E, d) with finite limit at ∞. Show that (PC) holds.

Hint. Let $x_n \to x$. It suffices to show $\{P^{x_n}\}$ is tight on $D(E)$ (why?). By Aldous' criterion (Walsh (1986), Thm. 6.8(a)) it suffices to consider a sequence of stopping times $\{T_n\}$, $T_n \le t_0 < \infty$, a sequence $\delta_n \downarrow 0$, and show

$$\lim_{n\to\infty} P^{x_n}(d(Y_{T_n+\delta_n}, Y_{T_n}) > \varepsilon) = 0 \qquad \forall \varepsilon > 0.$$

Exercise II.2.4. Give an example of a Markov process satisfying the hypotheses (II.2.1) and (II.2.2) of this Section but for which (PC) fails.

Hint. Take $E = \mathbb{R}_+$ and consider the generator

$$Af(y) = (f(1)-f(0))1(y=0)+(f(1/y)-f(y))1(0<y\le 1)+y(f(1)-f(y))1(y>1).$$

Thanks go to Tom Kurtz for suggesting this example.

3. Branching Particle Systems

Let Y be the E-valued Markov process from the previous Section and introduce a drift function $g \in C_b(E)$ and branching variance (or rate) function $\gamma \in C_b(E)_+$. Recall that G_+ denotes the non-negative elements in the set G. The state space for our limiting and approximating processes will be $M_F(E)$, the space of finite measures on E with the topology of weak convergence, and we choose an initial state $X_0 \in M_F(E)$.

For $N \in \mathbb{N}$ and $x \in E$, $\nu^N(x,\cdot) \in M_1(\mathbb{Z}_+)$ is the offspring law for a parent located at x. We assume $x \to \nu^N(x,\cdot)$ is measurable and satisfies ($\overset{ucb}{\to}$ denotes convergence on E which is uniform on compacts and bounded on E):

$$\text{(a)} \int k\nu^N(x,dk) = 1 + \frac{g_N(x)}{N} \quad \text{where} \quad g_N \overset{ucb}{\to} g \text{ as } N \to \infty,$$

$$(II.3.1) \qquad \text{(b) Var } (\nu^N(x,\cdot)) = \gamma_N(x) \quad \text{where} \quad \gamma_N \overset{ucb}{\to} \gamma \text{ as } N \to \infty,$$

$$\text{(c) } \exists \delta > 0 \quad \text{such that} \quad \sup_{N,x} \int k^{2+\delta}\nu^N(x,dk) < \infty.$$

Remarks II.3.1. (1) At the cost of complicating our arguments somewhat, (II.3.1) (c) could be weakened to the uniform integrability of k^2 with respect to $\{\nu^N(x,\cdot) : x \in E, N \in \mathbb{N}\}$.

(2) Given $\gamma \in C_b(E)_+$ and $g \in C_b(E)$ it is not hard to see there is a sequence $\{\nu^N\}$ satisfying (II.3.1). In fact there is a $k \in \mathbb{Z}$, $k \ge 2$ and functions p_N and q_N such that

$$\nu^N(x,\cdot) = \delta_0(1 - p_N(x) - q_N(x)) + \delta_1 p_N(x) + \delta_k q_N(x)$$

will satisfy (II.3.1) with $g_N \equiv g$ for $N \ge N_0$.

Exercise II.3.1. Prove this.

Hint. A simple calculation shows that if (II.3.1(a,b)) hold for ν^N as above with $g_N = g$, then $p_N(x) = 1 + g(x)/N - \alpha_N(x)/(k-1)$ and $q_N(x) = \alpha_N(x)/(k^2 - k)$, where

$$\alpha_N(x) = \gamma_N(x) + g(x)/N + g(x)^2/N^2.$$

Let $\eta_N = \|g\|_\infty/N + \|g\|_\infty^2/N^2 + \|g\|_\infty/\sqrt{N}$ and set $\gamma_N(x) = \gamma(x) \vee \eta_N$. Show that you can choose k sufficiently large so that $p_N(x), q_N(x) \ge 0$, $p_N(x) + q_N(x) \le 1$ for N large, and (II.3.1) is valid with $g_N = g$ for such an N.

We now describe a system of branching particles which undergo near critical branching at times k/N according to the laws $\nu_N(x,\cdot)$ where x is the location of the parent. In between branch times particles migrate as independent copies of the process Y from the previous section. It will be important to have a labeling scheme to refer to the branches of the resulting tree of Y-processes. We follow the arboreal labeling of Walsh (1986) – in fact this section is really the missing Ch. 10 of Walsh's SPDE notes. We have decided to work in a discrete time setting but could just as well work in the continuous time setting in which inter-branch intervals are exponential rate N random variables.

We label particles by multi-indices

$$\alpha \in I = \bigcup_{n=0}^{\infty} \mathbb{N}^{\{0,\dots,n\}} = \{(\alpha_0,\dots,\alpha_n) : \alpha_i \in \mathbb{N}, n \in \mathbb{Z}_+\}.$$

Let $|(\alpha_0,\dots,\alpha_n)| = n$ be the generation of α and write

$$\beta < \alpha \Leftrightarrow \beta = (\alpha_0,\dots,\alpha_i) \equiv \alpha|i \text{ for some } i \le |\alpha|,$$

i.e. if β is an ancestor of α. We let $\alpha \vee k = (\alpha_0,\dots\alpha_n,k)$ denote the kth offspring of α and $\alpha \wedge \beta$ denote the "greatest common ancestor" of α and β (set $\alpha\wedge\beta = \phi$ if $\alpha_0 \ne \beta_0$ and $|\phi| = -\infty$), and let $\pi\alpha = (\alpha_0,\dots,\alpha_{n-1})$ denote the parent of α if $n > 0$.

Adjoin Δ to E as a discrete point to form $E_\Delta = E \cup \{\Delta\}$ and let P^Δ be point mass at the constant path identical to Δ. Let $\Omega = E_\Delta^{\mathbb{N}} \times D(E_\Delta)^{\mathbb{N}} \times \mathbb{Z}_+^{\mathbb{N}}$ and let $\mathcal{F}$ denote its product σ-field. Sample points in Ω are denoted by

$$\omega = ((x_i, i \in \mathbb{N}), (Y^\alpha, \alpha \in I), (N^\alpha, \alpha \in I)).$$

Now fix $N \in \mathbb{N}$ and define a probability $P = P^N$ on $(\Omega, \mathcal{F})$ as follows:
$(II.3.2)$

(a) $(x_i, i \le M_N)$ is a Poisson point process with intensity $NX_0(\cdot)$ and $x_i = \Delta$ if $i > M_N$.

(b) Given $\mathcal{G}_n = \sigma(x_i, i \in \mathbb{N}) \vee \sigma(N^\beta, Y^\beta, |\beta| < n)$, $\{Y^\alpha : |\alpha| = n\}$ are (conditionally) independent and (for $|\alpha| = n$)

$$P(Y^\alpha \in A|\mathcal{G}_n)(\omega) = P^{x_{\alpha_0}(\omega)}\Big(Y\Big(\cdot\wedge\Big(\frac{n+1}{N}\Big)\Big) \in A \Big| Y\Big(\cdot \wedge \frac{n}{N}\Big) = Y^{\pi\alpha}_\cdot(\omega)\Big)$$

where $Y^{\pi\alpha} \equiv x_{\alpha_0}$ if $|\alpha| = 0$. That is, $Y^\alpha\Big(\cdot\wedge\dfrac{|\alpha|}{N}\Big) = Y^{\pi\alpha}(\cdot)$and given $\mathcal{G}_n$, $\{Y^\alpha|_{[|\alpha|/N,(|\alpha|+1)/N]} : |\alpha| = n\}$ evolve as independent copies of Y starting from $Y^{\pi\alpha}(|\alpha|/N)$, and stopped at $(|\alpha|+1)/N$.

(c) Given $\bar{\mathcal{G}}_n = \mathcal{G}_n \vee \sigma(Y^\alpha : |\alpha| = n)$, $\{N^\alpha : |\alpha| = n\}$ are (conditionally) independent and $P(N^\alpha \in \cdot|\bar{\mathcal{G}}_n)(\omega) = \nu^N(Y^\alpha((|\alpha|+1)/N,\omega),\cdot)$.

It should be clear from the above that $\{Y^\alpha : \alpha \in I\}$ is an infinite tree of branching Y processes, where $Y^\alpha_t = Y^\beta_t$ for $0 \le t < (|\alpha\wedge\beta|+1)/N$. Let $\underline{t} = [Nt]/N$ for $t \ge 0$ where $[x]$ is the integer part of x, set $T = T_N = \{kN^{-1} : k \in \mathbb{Z}_+\}$ and let $\tau = 1/N$. It will be convenient to work with respect to the right continuous filtration given by

$$\mathcal{F}_t = \mathcal{F}_t^N = \sigma\left((x_i)_{i\in\mathbb{N}}, (Y^\alpha, N^\alpha)_{|\alpha|<N\underline{t}}\right) \vee \left(\bigcap_{r>t} \sigma\left(Y_s^\alpha : |\alpha| = N\underline{t}, s \le r\right)\right)$$

It also will be useful to introduce the slightly larger σ-field

$$\overline{\mathcal{F}}_{\underline{t}} = \mathcal{F}_{\underline{t}} \vee \sigma(Y^\alpha : |\alpha| = N\underline{t}).$$

Here are some consequences of our definition of P for each $\alpha \in I$ and $\underline{t} = |\alpha|/N$:

(II.3.3) $\{(Y^\alpha, \mathcal{F}_s) : s \ge \underline{t}\}$ is a Markov process,and for all $s \in [\underline{t}, \underline{t}+\tau]$,
$P(Y^\alpha(s+\cdot) \in A|\mathcal{F}_s)(\omega) = P^{Y_s^\alpha(\omega)}(Y(\cdot \wedge (\underline{t}+\tau-s)) \in A)$ a.s.
for all $A \in \mathcal{D}$.

(II.3.4) $P(Y^\alpha \in A|\mathcal{F}_0)(\omega) = P^{x_{\alpha_0}(\omega)}(Y(\cdot \wedge (\underline{t}+\tau)) \in A)$ a.s. for all $A \in \mathcal{D}$.

(II.3.5) $P(N^\alpha \in \cdot|\bar{\mathcal{F}}_{\underline{t}}) = \nu^N(Y^\alpha_{\underline{t}+\tau}, \cdot)$ a.s.

Clearly (II.3.5) is a restatement of (II.3.2)(c). (II.3.4) should be clear from (II.3.3) and (II.3.2)(b) (one can, for example, induct on $|\alpha|$). To prove (II.3.3), it suffices to prove the stated equality for $s \in [\underline{t}, \underline{t}+\tau)$, so fix such an s. The stated result is an easy consequence of (II.3.2)(b) if $\mathcal{F}_s$ is replaced by the smaller σ-field $\mathcal{G}_{N\underline{t}} \vee \mathcal{F}_s^{Y^\alpha}$, where $\mathcal{F}_s^{Y^\alpha}$ is the right continuous filtration generated by Y^α. Now use the fact that $\mathcal{H}^\alpha \equiv \sigma(Y^\beta : |\beta| = N\underline{t}, \beta \ne \alpha)$ is conditionally independent of Y^α given $\mathcal{G}_{N\underline{t}}$ (by (II.3.2(b)) to see that the stated result is valid if $\mathcal{F}_s$ is replaced by the larger σ-field $\mathcal{G}_{N\underline{t}} \vee \mathcal{F}_s^{Y^\alpha} \vee \mathcal{H}^\alpha$. Now condition this last equality with respect to $\mathcal{F}_s$ to obtain (II.3.3).

Remark II.3.2. If $X_0^N = \frac{1}{N}\sum_1^{M_N} \delta_{x_i}$, an easy consequence of (II.3.2)(a) and the weak law of large numbers is

(II.3.6) $X_0^N(\phi) \xrightarrow{P} X_0(\phi)$ as $N \to \infty$ for any bounded measurable ϕ on E.

Note also that $E(X_0^N(\cdot)) = X_0(\cdot)$. From (II.3.6) it is easy to show that $X_0^N \xrightarrow{P} X_0$. For example, one could use the existence of a countable convergence determining class of functions on E (see the proof of Theorem 3.4.4 in Ethier and Kurtz (1986)).

Instead of assuming $\{x_i^N, i \le M_N\}$ is as in (II.3.2)(a), we could assume more generally that $\{x_i^N, i \le M_N\}$ are random points (M_N is also random) chosen so that $X_0^N = \frac{1}{N}\sum_1^{M_N} \delta_{x_i^N} \xrightarrow{P} X_0$, $\sup_N E(X_0^N(1)^2) < \infty$, and $E(X_0^N(\cdot)) \le c_0 X_0(\cdot)$ as measures. The only required change is to include c_0 as multiplicative factor in the upper bound in Lemma II.3.3 below. For example, we could assume $\{x_i, i \le M_N\}$ are i.i.d. with law $X_0(\cdot)/X_0(1)$ and $M_N = [NX_0(1)]$ and set $c_0 = 1$.

The interpretation of (II.3.2) (b,c) is that for $|\alpha| = N\underline{t}$, the individual labelled by α follows the trajectory Y^α on $[\underline{t}, \underline{t}+\tau]$ and at time $\underline{t}+\tau$ dies and is replaced by its N^α children. The next step is to use the N^α to prune the infinite tree of branching Y^α processes. The termination time of the αth branch is

$$\zeta^\alpha = \begin{cases} 0, & \text{if } x_{\alpha_0} = \Delta \\ \min\left\{(i+1)/N : i < |\alpha|, N^{\alpha|i} < \alpha_{i+1}\right\}, & \text{if this set is not } \emptyset \text{ and } x_{\alpha_0} \neq \Delta \\ (|\alpha|+1)/N, & \text{otherwise.} \end{cases}$$

Note in the first case, the αth particle was never born since $\alpha_0 > M_N$. In the second case, $\alpha_{i+1} > N^{\alpha|i}$ means the α_{i+1}st offspring of $\alpha|i$ doesn't exist. Finally in the last instance, the family tree of α is still alive at $(|\alpha|+1)/N$ but we have run out of data to describe its state beyond this final time.

We write $\alpha \overset{\omega}{\sim} t$ (or $\alpha \sim t$) iff $|\alpha|/N \leq t < (|\alpha|+1)/N = \zeta^\alpha$, i.e., iff α labels a particle alive at time t. Clearly $\alpha \sim t$ iff $\alpha \sim \underline{t}$. Note that we associate α with the particle alive on $[|\alpha|/N, (|\alpha|+1)/N]$, although of course Y^α_s, $s < (|\alpha|+1)/N$ describes the past history of its ancestors. From Feller's theorem (Theorem II.1.2) it is natural to assign mass $\frac{1}{N}$ to each particle alive at time t and define

$$X^N_t = \frac{1}{N}\sum_{\alpha \sim t} \delta_{Y^\alpha_t} = \frac{1}{N}\sum_{\alpha \sim \underline{t}} \delta_{Y^\alpha_t}, \quad \text{i.e.,} \quad X^N_t(A) = \#\{Y^\alpha_t \in A : \alpha \sim t\}/N, \quad A \in \mathcal{E}.$$

Since $N^\alpha < \infty$ for all $\alpha \in I$ a.s., clearly $X^N_t \in M_F(E)$ for all $t \geq 0$ a.s. Note also that $Y^\alpha \in D(E)$ for all α with $x_{\alpha_0} \neq \Delta$ a.s., and therefore $X^N_t = \frac{1}{N}\sum_{\alpha \sim \underline{t}} \delta_{Y^\alpha_t}$ has sample paths in $D(M_F(E))$ a.s. on each $[\underline{t}, \underline{t}+\tau)$, and hence on all of $\mathbb{R}_+$. The associated *historical process* is

$$H^N_t = \frac{1}{N}\sum_{\alpha \sim t} \delta_{Y^\alpha_{\cdot \wedge t}} \in M_F(D(E)).$$

Again $H^N_\cdot \in D(\mathbb{R}_+, M_F(D(E)))$. Therefore X^N_t is the (normalized) empirical measure of the particles alive at time t while H^N_t records the past trajectories of the ancestors of particles alive at time t. Clearly we have

$$X^N_t(\phi) = \int \phi(y_t) H^N_t(dy).$$

Exercise II.3.2. Show that
(i) $\{\alpha \sim t\} \in \mathcal{F}_t$.
(ii) X^N_t is $\mathcal{F}_t$-measurable.
(A trivial exercise designed only to convince you that $|\alpha| < N\underline{t}$ in the above definition is correct.)

Our goal is to show $X^N \overset{w}{\Rightarrow} X$ in $D(M_F(E))$ and characterize X as the unique solution of a martingale problem. The weak convergence of H^N to the associated historical process H will then follow easily by considering the special case in Example II.2.4 (c). As any student of Ethier and Kurtz (1986) knows, the proof proceeds in two steps:

1. Tightness of $\{X^N\}$ and derivation of limiting martingale problem.
2. Uniqueness of solutions to the martingale problem.

These are carried out in the next 3 sections.

We close this section with a simple bound for the first moments.

Notation. $P^\mu = \int P^x(\cdot)\mu(dx), \mu \in M_F(E)$. If $y \in C(E)$, let $y^t(s) = y(t \wedge s)$.

Lemma II.3.3. Let $g_\infty = \sup_N \|g_N\|_\infty$ and $\overline{g} = \sup\{g_N(x) : x \in E, N \in \mathbb{N}\}$.

(a) If $\psi : D(\mathbb{R}_+, E) \to \mathbb{R}_+$ is Borel, then for any $t \geq 0$

$$E\left(H_t^N(\psi)\right) \leq e^{\overline{g}\underline{t}} E^{X_0}(\psi(Y^t)) \leq e^{g_\infty \underline{t}} E^{X_0}(\psi(Y^t)).$$

In particular, $E(X_t^N(\phi)) \leq e^{\overline{g}t} E^{X_0}(\phi(Y_t)) \; \forall \phi \in \mathcal{E}_+$.

(b) For all $x, K > 0$ and for all $N \geq N_0(g_\infty)$,

$$P\big(\sup_{t \leq K} X_t^N(1) \geq x\big) \leq e^{3g_\infty K} X_0(1) x^{-1}.$$

Proof. (a) Let $P_{s,t}(\psi)(y) = E^{y(0)}(\psi(Y^t) \mid Y^s = y)$, $s \leq t$. We prove the result for $t \leq \underline{t}$ by induction on $\underline{t}$. If $\underline{t} = 0$, then one has equality in the above. Assume the result for $t \leq \underline{t}$. Then

$$H^N_{\underline{t}+\tau}(\psi) = \frac{1}{N} \sum_{\alpha \sim \underline{t}} \psi\left(Y^\alpha_{\cdot \wedge (\underline{t}+\tau)}\right) N^\alpha,$$

and so

$$\begin{aligned}
E\left(H^N_{\underline{t}+\tau}(\psi)\right) &= E\left(\frac{1}{N} \sum_{\alpha \sim \underline{t}} \psi\left(Y^\alpha_{\cdot \wedge (\underline{t}+\tau)}\right) E\left(N^\alpha \mid \mathcal{F}_{\underline{t}}\right)\right) \\
&= E\left(\frac{1}{N} \sum_{\alpha \sim \underline{t}} \psi\left(Y^\alpha_{\cdot \wedge (\underline{t}+\tau)}\right)\left(1 + g_N\left(Y^\alpha_{\underline{t}+\tau}\right)\tau\right)\right) \quad \text{(by (II.3.5))} \\
&\leq (1 + \overline{g}\tau) E\left(\frac{1}{N} \sum_{\alpha \sim \underline{t}} P_{\underline{t}, \underline{t}+\tau} \psi\left(Y^\alpha_{\cdot \wedge \underline{t}}\right)\right) \quad \text{(by (II.3.3))} \\
&\leq e^{\overline{g}\tau} e^{\overline{g}\underline{t}} E^{X_0}\left(P_{\underline{t}, \underline{t}+\tau} \psi(Y_{\cdot \wedge \underline{t}})\right) \quad \text{(by induction hypothesis)} \\
&= e^{\overline{g}(\tau + \underline{t})} E^{X_0}\left(\psi(Y_{\cdot \wedge (\underline{t}+\tau)})\right).
\end{aligned}$$

Finally it should be clear from the above that the result follows for $t \in (\underline{t}, \underline{t} + \tau)$.

(b) Claim $e^{2g_\infty \underline{t}} X^N_{\underline{t}}(1)$ is an $(\overline{\mathcal{F}}_{\underline{t}})$-submartingale for $N \geq N_0(g_\infty)$. From the above calculation we have

$$\begin{aligned}
&e^{2g_\infty(\underline{t}+\tau)} E\left(X^N_{\underline{t}+\tau}(1) \mid \overline{\mathcal{F}}_{\underline{t}}\right) \\
&\quad \geq e^{2g_\infty(\underline{t}+\tau)}(1 - g_\infty \tau) X^N_{\underline{t}}(1) \geq e^{2g_\infty \underline{t}} X^N_{\underline{t}}(1),
\end{aligned}$$

for $N \geq N_0(g_\infty)$. The weak L^1 inequality for non-negative submartingales and (a) now complete the proof. ∎

Remark II.3.4. It is clear from the above argument that if $g_N \equiv 0$, then equality holds in (a).

4. Tightness

We first specialize Theorem 3.1 of Jakubowski (1986), which gives necessary and sufficient conditions for tightness in $D(\mathbb{R}_+, S)$, to the case $S = M_F(E)$. As E is Polish, $M_F(E)$ is also Polish–see Theorem 3.1.7 of Ethier and Kurtz (1986) for the corresponding result for $M_1(E)$ from which the result follows easily. (An explicit complete metric is defined prior to Lemma II.7.5.) Therefore $D(M_F(E))$ is also Polish and Prohorov's theorem implies that a collection of laws on $D(M_F(E))$ is tight iff it is relatively compact.

Definition. A collection of processes $\{X^\alpha : \alpha \in I\}$ with paths in $D(S)$ is C-relatively compact in $D(S)$ iff it is relatively compact in $D(S)$ and all weak limit points are a.s. continuous.

Definition. $D_0 \subset C_b(E)$ is separating iff for any $\mu, \nu \in M_F(E)$, $\mu(\phi) = \nu(\phi)$ $\forall \phi \in D_0$ implies $\mu = \nu$.

Theorem II.4.1. Let D_0 be a separating class in $C_b(E)$ containing 1. A sequence of cadlag $M_F(E)$-valued processes $\{X^N, N \in \mathbb{N}\}$ is C-relatively compact in $D(M_F(E))$ iff the following conditions hold:

(i) $\forall \varepsilon, T > 0$ there is a compact set $K_{T,\varepsilon}$ in E such that

$$\sup_N P\left(\sup_{t\le T} X_t^N(K_{T,\varepsilon}^c) > \varepsilon\right) < \varepsilon.$$

(ii) $\forall \phi \in D_0$, $\{X_\cdot^N(\phi) : N \in \mathbb{N}\}$ is C-relatively compact in $D(\mathbb{R}_+, \mathbb{R})$.

If, in addition, D_0 is closed under addition, then the above equivalence holds when ordinary relatively compactness in D replaces C-relative compactness in both the hypothesis and conclusion.

Remark. A version of this result is already implicit in Kurtz (1975) (see the Remark after Theorem 4.20). A proof of the sufficiency of the above conditions in the C-relatively compact setting is given at the end of this Section. All the ideas of the proof may be found in Theorem 3.9.1 and Corollary 3.9.2 of Ethier and Kurtz (1986).

Although the C-relatively compact version is the result we will need, a few words are in order about the result in the general setting. (i) essentially reduces the result to the case when E is compact. In this case it is not hard to see there is a countable subset $D_0' \subset D_0$ closed under addition such that $\psi(\mu) = (\mu(f))_{f\in D_0'}$ is a homeomorphism from $M_F(E)$ onto its image in $\mathbb{R}^{D_0'}$. The same is true of the map $X_t \to (X_t(\phi))_{\phi\in D_0'}$ from $D(M_F(E))$ onto its image in $D(\mathbb{R}_+, \mathbb{R}^{D_0'})$. To complete the proof we must show $D(\mathbb{R}_+, \mathbb{R}^{D_0'})$ and $D(\mathbb{R}_+, \mathbb{R})^{D_0'}$ are homeomorphic. This is the step which requires D_0' to be closed under addition. As any scholar of the J_1-topology knows, $X^n \to X$ and $Y^n \to Y$ in $D(\mathbb{R}_+, \mathbb{R})$ need not imply $(X^n, Y^n) \to (X, Y)$ in $D(\mathbb{R}_+, \mathbb{R}^2)$, but it does if in addition $X^n + Y^n \to X + Y$. See Jakubowski (1986) for the details.

Notation. $\mathcal{F}_t^X = \bigcap_{u>t} \sigma(X_s : s \le u)$ denotes the right-continuous filtration generated by a process X. Let $A^f\phi = (A + f)\phi$ for $\phi \in \mathcal{D}(A), f \in C_b(E)$.

We will use Theorem II.4.1 to prove the following tightness result. Recall that our standing hypotheses (II.2.1), (II.2.2) and (II.3.1) are in force.

Proposition II.4.2. $\{X^N\}$ is C-relatively compact in $D(M_F(E))$. Each weak limit point, X, satisfies

$$(MP)_{X_0}\quad \begin{array}{l}\forall\phi\in\mathcal{D}(A),\ M_t(\phi)=X_t(\phi)-X_0(\phi)-\displaystyle\int_0^t X_s(A^g\phi)ds \text{ is a continuous}\\[2ex] (\mathcal{F}_t^X)\text{-martingale such that } M_0(\phi)=0 \text{ and } \langle M(\phi)\rangle_t=\displaystyle\int_0^t X_s(\gamma\phi^2)ds.\end{array}$$

Proof. We take $D_0=\mathcal{D}(A)$ in Theorem II.4.1, which is a separating class by Corollary II.2.3. We first check (ii) in II.4.1.

Let $\phi\in\mathcal{D}(A)$ and define

$$M_t^\alpha=\phi(Y_t^\alpha)-\phi(Y_{\underline{t}}^\alpha)-\int_{\underline{t}}^t A\phi(Y_s^\alpha)ds,\ t\in\left[\underline{t},\underline{t}+N^{-1}\right],\ \underline{t}=|\alpha|/N,\alpha\in I.$$

Note that if we group the population at $\underline{s}+N^{-1}$ according to their parents at $\underline{s}$ we get

$$X^N_{\underline{s}+N^{-1}}(\phi)=\frac{1}{N}\sum_{\alpha\sim\underline{s}}\phi\left(Y^\alpha_{\underline{s}+N^{-1}}\right)N^\alpha.$$

Therefore,

$(II.4.1)$

$$\begin{aligned}X^N_{\underline{s}+N^{-1}}(\phi)-X^N_{\underline{s}}(\phi)&=\frac{1}{N}\sum_{\alpha\sim\underline{s}}\left[\phi\left(Y^\alpha_{\underline{s}+N^{-1}}\right)N^\alpha-\phi\left(Y^\alpha_{\underline{s}}\right)\right]\\&=\frac{1}{N}\sum_{\alpha\sim\underline{s}}\phi\left(Y^\alpha_{\underline{s}+N^{-1}}\right)\left[N^\alpha-\left(1+g_N\left(Y^\alpha_{\underline{s}+N^{-1}}\right)N^{-1}\right)\right]+N^{-2}\sum_{\alpha\sim\underline{s}}\phi g_N\left(Y^\alpha_{\underline{s}+N^{-1}}\right)\\&\quad+\frac{1}{N}\sum_{\alpha\sim\underline{s}}M^\alpha_{\underline{s}+N^{-1}}+\int_{\underline{s}}^{\underline{s}+N^{-1}}\frac{1}{N}\sum_{\alpha\sim\underline{s}}A\phi\left(Y_s^\alpha\right)ds\end{aligned}$$

and

$$(II.4.2)\quad \begin{aligned}X^N_t(\phi)-X^N_{\underline{t}}(\phi)&=\frac{1}{N}\sum_{\alpha\sim\underline{t}}(\phi\left(Y^\alpha_t\right)-\phi(Y^\alpha_{\underline{t}}))\\&=\frac{1}{N}\sum_{\alpha\sim\underline{t}}M^\alpha_t+\int_{\underline{t}}^t\frac{1}{N}\sum_{\alpha\sim\underline{t}}A\phi\left(Y^\alpha_s\right)ds.\end{aligned}$$

Sum (II.4.1) over $\underline{s} < \underline{t}$ and then add (II.4.2) to arrive at

$$X_t^N(\phi) = X_0^N(\phi) + \frac{1}{N}\sum_{\underline{s}<\underline{t}}\sum_{\alpha\sim\underline{s}} \phi\left(Y^\alpha_{\underline{s}+N^{-1}}\right)\left(N^\alpha - \left(1 + g_N(Y^\alpha_{\underline{s}+N^{-1}})N^{-1}\right)\right)$$

$$+ \int_0^{\underline{t}} X_{\underline{s}}^N(\phi g_N)ds + N^{-2}\sum_{\underline{s}<\underline{t}}\sum_{\alpha\sim\underline{s}}\left(\phi g_N\left(Y^\alpha_{\underline{s}+N^{-1}}\right) - \phi g_N(Y^\alpha_{\underline{s}})\right)$$

$$+ \left[\frac{1}{N}\sum_{\underline{s}<\underline{t}}\sum_{\alpha\sim\underline{s}} M^\alpha_{\underline{s}+N^{-1}} + \frac{1}{N}\sum_{\alpha\sim\underline{t}} M^\alpha_t\right] + \int_0^t X_s^N(A\phi)ds,$$

and therefore

$$(MP)^N \qquad X_t^N(\phi) = X_0^N(\phi) + M_{\underline{t}}^{b,N}(\phi) + \int_0^{\underline{t}} X_{\underline{s}}^N(\phi g_N)ds + \delta_{\underline{t}}^N(\phi)$$

$$+ M_t^{s,N}(\phi) + \int_0^t X_s^N(A\phi)ds.$$

In the last line, the terms $M_{\underline{t}}^{b,N}(\phi)$, $\delta_{\underline{t}}^N(\phi)$ and $M_t^{s,N}(\phi)$ are defined to be the corresponding terms in the previous expression.

We start by handling the error term δ^N.

Lemma II.4.3. $\sup_{\underline{t}\le K} |\delta_{\underline{t}}^N(\phi)| \xrightarrow{L^1} 0$ as $N \to \infty$ $\forall K \in \mathbb{N}$.

Proof. If $h_N(y) \equiv E^y\left(|g_N\phi(Y_{N^{-1}}) - g_N\phi(Y_0)|\right)$ then

$$E\left(\sup_{\underline{t}\le K}|\delta_{\underline{t}}^N(\phi)|\right) \le \sum_{\underline{s}<K} N^{-1}E\left(\frac{1}{N}\sum_{\alpha\sim\underline{s}} E\left(\left|g_N\phi(Y^\alpha_{\underline{s}+N^{-1}}) - g_N\phi(Y^\alpha_{\underline{s}})\right| \mid \mathcal{F}_{\underline{s}}\right)\right)$$

$$= \int_0^K E\left(X_{\underline{s}}^N(h_N)\right)ds \qquad \text{(by (II.3.3))}$$

$$\le e^{g_\infty K}\int_0^K E^{X_0}\left(h_N(Y_{\underline{s}})\right)ds \qquad \text{(Lemma II.3.3(a))}$$

$$= e^{g_\infty K}E^{X_0}\left(\int_0^K |g_N\phi(Y_{\underline{s}+N^{-1}}) - g_N\phi(Y_{\underline{s}})|\,ds\right).$$

Now since $\{Y_s, Y_{s-} : s \le K\}$ is a.s. compact, (II.3.1)(a) and Dominated Convergence show that for some $\eta_N \to 0$,

$$\begin{aligned} E\Big(\sup_{t\le K} |\delta^N_{\underline{t}}(\phi)|\Big) &\le e^{g_\infty K} E^{X_0}\Big(\int_0^K |g\phi(Y_{\underline{s}+N^{-1}}) - g\phi(Y_{\underline{s}})|\,ds\Big) + \eta_N \\ &\to e^{g_\infty K} E^{X_0}\Big(\int_0^K |g\phi(Y_s) - g\phi(Y_{s-})|\,ds\Big) \quad \text{as } N\to\infty \\ &= 0. \quad \blacksquare \end{aligned}$$

Use (II.3.3) and argue as in the proof of Proposition II.2.1 to see that

$$(M^\alpha_t, \mathcal{F}_t)_{t\in[\underline{t}, \underline{t}+N^{-1}]} \text{ is a martingale.}$$

This and the fact that $\{\alpha \sim \underline{t}\} \in \mathcal{F}_{\underline{t}}$ (recall Exercise II.3.2) easily imply that $(M^{s,N}_t(\phi), \mathcal{F}_t)_{t\ge 0}$ is a martingale. Lemma II.3.3 (a) with $\phi \equiv 1$ implies integrability. Perhaps somewhat surprisingly, we now show that $M^{s,N}(\phi)$ will not contribute to the limit as $N \to \infty$. This is essentially a Strong Law effect. A moment's thought shows that the fact that Y has no fixed time discontinuities (by (QLC)) must play an implicit role in the proof as the following result fails without it.

Lemma II.4.4. $\sup_{t\le K} |M^{s,N}_t(\phi)| \xrightarrow{L^2} 0$ as $N \to \infty$ $\forall K > 0$.

Proof. Let $h_N(y) = E^y\big([\phi(Y_{1/N}) - \phi(Y_0)]^2\big)$. The definition of M^α_t and an easy orthogonality argument shows that for $K \in \mathbb{N}$

$$\begin{aligned} E\left(M^{s,N}_K(\phi)^2\right) &= N^{-2}\sum_{\underline{s}<K} E\left(\sum_{\alpha\sim\underline{s}} E\left((M^\alpha_{\underline{s}+N^{-1}})^2 \mid \mathcal{F}_{\underline{s}}\right)\right) \\ &\le 2N^{-2}\sum_{\underline{s}<K}\left[E\left(\sum_{\alpha\sim\underline{s}}\left(h_N(Y^\alpha_{\underline{s}}) + \|A\phi\|^2_\infty N^{-2}\right)\right)\right] \quad \text{(by (II.3.3))} \\ &\le 2E\left(\int_0^K X^N_{\underline{s}}(h_N) + \|A\phi\|^2_\infty N^{-2} X^N_{\underline{s}}(1)\,ds\right) \\ &\le 2e^{g_\infty K}\left[E^{X_0}\left(\int_0^K \big(\phi(Y_{\underline{s}+N^{-1}}) - \phi(Y_{\underline{s}})\big)^2 ds\right) + KN^{-2}\|A\phi\|^2_\infty X_0(1)\right]. \end{aligned}$$

In the last line we have used Lemma II.3.3(a) and argued as in the proof of Lemma II.4.3. As in that result the above expression approaches 0 as $N \to \infty$ by Dominated Convergence. This proves the above expectation goes to 0 and the result follows by the strong L^2 inequality for martingales. $\blacksquare$

Recall that $T = T_N = \{j/N : j \in \mathbb{Z}^+\}$. We claim $(M_{\underline{t}}^{b,N}(\phi), \overline{\mathcal{F}}_{\underline{t}})_{\underline{t}\in T}$ is also a martingale. To see this note that

$$\begin{aligned}
&E\left(M_{\underline{t}+N^{-1}}^{b,N}(\phi) - M_{\underline{t}}^{b,N}(\phi) \mid \overline{\mathcal{F}}_{\underline{t}}\right) \\
&= \frac{1}{N}\sum_{\alpha\sim\underline{t}} \phi\left(Y_{\underline{t}+N^{-1}}^{\alpha}\right) E\left(N^{\alpha} - \left(1 + g_N(Y_{\underline{t}+N^{-1}}^{\alpha})N^{-1}\right) \mid \overline{\mathcal{F}}_{\underline{t}}\right) \\
&= 0
\end{aligned}$$

by our definition of the conditional law of N^{α}. Integrability of this increment is again clear from Lemma II.3.3(a) and $E(|N^{\alpha}|) \le C$. In view of the fact that our spatial martingales vanish in the limit we expect that the martingale $M_t(\phi)$ in $(MP)_{X_0}$ must arise from the "branching martingales" $M^{b,N}(\phi)$.

To analyze $M^{b,N}$ we use the following "well-known" result.

Lemma II.4.5. Let $(M_{\underline{t}}^N, \overline{\mathcal{F}}_{\underline{t}}^N)_{\underline{t}\in T_N}$ be martingales with $M_0^N = 0$. Let $\langle M^N\rangle_{\underline{t}} = \sum_{0\le\underline{s}<\underline{t}} E((M_{\underline{s}+N^{-1}}^N - M_{\underline{s}}^N)^2 \mid \overline{\mathcal{F}}_{\underline{s}}^N)$, and extend $M_\cdot^N$ and $\langle M^N\rangle_\cdot$ to $\mathbb{R}_+$ as right-continuous step functions.

(a) If $\{\langle M^N\rangle_\cdot : N \in \mathbb{N}\}$ is C-relatively compact in $D(\mathbb{R})$ and

$$(II.4.3) \qquad \sup_{0\le\underline{t}\le K} |M^N(\underline{t} + N^{-1}) - M^N(\underline{t})| \xrightarrow{P} 0 \quad \text{as} \quad N \to \infty \quad \forall K > 0,$$

then $\{M^N\}$ is C-relatively compact in $D(\mathbb{R})$.

(b) If, in addition,

$$(II.4.4) \qquad \left\{\left(M_{\underline{t}}^N\right)^2 + \langle M^N\rangle_{\underline{t}} : N \in \mathbb{N}\right\} \quad \text{is uniformly integrable} \quad \forall \underline{t} \in T,$$

then $M^{N_k} \overset{w}{\Rightarrow} M$ in $D(\mathbb{R})$ implies M is a continuous L^2 martingale and $(M^{N_k}, \langle M^{N_k}\rangle) \overset{w}{\Rightarrow} (M, \langle M\rangle)$ in $D(\mathbb{R})^2$.

(c) Under (II.4.4), the converse to (a) holds.

Proof. (a) is immediate from Theorems VI.4.13 and VI.3.26 of Jacod-Shiryaev (1987). A nonstandard proof (and statement) of the entire result may also be found in Theorems 8.5 and 6.7 of Hoover-Perkins (1983). (b) remains valid without the $\langle M^N\rangle_{\underline{t}}$ term in (II.4.4) but the proof with this condition becomes completely elementary as the reader can easily check. ∎

The key ingredient in the above result is a predictable square function inequality of Burkholder (1973):

(PSF)

$\exists c = c(c_0)$ such that if $\phi : \mathbb{R}_+ \to \mathbb{R}_+$ is continuous, increasing, $\phi(0) = 0$ and $\phi(2\lambda) \le c_0\phi(\lambda)$ for all $\lambda \ge 0$, $(M_n, \mathcal{F}_n)$ is a martingale, $M_n^* = \sup_{k\le n}|M_k|$,

$$\langle M\rangle_n = \sum_{k=1}^{n} E\left((M_k - M_{k-1})^2 \mid \mathcal{F}_{k-1}\right) + E(M_0^2) \text{ and } d_n^* = \max_{1\le k\le n}|M_k - M_{k-1}|,$$

then $E\left(\phi(M_n^*)\right) \le c\left[E\left(\phi\left(\langle M\rangle_n^{1/2}\right) + \phi(d_n^*)\right)\right]$.

To apply the above Lemma to $M_{\underline{t}}^{b,N}(\phi)$, first note that by (II.3.2)(c), if $\alpha \sim \underline{s}$, $\beta \sim \underline{s}$, and $\alpha \neq \beta$, then N^α and N^β are conditionally independent given $\bar{\mathcal{F}}_{\underline{s}}$. The resulting orthogonality then shows that

$$E\left(\left(M_{\underline{s}+N^{-1}}^{b,N}(\phi) - M_{\underline{s}}^{b,N}(\phi)\right)^2 \Big| \overline{\mathcal{F}}_{\underline{s}}\right)$$

$$= \frac{1}{N^2}\sum_{\alpha\sim\underline{s}} \phi(Y_{\underline{s}+N^{-1}}^\alpha)^2 E\left(\left(N^\alpha - (1+g_N(Y_{\underline{s}+N^{-1}}^\alpha)N^{-1})\right)^2 \Big| \overline{\mathcal{F}}_{\underline{s}}\right)$$

$$= \frac{1}{N^2}\sum_{\alpha\sim\underline{s}} \phi(Y_{\underline{s}+N^{-1}}^\alpha)^2 \gamma_N(Y_{\underline{s}+N^{-1}}^\alpha).$$

Sum over $\underline{s} < \underline{t}$ to see that

$$(II.4.5) \qquad \langle M^{b,N}(\phi)\rangle_{\underline{t}} = \int_0^{\underline{t}} X_{\underline{s}}^N(\phi^2\gamma_N)ds + \varepsilon_{\underline{t}}^N(\phi) \leq c_\phi \int_0^t X_{\underline{s}}^N(1)ds,$$

where

$$\varepsilon_{\underline{t}}^N(\phi) = N^{-2}\sum_{\underline{s}<\underline{t}}\sum_{\alpha\sim\underline{s}} \phi^2\gamma_N(Y_{\underline{s}+N^{-1}}^\alpha) - \phi^2\gamma_N(Y_{\underline{s}}^\alpha).$$

Just as in Lemma II.4.3, for any $K > 0$,

$$(II.4.6) \qquad \sup_{\underline{t}\leq K} |\varepsilon_{\underline{t}}^N(\phi)| \xrightarrow{L^1} 0 \quad \text{as} \quad N\to\infty.$$

We also see from the above that for $0 \leq \underline{s} < \underline{t} \leq K$,

$$\langle M^{b,N}(\phi)\rangle_{\underline{t}} - \langle M^{b,N}(\phi)\rangle_{\underline{s}} \leq c \sup_{\underline{s}\leq K} X_{\underline{s}}^N(1)|\underline{t}-\underline{s}|,$$

which in view of Lemma II.3.3(b) and Arzela-Ascoli implies the C-relative compactness of $\{\langle M^{b,N}(\phi)\rangle_\cdot : N \in \mathbb{N}\}$ in $D(\mathbb{R})$. To verify (II.4.3) and (II.4.4) we will use the following result, whose proof we defer.

Lemma II.4.6. $\sup_N E\left(\sup_{t\leq K} X_t^N(1)^2\right) < \infty$ for any $K > 0$.

Exercise II.4.1 (a) If δ is as in (II.3.1(c)) use (PSF) to show

$$\lim_{N\to\infty} E\left(\sum_{\underline{t}\leq K} |\Delta M^{b,N}(\phi)(\underline{t})|^{2+\delta}\right) = 0 \qquad \forall K > 0.$$

Hint: Conditional on $\overline{\mathcal{F}}_{\underline{t}}$, $\Delta M^{b,N}(\phi)(\underline{t})$ is a sum of mean 0 independent r.v.'s to which one may apply (PSF).
(b) Use (a), (PSF), and Lemma II.4.6 to show that $\forall K > 0$

$$\sup_N E\left(\sup_{\underline{t}\leq K} |M_{\underline{t}}^{b,N}(\phi)|^{2+\delta}\right) < \infty$$

and, in particular, $\left\{\sup_{t\le K}\left|M_{\underline{t}}^{b,N}(\phi)\right|^2 : N\in\mathbb{N}\right\}$ is uniformly integrable.

The uniform integrability of $\left\{\langle M^{b,N}(\phi)\rangle_{\underline{t}} : N\in\mathbb{N}\right\}$ is clear from (II.4.5) and Lemma II.4.6, and so the above Exercise allows us to apply the relative compactness lemma for martingales (Lemma II.4.5) to conclude:

(M^b)

$\{M_{\cdot}^{b,N}(\phi) : N\in\mathbb{N}\}$ are C-relatively compact in D, and all limit points are continuous L^2 martingales. If $M_{\cdot}^{b,N_k}(\phi)\overset{w}{\Rightarrow}M_{\cdot}(\phi)$, then

$$\left(M_{\cdot}^{b,N_k}(\phi),\langle M^{b,N_k}(\phi)\rangle_{\cdot}\right)\overset{w}{\Rightarrow}\left(M_{\cdot}(\phi),\langle M(\phi)\rangle\right)\quad\text{in}\quad D^2.$$

Proof of Lemma II.4.6. Let $\phi=1$ in $(MP)^N$ and combine the terms $\int_0^t X_{\underline{s}}^N(g_N)ds+\delta_{\underline{t}}^N(1)$ to see that (recall that $g_\infty=\sup_N\|g_N\|$)

$$(II.4.7)\qquad X_{\underline{t}}^N(1)\le X_0^N(1)+M_{\underline{t}}^{b,N}(1)+g_\infty\int_0^t X_{\underline{s}}^N(1)ds.$$

Doob's strong L^2 inequality and Lemma II.3.3(a) imply

$$(II.4.8)\qquad \begin{aligned} E\left(\sup_{t\le K}M_{\underline{t}}^{b,N}(1)^2\right)&\le cE\left(\langle M^{b,N}(1)\rangle_K\right)\\ &\le cE\left(\int_0^K X_{\underline{s}}^N(1)ds\right)\qquad\text{(by (II.4.5))}\\ &\le c(K)X_0(1).\end{aligned}$$

Jensen's inequality and (II.4.7) show that for $u\le K$

$$\sup_{t\le u}X_t^N(1)^2\le cX_0^N(1)^2+c\sup_{t\le K}M_{\underline{t}}^{b,N}(1)^2+cg_\infty^2K\int_0^{\underline{u}}X_{\underline{s}}^N(1)^2ds.$$

Consider the first $\underline{u}\le K$ at which the mean of the left side is infinite. If such a $\underline{u}$ exists, the last integral has finite mean and so does the right-hand side. This contradiction shows $E\big(\sup_{t\le K}X_t^N(1)^2\big)<\infty$ and a simple Gronwall argument now gives a bound uniform in N, namely for all $u\le K$,

$$E\left(\sup_{t\le u}X_t^N(1)^2\right)\le c\left[\sup_N E\left(X_0^N(1)^2\right)+c(K)X_0(1)\right]e^{cg_\infty^2Ku}.\quad\blacksquare$$

To complete the verification of (ii) in Theorem II.4.1, return to $(MP)^N$. Recall from Remark II.3.2 that $X_0^N(\phi)\overset{P}{\to}X_0(\phi)$ as $N\to\infty$. The Arzela-Ascoli Theorem and Lemma II.3.3(b) show that $\int_0^{\underline{t}}X_{\underline{s}}^N(\phi g_N)ds$ and $\int_0^t X_s^N(A\phi)ds$ are C-relatively compact sequences in $D(\mathbb{R})$. Therefore $(MP)^N$, (M^b) and Lemmas II.4.3 and II.4.4 show that $\{X^N_\cdot(\phi)\}$ is C-relatively compact in $D(\mathbb{R})$ for each ϕ in $\mathcal{D}(A)$, and (ii) is verified.

We now give the

Proof of the Compact Containment Condition (i) in Theorem II.4.1. Let $\varepsilon, T > 0$ and $\eta = \eta(\varepsilon, T) > 0$ (η will be chosen below). As any probability on a Polish space is tight, we may choose a compact set $K_0 \subset D(E)$ so that $P^{X_0}(Y \in K_0^c) < \eta$. Let $K = \{y_t, y_{t-} : t \le T, y \in K_0\}$. It is easy to see K is compact in E (note that if $t_n \to t$ and $y_n \to y$, then $y_{n_k}(t_{n_k}) \to y(t)$ or $y(t-)$ for some subsequence $\{n_k\}$ and similarly for $y_{n_k}(t_{n_k}-)$). Clearly

$$P^{X_0}(Y_t \in K^c \quad \text{or} \quad Y_{t-} \in K^c \quad \exists t \le T) < \eta.$$

Let

$$\begin{aligned} R_t^N &= e^{2g_\infty \underline{t}} H_t^N\{y : y(s) \in K^c \quad \text{for some} \quad s \le t\}. \\ &= e^{2g_\infty \underline{t}} \frac{1}{N} \sum_{\alpha \sim t} \sup_{s \le t} 1_{K^c}(Y_s^\alpha). \end{aligned}$$

Claim R_t^N is an $\mathcal{F}_t^N$-submartingale for $N \ge N_0$. As $R^N_\cdot$ is increasing on $[\underline{t}, \underline{t}+\tau)$, it suffices to show that

$$(II.4.9) \qquad E\left(R_{\underline{t}}^N - R_{\underline{t}-}^N \mid \mathcal{F}_{\underline{t}-}^N\right) \ge 0 \quad \text{a.s.}$$

We have

$$\begin{aligned} R_{\underline{t}}^N - R_{\underline{t}-}^N &= \frac{1}{N} \sum_{\alpha \sim \underline{t} - N^{-1}} \left[e^{2g_\infty \underline{t}} \sup_{s \le \underline{t}} 1_{K^c}(Y_s^\alpha) N^\alpha - e^{2g_\infty(\underline{t} - N^{-1})} \sup_{s < \underline{t}} 1_{K^c}(Y_s^\alpha) \right] \\ &\ge \frac{1}{N} \sum_{\alpha \sim \underline{t} - N^{-1}} \left[e^{2g_\infty \underline{t}} N^\alpha - e^{2g_\infty(\underline{t} - N^{-1})} \right] \sup_{s < \underline{t}} 1_{K^c}(Y_s^\alpha) \\ &= \frac{1}{N} \sum_{\alpha \sim \underline{t} - N^{-1}} e^{2g_\infty \underline{t}} \left[N^\alpha - g_N(Y_{\underline{t}}^\alpha)/N - 1 \right] \sup_{s < \underline{t}} 1_{K^c}(Y_s^\alpha) \\ &\quad + \frac{1}{N} \sum_{\alpha \sim \underline{t} - N^{-1}} e^{2g_\infty \underline{t}} \left[g_N(Y_{\underline{t}}^\alpha)/N + 1 - e^{-2g_\infty/N} \right] \sup_{s < \underline{t}} 1_{K^c}(Y_s^\alpha). \end{aligned}$$

The conditional expectation of the first term with respect to $\mathcal{F}_{\underline{t}-}^N$ is 0. The second term is at least

$$\frac{1}{N} \sum_{\alpha \sim \underline{t} - N^{-1}} e^{2g_\infty \underline{t}} \left[-g_\infty/N + 1 - e^{-2g_\infty/N} \right] \sup_{s < \underline{t}} 1_{K^c}(Y_s^\alpha) \ge 0, \quad \text{for} \quad N \ge N_0(g_\infty),$$

and (II.4.9) is proved. Now use the weak L^1 inequality for submartingales and

Lemma II.3.3 (a) to see that for $N \geq N_0(g_\infty)$

$$\begin{aligned}P\left(\sup_{t\leq T} X_t^N(K^c) > \varepsilon\right) &\leq P\left(\sup_{t\leq T} R_t^N > \varepsilon\right)\\ &\leq \varepsilon^{-1}E(R_T^N)\\ &\leq e^{2g_\infty T}\varepsilon^{-1}e^{\bar{g}T}P^{X_0}(Y_s\in K^c \ \ \exists s\leq T)\\ &< \varepsilon\end{aligned}$$

by an appropriate choice of $\eta = \eta(\varepsilon, T)$. It is trivial to enlarge K if necessary to accommodate $N < N_0(g_\infty)$, e.g. by using the converse of Theorem II.4.1 and the fact that each P_N is trivially tight. ∎

Completion of Proof of Proposition II.4.2. By Theorem II.4.1 $\{X_\cdot^N\}$ is C-relatively compact in $D(M_F(E))$. To complete the proof of Proposition II.4.2 we must verify that all limit points satisfy $(MP)_{X_0}$. Assume $X^{N_k}\overset{w}{\Rightarrow}X$. By a theorem of Skorohod (see Theorem 3.1.8 of Ethier-Kurtz (1986)) we may assume X^{N_k}, X are defined on a common probability space and $X^{N_k} \to X$ in $D(M_F(E))$ a.s. Let $\phi \in \mathcal{D}(A)$. Note that $X \to \int_0^t X_r(A\phi)dr$ and $X \to \int_0^t X_r(g\phi)dr$ are continuous maps from $D(\mathbb{R}_+, M_F(E))$ to $C(\mathbb{R}_+, \mathbb{R})$, and $X \to X_\cdot(\phi)$ is a continuous function from $D(M_F(E))$ to $D(\mathbb{R})$. This, Lemmas II.4.3 and II.4.4, the convergence in probability of $X_0^{N_k}(\phi)$ to $X_0(\phi)$ (by (II.3.6)), the uniform convergence on compacts of g_{N_k} to g and condition (i) in Theorem II.4.1 allow us to take limits in $(MP)^{N_k}$ and conclude that $M_\cdot^{b,N_k}(\phi) \to M_\cdot(\phi)$ in $D(\mathbb{R}_+, \mathbb{R})$ in probability where

$$X_t(\phi) = X_0(\phi) + M_t(\phi) + \int_0^t X_s(A\phi + g\phi)ds.$$

By (M^b), $M_\cdot(\phi)$ is a continuous square integrable martingale. In addition, (M^b) together with (II.4.5), (II.4.6), $\gamma_{N_k}\overset{ucb}{\to}\gamma$ and the compact containment condition (i) allow one to conclude

$$\langle M(\phi)\rangle_t = P - \lim_{k\to\infty} \int_0^{\underline{t}} X_{\underline{s}}^{N_k}(\phi^2\gamma_{N_k})ds = \int_0^t X_s(\phi^2\gamma)ds.$$

To complete the derivation of $(MP)_{X_0}$ we must show that $M_t(\phi)$ is an $(\mathcal{F}_t^X)$-martingale and not just a martingale with respect to its own filtration. To see this let $s_1 \leq s_2 \ldots \leq s_n \leq s < t$, $\psi : M_F(E)^n \to \mathbb{R}$ be bounded and continuous and use Exercise II.4.1 (b) to take limits in

$$E\left(\left(M_{\underline{t}}^{b,N_k}(\phi) - M_{\underline{s}}^{b,N_k}(\phi)\right)\psi\left(X_{s_1}^{N_k}, \ldots, X_{s_n}^{N_k}\right)\right) = 0.$$

This shows that X satisfies $(MP)_{X_0}$ and thus completes the proof of Proposition II.4.2. ∎

Proof of Theorem II.4.1. We only prove the sufficiency of the two conditions as the necessity is quite easy. Let d be a complete metric on $M_F(E)$ and if $x \in D(M_F(E))$ and $\delta, T > 0$, set

$$w(x,\delta,T) = \sup\{d(x(t),x(s)) : s,t \le T,\ |s-t| \le \delta\}.$$

A standard result for general Polish state spaces states that $\{X^N : N \in \mathbb{N}\}$ is C-relatively compact if and only if

(*II*.4.10) $\forall \varepsilon, T > 0$ there is a compact set $K^0_{\varepsilon,T} \subset M_F(E)$ such that $\sup_N P(X^N(t) \notin K^0_{\varepsilon,T}$ for some $t \le T) \le \varepsilon$,

and

(*II*.4.11) $\forall \varepsilon, T > 0$ there is a $\delta > 0$ so that $\limsup_{N\to\infty} P(w(X^N,\delta,T) \ge \varepsilon) \le \varepsilon.$

For example, this follows from Corollary 3.7.4, Remark 3.7.3, and Theorem 3.10.2 of Ethier and Kurtz (1986).

We first verify the compact containment condition (II.4.10). Let $\varepsilon, T > 0$. By condition (i) there are compact subsets K_m of E so that

$$\sup_N P(\sup_{t\le T} X^N_t(K^c_m) > 2^{-m}) < \varepsilon 2^{-m-1}.$$

Take $\phi = 1$ in condition (ii) and use (II.4.10) for the real-valued processes $X^N_\cdot(1)$ to see there is an $R = R(\varepsilon, T)$ so that

$$\sup_N P(\sup_{t\le T} X^N_t(1) > R) < \varepsilon/2.$$

Define

$$C^0 = \{\mu \in M_F(E) : \mu(K^c_m) \le 2^{-m} \text{ for all } m \in \mathbb{N}, \text{ and } \mu(1) \le R\}.$$

Then the choice of R and K_m imply that

$$P(X^N_t \notin C^0 \text{ for some } t \le T) < \varepsilon.$$

To verify compactness of $\overline{C^0}$, let $\{\mu_n\}$ be a sequence in C^0. To find a weakly convergent subsequence we may assume that $\inf \mu_n(E) > \delta > 0$. Tightness of $\{\mu_n/\mu_n(E)\}$ is now clear, and so by Prohorov's theorem there is a subsequence $\{n_k\}$over which these normalized measures converge weakly. As the total masses are bounded by R, we may take a further subsequence to ensure convergence of the total masses and hence obtain weak convergence of $\{\mu_{n_k}\}$. It follows that $K_0 = \overline{C^0}$ is compact and so will satisfy (II.4.10).

The next step is to show

(*II*.4.12) $\forall f \in C_b(M_F(E))$, $\{f \circ X^N : N \in \mathbb{N}\}$ is C-relatively compact in $D(\mathbb{R})$.

Let $f \in C_b(M_F(E))$ and $\varepsilon, T > 0$. Choose K^0 as in (II.4.10) and define

$$A = \Big\{h : M_F(E) \to \mathbb{R} \ : \ h(\mu) = \sum_{i=1}^{k} a_i \prod_{j=1}^{m_i} \mu(f_{i,j}),\ a_i \in \mathbb{R},\ f_{i,j} \in D_0, k, m_i \in \mathbb{Z}_+\Big\}$$
$$\subset C_b(M_F(E)).$$

Then A is an algebra containing the constant functions and separating points in $M_F(E)$. By Stone-Weierstrass there is an $h \in A$ so that $\sup_{\mu \in K^0} |h(\mu) - f(\mu)| < \varepsilon$. If $\{Y^N\}$ and $\{Z^N\}$ are C-relatively compact in $D(\mathbb{R})$ then so are $\{aY^N + bZ^N\}$ and $\{Y^N Z^N\}$ for any $a, b \in \mathbb{R}$. This is easy to show using (II.4.10) and (II.4.11), for example (but is false for ordinary relative compactness in $D(\mathbb{R})$). Therefore condition (ii) of the Theorem implies that $\{h \circ X^N\}$ is C-relatively compact and by (II.4.11) there is a $\delta > 0$ so that

$$(II.4.13) \qquad \limsup_{N\to\infty} P(w(h \circ X^N, \delta, T) \geq \varepsilon) \leq \varepsilon.$$

If $s, t \leq T$ and $|t - s| \leq \delta$, then

$$|f(X_t^N) - f(X_s^N)| \leq 2\|f\|_\infty 1(X^N([0,T]) \not\subset K^0) + 2 \sup_{\mu \in K^0} |h(\mu) - f(\mu)| + |h(X_t^N) - h(X_s^N)|,$$

and so,

$$w(f \circ X^N, \delta, T) \leq 2\|f\|_\infty 1(X^N([0,T]) \not\subset K^0) + 2\varepsilon + w(h \circ X^N, \delta, T).$$

Therefore

$$\limsup_{N\to\infty} P(w(f \circ X^N, \delta, T) \geq 3\varepsilon)$$
$$\leq \limsup_{N\to\infty} P(X^N([0,T]) \not\subset K^0) + \limsup_{N\to\infty} P(w(h \circ X^N, \delta, T) \geq \varepsilon) \leq 2\varepsilon,$$

the last by (II.4.13) and the choice of K^0. We have verified (II.4.11) with $\{f \circ X^N\}$ in place of $\{X^N\}$, and as (II.4.10) is trivial for this process, (II.4.12) follows.

It remains to verify (II.4.11). We may assume d is bounded by 1. Let $\varepsilon, T > 0$, and K^0 is as in (II.4.10). Choose $\mu_i \in K^0$, $i \leq M$, so that $K^0 \subset \cup_{i=1}^M B(\mu_i, \varepsilon)$, and let $f_i(\mu) = d(\mu_i, \mu)$. Clearly $f_i \in C_b(M_F(E))$. We showed in the previous paragraph that there is a $\delta > 0$ so that

$$(II.4.14) \qquad \sum_{i=1}^{M} \limsup_{N\to\infty} P(w(f_i \circ X^N, \delta, T) \geq \varepsilon) \leq \varepsilon.$$

If $\mu, \nu \in K^0$, choose μ_j so that $d(\nu, \mu_j) < \varepsilon$. Then

$$\begin{aligned} d(\mu, \nu) &\leq d(\mu, \mu_j) + d(\mu_j, \nu) \\ &\leq |d(\mu, \mu_j) - d(\mu_j, \nu)| + 2d(\mu_j, \nu) \\ &\leq \max_i |f_i(\mu) - f_i(\nu)| + 2\varepsilon. \end{aligned}$$

Let $s,t \le T$, $|s-t| \le \delta$. Then the above inequality implies that

$$d(X_t^N, X_s^N) \le \max_i |f_i \circ X^N(t) - f_i \circ X^N(s)| + 2\varepsilon + 1(X^N([0,T]) \not\subset K^0),$$

and therefore,

$$w(X^N, \delta, T) \le \max_i w(f_i \circ X^N, \delta, T) + 2\varepsilon + 1(X^N([0,T]) \not\subset K^0).$$

It follows that

$$\begin{aligned}&\limsup_{N\to\infty} P(w(X^N,\delta,T) \ge 3\varepsilon)\\ &\quad\le \limsup_{N\to\infty} P(\max_i w(f_i \circ X^N, \delta, T) \ge \varepsilon) + \limsup_{N\to\infty} P(X^N([0,T]) \not\subset K^0) \le 2\varepsilon,\end{aligned}$$

the last by (II.4.14) and the choice of K^0. This gives (II.4.11) and so the proof is complete. ∎

5. The Martingale Problem

In order to prove convergence in Proposition II.4.2 it suffices to show solutions to $(MP)_{X_0}$ are unique in law. We will show this is the case and state the main result of this section in Theorem II.5.1 below. Let g, $\gamma \in C_b(E)$ with $\gamma \ge 0$ as before. A is the weak generator of our BSMP, Y, satisfying (II.2.1) and (II.2.2). Recall that $A^g\phi = A\phi + g\phi$. $(\Omega, \mathcal{F}, \mathcal{F}_t, \mathbb{P})$ will denote a filtered probability space with $(\mathcal{F}_t)$ right-continuous.

Definition. Let ν be a probability on $M_F(E)$. An adapted a.s. continuous $M_F(E)$-valued process, X, on $(\Omega, \mathcal{F}, \mathcal{F}_t, \mathbb{P})$ satisfies $(LMP)_\nu$ (or $(LMP)_\nu^{g,\gamma,A}$) iff

$$(LMP)_\nu \quad \begin{aligned}&X_0 \text{ has law } \nu \text{ and } \forall \phi \in \mathcal{D}(A)\ M_t(\phi) = X_t(\phi) - X_0(\phi) - \int_0^t X_s(A^g\phi)ds\\ &\text{is an } (\mathcal{F}_t)-\text{local martingale such that } \langle M(\phi)\rangle_t = \int_0^t X_s(\gamma\phi^2)ds.\end{aligned}$$

Remark. If $\int X_0(1)d\nu(X_0) = \infty$, then the integrability of $M_t(1)$ may fail and so we need to work with a local martingale problem. We let $(MP)_\nu$ denote the corresponding martingale problem (i.e. $M_t(\phi)$ is an $(\mathcal{F}_t)$-martingale), thus slightly abusing the notation in Proposition II.4.2. That result shows that if $X_0 \in M_F(E)$, then any limit point of $\{X^N\}$ satisfies $(MP)_{\delta_{X_0}}$ on the canonical space of measure-valued paths.

Definition. $(LMP)_\nu$ is well-posed if a solution exists on some $(\Omega, \mathcal{F}, \mathcal{F}_t, \mathbb{P})$ and the law of any solution (on $C(\mathbb{R}_+, M_F(E))$) is unique.

Notation. $\Omega_X = C(\mathbb{R}_+, M_F(E))$, $\mathcal{F}_X$ = Borel sets on Ω_X, $\Omega_D = D(\mathbb{R}_+, M_F(E))$.

Theorem II.5.1. (a) $(LMP)_\nu$ is well-posed $\forall \nu \in M_1(M_F(E))$.

(b) There is a family of probabilities $\{\mathbb{P}_{X_0} : X_0 \in M_F(E)\}$ on $(\Omega_X, \mathcal{F}_X)$ such that if $X_t(\omega) = \omega_t$, then

(i) $\mathbb{P}_\nu(\cdot) = \int \mathbb{P}_{X_0}(\cdot) d\nu(X_0)$ is the law of any solution to $(LMP)_\nu$ for any probability ν on $M_F(E)$.

(ii) $(\Omega_X, \mathcal{F}_X, \mathcal{F}_t^X, X, \mathbb{P}_{X_0})$ is a BSMP.

(iii) If $(Z_t)_{t\geq 0}$ satisfies $(LMP)_\nu$ on $(\Omega, \mathcal{F}, \mathcal{F}_t, \mathbb{P})$ and T is an a.s. finite $(\mathcal{F}_t)$-stopping time, then $\mathbb{P}(Z_{T+\cdot} \in A \mid \mathcal{F}_T)(\omega) = \mathbb{P}_{Z_T(\omega)}(A)$ a.s. $\forall A \in \mathcal{F}_X$.

(c) If (II.3.1) holds and $\{X^N\}$ are as in Proposition II.4.2, then

$$\mathbb{P}(X^N \in \cdot) \stackrel{w}{\Rightarrow} \mathbb{P}_{X_0} \quad \text{on} \quad \Omega_D.$$

(d) If $T_t F(X_0) = \mathbb{P}_{X_0}(F(X_t))$, then $T_t : C_b(M_F(E)) \to C_b(M_F(E))$.

The key step in the above is the uniqueness of solution to $(LMP)_\nu$. The remaining properties will be standard consequences of this and the method (duality) used to establish uniqueness (see, e.g., Theorem II.5.6 below). A process satisfying $(LMP)_\nu$ on $(\Omega, \mathcal{F}, \mathcal{F}_t, P)$ is called an $(\mathcal{F}_t)$-(A, γ, g)-Dawson-Watanabe superprocess, or $(\mathcal{F}_t)$-(Y, γ, g)-DW superprocess, with initial law ν, or, if $\nu = \delta_{X_0}$, starting at X_0.

The following standard monotone class lemma will be useful.

Lemma II.5.2. Let $\mathcal{H} \subset b\mathcal{F}$ be a linear class containing 1 and closed under $\stackrel{bp}{\to}$. Let $\mathcal{H}_0 \subset \mathcal{H}$ be closed under products. Then $\mathcal{H}$ contains all bounded $\sigma(\mathcal{H}_0)$-measurable functions.

Proof. See p. 497 of Ethier-Kurtz (1986). ∎

Let X satisfy $(LMP)_\nu$ on some $(\Omega, \mathcal{F}, \mathcal{F}_t, \mathbb{P})$. Let $\mathcal{M}_{\text{loc}}$ be the space of continuous $(\mathcal{F}_t)$-local martingales such that $M_0 = 0$. Here processes which agree off an evanescent set are identified. Let $\mathcal{P}$ be the σ-field of $(\mathcal{F}_t)$-predictable sets in $\mathbb{R}_+ \times \Omega$, and define

$$\mathcal{L}^2_{\text{loc}} = \Big\{\psi : \mathbb{R}_+ \times \Omega \times E \to \mathbb{R} : \psi \text{ is } \mathcal{P}\times\mathcal{E}\text{-measurable}, \int_0^t X_s(\psi_s^2\gamma)ds < \infty \quad \forall t > 0\Big\},$$

and

$$\mathcal{L}^2 = \{\psi \in \mathcal{L}^2_{\text{loc}} : E\Big(\int_0^t X_s(\psi_s^2\gamma)ds\Big) < \infty \ \forall t > 0\}.$$

A $\mathcal{P} \times \mathcal{E}$-measurable function ψ is simple (write $\psi \in \mathcal{S}$) iff

$$\psi(t, \omega, x) = \sum_{i=0}^{K-1} \psi_i(\omega)\phi_i(x) 1_{(t_i, t_{i+1}]}(t)$$

for some $\phi_i \in \mathcal{D}(A)$, $\psi_i \in b\mathcal{F}_{t_i}$ and $0 = t_0 < t_1 \ldots < t_K \leq \infty$. For such a ψ define

$$M_t(\psi) \equiv \int_0^t \int \psi(s, x) dM(s, x) = \sum_{i=0}^{K-1} \psi_i \left(M_{t\wedge t_{i+1}}(\phi_i) - M_{t\wedge t_i}(\phi_i)\right).$$

Then a standard argument shows that $M_t(\psi)$ is well-defined (i.e., independent of the choice of representation for ψ) and so $\psi \mapsto M(\psi)$ is clearly linear. If

$$\tilde{\psi}_i(s,\omega) = \psi_i(\omega)1_{(t_i,t_{i+1}]}(s),$$

then $\tilde{\psi}_i$ is $\mathcal{P}$-measurable and $M_t(\psi) = \sum_{i=0}^{K-1} \int_0^t \tilde{\psi}_i dM_s(\phi_i)$. Therefore $M_t(\psi)$ is in $\mathcal{M}_{\text{loc}}$ and a simple calculation gives

$$\langle M(\psi)\rangle_t = \int_0^t X_s\left(\gamma\psi_s^2\right) ds.$$

Lemma II.5.3. For any $\psi \in \mathcal{L}^2_{\text{loc}}$ there is a sequence $\{\psi_n\}$ in $\mathcal{S}$ such that

$$\mathbb{P}\left(\int_0^n \int (\psi_n - \psi)^2(s,\omega,x)\gamma(x)X_s(dx)ds > 2^{-n}\right) < 2^{-n}.$$

Proof. Let $\overline{\mathcal{S}}$ denote the set of bounded $\mathcal{P}\times\mathcal{E}$-measurable functions which can be approximated as above. $\overline{\mathcal{S}}$ is clearly closed under $\xrightarrow{bp}$. Since $\overline{D(A)}^{bp} = b\mathcal{E}$, $\overline{\mathcal{S}}$ contains

$$\psi(t,\omega,x) = \sum_{i=0}^{K-1} \psi_i(\omega,x)1_{(t_i,t_{i+1}]}(t)$$

where $0 = t_0 < \ldots < t_K \le \infty$ and $\psi_i(\omega,x) = f_i(\omega)\phi_i(x)$, $\phi_i \in b\mathcal{E}$, $f_i \in b\mathcal{F}_{t_i}$. Now apply Lemma II.5.2 to the class $\mathcal{H}$ of $\psi_i \in b(\mathcal{F}_{t_i}\times\mathcal{E})$ for which ψ as above is in $\overline{\mathcal{S}}$. Using $\mathcal{H}_0 = \{f_i(\omega)\phi_i(x) : f_i \in b\mathcal{F}_{t_i}, \phi_i \in b\mathcal{E}\}$, we see that $\psi(t,\omega,x) = \sum_{i=0}^{K-1} \psi_i(\omega,x)1_{(t_i,t_{i+1}]}(t)$ is in $\overline{\mathcal{S}}$ for any $\psi_i \in b(\mathcal{F}_{t_i}\times\mathcal{E})$.

If $\psi \in b(\mathcal{P}\times\mathcal{E})$, then

$$\psi_n(s,\omega,x) = 2^n \int_{(i-1)2^{-n}}^{i2^{-n}} \psi(r,\omega,x)dr \quad \text{if} \quad s \in (i2^{-n},(i+1)2^{-n}] \qquad i = 1,2,\ldots$$

satisfies $\psi_n \in \overline{\mathcal{S}}$ by the above. For each (ω,x), $\psi_n(s,\omega,x) \to \psi(s,\omega,x)$ for Lebesgue a.a. s by Lebesgue's differentiation theorem (e.g. Theorem 8.8 of Rudin (1974)) and it follows easily that $\psi \in \overline{\mathcal{S}}$. Finally if $\psi \in \mathcal{L}^2_{\text{loc}}$, the obvious truncation argument and Dominated Convergence (set $\psi_n = (\psi \wedge n) \vee (-n)$) completes the proof. ∎

Proposition II.5.4. There is a unique linear extension of $M : \mathcal{S} \to \mathcal{M}_{\text{loc}}$ to a map $M : \mathcal{L}^2_{\text{loc}} \to \mathcal{M}_{\text{loc}}$ such that $\langle M(\psi)\rangle_t = \int_0^t X_s(\gamma\psi_s^2)ds$ $\forall t \ge 0$ a.s. $\forall \psi \in \mathcal{L}^2_{\text{loc}}$. If $\psi \in \mathcal{L}^2$, then $M(\psi)$ is a square integrable $\mathcal{F}_t$-martingale.

Proof. Assume M satisfies the above properties and $\psi \in \mathcal{L}^2_{\text{loc}}$. Choose $\psi_n \in \mathcal{S}$ as in Lemma II.5.3. By linearity,

$$\langle M(\psi) - M(\psi_n)\rangle_n = \langle M(\psi - \psi_n)\rangle_n = \int_0^n X_s\left(\gamma(\psi(s) - \psi_n(s))^2\right) ds$$
$$\leq 2^{-n} \text{ w.p. } > 1 - 2^{-n}.$$

A standard square function inequality and the Borel-Cantelli Lemma imply

$$\sup_{t\leq n} \left|M_t(\psi) - M_t(\psi_n)\right| \to 0 \quad \text{a.s.} \quad \text{as} \quad n \to \infty.$$

This proves uniqueness and shows how we must define the extension. The details showing that this extension has the required properties are standard and left for the reader. Finally it is easy to use the dominated convergence theorem to see that if $M \in \mathcal{M}_{\text{loc}}$ satisfies $E(\langle M\rangle_t) < \infty$ for all $t > 0$ then M is a square integrable martingale. This proves the final assertion for $\psi \in \mathcal{L}^2$. ∎

Remarks. II.5.5. (1) By polarization if $\phi, \psi \in \mathcal{L}^2_{\text{loc}}$,

$$\langle M(\phi), M(\psi)\rangle_t = \int_0^t X_s(\gamma\phi_s\psi_s)ds.$$

In particular if A_1 and A_2 are disjoint sets in $\mathcal{E}$ and $M(A_i) = M(1_{A_i})$, then $\langle M(A_1), M(A_2)\rangle_t = 0$ and so M_t is an orthogonal (local) martingale measure in the sense of Ch. 2 of Walsh (1986) where the reader can find more general constructions of this type.

(2) If $\int X_0(1)d\nu(X_0) < \infty$, then take $\phi \equiv 1$ in $(LMP)_\nu$ and use Gronwall's and Fatou's Lemmas to see that $E(X_t(1)) \leq E(X_0(1))e^{\bar{g}t}$ where $\bar{g} = \sup_x g(x)$. If ψ is $\mathcal{P} \times \mathcal{E}$ measurable and bounded on $[0,T] \times \Omega \times E$ for all $T > 0$, then the above shows that $\psi \in \mathcal{L}^2$ and so $M_t(\psi)$ is an L^2 martingale.

Let $\mathcal{M}_F$ denote the Borel σ-field on $M_F(E)$.

Theorem II.5.6. Assume:

(H_1) $\forall X_0 \in M_F(E)$ there is a solution to $(LMP)_{\delta_{X_0}}$.

(H_2) $\forall t \geq 0$ there is a Borel map $p_t : M_F(E) \to M_1(M_F(E))$ such that if $\nu \in M_1(M_F(E))$ and X satisfies $(LMP)_\nu$, then

$$\mathbb{P}(X_t \in A) = \int p_t(X_0, A)d\nu(X_0) \quad \forall A \in \mathcal{M}_F.$$

Then (a) and (b) of Theorem II.5.1 hold.

Remark. This is a version of the well-known result that uniqueness of the one-dimensional distributions for solutions of a martingale problem implies uniqueness in law and the strong Markov property of any solution. The proof is a simple adaptation of Theorem 4.4.2 of Ethier-Kurtz (1986). Note that (H_1) has already been verified because any limit point in Proposition II.4.2 will satisfy $(LMP)_{\delta_{X_0}}$.

Proof. Let Z satisfy $(LMP)_\nu$ on $(\Omega, \mathcal{F}, \mathcal{F}_t, \mathbb{P})$ and T be an a.s. finite $(\mathcal{F}_t)$-stopping time. Choose $A \in \mathcal{F}_T$ so that $\mathbb{P}(A) > 0$ and define Q on $(\Omega, \mathcal{F})$ by $Q(B) = \mathbb{P}(B \mid A)$. Let $W_t = Z_{T+t}$ and $\mathcal{G}_t = \mathcal{F}_{T+t}$, $t \geq 0$. If $\nu_0 = Q(W_0 \in \cdot)$, we claim W satisfies $(LMP)_{\nu_0}$ on $(\Omega, \mathcal{F}, \mathcal{G}_t, Q)$. Define

$$S_k = \inf\Big\{t : \int_T^{t+T} Z_r(1)dr > k\Big\} \wedge k.$$

One easily checks that $S_k + T$ is an $(\mathcal{F}_t)$-stopping time and S_k is a $(\mathcal{G}_t)$-stopping time. Clearly $S_k \uparrow \infty$ $\mathbb{P}$-a.s., and hence Q-a.s. Let M^Z be the martingale measure associated with Z, and for $\phi \in \mathcal{D}(A)$ let

$$M_t^W(\phi) = W_t(\phi) - W_0(\phi) - \int_0^t W_s(A^g\phi)\, ds = M_{T+t}^Z(\phi) - M_T^Z(\phi).$$

Fix $0 \leq s < t$ and $D \in \mathcal{G}_s = \mathcal{F}_{s+T}$. The definition of S_k ensures that

$$N_t = \int_0^t \int 1(T \leq s \leq S_k + T)\phi(x) M^Z(ds, dx)$$

is an L^2 bounded martingale ($\langle N \rangle_t$ is bounded), and therefore,

$$\begin{aligned} Q\left(\left(M_{t\wedge S_k}^W(\phi) - M_{s\wedge S_k}^W(\phi)\right) 1_D\right) &= \mathbb{P}\left((N_{T+t} - N_{T+s})\, 1_{D\cap A}\right)/\mathbb{P}(A) \\ &= 0 \end{aligned}$$

by optional sampling, because $D \cap A \in \mathcal{F}_{s+T}$. This proves $M_t^W(\phi)$ is a $(\mathcal{G}_t)$-local martingale under Q and a similar argument shows the same is true of

$$M_t^W(\phi)^2 - \int_0^t W_r(\gamma\phi^2)dr.$$

This shows that W satisfies $(LMP)_{\nu_0}$ on $(\Omega, \mathcal{F}, \mathcal{G}_t, Q)$. (H_2) implies that for $t \geq 0$ and $C \in \mathcal{M}_F$,

$$Q(W_t \in C) = \int p_t(\mu, C)\nu_0(d\mu),$$

that is

$$\mathbb{P}(Z_{T+t} \in C \mid A) = \mathbb{E}(p_t(Z_T, C) \mid A),$$

and so

$$\mathbb{P}(Z_{T+t} \in C \mid \mathcal{F}_T) = p_t(Z_T, C) \quad \mathbb{P}\text{-a.s.} \tag{II.5.1}$$

Therefore $\{Z_t\}$ is $(\mathcal{F}_t)$-strong Markov with Borel transition kernel p_t and initial law ν, and hence the uniqueness in II.5.1 (a) is proved.

(H_1) and the above allow us to use the above Markov kernel to define the law, $\mathbb{P}_{X_0}$ (on Ω_X) of any solution to $(LMP)_{\delta_{X_0}}$. (II.5.1) implies $X_0 \to \mathbb{P}_{X_0}(A)$ is Borel for finite-dimensional A and hence for all A in $\mathcal{F}_X$. It also implies

$$\mathbb{P}(Z_{T+\cdot} \in A \mid \mathcal{F}_T)(\omega) = \mathbb{P}_{Z_T(\omega)}(A)$$

first for A finite-dimensional and hence for all A in $\mathcal{F}_X$.

Now consider the "canonical solution" to $(LMP)_\nu$, $X_t(\omega) = \omega_t$, on Ω_X under $\mathbb{P}_\nu(\cdot) = \int \mathbb{P}_{X_0}(\cdot)d\nu(X_0)$. It is easy to check that X solves $(LMP)_\nu$ under $\mathbb{P}_\nu$ for any $\nu \in M_1(M_F(E))$. (Note that if $S_k = \inf\{t : \int_0^t X_s(1)ds \geq k\} \wedge k$ then $M_{t\wedge S_k}(\phi)$ is a square integrable martingale under $\mathbb{P}_{X_0}$ and $\mathbb{P}_{X_0}(M_{t\wedge S_k}(\phi)^2) \leq \|\gamma\phi^2\|_\infty k$ for each $X_0 \in M_F(E)$ and so the same is true under $\mathbb{P}_\nu$.) This proves the existence part of Theorem II.5.1(a). By the above $(\Omega_X, \mathcal{F}_X, \mathcal{F}_t^X, X, \mathbb{P}_{X_0})$ is Borel strong Markov, and the proof is complete. ∎

To verify (H_2) we first extend $(LMP)_\nu$ to time dependent functions. Recall that X satisfies $(LMP)_\nu$.

Definition. Let $T > 0$. A function $\phi : [0,T] \times E \to \mathbb{R}$ is in $\mathcal{D}(\vec{A})_T$ iff

(1) For any x in E, $t \to \phi(t,x)$ is absolutely continuous and there is a jointly Borel measurable version of its Radon-Nikodym derivative $\dot{\phi}(t,x) = \frac{\partial\phi}{\partial t}(t,x)$ which is bounded on $[0,T] \times E$ and continuous in x for each $t \in [0,T]$.

(2) For any $t \in [0,T]$, $\phi(t,\cdot) \in \mathcal{D}(A)$ and $A\phi_t$ is bounded on $[0,T] \times E$.

Proposition II.5.7. If $\phi \in D(\vec{A})_T$, then

$$X_t(\phi_t) = X_0(\phi_0) + M_t(\phi) + \int_0^t X_s(\dot{\phi}_s + A^g\phi_s)ds \quad \forall t \in [0,T] \quad \text{a.s.}$$

Proof. Set $t_i^n = i2^{-n}$ and define $\phi^n(t,x) = 2^n \int_{t_{i-1}^n}^{t_i^n} \phi(t,x)dt$ if $t_{i-1}^n \leq t < t_i^n$, $i \geq 1$. Clearly $\phi^n \xrightarrow{bp} \phi$. It is easy to see $\phi^n(t,\cdot) \in \mathcal{D}(A)$ and

$$A\phi_t^n(x) = \int_{t_{i-1}^n}^{t_i^n} A\phi_u(x)du\, 2^n \quad \text{if} \quad t_{i-1}^n \leq t < t_i^n. \tag{II.5.2}$$

By the (local) martingale problem we have
(II.5.3)

$$X_t(\phi_t^n) = X_{t_{i-1}^n}(\phi_{t_{i-1}^n}^n) + \int_{t_{i-1}^n}^t X_s\left(A^g\phi_{t_{i-1}^n}^n\right)ds + \int_{t_{i-1}^n}^t \int \phi^n(s,x)dM(s,x), \; t \in [t_{i-1}^n, t_i^n).$$

By the continuity of X we have for $i \geq 2$

$$X_{t_{i-1}^n}\left(\phi_{t_{i-1}^n}^n\right) = X_{t_{i-1}^n-}\left(\phi_{t_{i-1}^n-}^n\right) + X_{t_{i-1}^n}\left(\phi_{t_{i-1}^n}^n - \phi_{t_{i-2}^n}^n\right),$$

and so for $i \geq 2$ and $t \in [t_{i-1}^n, t_i^n)$,

$$X_t(\phi_t^n) = X_{t_{i-1}^n-}\left(\phi_{t_{i-1}^n-}^n\right) + X_{t_{i-1}^n}\left(\phi_{t_{i-1}^n}^n - \phi_{t_{i-2}^n}^n\right) + \int_{t_{i-1}^n}^t X_s\left(A^g \phi_{t_{i-1}^n}^n\right) ds$$

$$+ \int_{t_{i-1}^n}^t \int \phi^n(s,x) dM(s,x).$$

If $t \uparrow t_i^n$, we get a telescoping sum which we may sum over $t_i^n \leq t$ $(i \geq 2)$ and add (II.5.3) with $i = 1$ and $t = t_1^n-$, together with the above expression for $X_t(\phi_t^n) - X_{t_i^n-}(\phi_{t_i^n-})$, where $t \in [t_i^n, t_{i+1}^n)$. If $C_t^n = \sum_{i=1}^{\infty} 1(t_i^n \leq t) X_{t_i^n}(\phi_{t_i^n}^n - \phi_{t_{i-1}^n}^n)$, we get

$$(II.5.4) \qquad X_t(\phi_t^n) = X_0(\phi_0^n) + C_t^n + \int_0^t X_s\left(A\phi_s^n + g\phi_s^n\right) ds + M_t(\phi^n).$$

Note that if $[t] = [2^n t]2^{-n}$, $[t]^+ = ([2^n t] + 1)2^{-n}$, then

$$C_t^n = \sum_{i=1}^{[2^n t]} \int_{t_{i-1}^n}^{t_i^n} X_{t_i^n}\left(\phi_{s+2^{-n}} - \phi_s\right) 2^n ds$$

$$= \sum_{i=1}^{[2^n t]} \int_{t_{i-1}^n}^{t_i^n} \int_s^{s+2^{-n}} X_{t_i^n}(\dot{\phi}_r) dr\, ds 2^n$$

$$= \sum_{i=1}^{[2^n t]} \int_{t_{i-1}^n}^{t_i^n} \int_s^{s+2^{-n}} X_{t_i^n}(\dot{\phi}_r) - X_r(\dot{\phi}_r)\, dr ds 2^n + \int_0^{2^{-n}} r X_r(\dot{\phi}_r)\, dr 2^n$$

$$+ \int_{2^{-n}}^{[t]} X_r(\dot{\phi}_r)\, dr + \int_{[t]}^{[t]^+} X_r(\dot{\phi}_r)([t]^+ - r)\, dr 2^n.$$

The sum of the last three terms approach $\int_0^t X_r(\dot{\phi}_r)\, dr$ for all $t \geq 0$ a.s. If

$$h_n(r) = \sup\{|X_u(\dot{\phi}_r) - X_r(\dot{\phi}_r)| : |u - r| \leq 2^{-n}, u \geq 0\},$$

then $h_n \xrightarrow{bp} 0$ a.s. by the continuity of $X_u(\dot{\phi}_r)$ in u and the first term is at most

$$\int_0^{[t]^+} h_n(r)\, dr \to 0 \qquad \forall t \geq 0 \quad \text{a.s.}$$

We have proved that $C_t^n \to \int_0^t X_r(\dot{\phi}_r)\,dr$ for all $t \geq 0$ a.s. By (II.5.2) we also have

$$\int_0^t X_s(A\phi_s^n)\,ds = \sum_{i=1}^{[2^n t]+1} \int_{t_{i-1}^n}^{t_i^n \wedge t} \int_{t_{i-1}^n}^{t_i^n} X_s(A\phi_u)\,du\,ds 2^n,$$

and an argument very similar to the above shows that

$$\lim_{n\to\infty} \int_0^t X_s(A\phi_s^n)\,ds = \int_0^t X_s(A\phi_s)ds \qquad \forall t \geq 0 \quad \text{a.s.}$$

By considering $\langle M(\phi) - M(\phi^n)\rangle_t$ we see that

$$\sup_{t\leq K} |M_t(\phi) - M_t(\phi^n)| \xrightarrow{\mathbb{P}} 0 \quad \text{as} \quad n \to \infty \quad \text{for all} \quad K > 0.$$

The other terms in (II.5.4) converge trivially by Dominated Convergence and so we may let $n \to \infty$ in (II.5.4) to complete the proof. ∎

Not surprisingly the above extension is also valid for the martingale problem for Y.

Notation. If $\phi \in \mathcal{D}(\vec{A})_T$, let $\vec{A}\phi(t,x) = \dot{\phi}(t,x) + A\phi_t(x)$ for $(t,x) \in [0,T] \times E$.

Proposition II.5.8. If $\phi \in \mathcal{D}(\vec{A})_T$, then

$$N_t = \phi(t, Y_t) - \phi(0, Y_0) - \int_0^t \vec{A}\phi(s, Y_s)\,ds \quad t \in [0,T]$$

is a bounded a.s. cadlag $\mathcal{D}_t$-martingale under P^x for all $x \in E$. Its jumps are contained in the jumps of Y a.s.

Proof. The continuity properties of ϕ imply that

$$\lim_{s\to t+} \phi(s, Y_s) = \phi(t, Y_t) \ \text{ for all } t \in [0,T) \ P^x - a.s.$$

and

$$\lim_{s\to t-} \phi(s, Y_s) = \phi(t, Y_{t-}) \ \text{ for all } t \in (0,T] \ P^x - a.s.$$

Therefore N is a.s. cadlag on $[0,T]$ and can only jump at the jump times of Y a.s. The definition of $\mathcal{D}(\vec{A})_T$ implies that ϕ and $\vec{A}\phi$ are bounded on $[0,T] \times E$ and hence N is also uniformly bounded.

Take mean values in Proposition II.5.7 with $g \equiv 0$ and $X_0 = \delta_x$ and use Remark II.5.5 (2) and Exercise II.5.2 (b)(i) below to see that

$$P_t\phi_t(x) = \phi_0(x) + \int_0^t P_s(\vec{A}\phi_s)(x)\,ds \ \text{ for all } (t,x) \in [0,T] \times E.$$

If $u \in [0,T)$ is fixed then $(t,x) \mapsto \phi(u+t,x)$ is in $\mathcal{D}(\vec{A})_{T-u}$ and so the above implies

$$P_t\phi_{u+t}(x) = \phi_u(x) + \int_0^t P_s(\vec{A}\phi_{s+u})\,ds\ \forall (t,x) \in [0,T-u]\times E \quad \forall u \in [0,T).$$

It is now a simple exercise using the Markov property of Y to see that the above implies that N_t is a $\mathcal{D}_t$-martingale under each P^x. ∎

Green Function Representation

Let $P_t^g\phi(x) = E^x\Big(\phi(Y_t)\exp\Big\{\int_0^t g(Y_s)\,ds\Big\}\Big)$.

Exercise II.5.1. (a) Show that $P_t^g : C_b(E) \to C_b(E)$.
Hint: One approach is to use a Taylor series for exp and recall $P_t : C_b(E) \to C_b(E)$.
(b) Show $\phi \in \mathcal{D}(A) \Leftrightarrow (P_t^g\phi - \phi)t^{-1} \overset{bp}{\to} \psi \in C_b(E)$ as $t \downarrow 0$, and in this case $\psi = A^g\phi$.
(c) Show that $P_t^g : \mathcal{D}(A) \to \mathcal{D}(A)$ and $\frac{d}{dt}P_t^g\phi = A^g P_t^g\phi = P_t^g A^g\phi \quad \forall \phi \in \mathcal{D}(A)$.

The next Exercise will be used extensively.

Exercise II.5.2. Assume X solves $(LMP)_\nu$.
(a) Prove that $\forall \phi \in b\mathcal{E}$

$$(GFR) \qquad X_t(\phi) = X_0\left(P_t^g\phi\right) + \int_0^t\int P_{t-s}^g\phi(x)dM(s,x) \qquad \text{a.s.} \quad \forall t \geq 0.$$

Hint: Assume first $\phi \in \mathcal{D}(A)$ and apply Proposition II.5.7 for an appropriate choice of $\phi(s,x)$.
(b) Assume $\nu = \delta_{X_0}$. (i) Show that $\mathbb{P}\big(X_t(\phi)\big) = X_0\big(P_t^g\phi\big)\ \forall t \geq 0$ and $\phi \in b\mathcal{E}$.
(ii) Show that if $0 \leq s \leq t$ and $\phi, \psi \in b\mathcal{E}$,

$$\mathbb{P}\left(X_s(\phi)X_t(\psi)\right) = X_0\left(P_s^g\phi\right)X_0\left(P_t^g\psi\right) + \int_0^s X_0(P_r^g\left(\gamma P_{s-r}^g\phi P_{t-r}^g\psi\right))\,dr.$$

Hint: Recall Remark II.5.5.

Definition. If W is an $M_F(E)$-valued random vector and $\phi \in b\mathcal{E}_+$, $L_W(\phi) = E(e^{-W(\phi)}) \equiv E\big(e_\phi(W)\big)$ is the Laplace functional of W, or of $P_W = P(W \in \cdot)$.

Lemma II.5.9. Assume $D_0 \subset (b\mathcal{E})_+$ satisfies $\overline{D_0}^{bp} = b\mathcal{E}_+$. Then $L_W(\phi) = L_{W'}(\phi)$ for all $\phi \in D_0$ implies $P_W = P_{W'}$.

Proof. Clearly equality of L_W and $L_{W'}$ on D_0 implies $L_W = L_{W'}$ on $b\mathcal{E}_+$. An elementary argument (see the proof of Proposition 3.4.4 of Ethier-Kurtz (1986)) shows that there is a countable convergence determining set $V \subset C_b(E)_+$ (i.e. $\nu_n \to \nu$ in $M_F(E) \Leftrightarrow \nu_n(\phi) \to \nu(\phi) \quad \forall \phi \in V$). For any $\phi \in V$, $\nu \to \nu(\phi)$ is measurable with respect to $\sigma(e_\phi : \phi \in V)$. This implies that the class of open sets in $M_F(E)$, and hence $\mathcal{M}_F$, is contained in $\sigma(e_\phi : \phi \in V)$. Apply the Monotone Class Lemma II.5.2 with $\mathcal{H} = \{\Phi \in b\mathcal{M}_F : E(\Phi(W)) = E(\Phi(W'))\}$ and $\mathcal{H}_0 = \{e_\phi : \phi \in C_b(E)_+\}$ to see that $L_W = L_{W'}$ on $C_b(E)_+$ implies $P_W = P_{W'}$. ∎

We will verify (H_2) by giving an explicit formula for L_{X_t}. If X and X' are independent solutions of $(LMP)_{\delta_{X_0}}$ and $(LMP)_{\delta_{X_0'}}$, respectively, then it is easy to check that $X+X'$ satisfies $(LMP)_{\delta_{X_0+X'_0}}$. This "additive property" is also clear for the approximating branching particle systems considered in Proposition II.4.2; the particles do not interact and so superimposing two such systems gives another such system. This leads to $L_{X_t+X'_t} = L_{X_t}\cdot L_{X'_t}$ and suggests the use of Laplace functionals to characterize the law of X_t. It also explains the "multiplicative" part of the terminology in "critical multiplicative branching measure diffusions", the catchy name for Dawson-Watanabe superprocesses prior to 1987.

Let $\psi\in\mathcal{D}(\vec{A})_t$ for a fixed $t>0$ and $f\in C_b(E)_+$. By Proposition II.5.7 and Itô's Lemma for $u\le t$,

$$\exp\Big\{-X_u(\psi_u)-\int_0^u X_s(f)\,ds\Big\}$$
$$=\exp\big(-X_0(\psi_0)\big)-\int_0^u\int\exp\Big\{-X_s(\psi_s)-\int_0^s X_r(f)\,dr\Big\}\psi(s,x)dM(s,x)$$
$$+\int_0^u\exp\Big\{-X_s(\psi_s)-\int_0^s X_r(f)\,dr\Big\}\Big[-X_s(\dot\psi_s+A^g\psi_s+f-\gamma\psi_s^2/2)\Big]\,ds.$$

Let N_u denote the stochastic integral on the righthand side. Let $\phi\in\mathcal{D}(A)_+$. Now choose a non-negative ψ so that the drift term vanishes, and $\psi_t=\phi$, i.e.,

$$(II.5.5)\qquad \dot\psi_s+A^g\psi_s+f-\gamma\psi_s^2/2=0,\quad 0\le s\le t\quad \psi_t=\phi.$$

The previous equation then becomes

$$(II.5.6)\qquad \exp\Big\{-X_u(\phi)-\int_0^u X_s(f)\,ds\Big\}=\exp\big(-X_0(\psi_0)\big)+N_u\quad u\le t.$$

This shows that the local martingale N is bounded and therefore is a martingale satisfying $E(N_t)=0$. Take expectations in (II.5.6) with $u=t$ to see that

$$(II.5.7)\qquad E\Big(\exp\Big\{-X_t(\phi)-\int_0^t X_s(f)\,ds\Big\}\Big)=\int e^{-X_0(\psi_0)}d\nu(X_0).$$

If $a(x,\lambda)=g(x)\lambda-\gamma(x)\lambda^2/2$ and $V_s\equiv V_s^f\phi\equiv\psi_{t-s}$, then $\psi\in\mathcal{D}(\vec{A})_t$ iff $V\in\mathcal{D}(\vec{A})_t$, and (II.5.5) and (II.5.7) (for all $t\ge0$) become, respectively:

$$(SE)_{\phi,f}\qquad \frac{\partial V_s}{\partial s}=AV_s+a(\cdot,V_s)+f,\quad s\ge0,\quad V_0=\phi,$$

and

$$(LE)\qquad E\Big(\exp\Big\{-X_t(\phi)-\int_0^t X_s(f)\,ds\Big\}\Big)=\int e^{-X_0(V_t^f\phi)}d\nu(X_0)\quad\forall t\ge0.$$

These arguments trivially localize to $t\in[0,T]$ and hence we have proved:

Proposition II.5.10. Let $\phi \in \mathcal{D}(A)_+$ and $f \in C_b(E)_+$. If $V \in \mathcal{D}(\vec{A})_T$ for some $T > 0$ and is a non-negative solution V to $(SE)_{\phi,f}$ on $[0,T] \times E$, then (LE) holds for all $0 \le t \le T$ for any X satisfying $(LMP)_\nu$.

The next result will extend this to a larger class of ϕ and f and solutions to the following mild form of (SE):

$$(ME)_{\phi,f} \qquad V_t = P_t\phi + \int_0^t P_{t-s}(f + a(\cdot, V_s))\, ds.$$

It is easy to check, e.g. by using Proposition II.5.8 to write $V_{t-s}(Y_s)$ as the sum of a martingale and process of bounded variation, that any solution to (SE) with $V|_{[0,t]\times E} \in D(\vec{A})_t$ for all $t \ge 0$ satisfies (ME). Conversely, a solution to (ME) will satisfy

$$(II.5.8) \qquad \frac{V_{t+h} - V_t}{h} = \frac{(P_h - I)}{h} V_t + \frac{1}{h}\int_0^h P_r\big(a(\cdot, V_{t+h-r}) + f\big)\, dr,$$

and hence should satisfy (SE) provided that these limits exist.

Theorem II.5.11. Let $\phi, f \in b\mathcal{E}_+$.

(a) There is a unique jointly Borel measurable solution $V_t^f\phi(x)$ of $(ME)_{\phi,f}$ such that $V^f\phi$ is bounded on $[0,T] \times E$ for all $T > 0$. Moreover $V^f\phi \ge 0$.

(b) If, in addition, $\phi \in \mathcal{D}(A)_+$ and f is continuous, then $V^f\phi\, |_{[0,T]\times E} \in \mathcal{D}(\vec{A})_T$ $\forall T > 0$, $\dot{V}_t^f\phi(x)$ and $AV_t^f(x)$ are continuous in t (as well as x), and $V^f\phi$ satisfies $(SE)_{\phi,f}$.

(c) If X is any solution of $(LMP)_\nu$ then (LE) holds.

In view of the locally Lipschitz nature of $a(x, \lambda)$ in λ, (a) and (b) of the above result are to be expected and will follow from a standard fixed point argument, although some care is needed as a does not have linear growth in λ. The regularity of the solutions in (b) will be the delicate part of the argument. Note that if the spatial motion is Brownian, the argument here may be simplified considerably because of the regularity of the Brownian transition density.

(c) follows immediately for $\phi \in \mathcal{D}(A)_+$ and $f \in C_b(E)_+$ by Proposition II.5.10. It is then not hard to derive (LE) for all $\phi, f \in b\mathcal{E}_+$ by taking bounded pointwise limits.

We defer the details of the proof until the next section and now complete the **Proof of Theorem II.5.1.** We first verify (H_2) of Theorem II.5.6. If X satisfies $(LMP)_{\delta_{X_0}}$ for $X_0 \in M_F(E)$, then (from Theorem II.5.11) for each $\phi \in \mathcal{D}(A)_+$

$$(II.5.9) \qquad \mathbb{P}\left(e^{-X_t(\phi)}\right) = e^{-X_0(V_t^0\phi)}.$$

This uniquely determines the law, $p_t(X_0, \cdot)$, of X_t by Lemma II.5.9. Let $X_0^n \to X_0$ in $M_F(E)$, then for any $\phi \in C_b(E)$, the mean measure calculation in Exercise II.5.2(b) shows that

$$\int \mu(\phi) p_t(X_0^n, d\mu) = X_0^n(P_t^g\phi) \to X_0(P_t^g\phi) = \int \mu(\phi) p_t(X_0, d\mu).$$

This weak convergence shows that if $\varepsilon > 0$ there is a compact subset of E, K, such that $\sup_n \int \mu(K^c)p_t(X_0^n, d\mu) < \varepsilon$. This shows that $\{p_t(X_0^n, \cdot) : n \in \mathbb{N}\}$ is tight on $M_F(E)$. For example, one can apply Theorem II.4.1 to the set of constant $M_F(E)$-valued processes. (II.5.9) shows that for $\phi \in \mathcal{D}(A)_+$

$$\int e_\phi(\mu)p_t(X_0^n, d\mu) = e^{-X_0^n(V_t^0\phi)} \to e^{-X_0(V_t^0\phi)} = \int e_\phi(\mu)p_t(X_0, d\mu),$$

and hence that $p_t(X_0, \cdot)$ is the only possible weak limit point. We have proved that $X_0 \to p_t(X_0, \cdot)$ is continuous, and in particular is Borel.

If X satisfies $(LMP)_\nu$, then Theorem II.5.11 (c) shows that

$$\mathbb{P}(e_\phi(X_t)) = \int e^{-X_0(V_t^0\phi)} d\nu(X_0) = \int\int e_\phi(\mu)p_t(X_0, d\mu)d\nu(X_0) \quad \forall \phi \in b\mathcal{E}_+,$$

and so (H_2) follows by Lemma II.5.9. This allows us to apply Theorem II.5.6 and infer (a) and (b). The above continuity of $p_t(X_0, \cdot)$ in X_0 implies (d). Finally the uniqueness in (a) shows that all the weak limit points in Proposition II.4.2 coincide with $\mathbb{P}_{X_0}$ and so the convergence in (c) follows. ∎

The following Feynman-Kac formula shows solutions to $(ME)_{\phi,f}$ for non-negative ϕ and f are necessarily non-negative and will be useful in Chapter III.

Proposition II.5.12. Suppose ϕ, $f \in b\mathcal{E}$ and $V : [0, T] \to \mathbb{R}$ is a bounded Borel function satisfying $(ME)_{\phi,f}$ for $t \le T$. For $u \le t \le T$ define

$$C_u = C_u^{(t)} = \int_0^u g(Y_r) - \frac{\gamma(Y_r)}{2}V(t-r, Y(r))\,dr.$$

Then for all $(t, x) \in [0, T] \times E$,

$$V(t,x) = E^x(\phi(Y_t)e^{C_t}) + \int_0^t E^x(f(Y_r)e^{C_r})\,dr.$$

Proof. Let $0 \le s \le t$. Use $(ME)_{\phi,f}$ with $t - s$ in place of t and apply P_s to both sides to derive

$$\begin{aligned} P_sV_{t-s} &= P_t\phi + \int_0^{t-s} P_{t-r}(f + gV_r - \gamma V_r^2/2)\,dr \\ &= V_t - \int_0^s P_r(f + gV_{t-r} - \gamma V_{t-r}^2/2)dr. \end{aligned}$$

The Markov Property now shows that

$$N_s = V_{t-s}(Y_s) - V_t(Y_0) + \int_0^s f(Y_r) + g(Y_r)V_{t-r}(Y_r) - \gamma(Y_r)V_{t-r}(Y_r)^2/2\,dr, \quad s \le t$$

is a bounded $\mathcal{D}_s$-martingale under P^x for all x. Itô's Lemma then implies

$$e^{C_s}V_{t-s}(Y_s) = V_t(x) - \int_0^s e^{C_r}f(Y_r)\,dr + \int_0^s e^{C_r}dN_r.$$

The stochastic integral is a mean zero L^2-martingale under each P^x and so we may set $s = t$ and take means to complete the proof. ∎

Extinction Probabilities

The Laplace functional equation (LE) is a powerful tool for the analysis of X. As a warm-up we use it to calculate extinction probabilities for X. Assume $g(\cdot) \equiv g \in \mathbb{R}$, $\gamma(\cdot) \equiv \gamma > 0$, and $X_0 \in M_F(E)$ is deterministic. Then, setting $\phi \equiv 1$ in $(LMP)_{\delta_{X_0}}$, we see that $X(1)$ satisfies the martingale problem characterizing the solution of

$$(II.5.10) \qquad X_t(1) = X_0(1) + \int_0^t \sqrt{\gamma X_s(1)}dB_s + \int_0^t gX_s(1)\, ds,$$

where B is a linear Brownian motion. An immediate consequence of Theorem II.5.1(c) is $X^n_\cdot(1) \overset{w}{\Rightarrow} X_\cdot(1)$ in $D(\mathbb{R})$, which when $g \equiv 0$ reduces to Feller's Theorem II.1.2.

Assume first $g = 0$ and for $\lambda > 0$ let $V_t = V_t^\lambda$ solve

$$\frac{\partial V_t}{\partial t} = AV_t - \frac{\gamma V_t^2}{2}, \quad V_0 \equiv \lambda.$$

Clearly the solution is independent of x and so one easily gets

$$V_t^\lambda = 2\lambda(2 + \lambda t\gamma)^{-1}.$$

(LE) implies that the Laplace functional of the total mass of the DW-superprocess is

$$(II.5.11) \qquad \mathbb{P}_{X_0}\left(e^{-\lambda X_t(1)}\right) = \exp\left\{-\frac{X_0(1)2\lambda}{2 + \lambda t\gamma}\right\}.$$

Let $\lambda \to \infty$ to see

$$(II.5.12) \qquad \mathbb{P}_{X_0}(X_t = 0) = \mathbb{P}_{X_0}(X_s = 0\, \forall s \geq t) = \exp\left\{\frac{-2X_0(1)}{t\gamma}\right\}.$$

In particular, by letting $t \to \infty$ we see that X becomes extinct in finite time $\mathbb{P}_{X_0}$-a.s. See Knight (1981, p. 100) for the transition density of this total mass process.

Exercise. II.5.3. Assume $\gamma(\cdot) \equiv \gamma > 0, g(\cdot) \equiv g$ are constants.

(a) Find $\mathbb{P}_{X_0}(X_s \equiv 0 \quad \forall s \geq t)$. $\left(\text{Answer} : \exp\left\{\frac{-2X_0(1)g}{\gamma(1-e^{-gt})}\right\} \text{ if } g \neq 0.\right)$

(b) Show that $\mathbb{P}_{X_0}$ (X becomes extinct in finite time) $= \begin{cases} 1 & \text{if } g \leq 0 \\ \exp\left\{\frac{-2X_0(1)g}{\gamma}\right\} & \text{if } g > 0 \end{cases}$.

(c) If $g > 0$ prove that $\mathbb{P}_{X_0}$-a.s.

$$X \text{ becomes extinct in finite time or } \lim_{t\to\infty} X_t(1) = \infty.$$

Hint. Show that $e^{-\lambda X_t(1)}$ is a supermartingale for sufficiently small $\lambda > 0$ and $\lim_{t\to\infty} \mathbb{E}_{X_0}\left(e^{-\lambda X_t(1)}\right) = \exp\left\{\frac{-2X_0(1)g}{\gamma}\right\}$.

Exercise II.5.4, Assume $X_0 \in M_F(E) - \{0\}$, $\nu_N(x, dk) \equiv \nu_N(dk)$ is independent of x and $g_N \equiv 0$. Prove that for ϕ bounded and measurable,

$$E(X_t^N(\phi)|X_s^N(1), s \geq 0) = \frac{X_0(P_t\phi)}{X_0(1)} X_t^N(1).$$

Conclude that if X is the $(Y, \gamma, 0)$-DW-superprocess ($\gamma \geq 0$ is constant), then

$$E(X_t(\phi)|X_s(1), s \geq 0) = \frac{X_0(P_t\phi)}{X_0(1)} X_t(1).$$

Hint. Condition first on the larger σ-field $\sigma(N^\alpha, \alpha \in I) \vee \sigma(M_N)$ (recall M_N is the Poisson number of initial points).

Remark II.5.13. Assume that X satisfies $(LMP)_\nu$ on $(\Omega, \mathcal{F}, \mathcal{F}_t, \mathbb{P})$, where A is the generator of the d-dimensional symmetric stable process of index $\alpha \in (0, 2]$ and the smaller class $C_b^\infty(\mathbb{R}^d)$ is used in place of $\mathcal{D}(A)$. Recalling that $C_b^\infty(\mathbb{R}^d)$ is a core for $\mathcal{D}(A)$ (see Example II.2.4), we may pass to the bounded pointwise closure of $\{(\phi, A\phi) : \phi \in C_b^\infty(\mathbb{R}^d)\}$ in $(LMP)_\nu$ by Dominated Convergence. Here note that if $T_k = \inf\{t : X_t(1) > k\} \wedge k$ and $\phi_n \xrightarrow{bp} \phi$, then $M_{t\wedge T_k}(\phi_n)$ is a bounded martingale for all n. Therefore X satisfies $(LMP)_\nu$ and so is an $(\mathcal{F})_t$-(A, γ, g)-DW superprocess with initial law ν.

Exercise II.5.5. Let X be the $(Y, \gamma, 0)$-DW superprocess starting at $X_0 \in M_F(E)$, where Y is the d-dimensional symmetric stable process of index $\alpha \in (0, 2]$ and $\gamma > 0$ is constant. For $\gamma_0, \lambda > 0$, let $\phi_\lambda(x) = \phi(x\lambda^{-1/\alpha})$ and

$$X_t^{(\lambda)}(\phi) = \frac{\gamma_0}{\lambda} X_{\lambda t}(\phi_\lambda), \quad t \geq 0.$$

Prove that $X^{(\lambda)}$ is a $(Y, \gamma\gamma_0, 0)$-DW superprocess (starting of course at $X_0^{(\lambda)}$).

6. Proof of Theorem II.5.11.

Step 1. If $\phi, f \in b\mathcal{E}$, there is a $t_{\max} \in (0, \infty]$ and a unique solution V to $(ME)_{\phi,f}$ on $[0, t_{\max})$ which is bounded on $[0, T] \times E$ for all $T < t_{\max}$ and satisfies $\varliminf_{t\uparrow t_{\max}} \|V_t\| = \infty$ if $t_{\max} < \infty$. If in addition $\phi, f \in C_b(E)$ and $\|P_t\phi - \phi\| \to 0$ as $t \downarrow 0$ (as is the case for $\phi \in \mathcal{D}(A)$), then $V : [0, t_{\max}) \to C_b(E)$ is continuous in norm.

This is a standard fixed point argument which only requires $a(\cdot, 0) = 0$ and

$$\forall K > 0, a\,|_{E\times[-K,K]} \in C_b(E \times [-K, K]) \text{ and } |a(x, \lambda) - a(x, \lambda')| \leq C_K|\lambda - \lambda'| \tag{II.6.1}$$

for all x in $E, \lambda, \lambda' \in [-K, K]$ for some increasing $\{C_K\}$ and $C_0 \geq 1$.

We start with ϕ, f as in the second part of the above assertion. We will view f as fixed but will let ϕ vary and will choose $\delta = \delta(\|\phi\|) > 0$ below. Define $\psi : C([0, \delta], C_b(E)) \to C([0, \delta], C_b(E))$ by

$$\psi(U)(t) = P_t\phi + \int_0^t P_{t-s}(f + a(U_s))ds.$$

Note that $a(U_s) \equiv a(\cdot, U_s(\cdot)) \in C_b(E)$. To see that ψ does map into the above space, note first that $t \to P_t\phi$ is in $C(\mathbb{R}_+, C_b(E))$ by our choice of ϕ and the semigroup property. If $0 \le t < t+h \le \delta$, then

$$\left| \int_0^{t+h} P_{t+h-s}(f + a(U_s))ds - \int_0^t P_{t-s}(f + a(U_s))ds \right|$$

$$\le \left| \int_0^h P_{t+h-s}(f + a(U_s))ds \right| + \int_0^t \|P_{t-s}(a(U_{s+h})) - P_{t-s}(a(U_s))\|ds$$

$$\le \left[\|f\| + \sup_{s\le\delta} \|a(U_s)\| \right] h + \int_0^t \|a(U_{s+h}) - a(U_s)\|ds$$

$$\to 0 \quad \text{as} \quad h \downarrow 0$$

by (II.6.1) and $\sup_{s\le\delta} \|U_s\| < \infty$. Hence ψ is as claimed above.

Take $K = 2\|\phi\| + 1$ and for $U \in C([0,\delta], C_b(E))$ define $\|U\| = \sup_{t\le\delta} \|U_t\|$ and let $\overline{B}(0,K)$ be the set of such U with $\|U\| \le K$. If $U \in \overline{B}(0,K)$ and

$$0 < \delta \le \varepsilon_1(\|\phi\|) = (K - \|\phi\|)/(\|f\| + KC_K),$$

then

$$\|\psi(U)\| \le \|\phi\| + \delta\|f\| + \delta \sup_{|\lambda|\le K} \|a(\cdot,\lambda)\| \le \|\phi\| + \delta[\|f\| + KC_K] \le K$$

and therefore $\psi : \overline{B}(0,K) \to \overline{B}(0,K)$. If, in addition, $0 < \delta \le \varepsilon_2(\|\phi\|) = 1/2C_K$, then an application of (6.1) shows that for $U, V \in \overline{B}(0,K)$

$$\|\psi(U) - \psi(V)\| \le \int_0^\delta \|a(U_s) - a(V_s)\|ds \le C_K\delta\|U - V\| \le \frac{1}{2}\|U - V\|.$$

Now let $\delta = \delta(\|\phi\|) = \min(\varepsilon_1(\|\phi\|), \varepsilon_2(\|\phi\|))$ and note that

$$(II.6.2) \qquad \inf_{0\le r\le M} \delta(r) > 0 \quad \text{for any} \quad M > 0.$$

Then ψ is a contraction on the complete metric space $\overline{B}(0,K)$ and so has a unique fixed point V_t which solves $(ME)_{\phi,f}$ for $t \le \delta$.

To repeat this construction with V_δ in place of ϕ we must check that $\|P_hV_\delta - V_\delta\| \to 0$ as $h \downarrow 0$. Use $(ME)_{\phi,f}$ at $t = \delta$ to see this reduces to

$$\| \int_0^\delta P_{h+\delta-s}a(V_s) - P_{\delta-s}a(V_s)ds\| \to 0 \quad \text{as} \quad h \downarrow 0.$$

The above norm is at most ($0 < h < \delta$)

$$\|\int_0^h P_{h+\delta-s}a(V_s)ds\| + \|\int_{\delta-h}^{\delta} P_{\delta-s}a(V_s)ds\| + \int_0^{\delta-h} \|P_{\delta-s}\big(a(V_{s+h}) - a(V_s)\big)\|ds$$

$$\leq 2\sup_{s\leq\delta}\|a(V_s)\|h + \int_0^{\delta-h} \|a(V_{s+h}) - a(V_s)\|ds$$

$$\to 0 \quad \text{as} \quad h \downarrow 0$$

by the norm-continuity of V and (II.6.1). By repeating the previous argument with V_δ in place of ϕ we can extend $V.$ to a norm-continuous solution to $(ME)_{\phi,f}$ on $[0, \delta_1 + \delta_2]$ where $\delta_1 = \delta(\|\phi\|)$ and $\delta_2 = \delta(\|V_{\delta_1}\|)$. Continue inductively to construct a norm-continuous solution to $(ME)_{\phi,f}$ on $[0, t_{\max})$, where $t_{\max} = \sum\limits_{n\geq 1}^{\infty} \delta_n$ and $\delta_{n+1} = \delta(\|V_{\delta_1+\cdots+\delta_n}\|)$. If $t_{\max} < \infty$, clearly $\lim \delta_n = 0$ and so (II.6.2) implies $\lim\limits_{n\to\infty} \|V_{\delta_1+\cdots+\delta_{n-1}}\| = \infty$ and hence $\overline{\lim\limits_{t\uparrow t_{\max}}} \|V_t\| = \infty$.

For $\phi, f \in b\mathcal{E}$ one can use the same existence proof with $L^\infty([0,\delta]\times E)$, the space of bounded Borel functions with the supremum norm, in place of $C_b\big([0,\delta], C_b(E)\big)$. We need the fact $P_t\phi(x)$ is jointly Borel which is clear for $\phi \in C_b(E)$ (because (II.2.2) and (QLC) imply continuity in each variable separately) and hence for all $\phi \in b\mathcal{E}$ by a monotone class argument. It follows easily that $\psi(U)(t,x)$ is Borel and the argument proceeds as above to give a Borel solution of $(ME)_{\phi,f}$ on $[0, t_{\max}) \times E$.

Turning to uniqueness, assume V and $\tilde{V}$ are solutions to $(ME)_{\phi,f}$ on $[0, t_{\max})$ and $[0, \tilde{t}_{\max})$, respectively so that V and $\tilde{V}$ are locally (in t) bounded and, in particular, $K = \sup\limits_{s\leq t} \|V_s\| \vee \|\tilde{V}_s\| < \infty$ for $t < t_{\max} \wedge \tilde{t}_{\max}$. Then for such a t and K,

$$\|V_t - \tilde{V}_t\| \leq C_K \int_0^t \|V_s - \tilde{V}_s\|ds$$

which implies $V = \tilde{V}$ on $[0, t_{\max} \wedge \tilde{t}_{\max})$ by Gronwall's Lemma ($s \to \|V_s - \overline{V}_s\|$ is universally measurable). Clearly $t_{\max} < \tilde{t}_{\max}$ is impossible because then $\overline{\lim\limits_{t\uparrow t_{\max}}} \|V_t\| = \infty$ would imply $\lim\limits_{t\uparrow t_{\max}} \|\tilde{V}_t\| = \infty$ which is impossible for $t_{\max} < \tilde{t}_{\max}$ by our local boundedness assumption on the solution $\tilde{V}$. Therefore $t_{\max} = \tilde{t}_{\max}$ and so $V = \tilde{V}$. This completes Step 1.

Step 2. If $\phi \in \mathcal{D}(A)$ and $f \in C_b(E)$, then the above solution satisfies the conclusions of (b) for $T, t < t_{\max}$.

The key step will be the existence of $\frac{\partial V}{\partial t}$. In addition to (II.6.1), the only property of a we will use is

$$(II.6.3) \qquad \begin{aligned} &a'(x,\lambda) \equiv \frac{\partial}{\partial\lambda}a(x,\lambda) \in C(E \times \mathbb{R}_+) \text{ and satisfies} \\ &\lim_{\delta\downarrow 0} \sup_{|\lambda|\leq K} \|a'(\lambda) - a'(\lambda+\delta)\| = 0 \text{ and } \sup_{|\lambda|\leq K} \|a'(\cdot,\lambda)\| < \infty \quad \forall K > 0. \end{aligned}$$

Fix $0 < T < t_{\max}$. Recall from (II.5.8) that for $h > 0$, if

$$R_t^h = h^{-1}\int_0^h P_r\big(f + a(V_{t+h-r})\big)dr,$$

then $V_t^h = (V_{t+h} - V_t)h^{-1}$ satisfies

$$(II.6.4) \qquad V_t^h = \frac{(P_h - I)}{h}V_t + R_t^h.$$

The norm continuity of V_t (from Step 1) and (II.6.1) show that as $h, r \downarrow 0$ ($r < h$), $\|a(V_{t+h-r}) - a(V_t)\| \to 0$ and so $P_r\big(f + a(V_{t+h-r})\big) \xrightarrow{bp} f + a(V_t)$ on $[0,T] \times E$. It follows that $R_t^h \xrightarrow{bp} f + a(V_t) \in C_b([0,T] \times E)$ as $h \to 0+$. Therefore it is clear from (II.6.4) that if

(II.6.5) $V_t^h \xrightarrow{bp} \dot{V}_t$ on $[0,T] \times E$ and the limit is continuous in each variable separately,

then the conclusions of (b) hold on $[0,T] \times E$ and Step 2 is complete.
To prove (II.6.5), write

$$V_t^h = \frac{P_{t+h}\phi - P_t\phi}{h} + h^{-1}\int_0^h P_{t+h-s}\big(f + a(V_s)\big)ds$$

$$+ \int_0^t P_{t-s}\left(\big(a(V_{s+h}) - a(V_s)\big)h^{-1} - a'(V_s)V_s^h\right)ds + \int_0^t P_{t-s}\big(a'(V_s)V_s^h\big)ds$$

$(II.6.6) \equiv 1_h + 2_h + 3_h + 4_h.$

The norm continuity of V_s, and hence of $a(V_s)$, together with $\phi \in \mathcal{D}(A)$ imply

$$(II.6.7) \qquad 1_h + 2_h \xrightarrow{bp} P_t(A\phi) + P_t\big(f + a(\phi)\big) \quad \text{on} \quad [0,T] \times E \quad \text{as} \quad h \downarrow 0.$$

Note also that the limit is continuous in each variable if the other is fixed. The mean value theorem shows there is a $\zeta_s^h(x)$ between $V_s(x)$ and $V_{s+h}(x)$ such that

$$\left[\big(a(V_{s+h}) - a(V_s)\big)h^{-1} - a'(V_s)V_s^h\right](x) = \big(a'\left(x, \zeta_s^h(x)\right) - a'\left(x, V_s(x)\right)\big)\, V_s^h(x).$$

This together with (II.6.3) and the norm continuity of V_s imply

$$(II.6.8) \qquad \sup_x |3_h| \le \eta_h \int_0^t \|V_s^h\|ds \quad \text{for some} \quad \eta_h \to 0 \quad \text{as} \quad h \to 0+.$$

Our local boundedness condition on a' (see (II.6.3)) and norm continuity of V imply

$$\sup_x |4_h| \le C\int_0^t \|V_s^h\|ds.$$

Use the above bounds in (II.6.6) to get

$$\|V_t^h\| \le C + C\int_0^t \|V_s^h\|ds, \quad t \le T$$

and hence

$$(II.6.9) \qquad \sup_{t\le T}\|V_t^h\| \le Ce^{CT}.$$

We now may conclude from (II.6.8) that

$$(II.6.10) \qquad \sup_{t\le T,x}|3_h| \to 0 \quad \text{as} \quad h\downarrow 0.$$

The above results and (II.6.6) suggest that $\dot{V}_t$ (if it exists) should solve

$$(II.6.11) \qquad W_t = P_t(A\phi + f + a(\phi)) + \int_0^t P_{t-s}(a'(V_s)W_s)ds.$$

A slight modification of Step 1 shows there is a unique solution of (II.6.11) in $L^\infty([0,T],C_b(E))$. To see this, set $\theta = A\phi + f + a(\phi) \in C_b(E)$ and define

$$\psi : L^\infty([0,T],C_b(E)) \to L^\infty([0,T],C_b(E))$$

by

$$\psi(W)(t) = P_t\theta + \int_0^b P_{t-s}(a'(V_s)W_s)ds.$$

Clearly $h(s,x,\lambda) = a'(x,V_s(1))\lambda$ is Lipschitz in λ uniformly in $(s,x)\in[0,T]\times E$ and so as in Step 1 we get the existence of a unique fixed point W first on $L^\infty([0,\delta],C_b(E))$ for appropriate $\delta>0$ and then on $L^\infty([0,T],C_b(E))$ by iteration because the linear growth of h in λ means the solution cannot explode. As $W_t(x)$ is continuous in x for each t and continuous in t for each x (see (II.6.11)), to prove (II.6.5) it suffices to show

$$(II.6.12) \qquad V_t^h \overset{bp}{\to} W_t \quad \text{on} \quad [0,T]\times E.$$

In view of (II.6.9) we only need establish pointwise convergence. For this we may fix $h_n \downarrow 0$ and define $r(t,x) = \overline{\lim}_{n\to\infty} |V_t^{h_n}(x) - W_t(x)|$ which is bounded on $[0,T]\times E$ because W is. Apply (II.6.6), (II.6.7), (II.6.10) and (II.6.11) to see that

$$\begin{aligned} r(t,x) &= \overline{\lim_{n\to\infty}}\left|\int_0^t P_{t-s}\left(a'(V_s)(V_s^{h_n} - W_s)\right)(x)ds\right| \\ &\le C\int_0^t P_{t-s}(r_s)(x)ds, \end{aligned}$$

and so

$$\|r_t\| \le C \int_0^t \|r_s\| ds.$$

This implies $r \equiv 0$ and hence (II.6.12). The proof of Step 2 is complete.

Step 3. If $\phi, f \in b\mathcal{E}_+$, then $t_{\max} = \infty$, $V_t = V_t^f \phi \ge 0$ and is bounded on $[0,T] \times E$ $\forall T > 0$ and (LE) holds if X is any solution of $(LMP)_\nu$.

The non-negativity is immediate from Proposition II.5.12. For the other assertions assume first $\phi \in \mathcal{D}(A)_+$, $f \in C_b(E)_+$. Step 2, the non-negativity of $V^f \phi$, and Proposition II.5.10 show that (LE) is valid for $t < t_{\max}$. If $\overline{g} = \sup_x g(x)$, $(ME)_{\phi,f}$ and the non-negativity of $V^f \phi$ imply

$$\|V_t^f \phi\| \le \|\phi\| + t\|f\| + \overline{g} \int_0^t \|V_s^f \phi\| ds, \quad t < t_{\max}$$

and therefore

$$(II.6.14) \qquad \|V_t^f \phi\| \le (\|\phi\| + t\|f\|)\, e^{\overline{g}t}, \quad t < t_{\max}.$$

This means $\|V_t^f \phi\|$ cannot explode at a finite $t_{\max}$ and so $t_{\max} = \infty$.

Turning to more general (ϕ, f), let

$$\mathcal{H} = \left\{ (\phi, f) \in (b\mathcal{E}_+)^2 : t_{\max} = \infty, \quad \text{(LE) holds} \right\}.$$

Assume $(\phi_n, f_n) \overset{bp}{\to} (\phi, f)$ and $(\phi_n, f_n) \in \mathcal{H}$. By (II.6.14) we have

$$\sup_{n, t \le T} \|V_t^{f_n} \phi_n\| < \infty \quad \forall T > 0.$$

Apply (LE) with $\nu = \delta_{\delta_x}$ and (ϕ_n, f_n) to see that $V_t^{f_n} \phi_n \overset{bp}{\to} V_t^\infty$ on $[0,T] \times E$ $\forall T > 0$ (the boundedness is immediate from the above). Now let $n \to \infty$ in $(ME)_{\phi_n, f_n}$ and use Dominated Convergence to see that $V_t^\infty = V_t^f \phi$ and for (ϕ, f), $t_{\max} = \infty$, and (LE) holds by taking limits in this equation for (ϕ_n, f_n). This shows $\mathcal{H}$ is closed under $\overset{bp}{\to}$. As $\mathcal{H} \supset \mathcal{D}(A)_+ \times C_b(E)_+$ (by the previous argument) and $\mathcal{D}(A)_+$ is *bp*-dense in $b\mathcal{E}_+$ (Corollary II.2.3) we may conclude that $\mathcal{H} = (b\mathcal{E}_+)^2$. This completes Step 3 because the boundedness claim is immediate from $t_{\max} = \infty$ and the local boundedness established in Step 1.

(a) is immediate from Steps 1 and 3. (c) follows from Step 3. (b) is clear from Step 2 and $t_{\max} = \infty$ in Step 3. ∎

7. Canonical Measures

Definition. A random finite measure, X, on E is infinitely divisible iff for any natural number n there are i.i.d. random measures $\{X_i : i \leq n\}$ such that X and $X_1 + \ldots + X_n$ have the same law on $M_F(E)$.

Example. Let $(X_t, t \geq 0)$ be a (Y, γ, g)-DW-superprocess starting at $X_0 \in M_F(E)$. If $\{X^i_\cdot : i \leq n\}$ are iid copies of the above DW-process but starting at X_0/n, then

$$(II.7.1) \qquad X_\cdot \overset{\mathcal{D}}{\equiv} X^1_\cdot + \ldots + X^n_\cdot \quad \text{as continuous} \quad M_F(E)\text{-valued processes.}$$

This follows from Theorem II.5.1, by noting that $X^1 + \ldots + X^n$ satisfies the martingale problem which characterizes the law of X (or by using the convergence theorem and the corresponding decomposition for the approximating branching particle systems). In particular for each fixed $t \geq 0$, X_t is an infinitely divisible random measure.

For our purposes, Chapter 3 of Dawson (1992) is a good reference for infinitely divisible random measures on a Polish space (see also Kallenberg (1983) for the locally compact case). The following canonical representation is essentially derived in Theorem 3.4.1 of Dawson (1992).

Theorem II.7.1. Let X be an infinitely divisible random measure on E such that $E\big(X(1)\big) < \infty$. There is a unique pair (M, R) such that $M \in M_F(E)$, R is a measure on $M_F(E) - \{0\}$ satisfying $\int \nu(1) R(d\nu) < \infty$, and

$$(II.7.2) \quad E\big(\exp\big(-X(\phi)\big)\big) = \exp\Big\{-M(\phi) - \int 1 - e^{-\nu(\phi)} R(d\nu)\Big\} \ \forall \phi \in (b\mathcal{E})_+.$$

Conversely if M and R are as above, then the right-hand side of (II.7.2) is the Laplace functional of an infinitely divisible random measure X satisfying $E\big(X(1)\big) < \infty$.

Definition. The measure R in (II.7.2) is the canonical measure associated with X.

We will give an independent construction of the canonical measure associated with X_t, a DW-superprocess evaluated at t, below (see Theorem II.7.2 and Exercise II.7.1). There are some slight differences between the above and Theorem 3.4.1 of Dawson (1992) and so we point out the necessary changes in the

Proof of Theorem II.7.1. A measure, μ, on E is locally finite $\big(\mu \in M_{LF}(E)\big)$ iff it is finite on bounded sets. Suppose X is infinitely divisible and $E\big(X(1)\big) < \infty$. Theorem 3.4.1 of Dawson (1992) shows there is a locally finite measure, M, on E and a measure, R, on $M_{LF}(E) - \{0\}$ such that (II.7.2) holds for all $\phi \in (b\mathcal{E})_+$ with bounded support. Fix such a ϕ. Then for $\lambda > 0$

$$(II.7.3) \qquad u(\lambda\phi) \equiv -\log E\left(e^{-X(\lambda\phi)}\right) = \lambda M(\phi) + \int 1 - e^{-\lambda\nu(\phi)} R(d\nu).$$

Since $\nu(\phi)e^{-\lambda\nu(\phi)} \leq C_\lambda(1 - e^{-\lambda\nu(\phi)})$ and $\int 1 - e^{-\lambda\nu(\phi)} R(d\nu) < \infty$ for $\lambda > 0$, it follows that $\int \nu(\phi)e^{-\lambda\nu(\phi)} \, R(d\nu) < \infty$ for $\lambda > 0$. An application of the Mean Value and Dominated Convergence Theorems allows us to differentiate (II.7.3) with respect to $\lambda > 0$ and conclude

$$E\left(X(\phi)e^{-\lambda X(\phi)}\right)\left[E\left(e^{-\lambda X(\phi)}\right)\right]^{-1} = M(\phi) + \int \nu(\phi)e^{-\lambda\nu(\phi)} R(d\nu).$$

Let $\lambda \to 0+$ and use Monotone Convergence to see

$$(II.7.4) \qquad E\big(X(\phi)\big) = M(\phi) + \int \nu(\phi) R(d\nu)$$

first for ϕ as above and then for all non-negative measurable ϕ by Monotone Convergence. Take $\phi = 1$ to see M is finite, $\int \nu(1)R(d\nu) < \infty$ and so R is supported by $M_F(E) - \{0\}$. We can also take monotone limits to see that (II.7.2) holds for all $\phi \in (b\mathcal{E})_+$.

For uniqueness note from (II.7.3) that for any ϕ in $(b\mathcal{E})_+$

$$\lim_{\lambda\to\infty} u(\lambda\phi)\lambda^{-1} = M(\phi) + \lim_{\lambda\to\infty} \int \left(1 - e^{-\lambda\nu(\phi)}\right) \lambda^{-1} R(d\nu) = M(\phi),$$

where in the last line we used $(1 - e^{-\lambda\nu(\phi)})\lambda^{-1} \le \nu(\phi)$ and Dominated Convergence. This shows that M and $\int h(\nu)R(d\nu)$ are determined by the law of X for h in

$$\mathcal{C} = \{h(\nu) = \sum_1^K b_i e^{-\nu(\phi_i)} : b_i \in \mathbb{R},\ \phi_i \in (b\mathcal{E})_+,\ h(0) = 0\}.$$

Note that for integration purposes $h(\nu) = -\sum_1^K b_i(1 - e^{-\langle \nu, \phi_i\rangle})$, and $\mathcal{C}$ is a vector space closed under multiplication. As in Lemma II.5.9, the Monotone Class Lemma 5.2 shows these integrals determine R.

Assume conversely that (II.7.2) holds for some M, R as in the Theorem. As in Theorem 3.4.1 of Dawson the right-hand side is the Laplace functional of some random measure which clearly must then be infinitely divisible. One then obtains (II.7.4) as above and this shows $E\big(X(1)\big) < \infty$. ∎

Assume now that X is the (Y, γ, g)-DW superprocess with $\gamma(\cdot) \equiv \gamma > 0$ constant, $g \equiv 0$, and law $\mathbb{P}_{X_0}$ if X starts at $X_0 \in M_F(E)$. Let $x_0 \in E$ and consider the approximating branching particle systems, X^N, in Theorem II.5.1 starting at δ_{x_0} (under $P^N_{\delta_{x_0}}$) and δ_{x_0}/N (under $P^N_{\delta_{x_0}/N}$), and with $g_N \equiv 0$ and $\nu_N(x, dk) = \nu(dk)$ independent of (x, N). In the former case we start N particles at x_0 (see Remark II.3.2) and in the latter we start a single particle at x_0. Let $\phi \in C_b(E)_+$ and write $V_t\phi$ for $V_t^0\phi$, the unique solution of $(ME)_{\phi,0}$. Lemma II.3.3 and Remark II.3.4 (the arguments go through unchanged for our slightly different initial conditions) show that

$$(II.7.5) \qquad N P^N_{\delta_{x_0}/N}\left(X_t^N(\phi)\right) = P^N_{\delta_{x_0}}\left(X_t^N(\phi)\right) = P_t\phi(x_0).$$

Theorem II.5.1 and (LE) imply that

$$\left[P^N_{\delta_{x_0}/N}\left(\exp\left(-X_t^N(\phi)\right)\right)\right]^N = P^N_{\delta_{x_0}}\left(\exp\left(-X_t^N(\phi)\right)\right) \to \exp\left(-V_t\phi(x_0)\right) \text{ as } N \to \infty.$$

Take logarithms and use $\log z \sim z - 1$ as $z \to 1$ (the expression under the Nth power must approach 1) to conclude

$$(II.7.6) \qquad \lim_{N\to\infty} \int \left(1 - e^{-X_t^N(\phi)}\right) N dP^N_{\delta_{x_0}/N} = V_t\phi(x_0).$$

Also note by Kolmogorov's Theorem (II.1.1(a)) that

$$(II.7.7) \qquad \lim_{N\to\infty} NP^N_{\delta_{x_0}/N}(X^N_t \neq 0) = 2/\gamma t.$$

(II.7.7) and (II.7.5) easily imply tightness of $NP^N_{\delta_{x_0}/N}(X^N_t \in \cdot, X^N_t \neq 0)$ and (II.7.6) shows the limit points coincide. The details are provided below.

Theorem II.7.2. For each $x_0 \in E$ and $t > 0$ there is a finite measure $R_t(x_0, \cdot)$ on $M_F(E) - \{0\}$ such that

(i) $NP^N_{\delta_{x_0}/N}(X^N_t \in \cdot, X^N_t \neq 0) \stackrel{w}{\Rightarrow} R_t(x_0, \cdot)$ on $M_F(E)$ and $x_0 \mapsto R_t(x_0, \cdot)$ is Borel measurable,

(ii) $\mathbb{P}_{X_0}(\exp(-X_t(\phi))) = \exp\left\{-\int\int 1 - e^{-\nu(\phi)} R_t(x_0, d\nu) dX_0(x_0)\right\} \ \forall \phi \in b\mathcal{E}_+$,

(iii) $R_t(x_0, M_F(E) - \{0\}) = 2/\gamma t$, $\int \nu(\phi) R_t(x_0, d\nu) = P_t\phi(x_0) \ \forall \phi \in b\mathcal{E}$,

$$R_t(x_0, \{\nu : \nu(1) \in A\}) = (2/\gamma t)^2 \int 1_A(x) \exp\{-2x/\gamma t\} dx \ \forall A \in \mathcal{B}((0,\infty)),$$

(iv) $\int \psi(\nu(1))\nu(\phi) R_t(x_0, d\nu) = \int_0^t \psi(\gamma t z/2) z e^{-z}\, dz\, P_t\phi(x_0) \quad \forall \phi \in b\mathcal{E}, \psi \in b\mathcal{B}(\mathbb{R}_+)$.

Proof. A sequence $\{\mu_N\}$ of finite, non-zero measures on $M_F(E)$ is tight if $\sup_N \mu_N(1) < \infty$ and for any $\varepsilon > 0$ there is a compact set $K_\varepsilon \subset E$ such that $\mu_N(\{\nu : \nu(K^c_\varepsilon) > \varepsilon\})/\mu_N(1) < \varepsilon$. For example, one may apply Theorem II.4.1 to the set of constant paths $\tilde{X}^N \equiv \tilde{X}^N_0$ with law $\mu_N/\mu_N(1)$. (II.7.5) and (II.7.7) easily imply these conditions for $\mu_N(\cdot) = NP^N_{\delta_{x_0}/N}(X^N_t \in \cdot, X^N_t \neq 0)$. Let μ_∞ be any weak limit point in $M_F(M_F(E))$. Then (II.7.6) implies

$$(II.7.8) \qquad \int 1 - e^{-\nu(\phi)} d\mu_\infty(\nu) = V_t\phi(x_0) \qquad \forall \phi \in C_b(E)_+.$$

Take $\phi \equiv \lambda > 0$ in the above, recall $V_t\lambda = 2\lambda(2 + \lambda t\gamma)^{-1}$ and let $\lambda \to \infty$ to see that

$$\begin{aligned} \mu_\infty(M_F(E) - \{0\}) &= 2/\gamma t \\ &= \lim_{N\to\infty} \mu_N(1) \qquad \text{(by (II.7.7))} \\ &= \mu_\infty(1). \end{aligned}$$

This shows $\mu_\infty(\{0\}) = 0$ and, together with (II.7.8), implies

$$(II.7.9) \qquad \int e^{-\nu(\phi)} d\mu_\infty(\nu) = 2/\gamma t - V_t\phi(x_0) \qquad \forall \phi \in C_b(E)_+.$$

As in Lemma II.5.9, this uniquely determines μ_∞ and shows $\mu_N \stackrel{w}{\Rightarrow} \mu_\infty$. The Borel measurability of $R_t(x_0)$ in x_0 is then clear from the Borel measurability of the approximating measures. The proof of (i), (ii) (by (LE)), and the first assertion in (iii) is complete. The second assertion in (iii) is a special case of (iv), proved below. The final assertion in (iii) is obtained by setting $\phi \equiv \lambda$ in (II.7.9), as was already done in the above.

For (iv) it suffices to consider ψ and ϕ bounded and continuous. For the branching particle system described above, $\mathcal{N} = \sigma(N^\alpha : \alpha \in I)$ is independent of $\sigma(Y^\alpha : \alpha \in I)$ and so

$$\begin{aligned} &NP^N_{\delta_{x_0}/N}(\psi(X^N_t(1))X^N_t(\phi)1(X^N_t(1) \neq 0)) \\ &= NP^N_{\delta_{x_0}/N}(\psi(X^N_t(1))1(X^N_t(1) \neq 0)\frac{1}{N}\sum_{\alpha\sim t} P^N_{\delta_{x_0}/N}(\phi(Y^\alpha_t)|\mathcal{N})) \\ &= NP^N_{\delta_{x_0}/N}(\psi(X^N_t(1))1(X^N_t(1) \neq 0)X^N_t(1))P_t\phi(x_0). \end{aligned}$$

Now let $N \to \infty$ in the above. Lemma II.4.6 and $NP^N_{\delta_{x_0}/N}(X^N_t(1)^2) \leq P^N_{\delta_{x_0}}(X^N_t(1)^2)$ give us the necessary uniform integrability to use (i) and conclude that

$$\int \psi(\nu(1))\nu(\phi)R_t(x_0, d\nu) = \int \psi(\nu(1))\nu(1)R_t(x_0, d\nu)P_t\phi(x_0),$$

and the last part of (iii) completes the proof of (iv). ∎

Clearly we have given a direct construction of the canonical measure, $R_t(x_0, \cdot)$, of X_t under $\mathbb{P}_{\delta_{x_0}}$. In this case $M \equiv 0$. For general γ, g it is not hard to modify the above to recover the canonical representation from our convergence theorem. We leave this as Exercise II.7.1 below. In general M will not be 0 as can readily be seen by taking $\gamma \equiv 0$.

Exercise II.7.1. Let X be a (Y, γ, g)-DW superprocess starting at δ_{x_0} where $\gamma \in C_b(E)_+$ and $g \in C_b(E)$. Extend the proof of Theorem II.7.2 to show there is an $M_t(x_0, \cdot) \in M_F(E)$ and a σ-finite measure $R_t(x_0, \cdot)$ on $M_F(E) - \{0\}$ such that $\int \nu(1)R_t(x_0, d\nu) < \infty$ and

$$\mathbb{E}_{\delta_{x_0}}(\exp(-X_t(\phi))) = \exp\left\{-M_t(x_0, \phi) - \int 1 - e^{-\nu(\phi)}R_t(x_0, d\nu)\right\} \quad \forall\phi \in (b\mathcal{E})_+.$$

Hint. Recalling (II.7.6), let $\varepsilon > 0$, $\phi \in C_b(E)_+$, and write

$$\begin{aligned} \int\left(1 - e^{-X^N_t(\phi)}\right)NdP^N_{\delta_{x_0}/N} &= N\int\left(1 - e^{-X^N_t(\phi)} - X^N_t(\phi)\right)1(X^N_t(1) \leq \varepsilon)dP^N_{\delta_{x_0}/N} \\ (*) \qquad &+ \int X^N_t(\phi)1\left(X^N_t(1) \leq \varepsilon\right)NdP^N_{\delta_{x_0}/N} \\ &+ \int(1 - e^{-X^N_t(\phi)})1(X^N_t(1) > \varepsilon)NdP^N_{\delta_{x_0}/N}. \end{aligned}$$

Show that the first term goes to 0 as $\varepsilon \downarrow 0$ uniformly in N and that

$$\left\{\int X^N_t(\cdot)1(X^N_t(1) \leq \varepsilon)NdP^N_{\delta_{x_0}/N} : N \in \mathbb{N}\right\}$$

and

$$\{NP^N_{\delta_{x_0}/N}(X^N_t \in \cdot, X^N_t(1) > \varepsilon) : N \in \mathbb{N}\}$$

are tight on E and $M_F(E)$, respectively. Now let $N \to \infty$ through an appropriate subsequence and then $\varepsilon = \varepsilon_k \downarrow 0$ in $(*)$ to obtain the desired decomposition.

Theorem II.7.2 and (II.7.7) imply

$$(II.7.10)\qquad P^N_{\delta_{x_0}/N}(X^N_t \in \cdot \mid X^N_t \neq 0) \overset{w}{\Rightarrow} R_t(x_0,\cdot)/R_t(x_0,1),$$

that is, $R_t(x_0,\cdot)$, when normalized, is the law of a cluster at time t of descendants of a common ancestor at $t=0$ conditioned on the existence of such descendants. Note that Yaglom's Theorem II.1.1(b) is immediate from (II.7.10) and Theorem II.7.2(iii).

Exercise II.7.2. (a) If X is a $(Y,\gamma,0)$-DW-superprocess under $\mathbb{P}_{X_0}$ with $\gamma>0$ constant, prove that $\varepsilon^{-1}\mathbb{P}_{\varepsilon\delta_{x_0}}(X_t\in\cdot, X_t\neq 0)\overset{w}{\Rightarrow}R_t(x_0,\cdot)$ as $\varepsilon\downarrow 0$ on $M_F(E)$.
Hint. Use Theorem II.7.2 (ii) to show convergence of the corresponding Laplace functionals.

(b) If T_t is the semigroup of X, show that $R_t(x_0,\psi)=\int T_{t-\tau}\psi(\nu)R_\tau(x_0,d\nu)$ for all $0<\tau\le t$, $x_0\in E$, and ψ bounded measurable on $M_F(E)-\{0\}$.
Hint. Use (a), first consider $\psi\in C_b(M_F(E))$ such that $\psi(0)=0$, and recall that $T_t: C_b(M_F(E))\to C_b(M_F(E))$.

If $X_0\in M_F(E)$, let Ξ^{t,X_0} be a Poisson point process on $M_F(E)-\{0\}$ with intensity $\int R_t(x_0,\cdot)dX_0(x_0)$. Theorem II.7.2(ii) implies that

$$(II.7.11)\qquad \int \nu\,\Xi^{t,X_0}(d\nu)\quad\text{is equal in law to}\quad \mathbb{P}_{X_0}(X_t\in\cdot).$$

In view of (II.7.7) and (II.7.10) we see that this representation decomposes X_t according to the Poisson number of ancestors at time 0 with descendants alive at time t. This perspective will allow us to extend Theorem II.7.2 and this Poisson decomposition to the sample paths of X. Indeed, (II.7.1) shows that infinite divisibility is valid on the level of sample paths.

Let $\zeta:\Omega_D\to[0,\infty]$ be given by $\zeta(X)=\inf\{t>0: X_t=0\}$ and define

$$\Omega^{Ex}=\{X\in\Omega_D: X_0=0,\zeta>0, X_t\equiv 0\quad \forall t\ge\zeta\},$$
$$\Omega^{Ex}_C=\{X\in\Omega^{Ex}: X_\cdot\ \text{ is continuous}\}\subset\Omega_X,$$

equipped with the subspace topologies they inherit from Ω_D and Ω_X, respectively. If $\{\mathbb{N}_k: k\in\mathbb{N}\cup\{\infty\}\}$ are measures on Ω^{Ex} we write $\mathbb{N}_k\overset{w}{\Rightarrow}\mathbb{N}_\infty$ on Ω^{Ex} if $\mathbb{N}_k(\zeta>t)<\infty$ for all $k\in\mathbb{N}\cup\{\infty\}$ and $t>0$, and

$$\mathbb{N}_k(X\in\cdot,\zeta>t)\overset{w}{\Rightarrow}\mathbb{N}_\infty(X\in\cdot,\zeta>t)\text{ as }k\to\infty,\text{ as finite measures on }\Omega_D\ \forall t>0.$$

Theorem II.7.3. (a) For each $x_0\in E$ there is a σ-finite measure, $\mathbb{N}_{x_0}$, on Ω^{Ex}_C such that $NP_{\delta_{x_0}/N}(X^N_\cdot\in\cdot)\overset{w}{\Rightarrow}\mathbb{N}_{x_0}$ on Ω^{Ex}.
(b) For all $t>0$, $\mathbb{N}_{x_0}(X_t\in\cdot,\zeta>t)=R_t(x_0,\cdot)$.
(c) Let Ξ be a Poisson point process on Ω^{Ex}_C with intensity $\mathbb{N}_{x_0}$. Then $X_t=\int\nu_t d\Xi(\nu)$, $t>0$, has the law of a $(Y,\gamma,0)$-DW-superprocess starting at δ_{x_0}.

Remark II.7.4. Note that (II.7.7) and the equality $\mathbb{N}_{x_0}(\zeta>t)=R_t(x_0,1)=2/\gamma t$ allow us to use (a) to conclude

$$P_{\delta_{x_0}/N}(X^N\in\cdot\mid X^N_t\neq 0)\overset{w}{\Rightarrow}\mathbb{N}_{x_0}(X\in\cdot\mid\zeta>t)\quad\text{on}\quad\Omega_D\quad\forall t>0.$$

In this way $\mathbb{N}_{x_0}$ describes the time evolution of a cluster starting from a single ancestor at x_0 given that it survives for some positive length of time. We call $\mathbb{N}_{x_0}$ the canonical measure of the process X. It has been studied by El Karoui and Roelly (1991) and Li and Shiga (1995). A particularly elegant construction of $\mathbb{N}_{x_0}$ in terms of Le Gall's snake may be found in Chapter IV of Le Gall (1999). The reader may want to skip the proof of Theorem II.7.3 on a first reading.

Proof. Let $X^N_\cdot$ be as before under $P^N_{\delta_{x_0}}$, and for $i \le N$ let $X^{N,i}_t = \frac{1}{N} \sum_{\alpha\sim t, \alpha_0=i} \delta_{Y^\alpha_t}$ be the portion of X^N_t descending from the ith initial ancestor. Fix $t > 0$ and set $\Lambda^N_t = \{i \le N : X^{N,i}_t \neq 0\}$. The mutual independence of $\mathcal{G}_i = \sigma(Y^\alpha, N^\alpha : \alpha > i)$, $i = 1 \dots, N$, shows that conditional on Λ^N_t, $\{X^{N,i}_\cdot : i \in \Lambda^N_t\}$ are iid with law $P^N_{\delta_{x_0}/N}(X^N_\cdot \in \cdot \mid X^N_t \neq 0)$. We also have

$$X^N_{t+\cdot} = \sum_{i\in\Lambda^N_t} X^{N,i}_{t+\cdot}. \tag{II.7.12}$$

Clearly $|\Lambda^N_t| = \text{card}(\Lambda^N_t)$ is binomial $(N, P_{\delta_{x_0}/N}(X^N_t \neq 0))$ and so by (II.7.7), converges weakly to a Poisson random variable Λ_t with mean $2/\gamma t$. The left side of (II.7.12) converges weakly on Ω_D to $\mathbb{P}_{\delta_{x_0}}(X_{t+\cdot} \in \cdot)$ (use the fact that the limit is a.s. continuous) and so for $\varepsilon > 0$ there is a compact set $K_\varepsilon \subset \Omega_D$ such that $P^N_{\delta_{x_0}}(X^N_{t+\cdot} \in K^c_\varepsilon) < \varepsilon$ for all N. Use (II.7.12) to see that this means that for all N

$$\varepsilon > P^N_{\delta_{x_0}}(X^N_{t+\cdot} \in K^c_\varepsilon, |\Lambda^N_t| = 1) = P^N_{\delta_{x_0}/N}(X^N_{t+\cdot} \in K^c_\varepsilon | X^N_t \neq 0)\, P(|\Lambda^N_t| = 1).$$

This proves tightness of $\{P^N_{\delta_{x_0}/N}(X^N_{t+\cdot} \in \cdot \mid X^N_t \neq 0) : N \in \mathbb{N}\}$ on Ω_D because $\lim_{N\to\infty} P(|\Lambda^N_t| = 1) = (2/\gamma t)e^{-2/\gamma t} > 0$. Let P^t be any limit point on Ω_D. If Λ_t is the above Poisson random variable and conditional on Λ_t, $\{X^i_\cdot : i \le \Lambda_t\}$ are iid with law P^t, then we may let $N \to \infty$ through an appropriate subsequence in (II.7.12) and conclude

$$\sum_{i=1}^{\Lambda_t} X^i_\cdot \quad \text{has law} \quad \mathbb{P}_{\delta_{x_0}}(X_{t+\cdot} \in \cdot). \tag{II.7.13}$$

Note that $P(\Lambda_t \ge 1) = 1 - e^{-2/\gamma t} = \mathbb{P}_{\delta_{x_0}}(X_t \neq 0)$ (recall (II.5.12)). From this and the above we may conclude that $\sum_{i=1}^{\Lambda_t} X^i_\cdot = 0$ iff $\Lambda_t = 0$ and therefore

$$P\left(\Lambda_t \ge 1, \sum_{i=1}^{\Lambda_t} X^i_\cdot \in \cdot\right) = \mathbb{P}_{\delta_{x_0}}(X_{t+\cdot} \in \cdot, X_t \neq 0).$$

The measure on the right is supported on

$$\Omega'_X = \{X \in \Omega_X : X_0 \neq 0, \zeta > 0, X_s = 0 \text{ all } s \ge \zeta\},$$

and so the same must be true of $P^t = \mathcal{L}(X^i_\cdot)$ as $P(\Lambda_t = 1, X^1_0 \notin \Omega'_X) = 0$ by the above.

If $0 \le t_1 < \ldots < t_k$ and $\phi_i \in C_b(E)_+$ for $1 \le i \le k$, (II.7.13) shows that

$$\mathbb{P}_{\delta_{x_0}}\Big(\exp\Big\{-\sum_1^k X_{t_j+t}(\phi_j)\Big\}\Big) = \exp\Big\{-\int 1-\exp\Big(-\sum_1^k \nu_{tj}(\phi_j)\Big)dP^t(\nu)2/\gamma t\Big\}.$$

This uniquely determines $\int \exp\big(-\sum_1^k \nu_{tj}(\phi_j)\big)dP^t(\nu)$ and hence the finite-dimensional distributions of P^t by a now familiar monotone class argument. We have shown (use (II.7.7))

$$\begin{aligned} &NP^N_{\delta_{x_0}/N}\left(X^N_{t+\cdot} \in \cdot, X^N_t \neq 0\right) \\ (II.7.14)\qquad &= NP^N_{\delta_{x_0}/N}\left(X^N_t \neq 0\right) P^N_{\delta_{x_0}/N}\left(X^N_{t+\cdot} \in \cdot \mid X^N_t \neq 0\right) \\ &\overset{w}{\Rightarrow} \frac{2}{\gamma t} P^t \quad \text{on } \Omega_D, \end{aligned}$$

where the limit is supported on Ω'_X.

To handle the small values of t we need a definition and a Lemma.

Notation. $\mathrm{Lip}_1 = \{\phi : E \to \mathbb{R} : \|\phi\| \le 1, |\phi(x) - \phi(y)| \le \rho(x,y)\}$ (ρ is a fixed complete metric on E).

Definition. The Vasershtein metric $d = d_\rho$ on $M_F(E)$ is

$$d(\mu,\nu) = \sup\{|\mu(\phi) - \nu(\phi)| : \phi \in \mathrm{Lip}_1\}.$$

Then d is a complete metric which induces the weak topology on $M_F(E)$ (e.g. see Ethier and Kurtz (1986), p. 150, Exercise 2). It only plays an incidental role here but will be important in treating models with spatial interactions in Chapter V.

Redefine $X^N_\cdot$ near $t = 0$ by

$$\tilde X^N_t = \begin{cases} X^N_t & \text{if } t \ge N^{-3} \\ tN^3 X^N_{N^{-3}} & \text{if } t \in [0, N^{-3}) \end{cases}.$$

Lemma II.7.5. (a) $NP^N_{\delta_{x_0}/N}\big(\sup_{s\le\delta} X^N_s(1) > \varepsilon\big) \le 4\gamma\delta\varepsilon^{-2}$ for all $\delta, \varepsilon > 0$ and $N \ge 2/\varepsilon$.

(b) There are $N_0 \in \mathbb{N}$ and $\delta_0 > 0$ such that

$$NP^N_{\delta_{x_0}/N}\left(\sup_{s\le\delta} \tilde X^N_s(1) > \delta^{1/5}\right) \le \delta^{1/2} \quad \text{for } 0 < \delta \le \delta_0 \quad \text{and} \quad N \ge N_0.$$

(c) $NP^N_{\delta_{x_0}/N}\left(\sup_t d\left(X^N_t, \tilde X^N_t\right) \ge 4/N\right) \le \gamma N^{-1}$.

Proof. (a) Use $(MP)^N$ in Section II.4 and (II.4.5) with $\phi = 1$, $g_N = 0$ and γ_N constant to see that under $P^N_{\delta_{x_0}/N}$, $X^N_t(1)$ is a martingale with predictable square function $\gamma \int_0^t X^N_s(1)ds$. The weak L^1 inequality and Lemma II.3.3 (a) imply that for

$N > 2/\varepsilon$

$$NP^N_{\delta_{x_0}/N}\left(\sup_{s\le\delta} X^N_s(1) > \varepsilon\right) \le NP^N_{\delta_{x_0}/N}\left(\sup_{s\le\delta} X^N_s(1) - X^N_0(1) \ge \varepsilon/2\right)$$
$$\le N4\varepsilon^{-2}P^N_{\delta_{x_0}/N}\left(\gamma\int_0^\delta X^N_s(1)ds\right) = 4\gamma\delta\varepsilon^{-2}.$$

(b) Assume first $\delta \ge N^{-3}$. Then for $N \ge N_0$, $2\delta^{-1/5} \le N$ and so by (a) for $\delta < \delta_0$,

$$NP^N_{\delta_{x_0}/N}\left(\sup_{s\le\delta} \tilde{X}^N_s(1) > \delta^{1/5}\right) = NP^N_{\delta_{x_0}/N}\left(\sup_{N^{-3}\le s\le\delta} X^N_s(1) > \delta^{1/5}\right)$$
$$\le 4\gamma\delta^{3/5} \le \delta^{1/2}.$$

Assume now $\delta < N^{-3}$. Then the above probability equals

$$NP^N_{\delta_{x_0}/N}\left(\delta N^3 X^N_{N^{-3}}(1) > \delta^{1/5}\right) = NP^N_{\delta_{x_0}/N}\left(X^N_{N^{-3}}(1) > \delta^{-4/5}N^{-3}\right).$$

Our assumption on δ implies $2\delta^{4/5}N^3 < 2N^{3/5} < N$ for $N \ge N_0$ and so (a) bounds the above by

$$4\gamma N^{-3}N^6\delta^{8/5} < 4\gamma\delta^{3/5} < \delta^{1/2} \quad \text{for} \quad \delta \le \delta_0.$$

(c) If $f \in \mathrm{Lip}_1$ and $t < N^{-3}$,

$$\left|X^N_t(f) - \tilde{X}^N_t(f)\right| \le \|f\|\left[X^N_t(1) + tN^3X^N_{N^{-3}}(1)\right] \le 2\sup_{t\le N^{-3}} X^N_t(1).$$

This shows that the right-hand side is an upper bound for $\sup_t d(X^N_t, \tilde{X}^N_t)$ and an application of (a) completes the proof. ∎

We now complete the

Proof of Theorem II.7.3. Lemma II.7.5 (c) and (II.7.14) show that if $t_n \downarrow 0$ ($t_n > 0$) is fixed and $\varepsilon > 0$, we may choose $\tilde{K}^\varepsilon_n$ compact in Ω_D such that

$$(II.7.15) \qquad \sup_N NP^N_{\delta_{x_0}/N}\left(\tilde{X}^N_{t_n+\cdot} \notin \tilde{K}^\varepsilon_n, X^N_{t_n} \ne 0\right) < \varepsilon 2^{-n}.$$

For $t > 0$ define

$$K^\varepsilon_t = \left\{X \in \Omega_D : X_{t_n+\cdot} \in \tilde{K}^\varepsilon_n \ \ \forall t_n \le t \text{ and } \sup_{s\le 2^{-2n}} X_s(1) \le 2^{-2n/5} \text{ for all } n \ge 1/\varepsilon\right\}.$$

Lemma II.7.5(b) and (II.7.15) show that for $\varepsilon < \varepsilon_0$ and $N \ge N_0$

$$\begin{aligned} &NP^N_{\delta_{x_0}/N}\left(\tilde X^N \notin K^\varepsilon_t, X^N_t \neq 0\right) \\ &\le \sum_n 1(t_n \le t) NP^N_{\delta_{x_0}/N}\left(\tilde X^N_{t_n+\cdot} \notin \tilde K^\varepsilon_n, X^N_{t_n} \neq 0\right) \\ &\qquad + \sum_{n\ge 1/\varepsilon} NP^N_{\delta_{x_0}/N}\Big(\sup_{s\le 2^{-2n}} \tilde X^N_s(1) > 2^{-2n/5}\Big) \\ &\le \sum_{n=1}^\infty \varepsilon 2^{-n} + \sum_{n\ge 1/\varepsilon} 2^{-n} \le 2\varepsilon. \end{aligned}$$

It is a routine Skorokhod space exercise to check that $\overline{K^\varepsilon_t}$ is a compact subset of $\Omega^0_D = \{X \in \Omega_D : X_0 = 0\}$. This together with Lemma II.7.5(c) shows that $\{NP^N_{\delta_{x_0}/N}(X^N \in \cdot, \zeta > t) : N \in \mathbb{N}\}$ is relatively compact in Ω_D and all limit points are supported on Ω^0_D.

Fix $t_0 > 0$. Choose $N_k \to \infty$ such that

$$(II.7.16)\qquad NP^N_{\delta_{x_0}/N}\left(X^N \in \cdot, \zeta > t_0\right) \stackrel{w}{\Rightarrow} \mathbb{N}^{t_0} \text{ on } \Omega_D \text{ as } N \to \infty \text{ through } \{N_k\}.$$

To ease our notation we write N for N_k and $\mathbb{Q}_N$ for $NP^N_{\delta_{x_0}/N}$. By taking a further subsequence we may assume (II.7.16) holds with t_m in place of t_0 (recall $t_m \downarrow 0$). Clearly $\mathbb{N}^{t_m}(\cdot)$ are increasing in m and so we may define a measure on Ω_D by $\mathbb{N}_{x_0}(A) = \lim_{m\to\infty} \mathbb{N}^{t_m}(A)$. Let $t_m < t$. Theorem II.7.2 implies that

$$\lim_{\varepsilon\downarrow 0} \overline{\lim_{N\to\infty}}\, \mathbb{Q}_N\left(X^N_t(1) \in (0,\varepsilon), \zeta > t_m\right) \le \lim_{\varepsilon\downarrow 0} R_t\left(x_0, \{\nu : \nu(1) \in (0,\varepsilon)\}\right) = 0.$$

A standard weak convergence argument now gives

$$\begin{aligned} \lim_{N\to\infty} \mathbb{Q}_N\left(X^N \in \cdot, \zeta > t\right) &= \lim_{N\to\infty} \mathbb{Q}_N\left(X^N \in \cdot, X^N_t(1) > 0, \zeta > t_m\right) \\ &= \mathbb{N}^{t_m}\left(\cdot\,, X_t(1) > 0\right). \end{aligned}$$

This shows that the measure on the right side is independent of m and so the above implies

$$(II.7.17)\qquad \mathbb{Q}_N\left(X^N \in \cdot, \zeta > t\right) \stackrel{w}{\Rightarrow} \mathbb{N}_{x_0}\left(\cdot\,, X_t(1) > 0\right) \qquad \forall t > 0,$$

and in particular (take $t = t_m$)

$$(II.7.18)\qquad \mathbb{N}_{x_0}\left(\cdot, X_{t_m}(1) > 0\right) = \mathbb{N}^{t_m}(\cdot\,), \quad m = 0, 1, 2, \ldots.$$

(II.7.17) shows that the measures $\mathbb{N}_{x_0}\big(\cdot\,, X_t(1) > 0\big)$ are decreasing in t and this implies for each $s < t$, $\mathbb{N}_{x_0}$ a.s. $X_s = 0$ implies $X_t = 0$. Right-continuity implies

$$(II.7.19)\qquad X_s = 0 \Rightarrow X_t = 0 \quad \forall s < t \quad \mathbb{N}_{x_0}\text{-a.s.}$$

This implies $\{\zeta = 0\} \subset \bigcap_m \{X_{t_m} = 0\}$ $\mathbb{N}_{x_0}$-a.s., and therefore $\mathbb{N}^{t_m}$-a.s. Therefore

$$\mathbb{N}^{t_m}(\zeta = 0) \le \mathbb{N}^{t_m}(X_{t_m} = 0) = 0,$$

the last by (II.7.18). It follows that $\mathbb{N}_{x_0}(\zeta = 0) = \lim_{m\to 0} \mathbb{N}^{t_m}(\zeta = 0) = 0$ which, together with (II.7.19) shows that $\mathbb{N}_{x_0}$ is supported by Ω^{Ex}. (II.7.17) may therefore be written as

$$(II.7.20) \qquad \mathbb{Q}_N\left(X^N \in \cdot, \zeta > t\right) \overset{w}{\Rightarrow} \mathbb{N}_{x_0}\left(\cdot, \zeta > t\right) \qquad \forall t > 0.$$

The convergence in (II.7.14) and the above together imply

$$(II.7.21) \qquad \mathbb{N}_{x_0}\left(X_{t+\cdot} \in \cdot, \zeta > t\right) = \frac{2}{\gamma t} P^t(\cdot)$$

(Note that (II.7.20) alone would not give this if t is a point of discontinuity, but as the limit in (II.7.14) exists we only need identify the finite-dimensional distributions in terms of $\mathbb{N}_{x_0}$ and (II.7.20) is enough for this.) This implies $\mathbb{N}_{x_0}$-a.s. continuity of $X_{t+\cdot}$ for all $t > 0$ (recall $P^t(\Omega_X'^c) = 0$) and hence shows $\mathbb{N}_{x_0}$ is supported on Ω_C^{Ex}. (II.7.21) also identifies the finite dimensional distributions of $\mathbb{N}_{x_0}$ and so by (II.7.18) with $m = 0$ we may conclude that all limit points in (II.7.16) equal $\mathbb{N}_{x_0}(\cdot, \zeta > t_0)$. This proves (a). (b) is then immediate from Theorem II.7.2(i).

Let Ξ be as in (c). Note that $\Xi(\{\nu : \nu_t \neq 0\})$ is Poisson with mean $\mathbb{N}_{x_0}(\zeta > t) = 2/\gamma t$ (by (II.7.21)) and so $\int \nu_{t+\cdot}\Xi(d\nu) \overset{\mathcal{D}}{=} \sum_{i=1}^{\Lambda_t} X^i_{t+\cdot}$, where Λ_t is Poisson $(2/\gamma t)$ and given Λ_t, $\{X^i_{t+\cdot} : i \le \Lambda_t\}$ are iid with law $\mathbb{N}_{x_0}(X_{t+\cdot} \mid \zeta > t) = P^t$ (by (II.7.21)). Compare this with (II.7.13) and let $t \downarrow 0$ to complete the proof of (c). ∎

8. Historical Processes.

We return to the path-valued setting of Example II.2.4(c) under the assumption

$$(PC) \qquad x \to P^x \quad \text{is continuous.}$$

In addition to the $\hat{E}$-valued BSMP W_t with laws $(\hat{P}_{\tau,y})$ described there, we introduce probabilities $\left\{P_{\tau,y} : (\tau, y) \in \hat{E}\right\}$ on $D(E)$ by

$$P_{\tau,y}(A) = P^{y(\tau)}\left(\{w : (y/\tau/w) \in A\}\right).$$

If $W_\cdot$ has law $\hat{P}_{\tau,y}$ and Y has law $P_{\tau,y}$ then

$$(II.8.1) \qquad (W_t)_{t\ge 0} \overset{\mathcal{D}}{=} \left(\tau + t, Y^{\tau+t}\right)_{t\ge 0}.$$

Let $\hat{g} \in C_b(\hat{E})$, $\hat{\gamma} \in C_b(\hat{E})_+$, and for $\tau \ge 0$ define

$$M_F^\tau(D) = \left\{m \in M_F(D(E)) : y^\tau = y \quad m - a.a.\ y\right\}.$$

Fix $m \in M_F^\tau(D)$ so that $\delta_\tau \times m \in M_F(\hat{E})$ and let $\hat{X}$ be the $(W, \hat{\gamma}, \hat{g})$-DW superprocess starting at $\delta_\tau \times m$ with law $\hat{\mathbb{P}}_{\tau,m} (\equiv \hat{\mathbb{P}}_{\delta_\tau \times m})$ on the canonical space of continuous $M_F(\hat{E})$-valued paths. Introduce

$$\Omega_H[\tau, \infty) = \left\{H_\cdot \in C\Big([\tau, \infty), M_F(D(E))\Big) : H_t \in M_F^t(D) \quad \forall t \ge \tau\right\},$$

and let $\Omega_H = \Omega_H[0,\infty)$ with its Borel σ-field $\mathcal{F}_H$. Let $\Pi : \hat{E} \to D(E)$ be the projection map and define an $M_F(D(E))$-valued process $(H_t, t \geq \tau)$ by

$$H_{\tau+t}(A) = \hat{X}_t(\Pi^{-1}(A)).$$

Lemma II.8.1. $\hat{X}_t = \delta_{\tau+t} \times H_{\tau+t}\ \forall t \geq 0$ and $H_t \in M_F^t(D)\ \forall t \geq \tau\ \hat{\mathbb{P}}_{\tau,m}$-a.s.

Proof. Let $\hat{P}_t^{\hat{g}} f(\tau, y) = \hat{P}_{\tau,y}\big(\exp\{\int_0^t \hat{g}(W_s)\,ds\} f(W_t)\big)$ and

$$\Lambda(t) = \{(u,y) \in \hat{E} : u \neq \tau + t\}.$$

Then by Exercise II.5.2(b)

$$\mathbb{P}_{\tau,m}\left(\hat{X}_t(\Lambda(t))\right) = \int \hat{P}_t^{\hat{g}}(1_{\Lambda(t)})(\tau,y)\,dm(y) = 0$$

because $W_t = (\tau + t, Y^{\tau+t})\ \hat{P}_{\tau,y}$-a.s. This shows $\hat{X}_t = \delta_{\tau+t} \times H_{\tau+t}\ \mathbb{P}_{\tau,m}$-a.s. for each $t \geq 0$ and hence for all $t \geq 0$ a.s. by the right-continuity of both sides. Since $\hat{X}_t \in M_F(\hat{E})\ \forall t \geq 0$ a.s., the second assertion follows immediately. ∎

The process of interest is the $M_F(D)$-valued process H. We abuse our notation and also use H_t to denote the coordinate variables on Ω_H and let

$$\mathcal{F}^H[s, t+] = \bigcap_{n=1}^{\infty} \sigma(H_r : s \leq r \leq t + 1/n).$$

Define $\mathbb{Q}_{\tau,m}$ on $(\Omega_H, \mathcal{F}^H[\tau,\infty))$ by $\mathbb{Q}_{\tau,m}(\cdot) = \hat{\mathbb{P}}_{\tau,m}(H \in \cdot)$, where H is as in Lemma II.8.1. The fact that $\hat{X}$ is a BSMP easily shows that

$$H \equiv (\Omega_H, \mathcal{F}_H, \mathcal{F}^H[\tau, t+], H_t, \mathbb{Q}_{\tau,m})$$

is an inhomogeneous Borel strong Markov process (IBSMP) with continuous paths in $M_F^t(D) \subset M_F(D)$. This means

(i) $\forall u > 0$ and $A \in \mathcal{F}^H[u,\infty)$ $(\tau,m) \to \mathbb{Q}_{\tau,m}(A)$ is Borel measurable on $\{(\tau,m) : m \in M_F^\tau(D), \tau \leq u\}$.

(ii) $H_\tau = m$, $H_t \in M_F^t(D)\ \forall t \geq \tau$, and $H_\cdot$ is continuous $\mathbb{Q}_{\tau,m}$-a.s.

(iii) If $m \in M_F^\tau(D)$, $\psi \in b\mathcal{B}([\tau,\infty] \times M_F(D))$ and $T \geq \tau$ is an $(\mathcal{F}^H[\tau,t+])_{t\geq\tau}$-stopping time, then

$$\mathbb{Q}_{\tau,m}\left(\psi(T, H_{T+\cdot}) \mid \mathcal{F}^H[\tau, T+]\right)(\omega) = \mathbb{Q}_{T(\omega),H_{T(\omega)}}\Big(\psi\big(T(\omega), H_{T(\omega)+\cdot}\big)\Big)$$
$$\mathbb{Q}_{\tau,m} - \text{a.s. on} \quad \{T < \infty\}.$$

This is a simple consequence of the fact that $\hat{X}$ is a BSMP (only (iii) requires a bit of work) and the routine proof is contained in Dawson-Perkins (1991) (Proof of Theorem 2.1.5 in Appendix 1). We call H the $(Y, \hat{\gamma}, \hat{g})$-historical process.

Of course it is now a simple matter to interpret the weak convergence theorem (Theorem II.5.1(c)), local martingale problem $(LMP)_\nu$, and Laplace equation (LE) for $\hat{X}$, in terms of H.

To link the weak convergence result with that for the (Y,γ,g)-superprocess consider the special case where $\hat{\gamma}(t,y) = \gamma(y(t))$, $\hat{g}(t,y) = g(y(t))$ for some $g \in$

$C_b(E)$ and $\gamma \in C_b(E)_+$, $\tau = 0$ and $m \in M_F^0(D)$. Note that we can, and shall, consider m as a finite measure on E. Note also that $\hat{\gamma}$ and $\hat{g}$ are continuous on $\hat{E}$ (but not necessarily on $\mathbb{R}_+ \times D$) – see Exercise II.8.1 below. It is then natural to assume our approximating branching mechanisms $\hat{\nu}^N((t,y),\cdot) = \nu^N(y(t),\cdot)$ where $\{\nu^N\}$ satisfy (II.3.1). Let $\{Y^\alpha : \alpha \sim t\}$ be the system of branching Y-processes constructed in Section II.3. If $W_t^\alpha = (t, Y^\alpha_{\cdot\wedge t})$, then $\{W^\alpha : \alpha \sim t\}$ is the analogous system of branching W-processes and so Theorem II.5.1(c) implies

$$(II.8.2) \qquad \hat{X}_t^N(\cdot) = \frac{1}{N}\sum_{\alpha\sim t} \delta_{W_t^\alpha} \overset{w}{\Rightarrow} \hat{X}_t \quad \text{on} \quad D(\mathbb{R}_+, M_F(\hat{E})),$$

where $\hat{X}$ has law $\mathbb{P}_{0,m}$.

Recall $H_t^N \in M_F(D(E))$ is defined by $H_t^N = \hat{X}_t^N(\Pi^{-1}(\cdot))$, i.e.,

$$H_t^N = \frac{1}{N}\sum_{\alpha\sim t} \delta_{Y^\alpha_{\cdot\wedge t}},$$

and so, taking projections in (II.8.2), we have

$$(II.8.3) \qquad \mathbb{P}(H^N_\cdot \in \cdot) \overset{w}{\Rightarrow} \mathbb{Q}_{0,m}(\cdot).$$

Remark II.8.2. It is possible to prove this weak convergence result without the continuity assumption (PC) and to prove Theorem II.5.1(c) with the assumption (II.2.2) $P_t : C_b \to C_b$ replaced by the weaker condition (QLC) (i.e. Y is a Hunt process). For γ constant and $g \equiv 0$ these results are proved in Theorems 7.15 and 7.13, respectively, of Dawson-Perkins (1991). Our proof of the compact containment condition (i) can be used to simplify the argument given there. Without our continuity assumptions one must work with the fine topology and use a version of Lusin's theorem. The processes of interest to us satisfy our continuity conditions and so we will not discuss these extensions.

It is a relatively simple matter to take projections in (II.8.2) (or (II.8.3)) and compare with Theorem II.5.1(c) to see that

$$(II.8.4) \qquad X_t \equiv H_t(y_t \in \cdot) \text{ is a } (Y,\gamma,g)\text{-DW superprocess starting at } m \text{ under } \mathbb{Q}_{0,m}.$$

We leave this as Exercise II.8.1. See Exercise II.8.3 for another approach.

Exercise II.8.1. (a) Define $\hat{\Pi} : \mathbb{R}_+ \times D(E) \to E$ by $\hat{\Pi}(t,y) = y(t)$. Show that $\hat{\Pi}$ is not continuous but its restriction to $\hat{E}$ is.
Hint. On $\hat{E}$, $\hat{\Pi}(t,y) = y(T)$ for all $T \ge t$.

(b) For $H \in \Omega_H$ define $\tilde{\Pi}(H)(t) = H_t(y_t \in \cdot) \in M_F(E)$. Show that $\tilde{\Pi} : \Omega_H \to \Omega_X$ and is continuous.
Hints. (i) Show

$$\begin{aligned} T :&\Omega_H \to C(\mathbb{R}_+, M_F(\hat{E})) \\ &H_t \to \delta_t \times H_t \end{aligned}$$

is continuous.
(ii) Show that $\tilde{\Pi}(H)_t = T(H)_t \circ \hat{\Pi}^{-1}$.

(c) Use either (II.8.2) and (a), or (II.8.3) and (b), to prove that under $\mathbb{Q}_{0,m}$, $\tilde{\Pi}(H.)$ is a (Y, γ, g)-DW superprocess starting at $m \in M_F(E)$.

Consider now the version of (MP) which characterizes $\mathbb{Q}_{\tau,m}$. For $s \le t$ and ϕ in $b\mathcal{D}\Big(\mathcal{D} = \mathcal{B}(D(E))\Big)$ let $P_{s,t}\phi(y) = P_{s,y}(\phi(Y^t))$ be the inhomogeneous semigroup associated with the path-valued process Y^t. If $\hat{A}$ is the weak generator of W and $\hat{A}\phi(s,y) \equiv \hat{A}_s\phi(y)$, then it is easy to see from Proposition II.2.1 that for $\phi \in C_b(\hat{E})$,

$$\phi \in \mathcal{D}(\hat{A}) \Leftrightarrow (P_{s,s+h}\phi_{s+h} - \phi_s)/h \xrightarrow{bp} \hat{A}_s\phi \text{ as } h \downarrow 0 \text{ for some } \hat{A}_s\phi(y) \text{ in } C_b(\hat{E})$$

$$\Leftrightarrow \text{ For some } \hat{A}_s\phi(y) \text{ in } C_b(\hat{E}),\ \phi(t, Y^t) - \phi(s, Y^s) - \int_s^t \hat{A}_r\phi(Y^r)\, dr,$$

$$t \ge s, \text{ is a } P_{s,y}\text{-martingale } \forall (s,y) \in \hat{E}.$$

If $m \in M_F^\tau(D)$, an $(\mathcal{F}_t)_{t\ge\tau}$-adapted process $(H_t)_{t\ge\tau}$ with sample paths in $\Omega_H[\tau,\infty)$ defined on $(\Omega, \mathcal{F}, (\mathcal{F}_t)_{t\ge\tau}, \mathbb{P})$ satisfies $(HMP)_{\tau,m}$ iff $H_\tau = m$ a.s. and

$$\forall \phi \in \mathcal{D}(\hat{A}) \quad M_t(\phi) = H_t(\phi_t) - H_\tau(\phi_\tau) - \int_\tau^t H_s(\hat{A}_s\phi + \hat{g}_s\phi_s)\, ds \quad \text{is a}$$

$$\text{continuous } (\mathcal{F}_t)\text{-martingale with } \quad \langle M(\phi)\rangle_t = \int_\tau^t H_s(\hat{\gamma}_s\phi_s^2)\, ds \quad \forall t \ge \tau \quad \text{a.s.}$$

The following is immediate from Theorem II.5.1, applied to $\hat{X}$. As we are considering deterministic initial conditions we may work with a martingale problem rather than a local martingale problem (recall Remark II.5.5(2)).

Theorem II.8.3. (a) $(HMP)_{\tau,m}$ is well-posed. $\mathbb{Q}_{\tau,m}$ is the law of any solution to $(HMP)_{\tau,m}$.
(b) If K satisfies $(HMP)_{\tau,m}$ on $(\Omega, \mathcal{F}, (\mathcal{F}_t)_{t\ge\tau}\mathbb{P})$ and $T \ge \tau$ is an $(\mathcal{F}_t)_{t\ge\tau}$-stopping time, then

$$\mathbb{P}(K_{T+\cdot} \in A \mid \mathcal{F}_T)(\omega) = \mathbb{Q}_{T(\omega),K_{T(\omega)}}(H_{T(\omega)+\cdot} \in A) \quad \text{a.s.} \quad \forall A \in \mathcal{F}_H. \ \blacksquare$$

We call a solution $(K_t, t \ge \tau)$ on $(\Omega, \mathcal{F}, (\mathcal{F}_t)_{t\ge\tau}, \mathbb{P})$, an $(\mathcal{F}_t)$-historical process (or $(\mathcal{F}_t) - (Y, \hat{\gamma}, \hat{g})$-historical process) starting at (τ, m).

The Feyman-Kac semigroup associated with Y^t and $\hat{g}$ is

$$P_{s,t}^{\hat{g}}\phi(y) = P_{s,y}\Big(\exp\Big\{\int_s^t \hat{g}(u, Y^u)\, du\Big\}\phi(Y^t)\Big) \qquad 0 \le s \le t, \quad \phi \in b\mathcal{D}.$$

The mean measure formula for DW-superprocesses (Exercise II.5.2 (b)) gives

$$\hat{\mathbb{P}}_{\tau,m}\Big(\hat{X}_t(\phi)\Big) = \int \hat{P}_{\tau,y}\Big(\exp\Big\{\int_0^t \hat{g}(W_s)\, ds\Big\}\phi(W_t)\Big)\, dm(y),$$

which by (II.8.1) and Lemma II.8.1 gives (set $\phi(t,y)=\psi(y^t)$ for $\psi\in b\mathcal{D}$)

$$(II.8.5)\qquad \mathbb{Q}_{\tau,m}\left(H_t(\psi)\right)=\int P^{\hat{g}}_{\tau,t}\psi(y)\,dm(y)\qquad t\ge\tau,\quad \psi\in b\mathcal{D}.$$

Let $\{\hat{P}_t\}$ denote the semigroup of W and let $\hat{\phi}$, $\hat{f}\in b\hat{\mathcal{E}}_+$ ($\hat{\mathcal{E}}$ is the Borel σ-field of $\hat{E}$). The Laplace equation for $\hat{X}$ (Theorem II.5.11) shows that if $\hat{V}_t$ is the unique solution of

$$(\widehat{ME})_{\hat{\phi},\hat{f}}\qquad \hat{V}_t=\hat{P}_t\hat{\phi}+\int_0^t\hat{P}_s\left(\hat{f}+\hat{g}\hat{V}_{t-s}-\frac{\hat{\gamma}(\hat{V}_{t-s})^2}{2}\right)ds,$$

then

$$(\widehat{LE})\qquad \hat{\mathbb{P}}_{\tau,m}\Big(\exp\Big(-\hat{X}_t(\hat{\phi})-\int_0^t\hat{X}_s(\hat{f})\,ds\Big)\Big)=\exp\Big\{-\int\hat{V}_t(\tau,y)\,dm(y)\Big\}.$$

Let $D^s=\{y\in D(E):y=y^s\}$. Defining $V_{s,t}(y)=\hat{V}_{t-s}(s,y)$ $(s\le t, y\in D^s)$ establishes a one-to-one correspondence between solutions of $(\widehat{ME})_{\hat{\phi},\hat{f}}$ and solutions of

$$(ME)_{\hat{\phi},\hat{f}}\qquad V_{\tau,t}(y)=P_{\tau,t}(\hat{\phi}_t)(y)+\int_\tau^t P_{\tau,s}\left(\hat{f}_s+\hat{g}_sV_{s,t}-\hat{\gamma}_sV^2_{s,t}/2\right)(y)\,ds.$$

Note that in $(ME)_{\hat{\phi},\hat{f}}$ we may fix t (and $\hat{\phi}_t$) and obtain an equation in $y\in D^\tau$, $\tau\le t$. Using Lemma II.8.1, and setting $\hat{\phi}(t,y)=\phi(y^t)$ for $\phi\in b\mathcal{D}_+$, and $\hat{f}(t,y)=f(t,y^t)$ for $f\in b(\mathcal{B}(\mathbb{R}_+)\times\mathcal{D})_+$, we readily translate $(\widehat{LE})$ into

Theorem II.8.4. Assume $\phi(y)$ and $f(t,y)$ are non-negative, bounded, Borel functions on $D(E)$ and $\mathbb{R}_+\times D(E)$, respectively. Let $V_{\tau,t}(y)$ be the unique solution of $(ME)_{\hat{\phi},\hat{f}}$ with $\hat{\phi}$ and $\hat{f}$ as above. Then

$$(HLE)\qquad \mathbb{Q}_{\tau,m}\Big(\exp\Big\{-H_t(\phi)-\int_\tau^t H_s(f_s)\,ds\Big\}\Big)=\exp\Big\{-\int V_{\tau,t}(y)\,dm(y)\Big\}.$$

Exercise II.8.2. Assume $(H_t,t\ge\tau)$ is an $(\mathcal{F}_t)-(Y,\hat{\gamma},0)$-historical process starting at (τ,m). Show that for any $\phi\in b\mathcal{D}$, $\int\phi(y^\tau)H_t(dy)$is a continuous $(\mathcal{F}_t)$-martingale.
Hint. The martingale property is easy (use (II.8.5)). For the continuity, start with ϕ continuous and then show the class of ϕ for which continuity holds is closed under $\xrightarrow{bp}$.

Exercise II.8.3. Let $m\in M^\tau_F(D)$, and assume $\hat{g}(s,y)=g(y(s))$ and $\hat{\gamma}(s,y)=\gamma(y(s))$ for some $g\in C_b(E)$ and $\gamma\in C_b(E)_+$. Let K satisfy $(HMP)_{\tau,m}$ on $(\Omega,\mathcal{F},(\mathcal{F}_t)_{t\ge\tau},P)$. Define $X_t\in M_F(E)$ by $X_t(A)=K_{\tau+t}(\{y:y_{\tau+t}\in A\})$. From the hint in Exercise II.8.1(a) it is easy to see that $X_\cdot$ is a.s. continuous.

(a) If $\phi \in \mathcal{D}(A)$, prove that $\hat{\phi}(s,y) = \phi(y(s))$, $(s,y) \in \hat{E}$, defines a function in $\mathcal{D}(\hat{A})$ and $\hat{A}\phi(s,y) = A\phi(y(s))$.
(b) Show that X solves $(MP)^{g,\gamma,A}_{X_0}$ and conclude that $P(X \in \cdot) = \mathbb{P}_{X_0}$ is the law of the (Y,γ,g)-superprocess starting at $X_0 = m(y_\tau \in \cdot)$.

Since $\hat{X}_t$ is an infinitely divisible random measure, the same is true of H_t under $\mathbb{Q}_{\tau,m}$ for each $t \geq \tau$. We can therefore introduce the canonical measures from Section 7 in the historical setting. Assume $\hat{\gamma} \equiv \gamma$ is constant and $\hat{g} \equiv 0$. If $(\tau, y) \in \hat{E}$, let $\hat{R}_t(\tau, y)$ denote the canonical measure of $\hat{X}_t$ from Theorem II.7.2. Then Lemma II.8.1 and (II.7.11), applied to $\hat{X}$, imply that

$$\hat{R}_t(\tau,y)(\cdot) = \delta_{\tau+t} \times R_{\tau,\tau+t}(y,\cdot),$$

where (by Theorem II.7.2)

(a) $R_{\tau,t}(y,\cdot)$ is a finite measure on $M_F^t(D) - \{0\}$ which is Borel in $y \in D^\tau$, and satisfies $R_{\tau,t}(M_F^t(D) - \{0\}) = \dfrac{2}{\gamma(t-\tau)}$,

$$\int \psi(\nu(1))\nu(\phi)R_{\tau,t}(y,d\nu) = \int_0^\infty \psi(\gamma(t-\tau)z/2)ze^{-z}dzP_{\tau,y}(\phi(Y^t))$$

for any bounded Borel $\phi : D(E) \to \mathbb{R}$ and $\psi : \mathbb{R}_+ \to \mathbb{R}$, and $R_{\tau,t}(y,\{w : w^\tau \neq y\}) = 0$.

(II.8.6) (b) $$\mathbb{Q}_{\tau,m}(\exp(-H_t(\phi))) = \exp\left(-\int\int 1 - e^{-\nu(\phi)}R_{\tau,t}(y,d\nu)m(dy)\right)$$

for any Borel $\phi : D(E) \to \mathbb{R}_+$.

(c) If $m \in M_F^\tau(D)$ and Ξ is a Poisson point process on $M_F(D) - \{0\}$ with intensity $\int R_{\tau,t}(y,\cdot)dm(y)$ then $\int \nu\Xi(d\nu)$ has law $\mathbb{Q}_{\tau,m}(H_t \in \cdot)$.

The fact that $w^\tau = y$ for $R_{\tau,t}(y)$-a.a. w (in (a)) is immediate from (c) and the corresponding property for H_t under $\mathbb{Q}_{\tau,\delta_y}$ (use (II.8.5) to see the latter).

The uniqueness of the canonical measure in Theorem II.7.1 and the fact that under $\mathbb{Q}_{\tau,\delta_y}$, $X_t(\cdot) = H_{\tau+t}(\{y' : y'_{\tau+t} \in \cdot\})$ is a $(Y,\gamma,0)$-superprocess starting at y_τ $(y \in D^\tau)$ by Exercise II.8.3, show that if $\hat{\Pi}_t(y) = y_t$ and $(R_t(x))_{x\in E}$ are the canonical measures for X, then

(II.8.7) $$R_{\tau,\tau+t}(y, \nu \circ \hat{\Pi}^{-1}_{\tau+t} \in \cdot) = R_t(y_\tau, \cdot).$$

III. Sample Path Properties of Superprocesses
1. Historical Paths and a Modulus of Continuity

Assume $(X, (\mathbb{P}_\mu)_{\mu\in M_F(E)})$ is a $(Y,\gamma,0)$-DW-superprocess with $\gamma(\cdot)\equiv\gamma>0$ constant, (PC) holds, and $(H, (\mathbb{Q}_{\tau,m})_{(\tau,m)\in\hat{E}})$ is the corresponding historical process on their canonical path spaces, Ω_X and Ω_H, respectively.

Theorem III.1.1. (Historical Cluster Representation). Let $m\in M_F^\tau(D)$ and $\tau\le s<t$. If $r_s(H_t)(\cdot)=H_t(\{y: y^s\in\cdot\})$, then

$$(III.1.1)\qquad \mathbb{Q}_{\tau,m}\left(r_s(H_t)\in\cdot\mid \mathcal{F}^H[\tau,s+]\right)(\omega)\overset{\mathcal{D}}{\equiv}\sum_{i=1}^M e_i\delta_{y_i},$$

where $(e_i,y_i)_{i\le M}$ are the points of a Poisson point process on $\mathbb{R}_+\times D$ with intensity $(\frac{2}{\gamma(t-s)})(\nu_{t-s}\times H_s(\omega))$ and ν_{t-s} is an exponential law with mean $\gamma(t-s)/2$. That is, the right-hand side of (III.1.1) defines a regular conditional distribution for the random measure on the left side given $\mathcal{F}^H[\tau,s+]$.

Proof. By the Markov property of H we may assume $s=\tau$. Fix $m\in M_F^\tau(D)$. Let $A_1,\dots,A_n$ be a Borel partition of D and define $m_i=m(\cdot\cap A_i)$. Let $H^1,\dots,H^n$ be independent $(\mathcal{F}_t)$-historical processes with H^i starting at (τ,m_i) on some $(\Omega,\mathcal{F},\mathcal{F}_t,\mathbb{P})$. Then by checking the historical martingale problem $(HMP)_{\tau,m}$ we can easily see that $H\equiv\sum_{i=1}^n H^i$ is an $(\mathcal{F}_t)$-historical process starting at (τ,m) and so has law $\mathbb{Q}_{\tau,m}$. For each i, the mean value property for historical processes (II.8.5) implies (recall $y=y^\tau$ m-a.e.) for each $t>\tau$

$$\mathbb{P}\left(H_t^i(\{y:y^\tau\in A_i^c\})\right)=\int 1_{A_i}(y^\tau)P_{\tau,y^\tau}(y^\tau\in A_i^c)dm(y)=0.$$

The process inside the expected value on the left side is a.s. continuous in t by Exercise II.8.2 and so is identically 0 for all $t\ge\tau$ a.s. This implies

$$(III.1.2)\qquad H_t^i(\cdot)=H_t\left(\cdot\cap\{y^\tau\in A_i\}\right)\qquad i=1\dots n\quad\text{for all}\quad t\ge\tau\quad\text{a.s.}$$

It follows that if we start with H. under $\mathbb{Q}_{\tau,m}$ and define H^i by (III.1.2), then

$$(III.1.3)\qquad \begin{array}{l}(H^i_\cdot, i\le n)\quad\text{are independent}\quad \mathcal{F}^H[\tau,t+]\text{-historical processes}\\ \text{starting at}\quad (\tau,m_i)_{i\le n},\quad\text{respectively.}\end{array}$$

In particular $\{H^i_{s+\tau}(1), s\ge 0\}_{i=1\dots n}$ are independent Feller diffusions (i.e. solutions of (II.5.10) with $g=0$) with initial values $\{m(A_i): i=1\dots n\}$, respectively. Recalling the Laplace transforms of $H_t^i(1)$ from (II.5.11), we have for any $\lambda_1,\dots,\lambda_n\ \ge 0$

Originally published in: *Ecole d'Eté de Probabilités de Saint-Flour XXIX – 1999*, Lecture Notes in Mathematics, Vol. **1781**, 193–246, DOI: 10.1007/3-540-47944-9_7,

and $f(y)=\sum_{i=1}^{n}\lambda_i 1_{A_i}(y)$ $(y\in D^\tau)$,

$$\begin{aligned}\mathbb{Q}_{\tau,m}\Big(\exp\Big\{-\int f(y^\tau)H_t(dy)\Big\}\Big)&=\mathbb{Q}_{\tau,m}\Big(\exp\Big\{-\sum_{i=1}^{n}\lambda_i H_t^i(1)\Big\}\Big)\\&=\exp\Big\{-\sum_{i=1}^{n}\frac{2\lambda_i m(A_i)}{2+\lambda_i(t-\tau)\gamma}\Big\}\\&=\exp\Big\{-\int 2f(y)(2+f(y)(t-\tau)\gamma)^{-1}dm(y)\Big\}\\&=E\Big(\exp\Big\{-\int ef(y)\Xi(de,dy)\Big\}\Big),\end{aligned}$$

where Ξ is a Poisson point process on $\mathbb{R}_+\times D$ with intensity $\frac{2}{\gamma(t-\tau)}(\nu_{t-\tau}\times m)$. As the above equation immediately follows for any Borel $f\geq 0$ on D, the proof is complete because we have shown $\int 1(y^\tau\in\cdot)H_t(dy)$ and $\int e1(y\in\cdot)\Xi(de,dy)$ have the same Laplace functionals. ∎

A consequence of the above argument is

Lemma III.1.2. If A is a Borel subset of $D(E)$, and $m\in M_F^\tau(D)$, then $X_s=H_{s+\tau}(\{y:y^\tau\in A\})$ is a Feller branching diffusion (a solution of (II.5.10) with $g=0$), and for $t>\tau$,

$$\mathbb{Q}_{\tau,m}\left(H_s\left(\{y:y^\tau\in A\}\right)=0\quad\forall s\geq t\right)=\exp\left\{-\frac{m(A)2}{\gamma(t-\tau)}\right\}.$$

Proof. This is immediate from (III.1.3) and the extinction probabilities found in (II.5.12). ∎

The above Theorem shows for $s<t$, $H_t(\{y:y^s\in\cdot\})$ is a purely atomic measure. The reader should be able to see that (conditionally) this measure is a Poisson superposition of exponential masses directly from Kolmogorov's and Yaglom's theorems (Theorem II.1.1). If $\tau=s=0$, the above may also be easily derived from the corresponding canonical measure representation (II.7.11) for H_t and projecting it down to y_0. Note that the exponential masses come from the last assertion of Theorem II.7.2(iii). An extension of the above cluster decomposition which describes the future evolution of the descendants of these finite number of clusters will be proved using similar ideas in Section III.6 (see Theorem III.6.1 and Corollary III.6.2).

Until otherwise indicated assume $((X_t)_{t\geq 0},(\mathbb{P}_\mu)_{\mu\in M_F(\mathbb{R}^d)})$ is a super-Brownian motion with branching rate $\gamma>0$ (we write X is SBM(γ)). This means X is a $(B,\gamma,0)$-DW-superprocess on its canonical space Ω_X, where B is a standard Brownian motion in $\mathbb{R}^d$ and γ is a positive constant. $H.$ will denote the corresponding historical process on Ω_H. We call H a historical Brownian motion (with branching rate γ).

The following result is essentially proved in Dawson-Iscoe-Perkins (1989) but first appeared in this form as Theorem 8.7 of Dawson-Perkins (1991). It gives a uniform modulus of continuity for all the paths in the closed support of H_t for all $t\geq 0$. The simple probabilistic proof given below seems to apply in a number of different settings. See, for example, Mueller and Perkins (1992) where it is applied to the supports of the solutions of a class of parabolic stochastic pde's. It also extends readily to more general branching mechanisms (see Dawson and Vinogradov (1994)) and to the interacting models considered in Chapter V (see Chapter 3 of Perkins (1995)).

Notation: $S(\mu)$ denotes the closed support of a measure μ on the Borel sets of a metric space. $h(r) = \left(r \log 1/r\right)^{1/2}$ is Lévy's modulus function.

Theorem III.1.3. (Historical Modulus of Continuity). If $\delta, c > 0$, let $K(\delta, c) = \{y \in C(\mathbb{R}^d) : |y(r) - y(s)| \le ch(|r-s|)\ \forall r, s \ge 0 \text{ satisfying } |r-s| \le \delta\}$. $\mathbb{Q}_{0,m}$ denotes the law of historical Brownian motion with branching rate $\gamma > 0$ starting at $(0, m)$.

(a) If $c > 2$, then $\mathbb{Q}_{0,m}$-a.s. there is a $\delta(c,\omega) > 0$ such that

$$S(H_t)(\omega) \subset K\big(\delta(c,\omega), c\big) \quad \forall t \ge 0.$$

Moreover there are constants $\rho(c) > 0$ and $C(d,c)$ such that

$$\mathbb{Q}_{0,m}(\delta(c) \le r) \le C(d,c)m(1)\gamma^{-1}r^{\rho} \text{ for } r \in [0,1].$$

(b) If $m \ne 0$ and $c < 2$, then $\mathbb{Q}_{0,m}$-a.s. for all $\delta > 0$ there is a t in $(0,1]$ such that $H_t(K(\delta,c)^c) > 0$.

Remark. This should be compared to Lévy's modulus of continuity for a simple Brownian path for which $c = \sqrt{2}$ is critical. This reflects the fact that the tree of Brownian paths underlying H. has infinite length. We prove (a) below only for a sufficiently large c.

Proof of (a) (for large c). Use Lemma III.1.2 and the Markov property to see that

$$\begin{aligned}
&\mathbb{Q}_{0,m}\Big(H_t\Big(\Big\{y : \Big|y\Big(\frac{j}{2^n}\Big) - y\Big(\frac{j-1}{2^n}\Big)\Big| > ch(2^{-n})\Big\}\Big) > 0 \quad \exists t \ge (j+1)/2^n\Big) \\
&= \mathbb{Q}_{0,m}\Big(1 - \exp\Big\{-\frac{2^{n+1}}{\gamma}H_{j/2^n}\Big(\Big\{y : \Big|y\Big(\frac{j}{2^n}\Big) - y\Big(\frac{j-1}{2^n}\Big)\Big| > ch(2^{-n})\Big\}\Big)\Big\}\Big) \\
&\le 2^{n+1}\gamma^{-1}\mathbb{Q}_{0,m}\left(H_{j/2^n}\left(|y(j/2^n) - y((j-1)/2^n)| > ch(2^{-n})\right)\right).
\end{aligned}$$

Now recall from (II.8.5) that the mean measure of H_t is just Wiener measure stopped at t. The above therefore equals

$$\begin{aligned}
&2^{n+1}\gamma^{-1}\int P_y\left(|B(j/2^n) - B((j-1)/2^n)| > ch(2^{-n})\right)dm(y) \\
&\le 2^{n+1}\gamma^{-1}m(1)c_d n^{d/2-1}2^{-nc^2/2}
\end{aligned}$$

by a Gaussian tail estimate. Sum over $1 \le j \le n2^n$ to see that

$$\begin{aligned}
&\mathbb{Q}_{0,m}\Big(H_t\Big(\Big\{y : \Big|y\Big(\frac{j}{2^n}\Big) - y\Big(\frac{j-1}{2^n}\Big)\Big| > ch(2^{-n})\Big\}\Big) > 0, \\
&\qquad\qquad \text{for some } t \ge (j+1)2^{-n} \text{ and } 1 \le j \le n2^n\Big) \\
&\le c_d\gamma^{-1}m(1)n^{d/2}2^{2n+1}2^{-nc^2/2},
\end{aligned}$$

which is summable if $c > 2$. Assuming the latter, we may use Borel-Cantelli to see $\exists N(\omega) < \infty$ a.s. such that $H_t \equiv 0$ for $t \geq N(\omega)$ and

$$(III.1.4) \quad n \geq N(\omega) \Rightarrow \left|y\Big(\frac{j}{2^n}\Big) - y\Big(\frac{j-1}{2^n}\Big)\right| \leq ch(2^{-n}) \quad \forall j \geq 1,\ (j+1)2^{-n} \leq t, \quad H_t\text{-a.a. } y \quad \forall t \leq n.$$

We now follow Lévy's proof for Brownian motion. Let $\delta(c_2, \omega) = 2^{-N(\omega)} > 0$ *a.s.*, where c_2 will be chosen large enough below. The required bound on $\mathbb{Q}_{0,m}(\delta(c_2) \leq r)$ is clear from the above bound and the extinction probability estimate formula (II.5.12). Let $N \geq t > 0$ and choose y outside of an H_t-null set so that (III.1.4) holds. Assume $r < s \leq t$ and $0 < s - r \leq 2^{-N}$ and choose $n \geq N$ so that $2^{-n-1} < s - r \leq 2^{-n}$. For $k \geq n$, choose $s_k \in \{j2^{-k} : j \in \mathbb{Z}_+\}$ such that $s_k + 2^{-k} \leq s$ and s_k is the largest such value (set $s_k = 0$ if $s < 2^{-k}$). One easily checks that

$$s_k \uparrow s,\ s_{k+1} = s_k + j_{k+1}2^{-(k+1)} \text{ for } j_{k+1} = 0, 1 \text{ or } 2 \text{ (and } j_{k+1} = 0 \text{ only arises if } s < 2^{-k-1}).$$

Note also that $s_k + (j_{k+1} + 1)2^{-k-1} = s_{k+1} + 2^{-k-1} \leq s \leq t$. Therefore the choice of y and (III.1.4) imply $|y(s_{k+1}) - y(s_k)| \leq j_{k+1}ch(2^{-k-1})$, and so for some $c_1 > 0$ ($c_1 = 5c/\log 2$ will do),

$$(III.1.5) \quad |y(s) - y(s_n)| \leq \sum_{k=n}^{\infty} j_{k+1}ch(2^{-k-1}) \leq 2c\sum_{k=n}^{\infty} h(2^{-k-1}) \leq c_1 h(2^{-n-1}) \leq c_1 h(s-r).$$

Similarly one constructs $r_k \uparrow r$ so that

$$(III.1.6) \quad |y(r) - y(r_n)| \leq c_1 h(s-r).$$

The restriction $s - r \leq 2^{-n}$ implies $s_n = r_n + j_n 2^{-n}$ where $j_n = 0$ or 1 and so $s_n \leq s \leq t$, which means by (III.1.4) that

$$(III.1.7) \quad |y(s_n) - y(r_n)| \leq ch(2^{-n}) \leq c\sqrt{2}h(s-r).$$

(III.1.5)-(III.1.7) imply $|y(s) - y(r)| \leq (2c_1 + c\sqrt{2})h(s-r) \equiv c_2 h(s-r)$. This proves $H_t(K(\delta(c_2), c_2)^c) = 0$ for $t \leq N(\omega)$ and so all $t \geq 0$ because $H_t = 0$ for $t > N$. $K(\delta, c_2)$ is closed and therefore $S(H_t) \subset K(\delta(c_2), c)$ for all $t \geq 0$.

To get (a) for any $c > 2$, one works with a finer set of grid points $\{(j + \frac{p}{M})\theta^n : j \in \mathbb{Z}_+,\ 0 \leq p < M\}$ ($\theta < 1$ sufficiently small and M large) in place of $\{j2^{-n} : j = 0, 1, 2 \ldots\}$ to get a better approximation to r, s, as in Lévy's proof for Brownian motion. For example, see Theorem 8.4 of Dawson-Perkins (1991).

(b) Let $c < 2$ and $1 > \eta > 0$. If

$$B_j^n = \{y \in C : |y(2j2^{-n}) - y((2j-1)2^{-n})| > ch(2^{-n})\},$$

it suffices to show
(*III*.1.8)

$$\sup_{j\in\mathbb{N},(2j+1)2^{-n}<\eta} H_{(2j+1)2^{-n}}(B_j^n) > 0 \text{ for large } n \text{ for a.a. } \omega \text{ satisfying } \inf_{t<\eta} H_t(1) \geq \eta.$$

This is because the first event implies $\sup_{t<\eta} H_t(K(\delta,c)^c) > 0\ \forall\delta > 0$, and $\mathbb{Q}_{0,m}(\inf_{t<\eta} H_t(1) \geq \eta) \uparrow 1$ as $\eta \downarrow 0$. If

$$A_j^n = \{\omega : H_{(2j+1)2^{-n}}(B_j^n) = 0,\ H_{(2j+1)2^{-n}}(1) \geq \eta\},$$

then we claim that

$$(III.1.9) \qquad \sum_{n=1}^{\infty} \mathbb{Q}_{0,m}(\cap_{1\leq j,(2j+1)2^{-n}\leq\eta} A_j^n) < \infty.$$

Assume this for the moment. The Borel-Cantelli Lemma implies

$$\text{w.p.1 for } n \text{ large enough } \exists (2j+1)2^{-n} < \eta \text{ such that}$$
$$H_{(2j+1)2^{-n}}(B_j^n) > 0 \text{ or } H_{(2j+1)2^{-n}}(1) < \eta,$$

which implies

$$\text{w.p.1 either } \inf_{t<\eta} H_t(1) < \eta \text{ or for large } n \sup_{(2j+1)2^{-n}<\eta} H_{(2j+1)2^{-n}}(B_j^n) > 0.$$

This gives (III.1.8) and so completes the proof.

Turning to (III.1.9), note that

$$\begin{aligned}\mathbb{Q}_{0,m}(A_j^n|\mathcal{F}^H[0,2j2^{-n}+]) &\leq \mathbb{Q}_{0,m}(H_{(2j+1)2^{-n}}(B_j^n) = 0|\mathcal{F}^H[0,2j2^{-n}+])\\ &\leq \exp\Big\{-2^{n+1}\gamma^{-1}H_{2j2^{-n}}(B_j^n)\Big\},\end{aligned}$$

the last by Theorem III.1.1 with $t = (2j+1)2^{-n}$ and $s = 2j2^{-n}$. Let $R_{s,t}(y,\cdot)$ be the canonical measures associated with H, introduced in (II.8.6). Condition the above with respect to $\mathcal{F}^H[0,(2j-1)2^{-n}+]$ and use (II.8.6)(b) and the Markov property to see that

$$\begin{aligned}&\mathbb{Q}_{0,m}(A_j^n|\mathcal{F}^H[0,(2j-1)2^{-n}+])\\ &\leq \exp\Big\{-\iint 1-\exp(-2^{n+1}\gamma^{-1}\nu(B_j^n))R_{(2j-1)2^{-n},2j2^{-n}}(y,d\nu)H_{(2j-1)2^{-n}}(dy)\Big\}\\ &\leq \exp\Big\{-\iint 1(\nu(1)\leq 2^{-n})[1-\exp(-2^{n+1}\gamma^{-1}\nu(B_j^n))]\\ &\qquad\qquad R_{(2j-1)2^{-n},2j2^{-n}}(y,d\nu)H_{(2j-1)2^{-n}}(dy)\Big\}\\ &\leq \exp\Big\{-c_0\iint 1(\nu(1)\leq 2^{-n})2^{n+1}\gamma^{-1}\nu(B_j^n)\\ &\qquad\qquad R_{(2j-1)2^{-n},2j2^{-n}}(y,d\nu)H_{(2j-1)2^{-n}}(dy)\Big\},\end{aligned}$$

where $c_0 = c_0(\gamma) > 0$ satisfies $1 - e^{-2x/\gamma} \geq c_0 x$ for all $x \in [0,1]$. (II.8.6)(a) shows that on A^n_{j-1}

$$\begin{aligned}
&\mathbb{Q}_{0,m}(A^n_j | \mathcal{F}^H[0,(2j-1)2^{-n}+]) \\
&\leq \exp\Big\{-c_0 2^{n+1}\gamma^{-1}\int_0^\infty 1(\gamma 2^{-n-1}z \leq 2^{-n}) z e^{-z} dz \\
&\quad \times \int P_{(2j-1)2^{-n},y}(|B(2j2^{-n}) - B((2j-1)2^{-n}| > ch(2^{-n}))H_{(2j-1)2^{-n}}(dy)\Big\} \\
&\leq \exp\Big\{-c_0 2^{n+1}\gamma^{-1} c_1 P^0(|B(1)| \geq c\sqrt{n\log 2}) H_{(2j-1)2^{-n}}(1)\Big\} \\
&\leq \exp\{-c_2(\gamma,c)\eta 2^{n(1-c^2/2)} n^{-1/2}\},
\end{aligned}$$

for some $c_2 > 0$, at least for $n \geq n_0(c)$ by a Gaussian tail estimate and the fact that $H_{(2j-1)2^{-n}}(1) \leq \eta$ on A^n_{j-1}. Therefore for $n \geq n_1(c,\eta)$,

$$\mathbb{Q}_{0,m}\Big(\cap_{j\geq 1,(2j+1)2^{-n}\leq\eta} A^n_j\Big) \leq \exp\Big\{\frac{-c_2\eta^2}{2}\frac{2^{n(2-c^2/2)}}{\sqrt{n}}\Big\},$$

which is summable over n since $c < 2$. This gives (III.1.9) and we are done. ∎

Remark. It is easy to obtain versions of the above result for continuous spatial motions other than Brownian motion (see Theorem 8.6 in Dawson-Perkins (1991)).

Notation. $\Pi_t : D(E) \to E$ is the projection map $\Pi_t(y) = y(t)$.

Recall X denotes SBM(γ) and H is the associated historical Brownian motion.

Corollary III.1.4. (a) $S(H_t)$ is compact in $C(\mathbb{R}^d)$ $\forall t > 0$ $\mathbb{Q}_{0,m}$-a.s.
(b) $S(X_t) = \Pi_t(S(H_t))$ and hence is compact in $\mathbb{R}^d$ $\forall t > 0$ $\mathbb{P}_m$-a.s.

Proof. (a) Lemma III.1.2 shows that for any $\eta > 0$,

$$\begin{aligned}
\mathbb{Q}_{0,m}\left(H_s(\{|y_0| > R\}) = 0 \text{ for all } s \geq \eta\right) &= \exp\left\{-\frac{2m(\{(y_0) > R\})}{\gamma\eta}\right\} \\
&\to 1 \text{ as } R \to \infty.
\end{aligned}$$

This and the previous theorem show that for $\mathbb{Q}_{0,m}$-a.a. ω there is a $\delta(3,\omega) > 0$ and an $R(\omega) < \infty$ such that

$$S(H_t)(\omega) \subset K(\delta(3,\omega),3) \cap \{y : |y_0| \leq R(\omega)\} \quad \forall t \geq \eta.$$

The set on the righthand side is compact by the Arzela-Ascoli theorem. Let $\eta \downarrow 0$ to complete the proof of (a).
(b) From (II.8.4) we may assume that $X_t = H_t \circ \Pi_t^{-1}$ for all $t \geq 0$. Note that $S(H_t) \subset \Pi_t^{-1}\Big(\Pi_t(S(H_t))\Big)$ and therefore

$$X_t\left(\Pi_t(S(H_t))^c\right) = H_t\left(\Pi_t^{-1}\left(\Pi_t(S(H_t))\right)^c\right) \leq H_t(S(H_t)^c) = 0.$$

This shows $\Pi_t(S(H_t))$ supports X_t and as it is compact $\forall t>0$ by (a), $S(X_t)\subset \Pi_t(S(H_t))$ $\forall t>0$ and is also compact a.s. If $w\in S(H_t)$, $H_t(\{y:|y_t-w_t|<\varepsilon\})>0$ for all $\varepsilon>0$ and so $w_t\in S(X_t)$. This shows the reverse inclusion. ∎

A measure-valued process $X.$ has the compact support property (CSP) iff $S(X_0)$ compact implies $S(X_t)$ is compact for all $t>0$ a.s. Corollary III.1.4 shows that super-Brownian motion has the (CSP) (in fact $S(X_0)$ need not be compact). Obviously this property fails for the heat kernel $P_t\phi$. The (CSP) for SBM was first proved by Iscoe (1988). The next result provides the natural rate of propagation for X suggested by the historical modulus of continuity.

Notation. $A\subset\mathbb{R}^d$, $\delta>0$, $A^\delta=\{x\in\mathbb{R}^d: d(A,x)\equiv\inf\{|y-x|:y\in A\}<\delta\}$.

Corollary III.1.5. With probability 1 for any $c>2$ there is a $\delta(c,\omega)>0$ such that if $0<t-s<\delta(c,\omega)$, then $S(X_t)\subset S(X_s)^{ch(t-s)}$.

To avoid an unexpected decline in $S(X_s)$ on the right side of this inclusion we need a lemma.

Lemma III.1.6. For $\mathbb{Q}_{0,m}$-a.a. ω if $0\le s\le t$ and $y\in S(H_t)$, then $y(s)\in S(X_s)$.

Proof. If $0\le s<s'$ are fixed, Theorem III.1.1 shows that conditional on $\mathcal{F}^H[0,s+]$, $H_{s'}(y(s)\in\cdot)$ is supported on a finite number of points $x_1\dots x_n$ in $S(X_s)$. The Markov property and (III.1.3) show that conditional on $\mathcal{F}^H[0,s'+]$,

$$\{H_t(y_s=x_i):t\ge s'\}_{i\le n}\text{ and }\{H_t(y_s\notin\{x_1\dots x_n\}):t\ge s'\}$$

are independent Feller diffusions. The latter is therefore a.s. identically 0 and so w.p.1 for all $t\ge s'$,

$$\begin{aligned}\{y(s):y\in S(H_t)\}&\subset S(H_t(y_s\in\cdot))\qquad\text{(trivial)}\\&\subset S(H_{s'}(y_s\in\cdot))\\&\subset S(X_s).\end{aligned}$$

Take the union over all $0\le s\le s'$ in $\mathbb{Q}$ to conclude

$$(III.1.8)\quad w.p.1\quad\text{for all}\quad s\in\mathbb{Q}^{\ge0}\quad\text{and all}\quad t>s\quad \{y(s):y\in S(H_t)\}\subset S(X_s).$$

A simple consequence of our modulus of continuity and $X_t=H_t(y_t\in\cdot)$ is that if $B=B(x,\varepsilon)$, $B'=B(x,\varepsilon/2)$ and $m(B)=0$, then $\mathbb{Q}_{\tau,m}$-a.s. $\exists\eta>0$ such that $X_t(B')=0$ for all $\tau\le t<\eta$. Use this and the strong Markov property at time $T_r(B)=\inf\{t\ge r: X_t(B)=0\}$ where $B=B(x,\varepsilon)$ is a rational ball ($x\in\mathbb{Q}^d,\varepsilon\in\mathbb{Q}^{>0}$) and B' is as above to conclude:

$$(III.1.9)\quad \begin{array}{l}w.p.1\text{ for all }r\in\mathbb{Q}^{\ge0}\text{ and rational ball }B\ \exists\eta>0\text{ such that}\\ X_s(B')=0\text{ for all }s\text{ in }[T_r(B),T_r(B)+\eta).\end{array}$$

Choose ω so that (III.1.8) and (III.1.9) hold. Let $y\in S(H_t)$, $s<t$ and suppose $y(s)\notin S(X_s)$ (the $s=t$ case is handled by Corollary III.1.4). Choose a rational ball B so that $y(s)\in B'$ and $X_s(B)=0$, $\eta'>0$, and a rational r in $(s-\eta',s]$. Then

$T_r(B) \le s$ because $X_s(B) = 0$ and so by (III.1.9) there is an open interval I in $(s-\eta', s+\eta')$ such that $X_u(B') = 0$ for all u in I. In particular there are rationals $u_n \to s$ such that $X_{u_n}(B') = 0$. On the other hand by (III.1.8) and the continuity of y, $y(u_n) \in B'$ and $y(u_n) \in S(X_{u_n})$ for n large which implies $X_{u_n}(B') > 0$, a contradiction. ∎

Proof of Corollary III.1.5. Apply Theorem III.1.3, Corollary III.1.4 and Lemma III.1.6 to see that w.p.1 if $0 < t - s < \delta(c, \omega)$,

$$S(X_t) = \Pi_t(S(H_t)) \subset \Pi_s(S(H_t))^{ch(t-s)} \subset S(X_s)^{ch(t-s)}. \quad ∎$$

Remark. Presumably $c = 2$ is also sharp in Corollary III.1.5 if $d \ge 2$, although this appears to be open. It would be of particular interest to find the best result in $d = 1$ as the behaviour of $\partial S(X_t)$ in $d = 1$ could shed some light on the SPDE for super-Brownian motion in one dimension.

Definition. For $I \subset \mathbb{R}_+$ we call $\mathcal{R}(I) = \bigcup_{t \in I} S(X_t)$, the range of X on I, and $\overline{\mathcal{R}}(I) = \overline{\mathcal{R}(I)}$ is the closed range of X on I. The range of X is $\mathcal{R} = \bigcup_{\delta > 0} \overline{\mathcal{R}}([\delta, \infty))$.

It is not hard to see that $\mathcal{R} - \mathcal{R}(0,\infty)$ is at most a countable set of "local extinction points" (see Proposition 4.7 of Perkins (1990) and the discussion in Section III.7 below). $\mathcal{R}$ is sometimes easier to deal with than $\mathcal{R}((0,\infty))$. The reason for not considering $\mathcal{R}([0,\infty))$ or $\overline{\mathcal{R}}([0,\infty))$ is that it will be $\mathbb{R}^d$ whenever $S(X_0) = \mathbb{R}^d$.

Corollary III.1.7. $\overline{\mathcal{R}}([\delta,\infty))$ is compact for all $\delta > 0$ a.s. $\overline{\mathcal{R}}([0,\infty))$ is a.s. compact if $S(X_0)$ is.

Proof. Immediate from Corollaries III.1.4 and III.1.5. ∎

In view of the increase in the critical value of c in Theorem III.1.3 from that for a single Brownian path, it is not surprising that there are diffusions Y for which the (CSP) fails for the associated DW-superprocess X. Example 8.16 in Dawson-Perkins (1991) gives a time-inhomogeneous $\mathbb{R}$-valued diffusion and $T > 0$ for which $S(X_T) = \phi$ or $\mathbb{R}$ a.s. For jump processes instantaneous propagation is to be expected as is shown in the next Section.

2. Support Propagation and Super-Lévy Processes

Let Y be a Poisson (rate $\lambda > 0$) process on $\mathbb{Z}_+$ and consider X, the $(Y,\gamma,0)$-DW superprocess with $\gamma > 0$ constant. Then $A\phi(i) = \lambda(\phi(i+1) - \phi(i))$, $i \in \mathbb{Z}_+$, and taking $\phi(i) = 1(i = j)$ in the martingale problem for X, we see that $X_t = (X_t(j))_{j \in \mathbb{Z}_+}$ may also be characterized as the pathwise unique solution of

$$(III.2.1)\quad X_t(j) = X_0(j) + \lambda \int_0^t (X_s(j-1) - X_s(j))ds + \int_0^t \sqrt{\gamma X_s(j)}dB_s^j,\ j \in \mathbb{Z}_+.$$

Here $\{B_\cdot^j : j \in \mathbb{Z}_+\}$ is a collection of independent linear Brownian motions, and $X_s(-1) \equiv 0$. Pathwise uniqueness holds by the method of Yamada and Watanabe (1971).

Let $X_0 = \alpha\delta_0$. Note that $X_t(\{0,\ldots,n\})$ is a non-negative supermartingale and so sticks at 0 when it first hits 0 at time ζ_n. Evidently

$$\zeta_n \uparrow \zeta = \inf\{t : X_t(\mathbb{Z}_+) = 0\} < \infty \quad \text{a.s.}$$

Clearly $m_t = \inf S(X_t) = n$ if $t \in [\zeta_{n-1}, \zeta_n)$ ($\zeta_{-1} = 0$) and so is increasing in t a.s., and X_t becomes extinct as the lower end of its support approaches infinity. On the other hand it is clear (at least intuitively) from (III.2.1) that the mass at m_t will immediately propagate to $m_t + 1$ which in turn immediately propagates to $m_t + 2$, and so on. Therefore we have

$$(III.2.2) \qquad S(X_t) = \{m_t, m_t + 1, \ldots\} \quad \text{a.s.} \quad \forall t > 0.$$

This result will be a special case of Theorem III.2.4 below. Note, however, that $S(X_t) = \{m_t\}$ at exceptional times is a possibility by a simple comparison argument with the square of a low-dimensional Bessel process (see Exercise III.2.1).

The above example suggests $S(X_t)$ will propagate instantaneously to any points to which Y can jump. This holds quite generally (Corollary 5.3 of Evans-Perkins (1991)) but for the most part we consider only d-dimensional Lévy processes here. Our first result, however, holds for a general (Y, γ, g)-DW superprocess X with law $\mathbb{P}_{X_0}$. Recall if $Z_t = Z_0 + M_t + V_t$ is the canonical decomposition of a semimartingale then its local time at 0 is

$$(LT) \qquad L_t^0(Z) = \lim_{\varepsilon \downarrow 0} \varepsilon^{-1} \int_0^t 1(0 \le Z_s \le \varepsilon) d\langle M \rangle_s \quad \text{a.s.}$$

and in particular

$$(III.2.3) \qquad \int_0^\infty 1(Z_s = 0) d\langle M \rangle_s = 0.$$

Theorem III.2.1. If $\phi \in \mathcal{D}(A)_+$, then with probability 1 for Lebesgue-a.a. s, $X_s(A\phi) > 0$ implies $X_s(\phi) > 0$.

Proof. This should be intuitively clear from $(MP)_{X_0}$ as the drift $X_s(A\phi)ds > 0$ should keep $X_s(\phi) > 0$ a.s. Since $X_t(\phi) = X_0(\phi) + M_t(\phi) + \int_0^t X_s(A^g\phi)ds$, we have with probability one,

$$L_t \equiv L_t^0(X(\phi)) = \lim_{\varepsilon\downarrow 0}\int_0^t 1(0 < X_s(\phi) \le \varepsilon) X_s(\gamma\phi^2) ds \varepsilon^{-1}$$

$$\le \|\gamma\phi\| \lim_{\varepsilon\downarrow 0}\int_0^t 1(0 < X_s(\phi) \le \varepsilon) X_s(\phi)\varepsilon^{-1} ds$$

$$\le \|\gamma\phi\| \lim_{\varepsilon\downarrow 0}\int_0^t 1(0 < X_s(\phi) \le \varepsilon) ds$$

$$= 0 \quad \text{a.s.}$$

Tanaka's Formula implies

$$X_t(\phi)^+ = X_0(\phi) + \int_0^t 1\left(X_s(\phi) > 0\right) dM_s(\phi) + \int_0^t 1\left(X_s(\phi) > 0\right) X_s(A\phi) ds$$
$$+ \int_0^t 1\left(X_s(\phi) > 0\right) X_s(g\phi) ds.$$

Clearly $\int_0^t 1(X_s(\phi) = 0)\, X_s(g\phi)\, ds = 0$, and by (III.2.3), $\int_0^t 1\left(X_s(\phi) = 0\right) dM_s(\phi) = 0$. The above therefore implies

$$X_t(\phi) = X_t(\phi)^+ = X_0(\phi) + M_t(\phi) + \int_0^t X_s(A^g\phi) ds - \int_0^t 1(X_s(\phi) = 0) X_s(A\phi) ds$$

$$= X_t(\phi) - \int_0^t 1(X_s(\phi) = 0) X_s(A\phi) ds.$$

We conclude that $\int_0^t 1(X_s(\phi) = 0) X_s(A\phi) ds = 0 \ \forall t \ge 0$ a.s. and the result follows. ∎

Assume now that Y is a Lévy process in $\mathbb{R}^d$ with Lévy measure ν. Then $\mathcal{D}(A)$ contains C_K^∞, the C^∞-functions on $\mathbb{R}^d$ with compact support. Let B be an open ball in $\mathbb{R}^d$ and choose $\phi \in (C_K^\infty)_+$ such that $\{\phi > 0\} = B$. Then for $x \notin B$, $A\phi(x) = \int \phi(x+y)\nu(dy)$ (see, e.g., Theorem IV.4.1 of Gihman and Skorokhod (1975), or Example II.2.4(b) when Y is an asymmetric α-stable process). This means that $X_s(B) = 0$ implies $X_s(A\phi) = X_s * \nu(\phi)$, where $*$ denotes convolution of measures. Theorem III.2.1 therefore implies w.p.1 $X_s * \nu(B) > 0$ implies $X_s(B) > 0$, for Lebesgue a.a. s. Taking a union over balls with rational radii and centers we conclude

$$(III.2.4) \qquad S(X_s * \nu) \subset S(X_s) \quad \text{for Lebesgue a.a. } s > 0 \quad \text{a.s.}$$

The "Lebesgue a.a. s" is a nuisance as we would like to verify this inclusion for a fixed $s > 0$ (it is false for all $s > 0$ simultaneously as Exercise III.2.1 shows). The following result allows us to do this and also has several other applications.

Theorem III.2.2. Let X be the (Y, γ, g)-DW-superprocess where $\gamma(\cdot) \equiv \gamma > 0$ is constant and $g \in C_b(E)$. Let $\mu_1, \mu_2 \in M_F(E)$. The following are equivalent:
(i) $\mu_1 P_s \ll \mu_2 P_t \quad \forall\, 0 < s \leq t$
(ii) $\mathbb{P}_{\mu_1}(X_s \in \cdot) \ll \mathbb{P}_{\mu_2}(X_t \in \cdot) \quad \forall\, 0 < s \leq t$
(iii) $\mathbb{P}_{\mu_1}(X_{s+\cdot} \in \cdot) \ll \mathbb{P}_{\mu_2}(X_{t+\cdot} \in \cdot)$ (on $C(\mathbb{R}_+, M_F(E))$) $\quad \forall 0 < s \leq t$.

The original proof in Evans-Perkins (1991) used exact moment measure calculations and Theorem II.7.2. A simpler argument using only the latter is given at the end of the Section.

Example III.2.3. Let X be a super-α-stable process, i.e. $g = 0$, $\gamma(\cdot) = \gamma > 0$ constant and Y is the symmetric α-stable process in Example II.2.4(b) (and so is Brownian motion if $\alpha = 2$). For any $\mu_1, \mu_2 \in M_F(\mathbb{R}^d) - \{0\}$, (i) is trivial as $\mu_1 P_s$ is equivalent to Lebesgue measure for all $s > 0$. Therefore $\mathbb{P}_{\mu_1}(X_s \in \cdot)$ and $\mathbb{P}_{\mu_2}(X_t \in \cdot)$ are equivalent measures on $M_F(\mathbb{R}^d)$ and $\mathbb{P}_{\mu_1}(X \in \cdot)$ and $\mathbb{P}_{\mu_2}(X \in \cdot)$ are equivalent measures on $(\Omega_X, \sigma(X_r : r \geq \delta))$ for any $\delta > 0$. For $0 < \alpha < 2$ the first equivalence allows us to consider a fixed s in (III.2.4) and conclude $S(X_s * \nu) \subset S(X_s)$ a.s. $\forall s > 0$. Recall that $\nu(dx) = c|x|^{-d-\alpha}dx$ and conclude

$$(III.2.5) \qquad S(X_s) = \phi \text{ or } \mathbb{R}^d \text{ a.s. } \forall s > 0.$$

A similar application of Theorem III.2.2 easily gives (III.2.2) for super-Poisson processes. More generally we have

Theorem III.2.4. Let Y be a Lévy process on $\mathbb{R}^d$ with Lévy measure ν, let $\gamma(\cdot) \equiv \gamma > 0$ be constant and let X be the $(Y, \gamma, 0)$-DW-superprocess starting at X_0 under $\mathbb{P}_{X_0}$. If ν^{*k} is the k-fold convolution of ν with itself then

$$\bigcup_{k=1}^{\infty} S(\nu^{*k} * X_t) \subset S(X_t)\ \mathbb{P}_{X_0} - \text{a.s.} \quad \forall t > 0, \quad X_0 \in M_F(\mathbb{R}^d).$$

Proof. Choose X_0 in $M_F(\mathbb{R}^d)$ so that $X_0(A) = 0$ iff A is Lebesgue null. Then $P^{X_0}(Y_t \in A) = 0$ iff A is Lebesgue null and so just as in the α-stable case above, Theorem III.2.2 and (III.2.4) imply that if $\Lambda = \{\mu \in M_F(E) : S(\nu * \mu) \subset S(\mu)\}$ then $\mathbb{P}_{X_0}(X_t \in \Lambda) = 1\ \forall t > 0$. The cluster decomposition (II.7.11) implies that for each $t > 0$,

$$(III.2.6) \qquad R_t(x_0, \Lambda^c) = 0 \quad \text{for Lebesgue a.a. } x_0.$$

Let $\tau_y : M_F(\mathbb{R}^d) \to M_F(\mathbb{R}^d)$ be the translation map $\tau_y(\mu)(A) = \int 1_A(x+y)d\mu(y)$. Then $R_t(x_0, \tau_y^{-1}(\cdot)) = R_t(x_0 + y, \cdot)$, e.g., by Theorem II.7.2(i) and the translation invariance of X_t^N. Clearly $\tau_y^{-1}(\Lambda) = \Lambda$ and so (III.2.6) implies $R_t(x_0, \Lambda^c) = 0$ for any $x_0 \in \mathbb{R}^d$. Another application of the cluster decomposition (II.7.11) (use $\bigcup_{i=1}^{n} S(\mu_i) = S(\sum_1^n \mu_i)$) shows $S(\nu * X_t) \subset S(X_t)$ $\mathbb{P}_{X_0}$-a.s. for any $X_0 \in M_F(\mathbb{R}^d)$. Iterate this to complete the proof. ∎

Remark. Dawson's Girsanov theorem (Theorem IV.1.6 below) immediately gives the above for a general non-zero drift g in $C_b(E)$.

It is interesting to compare Theorem III.2.4 with the following result of Tribe (1992).

Theorem III.2.5. Assume Y is a Feller process on a locally compact metric space E, X is the Y-DW-superprocess starting at $X_0 \in M_F(E)$ under $\mathbb{P}_{X_0}$, and $\zeta = \inf\{t : X_t(1) = 0\}$. Then there is a random point F in E such that

$$\mathbb{P}_{X_0}(F \in A \mid X_s(1), s \geq 0)(\omega) = P^{X_0}(Y_{\zeta(\omega)} \in A)/X_0(1) \quad \text{a.s.} \quad \forall A \in \mathcal{E}$$

and

$$\lim_{t \uparrow \zeta} \frac{X_t(\cdot)}{X_t(1)} = \delta_F \quad \text{a.s. in} \quad M_F(E).$$

Proof. See Exercise III.2.2 below. ∎

This says the final extinction occurs at a single "death point" F, which, in view of the independence assumptions underlying X, is to be expected. On the other hand Example IV.2.3 shows that for an α-stable superprocess $S(X_t) = \mathbb{R}^d$ for Lebesgue a.a. $t < \zeta$ a.s. because of the ability of the mass concentrating near F to propagate instantaneously. The following result of Tribe (1992) shows the support will collapse down to F at exceptional times at least for $\alpha < 1/2$.

Theorem III.2.6. Assume X is a α-stable-DW-superprocess with $\alpha < 1/2$. Let F be as in Theorem III.2.5. For a.a. ω there are sequences $\varepsilon_n \downarrow 0$ and $t_n \uparrow \zeta$ such that $S(X_{t_n}) \subset B(F, \varepsilon_n)$.

We close this Section with the

Proof of Theorem III.2.2. The implications (ii) $\Rightarrow$ (i) and (ii) $\Rightarrow$ (iii) are immediate by considering $\mathbb{P}_{\mu_1}(X_s(\cdot))$ and using the Markov property, respectively. As (iii) $\Rightarrow$ (ii) is trivial, only (i) $\Rightarrow$ (ii) requires a proof. Dawson's Girsanov Theorem (Theorem IV.1.6 below) reduces this implication to the case where $g \equiv 0$ which we now assume.

Assume (i) and choose $0 < s \leq t$. Write $R_u(m, \cdot)$ for $\int R_u(x_0, \cdot) dm(x_0)$ and set $R_u^*(m, \cdot) = R_u(m, \cdot)/R_u(m, M_F(E) - \{0\})$, where $R_u(x_0, \cdot)$ are the canonical measures of X from Theorem II.7.2 and $m \in M_F(E) - \{0\}$.

The first step is to reduce the problem to

$$(III.2.7) \qquad R_s(\mu_1, \cdot) \ll R_t(\mu_2, \cdot) \text{ on } M_F(E) - \{0\}.$$

By (II.7.11), $\mathbb{P}_{\mu_1}(X_s \in \cdot)$ and $\mathbb{P}_{\mu_2}(X_t \in \cdot)$ are the laws of $\sum_{i=1}^{N_1} \nu_i^1$ and $\sum_{i=1}^{N_2} \nu_i^2$, respectively, where N_1 and N_2 are Poisson with means $2\mu_1(1)/\gamma s$ and $2\mu_2(1)/\gamma t$, respectively, and conditional on N_1, N_2, $\{\nu_i^1 : i \leq N_1\}$ and $\{\nu_i^2 : i \leq N_2\}$ are i.i.d. with law $R_s^*(\mu_1, \cdot)$ and $R_t^*(\mu_2, \cdot)$, respectively. (III.2.7) implies the n-fold product of $R_s^*(\mu_1, \cdot)$ will be absolutely continuous to the n-fold product of $R_t^*(\mu_2, \cdot)$. Therefore we can sum over the values of N_1 and N_2 to obtain $\mathbb{P}_{\mu_1}(X_s \in \cdot) \ll \mathbb{P}_{\mu_2}(X_t \in \cdot)$ as required.

Let Ξ^{τ,ν_0} denote a Poisson point process on $M_F(E) - \{0\}$ with intensity $R_\tau(\nu_0, \cdot)$. If $0 < \tau < t$, then Exercise II.7.2 (b) and (II.7.11) show that

(III.2.8)

$$R_t(\mu_2, \cdot) = \int \mathbb{P}\Big(\int \nu \, \Xi^{t-\tau,\nu_0}(d\nu) \in \cdot\Big) R_\tau(\mu_2, d\nu_0) \text{ as measures on } M_F(E) - \{0\}.$$

This and the fact that $R_{t-\tau}(\nu_0,1)=\frac{2\nu_0(1)}{\gamma(t-\tau)}$, show that

$$R_t(\mu_2,\cdot)\geq \mathbb{P}\Big(\int \nu\,\Xi^{t-\tau,\nu_0}(d\nu)\in\cdot,\ \Xi^{t-\tau,\nu_0}(M_F(E)-\{0\})=1\Big)\,R_\tau(\mu_2,d\nu_0)$$

$$(III.2.9)\qquad =\int \exp\Big(-\frac{2\nu_0(1)}{\gamma(t-\tau)}\Big)R_{t-\tau}(\nu_0,\cdot)\,R_\tau(\mu_2,d\nu_0).$$

Assume now that B is a Borel subset of $M_F(E)-\{0\}$ such that $R_t(\mu_2,B)=0$. Then (III.2.9) implies that

$$0=\iint R_{t-\tau}(x_0,B)d\nu_0(x_0)\,R_\tau(\mu_2,d\nu_0).$$

Recall from Theorem II.7.2 (c) that the mean measure associated with $R_\tau(\mu_2,\cdot)$ is $\mu_2 P_\tau$. Therefore the above implies that

$$R_{t-\tau}(x_0,B)=0\quad \mu_2P_\tau-\text{a.a. } x_0.$$

Now apply (i) to see that for $0\leq h<\tau$,

$$R_{t-\tau}(x_0,B)=0\quad \mu_1P_{\tau-h}-\text{a.a. } x_0.$$

Now reverse the above steps to conclude

$$0=\int R_{t-\tau}(\nu_0,B)R_{\tau-h}(\mu_1,d\nu_0),$$

and for $s>\varepsilon>0$, set $\tau=t-s+\varepsilon$ and $h=t-s$ to get

$$\int R_{s-\varepsilon}(\nu_0,B)R_\varepsilon(\mu_1,d\nu_0)=0.$$

Use this result in (III.2.9) with our new parameter values to see that

$$0=\int \mathbb{P}\Big(\int \nu\,\Xi^{s-\varepsilon,\nu_0}(d\nu)\in B,\ \Xi^{s-\varepsilon,\nu_0}(M_F(E)-\{0\})=1\Big)R_\varepsilon(\mu_1,d\nu_0)$$

$$(III.2.10)\ \geq R_s(\mu_1,B)-\int \mathbb{P}\Big(\Xi^{s-\varepsilon,\nu_0}(M_F(E)-\{0\})\geq 2\Big)R_\varepsilon(\mu_1,d\nu_0),$$

where in the last line we used (III.2.8) with our new parameter values and also the fact that $0\notin B$ so that there is at least one point in the Poisson point process $\Xi^{s-\varepsilon,\nu_0}$. The elementary inequality $1-e^{-x}-xe^{-x}\leq x^2/2$ for $x\geq 0$ and Theorem II.7.2 (iii) show that if $\gamma'=\gamma(s-\varepsilon)/2$, then the last term in (III.2.10) is

$$\begin{aligned}
&\int\Big(1-\exp(-\nu_0(1)/\gamma')-(\nu_0(1)/\gamma')\exp(-\nu_0(1)/\gamma')\Big)R_\varepsilon(\mu_1,d\nu_0)\\
&=\int_0^\infty\Big(1-e^{-x/\gamma'}-\frac{x}{\gamma'}e^{-x/\gamma'}\Big)\Big(\frac{2}{\gamma\varepsilon}\Big)^2e^{-2x/\gamma\varepsilon}\,dx\\
&\leq\int_0^\infty\frac{x^2}{2\gamma'^2}\Big(\frac{2}{\gamma\varepsilon}\Big)^2e^{2x/\gamma\varepsilon}\,dx\\
&=\gamma'^{-2}\gamma\varepsilon/2.
\end{aligned}$$

Use this bound in (III.2.10) to conclude that $R_s(\mu_1, B) \le \gamma'^{-2}\gamma\varepsilon/2$ and for any ε as above and hence $R_s(\mu_1, B) = 0$, as required. ∎

Exercise III.2.1. Let X_t be the super-Poisson process which satisfies (III.2.1) with $X_0 = \alpha\delta_0$ ($\alpha > 0$). Prove that $P\big(S(X_t) = \{0\}\ \exists t > 0\big) > 0$ and conclude that (III.2.2) is not valid for all $t > 0$ a.s.
Hints. By a simple scaling argument we may take $\gamma = 4$.
(i) Show that

$$X_t(0) = \alpha - \int_0^t \lambda X_s(0)ds + \int_0^t 2\sqrt{X_s(0)}dB_s^0$$

$$X'_t \equiv X_t(\{0\}^c) = \int_0^t \lambda X_s(0)ds + \int_0^t 2\sqrt{X'_s}dB'_s,$$

where (B^0, B') are independent linear Brownian motions.
(ii) Let $T_1 = \inf\{t : X_t(0) \le \lambda^{-1}\}$, $T_2 = \inf\{t : t > T_1, X_t(0) \notin (\frac{1}{2}\lambda^{-1}, \frac{3}{2}\lambda^{-1})\} - T_1$ and let

$$Y_t = X'_{T_1} + \frac{3}{2}t + \int_{T_1}^{t+T_1} 2\sqrt{Y_s}dB'_s.$$

Y is the square of a 3/2-dimensional Bessel process and hits 0 a.s. (see V.48 of Rogers and Williams (1987)), $Y_\cdot$ and $X_{\cdot+T_1}(0)$ are conditionally independent given $\mathcal{F}_{T_1}$. Argue that Y hits 0 before T_2 with positive probability and use this to infer the result.

Exercise III.2.2. (Tribe (1992)). Let X be a Y-DW-superprocess (hence $g \equiv 0, \gamma$ constant) on $(\Omega, \mathcal{F}, \mathcal{F}_t, \mathbb{P})$, where Y is a Feller process on a locally compact metric space E and $X_0 \in M_F(E) - \{0\}$. This means P_t is norm continuous on $C_\ell(E)$ the space of bounded continuous functions with a finite limit at ∞. Let $E_\infty = E \cup \{\infty\}$ be the one-point compactification of E.
(a) Let $C_t = \int_0^t \frac{1}{X_s(1)}\,ds$ for $t < \zeta$. It is well-known and easy to show that $C_{\zeta-} = \infty$ (see Shiga (1990), Theorem 2.1). Let $D_t = \inf\{s : C_s > t\}$ for $t \ge 0$, $\tilde{Z}_t(\cdot) = X_{D_t}(\cdot)$ and $Z_t(\cdot) = \tilde{Z}_t(\cdot)/\tilde{Z}_t(1)$. Show that if $\phi \in \mathcal{D}(A)$, then

$$Z_t(\phi) = X_0(\phi)/X_0(1) + \int_0^t \tilde{Z}_s(A\phi)ds + N_t(\phi),$$

where $N_t(\phi)$ is a continuous $(\mathcal{F}_{D_t})$-martingale such that

$$\langle N(\phi)\rangle_t = \int_0^t Z_s(\phi^2) - Z_s(\phi)^2 ds.$$

(b) Show that $\int_0^\infty \tilde{Z}_s(|A\phi|)ds < \infty$ a.s. and then use this to prove that $N_t(\phi)$ converges a.s. as $t \to \infty$ for all ϕ in $\mathcal{D}(A)$. Conclude that $Z_t(\cdot) \overset{a.s.}{\to} Z_\infty(\cdot)$ in $M_F(E_\infty)$ for some $Z_\infty \in M_F(E_\infty)$.
(c) Prove that $Z_\infty = \delta_F$ a.s. for some random point F in E_∞ and hence conclude that $X_t(\cdot)/X_t(1) \overset{a.s.}{\to} \delta_F(\cdot)$ as $t \uparrow \zeta$.
Hint. Prove $\lim_{t\to\infty} Z_t(\phi^2) - Z_t(\phi)^2 = 0$ for all ϕ in $\mathcal{D}(A)$, by using the a.s. convergence of $N_t(\phi)$.
(d) Use Exercise II.5.4 to see that

$$\mathbb{P}_{X_0}(F \in A \mid X_s(1), s \geq 0)(\omega) = P^{X_0}(Y_{\zeta(\omega)} \in A)/X_0(1) \text{ a.s. for all } A \in \mathcal{B}(E_\infty),$$

and in particular $F \in E$ a.s.

3. Hausdorff Measure Properties of the Supports.

Definition. If $h : \mathbb{R}_+ \to \mathbb{R}$ is strictly increasing and continuous near 0 and $h(0) = 0$ (write $h \in \mathcal{H}$), the Hausdorff h-measure of $A \subset \mathbb{R}^d$ is

$$h - m(A) = \liminf_{\delta \downarrow 0} \Big\{ \sum_{i=0}^{\infty} h(\text{diam}(B_i)) : A \subset \bigcup_1^\infty B_i,\ B_i \text{ balls with } \text{diam}(B_i) \leq \delta \Big\}$$

The Hausdorff dimension of A is $\dim(A) = \inf\{\alpha : x^\alpha - m(A) < \infty\} (\leq d)$.

The first result gives a global (in time) upper bound on $S(X_t)$ for $d \geq 3$ which allows one to quickly understand the 2-dimensional nature of $S(X_t)$. Until otherwise indicated $(X_t, t \geq 0)$ is a SBM(γ) starting at $X_0 \in M_F(\mathbb{R}^d)$ under $\mathbb{P}_{X_0}$, P_t is the Brownian semigroup and A is its (weak) generator.

Proposition III.3.1. Let $\psi(r) = r^2(\log^+ 1/r)^{-1}$ and $d \geq 3$. Then

$$\psi - m(S(X_t)) < \infty\ \forall t > 0 \text{ a.s.}$$

and, in particular, $\dim S(X_t) \leq 2\ \forall t > 0$ a.s.
Proof. By the Historical Cluster Representation (Theorem III.1.1)

$$S(H_{j/2^n}(\{y : y((j-1)2^{-n}) \in \cdot\}) = \{x_1^{j,n} \ldots, x_{M(j,n)}^{j,n}\},$$

where conditional on $\mathcal{F}^H_{(j-1)2^{-n}}$, $M(j,n)$ has a Poisson distribution with mean $2^{n+1}\gamma^{-1}H_{(j-1)2^{-n}}(1)$ $(j, n \in \mathbb{N})$. The Historical Modulus of Continuity (Theorem III.1.3) implies that for a.a. ω if $2^{-n} < \delta(\omega, 3)$ and $t \in [j/2^n, (j+1)/2^n]$, then

$$\begin{aligned} S(X_t) &\subset S(X_{j2^{-n}})^{3h(2^{-n})} \quad &\text{(Corollary III.1.5)} \\ &= \left[\Pi_{j2^{-n}}(S(H_{j2^{-n}}))\right]^{3h(2^{-n})} \quad &\text{(Corollary III.1.4(b))} \\ &\subset \bigcup_{i=1}^{M(j,n)} B\left(x_i^{j,n}, 6h(2^{-n})\right). \end{aligned} \tag{III.3.1}$$

(The historical modulus is used in the last line.) A simple tail estimate for the Poisson distribution gives

$$P\left(\bigcup_{j=1}^{n2^n}\{M(j,n)>2^{n+2}\gamma^{-1}(H_{(j-1)2^{-n}}(1)+1)\}\right)$$
$$\le\sum_{j=1}^{n2^n}E\left(\exp\{-2^{n+2}\gamma^{-1}(H_{(j-1)2^{-n}}(1)+1)\}\exp\{H_{(j-1)2^{-n}}(1)2^{n+1}\gamma^{-1}(e-1)\}\right)$$
$$\le n2^n\exp\left(-2^{n+2}\gamma^{-1}\right).$$

By the Borel-Cantelli Lemma w.p.1 for large enough n (III.3.1) holds, and for all $j2^{-n}\le n$,

$$\sum_{i=1}^{M(j,n)}\psi(12h(2^{-n}))\le\sup_t\left(H_t(1)+1\right)2^{n+2}\gamma^{-1}\psi(12h(2^{-n}))\le c\sup_t\left(H_t(1)+1\right).$$

This implies $\psi-m(S(X_t))<\infty$ $\forall t>0$ a.s. because $X_t=0$ for t large enough. ∎

Remark III.3.2. By taking unions over $j\le 2^nK$ in (III.3.1) we see from the above argument that for $K\in\mathbb{N}$ and $2^{-n}<\delta(\omega,3)$

$$\overline{\bigcup_{t\in[2^{-n},K]}S(X_t)}\subset\bigcup_{j=1}^{K2^n}\bigcup_{i=1}^{M(j,n)}\overline{B(x_i^{j,n},6h(2^{-n}))}$$

and for n large enough this is a union of at most $K_0 2^{2n+2}\gamma^{-1}\sup_t\left(H_t(1)+1\right)$ balls of radius $6h(2^{-n})$. As $X_t=0$ for $t>K(\omega)$, this shows that $f-m(\overline{\mathcal{R}}([\delta,\infty)))<\infty$ $\forall\delta>0$ where $f(r)=r^4(\log 1/r)^{-2}$, and so $\dim\mathcal{R}\le 4$. A refinement of this result which in particular shows that $\mathcal{R}$ is Lebesgue null in the critical 4-dimensional case is contained in Exercise III.5.1 below. The exact results are described below in Theorem III.3.9.

In order to obtain more precise Hausdorff measure functions one must construct efficient coverings by balls of variable radius (unlike those in Proposition III.3.1). Intuitively speaking, a covering is efficient if the balls contain a maximal amount of X_t-mass for a ball of its radius. This suggests that the lim sup behaviour of $X_t(B(x,r))$ as $r\downarrow 0$ is critical. The following result of Rogers and Taylor (1961) (see Perkins (1988) for this slight refinement) plays a central role in the proof of the exact results described below (see Theorems III.3.8 and III.3.9).

Proposition III.3.3. There is a $c(d)>0$ such that for any $h\in\mathcal{H}$, $K>0$ and $\nu\in M_F(\mathbb{R}^d)$,
(a) $\nu(A)\le Kh-m(A)$ whenever A is a Borel subset of

$$E_1(\nu,h,K)=\{x\in\mathbb{R}^d:\overline{\lim_{r\downarrow 0}}\,\nu(B(x,r))/h(r)\le K\}.$$

(b) $\nu(A) \geq c(d)Kh - m(A)$ whenever A is a Borel subset of

$$E_2(\nu, h, K) = \{x \in \mathbb{R}^d : \overline{\lim_{r\downarrow 0}} \nu(B(x,r))/h(r) \geq K\}.$$

We can use Proposition III.3.3 (a) to get a lower bound on $S(X_t)$ which complements the upper bound in Proposition III.3.1.

Notation. If $d \geq 3$, let $\overline{h}_d(r) = r^2 \log^+ \frac{1}{r}$ and define $\overline{h}_2(r) = r^2(\log^+ \frac{1}{r})^2$.

Theorem III.3.4. If $d \geq 2$ there is a $c(d) > 0$ such that for all $X_0 \in M_F(\mathbb{R}^d)$, $\mathbb{P}_{X_0}$-a.s.

$$\forall\delta > 0 \text{ there is an } r_0(\delta,\omega) > 0 \text{ so that } \sup_{x,t\geq\delta} X_t\big(B(x,r)\big) \leq \gamma c(d)\overline{h}_d(r)\ \forall r \in (0, r_0).$$

This result is very useful when handling singular integrals with respect to X_t (e.g., see the derivation of the Tanaka formula in Barlow-Evans-Perkins (1991)). Before proving this, here is the lower bound on $S(X_t)$ promised above.

Corollary III.3.5. If $d \geq 2$, $X_t(A) \leq \gamma c(d)\overline{h}_d - m(A \cap S(X_t))$ $\forall$ Borel set A and $t > 0$ $\mathbb{P}_{X_0}$-a.s. $\forall X_0 \in M_F(\mathbb{R}^d)$. In addition, if $\zeta = \inf\{t : X_t = 0\}$, then $\dim S(X_t) = 2$ for $0 < t < \zeta$ $\mathbb{P}_{X_0}$-a.s. $\forall X_0 \in M_F(\mathbb{R}^d)$.

Proof. By the previous result we may apply Proposition III.3.3 (a) to the sets $A \cap S(X_t)$ for all Borel A, $t > 0$ to get the first inequality. This with Proposition III.3.1 together imply that for $d \geq 2$, $\dim S(X_t) = 2$ $\forall 0 < t < \zeta$ a.s. ∎

Notation. If $f \geq 0$ is Borel measurable, let $G(f,t) = \int_0^t \sup_x P_s f(x) ds$.

Lemma III.3.6. If f is a non-negative Borel function such that $G(f,t)\gamma/2 < 1$, then

$$\mathbb{P}_{X_0}\big(\exp(X_t(f))\big) \leq \exp\Big\{X_0(P_t f)(1 - \frac{\gamma}{2}G(f,t))^{-1}\Big\} < \infty.$$

Proof. Let $k(s) = \big(1 - \frac{\gamma}{2}G(f, t-s)\big)^{-1}$ and $\phi(s,x) = P_{t-s}f(x)k(s)$ $(t > 0, s \in (0,t])$. If $\varepsilon > 0$, we claim $\phi\,|_{[0,t-\varepsilon]\times\mathbb{R}^d} \in \mathcal{D}(\vec{A})_{t-\varepsilon}$. To see this note that $P_\varepsilon f \in \mathcal{D}(A)$ so that $\frac{\partial}{\partial s}P_{t-s}f = \frac{\partial}{\partial s}P_{t-\varepsilon-s}(P_\varepsilon f) = -P_{t-\varepsilon-s}(AP_\varepsilon f)$ is continuous on $\mathbb{R}^d$ and is bounded on $[0, t-\varepsilon] \times \mathbb{R}^d$. The same is true of $\frac{\partial\phi}{\partial s}(s,x)$, and clearly

$$P_{t-s}f = P_{t-s-\varepsilon}(P_\varepsilon f) \in \mathcal{D}(A)$$

implies $\phi(s,\cdot) \in \mathcal{D}(A)$ and $A\phi_s = k(s)P_{t-s-\varepsilon}(AP_\varepsilon f)$ is bounded on $[0, t-\varepsilon] \times \mathbb{R}^d$. By Proposition II.5.7, for $s < t$

$$\begin{aligned} Z_s &\equiv X_s(P_{t-s}f)k_s \\ &= X_0(P_t f)k_0 + \int_0^s\int P_{t-r}f(x)k_r M(dr,dx) + \int_0^s X_r(P_{t-r}f)\dot{k}_r dr. \end{aligned}$$

By Itô's lemma there is a continuous local martingale N_s with $N_0 = 0$ so that for $s < t$,

$$e^{Z_s} = \exp\left(X_0(P_t f)k_0\right) + N_t + \int_0^s e^{Z_r} X_r\left[(P_{t-r}f)\dot{k}_r + \frac{\gamma}{2}(P_{t-r}f)^2 k_r^2\right] dr.$$

Our choice of k shows the quantity in square brackets is less than or equal to 0. This shows that e^{Z_s} is a non-negative local supermartingale, and therefore a supermartingale by Fatou's Lemma. Fatou's lemma also implies

$$E(e^{Z_t}) \le \liminf_{s\uparrow t} E(e^{Z_s}) \le e^{Z_0},$$

which gives the result. ∎

Remark III.3.7. The above proof shows the Lemma holds for $f \in \mathcal{D}(A)_+$ for any BSMP Y satisfying (II.2.2), or for $f \in b\mathcal{B}(\mathbb{R}^d)_+$ and any BSMP Y satisfying (II.2.2) and $P_t : b\mathcal{B}(\mathbb{R}^d)_+ \to \mathcal{D}(A)$ for $t > 0$. (We still assume $g = 0$, γ constant.) In particular Lemma III.3.6 is valid for the α-stable-DW-superprocess. Schied (1996) contains more on exponential inequalities as well as the idea underlying the above argument.

Proof of Theorem III.3.4. Recall $h(r) = (r\log 1/r)^{1/2}$. We first control $X_t(B(x, h(2^{-n})))$ using balls in

$$\mathcal{B}_n = \left\{B\big(x_0, (\sqrt{d}+4)h(2^{-n})\big) : x_0 \in 2^{-n/2}\mathbb{Z}^d \cap [-n,n]^d\right\}.$$

Here $\varepsilon\mathbb{Z}^d = \{\varepsilon n : n \in \mathbb{Z}^d\}$. Assume $2^{-n} < \delta(\omega, 3)$, where $\delta(\omega,3)$ is as in the Historical Modulus of Continuity. If $x \in [-n,n]^d$, choose $x_0 \in 2^{-n/2}\mathbb{Z}^d \cap [-n,n]^d$ such that $|x_0 - x| < \sqrt{d}\,2^{-n/2} \le \sqrt{d}\,h(2^{-n})$ if $n \ge 3$. Assume $j \in \mathbb{Z}_+$ and $j2^{-n} \le t \le (j+1)2^{-n}$. For H_t-a.a. y if $y(t) \in B(x, h(2^{-n}))$, then

$$\begin{aligned}|y(j2^{-n}) - x_0| &\le |y(j2^{-n}) - y(t)| + |y(t) - x| + |x - x_0|\\ &\le (4+\sqrt{d})h(2^{-n}).\end{aligned}$$

If $B = B(x_0, (4+\sqrt{d})h(2^{-n})) \in \mathcal{B}_n$, this shows that

$$\begin{aligned}X_t(B(x, h(2^{-n}))) = H_t(y(t) \in B(x, h(2^{-n}))) &\le H_t\left(\{y(j2^{-n}) \in B\}\right)\\ &\equiv M_{j2^{-n},B}(t - j2^{-n}).\end{aligned}$$

Lemma III.1.2 and the Markov property show that $M_{j2^{-n},B}(t)$ is a martingale and so $M_{j2^{-n}}(t) = \sup_{B\in\mathcal{B}_n} M_{j2^{-n},B}(t)$ is a non-negative submartingale. The above bound implies

$$(III.3.2)\quad \sup_{j2^{-n}\le t\le (j+1)2^{-n}} \ \sup_{x\in[-n,n]^d} X_t(B(x, h(2^{-n}))) \le \sup_{t\le 2^{-n}} M_{j2^{-n}}(t)\ \forall\, j \in \mathbb{Z}_+$$
$$\text{whenever } 2^{-n} \le \delta(\omega, 3).$$

We now use the exponential bound in Lemma III.3.6 to bound the right side of the above. Let $\beta = \beta(d) = 1 + 1(d = 2)$ and set $\varepsilon_n = c_1 2^{-n} n^{1+\beta(d)}$ where c_1 will be chosen (large enough) below. An easy calculation shows that

$$(III.3.3) \quad G(B(x,r),t) \equiv G\left(1_{B(x,r)},t\right) \le c_2(d) r^2 \left[1 + 1(d=2)\left(\log(\sqrt{t}/r)\right)^+\right]$$

from which it follows that

$$(III.3.4) \qquad \sup_{B\in\mathcal{B}_n, t\le n} G(B,t) \le c_3(d) 2^{-n} n^{\beta(d)}.$$

If $\delta > 0$ and $\lambda_n = c_3^{-1}\gamma^{-1}2^{n-1}n^{-\beta(d)}$, then the weak maximal inequality for submartingales implies

(III.3.5)

$$\begin{aligned}
&\mathbb{P}_{X_0}\left(\sup_{\delta 2^n \le j < n2^n} \sup_{t\le 2^{-n}} M_{j2^{-n}}(t) \ge \varepsilon_n\right) \\
&\le n2^n \sup_{\delta 2^n \le j < n2^n} e^{-\lambda_n\varepsilon_n}\mathbb{P}_{X_0}\left(e^{\lambda_n M_{j2^{-n}}(2^{-n})}\right) \\
&\le n2^n|\mathcal{B}_n| \sup_{\delta 2^n\le j<n2^n, B\in\mathcal{B}_n} e^{-\lambda_n\varepsilon_n}\mathbb{P}_{X_0}\Big(\exp\left(X_{j/2^n}(B)2\lambda_n(2-\lambda_n\gamma 2^{-n})^{-1}\right)\Big),
\end{aligned}$$

where we have used Lemma III.1.2, the Markov property and Lemma III.3.6 with $f \equiv \lambda_n$. The latter requires $\lambda_n 2^{-n}\gamma/2 < 1$ which is certainly true for $n \ge n_0$. In fact for $n \ge n_0$ (III.3.5) is no more than

$$(III.3.6) \quad \begin{aligned}
& n2^n|\mathcal{B}_n| \sup_{\delta 2^n\le j<n2^n} \sup_{B\in\mathcal{B}_n} e^{-\lambda_n\varepsilon_n}\mathbb{P}_{X_0}\left(\exp\left(2\lambda_n X_{j/2^n}(B)\right)\right) \\
&\le n2^n|\mathcal{B}_n|e^{-\lambda_n\varepsilon_n} \sup_{\delta 2^n\le j<n2^n} \sup_{B\in\mathcal{B}_n} \exp\left(4\lambda_n P^{X_0}(B_{j/2^n}\in B)\right)
\end{aligned}$$

by another application of Lemma III.3.6, this time with $f = 2\lambda_n 1_B$. We also use (III.3.4) here to see that

$$2\lambda_n G(B, j/2^{-n}) \le 2\lambda_n c_3 2^{-n} n^{\beta(d)} = \gamma^{-1}$$

so that Lemma III.3.6 may be used and the upper bound given there simplifies to the expression in (III.3.6). An elementary bound shows that the right side of (III.3.6) is at most

$$\begin{aligned}
&cn^{1+d}2^{n+nd/2}e^{-\lambda_n\varepsilon_n}\exp\left(4\lambda_n c(\delta, X_0(1))h(2^{-n})^d\right) \\
&\le cn^{1+d}2^{n(1+d/2)}\exp\left(-c_1c_3^{-1}\gamma^{-1}2^{-1}n\right)c'(\delta, X_0(1)),
\end{aligned}$$

which is summable if $c_1 = \gamma 2c_3(1+\frac{d}{2}) \equiv \gamma c'(d)$. Borel-Cantelli and (III.3.2) imply that a.s. for large enough n,

$$\sup_{\delta\le t\le n} \sup_{x\in[-n,n]^d} X_t\left(B(x, h(2^{-n}))\right) < \gamma c' 2^{-n} n^{1+\beta(d)} \le \gamma c'' \overline{h}_d(h(2^{-n})).$$

An elementary interpolation completes the proof. ∎

In view of Proposition III.3.1 and Corollary III.3.5 it is natural to ask if there is an exact Hausdorff measure function associated with $S(X_t)$.

Notation.

$$h_d(r) = \begin{cases} r^2 \log^+ \log^+ 1/r & d \geq 3 \\ r^2(\log^+ 1/r)(\log^+ \log^+ \log^+ 1/r) & d = 2. \end{cases} \qquad (\log^+ x = (\log x) \vee e^e)$$

Theorem III.3.8. Assume X is a SBM(γ) starting at μ under $\mathbb{P}_\mu$.
(a) $[d \geq 2, t \text{ fixed}]$ There is a universal constant $c(d) > 0$ such that $\forall \mu \in M_F(\mathbb{R}^d)$, $t > 0$

$$X_t(A) = \gamma c(d) h_d - m(A \cap S(X_t)) \quad \forall A \in \mathcal{B}(\mathbb{R}^d) \quad \mathbb{P}_\mu\text{-a.s.}$$

(b) $[d \geq 2, t \text{ variable}]$ There are universal constants $0 < c(d) \leq C(d) < \infty$ such that for any $\mu \in M_F(\mathbb{R}^d)$

(i) If $d \geq 3$, $\gamma c(d) h_d - m(A \cap S(X_t)) \leq X_t(A) \leq \gamma C(d) h_d - m(A \cap S(X_t))$ $\forall A \in \mathcal{B}(\mathbb{R}^d)\ \forall t > 0\ \mathbb{P}_\mu$-a.s.

(ii) If $d = 2$, $\gamma c(2) h_3 - m(A \cap S(X_t)) \leq X_t(A) \leq \gamma C(2) \bar{h}_2 - m(A \cap S(X_t))$ $\forall A \in \mathcal{B}(\mathbb{R}^d)\ \forall t > 0\ \mathbb{P}_\mu$-a.s.

(c) $[d = 1]$ There is a jointly continuous process $\{u(t, x) : t > 0, x \in \mathbb{R}\}$ such that

$$X_t(dx) = u(t, x)dx \quad \forall t > 0 \quad \mathbb{P}_\mu\text{-a.s.}$$

Remarks. (1) (b) shows that if $d \geq 2$ then w.p.1 for any $t > 0$ $S(X_t)$ is a singular set of Hausdorff dimension 2 whenever it is non-empty. This fact has already been proved for $d \geq 3$ (Corollary III.3.5).

(2) (a) and (b) state that X_t distributes its mass over $S(X_t)$ in a deterministic manner. This extreme regularity of the local structure of $S(X_t)$ is due to the fact that for $d \geq 2$ the local density of mass at x is due entirely to close cousins of "the particle at x" and so will exhibit strong independence in x. The strong recurrence for $d = 1$ means this will fail in $\mathbb{R}^1$ and a non-trivial density, u, exists.

(3) We conjecture that one may take $c(d) = C(d)$ in (b)(i) for $d \geq 3$. The situation in the plane is much less clear.

(4) Curiously enough the exact Hausdorff measure functions for $S(X_t)$ are exactly the same as those for the range of a Brownian path (see Ciesielski-Taylor (1962), Taylor (1964)) although these random sets certainly look quite different. The two sets behave differently with respect to packing measure: $h(s) = s^2(\log\log 1/s)^{-1}$ is an exact packing measure function for the range of a Brownian path for $d \geq 3$ while $h(s) = s^2(\log 1/s)^{-1/2}$ is critical for $S(X_t)$ for $d \geq 3$ (see Le Gall-Perkins-Taylor (1995)).

(5) (b) is proved in Perkins (1988, 1989). In Dawson-Perkins (1991), (a) was then proved for $d \geq 3$ by means of a 0-1 law which showed that the Radon-Nikodym derivative of X_t with respect to $h_d - m(\,\cdot \cap S(X_t))$ is a.s. constant. The more delicate 2-dimensional case in (a) was established in Le Gall-Perkins (1995) using the Brownian snake. This approach has proved to be a very powerful tool in the study of path properties of (and other problems associated with) DW-superprocesses. The $d = 1$ result was proved independently by Reimers (1989) and Konno-Shiga (1988) and will be analyzed in detail in Section III.4 below. The existence of a density at a fixed time was proved by Roelly-Coppoletta (1986). The first Hausdorff measure result, $\dim S(X_t) \leq 2$, was established by Dawson and Hochberg (1979).

Here are the exact results for the range of super-Brownian motion, promised after Proposition III.3.1.

Theorem III.3.9. (a) $d \geq 4$. Let

$$\psi_d(r) = \begin{cases} r^4 \log^+ \log^+ 1/r & d > 4 \\ r^4(\log^+ 1/r)(\log^+ \log^+ \log^+ 1/r) & d = 4 \end{cases}.$$

There is a $c(d) > 0$ such that for all $X_0 \in M_F(\mathbb{R}^d)$,

$$\int_0^t X_s(A)ds = \gamma c(d)\psi_d - m(A \cap \mathcal{R}((0,t])) \ \forall A \in \mathcal{B}(\mathbb{R}^d) \ \forall t \geq 0 \ \mathbb{P}_{X_0} - \text{a.s.}$$

(b) $d \leq 3$. Assume X_0 has a bounded density if $d = 2, 3$. Then there is a jointly continuous density $\{v(t,x) : t \geq 0, x \in \mathbb{R}^d\}$ such that

$$\int_0^t X_s(A)ds = \int_A v(t,x)dx \quad \forall A \in \mathcal{B}(\mathbb{R}^d) \quad \forall t \geq 0 \quad \mathbb{P}_{X_0} - \text{a.s.}$$

Discussion. (a) for $d > 4$ is essentially proved in Dawson-Iscoe-Perkins (1989). There upper and lower bounds with differing constants were given. Le Gall (1999) showed that by a $0-1$ law the above constants were equal and used his snake to derive the critical 4-dimensional result.

(b) is proved in Sugitani (1989). ∎

Consider now the analogue of Theorem III.3.8 when X is an α-stable-DW-superprocess ($g \equiv 0, \gamma(\cdot) \equiv \gamma > 0$). That is Y is the symmetric α-stable process in $\mathbb{R}^d$ considered in Example II.2.4 (b). Let

$$h_{d,\alpha}(r) = r^\alpha \log^+ \log^+ 1/r \quad \text{if} \quad d > \alpha$$
$$\overline{h}_{d,\alpha}(r) = r^\alpha (\log^+ 1/r)^2 \quad \text{if} \quad d = \alpha$$
$$\psi_{d,\alpha}(r) = \begin{cases} r^\alpha & \text{if } d > \alpha \\ r^\alpha \log^+ 1/r & \text{if } d = \alpha \end{cases}.$$

In Perkins (1988) it is shown that for $d > \alpha$ there are real constants $0 < c_1 \leq c_2$, depending only on (d, α) so that

$$(III.3.7) \qquad \gamma c_1 \leq \overline{\lim_{r \downarrow 0}} \frac{X_t(B(x,r))}{h_{d,\alpha}(r)} \leq \gamma c_2 \quad X_t\text{-a.a. } x \quad \forall t > 0 \quad \mathbb{P}_{X_0}\text{-a.s.}$$

Let $\Lambda_t(\omega)$ be the set of x for which the above inequalities hold. Then $\Lambda_t(\omega)$ is a Borel set supporting $X_t(\omega)$ for all $t > 0$ a.s. and Proposition III.3.3 shows that (for $d > \alpha$)

$$(III.3.8) \qquad \begin{aligned} \gamma c(d) c_1 h_{d,\alpha} - m(A \cap \Lambda_t) \leq X_t(A) \leq & \gamma c_2 h_{d,\alpha} - m(A \cap \Lambda_t) \\ & \forall A \in \mathcal{B}(R^d), \ t > 0 \ \mathbb{P}_{X_0}\text{-a.s.} \end{aligned}$$

For t fixed a $0-1$ law then shows the $\overline{\lim}$ in (III.3.7) is γc_3 for X_t-a.a. x, say for $x \in \Lambda_t'$, a.s. and $X_t(\cdot) = c_4 \gamma h_{d,\alpha} - m(\cdot \cap \Lambda_t')$ a.s. for some $c_3, c_4 > 0$ (see Theorem

5.5 of Dawson-Perkins (1991)). Analogous results are also shown for $d = \alpha$ (=1 or 2) with $\psi_{d,\alpha} - m$ in the lower bound and $\overline{h}_{d,\alpha} - m$ in the upper bound. Such results are clearly false for $\alpha < 2$ if Λ_t is replaced by $S(X_t)(= \phi$ or $\mathbb{R}^d$ a.s. by Example III.2.3).

(III.3.7) suggests we define the *Campbell measure*, $Q_t \in M_F(M_F(\mathbb{R}^d) \times \mathbb{R}^d)$, associated with X_t by

$$Q_t(A \times B) = \int 1_A(X_t)X_t(B)d\mathbb{P}_{X_0}.$$

If $(X, Z) \in M_F(\mathbb{R}^d) \times \mathbb{R}^d$ are coordinate variables on $M_F(\mathbb{R}^d) \times \mathbb{R}^d$,then under Q_t, Z is chosen at random according to X. The regular conditional probabilities $Q_t(X \in \cdot \mid Z = x)$ are the Palm measures of X_t and describe X_t from the perspective of a typical point x in the support of X_t (see Dawson-Perkins (1991), Chapter 4, and Dawson (1992), Chapter 6, for more on Palm measures of superprocesses).

The first step in deriving (III.3.7) is to find the asymptotic mean size of $X_t(B(x,r))$ when x is chosen according to X_t. (a) of the following Exercise is highly recommended.

Exercise III.3.1. Let X be the α-stable-DW-superprocess and $Q_t(dX, dZ)$ be the Campbell measure defined above. If Y is the α-symmetric stable process in $\mathbb{R}^d$, Y_t has a smooth symmetric density $p_t(y) = p_t(|x|)$ such that $p_1(\cdot)$ is decreasing on $[0, \infty)$ and $p_1(r) \le c(1+r)^{-(d+\alpha)}$.

(a) If $d \ge \alpha$, show there is a constant $k_{d,\alpha} > 0$ such that

$$(III.3.9) \qquad \lim_{r\downarrow 0} \frac{E\left(\int X_t(B(x,r))dX_t(x)\right)}{\psi_{d,\alpha}(r)} \equiv \lim_{r\downarrow 0} \frac{\int X(B(Z,r))Q_t(dX,dZ)}{\psi_{d,\alpha}(r)} = k_{d,\alpha}\gamma X_0(1) \quad \forall t > 0.$$

Also show there is a $c = c(d, \alpha, \gamma, X_0(1))$ such that for any $\delta > 0$ there is an $r_0(\delta) > 0$ satisfying

$$(III.3.10) \qquad E\left(\int X_t(B(x,r))dX_t(x)\right) \le c\psi_{d,\alpha}(r) \quad \forall t \in [\delta, \delta^{-1}], \quad r \le r_0(\delta).$$

(b) Show the above results remain valid for $d < \alpha$ if $\psi_{d,\alpha}(r) = r$, $k_{d,\alpha}$ may depend on (t, γ, X_0), and c may also depend on δ.

(c) [Palm measure version] If $X_0P_t(x) = \int p_t(y-x)X_0(dy)$, show that

$$Q_t\left(X\left(B(Z,r)\right) \mid Z = x_0\right) = P^{X_0}\left(Y_t \in B(x_0,r)\right) + \frac{\gamma E^{X_0}\left(\int_0^t P^{Y_s}\left(|Y_{t-s} - x_0| < r\right)p_{t-s}(x_0 - Y_s)ds\right)}{X_0P_t(x_0)}.$$

*(d) Show there is a constant $k_{d,\alpha}$ (which may also depend on (γ, t, x_0, X_0) if $d < \alpha$) such that for any $t > 0$, $x_0 \in \mathbb{R}^d$,

$$(III.3.11) \qquad \lim_{r\downarrow 0} \frac{Q_t\left(X(B(Z,r)) \mid Z = x_0\right)}{\psi_{d,\alpha}(r)} = \gamma k_{d,\alpha}.$$

Recall that $\psi_{d,\alpha}(r) = r$ if $d < \alpha$.

The above Exercise shows that if $d > \alpha$, the mean of $X_t(B(x,r))$ when x is chosen according to X_t behaves like $k_{d,\alpha}\psi_{d,\alpha}(r)$ as $r \downarrow 0$. This explains the r^α part of $h_{d,\alpha}$ in (III.3.7). The $\log\log 1/r$ then comes from the exponential tail of $X_t(B(x,r))$ and the fact that it suffices to consider $r \downarrow 0$ through a geometric sequence (each exponentiation produces a log). It is an easy Borel-Cantelli Exercise to use the above mean results to obtain weaker versions of the upper bounds in (III.3.7) and (III.3.8).

Exercise III.3.2. (a) Use (III.3.10) to show that if $d \geq \alpha$, then for any $\varepsilon > 0$, $t > 0$

$$\lim_{r\downarrow 0} \frac{X_t(B(x,r))}{\psi_{d,\alpha}(r)(\log^+ 1/r)^{1+\varepsilon}} = 0 \quad X_t\text{-a.a. } x \quad \mathbb{P}_{X_0}\text{-a.s.}$$

(b) If $\phi^\varepsilon_{d,\alpha}(r) = \psi_{d,\alpha}(r)(\log^+ 1/r)^{1+\varepsilon}$, show that $\forall t > 0$,

$$\phi^\varepsilon_{d,\alpha} - m(A) = 0 \quad \text{implies} \quad X_t(A) = 0 \;\; \forall A \in \mathcal{B}(\mathbb{R}^d),\; \varepsilon > 0 \quad \mathbb{P}_{X_0}\text{-a.s.}$$

4. One-dimensional Super-Brownian motion and Stochastic PDE's

In this section we study super-Brownian motion in one dimension with constant branching rate $\gamma > 0$. In particular, we will prove Theorem III.3.8(c) and establish a one-to-one correspondence between the density of super-Brownian motion in one dimension and the solution of a parabolic stochastic pde driven by a white noise on $\mathbb{R}_+ \times \mathbb{R}$.

Let $\mathcal{F}_t$ be a right continuous filtration on $(\Omega, \mathcal{F}, P)$ and let $\mathcal{P} = \mathcal{P}(\mathcal{F}_\cdot)$ denote the σ-field of $\mathcal{F}_t$-predictable sets in $\mathbb{R}_+ \times \Omega$. Let $|A|$ denote the Lebesgue measure of a Borel set in $\mathbb{R}^d$ and let $\mathcal{B}_F(\mathbb{R}^d) = \{A \in \mathcal{B}(\mathbb{R}^d) : |A| < \infty\}$.

Definition. An $(\mathcal{F}_t)$-white noise, W, on $\mathbb{R}_+ \times \mathbb{R}^d$ is a random process $\{W_t(A) : t > 0, A \in \mathcal{B}_F(\mathbb{R}^d)\}$ such that

(*i*) $W_t(A \cup B) = W_t(A) + W_t(B)$ a.s. for all disjoint $A, B \in \mathcal{B}_F(\mathbb{R}^d)$ and $t > 0$.
(*ii*) $\forall A \in \mathcal{B}_F(\mathbb{R}^d)$, $t \mapsto W_t(A)$ is an $(\mathcal{F}_t)$-Brownian motion starting at 0 with diffusion parameter $|A|$.

See Chapters 1 and 2 of Walsh (1986) for more information on white noises. Note that if W is above and A and B are disjoint sets in $\mathcal{B}_F$, then

$$2W_t(A)W_t(B) = W_t(A \cup B)^2 - W_t(A)^2 - W_t(B)^2$$

is an $(\mathcal{F}_t)$-martingale. It follows that W is an orthogonal martingale measure in the sense of Chapter 2 of Walsh (1986) and Proposition 2.10 of Walsh (1986) shows that the above definition is equivalent to that given in Walsh (1986). As in Section II.5 (see Chapter 2 of Walsh (1986)) one can define a stochastic integral $W_t(\psi) = \int_0^t \int \psi(s,\omega,x)dW(s,x)$ for

$$\psi \in \mathcal{L}^2_{\mathrm{loc}}(W) \equiv \{\psi : \mathbb{R}_+ \times \Omega \times \mathbb{R}^d \to \mathbb{R} : \psi \text{ is } \mathcal{P} \times \mathcal{B}(\mathbb{R}^d) - \text{measurable}, \int_0^t \int \psi(s,\omega,x)^2 dx\, ds < \infty\ \forall t > 0 \text{ a.s.}\}.$$

The map $\psi \mapsto W(\psi)$ is a linear map from $\mathcal{L}^2_{\mathrm{loc}}$ to the space of continuous $\mathcal{F}_t$-local martingales and $W_t(\psi)$ has predictable square function

$$\langle W(\psi)\rangle_t = \int_0^t \int \psi(s,\omega,x)^2 dx\, ds \quad \text{for all } t > 0 \text{ a.s.}$$

Notation. If f, g are measurable functions on $\mathbb{R}^d$, let $\langle f, g\rangle = \int f(x)g(x)\, dx$ if the integral exists.

Definition. Let W be an $(\mathcal{F}_t)$-white noise on $\mathbb{R}_+ \times \mathbb{R}^d$ defined on $\bar\Omega = (\Omega, \mathcal{F}, \mathcal{F}_t, P)$, let $m \in M_F(\mathbb{R}^d)$ and let $f : \mathbb{R}_+ \to \mathbb{R}$. We say that an adapted continuous process $u : (0,\infty) \times \Omega \to C_K(\mathbb{R}^d)_+$ is a solution of

$$(SPDE)^f_m \qquad \frac{\partial u}{\partial t} = \frac{\Delta u}{2} + \sqrt{\gamma u}\dot W + f(u), \qquad u_{0+}(x)dx = m$$

on $\bar\Omega$ iff for every $\phi \in C^2_b(\mathbb{R}^d)$,

$$\langle u_t, \phi\rangle = m(\phi) + \int_0^t \langle u_s, \frac{\Delta\phi}{2}\rangle\, ds + \int_0^t \phi(x)\sqrt{\gamma u(s,x)}dW(s,x) + \int_0^t \int f(u(s,x))\phi(x)dx\, ds \quad \text{for all } t > 0 \text{ a.s.}$$

Remark III.4.1. Use the fact that $C^2_b(\mathbb{R}^d)$ contains a countable convergence determining class (such as $\{\sin(u\cdot x), \cos(u\cdot x) : u \in \mathbb{Q}^d\}$) to see that $(\mathrm{SPDE})^f_m$ implies that $\lim_{t\to 0+} u_t(x)dx = m$ a.s. in $M_F(\mathbb{R}^d)$. We have been able to choose a rather restrictive state space for u_t because the Compact Support Property for SBM (Corollary III.1.4) will produce solutions with compact support. This property will persist for stochastic pde's in which the square root in the white noise integrand is replaced by any positive power less than 1, but fails if this power equals 1 (see Mueller and Perkins (1992) and Krylov (1997)).

We write $(\mathrm{SPDE})_m$ for $(\mathrm{SPDE})^0_m$. The one-dimensional Brownian density and semigroup are denoted by p_t and P_t, respectively. We set $p_t(x) = 0$ if $t \le 0$.

Theorem III.4.2. (a) Let X be an $(\mathcal{F}'_t)$-SBM(γ) in one spatial dimension, starting at $X_0 \in M_F(\mathbb{R})$ and defined on $\bar\Omega' = (\Omega', \mathcal{F}', \mathcal{F}'_t, P')$. There is an adapted continuous $C_K(\mathbb{R})$-valued process $\{u_t : t > 0\}$ such that $X_t(dx) = u_t(x)\, dx$ for all $t > 0$ P'-a.s. Moreover for all $t > 0$ and $x \in \mathbb{R}$,

$$(III.4.1) \quad u_t(x) = P_tX_0(x) + \int_0^t \int p_{t-s}(y-x)dM(s,y) \quad P' - \text{a.s., and}$$

$$E'\Big(\sup_{v\le t}\Big[\int_0^v \int p_{t-s}(y-x)dM(s,y)\Big]^2\Big) < \infty.$$

(b) Let X and u be as above. There is a filtered space $\bar{\Omega}'' = (\Omega'', \mathcal{F}'', \mathcal{F}_t'', P'')$ so that

$$\bar{\Omega} = (\Omega, \mathcal{F}, \mathcal{F}_t, P) \equiv (\Omega' \times \Omega'', \mathcal{F}' \times \mathcal{F}'', (\mathcal{F}'_\cdot \times \mathcal{F}''_\cdot)_{t+}, P' \times P'')$$

carries an $\mathcal{F}_t$-white noise, W, on $\mathbb{R}_+ \times \mathbb{R}$ and $u \circ \Pi$ satisfies $(\mathrm{SPDE})_{X_0}$ on $\bar{\Omega}$, where $\Pi : \Omega \to \Omega'$ is the projection map.
(c) Assume u satisfies $(\mathrm{SPDE})_m$ (d-dimensional) on some $\bar{\Omega} = (\Omega, \mathcal{F}, \mathcal{F}_t, P)$. Then

$$X_t(dx) = \begin{cases} u(t,x)dx & \text{if } t > 0 \\ m(dx) & \text{if } t = 0 \end{cases}$$

defines an $(\mathcal{F}_t)$-SBM(γ) on $\bar{\Omega}$ starting at m.

A proof is given below. Clearly Theorem III.3.8(c) is immediate from (a).

Corollary III.4.3. (a) If $d = 1$, then for any $m \in M_F(\mathbb{R})$ there is a solution to $(\mathrm{SPDE})_m$ and the law of u on $C((0,\infty), C_K(\mathbb{R}))$ is unique.
(b) If $d \geq 2$ and $m \neq 0$, there is no solution to $(\mathrm{SPDE})_m$.
Proof. (a) The existence is included in Theorem III.4.2(b). The Borel subsets of $C_K(\mathbb{R})$ are generated by the coordinate maps. To prove uniqueness in law it therefore suffices to show that if u satisfies $(\mathrm{SPDE})_m$, then

$(III.4.2)$ $P((u_{t_i}(x_i))_{i \leq n} \in \cdot)$ is unique on $\mathcal{B}(\mathbb{R}^n)$ for any $0 < t_i$, $x_i \in \mathbb{R}$, $n \in \mathbb{N}$.

If X is the SBM(γ) in Theorem III.4.2(c), then $u_{t_i}(x_i) = \lim_{\varepsilon \to 0} X_{t_i}(p_\varepsilon(\cdot - x_i))$, and the uniqueness in law of the super-Brownian motion X clearly implies (III.4.2).

(b) If u is a solution to $(\mathrm{SPDE})_m$ for $d \geq 2$, then by Theorem III.4.2(c), $X_t(dx) = u(t,x)dx$ for $t > 0$ and $X_0 = m$ defines a super-Brownian motion starting at m which is absolutely continuous for $t > 0$. This contradicts the a.s. singularity of super-Brownian motion in dimensions greater than 1 (Theorem III.3.8(a), but note that Proposition III.3.1 suffices if $d \geq 3$). ∎

The uniqueness in law for the above stochastic pde does not follow from the standard theory since the square root function is not Lipschitz continuous. For Lipschitz continuous functions of the solution (as opposed to $\sqrt{\gamma u}$) solutions to (SPDE) are unique in law when the initial condition has a nice density (see Chapter 3 of Walsh (1986)). This needs naturally to
Open Problem. Does pathwise uniqueness hold in $(\mathrm{SPDE})_m$? That is, if u, v are solutions of $(\mathrm{SPDE})_m$ with the same white noise W, is it true that $u = v$ a.s.?

Note that the finite-dimensional version of this problem is true by Yamada-Watanabe (1971).

Recently Mytnik (1998) proved uniqueness in law for solutions of $(\mathrm{SPDE})_m$ when the square root in front of the white noise is replaced by a power between 1/2 and 1. His argument may be viewed as an extension of the exponential duality used to prove uniqueness for superprocesses but the dual process is now random. It does apply for slightly more general functions than powers but, as is often the case with duality arguments, the restriction on the functions is severe and artificial. This is one reason for the interest in the above problem as a pathwise uniqueness argument would likely be quite robust.

To prove Theorem III.4.2 we use the following version of Kolmogorov's continuity criterion for two-parameter processes.

Proposition III.4.4. Let $I:(t_0,\infty)\times\mathbb{R}\to\mathbb{R}$ be a process on $(\Omega,\mathcal{F},P)$ such that for some $p>1$, $a,b>2$, for any $T>t_0$, there is a $c=c(T)$ so that

$$E(|I(t',x')-I(t,x)|^p)\le c(T)[|t'-t|^a+|x'-x|^b]\ \forall t,t'\in(t_0,T],\ x,x'\in[-T,T].$$

Then I has a continuous version.

Proof. See Corollary 1.2 of Walsh (1986) where a modulus of continuity of the continuous version is also given.

Lemma III.4.5. (a) If $0\le\delta\le p$, then

$$|p_{t+\varepsilon}(z)-p_t(z)|^p\le(\varepsilon t^{-3/2})^\delta[p_{t+\varepsilon}(z)^{p-\delta}+p_t(z)^{p-\delta}]\quad\forall z\in\mathbb{R},\ t>0,\ \varepsilon\ge 0.$$

(b) If $0<\delta<1/2$, there is a $c(\delta)>0$ so that for any $0\le t\le t'\le T$ and $x,x'\in\mathbb{R}$,

$$\int_0^{t'}\int(p_{t'-s}(y-x')-p_{t-s}(y-x))^2\,dy\,ds\le|x'-x|+c(\delta)T^{1/2-\delta}|t'-t|^\delta.$$

Proof. (a) By the mean value theorem there is a $u\in[t,t+\varepsilon]$ such that

$$\begin{aligned}|p_{t+\varepsilon}(z)-p_t(z)|&=\varepsilon\left|\frac{\partial p_u}{\partial u}(z)\right|=\varepsilon\frac{p_u(z)}{2u}\left|\frac{z^2}{u}-1\right|\\&\le\varepsilon u^{-3/2}\le\varepsilon t^{-3/2},\end{aligned}$$

where a calculus argument has been used to see that

$$\sqrt{u}p_u(z)/2\left|\frac{z^2}{u}-1\right|\le(2\pi)^{-1/2}\left[\sup_{x>1/2}xe^{-x}\vee\frac{1}{2}\right]\le 1.$$

Therefore for $0\le\delta\le p$,

$$\begin{aligned}|p_{t+\varepsilon}(z)-p_t(z)|^p&\le(\varepsilon t^{-3/2})^\delta|p_{t+\varepsilon}(z)-p_t(z)|^{p-\delta}\\&\le(\varepsilon t^{-3/2})^\delta\left[p_{t+\varepsilon}(z)^{p-\delta}+p_t(z)^{p-\delta}\right].\end{aligned}$$

(b) Note that if $0\le t\le t'\le T$, then

$$\begin{aligned}&\int_0^{t'}\int(p_{t'-s}(y-x')-p_{t-s}(y-x))^2\,dy\,ds\\&=\int_t^{t'}\int p_{t'-s}(y-x')^2\,dy\,ds+\int_0^t\int(p_{t'-s}(y-x')-p_{t-s}(y-x))^2\,dy\,ds\end{aligned}$$

$(III.4.3)\equiv I_1+I_2.$

By Chapman-Kolmogorov,

$(III.4.4)$ $$I_1=\int_t^{t'}p_{2(t'-s)}(0)\,ds=\pi^{-1/2}(t'-t)^{1/2}.$$

If we expand the integrand in I_2, let $\Delta = x' - x$, and use Chapman-Kolmogorov, we get

$$\begin{aligned}I_2 &= \int_0^t p_{2(t'-s)}(0) + p_{2(t-s)}(0) - 2p_{t'-s+t-s}(\Delta)\,ds\\ &= \int_0^t p_{2(t'-s)}(0) - p_{t'-s+t-s}(\Delta)\,ds + \int_0^t p_{2(t-s)}(0) - p_{t'-s+t-s}(\Delta)\,ds\\ &\equiv I_3 + I_4.\end{aligned}$$

Consider I_3 and use (a) with $p=1$ and $0<\delta<1/2$ to see that

$$\begin{aligned}I_3 &= \int_0^t (p_{2(t'-s)}(0) - p_{2(t'-s)}(\Delta)) + (p_{2(t'-s)}(\Delta) - p_{t'-s+t-s}(\Delta))\,ds\\ &\le \frac{1}{2}\int_0^{2t'} p_s(0) - p_s(\Delta)\,ds\\ &\quad + (t'-t)^\delta \int_0^t (t'-s)^{-\delta 3/2}\left[p_{2(t'-s)}(\Delta)^{1-\delta} + p_{t'+t-2s}(\Delta)^{1-\delta}\right]ds\\ &\le 2^{-3/2}\pi^{-1/2}\int_0^{2t'} s^{-1/2}[1-\exp(-\Delta^2/(2s))]\,ds\\ &\quad + (t'-t)^\delta \int_0^t (t-s)^{-\delta 3/2-(1-\delta)/2}(4\pi)^{-(1-\delta)/2}2\,ds\\ &\le (4\pi)^{-1}|\Delta|\int_{\Delta^2/2t'}^\infty u^{-3/2}(1-e^{-u})\,du + 1.1(t'-t)^\delta \int_0^t s^{-1/2-\delta}\,ds\\ &\le (4\pi)^{-1}|\Delta|\int_0^\infty u^{-3/2}(1\wedge u)\,du + 1.1(1/2-\delta)^{-1}t^{1/2-\delta}(t'-t)^\delta\\ &\le \pi^{-1}|\Delta| + c'(\delta)T^{1/2-\delta}(t'-t)^\delta.\end{aligned}$$

Use this, the analogous bound for I_4, and (III.4.4) in (III.4.3) to conclude that the left-hand side of (III.4.3) is bounded by

$$\pi^{-1/2}|t'-t|^{1/2} + |x'-x| + c''(\delta)T^{1/2-\delta}(t'-t)^\delta \le |x'-x| + c(\delta)T^{1/2-\delta}(t'-t)^\delta. \quad \blacksquare$$

We let $p_\varepsilon^x(y) = p_\varepsilon(y-x)$. To use Proposition III.4.4 we will need the following bound on the moments of X_t.

Lemma III.4.6. If X is as in Theorem III.4.2 (a), then

$$E'(X_t(p_\varepsilon^x)^p) \le p!\gamma^p t^{p/2}\exp(X_0(1)/\gamma t) \quad \forall t,\varepsilon>0,\ x\in\mathbb{R},\ \text{and } p\in\mathbb{N}.$$

Proof. We apply Lemma III.3.6 with $f = \theta p_\varepsilon^x$, where $\theta = \gamma^{-1}t^{-1/2}$ and $\varepsilon, t>0$ and $x\in\mathbb{R}$ are fixed. Then

$$\frac{\gamma}{2}G(f,t) \equiv \frac{\gamma}{2}\int_0^t \sup_{x'} P_s f(x')\,ds = \frac{\gamma\theta}{2}\int_0^t p_{s+\varepsilon}(0)\,ds \le 1/2,$$

and so Lemma III.3.6 implies that

$$E'(\exp(\theta X_t(p_\varepsilon^x))) \le \exp(2\theta X_0(p_{t+\varepsilon}^x)) \le \exp(X_0(1)/\gamma t).$$

This shows that for any $p \in \mathbb{N}$,

$$E'(X_t(p_\varepsilon^x)^p) \le p!\theta^{-p}\exp(X_0(1)/\gamma t),$$

as is required. ∎

Proof of Theorem III.4.2. (a) We adapt the argument of Konno and Shiga (1988). From (GFR) (see Exercise II.5.2) we see that for each fixed $\varepsilon, t > 0$ and $x \in \mathbb{R}$,

$$(III.4.5) \qquad X_t(p_\varepsilon^x) = X_0(p_{t+\varepsilon}^x) + \int_0^t \int p_{t+\varepsilon-s}(y-x)dM(s,y) \text{ a.s.}$$

Lemma III.4.5 (a) with $\delta = p = 1$ implies that

$$(III.4.6) \qquad \lim_{\varepsilon\downarrow 0} \sup_{x\in\mathbb{R}, t\ge\eta} |X_0(p_{t+\varepsilon}^x) - X_0(p_t^x)| \le 2\varepsilon\eta^{-3/2}X_0(1) \quad \forall\eta > 0.$$

To take L^2 limits in the stochastic integral in (III.4.5) apply Lemma III.4.5(a) with $p = 2$ and $0 < \delta < 1/2$ to see that

$$\begin{aligned}
&E'\left(\int_0^t \int (p_{t+\varepsilon-s}(y-x) - p_{t-s}(y-x))^2 X_s(dy)\,ds\right)\\
&= \int \left[\int_0^t \int (p_{t+\varepsilon-s}(y-x) - p_{t-s}(y-x))^2 p_s(y-z)\,dy\,ds\right] m(dz)\\
&\le \int \left[\int_0^t \int (\varepsilon(t-s)^{-3/2})^\delta [p_{t+\varepsilon-s}(y-x)^{2-\delta} + p_{t-s}(y-x)^{2-\delta}]\right.\\
&\qquad\qquad \left.\times\, p_s(y-z)dy\,ds\right] m(dz)\\
&\le \varepsilon^\delta \int \left[\int_0^t \int |t-s|^{-3\delta/2}[(t+\varepsilon-s)^{(\delta-1)/2} p_{(t+\varepsilon-s)/(2-\delta)}(y-x)\right.\\
&\qquad\qquad \left. + (t-s)^{(\delta-1)/2} p_{(t-s)/(2-\delta)}(y-x)] p_s(y-z)\,dy\,ds\right] m(dz)\\
&\le 2\varepsilon^\delta \int \left[\int_0^t (t-s)^{-1/2-\delta}\Big(p_{(t+\varepsilon-s)/(2-\delta)+s}(x-z)\right.\\
&\qquad\qquad \left. + p_{(t-s)/(2-\delta)+s}(x-z)\Big)\,ds\right] m(dz)\\
&\le 4(2\pi)^{-1/2} m(1)\varepsilon^\delta \left[\int_0^t (t-s)^{-1/2-\delta} s^{-1/2}\,ds\right]\\
&= m(1)c(\delta)t^{-\delta}\varepsilon^\delta.
\end{aligned}$$

This implies that
(III.4.7)

$$\lim_{\varepsilon\downarrow 0} \sup_{x\in\mathbb{R}, t\ge\eta} E'\left(\left(\int_0^t \int p_{t+\varepsilon-s}(y-x) - p_{t-s}(y-x)dM(s,y)\right)^2\right) = 0 \;\forall\eta > 0,$$

and also shows that

(III.4.8)
$$\int_0^u \int p_{t-s}(y-x)dM(s,y), \quad u \le t \text{ is a continuous } L^2\text{-bounded martingale.}$$

By (III.4.6) and (III.4.7) we may take L^2 limits in (III.4.5) as $\varepsilon \downarrow 0$ and also choose $\varepsilon_n \downarrow 0$ so that for any $t > 0$ and $x \in \mathbb{R}$,

$$\lim_{n\to\infty} X_t(p^x_{\varepsilon_n}) = X_0(p^x_t) + \int_0^t \int p_{t-s}(y-x)dM(s,y) \text{ a.s. and in } L^2.$$

Therefore if we define $u(t,x) = \liminf_{n\to\infty} X_t(p^x_{\varepsilon_n})$ for all $t > 0$, $x \in \mathbb{R}$, then

$$(III.4.9) \quad u(t,x) = X_0(p^x_t) + \int_0^t \int p_{t-s}(y-x)dM(s,y) \text{ a.s. for all } t > 0 \text{ and } x \in \mathbb{R}.$$

Also standard differentiation theory shows that for each $t > 0$ with probability 1, $X_t(dx) = u(t,x)dx + X^s_t(dx)$, where X^s_t is a random measure such that $X^s_t \perp dx$. Now (III.4.8) and (III.4.9) imply that

$$E'\Big(\int u(t,x)\,dx\Big) = \int X_0(p^x_t)\,dx = X_0(1) = E'(X_t(1)).$$

This shows that $E'(X^s_t(1)) = 0$ and so

$$(III.4.10) \qquad X_t(dx) = u(t,x)dx \text{ a.s. for all } t > 0$$

Now fix $t_0 > 0$. Apply (III.4.5) with t replaced by t_0 and ε replaced by $t - t_0 > 0$ to obtain

$$X_{t_0}(p^x_{t-t_0}) = X_0(p^x_t) + \int_0^{t_0} \int p_{t-s}(y-x)dM(s,y) \text{ a.s. } \forall t > t_0, \ x \in \mathbb{R}.$$

This and (III.4.9) show that

$$u(t,x) = X_{t_0}(p^x_{t-t_0}) + \int_{t_0}^t \int p_{t-s}(y-x)dM(s,y) \text{ a.s. } \forall t > t_0, \ x \in \mathbb{R}$$

$$(III.4.11) \qquad \equiv X_{t_0}(p^x_{t-t_0}) + I(t,x).$$

Proposition III.4.4 is now used to obtain a continuous version of I as follows. Let $0 < t_0 < t \le t' \le T$, $x, x' \in \mathbb{R}$, and $p > 1$. Then Burkholder's inequality and

(III.4.10) show that (recall that $p_t(x) = 0$ if $t \le 0$)

$$\begin{aligned} &E'(|I(t',x') - I(t,x)|^{2p}) \\ &\le c_p E'\left(\left(\int_{t_0}^{t'} \gamma \int (p_{t'-s}(y-x') - p_{t-s}(y-x))^2 u(s,y)\, dy\, ds\right)^p\right) \\ &\le \gamma^p \left(\int_{t_0}^{t'} \int (p_{t'-s}(y-x') - p_{t-s}(y-x))^2\, dy\, ds\right)^{p-1} \\ &\quad \times \int_{t_0}^{t'} \int (p_{t'-s}(y-x') - p_{t-s}(y-x))^2 E'(u(s,y)^p)\, dy\, ds, \end{aligned}$$

by Jensen's inequality. Lemma III.4.6 and Fatou's Lemma show that

$$(III.4.12) \qquad E'(u(t,y)^p) \le p!\gamma^p t^{p/2} \exp(X_0(1)/\gamma t) \quad \forall t > 0,\ x \in \mathbb{R},\ p \in \mathbb{N}.$$

This, together with Lemma III.4.5(b) and the previous inequality, show that for any $0 < \delta < 1/2$,

$$\begin{aligned} E'\left(|I(t',x') - I(t,x)|^{2p}\right) &\le c_p\gamma^p[|x-x'| + c(\delta)T^{1/2-\delta}(t'-t)^\delta]^{p-1} \\ &\quad \times p!T^{p/2}\exp(X_0(1)/\gamma t_0)[|x-x'| + c(\delta)T^{1/2-\delta}(t'-t)^\delta] \\ (III.4.13) \qquad &\le c(p,\gamma,X_0(1),t_0,T)[|x-x'|^p + |t'-t|^{p\delta}]. \end{aligned}$$

Proposition III.4.4 shows there is a continuous version of I on $(t_0,\infty)\times\mathbb{R}$. Dominated Convergence shows that $(t,x) \mapsto X_{t_0}(p^x_{t-t_0})$ is also continuous on $(t_0,\infty)\times\mathbb{R}$ a.s., and so (III.4.11) shows there is continuous version of $u(t,x)$ on $(t_0,\infty)\times\mathbb{R}$ for all $t_0 > 0$ and hence on $(0,\infty)\times\mathbb{R}$. We also denote this continuous version by u. Clearly $u(t,x) \ge 0$ for all $(t,x) \in (0,\infty)\times\mathbb{R}$ a.s. Define a measure-valued process by $\tilde{X}_t(dx) = u(t,x)dx$. Then (III.4.10) shows that

$$(III.4.14) \qquad \tilde{X}_t = X_t \text{ a.s. for each } t > 0.$$

If $\phi \in C_K(\mathbb{R})$, then $t \mapsto \tilde{X}_t(\phi)$ is continuous on $(0,\infty)$ a.s. by Dominated Convergence and the continuity of u. Therefore the weak continuity of X and (III.4.14) imply

$$\tilde{X}_t(\phi) = X_t(\phi)\ \forall t > 0 \text{ a.s.},$$

and hence

$$X_t(dx) = \tilde{X}_t(dx) \equiv u(t,x)dx\ \forall t > 0 \text{ a.s.},$$

where u is jointly continuous. Since $\cup_{t\ge\eta}S(X_t)$ is compact for all $\eta > 0$ a.s. by Corollary III.1.7, and u is uniformly continuous on compact sets, it follows easily that $t \mapsto u(t,\cdot)$ is a continuous function from $(0,\infty)$ to $C_K(\mathbb{R})$. (III.4.1) holds by (III.4.8) and (III.4.9), and so the proof of (a) is complete.

(b) Let $(\Omega'', \mathcal{F}'', \mathcal{F}''_t, P'')$ carry an $(\mathcal{F}''_t)$-white noise W'' on $\mathbb{R}_+ \times R$, and define W on $\bar{\Omega}$ by

$$W_t(A) = \int_0^t \int 1_A(x) 1(u(s,x) > 0)(\gamma u(s,x))^{-1/2} dM(s,x)$$
$$+ \int_0^t \int 1_A(x) 1(u(s,x) = 0) dW''(s,x),$$

for $t > 0$ and $A \in \mathcal{B}_F(\mathbb{R})$. The first stochastic integral is a continuous square integrable $(\mathcal{F}'_t)$-martingale with square function $\int_0^t \int 1_A(x) 1(u(s,x) > 0)\, dx\, ds$ because the integrand is in the space $\mathcal{L}^2$ from Remark II.5.5(b). It is now easy to check that $W_t(A)$ is a continuous $(\mathcal{F}_t)$-martingale on $\bar{\Omega}$ with square function $|A|t$ and so is an $(\mathcal{F}_t)$-Brownian motion with diffusion parameter $|A|$. Clearly if A, B are disjoint in $\mathbb{B}_F(\mathbb{R})$, then $W_t(A \cup B) = W_t(A) + W_t(B)$ a.s. and so W is an $(\mathcal{F}_t)$-white noise on $\mathbb{R}_+ \times \mathbb{R}$. We may extend the stochastic integrals with respect to M and W'' to $\mathbb{P}(\mathcal{F}.) \times \mathcal{B}(\mathbb{R})$-measurable integrands because M and W'' are both orthogonal martingale measures with respect to $\mathcal{F}_t$ (we suppress the projection maps in our notation). It follows easily from the definition of W that if $\psi \in \mathcal{L}^2_{\text{loc}}(W)$, then

$$W_t(\psi) = \int_0^t \int \psi(s,\omega,x) 1(u(s,x) > 0)(\gamma u(s,x))^{-1/2} dM(s,x)$$
$$+ \int_0^t \int \psi(s,\omega,x) dW''(s,x).$$

Therefore if $\phi \in C_b^2(\mathbb{R})$, then $\int_0^t \int \phi(x)(\gamma u(s,x))^{1/2} dW(s,x) = M_t(\phi)$. The martingale problem for $X_t(dx)$ $(= u(t,x)dx$ if $t > 0)$ now shows that u satisfies $(\text{SPDE})_m$.

(c) The fact that $t \mapsto u(t,\cdot)$ is a continuous $C_K(\mathbb{R}^d)$-valued process shows that X_t is a continuous $M_F(\mathbb{R}^d)$-valued process for $t > 0$. As was noted in Remark III.4.1, $(\text{SPDE})_m$ implies that $\lim_{t\to 0+} X_t = m$ a.s. and so $X.$ is a.s. continuous on $[0,\infty)$. If $\phi \in C_b^2(\mathbb{R}^d)$ and $M_t(\phi) = \int_0^t \int \phi(x)(\gamma u(s,x))^{1/2} dW(s,x)$, then $M_t(\phi)$ is an $\mathcal{F}_t$-local martingale satisfying

$$\langle M(\phi)\rangle_t = \int_0^t \int \gamma\phi(x)^2 u(s,x)\, dx\, ds = \int_0^t X_s(\gamma\phi^2)\, ds.$$

Therefore $(\text{SPDE})_m$ and Remark II.5.11 imply that X_t satisfies $(\text{LMP})_{\delta_m}$ and so X is an $(\mathcal{F}_t)$-SBM(γ) starting at m. ∎

5. Polar Sets

Throughout this section X is a $SBM(\gamma)$ under $\mathbb{P}_{X_0}$, $X_0 \neq 0$. Recall the range of X is $\mathcal{R} = \bigcup_{\delta>0} \overline{\mathcal{R}}([\delta,\infty))$.

Definition. If $A \subset \mathbb{R}^d$ we say $X(\omega)$ charges A iff $X_t(\omega)(A) > 0$ for some $t > 0$, and $X(\omega)$ hits A iff $A \cap \mathcal{R}(\omega) \neq \phi$.

Theorem III.5.1. (a) If $\phi \in b\mathcal{B}(\mathbb{R}^d)$, $X_t(\phi)$ is a.s. continuous on $(0,\infty)$.
(b) $\mathbb{P}_{X_0}$ (X charges A) $> 0 \Leftrightarrow A$ has positive Lebesgue measure.
Proof. (a) See Reimers (1989b) or Perkins (1991), and the Remark below.

(b) Since $X_t(A)$ is a.s. continuous by (a), X charges A iff $X_t(A) > 0$ for some rational $t > 0$. The probability of the latter event is positive iff

$$\mathbb{P}_{X_0}(X_t(A)) > 0\ \exists t \in \mathbb{Q}^{>0} \Leftrightarrow P^{X_0}(B_t \in A) > 0\ \exists t \in \mathbb{Q}^{>0}$$
$$\Leftrightarrow A \text{ has positive Lebesgue measure.} \quad \blacksquare$$

Remark. In Theorem 4 of Perkins (1991) (a) is proved for a large class of Y – DW – superdiffusions whose semigroup satisfies a strong continuity condition. The simple idea of the proof is to first use the Garsia-Rodemich-Rumsey Lemma to obtain an explicit modulus of continuity for $X_t(\phi) - X_0(P_t\phi)$ for $\phi \in C_b(E)$ and then to show this modulus is preserved under $\overset{bp}{\to}$ of ϕ. The strong continuity condition on P_t implies

$$\int_0^1 \|P_{t+r}\phi - P_t\phi\| r^{-1}\, dr < \infty\ \forall t > 0,\ \phi \in b\mathcal{E},$$

which is already stronger than the continuity of $t \to P_t\phi(x)$ for each $\phi \in b\mathcal{E}$, $x \in E$. The latter is clearly a necessary condition for a.s. continuity of $X_t(\phi)$ on $(0,\infty)$ (take means). Whether or not this, or even norm continuity of $P_t\phi\ \forall\phi \in b\mathcal{E}$, is also sufficient for continuity of $X_t(\phi)$, $t > 0$ remains open.

The notion of hitting a set will be probabilistically more subtle and more important analytically.

Definition. A Borel set is polar for X iff $\mathbb{P}_{X_0}(X \text{ hits } A) = 0$ for all $X_0 \in M_F(\mathbb{R}^d)$ or equivalently iff $\mathbb{P}_{X_0}(X \text{ hits } A) = 0$ for some non-zero X_0 in $M_F(\mathbb{R}^d)$.

The above equivalence is a consequence of the equivalence of $\mathbb{P}_{X_0}$ and $\mathbb{P}_{X_0'}$ on the field $\bigcup_{\delta>0} \sigma(X_r,\ r \ge \delta)$ for any non-zero finite measures X_0 and X_0' (see Example III.2.3). We would like to find an analytic criterion for polarity. For ordinary Brownian motion Kakutani (1944) did this using Newtonian capacity.

Definition. Let g be a decreasing non-negative continuous function on $(0,\infty)$ with $g(0+) > 0$. If $\mu \in M_F(\mathbb{R}^d)$, $\langle\mu\rangle_g = \int\int g(|x-y|)\mu(dx)\mu(dy)$ is the g-energy of μ. If $A \in \mathcal{B}(\mathbb{R}^d)$, let

$$I(g)(A) = \inf\{\langle\mu\rangle_g : \mu \text{ a probability, } \mu(A) = 1\}$$

and let the g-capacity at A be $C(g)(A) = I(g)(A)^{-1} \in [0,\infty)$.

Note that the g-capacity of A is positive iff A is large enough to support a probability of finite g-energy.

Notation.

$$g_\beta(r) = \begin{cases} r^{-\beta} & \beta > 0 \\ 1 + (\log 1/r)^+ & \beta = 0 \\ 1 & \beta < 0 \end{cases}$$

If $\phi \in \mathcal{H}$ there is a close connection between sets of zero Hausdorff ϕ-measure and sets of zero ϕ^{-1}-capacity:
(III.5.1) $\phi - m(A) < \infty \Rightarrow C(\phi^{-1})(A) = 0$.
(III.5.2) $C(g_\beta)(A) = 0 \Rightarrow x^\beta(\log 1/x)^{-1-\varepsilon} - m(A) = 0\ \forall\varepsilon > 0, \beta > 0$.
(III.5.3) $C(g_0)(A) = 0 \Rightarrow (\log^+ 1/x)^{-1}(\log^+\log^+ 1/x)^{-1-\varepsilon} - m(A) = 0\ \forall\varepsilon > 0$.

Moreover these implications are essentially best possible. See Taylor (1961) for a discussion. In particular the capacitory dimension (defined in the obvious way) and Hausdorff dimension coincide.

For $d \geq 2$ the range of a Brownian motion $\{B_t : t \geq 0\}$ is two-dimensional and so should hit sets of dimension greater than $d-2$. Kakutani (1944) showed for $d \geq 2$ and $A \in \mathcal{B}(\mathcal{R}^d)$,

$$P(B_t \in A\ \exists t > 0) > 0 \Leftrightarrow C(g_{d-2})(A) > 0.$$

Recall from Theorem III.3.9 (see also Remark III.3.2) that $\mathcal{R}$ is a 4-dimensional set if $d \geq 4$ and hence should hit sets of dimension greater than $d-4$.

Theorem III.5.2. Let $A \in \mathcal{B}(\mathbb{R}^d)$. A is polar for X iff $C(g_{d-4})(A) = 0$. In particular, points are polar for X iff $d \geq 4$.

Remark III.5.3. The inner regularity of the Choquet capacities

$$A \to \mathbb{P}_{\delta_x}(A \cap \bar{\mathcal{R}}([\delta, \infty)) \neq 0) \ (\delta > 0) \text{ and } A \to C(g_{d-4})(A)$$

(see III.29 of Dellacherie and Meyer (1978)) allows one to consider only $A = K$ compact. The necessity of the zero capacity condition for polarity was proved in Perkins (1990) by a probabilistic inclusion-exclusion argument. The elegant proof given below is due to Le Gall (1999) (Section VI.2). The more delicate sufficiency was proved by Dynkin (1991). His argument proceeded in two steps:

1. $V(x) = -\log \mathbb{P}_{\delta_x}(\mathcal{R} \cap K = \phi)$ is the maximal non-negative solution of $\Delta V = \gamma V^2$ on K^c.
2. The only non-negative solution of $\Delta V = \gamma V^2$ on K^c is $V \equiv 0$ iff $C(g_{d-4})(K) = 0$.

Step 1 uses a probabilistic representation of solutions to the non-linear boundary value problem

$$\Delta V = \gamma V^2 \text{ in D, } V|_{\partial D} = g,$$

where D is a regular domain in $\mathbb{R}^d$ for the classical Dirichlet problem, and g is a bounded, continuous, non-negative function on ∂D. The representation is in terms of the exit measure X_D of X on D. X_D may be constructed as the weak limit of the of the sequence of measures obtained by stopping the branching particles in Section II.4 when they exit D and assigning mass $1/N$ to each exit location.

Step 2 is the analytical characterization of the sets of removable singularities for $\Delta V = \gamma V^2$ due to Baras and Pierre (1984). A self-contained description of both steps may be found in Chapter VI of Le Gall (1999). A proof of a slightly weaker result is given below (Corollary III.5.10).

The proof of the necessity of the zero capacity condition in Theorem III.5.2 will in fact show

Theorem III.5.4. For any $M > 0$ and $X_0 \in M_F(\mathbb{R}^d) - \{0\}$, there is a $c(M, X_0) > 0$ such that

$$\mathbb{P}_{X_0}(X \text{ hits } A) \geq c(M, X_0) C(g_{d-4})(A) \quad \text{for any Borel subset } A \text{ of } B(0, M).$$

In particular points are not polar if $d \leq 3$.

Proof. We may assume $A = K$ is a compact subset of $B(0, M)$ of positive g_{d-4}-capacity by the inner regularity of $C(g_{d-4})$. Choose a probability ν supported by K so that $\mathcal{E} \equiv \int\int g_{d-4}(x-y)d\nu(x)d\nu(y) < \infty$. Let $f : \mathbb{R}^d \to \mathbb{R}$ be a continuous, non-negative, radially symmetric function such that $\{f > 0\} = B(0,1)$ and

$\int f(y)\,dy = 1$. Define $f_\varepsilon(y) = \varepsilon^{-d} f(\varepsilon^{-1}y)$ and $\phi_\varepsilon(y) = f_\varepsilon * \nu(y) \equiv \int f_\varepsilon(y-z)\nu(dz)$. Note that

$$\{\phi_\varepsilon > 0\} \subset K^\varepsilon \equiv \{x : d(x,K) < \varepsilon\}. \tag{III.5.4}$$

We will use the following elementary consequence of the Cauchy-Schwarz inequality.

Lemma III.5.5. Assume $Z \geq 0$ has mean μ and variance $\sigma^2 < \infty$. Then

$$P(Z > 0) \geq \mu^2/(\mu^2 + \sigma^2).$$

Proof. $E(Z) = E(Z1(Z > 0)) \leq E(Z^2)^{1/2} P(Z > 0)^{1/2}$. Rearrange to get a lower bound on $P(Z > 0)$. ∎

Apply the above lemma to $Z = \int_1^2 X_s(\phi_\varepsilon) ds$. If $g_{1,2}(x) = \int_1^2 p_s(x)\,ds$, where p_s is the Brownian transition density, then

$$\begin{aligned}
\mathbb{P}_{X_0}\Big(\int_1^2 X_s(\phi_\varepsilon)\,ds\Big) &= \int_1^2 P^{X_0}(\phi_\varepsilon(B_s))\,ds \\
&= \int\Big[\int g_{1,2} * X_0(y) f_\varepsilon(y-z)\,dy\Big]\nu(dz) \\
&\to \int g_{1,2} * X_0(z)\nu(dz) \text{ as } \varepsilon \downarrow 0 \\
&> 0.
\end{aligned}$$

The above shows that for $\varepsilon < \varepsilon_0$,

$$\mathbb{P}_{X_0}\Big(\int_1^2 X_s(\phi_\varepsilon)\,ds\Big) \geq \frac{1}{2}\inf_{|z|\leq M} g_{1,2} * X_0(z) \equiv C_1(X_0, M). \tag{III.5.5}$$

By our second moment formula (see Exercise II.5.2(b))

$$\begin{aligned}
\mathrm{Var}\Big(\int_1^2 X_s(\phi_\varepsilon)\,ds\Big) &= \int_1^2\int_1^2 \mathrm{Cov}\,(X_s(\phi_\varepsilon), X_t(\phi_\varepsilon))\,dsdt \\
&= 2\gamma\int_1^2 dt\int_1^t ds\int_0^s dr X_0 P_r(P_{s-r}\phi_\varepsilon P_{t-r}\phi_\varepsilon) \\
&\leq 2\gamma\Big[\int_0^{1/2} dr\int_1^2 ds\int_s^2 dt X_0 P_r(P_{s-r}\phi_\varepsilon P_{t-r}\phi_\varepsilon) \\
&\quad + \int_{1/2}^2 dr\int_r^2 ds\int_s^2 dt X_0 P_r(P_{s-r}\phi_\varepsilon P_{t-r}\phi_\varepsilon)\Big] \\
&\equiv 2\gamma[I_1 + I_2].
\end{aligned} \tag{III.5.6}$$

In I_1, $s - r \geq 1/2$, $t - r \geq 1/2$ and so

$$P_{t-r}\phi_\varepsilon \vee P_{s-r}\phi_\varepsilon \leq c\int\int f_\varepsilon(y-z)\,dyd\nu(z) = c.$$

This implies (the value of c may change)

$$I_1 \leq cX_0(1). \tag{III.5.7}$$

If $G_2\phi_\varepsilon(y)=\int_0^2 P_t\phi_\varepsilon(y)\,dt$, then

$$(III.5.8)\qquad I_2\le\int_{1/2}^2 X_0P_r(G_2\phi_\varepsilon^2)\,dr\le cX_0(1)\int G_2\phi_\varepsilon(y)^2\,dy.$$

Lemma III.5.6. (a) $G_2\phi_\varepsilon(y)\le C\int g_{d-2}(y-z)\,d\nu(z)$.
(b) $\int g_{d-2}(z_1-y)g_{d-2}(z_2-y)\,dy\le Cg_{d-4}(z_1-z_2)$.
Proof. (a) Since $\int_0^2 p_t(x)dt\le cg_{d-2}(x)$, we have

$$G_2\phi_\varepsilon(y)\le c\int\int g_{d-2}(y-x)f_\varepsilon(x-z)\,dxd\nu(z).$$

The superharmonicity of g_{d-2} implies that the spherical averages of $g_{d-2}(y-x)f_\varepsilon(x-z)$ over $\{x:|x-z)|=r\}$ are at most $g_{d-2}(y-z)f_\varepsilon(r)$. This and the fact that $\int f_\varepsilon(y)\,dy=1$ allow us to conclude from the above that

$$G_2\phi_\varepsilon(y)\le c\int g_{d-2}(y-z)d\nu(z).$$

(b) Exercise. One approach is to use $g_{d-2}(x)\le c\int_0^1 p_t(x)\,dt$ and Chapman-Kolmogorov. ∎

Use (a) and (b) in (III.5.8) to see that

$$I_2\le cX_0(1)\int\int g_{d-4}(z_1-z_2)\,d\nu(z_1)\,d\nu(z_2)=cX_0(1)\mathcal{E}.$$

Now use the above with (III.5.5)–(III.5.8) in Lemma III.5.5 to see that

$$\mathbb{P}_{X_0}\Big(\int_1^2 X_s(\phi_\varepsilon)\,ds>0\Big)\ge\frac{C_1(X_0,M)^2}{C_1(X_0,M)^2+cX_0(1)(1+\mathcal{E})}\ge\frac{c(X_0,M)}{\mathcal{E}},$$

where we use $\mathcal{E}\ge c_M>0$ if $K\subset B(0,M)$ in the last line. Now minimize $\mathcal{E}$ to see that

$$\mathbb{P}_{X_0}\Big(\int_1^2 X_s(\phi_\varepsilon)\,ds>0\Big)\ge c(X_0,M)C(g_{d-4})(K).$$

This implies

$$\begin{aligned}\mathbb{P}_{X_0}(\overline{\mathcal{R}}([1,2])\cap K\ne\phi)&=\lim_{\varepsilon\downarrow0}\mathbb{P}_{X_0}(\overline{\mathcal{R}}([1,2])\cap K^\varepsilon\ne\phi)\\&\ge\lim_{\varepsilon\downarrow0}\mathbb{P}_{X_0}\Big(\int_1^2 X_s(\phi_\varepsilon)\,ds>0\Big),\end{aligned}$$

the last because $S(\phi_\varepsilon)\subset K^\varepsilon$. The above two inequalities complete the proof. ∎

Upper bounds on hitting probabilities appear to require a greater analytic component. We now obtain precise asymptotics for hitting small balls using the Laplace functional equation (LE) from Section II.5. Recall from Theorem II.5.11 that if $\Delta/2$

denotes the generator of Brownian motion, $f \in C_b(\mathbb{R}^d)_+$, and V_t is the unique solution of

$$(SE)_{0,f} \qquad \frac{\partial V}{\partial t} = \frac{\Delta}{2}V_t - \frac{\gamma}{2}V_t^2 + f,\ V_0 = 0,$$

then

$$(LE) \qquad \mathbb{P}_{X_0}(\exp\{-\int_0^t X_s(f)\,ds\}) = \exp(-X_0(V_t)).$$

Recall also that $f(\varepsilon) \sim g(\varepsilon)$ as $\varepsilon \downarrow 0$ means $\lim_{\varepsilon\downarrow 0} f(\varepsilon)/g(\varepsilon) = 1$.

Theorem III.5.7.(a) If $d \le 3$, then

$$\mathbb{P}_{X_0}(X \text{ hits } \{x\}) = 1 - \exp\Big\{-\frac{2(4-d)}{\gamma}\int |y-x|^{-2}\,dX_0(y)\Big\}.$$

(b) There is a $c(d) > 0$ such that if $x \notin S(X_0)$, then as $\varepsilon \downarrow 0$

$$\mathbb{P}_{X_0}(X \text{ hits } B(x,\varepsilon)) \sim \begin{cases} \frac{2}{\gamma}\int |y-x|^{-2}dX_0(y)(\log 1/\varepsilon)^{-1} & \text{if } d = 4 \\ \frac{c(d)}{\gamma}\int |y-x|^{2-d}\,dX_0(y)\varepsilon^{d-4} & \text{if } d > 4. \end{cases}$$

(c) There is a $K_d > 0$ so that if $d(x, S(X_0)) \ge 2\varepsilon_0$, then

$$\mathbb{P}_{X_0}(X \text{ hits } B(x,\varepsilon)) \le \frac{K_d}{\gamma}X_0(1)\varepsilon_0^{2-d}\begin{cases} (\log 1/\varepsilon)^{-1} & \forall \varepsilon \in (0, \varepsilon_0 \wedge \varepsilon_0^2) & \text{if } d = 4 \\ \varepsilon^{d-4} & \forall \varepsilon \in (0, \varepsilon_0) & \text{if } d > 4. \end{cases}$$

Remark. The constant $c(d)$ is the one arising in Lemma III.5.9 below.

Proof. By translation invariance we may assume $x = 0$. Choose $f \in C_b(\mathbb{R}^d)_+$, radially symmetric so that $\{f > 0\} = B(0,1)$. Let $f_\varepsilon(x) = f(x/\varepsilon)$, and let $u^{\lambda,\varepsilon}(t,x)$ be the unique solution of

$$(SE)_\varepsilon \qquad \frac{\partial u}{\partial t} = \frac{\Delta u}{2} - \frac{\gamma u^2}{2} + \lambda f_\varepsilon,\ u_0 = 0.$$

By scaling, $u^{\lambda,\varepsilon}(t,x) = \varepsilon^{-2}u^{\lambda\varepsilon^4,1}(t\varepsilon^{-2}, x\varepsilon^{-1}) \equiv \varepsilon^{-2}u^{\lambda\varepsilon^4}(t\varepsilon^{-2}, x\varepsilon^{-1})$. We have
$(III.5.9)$

$$\mathbb{P}_{X_0}\Big(\exp\Big\{-\lambda\int_0^t X_s(f_\varepsilon)\,ds\Big\}\Big) = \exp\Big\{-\int \varepsilon^{-2}u^{\lambda\varepsilon^4}(t\varepsilon^{-2}, x\varepsilon^{-1})\,dX_0(x)\Big\}.$$

The left side is decreasing in t and λ and so by taking $X_0 = \delta_x$ we see that $u^\lambda(t,x) \uparrow u(x) = u(|x|)$ $(\le \infty)$ as $t, \lambda \uparrow \infty$. Take limits in (III.5.9) to get

$$\begin{aligned} \mathbb{P}_{X_0}(X_s(B(0,\varepsilon)) = 0\ \forall s > 0) &= \mathbb{P}_{X_0}\Big(\int_0^\infty X_s(f_\varepsilon)\,ds = 0\Big) \\ &= \lim_{\lambda,t\to\infty} \mathbb{P}_{X_0}\Big(\exp\Big\{-\lambda\int_0^t X_s(f_\varepsilon)\,ds\Big\}\Big). \end{aligned}$$

$$(III.5.10) \qquad = \exp\{-\int \varepsilon^{-2}u(x\varepsilon^{-1})\,dX_0(x)\}.$$

The left-hand side increases as $\varepsilon \downarrow 0$ and so (take $X_0 = \delta_x$),

$(III.5.11)$ $\qquad \varepsilon \to \varepsilon^{-2}u(x/\varepsilon)$ decreases as $\varepsilon \downarrow 0$, and in particular the radial function $u(x)$ is decreasing in $|x|$.

By taking $\varepsilon = 1$ and $X_0 = \delta_x$ in (III.5.10), we see that

$$u(x) = -\log \mathbb{P}_{\delta_x}(X_s(B(0,1)) = 0\ \forall s > 0). \tag{III.5.12}$$

Suppose $|x| > 1$ and $\mathbb{P}_{\delta_x}(X_s(B(0,1)) = 0\ \forall s > 0) = 0$. Then by the multiplicative property

$$\mathbb{P}_{\frac{1}{n}\delta_x}(X_s(B(0,1)) = 0 \quad \forall s > 0) = 0 \ \forall n \in \mathbb{N}. \tag{III.5.13}$$

The extinction probability formula (II.5.12) and Historical Modulus of Continuity (Theorem III.1.3) imply ($\delta(3,\omega)$ is as in the latter result) for some C, $\rho > 0$,

$$\mathbb{Q}_{0,\frac{1}{n}\delta_x}(X_{\frac{1}{n}} = 0,\ \delta(3) > 1/n) \geq e^{-2/\gamma} - \frac{C}{\gamma}\frac{1}{n}\frac{1}{n^\rho} \geq \frac{1}{2}e^{-2/\gamma} \text{ if } n \geq n_0.$$

The above event implies $\mathcal{R} \subset B(x, 3h(1/n))$ and so $X_s(B(0,1)) = 0\ \forall s > 0$ providing $3h(1/n) < |x| - 1$. This contradicts (III.5.13) and so we have proved $u(x) < \infty$ for all $|x| > 1$, and so is bounded on $\{|x| \geq 1+\varepsilon\}\ \forall \varepsilon > 0$ by (III.5.11). Letting $\lambda, t \to \infty$ in $(SE)_1$, leads one to believe that u solves

$$\Delta u = \gamma u^2 \text{ on } \{|x| > 1\},\ \lim_{|x|\downarrow 1} u(x) = \infty,\ \lim_{|x|\to\infty} u(x) = 0. \tag{III.5.14}$$

Lemma III.5.8. As $\lambda, t \uparrow \infty$, $u^\lambda(t,x) \uparrow u(x)$, where for $|x| > 1$, u is the unique non-negative solution of (III.5.14). Moreover u is C^2 on $\{|x| > 1\}$.

Proof. The mild form of (SE) (recall $(ME)_{0,f}$ from Section II.5) gives (B_t is a Brownian motion under P^x)

$$u_t^\lambda(x) = -E^x\Big(\int_0^t \frac{\gamma}{2}u_{t-s}^\lambda(B_s)^2\,ds\Big) + \lambda E^x\Big(\int_0^t f(B_s)\,ds\Big). \tag{III.5.15}$$

Let C be an open ball with $\overline{C} \subset \{|x| > 1\}$ and let $T_C = \inf\{t : B_t \notin C\}$. If $x \in C$, the strong Markov property shows that

$$\begin{aligned} u_t^\lambda(x) =& E^x\Big(\int_0^{t\wedge T_C} -\frac{\gamma}{2}u_{t-s}^\lambda(B_s)^2\,ds\Big) \\ &+ E^x\Big(E^{B(T_C)}\Big(\int_0^{(t-T_C)^+} -\frac{\gamma}{2}u_{(t-T_C)^+-s}^\lambda(B_s)^2\,ds\Big) \\ &+ \lambda E^{B(T_C)}\Big(\int_0^{(t-T_C)^+} f(B_s)\,ds\Big)\Big), \end{aligned}$$

and so by (III.5.15), with $((t-T_C)^+, B(T_C))$ in place of (t,x),

$$u_t^\lambda(x) + E^x\Big(\int_0^{t\wedge T_C} \frac{\gamma}{2}u_{t-s}^\lambda(B_s)^2\,ds\Big) = E^x(u_{(t-T_C)^+}^\lambda(B_{T_C})).$$

Now let $t, \lambda \to \infty$ and use Monotone Convergence to see

$$(III.5.16) \qquad u(x) + E^x\Big(\int_0^{T_C} \frac{\gamma}{2} u(B_s)^2\, ds\Big) = E^x(u(B_{T_C}))\ \forall x \in C.$$

The righthand side is harmonic and therefore C^2 on C. The second term on the left is $\frac{\gamma}{2}\int_C g_C(x,y)u(y)^2\,dy$, where g_C is the Green function for C. This is C^2 on C by Theorem 6.6 of Port-Stone (1978). Itô's Lemma and (III.5.16) gives

$$E^x\Big(\int_0^{T_C} \frac{\Delta}{2} u(B_s)\, ds\Big)/E^x(T_C) = \frac{\gamma}{2} E^x\Big(\int_0^{T_C} u(B_s)^2\, ds)/E^x(T_C).$$

Now let $C = B(x, 2^{-n}) \downarrow \{x\}$ to see $\Delta u = \gamma u^2$ on $\{|x| > 1\}$.

Let $\mu(t,x)$ and $\sigma^2(t,x)$ be the mean and variance of $X_t(B(0,1))$ under $\mathbb{P}_{\delta_x}$. Then $\mu(|x|-1, x) \to \mu > 0$ as $|x| \downarrow 1$ and $\sigma^2(|x|-1,x) \le \gamma(|x|-1) \to 0$ as $|x| \downarrow 1$, by our moment formulae (Exercise II.5.2). Therefore Lemma III.5.5 and (III.5.12) show that for $|x| > 1$

$$\begin{aligned} e^{-u(x)} &= \mathbb{P}_{\delta_x}(X_s(B(0,1) = 0\ \forall s > 0) \\ &\le 1 - \mathbb{P}_{\delta_x}(X_{|x|-1}(B(0,1)) > 0) \\ &\le \sigma^2(|x|-1,x)/(\mu(|x|-1,x)^2 + \sigma^2(|x|-1,x)) \\ &\to 0 \text{ as } |x| \downarrow 1. \end{aligned}$$

Therefore $\lim_{|x|\downarrow 1} u(x) = \infty$. (III.5.11) shows that for $|x| \ge 2$,

$$(III.5.17) \qquad u(x) = \frac{4}{|x|^2}\Big(\frac{2}{|x|}\Big)^{-2} u\Big(\frac{2x}{|x|}\Big/\frac{2}{|x|}\Big) \le \frac{4}{|x|^2} u(2) \to 0 \text{ as } |x| \to \infty.$$

It remains only to show uniqueness in (III.5.14). This is an easy application of the classical maximum principle. Let u, v be solutions and $c > 1$. Then $w(x) = c^2 u(cx)$ solves $\Delta w = \gamma w^2$ on $\{|x| > c^{-1}\}$. Therefore

$$\lim_{|x|\downarrow 1} w(x) - v(x) = -\infty, \quad \lim_{|x|\to\infty} (w - v)(x) = 0$$

and the usual elementary argument shows that $w - v$ cannot have a positive local maximum. Therefore $w \le v$ and so (let $c \downarrow 1$) $u \le v$. By symmetry $u = v$. ∎

In view of (III.5.10) to obtain the required estimates for Theorem III.5.7 we need to know the asymptotic behavior of $u(x) = u(|x|)$ as $|x| \to \infty$. By the radial symmetry (III.5.14) is an ordinary differential equation in the radial variable and so precise results are known.

Lemma III.5.9. As $r \to \infty$

$$u(r) \sim \begin{cases} \frac{2}{\gamma}(4-d)r^{-2} & d \le 3 \\ \frac{2}{\gamma} r^{-2}/\log r & d = 4 \\ \frac{c(d)}{\gamma} r^{2-d},\ c(d) > 0 & d > 4 \end{cases}.$$

Proof. Iscoe (1988) gives just enough hints to reduce this to a calculus exercise (albeit a lengthy one if $d = 4$) before referring the reader to the differential equations literature. We will carry out this exercise later in this Section to give a self-contained proof. ∎

We are ready to complete the
Proof of Theorem III.5.7. Consider first the proof of (b) and (c) when $d=4$ (a similar argument works if $d>4$). From (III.5.10), we have

$$(III.5.18)\quad \mathbb{P}_{X_0}(X \text{ hits } B(0,\varepsilon)) = \mathbb{P}_{X_0}(X_s(B(0,\varepsilon))>0\ \exists s>0) = 1-\exp\{-\varepsilon^{-2}\int u(x/\varepsilon)\,dX_0(x)\}.$$

Let $2\varepsilon_0 = d(0,S(X_0))>0$. If $x\in S(X_0)$ and $0<\varepsilon<\varepsilon_0\wedge\varepsilon_0^2$, then the monotonicity of u and Lemma III.5.9 show that

$$\begin{aligned}(III.5.19)\quad (\log 1/\varepsilon)\varepsilon^{-2}u(x/\varepsilon) &\le (\log 1/\varepsilon)\varepsilon^{-2}u(2\varepsilon_0/\varepsilon)\\ &\le C(\log 1/\varepsilon)\varepsilon^{-2}(\varepsilon_0/\varepsilon)^{-2}(\log(\varepsilon_0/\varepsilon))^{-1}\\ &\le C2\varepsilon_0^{-2}.\end{aligned}$$

This proves the left side is uniformly bounded on $S(X_0)$ and so (III.5.18), Lemma III.5.9 and Dominated Convergence imply

$$\begin{aligned}\lim_{\varepsilon\downarrow 0}(\log 1/\varepsilon)\mathbb{P}_{X_0}(X \text{ hits } B(0,\varepsilon)) &= \lim_{\varepsilon\downarrow 0}\int(\log 1/\varepsilon)\varepsilon^{-2}u(x/\varepsilon)\,dX_0(x)\\ &= \int \lim_{\varepsilon\downarrow 0}(\log 1/\varepsilon)\varepsilon^{-2}u(x/\varepsilon)\,dX_0(x)\\ &= \frac{2}{\gamma}\int |x|^{-2}dX_0(x).\end{aligned}$$

This proves (b). To prove (c) use (III.5.18) and then (III.5.19) to see that for $\varepsilon<\varepsilon_0\wedge\varepsilon_0^2$

$$\begin{aligned}\mathbb{P}_{X_0}(X \text{ hits } B(0,\varepsilon)) &\le \varepsilon^{-2}\int u(x/\varepsilon)\,dX_0(x)\\ &\le 2C\varepsilon_0^{-2}X_0(1)(\log 1/\varepsilon)^{-1}.\end{aligned}$$

For (a) we consider 3 cases.
Case 1. $0\notin S(X_0)$.
Let $2\varepsilon_0 = d(0,S(X_0))$. By (III.5.11), and as in the proof of (b), if $x\in S(X_0)$ and $0<\varepsilon<\varepsilon_0$,

$$\varepsilon^{-2}u(x/\varepsilon)\le\varepsilon^{-2}u(2\varepsilon_0/\varepsilon)\le\varepsilon_0^{-2}u(2).$$

This allows us to use Dominated Convergence and Lemma III.5.9 to let $\varepsilon\downarrow 0$ in (III.5.10) and conclude

$$\begin{aligned}(III.5.20)\quad &\mathbb{P}_{X_0}\Big(\cup_{\varepsilon>0}\{X_s(B(0,\varepsilon))=0\ \forall s>0\}\Big)\\ &= \exp\Big\{-\int\frac{2}{\gamma}(4-d)|x|^{-2}\,dX_0(x)\Big\}.\end{aligned}$$

The event on the left hand side clearly implies $0\notin\mathcal{R}$. Conversely suppose $0\notin\mathcal{R}$. The historical modulus of continuity and $0\notin S(X_0)$ imply $0\notin\overline{\mathcal{R}}([0,\delta])$ for some $\delta>0$ w.p. 1. Therefore $0\notin\overline{\mathcal{R}}([0,\infty))$ and so $d(0,\overline{\mathcal{R}}([0,\infty)))>0$ which implies for some $\varepsilon > 0$ $X_s(B(0,\varepsilon)) = 0$ for all $s \geq 0$. Therefore (III.5.20) is the required equation.

Case 2. $0 \in S(X_0)$, $X_0(\{0\}) = 0$.

For $\eta > 0$, let $X_0 = X_0^{1,\eta} + X_0^{2,\eta}$, where $dX_0^{1,\eta}(x) = 1(|x| > \eta)\, dX_0(x)$. If $\delta > 0$,

$$\begin{aligned}\mathbb{P}_{X_0}(0 \notin \overline{\mathcal{R}}([\delta,\infty))) &= \mathbb{P}_{X_0^{1,\eta}}(0 \notin \overline{\mathcal{R}}([\delta,\infty)))\mathbb{P}_{X_0^{2,\eta}}(0 \notin \bar{\mathcal{R}}([\delta,\infty))) \\ &\geq \mathbb{P}_{X_0^{1,\eta}}(0 \notin \overline{\mathcal{R}})\mathbb{P}_{X_0^{2,\eta}}(X_\delta = 0) \\ &= \exp\Big\{\frac{-2(4-d)}{\gamma}\int 1(|x| > \eta)|x|^{-2} dX_0(x)\Big\}\exp\Big\{\frac{-2X_0^{2,\eta}(1)}{\gamma\delta}\Big\},\end{aligned}$$

where we have applied Case 1 to $X_0^{1,\eta}$ and used the extinction probability formula (II.5.12) in the last line. Let $\eta \downarrow 0$ and then $\delta \downarrow 0$ to get the required lower bound. For the upper bound just use $\mathbb{P}_{X_0}(0 \notin \mathcal{R}) \leq \mathbb{P}_{X_0^{1,\eta}}(0 \notin \mathcal{R})$, apply case 1 to $X_0^{1,\eta}$, and let $\eta \downarrow 0$.

Case 3. $X_0(\{0\}) > 0$.

Note that $\mathbb{P}_{X_0}(X_\delta(\{0\})) = P^{X_0}(B_\delta = 0) = 0$ and so we may use the Markov property and apply the previous cases a.s to X_δ to conclude

$$\begin{aligned}\mathbb{P}_{X_0}(X \text{ misses } \{0\}) &\leq \mathbb{P}_{X_0}(\mathbb{P}_{X_\delta}(X \text{misses } \{0\})) \\ &= \mathbb{P}_{X_0}\Big(\exp\Big\{-\frac{2(4-d)}{\gamma}\int |x|^{-2} dX_\delta(x)\Big\}\Big) \to 0 \text{ as } \delta \downarrow 0\end{aligned}$$

because the weak continuity and Fatou's lemma shows

$$\liminf_{\delta\downarrow 0}\int |x|^{-2} dX_\delta(x) = \infty \text{ a.s.}$$

Therefore X hits $\{0\}$ $\mathbb{P}_{X_0}$-a.s. and the result holds in this final case. ∎

Corollary III.5.10. Let

$$f_\beta(r) = \begin{cases} r^\beta & \text{if } \beta > 0 \\ (\log 1/r)^{-1} & \text{if } \beta = 0 \end{cases},$$

and $d \geq 4$. If $A \in \mathcal{B}(\mathbb{R}^d)$ and $f_{d-4} - m(A) = 0$, then A is polar for X. In particular, points are polar for X.

Proof. Assume without loss of generality that A is bounded. Let $A \subset \cup_{i=1}^\infty B(x_i^n, r_i^n)$ where $\lim_{n\to\infty}\sum_{i=1}^\infty f_{d-4}(2r_i^n) = 0$. Choose $x_0 \notin \overline{A}$ and n large so that $d(x_0, \cup_1^\infty B(x_i^n, r_i^n)) = \varepsilon_0 > 0$. Then Theorem III.5.7(c) implies that for n large,

$$\begin{aligned}\mathbb{P}_{\delta_{x_0}}(\mathcal{R}\cap A \neq \phi) &\leq \sum_{i=1}^\infty \mathbb{P}_{\delta_{x_0}}(\mathcal{R}\cap B(x_i^n, r_i^n) \neq \phi) \\ &\leq \frac{K_d}{\gamma}\varepsilon_0^{2-d}\sum_{i=1}^\infty f_{d-4}(r_i^n) \\ &\to 0 \text{ as } n \to \infty.\end{aligned}$$ ∎

Remark. (III.5.1) shows that this result is also a consequence of Theorem III.5.2 as the hypothesis on A is stronger than that in Theorem III.5.2. However, in practice it is often easier to verify the Hausdorff measure condition than the capacity condition, and (III.5.2)–(III.5.3) show the conditions are close.

Exercise III.5.1. Let

$$\tilde{\psi}_d(r) = \begin{cases} r^4 \log^+ 1/r & \text{if } d = 4 \\ r^4 & \text{if } d > 4 \end{cases}.$$

Use Theorem III.5.7(c) to show that $\tilde{\psi}_d - m(\overline{\mathcal{R}}([\delta,\infty))) < \infty \ \forall \delta > 0 \ \mathbb{P}_{X_0}$ – a.s. Conclude that $\mathcal{R}$ is $\mathbb{P}_{X_0}$ a.s. Lebesgue null if $d \geq 4$.

Hint. You may assume without loss of generality that $X_0 = \delta_0$. Why?

Proof of Lemma III.5.9. By considering $v = \gamma u$, we may assume without loss of generality that $\gamma = 1$. Radial symmetry shows that (III.5.14) is equivalent to the ordinary differential equation

$$(III.5.21) \qquad u''(r) + \frac{d-1}{r}u'(r) = u(r)^2 \text{ for } r > 1$$

with the associated boundary conditions, or equivalently

$$(III.5.22) \qquad (r^{d-1}u')' = r^{d-1}u(r)^2 \text{ for } r > 1, \quad \lim_{r\to\infty} u(r) = 0, \quad \lim_{r\to 1+} u(r) = \infty.$$

This shows that $r^{d-1}u'$ is non-decreasing and, as it is non-positive by (III.5.11), we conclude that

$$(III.5.23) \qquad -c_0(d) = \lim_{r\to\infty} r^{d-1}u'(r) \leq 0.$$

If $u(r_0) = 0$ for some $r_0 > 1$, then (III.5.12) implies that $X_1(B(0,1)) = 0 \ \mathbb{P}_{\delta_{r_0}}$-a.s. This contradicts the fact that $\mathbb{P}_{\delta_{r_0}}(X_1(B(0,1)) = P^{r_0}(B_1 \in B(0,1)) > 0$ and we have proved that

$$(III.5.24) \qquad u(r) > 0 \text{ for all } r > 1.$$

Integrate (III.5.22) twice, and use (III.5.23) and $u(\infty) = 0$ to obtain the integral equation

$$(III.5.25) \qquad u(r) = c_0(d)\int_r^\infty t^{1-d}dt + \int_r^\infty t^{1-d}\Big(\int_t^\infty s^{d-1}u(s)^2 ds\Big)dt.$$

This shows that $c_0(d) = 0$ if $d \leq 2$ (or else the first term would be infinite), and for $d \geq 3$ the above gives

$$(III.5.26) \qquad u(r) = \frac{c_0(d)}{d-2}r^{2-d} + \int_r^\infty \Big(\frac{r^{2-d} - s^{2-d}}{d-2}\Big)s^{d-1}u(s)^2 ds.$$

We claim that

$$(III.5.27) \qquad c_0(d) > 0 \text{ iff } d \geq 5.$$

Assume first that $d = 3$ or 4. (III.5.26) implies $u(r) \geq \frac{c_0(d)}{d-2} r^{2-d}$, and therefore for some $c' > 0$,

$$u(r) \geq c' \int_{2r}^{\infty} r^{2-d} s^{d-1} \frac{c_0(d)^2}{(d-2)^2} s^{4-2d} ds \geq c' c_0(d)^2 r^{2-d} \int_{2r}^{\infty} s^{3-d} ds.$$

As the last integral is infinite, this shows $c_0(d)$ must be 0. Assume now that $d \geq 5$ and $c_0(d) = 0$. Then (III.5.26) implies

$$(III.5.28) \qquad u(r) \leq \frac{1}{d-2} r^{2-d} \int_r^{\infty} s^{d-1} u(s)^2 ds \ll r^{2-d} \text{ as } r \to \infty,$$

because (III.5.26) implies the above integral is finite. Use this in (III.5.26) to see there is an $r_0 > 1$ such that $r \geq r_0$ implies

$$\begin{aligned} ru(r) &\leq \frac{1}{d-2} r^{3-d} \int_r^{\infty} s^{d-1} u(s)^2 ds \\ &\leq r^{3-d} \int_r^{\infty} su(s) ds \\ &\leq \int_r^{\infty} su(s) ds, \end{aligned}$$

where the above integral is finite by (III.5.28). Now iterate the above inequality as in Gronwall's lemma to see that $ru(r) = 0$ for $r \geq r_0$, contradicting (III.5.24). This completes the proof of (III.5.27).

Assume $d \geq 5$. Then (III.5.26) implies

$$\begin{aligned} r^{d-2} u(r) &= \frac{c_0(d)}{d-2} + \frac{1}{d-2} \int_r^{\infty} r^{d-2} [r^{2-d} - s^{2-d}] s^{d-1} u(s)^2 ds \\ &\equiv \frac{c_0(d)}{d-2} + h(r). \end{aligned}$$

Clearly $h(r) \leq c' \int_r^{\infty} s^{d-1} u(s)^2 ds \to 0$ as $r \to \infty$ because (III.5.26) implies the above integrals are finite. This proves the required result with $c(d) = \frac{c_0(d)}{d-2}$.

Assume $d \leq 4$ so that (III.5.25) and (III.5.27) give

$$(III.5.29) \qquad u(r) = \int_r^{\infty} t^{1-d} \int_t^{\infty} s^{d-5} (s^2 u(s))^2 ds dt.$$

Recall from (III.5.11) that $r^2 u(r) \downarrow L \geq 0$ as $r \to \infty$. Assume $d \leq 3$ and $L = 0$. If $\varepsilon \in (0,1)$, there is an $r_0(\varepsilon) > 1$ so that $r^2 u(r) \leq \varepsilon$ whenever $r \geq r_0$. Now use (III.5.29) to see that for $r \geq r_0$,

$$r^2 u(r) \leq \varepsilon^2 r^2 \int_r^{\infty} \frac{t^{-3}}{4-d} dt = \frac{\varepsilon^2}{(4-d)2} \leq \frac{\varepsilon}{2}.$$

Iterate the above to see that $u(r) = 0$ for all $r \geq r_0$, which contradicts (III.5.24) and hence proves $L > 0$. The fact that $r^2 u(r) \downarrow L$ and (III.5.29) together imply

$$r^2 \int_r^{\infty} t^{1-d} \int_t^{\infty} s^{d-5} ds dt \, (r^2 u(r))^2 \geq r^2 u(r) \geq L^2 r^2 \int_r^{\infty} t^{1-d} \int_t^{\infty} s^{d-5} ds dt,$$

and therefore

$$\frac{1}{2(4-d)}(r^2u(r))^2 \geq r^2u(r) \geq \frac{L^2}{2(4-d)}.$$

Let $r \to \infty$ to see that $L = \frac{L^2}{2(4-d)}$. This implies that $L = 2(4-d)$ (because $L > 0$) and so the result follows for $d \leq 3$.

It remains to consider the 4-dimensional case which appears to be the most delicate. In this case

$$(III.5.30) \qquad w(r) \equiv r^2u(r) \downarrow L = 0$$

because if L were positive the inner integral in (III.5.29) would be infinite. (III.5.26) shows that (recall $c_0(4) = 0$)

$$\begin{aligned} w(r) &= \frac{1}{2}\int_r^\infty s^3u(s)^2ds - \frac{1}{2}\int_r^\infty su(s)^2dsr^2 \\ &= \frac{1}{2}\Big[\int_r^\infty s^{-1}w(s)^2ds - \int_r^\infty s^{-3}w(s)^2dsr^2\Big] \\ (III.5.31) \qquad &\equiv \frac{1}{2}\Big[\int_r^\infty s^{-1}w(s)^2ds - g(r)\Big]. \end{aligned}$$

The monotonicity of w shows that

$$(III.5.32) \qquad g(r) \leq w(r)^2\int_r^\infty s^{-3}dsr^2 = w(r)^2/2 \downarrow 0 \text{ as } r \to \infty.$$

Let

$$\begin{aligned} v(r) \equiv (\log r)w(r) &= \frac{1}{2}\log r\int_r^\infty \frac{w(s)^2}{s}ds - \frac{1}{2}(\log r)g(r) \quad \text{by (III.5.31)} \\ (III.5.33) \qquad &\equiv \frac{1}{2}h(r) - \frac{1}{2}(\log r)g(r) \end{aligned}$$

Note that by (III.5.32),

$$(III.5.34) \qquad \frac{\frac{1}{2}(\log r)g(r)}{v(r)} = \frac{1}{2}\frac{g(r)}{w(r)} \leq \frac{1}{4}w(r) \to 0 \text{ as } r \to \infty,$$

and so

$$(III.5.35) \qquad \lim_{r\to\infty} \frac{\frac{1}{2}h(r)}{v(r)} = 1.$$

Now

$$\begin{aligned} h'(r) &= \frac{1}{r}\int_r^\infty \frac{w(s)^2}{s}ds - \frac{(\log r)w(r)^2}{r} \\ &= \frac{1}{r}\Big[2w(r) + g(r) - (\log r)w(r)^2\Big] \quad (\text{by } (III.5.31)) \\ &= \frac{w(r)}{r}\Big[2 + \frac{g(r)}{w(r)} - \frac{1}{2}h(r) + \frac{1}{2}(\log r)g(r)\Big]. \end{aligned}$$

(III.5.34) and (III.5.35) imply $\frac{1}{2}(\log r)g(r) = \varepsilon(r)h(r)$, where $\lim_{r\to\infty} \varepsilon(r) = 0$, and (III.5.32) shows that $d(r) = \frac{g(r)}{w(r)} \to 0$ as $r \to \infty$. We can therefore rewrite the above as

$$(III.5.36) \qquad h'(r) = \frac{w(r)}{r} a(r)[b(r) - h(r)],$$

where $\lim_{r\to\infty} a(r) = \frac{1}{2}$, and $\lim_{r\to\infty} b(r) = 4$.

We claim that $\lim_{r\to\infty} h(r)$ exists in $[0, \infty)$. If $h(r) > 4$ for large enough r this is clear since h is eventually decreasing (by (III.5.36)) and bounded below. In a similar way the claim holds if $h(r) < 4$ for large enough r. Assume therefore that $h(r) \le 4$ for some arbitrarily large r and $h(r) \ge 4$ for some arbitrarily large values of r. We claim that $\lim_{r\to\infty} h(r) = 4$. Let $\varepsilon > 0$ and suppose $\limsup_{r\to\infty} h(r) > 4+\varepsilon$. We may choose $r_n \uparrow \infty$ and $s_n \in (r_n, r_{n+1})$ so that $h(r_n) \ge 4+\varepsilon$ and $h(s_n) \le 4$ and then choose $u_n \in [s_{n-1}, s_n]$ so that h has a local maximum at u_n and $h(u_n) \ge 4+\varepsilon$. This implies $h'(u_n) = 0$ which contradicts (III.5.36) for n sufficiently large. We have proved that $\limsup_{r\to\infty} h(r) \le 4$. A similar argument shows that $\liminf_{r\to\infty} h(r) \ge 4$. In this way the claim is established. This together with (III.5.33) and (III.5.34) shows that

$$(III.5.37) \qquad L = \lim_{r\to\infty} v(r) \text{ exists in } \mathbb{R}_+.$$

An argument similar to that for $d \le 3$, and using (III.5.29), shows that $L > 0$.

We can write (III.5.33) as

$$(III.5.38) \qquad v(r) = \frac{1}{2} \log r \int_r^\infty \frac{v(s)^2}{(\log s)^2 s} ds - \varepsilon(r),$$

where $\lim_{r\to\infty} \varepsilon(r) = 0$ ($\varepsilon(r) \ge 0$) by (III.5.34) and (III.5.37). If $L > \varepsilon > 0$, there is an $r_0 > 1$ such that for $r \ge r_0$,

$$\frac{1}{2}(\log r)(L-\varepsilon)^2 \int_r^\infty (\log s)^{-2} s^{-1} ds - \varepsilon \le v(r) \le \frac{1}{2}(\log r)(L+\varepsilon)^2 \int_r^\infty (\log s)^{-2} s^{-1} ds.$$

Let $r \to \infty$ and then $\varepsilon \downarrow 0$ to see that $L = \frac{1}{2}L^2$ and so $L = 2$. The result for $d = 4$ follows. ∎

We next consider the fixed time analogue of Theorem III.5.7. Let $\frac{\Delta}{2}$ continue to denote the generator of Brownian motion. Recall from Theorem II.5.9 and Example II.2.4(a) that if $\phi \in C_b^2(\mathbb{R}^d)_+$ and $V_t\phi \ge 0$ is the unique solution of

$$(SE)_{\phi,0} \qquad \frac{\partial V}{\partial t} = \frac{\Delta V_t}{2} - \frac{\gamma}{2} V_t^2, \quad V_0 = \phi,$$

then

$$(LE) \qquad \mathbb{P}_{X_0}(\exp(X_t(\phi))) = \exp(-X_0(V_t\phi)).$$

Theorem III.5.11. Let $d \geq 3$. There is a constant C_d such that for all $X_0 \in M_F(\mathbb{R}^d)$, all $t \geq \varepsilon^2 > 0$, and all $x \in \mathbb{R}^d$,

$$\begin{aligned}\mathbb{P}_{X_0}(X_t(B(x,\varepsilon)) > 0) &\leq \frac{C_d}{\gamma}\int p_{\varepsilon^2+t}(y-x)X_0(dy)\varepsilon^{d-2}\\ &\leq \frac{C_d}{\gamma}t^{-d/2}X_0(\mathbb{R}^d)\varepsilon^{d-2}.\end{aligned}$$

Proof. Since $\frac{X_t}{\gamma}$ is a SBM(1) starting at X_0/γ (check the martingale problem as in Exercise II.5.5) we clearly may assume $\gamma = 1$. By translation invariance we may assume $x = 0$. Let $\phi \in C_b^2(\mathbb{R}^d)_+$ be a radially symmetric function such that $\{\phi > 0\} = B(0,1)$, let $\phi_\varepsilon(x) = \phi(x/\varepsilon)$ and let $v^{\lambda,\varepsilon}(t,x) \geq 0$ be the unique solution of $(SE)_{\lambda\phi_\varepsilon,0}$ from Theorem II.5.11. By scaling we have

$$v^{\lambda,\varepsilon}(t,x) = \varepsilon^{-2}v^{\lambda\varepsilon^2,1}(t\varepsilon^{-2}, x\varepsilon^{-1}) \equiv \varepsilon^{-2}v^{\lambda\varepsilon^2}(t\varepsilon^{-2}, x\varepsilon^{-1}).$$

By (LE),

$$(III.5.39)\quad 1-\mathbb{P}_{X_0}(\exp(-\lambda X_t(\phi_\varepsilon))) = 1-\exp\left(-\int \varepsilon^{-2}v^{\lambda\varepsilon^2}(t\varepsilon^{-2}, x\varepsilon^{-1})X_0(dx)\right).$$

The left-hand side is increasing in λ, and so by taking $X_0 = \delta_x$ we see that $v^\lambda(t,x) \uparrow v^\infty(t,x) \leq \infty$. Let $\lambda \to \infty$ in (III.5.39) to conclude that

$$(III.5.40)\quad \mathbb{P}_{X_0}(X_t(B(0,\varepsilon)) > 0) = 1-\exp\left(-\int \varepsilon^{-2}v^\infty(t\varepsilon^{-2}, x\varepsilon^{-1})X_0(dx)\right)$$

$$(III.5.41)\quad \leq \varepsilon^{-2}\int v^\infty(t\varepsilon^{-2}, x\varepsilon^{-1})X_0(dx).$$

We therefore require a good upper bound on v^∞. A comparison of (III.5.40), with $X_0 = \delta_x$, $\varepsilon = 1$, and the extinction probability (II.5.12) shows that

$$(III.5.42)\quad v^\infty(t,x) \leq \frac{2}{t}.$$

To get a better bound for small t and $|x| > 1$ we let $r > 1$ and suppose

$$(III.5.43)\quad \text{there exist } t_n \downarrow 0 \text{ such that } \sup_{|x|\geq r} v^\infty(t_n, x) \to \infty.$$

Then (III.5.40) implies that $\lim_{n\to\infty}\mathbb{P}_{\delta_{x_n}}(X_{t_n}(B(0,1)) > 0) = 1$ for some $|x_n| \geq r$. Therefore by translation invariance,

$$\begin{aligned}\liminf_{n\to\infty}\mathbb{P}_{\delta_0}(X_{t_n}(B(0,r-1)^c) > 0) &= \liminf_{n\to\infty}\mathbb{P}_{\delta_{x_n}}(X_{t_n}(B(x_n,r-1)^c) > 0)\\ &\geq \liminf_{n\to\infty}\mathbb{P}_{\delta_{x_n}}(X_{t_n}(B(0,1)) > 0) = 1.\end{aligned}$$

On the other hand our historical modulus of continuity (recall Corollary III.1.5) shows the left-hand side of the above is 0. Therefore (III.5.43) must be false. This with (III.5.42) proves that

$$(III.5.44) \qquad M(r) \equiv \sup_{|x|\geq r, t>0} v^\infty(t,x) < \infty \quad \forall r > 1.$$

Proposition II.5.12 gives the Feynman-Kac representation

$$(III.5.45) \qquad v^\lambda(t,x) = E^x\Big(\lambda\phi(B_t)\exp\Big(\frac{-1}{2}\int_0^t v^\lambda(t-s,B_s)\,ds\Big)\Big).$$

Use the strong Markov property at $T_r = \inf\{t : |B_t| = r\}$ to see that (III.5.45) implies that if $|x| \geq r > 1$, then

$$\begin{aligned} v^\lambda(t,x) =& E^x\Big(1(T_r < t)\exp\Big(\frac{-1}{2}\int_0^{T_r} v^\lambda(t-s,B_s)\,ds\Big) \\ & \times E^{B_{T_r}}\Big(\lambda\phi(B_{t-T_r})\exp\Big(\frac{-1}{2}\int_0^{t-T_r} v^\lambda(t-T_r-s,B_s)\,ds\Big)\Big)\Big) \\ \leq & E^x(1(T_r < t)v^\lambda(t-T_r,B_{T_r})). \end{aligned}$$

Let $\lambda \to \infty$ and use Monotone Convergence in the above to see that

$$(III.5.46) \qquad v^\infty(t,x) \leq E^x(1(T_r < t)v^\infty(t-T_r,B_{T_r})), \quad |x| \geq r > 1.$$

If we replace T_r by the deterministic time $t-s$ $(0 \leq s \leq t)$ in the above argument we get

$$(III.5.47) \qquad v^\infty(t,x) \leq P_{t-s}(v^\infty(s,\cdot))(x).$$

Combine (III.5.46) with (III.5.44) to see that for $|x| \geq 7 > r = 2$,

$$\begin{aligned} v^\infty(1,x) &\leq M(2)P^x(T_2 < 1) \\ &\leq 2M(2)P^0(|B_1| > |x| - 2) \text{ (by a } d\text{-dimensional reflection principle)} \\ &\leq c_1\exp(-|x|^2/4), \end{aligned}$$

where we used our bound $|x| > 7$ in the last line. Together with (III.5.42), this gives

$$v^\infty(1,x) \leq c_2 p(2,x) \text{ for all } x \in \mathbb{R}^d,$$

and so (III.5.47) with $s = 1$ implies

$$v^\infty(t,x) \leq c_2 p(t+1,x) \text{ for all } t \geq 1 \text{ and } x \in \mathbb{R}^d.$$

Use this in (III.5.41) to conclude that for $t \geq \varepsilon^2$,

$$\begin{aligned} \mathbb{P}_{X_0}(X_t(B(0,\varepsilon)) > 0) &\leq c_2\varepsilon^{-2}\int p(t\varepsilon^{-2}+1, x\varepsilon^{-1})X_0(dx) \\ &= c_2\varepsilon^{d-2}\int p(t+\varepsilon^2,x)X_0(dx). \end{aligned}$$

This gives the first inequality and the second inequality is then immediate. ∎

A corresponding lower bound is now left as an exercise.

Exercise III.5.2. Use Lemma III.5.5 and our first and second moment formulae (Exercise II.5.2) to prove:
(a) If $d \geq 3$, for any $K \in \mathbb{N}$, $\delta > 0$ $\exists \varepsilon_0(\delta, K) > 0$ and a universal constant $c_d > 0$ so that whenever $\frac{X_0(1)}{\gamma} \leq K$,

$$\mathbb{P}_{X_0}(X_t(B(x,\varepsilon)) > 0) \geq \frac{c_d}{\gamma} \int p_t(y-x) X_0(dy) \varepsilon^{d-2} \ \forall 0 < \varepsilon < \varepsilon_0(\delta, K),\ t \geq \delta.$$

(b) If $d = 2$, show the conclusion of (a) holds with $(\log 1/\varepsilon)^{-1}$ in place of ε^{d-2} and the additional restriction $t \leq \delta^{-1}$.
Hints: (1) Draw a picture or two to convince yourself that

$$P^y(B_t \in B(x,\varepsilon)) \geq c'_d \varepsilon^d p_t(y-x) \quad \forall t \geq \varepsilon^2.$$

(2) If $B = B(0,\varepsilon)$, then $P_s((P_{t-s}1_B)^2) \leq c''_d \left(\frac{\varepsilon^d}{(t-s)^{d/2}} \wedge 1\right) P_t 1_B$.

Remark III.5.12. (a) If $d \geq 3$, then Theorem III.5.11 and the above Exercise give sharp bounds on $\mathbb{P}_{X_0}(X_t(B(x,\varepsilon)) > 0$ as $\varepsilon \downarrow 0$ except for the value of the constants C_d and c_d. Theorem 3.1 of Dawson-Iscoe-Perkins (1989) shows that there is a universal constant $c(d) > 0$ such that

$$(III.5.48) \qquad \lim_{\varepsilon \downarrow 0} \varepsilon^{2-d} \mathbb{P}_{X_0}(X_t(B(x,\varepsilon)) > 0) = c(d) \int p_t(y-x) X_0(dy).$$

If $d = 2$, a companion upper bound to Example III.5.2(b)

$$\mathbb{P}_{X_0}(X_t(B(x,\varepsilon)) > 0) \leq C_2 t^{-1} X_0(1) |\log \varepsilon|^{-1} \quad \forall t \in [\varepsilon, \varepsilon^{-1}]\ \forall \varepsilon \in (0, 1/2)$$

is implied by Corollary 3 of Le Gall (1994). A version of (III.5.48) has not been obtained in this more delicate case.
(b) The analogue of Theorem III.5.2 for fixed times is
$(III.5.49)$
$A \cap S(X_t) = \emptyset\ \mathbb{P}_{X_0} - \text{a.s. iff } C(g_{d-2})(A) = 0, \quad \forall A \in \mathcal{B}(\mathbb{R}^d),\ t > 0,\ d \geq 1,\ X_0 \neq 0.$

The reader may easily prove the necessity of the capacity condition for fixed time polarity by means of a straightforward adaptation of the proof of Theorem III.5.4. This result was first proved in Perkins (1990) (see Theorem 6.1). As in Corollary III.5.10, Theorem III.5.11, and Le Gall's companion upper bound if $d = 2$, readily show that for $d \geq 2$, and all $t > 0$ and $A \in \mathcal{B}(\mathbb{R}^d)$,

$$(III.5.50) \qquad f_{d-2} - m(A) = 0 \Rightarrow A \cap S(X_t) = \emptyset\ \mathbb{P}_{X_0} - \text{a.s.}\ .$$

This is of course weaker than the sufficiency of the capacity condition in (III.5.49) which is again due to Dynkin (1992).

Parts (a) and (b) of the following Exercise will be used in the next Section.

Exercise III.5.3. Let X^1, X^2 be independent SBM's with branching rate γ which start at $X_0^1, X_0^2 \in M_F(\mathbb{R}^d) - \{0\}$.
(a) If $d \geq 5$ use Theorem III.5.11 to show that $S(X_t^1) \cap S(X_t^2) = \emptyset$ a.s.
(b) Prove that the conclusion of (a) remains valid if $d = 4$.

Hint. One way to handle this critical case is to first apply Theorem III.3.8 (which has not been proved in these notes) to X^1.

(c) If $d \le 3$ show that $P(S(X_t^1) \cap S(X_t^2) \neq \emptyset) > 0$.

Hint. Use the necessity of the capacity condition in (III.5.49) together with (III.5.2) (the connection between capacity and Hausdorff measure) and Corollary III.3.5 (the latter results are only needed if $d > 1$).

6. Disconnectedness of the Support

Our goal in this Section is to study the disconnectedness properties of the support of super-Brownian motion. The results seem to be scattered in the literature, often with proofs that are only sketched, and we will try to collect them in this section and give a careful proof of the main result (Theorem 6.3). This also gives us the opportunity to advertise an intriguing open problem. The historical clusters of Theorem III.1.1 will be used to disconnect the support and we start by refining that result.

Assume we are in the setting of Section II.8: H is the $(Y, \gamma, 0)$-historical process on the canonical space of paths $\Omega_H[\tau, \infty)$ with law $\mathbb{Q}_{\tau,m}$, and γ is a positive constant. Recall the canonical measures $\{R_{\tau,t}(y,\cdot) : t > \tau, y \in D^\tau\}$ associated with H from (II.8.6) and set

$$P^*_{\tau,t}(y, A) = \frac{R_{\tau,t}(y, A)}{R_{\tau,t}(y, M_F(D) - \{0\})} = \frac{\gamma(t-\tau)}{2} R_{\tau,t}(y, A).$$

From Exercise II.7.2 we may interpret $P^*_{\tau,t}$ as the law of H_t starting from an infinitesimal point mass on y at time τ and conditioned on non-extinction at time t.

If ν is a probability on $M_F^\tau(D)$ we abuse our notation slightly and write $\mathbb{Q}_{\tau,\nu}$ for $\int \mathbb{Q}_{\tau,m}\nu(dm)$ and also adopt a similar convention for the laws $\mathbb{P}_\mu$ (μ a probability on $M_F(E)$) of the corresponding superprocess. If $\tau \le s \le t$, define

$$r_{s,t}(H_t) \in M_F(D^s \times D) \text{ by } r_{s,t}(H_t)(A) = H_t(\{y : (y^s, y) \in A\}).$$

Theorem III.6.1. Let $m \in M_F^\tau(D) - \{0\}$, $t > \tau$ and let Ξ be a Poisson point process on $D^\tau \times \Omega_H[t, \infty)$ with intensity

$$\mu(A \times B) = \int 1_A(y) \mathbb{Q}_{t, P^*_{\tau,t}(y,\cdot)}(B) \frac{2}{\gamma(t-\tau)} m(dy).$$

Then (under $\mathbb{Q}_{\tau,m}$)

(a) $r_{\tau,t}(H_t) \overset{\mathcal{D}}{\equiv} \int \delta_y \times \nu_t \Xi(dy, d\nu)$,

(b) $(H_u)_{u \ge t} \overset{\mathcal{D}}{\equiv} \left(\int \nu_u \Xi(dy, d\nu) \right)_{u \ge t}$.

Proof. (a) Let $A_1, \ldots, A_n$ be a Borel partition of D, let $\phi_i : D \to \mathbb{R}_+$ be bounded Borel maps $i = 1, \ldots, n$, and define $H_t^i(\cdot) = H_t(\cdot \cap \{y^\tau \in A_i\})$ and $f : D^\tau \times D \to \mathbb{R}_+$

by $f(y,w)=\sum_{i=1}^n 1_{A_i}(y)\phi_i(w)$. Then (III.1.3) and (II.8.6)(b) imply

$$\begin{aligned}
&\mathbb{Q}_{\tau,m}(\exp(-\int f(y^\tau,y)H_t(dy))\\
&=\mathbb{Q}_{\tau,m}\left(\exp\left(-\sum_{i=1}^n H_t^i(\phi_i)\right)\right)\\
&=\prod_{i=1}^n \mathbb{Q}_{\tau,m}(\exp(-H_t^i(\phi_i)))
\end{aligned}$$

$$(III.6.1)\qquad =\exp\left(-\sum_{i=1}^n\int\int 1-e^{-\nu(\phi_i)}\frac{2}{\gamma(t-\tau)}P^*_{\tau,t}(y,d\nu)1_{A_i}(y)dm(y)\right).$$

On the other hand

$$\begin{aligned}
&E\left(\exp\left(-\int\int f(y,w)\nu_t(dw)\Xi(dy,d\nu)\right)\right)\\
&=E(\exp\left(-\int(1-\exp(-\int f(y,w)\nu_t(dw)))\mu(dy,d\nu)\right)\\
&=\exp\left(-\int\int 1-\exp(-\sum_{i=1}^n 1_{A_i}(y)\nu_t(\phi_i))\frac{2}{\gamma(t-\tau)}P^*_{\tau,t}(y,d\nu_t)m(dy)\right)
\end{aligned}$$

$$(III.6.2)\ =\exp\left(-\sum_{i=1}^n\int\int 1-e^{-\nu(\phi_i)}\frac{2}{\gamma(t-\tau)}P^*_{\tau,t}(y,d\nu)1_{A_i}(y)dm(y)\right).$$

As (III.6.1) and (III.6.2) are equal, the two random measures in (a) are equal in law by Lemma II.5.9.

(b) Let $\{y_i,H^i):i\le N\}$ be the points of Ξ. Then N is Poisson with mean $\frac{2m(D)}{\gamma(t-\tau)}$, given N, $\{y_i:i\le N\}$ are i.i.d with law $m/m(D)$, and

(III.6.3) given $\sigma(N,(y_i)_{i\le N})$, $\{H^i:i\le N\}$ are independent random processes with H^i having law $\mathbb{Q}_{t,P^*_{\tau,t}(y_i,\cdot)}$.

Therefore conditional on $\sigma(N,(y_i)_{i\le N},(H_t^i)_{i\le N})$, $\bar{H}_\cdot=\sum_{i=1}^N H^i_\cdot$ is a sum of N independent historical processes. Such a sum will clearly satisfy $(HMP)_{t,\bar{H}_t(\omega)}$ and therefore be a historical process itself (this is the multiplicative property of superprocesses). Therefore the conditional law of $(\bar{H}_u)_{u\ge t}$ given $\sigma(N,(y_i)_{i\le N},(H_t^i)_{i\le N})$ is $\mathbb{Q}_{t,\bar{H}_t}$. Part (a) implies that $\bar{H}_t$ has law $\mathbb{Q}_{\tau,m}(H_t\in\cdot)$. The Markov property of H under $\mathbb{Q}_{\tau,m}$ now shows that $(\bar{H})_{u\ge t}$ has law $\mathbb{Q}_{\tau,m}((H_u)_{u\ge t}\in\cdot)$ which is what we have to prove. ∎

We now reinterpret the above result directly in terms of the historical process H on its canonical space $(\Omega_H[\tau,\infty),\mathbb{Q}_{\tau,m})$. We in fact assume that H is the historical process of a super-Brownian motion with constant branching rate γ, although the result and its proof remain valid for any historical process such that H_s has no atoms $\mathbb{Q}_{\tau,m}$-a.s. for any $s>\tau$.

If $\tau\le s<t$ and $y\in D^s$, let $H_t^{s,y}(\cdot)=H_t(\{w\in\cdot:w^s=y\})$, i.e., $H_t^{s,y}$ is the contribution to H_t from descendants of y at time s.

Corollary III.6.2. Let $m \in M_F^\tau(D)$ and $\tau \le s < t$. Assume either $\tau < s$ or m is non-atomic. Then under $\mathbb{Q}_{\tau,m}$:
(a) Conditional on $\mathcal{F}^H[\tau, s+]$, $S(r_s(H_t))$ is the range of a Poisson point process with intensity $\frac{2H_s(\omega)}{\gamma(t-s)}$.
(b) $H_u = \sum_{y \in S(r_s(H_t))} H_u^{s,y}$ for all $u \ge t$ a.s.
(c) Conditional on $\mathcal{F}^H[\tau, s+] \vee \sigma(S(r_s(H_t)))$, $\{(H_u^{s,y})_{u\ge t} : y \in S(r_s(H_t))\}$ are independent processes and for each $y \in S(r_s(H_t))$, $(H_u^{s,y})_{u \ge t}$ has (conditional) law $\mathbb{Q}_{t,P_{s,t}^*(y,\cdot)}$.

Proof. (a) is included in Theorem III.1.1.
(b) Lemma III.1.2 and the Markov property show that $H_u(S(r_s(H_t))^c)_{u\ge t}$ is a continuous martingale starting at 0 and so is identically 0 a.s. (b) now follows from the definition of $H_u^{s,y}$.
(c) Theorem III.3.4 and Exercise II.8.3 show that H_s is non-atomic a.s (use Theorem III.3.8(c) if $d = 1$). Therefore by the Markov property we may assume without loss of generality that $s = \tau$ and m is non-atomic. We must show that conditional on $S(r_\tau(H_t))$,

$$\{(H_u^{\tau,y})_{u\ge t} : y \in S(r_\tau(H_t))\}$$

are independent, and, for each $y \in S(r_\tau(H_t))$, $(H_u^{\tau,y})_{u\ge t}$ has (conditional) law $\mathbb{Q}_{t,P_{\tau,t}^*(y,\cdot)}$. As this will depend only on the law of $(H_u)_{u\ge t}$, by Theorem III.6.1(b) we may assume that Ξ is as in that result and

$$(III.6.4) \qquad H_u = \int \nu_u \Xi(dy, d\nu) \text{ for all } u \ge t.$$

Let $\{(y_i, H^i) : i \le N\}$ be the points of Ξ (as in the proof of Theorem III.6.1(b)) so that (III.6.4) may be restated as

$$(III.6.5) \qquad H_u = \sum_{i=1}^N H_u^i \text{ for all } u \ge t.$$

Theorem III.6.1(a) implies that $\{y_i : i \le N\} = S(r_\tau(H_t))$ and the fact that m is non-atomic means that all the y_i's are distinct a.s. By (II.8.6)(a), $\nu(y^\tau \ne y_i) = 0$ $P_{\tau,t}^*(y_i,\cdot)$-a.a. ν and so $H_t^i(\{y : y^\tau \ne y_i\}) = 0$ a.s. As in the proof of (b) we may conclude that $H_u^i(\{y : y^\tau \ne y_i\}) = 0$ for all $u \ge t$ a.s. This shows that

$$\begin{aligned} H_u^i(\cdot) &= H_u^i(\cdot, y^\tau = y_i) \text{ for all } u \ge t \text{ a.s.} \\ &= H_u(\cdot, y^\tau = y_i) \text{ for all } u \ge t \text{ a.s. (by (III.6.5) and } y_i \ne y_j \text{ if } i \ne j \text{ a.s.)} \\ &= H_u^{\tau, y_i}(\cdot) \text{ for all } u \ge t \text{ a.s.} \end{aligned}$$

The required result now follows from (III.6.3). ∎

We are ready for the main result of this section (a sketch of this proof was given in Perkins (1995b)). In the rest of this Section we assume $\big(X, (\mathbb{P}_\mu)_{\mu\in M_F(E)}\big)$ is a super-Brownian motion with constant branching rate γ and $\big(H, (\mathbb{Q}_{\tau,m})_{(\tau,m)\in\hat{E}}\big)$ is the corresponding historical process.

Theorem III.6.3. If $d \ge 4$, $S(X_t)$ is totally disconnected $\mathbb{P}_{X_0}$-a.s. for each $t > 0$.

Proof. By Exercise II.5.1 (or II.5.3) we may work with $X_t = \tilde{\Pi}(H)_t$ under $\mathbb{Q}_{0,X_0}$. If $t > 0$ is fixed, $0 \le s < t$ and $y \in S(r_s(H_t))$, let $X_t^{s,y}(A) = H_t^{s,y}(y_t \in A) \equiv H_t^{s,y} \circ \hat{\Pi}_t^{-1}(A)$, where $\hat{\Pi}_t$ is the obvious projection map. Let $\varepsilon_n \downarrow 0$, where $0 < \varepsilon_n < t$. Corollary III.6.2(b) implies

$$(III.6.6) \qquad S(X_t) = \cup_{y \in S(r_{t-\varepsilon_n}(H_t))} S(X_t^{t-\varepsilon_n,y}) \ \forall n \ \text{a.s.}$$

By Corollary III.6.2(c) and (II.8.7), conditional on $\mathcal{F}^H([0,t-\varepsilon_n]) \vee \sigma(S(r_{t-\varepsilon_n}(H_t)))$, $\{X_t^{t-\varepsilon_n,y} : y \in S(r_{t-\varepsilon_n}(H_t))\}$ are independent and $X_t^{t-\varepsilon_n,y}$ has (conditional) law

$$P^*_{t-\varepsilon_n,t}(y, \nu \circ \hat{\Pi}_t^{-1} \in \cdot) = \frac{R_{\varepsilon_n}(y_{t-\varepsilon_n}, \cdot)}{R_{\varepsilon_n}(y_{t-\varepsilon_n}, 1)}.$$

A Poisson superposition of independent copies of $X_t^{t-\varepsilon_n,y}$ has law $\mathbb{P}_{y_{t-\varepsilon_n}}(X_{\varepsilon_n} \in \cdot)$ by (II.7.11) and so Exercise III.5.3 shows that

$$(III.6.7) \qquad \{S(X_t^{t-\varepsilon_n,y}) : y \in S_{t-\varepsilon_n}(H_t)\} \text{ are disjoint for all } n \text{ a.s.}$$

Let $\delta(3,\omega) > 0$ $\mathbb{Q}_{0,X_0}$-a.s. be as in the Historical Modulus of Continuity (Theorem III.1.3). Then that result and Corollary III.6.2(b) show that

$$S(H_t^{t-\varepsilon_n,y}) \subset S(H_t) \subset K(\delta(3,\omega),3)$$

and so by the definition of $X_t^{t-\varepsilon_n,y}$,

$$(III.6.8) \qquad S(X_t^{t-\varepsilon_n,y}) \subset B(y_{t-\varepsilon_n}, 3h(\varepsilon_n)) \ \text{ if } \varepsilon_n \le \delta(3,\omega).$$

Fix ω outside a null set so that (III.6.6), (III.6.7), and (III.6.8) hold, and $\delta(3,\omega) > 0$. Then $S(X_t)$ can be written as the disjoint union of a finite number of closed sets of arbitrarily small diameter and hence is totally disconnected. ∎

If X^1 and X^2 are independent super-Brownian motions starting from X_0^1 and X_0^2, respectively, then (see Theorem IV.3.2(b) below)

$$(III.6.9) \qquad \text{if } d \ge 6,\ S(X_u^1) \cap S(X_u^2) = \emptyset \ \text{ for all } u > 0 \text{ a.s.}$$

Using this in place of Exercise III.5.3 in the above proof we get a version of the above result which holds for all times simultaneously.

Theorem III.6.4. If $d \ge 6$, then $S(X_t)$ is totally disconnected for all $t > 0$ $\mathbb{P}_{X_0}-$ a.s.

Proof. Our setting and notation is that of the previous argument. Corollary III.6.2 implies that

(III.6.10)

$$S(X_u) = \cup_{y \in S(r_{(j-1)\varepsilon_n}(H_{j\varepsilon_n}))} S(X_u^{(j-1)\varepsilon_n,y}) \ \ \forall u \in [(j\varepsilon_n, (j+1)\varepsilon_n] \ \forall j,n \in \mathbb{N} \ \text{ a.s.}$$

Corollary III.6.2(c), Exercise II.8.3 and (II.8.7) show that, conditional on

$$\mathcal{F}^H([0,(j-1)\varepsilon_n]) \vee \sigma(S(r_{(j-1)\varepsilon_n}(H_{j\varepsilon_n}))),$$

$\{(X_u^{(j-1)\varepsilon_n,y})_{u\geq j\varepsilon_n} : y \in S(r_{(j-1)\varepsilon_n}(H_{j\varepsilon_n})\}$ are independent and $(X_{u+j\varepsilon_n})_{u\geq 0}$ has law $\mathbb{P}_{\mu_n(y_{(j-1)\varepsilon_n})}$ where $\mu_n(x,\cdot) = R_{\varepsilon_n}(x,\cdot)/R_{\varepsilon_n}(x,1)$. It therefore follows from (III.6.9) that

$$(III.6.11) \qquad \{S(X_u^{(j-1)\varepsilon_n,y}) : y \in S(r_{(j-1)\varepsilon_n}(H_{j\varepsilon_n}))\} \text{ are disjoint for all } u \in (j\varepsilon_n, (j+1)\varepsilon_n]\ \forall j, n \in \mathbb{N} \text{ a.s.}$$

Finally use the Historical Modulus of Continuity and Corollary III.6.2(b), as in the proof of Theorem III.6.3 to see that

$$(III.6.12)\ S(X_u^{(j-1)\varepsilon_n,y}) \subset B(y_{(j-1)\varepsilon_n}, 6h(\varepsilon_n)) \text{ forall } u \in [j\varepsilon_n, (j+1)\varepsilon_n] \ \forall j \in \mathbb{N} \text{ and } n \text{ such that } 2\varepsilon_n < \delta(3,\omega).$$

As before (III.6.10)-(III.6.12) show that with probability one, for all $u > 0$, $S(X_u)$ may be written as a finite disjoint union of closed sets of arbitrarily small diameter and so is a.s. totally disconnected. ∎

In one spatial dimension the existence of a jointly continuous density for X (see Section III.3.4) shows that the closed support cannot be totally disconnected for any positive time with probability one. This leaves the

Open Problem. In two or three dimensions, is the support of super-Brownian motion a.s. totally disconnected at a fixed time?

Nothing seems to be known in two dimensions and the only result in this direction for three dimensions is

Theorem 6.5. (Tribe (1991)) Let Comp(x) denote the connected component of $S(X_t)$ containing x. If $d \geq 3$, then Comp(x) $= \{x\}$ for X_t-a.a. x $\mathbb{P}_{X_0}$-a.s. for each $t > 0$.

Tribe's result leaves open the possibility that there is a non-trivial connected component in $S(X_t)$ having mass 0. The proof considers the history of a particle x chosen according to X_t and decomposes the support at time t into the cousins which break off from this trajectory in $[t-\varepsilon, t]$ and the rest of the population. He then shows that with positive probability these sets can be separated by an annulus centered at x. By taking a sequence $\varepsilon_n \downarrow 0$ and using a zero-one law he is then able to disconnect x from the rest of the support a.s. The status of Theorem 6.5 in two dimensions remains unresolved.

The critical dimension for Theorem III.6.4, i.e., above which the support is totally disconnected for all positive times, is also not known.

7. The Support Process

In this section we give a brief survey of some of the properties of the set-valued process $S(X_t)$. Let $\mathcal{K}$ be the set of compact subsets of $\mathbb{R}^d$. For non-empty K_1, $K_2 \in \mathcal{K}$, let

$$\rho_1(K_1,K_2) = \sup_{x\in K_1} d(x,K_2) \wedge 1,$$

$$\rho(K_1,K_2) = \rho_1(K_1,K_2) + \rho_1(K_2,K_1),$$

and set $\rho(K,\phi) = 1$ if $K \neq \phi$. ρ is the Hausdorff metric on $\mathcal{K}$ and $(\mathcal{K},\rho)$ is a Polish space (see Dugundji (1966), p. 205,253).

Assume X is SBM(γ) under $\mathbb{P}_{X_0}$ and let $S_t = S(X_t)$, $t \geq 0$. By Corollary III.1.4, $\{S_t : t > 0\}$ takes values in $\mathcal{K}$ a.s. Although the support map $S(\cdot)$ is not continuous on $M_F(\mathbb{R}^d)$, an elementary consequence of the weak continuity of X is that

$$(III.7.1) \qquad \lim_{t \to s} \rho_1(S_s, S_t) = 0 \quad \forall s > 0 \quad \text{a.s.}$$

On the other hand the Historical Modulus of Continuity (see Corollary III.1.5) shows that if $0 < t - s < \delta(\omega, 3)$

$$\rho_1(S_t, S_s) = \sup_{x \in S_t} d(x, S_s) \wedge 1 \leq 3h(t-s)$$

and so

$$(III.7.2) \qquad \lim_{t \downarrow s} \rho_1(S_t, S_s) = 0 \quad \forall s > 0 \quad \text{a.s.}$$

(III.7.1) and (III.7.2) show that $\{S_t : t > 0\}$ is a.s. right-continuous in $\mathcal{K}$. The a.s. existence of left limits is immediate from Corollary III.1.5 and the following simple deterministic result (see Lemma 4.1 of Perkins (1990)):

If $f : (0, \infty) \to \mathcal{K}$ is such that $\forall \varepsilon > 0 \; \exists \delta > 0$ so that
$0 \leq t - u < \delta$ implies $f(u) \subset f(t)^\varepsilon \equiv \{x : d(x, f(t)) < \varepsilon\}$,
then f possesses left and right limits at all $t > 0$.

(III.7.1) shows that $S_s \subset S_{s-}$ for all $s > 0$ a.s. When an "isolated colony" becomes extinct at time s at location F one expect $F \in S_{s-} - S_s$. These extinction points are the only kind of discontinuities which arise. Theorem 4.6 of Perkins (1990) shows that

$$\text{card}(S_{t-} - S_t) = 0 \quad \text{or} \quad 1 \quad \text{for all} \quad t > 0 \quad \text{a.s.}$$

The nonstandard proof given there may easily be translated into one use the historical process. For $d \geq 3$ the countable space-time locations of these extinction points are dense in the closed graph of X,

$$G_0(X) = \overline{\{(t, x) : x \in S_t, t \geq 0\}} = \{(t, x) : x \in S_{t-}, t > 0\} \cup \{0\} \times S_0$$

(see Theorem 4.8 of Perkins (1990)).

Of course if $S_0 \in \mathcal{K}$, the above arguments show that $\{S_t : t \geq 0\}$ is cadlag in $\mathcal{K}$ a.s. Assume now $S(X_0) \in \mathcal{K}$. Theorem III.3.8 suggests that in 2 or more dimensions the study of the measure-valued process X reduces to the study of the $\mathcal{K}$-valued process $S_t = S(X_t)$, as X_t is uniformly distributed over S_t according to a deterministic Hausdorff measure at least for Lebesgue a.a. t a.s. If

$$F = \{S \in \mathcal{K} : h_d - m(S) < \infty\},$$

then, as one can use finite unions of "rational balls" in the definition of $h_d - m$ on compact sets, it is clear that F is a Borel subset of $\mathcal{K}$ and $\Psi : F \to M_F(\mathbb{R}^d)$, given by $\Psi(S)(A) = h_d - m(S \cap A)$, is Borel. The support mapping $S(\cdot)$ is also a Borel function from $M_F^K(\mathbb{R}^d)$, the set of measures with compact support, to $\mathcal{K}$. Theorem III.3.8 implies that for $s, t > 0$ and $A \in \mathcal{B}(F)$

$$\begin{aligned}\mathbb{P}_{X_0}(S_{t+s} \in A \mid \mathcal{F}_t^X) &= \mathbb{P}_{X_t}(S_s \in A) \\ &= \mathbb{P}_{\Psi(S_t)}(S_s \in A)\end{aligned}$$

and so $\{S_t : t > 0\}$ is a cadlag F-valued Markov process. This approach however, does not yield the strong Markov property. For this we would need a means of recovering X_t from $S(X_t)$ that is valid for all $t > 0$ a.s. and although Theorem III.3.8 (b) comes close in $d \geq 3$, its validity for all $t > 0$ remains unresolved. Another approach to this question was initiated by Tribe (1994).

Notation. $d \geq 3$ $\quad X_t^\varepsilon(A) = |S(X_t)^\varepsilon \cap A|\varepsilon^{2-d}$, $\quad A \in \mathcal{B}(\mathbb{R}^d)$. Here $|\cdot|$ is Lebesgue measure.

The a.s. compactness of $S(X_t)$ shows $X_t^\varepsilon \in M_F(\mathbb{R}^d)$ $\forall t > 0$ a.s.

Theorem III.7.1. (Perkins (1994)) Assume $d \geq 3$. There is a universal constant $c(d) > 0$ such that $\lim_{\varepsilon \downarrow 0} X_t^\varepsilon = c(d)X_t$ $\forall t > 0$ $\mathbb{P}_{X_0}$-a.s. In fact if ϕ is a bounded Lebesgue-integrable function and $r < \frac{2}{d+2}$ then $\mathbb{P}_{X_0}$-a.s. there is an $\varepsilon_0(\omega) > 0$ so that $\sup_{t \geq \varepsilon^{1/4}} |X_t^\varepsilon(\phi) - c(d)X_t(\phi)| \leq \varepsilon^r$ for $0 < \varepsilon < \varepsilon_0$.

Remark. $c(d)$ is the constant given in (III.5.21) below which determines the asymptotic behaviour of $\mathbb{P}_{X_0}(X_t(B(x,\varepsilon)) > 0)$ as $\varepsilon \downarrow 0$.

It is now easy to repeat the above reasoning with the above characterization of X_t in place of the Hausdorff measure approach to see that $t \to S(X_t)$ is a Borel strong Markov process with cadlag paths.

Notation. $\Phi_\varepsilon : \mathcal{K} \to M_F(\mathbb{R}^d)$ is given by $\Phi_\varepsilon(S)(A) = |A \cap S^\varepsilon|\varepsilon^{2-d}$.

Define $\Phi : \mathcal{K} \to M_F(\mathbb{R}^d)$ by $\Phi(S) = \begin{cases} \lim_{n\to\infty} \Phi_{1/n}(S) & \text{if it exists} \\ 0 & \text{otherwise} \end{cases}$.

$E = \{S' \in \mathcal{K} : S(\Phi(S')) = S'\}$; $\Omega_E = D([0,\infty), E)$ with its Borel σ-field $\mathcal{F}^E$, canonical filtration $\mathcal{F}_t^E$, and coordinate maps S_t.

It is easy to check that Φ_ε is Borel and hence so are Φ and E. If $d \geq 3$, Theorem III.7.1 implies $S(X_t) \in E$ $\forall t > 0$ $\mathbb{P}_{X_0}$-a.s. and so for $S' \in E$ we may define a probability $\mathbb{Q}_{S'}$ on Ω_E by

$$\mathbb{Q}_{S'}(A) = \mathbb{P}_{\Phi(S')}(S(X_\cdot) \in A).$$

Corollary III.7.2. Assume $d \geq 3$. $(\Omega_E, \mathcal{F}^E, \mathcal{F}_t^E, S_t, \mathbb{Q}_{S'})$ is a Borel strong Markov process with right-continuous E-valued paths.

Proof. See Theorem 1.4 of Perkins (1994). ∎

Note. At a jump time of $S(X_t)$ the left limit of the support process will not be in E because

$$S(\Phi(S(X)_{t-})) = S(X_t) \neq S(X)_{t-}.$$

Open Problem. Is $S(X_t)$ strong Markov for $d = 2$?

The potential difficulty here is that $S(X_t)$ could fold back onto itself on a set of positive X_t measure at an exceptional time $t(\omega)$.

IV. Interactive Drifts

1. Dawson's Girsanov Theorem

Our objective is to study measure-valued diffusions which locally behave like DW-superprocesses, much in the same way as solutions to Itô's stochastic differential equations behave locally like Brownian motions. This means that we want to consider processes in which the branching rate, γ, the spatial generator, A, and the drift, g, all depend on the current state of the system, X_t, or more generally on the past behaviour of the system, $X|_{[0,t]}$. One suspects that these dependencies are listed roughly in decreasing order of difficulty. In this Chapter we present a general result of Dawson which, for a large class of interactive drifts, will give an explicit formula for the Radon-Nikodym derivative of law of the interactive model with respect to that of a driftless DW-superprocess.

We will illustrate these techniques with a stochastic model for two competing populations and hence work in a bivariate setting for most of the time. The models will also illustrate the limitations of the method as the interactions become singular. These singular interactions will be studied in the next Section.

Let E_i, Y_i, A_i, and γ_i, $i = 1, 2$ each be as in Theorem II.5.1 and set $\mathcal{E}_i = \mathcal{B}(E_i)$. Let $\Omega_X^i = C(\mathbb{R}_+, M_F(E_i))$ with its Borel σ-field $\mathcal{F}^{X,i}$ and canonical filtration $\mathcal{F}_t^{X,i}$ and introduce the canonical space for our interacting populations,

$$(\Omega^2, \mathcal{F}^2, \mathcal{F}_t^2) = \Big(\Omega_X^1 \times \Omega_X^2, \mathcal{F}^{X,1} \times \mathcal{F}^{X,2}, (\mathcal{F}_\cdot^{X,1} \times \mathcal{F}_\cdot^{X,2})_{t+}\Big).$$

The coordinate maps on Ω^2 will be denoted by $X = (X^1, X^2)$ and $\mathcal{P}$ will be the σ-field of $(\mathcal{F}_t^2)$-predictable sets in $\mathbb{R}_+ \times \Omega^2$. For $i = 1, 2$, let $m_i \in M_F(E_i)$, and g_i denote a $\mathcal{P} \times \mathcal{E}_i$-measurable map from $\mathbb{R}_+ \times \Omega^2 \times E_i$ to $\mathbb{R}$. A probability $\mathbb{P}$ on $(\Omega^2, \mathcal{F}^2)$ will satisfy $(MP)_g^m$ iff

$$\forall \phi_i \in \mathcal{D}(A_i) \quad X_t^i(\phi_i) = m_i(\phi_i) + \int_0^t X_s^i(A_i\phi - g_i(s, X, \cdot)\phi_i)ds + M_t^{i,g_i}(\phi_i)$$

defines continuous $(\mathcal{F}_t^2)$ – martingales $M_t^{i,g_i}(\phi)$ $(i = 1, 2)$ under $\mathbb{P}$ such that

$$M_0^{i,g_i}(\phi_i) = 0 \text{ and } \langle M^{i,g_i}(\phi_i), M^{j,g_j}(\phi_j)\rangle_t = \delta_{ij} \int_0^t X_s^i(\gamma_i \phi_i^2)ds.$$

Implicit in (MP) is the fact that $\int_0^t X_s^i(|g_i(s, X, \cdot)|)ds < \infty$ for all $t > 0$ $\mathbb{P}$-a.s. We have inserted a negative sign in front of g_i only because our main example will involve a negative drift.

Example IV.1.1. (Competing Species) Take $E_i = \mathbb{R}^d$, $A_i = \Delta/2$, $\gamma_i \equiv 1$. If p_t denotes the Brownian density, $\varepsilon > 0$ and $\lambda_i \geq 0$, let

$$g_i^\varepsilon(x, \mu) = \lambda_i \int p_\varepsilon(x - y)\mu(dy).$$

Consider two branching particle systems, $X^N = (X^{1,N}, X^{2,N})$, as in Section II.3 with independent spatial (Brownian) motions and independent critical binary branching mechanisms but with one change. At $t = i/N$ a potential parent of a 1-particle located at x_1 dies before it can reproduce with probability $g_1^\varepsilon(x_1, X_{t-}^{2,N})/N$. Similarly a potential parent in the 2-population located at x_2 dies with probability

Originally published in: *Ecole d'Eté de Probabilités de Saint-Flour XXIX – 1999*, Lecture Notes in Mathematics, Vol. **1781**, 247–280, DOI: 10.1007/3-540-47944-9_8,

$g_2^\varepsilon(x_2, X_{t-}^{2,N})/N$ before reaching child-bearing age. This means that the effective branching distribution for the i population is

$$\nu^{i,N}(x_i, X^N) = \delta_0 \frac{1}{2}(1 + g_i^\varepsilon(x_i, X_{t-}^{j,N})/N) + \delta_2 \frac{1}{2}(1 - g_i^\varepsilon(x_i, X_{t-}^{j,N})/N) \ (j \neq i)$$

and so depends on the current state of the population X_{t-}^N as well as the location of the parent. Note that $\int k\nu^{i,N}(x_i, X_{t-}^N)(dk) = 1 - g_i^\varepsilon(x_i, X_{t-}^{j,N})/N \ (j \neq i)$ and so $g_i^\varepsilon(x_i, X_{t-}^N)$ plays the role of g_N in (II.3.1).

The two populations are competing for resources and so a high density of 1's near a 2-particle decreases the likelihood of the successful reproduction of the 2 and a high density of 2's has similar detrimental effect on a 1-particle. λ_i is the susceptibility of the ith population and $\sqrt{\varepsilon}$ is the range of the interaction. The method of Section II.4 will show that if $X_0^i = m_i$ is the initial measure of the ith population, then $\{X^N\}$ is tight in $D(\mathbb{R}_+, M_F(\mathbb{R}^d)^2)$ and all limit points are in Ω^2 and satisfy

$$\forall \phi_i \in \mathcal{D}(\Delta/2) \ X_t^i(\phi_i) = m_i(\phi_i) + \int_0^t X_s^i(\Delta\phi_i/2)ds$$

$$(CS)_m^{\varepsilon,\lambda} \qquad - \int_0^t \int g_i^\varepsilon(x_i, X_s^j)\phi_i(x_i)X_s^i(dx_i)ds + M_t^i(\phi_i) \quad (i = 1, 2, \ j \neq i),$$

where $M_t^i(\phi_i)$ are continuous $(\mathcal{F}_t^2)$ – martingales such that

$$M_0^i = 0 \text{ and } \langle M^i(\phi_i), M^j(\phi_j)\rangle_t = \delta_{ij}\int_0^t X_s^i(\phi_i^2)ds.$$

The only technical point concerns the uniform (in N) bound required on $E(X_t^{i,N}(\phi))$ in the analogue of Lemma II.3.3. However, it is easy to couple X^N with branching particle systems with $\lambda_i = 0$, $Z^{i,N}$, $i = 1, 2$, (ignore the interactive killing) so that $X^{i,N} \leq Z^{i,N}$ and so the required bound is immediate from the $\lambda_i = 0$ case. Clearly $(CS)_m^{\varepsilon,\lambda}$ is a special case of $(MP)_g^m$ with

$$g_i(s, X, x) = g_i^\varepsilon(x, X_s^j), \ \ j \neq i.$$

First consider $(MP)_g^m$ in what should be a trivial case: $g_i(s, X, x) = g_i^0(x)$ for some $g_i^0 \in C_b(E_i)$. We let $(MP)_{g^0}^m$ denote this martingale problem. Let $\mathbb{P}_{m_i}^{i,g_i}$ be the law of the (A_i, γ_i, g_i^0)-DW-superprocess starting at m_i. If $g_i^0 \equiv \theta_i$ is constant, write $\mathbb{P}_{m_i}^{i,\theta_i}$ for this law and write $\mathbb{P}_{m_i}^i$ for $\mathbb{P}_{m_i}^{i,0}$. Clearly $\mathbb{P}_{m_1}^{1,g_1^0} \times \mathbb{P}_{m_2}^{2,g_2^0}$ satisfies $(MP)_{g^0}^m$ but it remains to show that it is the only solution. It is easy to extend the Laplace function equation approach in Section II.5 (see Exercise IV.1.1 below) but another approach is to use the following result which has a number of other interesting applications.

Theorem IV.1.2. (Predictable Representation Property). Let $\mathbb{P}_m$ be the law of the (Y, γ, g)-DW-superprocess starting at m on the canonical space of $M_F(E)$-

valued paths $(\Omega_X, \mathcal{F}_X, \mathcal{F}_t^X)$. If $V \in L^2(\mathcal{F}_X, \mathbb{P}_m)$, there is an f in

$$\mathcal{L}^2 = \{f : \mathbb{R}_+ \times \Omega_X \times E \to \mathbb{R} : f \text{ is } \mathcal{P}(\mathcal{F}_t^X) \times \mathcal{E} - \text{measurable and}$$
$$\mathbb{E}\Big(\int_0^t \int f(s,X,x)^2 \gamma(x) X_s(dx) ds\Big) < \infty \ \forall t > 0\}$$

such that

$$V = \mathbb{P}_m(V) + \int_0^\infty \int f(s,X,x) dM(s,x).$$

Proof. Let N_t be a square integrable $(\mathcal{F}_t^X)$-martingale under $\mathbb{P}_m$. As the martingale problem $(MP)_m$ for the superprocess X is well-posed, we see from Theorem 2 and Proposition 2 of Jacod (1977) that for each $n \in \mathbb{N}$ there is a finite set of functions, $\phi_n^1, \ldots \phi_n^{N(n)} \in \mathcal{D}(A)$, and a finite set of $(\mathcal{F}_t^X)$-predictable processes, $h_n^1, \ldots, h_n^{N(n)}$ such that $f_n(s,X,x) = \sum_i h_n^i(s,X)\phi_n^i(x) \in \mathcal{L}^2$ and

$$N_t = \mathbb{P}_m(N_0) + \lim_{n\to\infty} \int_0^t \int f_n(s,X,x) dM(s,x)$$

in $L^2(\Omega_X, \mathcal{F}_X, \mathbb{P}_m)$ for each $t \geq 0$. Hence for each such t,

$$\begin{aligned}
&\lim_{n,n'\to\infty} \mathbb{P}_m\Big(\int_0^t \int [f_n(s,X,x) - f_{n'}(s,X,x)]^2 \gamma(x) X_s(dx) ds\Big) \\
&= \lim_{n,n'\to\infty} \mathbb{P}_m\Big(\Big[\int_0^t \int f_n(s,X,x) dM(s,x) - \int_0^t \int f_{n'}(s,X,x) dM(s,x)\Big]^2\Big) \\
&= 0.
\end{aligned}$$

The completeness of $\mathcal{L}^2$ shows that there is an f in $\mathcal{L}^2$ so that

$$\lim_{n\to\infty} \int_0^t \int f_n(s,X,x) dM(s,x) = \int_0^t \int f(s,X,x) dM(s,x)$$

in $L^2(\Omega_X, \mathcal{F}_X, \mathbb{P}_m)$. This shows that any square integrable $(\mathcal{F}_t^X)$-martingale under $\mathbb{P}_m$ is a constant plus the stochastic integral of a process in $\mathcal{L}^2$ with respect to the martingale measure M. This is of course equivalent to the stated result. ∎

Corollary IV.1.3. $\mathbb{P}_{m_1}^{1,g_1^0} \times \mathbb{P}_{m_2}^{2,g_2^0}$ is the unique solution of $(MP)_{g^0}^m$.

Proof. Let $\mathbb{P}$ be any solution of $(MP)_{g^0}^m$. By the uniqueness of the martingale problem for the DW-superprocess (Theorem II.5.1) we know that $\mathbb{P}(X^i \in \cdot) = \mathbb{P}_{m_i}^{i,g_i^0}(\cdot)$. If ϕ_i is a bounded measurable function on Ω_X^i then by the above predictable representation property

$$\phi_i(X^i) = \mathbb{P}_{m_i}^{i,g_i^0}(\phi_i) + \int_0^\infty \int f_i(s,X_i,x) dM^i(s,x) \ \mathbb{P} - a.s.,\ i = 1,2.$$

(Note that the martingale measure arising in the martingale problem for X_i alone agrees with the martingale measure in $(MP)_{g^0}^m$ by the usual bootstrapping argument starting with simple functions.) The orthogonality of M^1 and M^2 implies that

$$\mathbb{P}(\phi_1(X_1)\phi_2(X_2)) = \mathbb{P}^{1,g_1^0}_{m_1}(\phi_1)\mathbb{P}^{2,g_2^0}_{m_2}(\phi_2). \quad \blacksquare$$

Exercise IV.1.1. Let $V_t^i\phi_i$ be the unique solution of

$$\frac{\partial V_t^i}{\partial t} = A_i V_t^i - \frac{\gamma_i (V_t^i)^2}{2} - g_i^0 V_t^i \quad V_0^i = \phi_i, \quad \phi_i \in \mathcal{D}(A_i).$$

Let ν be a probability on $M_F(E_1) \times M_F(E_2)$ and define $(LMP)^\nu_{g^0}$ in the obvious manner (ν is the law of X_0 and the martingale terms are now local martingales in general). Show that any solution $\mathbb{P}$ of $(LMP)^\nu_{g^0}$ satisfies

$$\mathbb{P}(\exp\{-X_t^1(\phi_1) - X_t^2(\phi_2)\}) = \int \exp\{-X_0^1(V_t^1\phi_1) - X_0^2(V_t^2\phi_2)\} d\nu(X_0).$$

Conclude that

$$\mathbb{P}(X_t \in \cdot) = \int \mathbb{P}^{1,g_1^0}_{m_1}(X_t^1 \in \cdot) \times \mathbb{P}^{2,g_2^0}_{m_2}(X_t^2 \in \cdot) d\nu(m),$$

and then convince yourself that the appropriate version of Theorem II.5.6 shows that $(LMP)^\nu_{g^0}$ is well-posed.

Consider now a more general martingale problem than $(MP)^m_g$ on a general filtered space $\bar{\Omega}' = (\Omega', \mathcal{F}', \mathcal{F}'_t, \mathbb{P}')$. If $m_i \in M_F(E_i)$, $i = 1, 2$, a pair of stochastic processes $(X^1, X^2) \in \Omega^2$ satisfies $(MP)^m_{C,D}$ iff

$$\forall \phi_i \in \mathcal{D}(A_i) \quad X_t^i(\phi_i) = m_i(\phi_i) + \int X_s^i(A_i\phi_i) ds - C_t^i(\phi_i) + D_t^i(\phi_i) + M_t^i(\phi_i),$$

where $M_t^i(\phi_i)$ is a continuous $(\mathcal{F}'_t)$ – martingale such that

$\langle M^i(\phi_i), M^j(\phi_j)\rangle_t = \delta_{ij} \int_0^t X_s^i(\gamma_i\phi_i^2) ds, M_0^i(\phi_i) = 0,$ and C^i, D^i are continuous, non-decreasing, adapted $M_F(E_i)$ – valued processes, starting at 0.

If $\bar{\Omega}'$ is as above, introduce

$$\Omega'' = \Omega' \times \Omega^2,\ \mathcal{F}'' = \mathcal{F}' \times \mathcal{F}^2,\ \mathcal{F}''_t = (\mathcal{F}'_\cdot \times \mathcal{F}^2_\cdot)_{t+},\ \bar{\Omega}'' = (\Omega'', \mathcal{F}'', \mathcal{F}''_t),$$

let $\omega'' = (\omega', \tilde{X}^1, \tilde{X}^2)$ denote points in Ω'' and let $\Pi : \Omega'' \to \Omega'$ be the projection map.

Proposition IV.1.4. (Domination Principle) Assume X satisfies $(MP)^m_{C,D}$ on $\bar{\Omega}'$ and for some $\theta_i \in C_b(E_i)_+$,

$$(DOM) \quad (D_t^i - D_s^i)(\cdot) \le \int_s^t X_r^i(\theta_i 1(\cdot)) dr \text{ (as measures on } E_i) \ \forall s < t,\ i = 1, 2.$$

There is a probability $\mathbb{P}$ on $(\Omega'', \mathcal{F}'')$ and processes $(Z^1, Z^2) \in \Omega^2$ such that
(a) If $W \in b\mathcal{F}'$, then $\mathbb{P}(W \circ \Pi | \mathcal{F}''_t) = \mathbb{P}'(W|\mathcal{F}'_t) \circ \Pi \ \ \mathbb{P} - a.s.$

(b) $X \circ \Pi$ satisfies $(MP)^m_{C\circ\Pi, D\circ\Pi}$ on $\bar{\Omega}''$.
(c) Z^1, Z^2 are independent, Z^i is an $(\mathcal{F}''_t) - (Y^i, \gamma_i, \theta_i)$-DW superprocess starting at m_i, and $Z^i_t \geq X^i_t \circ \Pi$ on Ω'' $\forall t \geq 0$, $i = 1, 2$.

Remark IV.1.5. Clearly (a) implies that $W = (X, D, C, M)$ on $\bar{\Omega}'$ and $W \circ \Pi$ on $\bar{\Omega}''$ have the same law. More significantly they have the same adapted distribution in the sense of Hoover and Keisler (1984). This means that all random variables obtained from W, respectively $W \circ \Pi$, by the operations of compositions with bounded continuous functions and taking conditional expectation with respect to $\mathcal{F}''_t$, respectively $\mathcal{F}'_t$, have the same laws. Therefore in studying (X, C, D, M) on $\bar{\Omega}'$ we may just as well study $(X, C, D, M) \circ \Pi$ on $\bar{\Omega}''$ and hence may effectively assume (X^1, X^2) is dominated by a pair of independent DW-superprocesses as above. We will do this in what follows without further ado.

Sketch of Proof. The proof of Theorem 5.1 in Barlow, Evans and Perkins (1991) goes through with only minor changes. We sketch the main ideas.
Step 1. DW-superprocesses with immigration.
Assume

$$\mu_i \in M^i_{LF} = \{\mu : \mu \text{ is a measure on } \mathbb{R}_+ \times E_i,\ \mu([0,T] \times E_i) < \infty\ \forall T > 0,$$
$$\mu(\{t\} \times E_i) = 0\ \forall t \geq 0\},$$

and $\tau \geq 0$. Consider the following martingale problem, denoted $(MP)^i_{\tau, m_i, \mu}$, for a DW-superprocess with immigration μ on some $(\Omega, \mathcal{F}, \mathcal{F}_t, \mathbb{P})$:

$$\forall \phi \in \mathcal{D}(A_i) \quad X_t(\phi) = m_i(\phi) + \int_\tau^t \int \phi(x) d\mu(r, x) + \int_\tau^t X_s(A_i^{\theta_i} \phi) ds + M_t(\phi),$$

$t \geq \tau$, where $M_t(\phi)$, $t \geq \tau$ is a continuous $(\mathcal{F}_t)$-martingale such that

$$M_\tau(\phi) = 0, \text{ and } \langle M(\phi) \rangle_t = \int_\tau^t X_s(\gamma_i \phi^2) ds.$$

Then $(MP)^i_{\tau, m_i, \mu}$ is well-posed and the law $\mathbb{P}^i_{\tau, m_i, \mu}$ of any solution on Ω_X satisfies

$$\mathbb{P}^i_{\tau, m_i, \mu}(\exp(-X_t(\phi)) = \exp\Big\{ - m_i(V^i_{t-\tau}\phi) - \int_\tau^t \int V^i_{t-s}\phi(x) d\mu(s, x)\Big\},$$

where $V^i_t \phi$ is as in Exercise IV.1.1. Moreover $(\Omega_X, \mathcal{F}_X, \mathcal{F}^X_t, X_t, \mathbb{P}^i_{\tau, m_i, \mu})$ is an inhomogeneous Borel strong Markov process and $(\tau, m_i, \mu) \to \mathbb{P}^i_{\tau, m_i, \mu}$ is Borel measurable. The existence of a solution may be seen by approximating μ by a sequence of measures each supported by $\{t_0, \ldots, t_m\} \times E_i$ for some finite set of points, and taking the weak limit through an appropriate sequence of the corresponding DW-superprocesses. For any solution to $(MP)^i$, the formula for the Laplace functional and other properties stated above may then be derived just as in Section II.5. Note that the required measurability is clear from the Laplace functional equation, the Markov property and a monotone class argument. (Alternatively, the existence of a unique Markov process satisfying this Laplace functional equation is a special case of Theorem 1.1 of Dynkin and the corresponding martingale problem may then be derived as in Fitzsimmons (1988,1989).)

Step 2. Definition of $\mathbb{P}$.

Set $\mathbb{Q}^i_\mu = \mathbb{P}^i_{0,0,\mu}$ and define

$$F^i_t(\cdot) = \int_0^t X^i_s(\theta^i 1(\cdot))ds - D^i_t(\cdot) + C^i_t(\cdot).$$

Then $F^i(\omega') \in M^i_{LF}$ $\mathbb{P}'$-a.s. and we can define $\mathbb{P}$ on Ω'' by

$$\mathbb{P}(A \times B_1 \times B_2) = \int_{\Omega'} 1_A(\omega')\mathbb{Q}^1_{F^1(\omega')}(B_1)\mathbb{Q}^2_{F^2(\omega')}(B_2)d\mathbb{P}(\omega').$$

This means that under $\mathbb{P}$, conditional on ω', $\tilde{X}^1$ and $\tilde{X}^2$ are independent, and $\tilde{X}^i$ is a $(Y_i, \gamma_i, \theta_i)$-DW-superprocess with immigration F_i. Define $Z^i_t(\omega', \tilde{X}) = X^i_t(\omega') + \tilde{X}^i_t$. For example if θ_i and D_i are both 0, then we can think of $\tilde{X}^i$ as keeping track of the "ghost particles" (and their descendants) killed off by C^i in the X^i population. When it is added to X^i one should get an ordinary DW-superprocess. (a) is a simple consequence of this definition and (b) is then immediate. To prove (c) we show Z satisfies the martingale problem on $\bar{\Omega}''$ corresponding to $(MP)^m_\theta$ and then use Corollary IV.1.3. This is a straightforward calculation (see Theorem 5.1 in Barlow-Evan-Perkins(1991)). The fact that Z^i dominates X^i is obvious. ∎

We now state and prove a bivariate version of Dawson's Girsanov Theorem for interactive drifts (Dawson (1978). The version given here is taken from Evans-Perkins (1994).

Theorem IV.1.6. Assume $\gamma_i(x) > 0$ for all x in E_i, $i = 1, 2$ and $\mathbb{P}_{m_1} \times \mathbb{P}_{m_2}$-a.s.,

$$(IV.1.1) \qquad \sum_{i=1}^2 \int_0^t \int \frac{g_i(s, X, x)^2}{\gamma_i(x)} X^i_s(dx)ds < \infty, \ \forall t > 0,$$

so that we can define a continuous local martingale under $\mathbb{P}_{m_1} \times \mathbb{P}_{m_2}$ by

$$R^g_t = \exp\Big\{ \sum_{i=1}^2 \int_0^t \frac{-g_i(s, X, x)}{\gamma_i(x)} dM^{i,0}(s, x) - \frac{1}{2}\int_0^t \int \frac{g_i(s, X, x)^2}{\gamma_i(x)} X^i_s(dx)ds \Big\}.$$

(a) If $\mathbb{P}$ satisfies $(MP)^m_g$ and (IV.1.1) holds $\mathbb{P}$-a.s., then

$$(IV.1.2) \qquad \frac{d\mathbb{P}}{d\mathbb{P}^1_{m_1} \times \mathbb{P}^2_{m_2}}\Big|_{\mathcal{F}^2_t} = R^g_t,$$

and in particular there is at most one law $\mathbb{P}$ satisfying $(MP)^m_g$ such that (IV.1.1) holds $\mathbb{P}$-a.s.

(b) If $|g_i|^2/\gamma_i(x)$ and $|g_i|$ are uniformly bounded for $i = 1, 2$ then R^g_t is an $(\mathcal{F}^2_t)$-martingale under $\mathbb{P}_{m_1} \times \mathbb{P}_{m_2}$ and (IV.1.2) defines the unique law $\mathbb{P}$ which satisfies $(MP)^m_g$.

(c) If $X_\cdot \le X'_\cdot$ (pointwise inequality of measures) implies

$$(IV.1.3) \qquad -\theta\sqrt{\gamma_i(x)} \le g_i(t, X, x) \le g_i(t, X', x), \ i = 1, 2 \ \text{ for all } (t, x),$$

for some constant $\theta \ge 0$, then the conclusion of (b) holds.

Proof. (a) Let

$$T_n = \inf\{t : \sum_{i=1}^{2}\int_0^t \Big[\int\Big(\frac{g_i(s,X,x)^2}{\gamma_i(x)}+1\Big)X_s^i(ds)+1\Big]ds \geq n\} \ (\leq n).$$

Assume $\mathbb{P}$ satisfies $(MP)_g^m$, (IV.1.1) holds $\mathbb{P}$-a.s., and define

$$\tilde{R}^g_{t\wedge T_n} = \exp\Big\{\sum_{i=1}^{2}\int_0^{t\wedge T_n}\int \frac{g_i(s,X,x)}{\gamma_i(x)}M^{i,g_i}(ds,dx) - \frac{1}{2}\int_0^{t\wedge T_n}\int\frac{g_i(s,X,x)^2}{\gamma_i(x)}X_s^i(dx)ds\Big\}.$$

Then $\tilde{R}^g_{\cdot\wedge T_n}$ is a uniformly integrable $(\mathcal{F}_t^2)$-martingale under $\mathbb{P}$ (e.g., by Theorem III.5.3 of Ikeda-Watanabe (1981)) and so $d\mathbb{Q}_n = \tilde{R}^g_{T_n}d\mathbb{P}$ defines a probability on $(\Omega^2,\mathcal{F}_t^2)$. If $\overset{m}{=}$ denotes equality up to local martingales and $\phi_i \in C_b^2(E_i)$, then integration by parts shows that under $\mathbb{P}$,

$$\begin{aligned} M^{i,0}_{t\wedge T_n}(\phi)\tilde{R}^g_{t\wedge T_n} &= \Big[M^{i,g_i}_{t\wedge T_n}(\phi_i) - \int_0^{t\wedge T_n}\int g_i(s,X,x)\phi_i(x)X_s^i(dx)ds\Big] \\ &\quad\times\Big[1+\int_0^{t\wedge T_n}\int \tilde{R}^g_s\frac{g_i(s,X,x)}{\gamma_i(x)}M^{i,g_i}(ds,dx)\Big] \\ &\overset{m}{=} -\int_0^{t\wedge T_n}\int \tilde{R}^g_s g_i(s,X,x)\phi_i(x)X_s^i(dx)ds \\ &\quad+\int_0^{t\wedge T_n}\tilde{R}^g_s d\langle M^{i,g_i}(\phi_i), M^{i,g_i}(g_i/\gamma_i)\rangle_s \\ &= 0. \end{aligned}$$

Therefore under $\mathbb{Q}_n$, $M^{i,0}_{t\wedge T_n}$ is an $(\mathcal{F}_t^2)$-local martingale. As $\mathbb{Q}_n \ll \mathbb{P}$ and quadratic variation is a path property, we also have

$$\langle M^{i,0}_{\cdot\wedge T_n}(\phi_i), M^{j,0}_{\cdot\wedge T_n}(\phi_j)\rangle_t = \delta_{ij}\int_0^{t\wedge T_n} X_s^i(\gamma_i\phi_i^2)ds \quad \forall t\geq 0 \ \mathbb{Q}_n - a.s.$$

which is uniformly bounded and hence shows $M^{i,0}_{t\wedge T_n}(\phi_i)$ is a $\mathbb{Q}_n$-martingale. Let $\tilde{\mathbb{Q}}_n$ denote the unique law on $(\Omega^2,\mathcal{F}^2)$ such that $\tilde{\mathbb{Q}}_n\big|_{\mathcal{F}^2_{T_n}} = \mathbb{Q}_n\big|_{\mathcal{F}^2_{T_n}}$ and the conditional law of $X_{T_n+\cdot}$ given $\mathcal{F}^2_{T_n}$ is $\mathbb{P}^1_{X^1_{T_n}}\times\mathbb{P}^2_{X^2_{T_n}}$. Then $\tilde{\mathbb{Q}}_n$ satisfies $(MP)_0^m$ and so, by Corollary IV.1.3, $\tilde{\mathbb{Q}}_n = \mathbb{P}^1_{m_1}\times\mathbb{P}^2_{m_2}$. Therefore (IV.1.1) implies

$$\mathbb{Q}_n(T_n < t) = \mathbb{P}^1_{m_1}\times\mathbb{P}^2_{m_2}(T_n<t)\to 0 \text{ as } n\to\infty.$$

Since (IV.1.1) holds $\mathbb{P}$-a.s., $\tilde{R}^g_t$ is an $(\mathcal{F}_t^2)$-local martingale under $\mathbb{P}$ and

$$\begin{aligned} \mathbb{P}(\tilde{R}^g_t) &\geq \mathbb{P}(\tilde{R}^g_{t\wedge T_n}1(T_n\geq t)) \\ &= \mathbb{P}(\tilde{R}^g_{t\wedge T_n}) - \mathbb{P}(\tilde{R}^g_{t\wedge T_n}1(T_n<t)) \end{aligned} \qquad (IV.1.4)$$

$$= 1 - \mathbb{Q}_n(T_n < t) \to 1 \text{ as } n \to \infty.$$

Therefore $\tilde{R}_t^g$ is a a $\mathbb{P}$-martingale and we may define a unique law, $\mathbb{Q}$, on $(\Omega^2, \mathcal{F}^2)$ by $d\mathbb{Q}\big|_{\mathcal{F}_t^2} = \tilde{R}_t^g d\mathbb{P}\big|_{\mathcal{F}_t^2}$ for all $t > 0$. Now repeat the above argument, but without the T_n's, to see that $\mathbb{Q} = \mathbb{P}_{m_1}^1 \times \mathbb{P}_{m_2}^2$. Note here that it suffices to show $M^{i,0}(\phi_i)$ are local martingales as the proof of Corollary IV.1.3 shows the corresponding local martingale problem is well-posed. Therefore

$$d\mathbb{P}\big|_{\mathcal{F}_t^2} = \left(\tilde{R}_t^g\right)^{-1} d(\mathbb{P}_{m_1}^1 \times \mathbb{P}_{m_2}^2)\big|_{\mathcal{F}_t^2} = R_t^g d(\mathbb{P}_{m_1}^1 \times \mathbb{P}_{m_2}^2)\big|_{\mathcal{F}_t^2} \quad \forall t > 0.$$

(b) Uniqueness is immediate from (a). Let T_n be as in (a), let

$$g^n(s, X, x) = 1(s \le T_n)(g_1(s, X, x), g_2(s, X, x)),$$

and define a probability on $(\Omega^2, \mathcal{F}^2)$ by $d\mathbb{Q}_n = R_{T_n}^g d(\mathbb{P}_{m_1}^1 \times \mathbb{P}_{m_2}^2)$. Now argue just as in the proof of (a) to see that $\mathbb{Q}_n$ solves $(MP)_{g^n}^m$. This martingale problem shows that

$$\begin{aligned}\mathbb{Q}_n(X_t^i(1)) &= m_i(1) + \mathbb{Q}_n\Big(\int_0^{t\wedge T_n} X_s^i(g_i(s, X, \cdot))ds\Big)\\ &\le m_i(1) + c\mathbb{Q}_n\Big(\int_0^{t\wedge T_n} X_s^i(1)ds\Big).\end{aligned}$$

The righthand side is finite by the definition of T_n and hence so is the lefthand side. A Gronwall argument now shows that $\mathbb{Q}_n(X_t^i(1)) \le m_i(1)e^{ct}$ and therefore

$$\begin{aligned}&\mathbb{Q}_n\Big(\sum_{i=1}^2 \int_0^t \int \Big[\frac{g_i(s, X, x)^2}{\gamma_i(x)} + 1\Big] X_s^i(dx) + 1\, ds\Big)\\ &\le (c^2 + 1)(m_1(1) + m_2(1))e^{ct}t + 2t \equiv K(t).\end{aligned}$$

This shows that $\mathbb{Q}_n(T_n < t) \le K(t)/n \to 0$ as $n \to \infty$. Argue exactly as in (IV.1.4) to see that $\mathbb{P}_{m_1}^1 \times \mathbb{P}_{m_2}^2(R_t^g) = 1$ and therefore R_t^g is a martingale under this product measure. A simple stochastic calculus argument as in the proof of (a) shows that (IV.1.2) does define a solution of $(MP)_g^m$. Note that, as for $\mathbb{Q}_n$, one sees that $\mathbb{P}(X_t^i(1)) \le m_i(1)e^{ct}$ and so $M^{i,g_i}(\phi)$ is a martingale (and not just a local martingale) because its square function is integrable.

(c) Define T_n, g^n and $\mathbb{Q}_n$ as in the proof of (b). As before, $\mathbb{Q}_n$ satisfies $(MP)_{g^n}^m$. The upper bound on $-g_i$ allows us to apply Proposition IV.1.4 and define processes $Z^i \ge X^i$, $i = 1, 2$ on the same probability space such that (X^1, X^2) has law $\mathbb{Q}_n$ and (Z^1, Z^2) has law $\mathbb{P}_{m_1}^{1,\theta\sqrt{\gamma_1}} \times \mathbb{P}_{m_2}^{2,\theta\sqrt{\gamma_2}}$. The conditions on g_i show that

$$\begin{aligned}\int_0^t \int \Big[\frac{g_i(s, X, x)^2}{\gamma_i(x)} + 1\Big] X_s^i(dx)ds &\le \int_0^t \int \Big[\theta^2 + \frac{g_i^+(s, X, x)^2}{\gamma_i(x)} + 1\Big] X_s^i(dx)ds\\ &\le \int_0^t \int \Big[\theta^2 + \frac{g_i^+(s, Z, x)^2}{\gamma_i(x)} + 1\Big] Z_s^i(dx)ds \qquad (IV.1.5)\\ &\le (\theta^2 + 1)\int_0^t \int \Big[\frac{g_i(s, Z, x)^2}{\gamma_i(x)} + 1\Big] Z_s^i(dx)ds.\end{aligned}$$

This implies that

$$(IV.1.6) \qquad \mathbb{Q}_n(T_n < t) \le \mathbb{P}^{1,\theta\sqrt{\gamma_1}}_{m_1} \times \mathbb{P}^{2,\theta\sqrt{\gamma_2}}_{m_2}(T_{n/(\theta^2+1)} < t).$$

Now (IV.1.1) and the fact that $\mathbb{P}^{1,\theta\sqrt{\gamma_1}}_{m_1} \times \mathbb{P}^{2,\theta\sqrt{\gamma_2}}_{m_2} \ll \mathbb{P}^1_{m_1} \times \mathbb{P}^2_{m_2}$ on $\mathcal{F}^2_t$ (from (b)) show that (IV.1.1) holds $\mathbb{P}^{1,\theta\sqrt{\gamma_1}}_{m_1} \times \mathbb{P}^{2,\theta\sqrt{\gamma_2}}_{m_2}$-a.s. and therefore the expression on the righthand side of (IV.1.6) approaches 0 as $n \to \infty$. Therefore the same is true for the lefthand side of (IV.1.6) and we can argue as in (IV.1.4) to see that R^g_t is an $(\mathcal{F}^2_t)$-martingale under $\mathbb{P}^1_{m_1} \times \mathbb{P}^2_{m_2}$. A simple stochastic calculus argument, as in (a), shows that (IV.1.2) defines a law $\mathbb{P}$ which satisfies $(MP)^m_g$. Note that initially one gets that the martingale terms in $(MP)^m_g$ are local martingales. As in (b) they are martingales because a simple Gronwall argument using the upper bound on $-g^i$ (use the above stopping times T_n and Fatou's Lemma) shows that $\mathbb{P}(X^i_t(1)) \le m_i(1)e^{ct}$.

For uniqueness assume $\mathbb{P}$ satisfies $(MP)^m_g$. As above we may use Proposition IV.1.4 to define processes X and Z on a common probability space such that $X^i \le Z^i$ $i = 1, 2$, X has law $\mathbb{P}$, and Z has law $\mathbb{P}^{1,\theta\sqrt{\gamma_1}}_{m_1} \times \mathbb{P}^{2,\theta\sqrt{\gamma_2}}_{m_2}$. Recall that we saw that (IV.1.1) holds a.s. with respect to this latter law and so the calculation in (IV.1.5) shows that it holds $\mathbb{P}$-a.s. as well. The uniqueness is therefore consequence of (a). ∎

Remark IV.1.7. (a) Simply take $g_2 = 0$ in the above to get the usual univariate form of Dawson's Girsanov theorem.

(b) In Theorem 2.3 (b) of Evans-Perkins (1994) this result is stated without the monotonicity part of (IV.1.3). This is false as one can easily see by taking $g_1 = 1/X^1_s(1)$ and noting that the total mass of the solution of $(MP)^m_g$ (if it existed) could now become negative because of the constant negative drift. Fortunately all the applications given there are valid because (IV.1.3) holds in each of them.

(c) If $-g_i \le c$ $i = 1, 2$ for some constant c, then $(MP)^m_g$ is equivalent to $(LMP)^m_g$, i.e., $(MP)^m_g$ but now $M^{i,g_i}_t(\phi_i)$ need only be a continuous local martingale. To see this, assume $\mathbb{P}$ satisfies $(LMP)^m_g$ and let $T^i_n = \inf\{t : X^i_t(1) \ge n\}$ $(n > m_i(1))$. Then $M^{i,g_i}_{t\wedge T^i_n}(1)$ is a square integrable martingale because $\langle M^{i,g_i}(1)\rangle_{t\wedge T^i_n} \le \|\gamma_i\|_\infty nt$. We have

$$X^i_t(1) \le m_i(1) + c\int_0^t X^i_s(1)\,ds + M^{i,g_i}_t(1).$$

Take mean values in the above inequality at time $t \wedge T^i_n$ to see that

$$E(X^i_{t\wedge T^i_n}(1)) \le m_i(1) + c\int_0^t E(X^i_{s\wedge T^i_n}(1))\,ds,$$

and so $E(X^i_{t\wedge T^i_n}(1)) \le m_i(1)e^{ct}$. By Fatou's Lemma this implies $E(X^i_t(1)) \le m_i(1)e^{ct}$. Therefore for each $\phi_i \in \mathcal{D}(\frac{\Delta}{2})$, $M^{i,g_i}_t(\phi_i)$ is an L^2-martingale since its square function is integrable.

As a first application of Theorem IV.1.6 we return to

Example IV.1.1. Recall that $(CS)_m^{\varepsilon,\lambda}$ was a special case of $(MP)_g^m$ with $\gamma_i \equiv 1$ and

$$g_i(s,X,x_i) = \lambda_i \int p_\varepsilon(x_i - x_j) X_s^j(dx_j) \; (j \neq i).$$

Clearly the monotonicity condition (IV.1.3) holds with $\theta = 0$ and (IV.1.1) is clear because $g_i(s,X,x_i) \le \lambda_i \varepsilon^{-d/2} X_s^j(1)$. Part (c) of the above theorem therefore shows that the unique solution of $(CS)_m^{\varepsilon,\lambda}$ is $\mathbb{P}_m^\varepsilon$, where, if $\mathbb{P}_m$ is the law of SBM ($\gamma \equiv 1$), then

$$\frac{d\mathbb{P}_m^\varepsilon}{d(\mathbb{P}_{m_1} \times \mathbb{P}_{m_2})}\Big|_{\mathcal{F}_t^2} = \exp\Big\{\sum_{i=1}^2 \Big[-\lambda_i \int_0^t \int\int p_\varepsilon(x_i - x_{3-i}) X_s^{3-i}(dx_{3-i}) dM^{i,0}(s,x_i)$$

$$(IV.1.7) \qquad -\frac{1}{2}\lambda_i^2 \int_0^t \int \Big[\int p_\varepsilon(x_i - x_{3-i}) X_s^{3-i}(dx_{3-i})\Big]^2 X_s^i(dx_i) ds\Big]\Big\}.$$

(IV.1.7) defines a collection of laws $\{\mathbb{P}_m^\varepsilon : m \in M_F(\mathbb{R}^d)\}$ on $(\Omega^2, \mathcal{F}^2)$. If ν is a probability on $M_F(\mathbb{R}^d)^2$ and $\mathbb{P}$ satisfies $(CS)_\nu^{\varepsilon,\lambda}$, that is the analogue of (CS) but with $\mathcal{L}(X_0) = \nu$ and $M_t^i(\phi_i)$ now a local martingale, then one easily sees that the regular conditional probability of X given X_0 satisfies $(CS)_{X_0}^{\varepsilon,\lambda}$ for ν-a.a. X_0. Therefore this conditional law is $\mathbb{P}_{X_0}^\varepsilon$ ν-a.s. and one can argue as in Theorem II.5.6 to see that $(\Omega^2, \mathcal{F}^2, \mathcal{F}_t^2, X_t, \mathbb{P}_m^\varepsilon)$ is a Borel strong Markov process. The Borel measurability is in fact clear from (IV.1.7).

Exercise IV.1.2 Assume Y is a Feller process on a locally compact separable metric space E with strongly continuous semigroup and fix $\gamma > 0$. Let $V_s(\omega) = \omega_s$ be the coordinate maps on $\Omega_V = C(\mathbb{R}_+, M_1(E))$. For each $V_0 \in M_(E)$ there is a unique law $\bar{\mathbb{P}}_{V_0}$ on Ω_V such that under $\bar{\mathbb{P}}_{V_0}$

$\forall \phi \in \mathcal{D}(A)$ $\quad V_t(\phi) = V_0(\phi) + \int_0^t V_s(A\phi)\, ds + M_t(\phi)$, where $M(\phi)$ is a continuous $(\mathcal{F}_t^V)$-martingale such that $M_0(\phi) = 0$ and $\langle M(\phi)\rangle_t = \gamma \int_0^t V_s(\phi^2) - V_s(\phi)^2\, ds.$

$\bar{\mathbb{P}}_{V_0}$ is the law of the Fleming-Viot process with mutation operator A (see Section 10.4 of Ethier-Kurtz (1986)).

For $c \ge 0$ and $m \in M_1(E)$ consider the following martingale problem for a law $\mathbb{P}$ on Ω_X:

$\forall \phi \in \mathcal{D}(A)$ $\quad X_t(\phi) = m(\phi) + \int_0^t X_s(A\phi) + c(1 - X_s(1))X_s(\phi)\, ds + M_t^c(\phi)$, where $M^c(\phi)$ is an $\mathcal{F}_t^X$-martingale such that $M_0^c(\phi) = 0$ and $\langle M^c(\phi)\rangle_t = \gamma \int_0^t X_s(\phi^2)\, ds.$

(a) Show there is a unique law $\mathbb{P}^c$ satisfying this martingale problem and find $\frac{d\mathbb{P}^c}{d\mathbb{P}^0}\big|_{\mathcal{F}_t^X}$.
(b) Show that for any $T, \varepsilon > 0$, $\lim_{c\to\infty} \mathbb{P}^c(\sup_{t\le T} |X_t(1) - 1| > \varepsilon) = 0$.

Hint. This is an exercise in one-dimensional diffusion theory–here is one approach. By a time change it suffices to show the required convergence for

$$W_t = 1 + \sqrt{\gamma}B_t + \int_0^t c(1 - W_s)ds.$$

Itô's Lemma implies that for any integer $p \geq 2$,

$$(W_t-1)^p+cp\int_0^t (W_s-1)^p ds = p\sqrt{\gamma}\int_0^t (W_s-1)^{p-1}dB_s+\frac{p(p-1)\gamma}{2}\int_0^t (W_s-1)^{p-2}ds.$$

Use induction and the above to conclude that for each even $p \geq 2$, $\lim_{c\to\infty} E(\int_0^t (W_s - 1)^p ds) = 0$. Now note that the left side of the above display is a nonnegative submartingale. Take $p = 4$ and use a maximal inequality.

(c) Define $S = \inf\{t : X_t(1) \leq 1/2\}$ and $Z_t(\cdot) = \frac{X_{t\wedge S}(\cdot)}{X_{t\wedge S}(1)} \in M_1(E)$. If $\phi \in \mathcal{D}(A)$, prove that $Z_t(\phi) = m(\phi) + \int_0^{t\wedge S} Z_s(A\phi)\,ds + N_t^c(\phi)$, where $N^c(\phi)$ is an $(\mathcal{F}_t^X)$-martingale under $\mathbb{P}^c$ starting at 0 and satisfying

$$\langle N^c(\phi)\rangle_t = \gamma\int_0^{t\wedge S} (Z_s(\phi^2) - Z_s(\phi)^2)X_s(1)^{-1}\,ds.$$

Show this implies $\lim_{c\to\infty} \mathbb{P}^c(Z_t(\phi)) = m(P_t\phi)$.

(d) Show that $\mathbb{P}^c(Z \in \cdot)\overset{w}{\Rightarrow}\tilde{\mathbb{P}}_m$ on Ω_V as $c \to \infty$ and conclude from (b) that $\mathbb{P}^c\overset{w}{\Rightarrow}\tilde{\mathbb{P}}_m$ on Ω_X (we may consider $\tilde{\mathbb{P}}_m$ as a law on Ω_X because $\Omega_V \in \mathcal{F}_X$).

Hint. Use Theorem II.4.1 to show that $\{\mathbb{P}^{c_n}(Z \in \cdot)\}$ is tight for any $c_n \uparrow \infty$. One approach to the compact containment is as follows:

Let d be a bounded metric on $E \cup \{\infty\}$, the one-point compactification of E, let $h_p(x) = e^{-pd(x,\infty)}$ and $g_p(x) = \int_0^1 P_s h_p(x)\,ds$. Then $Ag_p(x) = P_1 h_p(x) - h_p(x)$ and (c) gives

$$\sup_{t\leq T} Z_t(g_p) \leq m(g_p) + \sup_{t\leq T}|N_t^c(g_p)| + \int_0^T Z_s(P_1 h_p)ds.$$

Now use the first moment result in (c) and a square function inequality to conclude that

$$\lim_{p\to\infty}\limsup_{n\to\infty}\mathbb{P}^{c_n}\Big(\sup_{t\leq T} Z_t(g_p)\Big) = 0.$$

2. A Singular Competing Species Model–Dimension One

Consider $(CS)_m^{\varepsilon,\lambda}$ as the interaction range $\sqrt{\varepsilon} \downarrow 0$. In this limiting regime it is only the local density of the "2-population" at x that has an adverse effect on the "1-population" at x and conversely. It would seem simplest to first study this limiting model in the one-dimensional case where according to the results of Section III.4 we can expect these densities to exist. Throughout this Section $\mathbb{P}_m$ is the law of SBM ($\gamma \equiv 1$) and we continue to use the notation from the last Section with $E_i = \mathbb{R}$, $i = 1, 2$.

Define a Borel map $U : M_F(\mathbb{R}) \times \mathbb{R} \to [0, \infty]$ by

$$U(\mu, x) = \limsup_{n\to\infty} \frac{n}{2}\mu((x - \frac{1}{n}, x + \frac{1}{n}]),$$

and introduce the $\mathcal{P} \times \mathcal{B}(\mathbb{R})$-measurable canonical densities on Ω^2,

$$u_i(t, X, x) = U(X_t^i, x).$$

Then

$$\begin{aligned}\Omega_{ac} &= \{X \in \Omega^2 : X_t^i \ll dx \ \forall t > 0, \ i = 1, 2\} \\ &= \{X \in \Omega^2 : X_t^i(1) = \int u_i(t, x) dx \ \forall t > 0, \ i = 1, 2\}\end{aligned}$$

is a universally measurable subset of Ω^2 (e.g. by Theorem III.4.4 (a) of Dellacherie and Meyer (1978)).

Letting $\varepsilon \downarrow 0$ in $(CS)_m^{\lambda,\varepsilon}$ suggests the following definition: A probability $\mathbb{P}$ on $(\Omega^2, \mathcal{F}^2)$ satisfies $(CS)_m^\lambda$ iff

$$\begin{aligned}\text{For } i = 1, 2 \ \forall \phi_i \in \mathcal{D}(\Delta/2) \quad X_t^i(\phi_i) &= m_i(\phi_i) + \int_0^t X_s^i\Big(\frac{\Delta\phi_i}{2}\Big) ds \\ &\quad - \lambda_i \int_0^t \int \phi_i(x) u_1(s, x) u_2(s, x) \, dx \, ds + M_t^i(\phi_i),\end{aligned}$$

where $M_t^i(\phi_i)$ are continuous $(\mathcal{F}_t^2)$ – martingales under $\mathbb{P}$ such that $M_0^i(\phi_i) = 0$,

and $\langle M^i(\phi_i), M^j(\phi_j)\rangle_t = \delta_{ij} \int_0^t X_s^i(\phi_i^2) ds.$

Recall that $M^{i,0}$ $(i = 1, 2)$ are the orthogonal martingale measures on Ω^2 under $\mathbb{P}_{m_1} \times \mathbb{P}_{m_2}$–see the notation introduced at the beginning of this Chapter.

Theorem IV.2.1. Assume $d = 1$ and let

$$F = \Big\{(m_1, m_2) \in M_F(\mathbb{R})^2 : \int\int \log\Big(\frac{1}{|x_1 - x_2|}\Big)^+ dm_1(x_1) dm_2(x_2) < \infty\Big\}.$$

(a) For each $m \in F$, $(CS)_m^\lambda$ has a unique solution $\mathbb{P}_m^0$ given by

$$\begin{aligned}\frac{d\mathbb{P}_m^0}{d(\mathbb{P}_{m_1} \times \mathbb{P}_{m_2})}\Big|_{\mathcal{F}_t^2} = \exp\Big\{\sum_{i=1}^2 \Big[&- \lambda_i \int_0^t \int u_{3-i}(s, x) dM^{i,0}(s, x) \\ &- \frac{\lambda_i^2}{2} \int_0^t \int u_{3-i}(s, x)^2 u_i(s, x) dx ds\Big]\Big\}.\end{aligned}$$

In particular $\mathbb{P}_m^0(\Omega_{ac}) = 1$.

(b)$(\Omega^2, \mathcal{F}^2, \mathcal{F}_t^2, X_t, (\mathbb{P}_m^0)_{m \in F})$ is a continuous Borel strong Markov process taking values in F. That is, for each $m \in F$, $\mathbb{P}_m^0(X_t \in F \ \forall t \geq 0) = 1$, $m \to \mathbb{P}_m^0$ is Borel measurable, and the $(\mathcal{F}_t^2)$-strong Markov property holds.

(c) For each $m \in F$, $\mathbb{P}_m^\varepsilon \overset{w}{\Rightarrow} \mathbb{P}_m^0$ as $\varepsilon \downarrow 0$.

Proof. (a) Note first that $(CS)_m^\lambda$ is a special case of $(MP)_g^m$ with

$$g_i(s, X, x) = \lambda_i u_{3-i}(s, X, x).$$

To see this, note that if $\mathbb{P}$ satisfies $(CS)^\lambda_m$, then by Proposition IV.1.4 (with $D^i = \theta_i = 0$) we can define a process X with law $\mathbb{P}$ and a pair of independent super-Brownian motions ($\gamma = 1$), (Z^1, Z^2) on the same space so that $Z^i \geq X^i$. As $Z^i \ll dx$ by Theorem III.3.8(c), the same is true of X^i, and so in $(CS)^\lambda_m$, $u_i(s,x)dx = X^i_s(dx)$, and $\mathbb{P}$ satisfies $(MP)^m_g$ as claimed. The converse implication is proved in the same way. The fact that g_i can now take on the value ∞ will not alter any of the results (or proofs) in the previous section.

Now check the hypotheses of Theorem IV.1.6(c) for the above choice of g_i. Condition (IV.1.3) is obvious (with $\theta \equiv 0$). For (IV.1.1), by symmetry it suffices to show that

$$(IV.2.1) \qquad \mathbb{P}_{m_1} \times \mathbb{P}_{m_2}\Big(\int_0^t \int u_1(s,x)^2 u_2(s,x)\,dx\,ds\Big) < \infty \ \forall t > 0.$$

Recall from (III.4.1) that if $mP_t(x) = \int p_t(y-x)dm(y)$ then under $\mathbb{P}_{m_1} \times \mathbb{P}_{m_2}$,

$$u_i(t,x) = m_i P_t(x) + \int_0^t \int p_{t-s}(y-x)dM^{i,0}(s,y) \ \ a.s. \ \text{ for each } t, x,$$

where the stochastic integral is square integrable. This shows that

$$(IV.2.2) \qquad \mathbb{P}_{m_i}(u_i(t,x_i)) = m_i P_t(x_i),$$

and

$$\begin{aligned}
\mathbb{P}_{m_i}(u_i(t,x_i)^2) &= m_i P_t(x_i)^2 + \int_0^t \int p_{t-s}(y-x_i)^2 m_i P_s(y)\,dy\,ds \\
&\leq m_i P_t(x_i)^2 + \int_0^t (2\pi(t-s))^{-1/2}\,ds\, m_i P_t(x_i) \\
&\leq m_i P_t(x_i)^2 + \sqrt{t}\, m_i P_t(x_i) \\
(IV.2.3) \qquad &\leq m_i P_t(x_i)^2 + m_i(1).
\end{aligned}$$

Now use these estimates to bound the lefthand side of (IV.2.1) by

$$\int_0^t \int m_1 P_s(x)^2 m_2 P_s(x)\,dx\,ds + \int_0^t \int m_1(1) m_2 P_s(x)\,dx\,ds.$$

The second term is $m_1(1)m_2(1)t$ and so is clearly finite for all $t > 0$ for any pair of finite measures m. Bound $m_1 P_s(x)^2$ by $m_1(1)s^{-1/2} m_1 P_s(x)$ and use Chapman-Kolmogorov to see that the first term is at most

$$\begin{aligned}
&m_1(1)\int_0^t \int s^{-1/2} p_{2s}(y_1 - y_2) m_1(dy_1) m_2(dy_2)\,ds \\
&\leq m_1(1) \int \left(1 + \log\left(\frac{4t}{|y_1 - y_2|^2}\right)^+\right) dm_1(y_1) dm_2(y_2) \\
&< \infty \ \text{ if } m \in F.
\end{aligned}$$

(b) Let Z be the pair of independent dominating SBM's constructed in (a). Since Z^i_t has a continuous density on compact support for all $t > 0$ a.s. (Theorem III.4.2(a)

and Corollary III.1.4), clearly $Z_t \in F$ for all $t > 0$ a.s. and hence the same is true for X $\mathbb{P}^0_m$ a.s. The Borel measurability in m is clear from the Radon-Nikodym derivative provided in (a) and the strong Markov property is then a standard consequence of uniqueness (see, e.g. the corresponding discussion for $\mathbb{P}^\varepsilon_m$ at the end of the last section).
(c) Write $\mathbb{P}_m$ for $\mathbb{P}_{m_1} \times \mathbb{P}_{m_2}$. Let R^ε_t be the Radon-Nikodym derivative in (IV.1.7) and R^0_t be that in (a) above. It suffices to show $R^\varepsilon_t \to R^0_t$ in $\mathbb{P}_m$-probability because as these non-negative random variables all have mean 1, this would imply L^1 convergence. To show this convergence, by symmetry it clearly suffices to prove

$$\int_0^t \int [X^2_s P_\varepsilon(x) - u_2(s,x)]^2 u_1(s,x)\,dx\,ds \to 0 \text{ in } \mathbb{P}_m\text{-probability as } \varepsilon \downarrow 0.$$

If $\delta > 0$ is fixed, the fact that $(X^2_s,\ s \geq \delta)$ has a jointly continuous uniformly bounded density shows that

$$\int_\delta^t \int [X^2_s P_\varepsilon(x) - u_2(s,x)]^2 u_1(s,x)\,dx\,ds \to 0 \text{ in } \mathbb{P}_m\text{-probability as } \varepsilon \downarrow 0.$$

Therefore it suffices to show

$$\lim_{\delta \downarrow 0} \sup_{0<\varepsilon<1} \mathbb{P}_m\Big(\int_0^\delta \int [X^2_s P_\varepsilon(x)^2 + u_2(s,x)^2] u_1(s,x)\,dx\,ds\Big) = 0.$$

The argument in (a) easily handles the $u_2(s,x)^2$ term, so we focus on the $X^2_s P_\varepsilon(x)^2$ term. Use (IV.2.2) and (IV.2.3) to see that

$$\begin{aligned}
&\mathbb{P}_m\Big(\int_0^\delta \int X^2_s P_\varepsilon(x)^2 u_1(s,x)\,dx\,ds\Big)\\
&\leq \int_0^\delta \int\int p_\varepsilon(y-x) m_2 P_s(y)^2\,dy\, m_1 P_s(x)\,dx\,ds + m_2(1) m_1(1)\delta\\
&\leq \int_0^\delta m_2(1) s^{-1/2} \int m_2 P_{s+\varepsilon}(x) m_1 P_s(x)\,dx\,ds + m_1(1) m_2(1)\delta\\
&\to 0 \text{ as } \delta \downarrow 0,
\end{aligned}$$

by the same argument as that at the end of the proof of (a). ∎

Remark IV.2.2. In $(CS)^\lambda_m$ we may restrict the test functions ϕ_i to $C^\infty_b(\mathbb{R})$. To see this, first recall from Examples II.2.4 that this class is a core for $\mathcal{D}(\Delta/2)$. Now suppose the conclusion of $(CS)^\lambda_m$ has been verified for $C^\infty_b(\mathbb{R})$ and for a sequence of functions $\{(\phi^n_1, \phi^n_2)\}$ in $\mathcal{D}(\Delta/2)^2$ such that $(\phi^n_i, \frac{\Delta}{2}\phi^n_i) \xrightarrow{bp} (\phi_i, \frac{\Delta}{2}\phi_i)$ as $n \to \infty$ for $i = 1, 2$. It follows from $(CS)^\lambda_m$ for $\phi_i \equiv 1$ that $E(X^i_s(1)) \leq m_i(1)$. Therefore by Dominated Convergence

$$E(\sup_{t\leq T}(M^i_t(\phi^n_i) - M^i_t(\phi_i))^2) \leq cE\left(\int_0^T X^i_s((\phi^n_i - \phi_i)^2)\,ds\right) \to 0 \text{ as } n \to \infty.$$

By Dominated Convergence it is now easy to take limits in $(CS)^\lambda_m$ to see that this conclusion persists for the limiting functions (ϕ_1, ϕ_2). This establishes the claim.

We also showed in the proof of (a) above that $(CS)^\lambda_m$ is equivalent to $(MP)^m_g$ with $g_i = \lambda_i u_{3-i} \ge 0$. Hence by Remark IV.1.7(c), $(CS)^\lambda_m$ remains unchanged if we only assumed that $M^i_\cdot(\phi_i)$ are continuous $(\mathcal{F}^2_t)$-local martingales.

One can easily reformulate $(CS)^\lambda_m$ as a stochastic pde. Assume W_1, W_2 are independent white noises on $\bar\Omega = (\Omega, \mathcal{F}, \mathcal{F}_t, \mathbb{P})$. Recall that $C_K(\mathbb{R})$ is the space of continuous functions on $\mathbb{R}$ with compact support equipped with the sup norm.

A pair of non-negative processes $\{u_1(t,x), u_2(t,x) : t > 0, x \in \mathbb{R}\}$, is a solution of

$$(\mathrm{SPDE})^\lambda_m \qquad \frac{\partial u_i}{\partial t} = \frac{\Delta u_i}{2} - \lambda_i u_1 u_2 + \sqrt{u_i}\dot{W}_i, \quad u_i(0+, x)dx = m_i.$$

iff for $i = 1, 2$,

(*i*) $\{u_i(t,\cdot) : t > 0\}$ is continuous and $(\mathcal{F}_t)$ – adapted with values in $C_K(\mathbb{R})$.

$$(ii)\langle u_i(t), \phi\rangle \equiv \int u_i(t,x)\phi(x)dx = m_i(\phi) + \int_0^t \langle u_i(s), \frac{\phi''}{2}\rangle - \lambda_i\langle u_1(s)u_2(s), \phi\rangle\, ds$$
$$+ \int_0^t \int \phi(x)\sqrt{u_i(s,x)}dW_i(s,x), \ \forall t > 0 \text{ a.s. } \forall \phi \in C^2_b(\mathbb{R}).$$

As in Remark III.4.1 this implies

$$(IV.2.4) \qquad X^i_t(dx) \equiv u_i(t,x)dx \xrightarrow{\text{a.s.}} m_i \text{ as } t \downarrow 0.$$

Proposition IV.2.3. Assume $m \in F$.
(a) If (u_1, u_2) satisfies $(\mathrm{SPDE})^\lambda_m$, and X is given by (IV.2.4), then $\mathcal{L}(X) = \mathbb{P}^0_m$. In particular, the law of u on $C((0,\infty), C_K(\mathbb{R})^2)$ is unique.
(b) There is an $\bar\Omega' = (\Omega', \mathcal{F}', \mathcal{F}'_t, \mathbb{P}')$ such that if

$$\bar\Omega = (\Omega^2 \times \Omega', \mathcal{F}^2 \times \mathcal{F}', (\mathcal{F}^2_\cdot \times \mathcal{F}'_\cdot)_{t+}, \mathbb{P}^0_m \times \mathbb{P}')$$

and $\Pi : \Omega^2 \times \Omega' \to \Omega'$ is the projection map, then there is a pair of independent white noises, $\dot W_1, \dot W_2$ on $\bar\Omega$ such that $(u_1, u_2) \circ \Pi$ solves $(\mathrm{SPDE})^\lambda_m$ on $\bar\Omega$.

Proof. (a) The weak continuity of X follows from (IV.2.4) as in the proof of Theorem III.4.2(c). It now follows from Remark IV.2.2 that X satisfies $(CS)^\lambda_m$ and hence has law $\mathbb{P}^0_m$ by Theorem IV.2.1. The second assertion now follows as in the univariate case (Corollary III.4.3(c)).

(b) Let $u_{n,i}(t, X, x) = \frac{n}{2} X^i_t((x - \frac{1}{n}, x + \frac{1}{n}])$. We know $\mathbb{P}^0_m \ll \mathbb{P}_{m_1} \times \mathbb{P}_{m_2}$ and under the latter measure $u_i(t,x)$ is the jointly continuous density of X^i_t on $(0,\infty) \times \mathbb{R}$ (Theorem III.4.2(a)), and $\{(t,x) : u_i(t,x) > 0, t \ge \delta\}$ is bounded for every $\delta > 0$ (Corollary III.1.7). It follows that $\mathbb{P}_{m_1} \times \mathbb{P}_{m_2}$-a.s. and therefore $\mathbb{P}^0_m$-a.s. for every $\delta > 0$, and $i = 1, 2$,

$$\sup_{x\in\mathbb{R}} \sup_{t\in[\delta,\delta^{-1}]} |u_{n,i}(t,x) - u_{n',i}(t,x)|$$
$$= \sup_{x\in\mathbb{Q}} \sup_{t\in\mathbb{Q}\cap[\delta,\delta^{-1}]} |u_{n,i}(t,x) - u_{n',i}(t,x)| \to 0 \text{ as } n, n' \to \infty,$$

and $\exists R$ such that $\sup_{t\in[\delta,\delta^{-1}]} X_t^i(B(0,R)^c) = 0$. It follows that (i) holds $\mathbb{P}_m^0$-a.s. It remains to show that (ii) holds on this larger space. Choose Ω' carrying two independent white noises, W_1', W_2' on $\mathbb{R}_+ \times \mathbb{R}$. Define W_i on $\bar{\Omega}$ by

$$W_i(\omega', X)_t(A) = \int_0^t \int 1_A(x) \frac{1(u_i(s,X,x) > 0)}{\sqrt{u_i(s,X,x)}} dM^i(X)(s,x) + \int_0^t \int 1_A(x) 1(u_i(s,X,x) = 0) dW_i'(\omega')(s,x).$$

As in Theorem III.4.2(b), (W_1, W_2) are independent white noises on $\bar{\Omega}$ and $(u^1, u^2) \circ \Pi$ satisfies $(\mathrm{SPDE})_m^\lambda$ on $\bar{\Omega}$. Note the independence follows from the orthogonality of the martingales $W_1(t)(A)$ and $W_2(t)(B)$ for each A and B because these are Gaussian processes in (t, A). ∎

Here is a univariate version of the above result which may be proved in the same manner. If $\sigma^2, \gamma > 0$, $\lambda \geq 0$, and $\theta \in \mathbb{R}$, consider

$$(SPDE) \qquad \frac{\partial u}{\partial t} = \frac{\sigma^2 \Delta u}{2} + \sqrt{\gamma u}\dot{W} + \theta u - \lambda u^2, \quad u_{0+}(x)ds = m(dx),$$

where $m \in M_F(\mathbb{R})$, and the above equation is interpreted as before.

In the next result we also use $u_t(x) = \frac{dX_t}{dx}(x)$, to denote the canonical density of the absolutely continuous part of X_t on the canonical space of paths Ω_X (defined as before).

Proposition IV.2.4. Assume

$$\int\int \left(\log \frac{1}{|x_1 - x_2|}\right)^+ dm(x_1)dm(x_2) < \infty.$$

(a) There is a filtered space $(\Omega, \mathcal{F}, \mathcal{F}_t, \mathbb{P})$ carrying a solution of (SPDE).
(b) If u is any solution of (SPDE) and $\mathbb{P}$ is the law of $t \to u_t(x)dx$ on Ω_X, then

$$\frac{d\mathbb{P}}{d\mathbb{P}_m}\Big|_{\mathcal{F}_t^X} = \exp\left\{\int_0^t \int (\theta - \lambda u(s,x)) dM(s,x) - \frac{1}{2}\int_0^t \int (\theta - \lambda u(s,x))^2 X_s(dx)ds\right\}.$$

Here $\mathbb{P}_m$ is the law of super-Brownian motion starting at m with spatial variance σ^2, 0 drift and branching rate γ, and $dM(s,x)$ is the associated martingale measure. In particular the law of u on $C((0,\infty), C_K(\mathbb{R})^2)$ is unique.

The above result was pointed out by Don Dawson in response to a query of Rick Durrett. Durrett's question was prompted by his conjecture that the above SPDE arises in the scaling limit of a contact process in one dimension. The conjecture was confirmed by Mueller and Tribe (1994).

3. Collision Local Time

To study $(CS)_m^\lambda$ in higher dimensions we require an analogue of $u_1(s,x)u_2(s,x)dsdx$ which will exist in higher dimensions when the measures in question will not have densities. This is the collision local time of a pair of measure-valued processes which we now define.

Definition. Let $X = (X^1, X^2)$ be a pair of continuous $M_F(\mathbb{R}^d)$-valued processes on a common probability space and let p_t denote the standard Brownian transition density. The collision local time (COLT) of X is a continuous non-decreasing $M_F(\mathbb{R}^d)$-valued process $L_t(X)$ such that for any $\phi \in C_b(\mathbb{R}^d)$ and $t \geq 0$,

$$L_t^\varepsilon(X)(\phi) \equiv \int_0^t \int \phi\left(\frac{x_1 + x_2}{2}\right) p_\varepsilon(x_1 - x_2) X_s^1(dx_1) X_s^2(dx_2) ds \xrightarrow{P} L_t(X)(\phi)$$

as $\varepsilon \to 0$.

Definition. The graph of an $M_F(\mathbb{R}^d)$-valued process $(X_t, t \geq 0)$ is

$$G(X) = \cup_{\delta>0} \mathrm{cl}\{(t,x) : t \geq \delta,\ x \in S(X_t)\} \equiv \cup_{\delta>0} G_\delta(X) \subset \mathbb{R}_+ \times \mathbb{R}^d.$$

Remarks IV.3.1. (a) Clearly the process $L(X)$ is uniquely defined up to null sets. It is easy to check that $L(X)(ds, dx)$ is supported by $G(X^1) \cap G(X^2)$. This random measure gauges the intensity of the space-time collisions between the populations X^1 and X^2 and so can be used as a means of introducing local interactions between these populations. See the next section and Dawson et al (2000a) for examples.

(b) If $X_s^i(dx) = u_i(s,x)dx$, where u_i is a.s. bounded on $[0,t] \times \mathbb{R}^d$, then an easy application of Dominated Convergence, shows that

$$L_t(X)(dx) = \Big(\int_0^t u_1(s,x) u_2(s,x) ds\Big) dx.$$

However $L_t(X)$ may exist even for singular measures as we will see in Theorem IV.3.2 below.

(c) The definition of collision local time remains unchanged if $L_t^\varepsilon(X)(\phi_i)$ is replaced with $L_t^{\varepsilon,i}(X)(\phi_i) = \int_0^t \int p_\varepsilon(x_1 - x_2) X_s^j(dx_j) \phi_i(x_i) X_s^i(dx_i) ds$ $(i \neq j)$. This is easy to see by the uniform continuity of ϕ on compact sets.

Throughout this Section we will assume

(H_1) Z^i is an $(\mathcal{F}_t) - (SBM)(\gamma_i)$ starting at $m_i \in M_F(\mathbb{R}^d)$, $i = 1, 2$, defined on $(\Omega, \mathcal{F}, \mathcal{F}_t, \mathbb{P})$, and (Z^1, Z^2) are independent.

Let $Z_t = Z_t^1 \times Z_t^2$. Recall from Section III.5 that $g_\beta(r) = \begin{cases} r^{-\beta} & \text{if } \beta > 0 \\ 1 + \log^+ \frac{1}{r} & \text{if } \beta = 0 \\ 1 & \text{if } \beta < 0 \end{cases}$.

Theorem IV.3.2. (a) If $d \leq 5$, $m_i \neq 0$, and

$$(IC) \qquad \int g_{d-2}(|z_1 - z_2|) m_1(dz_1) m_2(dz_2) < \infty \quad \text{if } d \leq 4$$

$$\int g_{d-1}(|z_1 - z_2|) m_1(dz_1) m_2(dz_2) < \infty \quad \text{if } d = 5,$$

then $L_t(Z)$ exists, is not identically 0 and satisfies

$$\lim_{\varepsilon \downarrow 0} \| \sup_{t \leq T} |L_t^\varepsilon(Z)(\phi) - L_t(Z)(\phi)| \|_2 = 0 \quad \forall T > 0,\ \phi \in b\mathcal{B}(\mathbb{R}^d).$$

In particular, $\mathbb{P}(G(Z^1) \cap G(Z^2) \neq \emptyset) > 0$.
(b) If $d \geq 6$, then $G(Z^1) \cap G(Z^2) = \emptyset$ a.s.

We will prove this below except for the critical 6-dimensional case whose proof will only be sketched.

Lemma IV.3.3. If $d \geq 2$, there is a constant $C = C(d, \gamma_1, \gamma_2)$, and for each $\delta > 0$, a random $r_1(\delta, \omega) > 0$ a.s. such that for all $0 \leq r \leq r_1(\delta)$,

$$\sup_{t \geq \delta} \iint 1(|z_1 - z_2| \leq r) Z_t^1(dz_1) Z_t^2(dz_2) \leq C(\sup_t Z_t^1(1) + 1) r^{4-4/d} \Big(\log \frac{1}{r}\Big)^{2+2/d}.$$

Proof. We defer this to the end of this Section. It is a nice exercise using the results of Chapter III but the methods are not central to this Section. Clearly if $d = 1$ the above supremuim is a random multiple of r by Theorem III.4.2.

Corollary IV.3.4. If $d \geq 2$ and $0 \leq \beta < 4 - 4/d$, then with probability 1,

$$\limsup_{\varepsilon \downarrow 0} \sup_{t \geq \delta} \int g_\beta(|z_1 - z_2|) 1(|z_1 - z_2| \leq \varepsilon) Z_t^1(dz_1) Z_t^2(dz_2) = 0 \quad \forall \delta > 0$$

and

$$t \mapsto \iint g_\beta(|z_1 - z_2|) Z_t^1(dz_1) Z_t^2(dz_2) \text{ is continuous on } (0, \infty).$$

In particular, this is the case for $\beta = d - 2$ and $d \leq 5$.
Proof. Define a random measure on $[0, \infty)$ by

$$D_t(A) = Z_t(\{(z_1, z_2) : |z_1 - z_2| \in A\}).$$

If $0 < \beta < 4 - 4/d$ and $\varepsilon < r(\delta, \omega)$, then an integration by parts and Lemma IV.3.3 give

$$\begin{aligned}
&\sup_{t \geq \delta} \iint g_\beta(|z_1 - z_2|) 1(|z_1 - z_2| \leq \varepsilon) Z_t^1(dz_1) Z_t^2(dz_2) \\
&= \sup_{t \geq \delta} \Big[g_\beta(r) D_t([0, r])|_{0+}^{\varepsilon} + \beta \int_0^\varepsilon r^{-1-\beta} D_t([0, r])\, dr \Big] \\
&\leq C(\sup_t Z_t^1(1) + 1) \Big[\varepsilon^{-\beta + 4 - 4/d} (\log^+ \frac{1}{\varepsilon})^{2+2/d} + \beta \int_0^\varepsilon r^{3-\beta-4/d} (\log^+ \frac{1}{r})^{2+2/d}\, dr \Big] \\
&\to 0 \text{ as } \varepsilon \downarrow 0,
\end{aligned}$$

by our choice of β. It follows that for all $0 \leq \beta < 4 - 4/d$,

$$\lim_{M \to \infty} \sup_{t \geq \delta} \iint (g_\beta(|z_1 - z_2|) - g_\beta(|z_1 - z_2|) \wedge M)\, dZ_t^1 dZ_t^2 = 0 \quad \text{a.s.}$$

(if $\beta = 0$ we simply compare with a $\beta > 0$). The weak continuity of Z_t shows that $t \mapsto \int g_\beta(|z_1 - z_2|) \wedge M\, dZ_t$ is a.s. continuous and the second result follows. ∎

Throughout the rest of this Section we assume

(H_2) $X = (X^1, X^2)$ satisfies $(MP)_{C,0}^m$ for some C with $E_i = \mathbb{R}^d$ and $A_i = \Delta/2$ on $\bar{\Omega} = (\Omega, \mathcal{F}, \mathcal{F}_t, \mathbb{P})$.

Apply Proposition IV.1.4 with $D^i = 0$ to see that by enlarging the space we may assume there is a pair of independent SBM's (Z^1, Z^2) as in (H_1) defined on $\bar{\Omega}$ such that $X_t^i \leq Z_t^i$ for all $t \geq 0$ and $i = 1, 2$. Set $X_t = X_t^1 \times X_t^2$. We first derive a martingale problem for X and then construct $L_t(X)$ by means of a Tanaka formula.

Notation. $\frac{\vec{\Delta}}{2}$ is the generator of the standard $2d$-dimensional Brownian motion and $\vec{P}_t$ is its semigroup.

Lemma IV.3.5. For any $\phi \in \mathcal{D}(\vec{\Delta}/2)$,

$$\begin{aligned} X_t(\phi) =& X_0(\phi) + \int_0^t \int\int \phi(x_1, x_2)[X_s^1(dx_1)M^2(ds, dx_2) + X_s^2(dx_2)M^1(ds, dx_1)] \\ (IV.3.1) \quad & - \int_0^t \int\int \phi(x_1, x_2)[X_s^1(dx_1)C^2(ds, dx_2) + X_s^2(dx_2)C^1(ds, dx_1)] \\ & + \int_0^t X_s(\frac{\vec{\Delta}\phi}{2})\, ds. \end{aligned}$$

Proof. *Step 1.* $\phi(x_1, x_2) = \phi_1(x_1)\phi_2(x_2)$, $\phi_i \in \mathcal{D}(\vec{\Delta}/2)$.
Then $X_t(\phi) = X_t^1(\phi_1)X_t^2(\phi_2)$ and the result follows from $(MP)_{C,0}^m$ by an integration by parts.
Step 2. $\phi(x) = \vec{P}_\varepsilon\psi(x)$, where $\psi \in \mathcal{D}(\vec{\Delta}/2)$.
Then there is a sequence of finite Riemann sums of the form

$$\phi_n(x_1, x_2) = \sum_{y_1^{i,n}, y_2^{i,n}} p_\varepsilon(y_1^{i,n} - x_1)p_\varepsilon(y_2^{i,n} - x_2)\psi(y_1^{i,n}, y_2^{i,n})\Delta_n,$$

such that $\phi_n \overset{bp}{\to} \phi$ and

$$\begin{aligned} \frac{\vec{\Delta}}{2}\phi_n(x_1, x_2) &= \sum_{y_1^{i,n}, y_2^{i,n}} \frac{\vec{\Delta}}{2}(p_\varepsilon(y_1^{i,n} - \cdot)p_\varepsilon(y_2^{i,n} - \cdot))(x_1, x_2)\psi(y_1^{i,n}, y_2^{i,n})\Delta_n \\ &\overset{bp}{\to} \frac{\vec{\Delta}}{2}\vec{P}_\varepsilon\psi(x_1, x_2). \end{aligned}$$

By Step 1, (IV.3.1) holds for each ϕ_n. Now let $n \to \infty$ and use Dominated Convergence to obtain this result for ϕ.
Step 3. $\phi \in \mathcal{D}(\vec{\Delta}/2)$.
Let $\varepsilon_n \downarrow 0$ and note that $\vec{P}_{\varepsilon_n}\phi \overset{bp}{\to} \phi$ and $\frac{\vec{\Delta}}{2}\vec{P}_{\varepsilon_n}\phi = \vec{P}_{\varepsilon_n}\left(\frac{\vec{\Delta}}{2}\phi\right) \overset{bp}{\to} \frac{\vec{\Delta}}{2}\phi$ as $n \to \infty$. Now use (IV.3.1) for $\vec{P}_{\varepsilon_n}\phi$ (from Step 2) and let $n \to \infty$ to derive it for ϕ. ∎

Let $\phi \in C_K(\mathbb{R}^d)$ and apply the above result to $\phi_\varepsilon \in \mathcal{D}(\vec{\Delta}/2)$, which is chosen so that

$$(IV.3.2) \qquad \int_0^t X_s\left(\frac{\vec{\Delta}}{2}\phi_\varepsilon\right)ds = -L_t^\varepsilon(\phi).$$

This will be the case if

$$(IV.3.3)\qquad \frac{\vec{\Delta}}{2}\phi_\varepsilon(x_1,x_2)=-p_\varepsilon(x_1-x_2)\phi(\frac{x_1-x_2}{2})\equiv\psi_\varepsilon(x_1,x_2).$$

Let $\vec{U}_\lambda$ denote the $2d$-dimensional Brownian resolvent for $\lambda\ge 0$ and assume $d>2$. By Exercise II.2.2, $\phi_\varepsilon(x)=\vec{U}_0\psi_\varepsilon(x)\in\mathcal{D}(\vec{\Delta}/2)$ satisfies (IV.3.3). If $B_s=(B_s^1,B_s^2)$ is a $2d$-dimensional Brownian motion, then $\frac{B^1+B^2}{\sqrt{2}}$ and $\frac{B^1-B^2}{\sqrt{2}}$ are independent d-dimensional Brownian motions and so a simple calculation yields

$$\begin{aligned}\phi_\varepsilon(x_1,x_2)&=E^{x_1,x_2}\Big(\int_0^\infty p_\varepsilon(B_s^1-B_s^2)\phi\Big(\frac{B_s^1+B_s^2}{2}\Big)\,ds\Big)\\ (IV.3.4)\qquad &=2^{1-d}\int_0^\infty p_{\varepsilon/4+u}\Big(\frac{x_1-x_2}{2}\Big)P_u\phi(\frac{x_1+x_2}{2}\Big)\,du\\ &\equiv G_\varepsilon\phi(x_1,x_2).\end{aligned}$$

We may use (IV.3.2) in Lemma IV.3.5 and conclude that

$$\begin{aligned}X_t(G_\varepsilon\phi)&=X_0(G_\varepsilon\phi)\\ &\quad+\int_0^t\iint G_\varepsilon\phi(x_1,x_2)[X_s^1(dx_1)M^2(ds,dx_2)+X_s^2(dx_2)M^1(ds,dx_1)]\\ (T)_\varepsilon\qquad &\quad-\int_0^t\iint G_\varepsilon\phi(x_1,x_2)[X_s^1(dx_1)C^2(ds,dx_2)+X_s^2(dx_2)C^1(ds,dx_1)]\\ &\quad-L_t^\varepsilon(\phi)\qquad\forall t>0,\quad\text{for } d>2.\end{aligned}$$

(IV.3.4) shows that $G_\varepsilon\phi$ is defined for any $\phi\in b\mathcal{B}(\mathbb{R}^d)$ and that $\phi_n\xrightarrow{bp}\phi$ implies $G_\varepsilon\phi_n\xrightarrow{bp}G_\varepsilon\phi$. Now use Dominated Convergence to extend $(T)_\varepsilon$ to all $\phi\in b\mathcal{B}(\mathbb{R}^d)$. A similar argument with

$$G_{\lambda,\varepsilon}\phi(x_1,x_2)\equiv\vec{U}_\lambda\psi_\varepsilon(x_1,x_2)=2^{1-d}\int_0^\infty e^{-2\lambda u}p_{\varepsilon/4+u}\Big(\frac{x_1-x_2}{2}\Big)P_u\phi\Big(\frac{x_1+x_2}{2}\Big)\,du$$

in place of $G_\varepsilon\phi=G_{0,\varepsilon}\phi$ shows that for any $\phi\in b\mathcal{B}(\mathbb{R}^d)$,

$$\begin{aligned}X_t(G_{\lambda,\varepsilon}\phi)&=X_0(G_{\lambda,\varepsilon}\phi)\\ &\quad+\int_0^t\iint G_{\lambda,\varepsilon}\phi(x_1,x_2)[X_s^1(dx_1)M^2(ds,dx_2)+X_s^2(dx_2)M^1(ds,dx_1)]\\ (T)_{\lambda,\varepsilon}\qquad &\quad-\int_0^t\iint G_{\lambda,\varepsilon}\phi(x_1,x_2)[X_s^1(dx_1)C^2(ds,dx_2)+X_s^2(dx_2)C^1(ds,dx_1)]\\ &\quad+\lambda\int_0^t X_s(G_{\lambda,\varepsilon}\phi)\,ds-L_t^\varepsilon(\phi)\qquad\forall t>0,\quad\text{for } d\ge 1.\end{aligned}$$

As we want to let $\varepsilon\downarrow 0$ in the above formulae, introduce

$$G_{\lambda,0}\phi(x_1,x_2)=2^{1-d}\int_0^\infty e^{-2\lambda u}p_u\Big(\frac{x_1-x_2}{2}\Big)P_u\phi\Big(\frac{x_1+x_2}{2}\Big)\,du,\quad G_0\phi=G_{0,0}\phi,$$

when this integral is well-defined, as is the case if $\phi \geq 0$. A simple integration shows that for any $\varepsilon \geq 0$,

$$(IV.3.5) \quad G_\varepsilon|\phi|(x_1,x_2) \leq \|\phi\|_\infty G_0 1(x_1,x_2) = \|\phi\|_\infty k_d g_{d-2}(|x_1-x_2|) \quad \text{if } d>2,$$

where $k_d = \Gamma(d/2-1)2^{-1-d/2}\pi^{-d/2}$. Therefore $G_0\phi(x_1,x_2)$ is finite when ϕ is bounded, $x_1 \neq x_2$, and $d>2$.

Lemma IV.3.6. Let $\phi \in b\mathcal{B}(\mathbb{R}^d)$ and $d>2$. Then

$$|G_\varepsilon\phi(x_1,x_2) - G_0\phi(x_1,x_2)| \leq \|\phi\|_\infty c_d \min(|x_1-x_2|^{2-d}, \varepsilon|x_1-x_2|^{-d}).$$

If $\phi \geq 0$, $\lim_{\varepsilon\downarrow 0} G_\varepsilon\phi(x) = G_0\phi(x)$ ($\leq \infty$) for all x and $G_0\phi$ is lower semicontinuous.

Proof.

$$\begin{aligned}
&|G_\varepsilon\phi(x_1,x_2) - G_0\phi(x_1,x_2)| \\
&\leq \|\phi\|_\infty \int_0^\infty |p_{\varepsilon/4+u}((x_1-x_2)/2) - p_u((x_1-x_2)/2)|\,du \\
&\leq \|\phi\|_\infty \int_0^\infty \int_u^{u+\varepsilon/4} \left|\frac{\partial p_v}{\partial v}((x_1-x_2)/2)\right| dv\,du \\
&\leq \|\phi\|_\infty \Big[\int_{\varepsilon/4}^\infty \frac{\varepsilon}{4} p_v((x_1-x_2)/2)[(x_1-x_2)^2 v^{-2}/8 + d(2v)^{-1}]\,dv \\
&\qquad\qquad + \int_0^{\varepsilon/4} p_v((x_1-x_2)/2)[(x_1-x_2)^2(8v)^{-1} + d/2]dv\Big] \\
&\leq \|\phi\|_\infty c'_d \varepsilon \Big[\int_0^{(x_1-x_2)^2/2\varepsilon} e^{-y} y^{d/2-1}(y+d)\,dy |x_1-x_2|^{-d} \\
&\qquad\qquad + \int_{(x_1-x_2)^2/2\varepsilon}^\infty e^{-y} y^{d/2-2}[y+(d/2)]dy |x_1-x_2|^{2-d}\Big],
\end{aligned}$$

where we substituted $y=(x_1-x_2)^2(8v)^{-1}$ in the last line . The integrand in the first term of the last line is both bounded and integrable and so the first term is at most

$$c''_d \varepsilon \|\phi\|_\infty |x_1-x_2|^{-d} \min((x_1-x_2)^2(2\varepsilon)^{-1}, 1).$$

The integrand in the second term is at most $c(y^{-2}\wedge 1)$ and so the second term is bounded by

$$c''_d \min(|x_1-x_2|^{2-d}, \varepsilon|x_1-x_2|^{-d}).$$

This gives the first inequality and so for the second result we need only consider $x=(x_1,x_1)$. This is now a simple consequence of Monotone Convergence. The lower semicontinuity of $G_0\phi$ follows from the fact that it is the increasing pointwise limit of the sequence of continuous functions

$$\int_{2^{-n}}^\infty p_u((x_1-x_2)/2) P_u\phi((x_1+x_2)/2)\,du. \quad ∎$$

Lemma IV.3.7. If $3 \le d \le 5$, then for each $t > 0$ there is a $c_d(t)$ so that

$$\mathbb{E}\Big(\int_0^t \int \Big[\int g_{d-2}(|z_1 - z_2|) Z_s^1(dz_1)\Big]^2 Z_s^2(dz_2) ds\Big)$$
$$\le c_d(t) \iint [g_{2(d-3)}(|z_1 - z_2|) + 1]\, dm_1(z_1) dm_2(z_2).$$

Proof. We may assume $t \ge 1$. Recall that $m_2 P_s(x) = \int p_s(y - x) m_2(dy)$. Use the first and second moment calculations in Exercise II.5.2 to see that the above expectation is

$$\int_0^t \int \Big[\int g_{d-2}(|z_1 - z_2|) m_1 P_s(z_1) dz_1\Big]^2 m_2 P_s(z_2)\, dz_2 ds$$
$$+ \int_0^t \int \Big[\int_0^s \int P_{s-u}(g_{d-2}(|\cdot - z_2|))(z_1)^2 m_1 P_u(z_1) dz_1 du\Big] m_2 P_s(z_2)\, dz_2 ds$$
$$(IV.3.6) \quad \equiv I_1 + I_2.$$

Use

$$(IV.3.7) \qquad \int_0^\infty p_u(x)\, du = k(d) g_{d-2}(|x|)$$

and Chapman-Kolmogorov to see that

$$I_1 = c_d \int_0^t \int \Big[\int_0^\infty \int_u^\infty m_1 P_{s+u}(z_2) m_1 P_{s+u'}(z_2) du' du\Big] m_2 P_s(z_2) dz_2 ds$$
$$\le c_d \int_0^t \int_0^\infty \Big[\int_u^\infty (s + u')^{-d/2} du'\Big] m_1(1) \iint p_{2s+u}(z_1 - z_2) dm_1(z_1) dm_2(z_2) du ds$$
$$\le c_d m_1(1) \iint \Big[\int_0^t \int_{2s}^\infty (v - s)^{1-d/2} p_v(z_1 - z_2) dv ds\Big] dm_1(z_1) dm_2(z_2)$$
$$\le c_d m_1(1) \iint \Big[\int_0^\infty v^{1-d/2}(v \wedge t) p_v(z_1 - z_2) dv\Big] dm_1(z_1) dm_2(z_2)$$
$$\le c_d m_1(1) \iint \Big[\int_0^t v^{2-d/2} p_v(z_1 - z_2) dv + t \int_t^\infty v^{1-d} dv\Big] dm_1(z_1) dm_2(z_2).$$

A routine calculation now shows that (recall $t \ge 1$ to handle the second term)

$$(IV.3.8) \qquad I_1 \le c_d m_1(1) \begin{cases} \int (|z_1 - z_2|^{6-2d} + 1)\, dm_1(z_1) dm_2(z_2) & \text{if } d > 3 \\ \int (\log^+\big(\frac{2t}{|z_1 - z_2|}\big) + 1)\, dm_1(z_1) dm_2(z_2) & \text{if } d = 3. \end{cases}$$

For I_2, note first that (IV.3.7) implies

$$P_{s-u}(g_{d-2}(|\cdot - z_2|))(z_1) = k(d) \int_{s-u}^\infty p_v(z_1 - z_2)\, dv,$$

and so

$$I_2 = 2k(d)^2 \int_0^t ds \int_0^s du \int_{s-u}^\infty dv \int_v^\infty dv' \Big[\iint p_v(z_1 - z_2) p_{v'}(z_1 - z_2) \\ m_1 P_u(z_1) m_2 P_s(z_2) dz_1 dz_2 \Big]$$

$$\leq c_d \int_0^t ds \int_0^s du \int_{s-u}^\infty dv\, v^{1-d/2} \Big[\iint p_{u+v+s}(z_1 - z_2) m_1(dz_1) m_2(dz_2) \Big].$$

Use the fact that $p_{u+v+s}(x) \leq 2^{d/2} p_{2(u+v)}(x)$ for $s \leq u + v$ and integrate out $s \in [u, (u+v) \wedge t]$ in the above to get

$$I_2 \leq c_d \iint \Big[\int_0^t \int_0^\infty v^{1-d/2} \min(v, t-u) p_{2(u+v)}(z_1 - z_2)\, dv du \Big] m_1(dz_1) m_2(dz_2)$$

$$\leq c_d \iint \Big[\int_0^\infty \int_0^{w \wedge t} (w-u)^{1-d/2} ((w \wedge t) - u)\, du\, p_{2w}(z_1 - z_2)\, dw \Big] m_1(dz_1) m_2(dz_2)$$

$$(IV.3.9) \leq c_d \iint \Big[\int_0^\infty (w \wedge t)^{3-d/2} p_{2w}(z_1 - z_2)\, dw \Big] dm_1(z_1) dm_2(z_2).$$

A change of variables now gives (recall $t \geq 1$)

$$\int_0^\infty (w \wedge t)^{3-d/2} p_{2w}(\Delta)\, dw \leq c_d \Big[\Delta^{8-2d} \int_{\Delta^2/4t}^\infty x^{d-5} e^{-x} dx + t^{3-d/2} \int_t^\infty w^{-d/2} dw \Big]$$

$$\leq c_d \begin{cases} \Delta^{-2} + 1 & \text{if } d = 5 \\ \log^+\left(\frac{4t}{\Delta^2}\right) + 1 & \text{if } d = 4 \\ t & \text{if } d = 3. \end{cases}$$

Use this in (IV.3.9) to see that

$$I_2 \leq c_d(t) \begin{cases} \iint (|z_1 - z_2|^{-2} + 1)\, m_1(dz_1) m_2(dz_2) & \text{if } d = 5 \\ \iint (\log^+\left(\frac{1}{|z_1 - z_2|}\right) + 1)\, m_1(dz_1) m_2(dz_2) & \text{if } d = 4 \\ m_1(1) m_2(1) & \text{if } d = 3. \end{cases}$$

Combine this with (IV.3.8) and (IV.3.6) to complete the proof. ∎

Theorem IV.3.8. Assume X satisfies (H_2) where $d \leq 5$ and m_1, m_2 satisfy (IC).
(a) $L_t(X)$ exists and for any $\phi \in b\mathbb{B}(\mathbb{R}^d)$,

$$(IV.3.10) \qquad \sup_{t \leq T} |L_t^\varepsilon(X)(\phi) - L_t(X)(\phi)| \xrightarrow{L^2} 0 \text{ as } \varepsilon \downarrow 0 \quad \text{for all } T > 0.$$

(b) If $\lambda = 0$ and $d \geq 3$, or $\lambda > 0$ and $d \geq 1$, then for any $\phi \in b\mathcal{B}(\mathbb{R}^d)$,

$$X_t(G_{\lambda,0}\phi) = X_0(G_{\lambda,0}\phi) + \int_0^t \iint G_{\lambda,0}\phi(x_1,x_2)[X_s^1(dx_1)M^2(ds,dx_2) + X_s^2(dx_2)M^1(ds,dx_1)]$$

$$(T) \qquad - \int_0^t \iint G_{\lambda,0}\phi(x_1,x_2)[X_s^1(dx_1)C^2(ds,dx_2) + X_s^2(dx_2)C^1(ds,dx_1)]$$

$$+ \lambda \int_0^t X_s(G_{\lambda,0}\phi)\,ds - L_t(X)(\phi) \quad \forall t \geq 0 \quad \text{a.s.}$$

Each of the above processes are a.s. continuous in $t \geq 0$. The second term on the right-hand side is an L^2 $(\mathcal{F}_t)$-martingale and each of the other processes on the right-hand side has square integrable total variation on bounded time intervals.

Proof. We give the proof for $\lambda = 0$ and $d \geq 3$. The extra term involved when $\lambda > 0$ is very easy to handle and for $d \leq 3$ the entire proof simplifies considerably by means of a systematic use of Theorem III.3.4 (the reader may want to consider only this case, especially as the interactive models discussed in the next Section will only exist in these dimensions).

Let $\phi \in b\mathcal{B}(\mathbb{R}^d)_+$ and note that it suffices to prove the Theorem for such a non-negative ϕ. Consider the limit of each of the terms in $(T)_\varepsilon$ as $\varepsilon \downarrow 0$. (IC) and (IV.3.5) allow us to use Dominated Convergence and conclude from Lemma IV.3.6 that

$$(IV.3.11) \qquad \lim_{\varepsilon \downarrow 0} \iint G_\varepsilon \phi \, dm_1 dm_2 = \iint G_0 \phi \, dm_1 dm_2.$$

Let

$$N_t^\varepsilon(\phi) = \int_0^t \iint G_\varepsilon\phi(x_1,x_2)[X_s^1(dx_1)M^2(ds,dx_2) + X_s^2(dx_2)M^1(ds,dx_1)], \quad \varepsilon, t \geq 0.$$

Note that Lemma IV.3.7, (IV.3.5) and the domination $X^i \leq Z^i$ show that $N_t^\varepsilon(\phi)$ is a well-defined continuous square-integrable martingale even for $\varepsilon = 0$. Similarly, Lemmas IV.3.6 and IV.3.7, this domination, and Dominated Convergence show that for any $T > 0$,

$$\mathbb{E}(\sup_{t \leq T}(N_t^\varepsilon(\phi) - N_t^0(\phi))^2)$$

$$(IV.3.12) \qquad \leq c\mathbb{E}\Big(\int_0^T \gamma_2\Big(\int |G_\varepsilon\phi - G_0\phi|(x_1,x_2) Z_s^1(dx_1)\Big)^2 Z_s^2(dx_2)ds + \int_0^T \gamma_1\Big(\int |G_\varepsilon\phi - G_0\phi|(x_1,x_2) Z_s^2(dx_2)\Big)^2 Z_s^1(dx_1)ds\Big)$$

$$\to 0 \text{ as } \varepsilon \downarrow 0.$$

If $C(ds,dx_1,dx_2) = X_s^1(dx_1)C^2(ds,dx_2) + X_s^2(dx_2)C^1(ds,dx_1)$ then $(T)_\varepsilon$ implies for any $t > 0$,

$$\int_0^t \int G_\varepsilon 1(x_1,x_2)C(ds,dx_1,dx_2) \le m_1 \times m_2(G_\varepsilon 1) + N_t^\varepsilon(1)$$

$$\xrightarrow{L^2} m_1 \times m_2(G_0 1) + N_t^0(1),$$

the last by (IV.3.11) and (IV.3.12). Fatou's lemma and the equality in (IV.3.5) now show that

$$(IV.3.13) \qquad \mathbb{E}\Big(\Big(\int_0^t \iint g_{d-2}(|x_1 - x_2|)C(ds,dx_1,dx_2)\Big)^2\Big) < \infty \quad \forall t > 0.$$

This allows us to apply Lemma IV.3.6 and Dominated Convergence to conclude

$$(IV.3.14) \qquad \lim_{\varepsilon\downarrow 0}\mathbb{E}\Big(\int_0^t \iint |G_\varepsilon\phi - G_0\phi(x_1,x_2)|C(ds,dx_1,dx_2)^2\Big) = 0.$$

$(T)_\varepsilon$ shows that $X_t(G_\varepsilon\phi) \le X_0(G_\varepsilon\phi) + N_t^\varepsilon(\phi)$ for all $t \ge 0$ a.s. Let $\varepsilon \downarrow 0$, use Lemma IV.3.6 and Fatou's Lemma on the left-hand side, and (IV.3.11) and (IV.3.12) on the right-hand side to see that

$$(IV.3.15) \qquad X_t(G_0\phi) \le X_0(G_0\phi) + N_t^0(\phi) < \infty \quad \forall t \ge 0 \text{ a.s.}$$

Take $\phi = 1$ in the above inequality, recall that $N_t^0(1)$ is an L^2-martingale, and use the equality in (IV.3.5) to get

$$(IV.3.16) \qquad \mathbb{E}\Big(\Big(\sup_{t\le T}\iint g_{d-2}(|x_1 - x_2|)X_t^1(dx_1)X_t^2(dx_2)\Big)^2\Big) < \infty \quad \forall T > 0.$$

The bound in Lemma IV.3.6 shows that for any $T,\ \delta,\ \eta > 0$, if

$$S_\delta = \{(x_1,x_2) \in \mathbb{R}^{2d} : |x_1 - x_2| \le \delta\},$$

then

$$\begin{aligned}
&\sup_{t\le T} X_t(|G_\varepsilon\phi - G_0\phi|)\\
&\quad\le \sup_{t\le T} X_t(|G_\varepsilon\phi - G_0\phi|1_{S_\delta^c})\\
(IV.3.17) &\quad\quad + c_d\|\phi\|_\infty \sup_{t\le T}\int g_{d-2}(|x_1 - x_2|)1_{S_\delta}(x_1,x_2)X_t(dx_1,dx_2)\\
&\quad\le c_d\|\phi\|\Big[\varepsilon\delta^{-d}\sup_{t\le T} X_t(1) + \sup_{\eta\le t\le T}\int g_{d-2}(|x_1 - x_2|)1_{S_\delta}(x_1,x_2)X_t(dx_1,dx_2)\\
&\quad\quad + \sup_{t<\eta}\int g_{d-2}(|x_1 - x_2|)1_{S_\delta}(x_1,x_2)X_t(dx_1,dx_2)\Big].
\end{aligned}$$

Write $X_t(g_{d-2})$ for $\int g_{d-2}(|x_1 - x_2|)X_t(dx_1,dx_2)$. The lower semicontinuity of $(x_1,x_2) \to g_{d-2}(|x_1 - x_2|)$ (take $\phi = 1$ in Lemma IV.3.6 and use the equality in (IV.3.5)) and the weak continuity of X show that $\liminf_{t\downarrow 0} X_t(g_{d-2}) \ge X_0(g_{d-2})$.

On the other hand (IV.3.15) with $\phi = 1$ implies $\limsup_{t\downarrow 0} X_t(g_{d-2}) \le X_0(g_{d-2})$ a.s., and so

$$(IV.3.18) \qquad \lim_{t\downarrow 0} X_t(g_{d-2}) = X_0(g_{d-2}) \text{ a.s.}$$

Choose $\delta_n \downarrow 0$ so that $X_0(\{(x_1,x_2) : |x_1 - x_2| = \delta_n\}) = 0$. Weak continuity then implies $\lim_{t\downarrow 0} X_t(g_{d-2}1_{S^c_{\delta_n}}) = X_0(g_{d-2}1_{S^c_{\delta_n}})$ and so (IV.3.18) gives

$$(IV.3.19) \qquad \lim_{t\downarrow 0} X_t(g_{d-2}1_{S_{\delta_n}}) = X_0(g_{d-2}1_{S_{\delta_n}}) \text{ a.s.}$$

Let $\varepsilon_0 > 0$ and first choose an natural number N_0 so that the right-hand side is at most ε_0 for $n \ge N_0$. Next use (IV.3.19) to choose $\eta = \eta(\varepsilon_0)$ so that

$$(IV.3.20) \qquad \forall n \ge N_0 \quad \sup_{t<\eta} X_t(g_{d-2}1_{S_{\delta_n}}) \le \sup_{t<\eta} X_t(g_{d-2}1_{S_{\delta_{N_0}}}) < 2\varepsilon_0.$$

By Corollary IV.3.4 we may omit a $\mathbb{P}$-null set and then choose $N_1(\eta) \ge N_0$ so that

$$(IV.3.21) \qquad \sup_{t\ge\eta} X_t(g_{d-2}1_{S_{\delta_{N_1}}}) < \varepsilon_0.$$

Now take $\delta = \delta_{N_1}$ and $\eta = \eta(\varepsilon_0)$ in (IV.3.17). By (IV.3.20) and (IV.3.21) we see that outside a null set for $\varepsilon < \varepsilon(\varepsilon_0)$, the right-hand side of (IV.3.17) will be at most $\|\phi\|_\infty c_d 4\varepsilon_0$. We have proved

$$(IV.3.22) \qquad \limsup_{\varepsilon\downarrow 0} \sup_{t\le T} X_t(|G_\varepsilon\phi - G_0\phi|) = 0 \quad \forall T > 0 \text{ a.s. and in } L^2,$$

where Dominated Convergence, Lemma IV.3.6, and (IV.3.16) are used for the L^2-convergence.

(IV.3.11), (IV.3.12), (IV.3.14) and (IV.3.22) show that each term in $(T)_\varepsilon$, except perhaps for $L^\varepsilon_t(\phi)$, converges uniformly in compact time intervals in L^2. Therefore there is an a.s. continuous process $\{\tilde{L}_t(\phi) : t \ge 0\}$, so that

$$(IV.3.23) \qquad \lim_{\varepsilon\downarrow 0} \| \sup_{t\le T} |L^\varepsilon_t(\phi) - \tilde{L}_t(\phi)| \|_2 = 0 \quad \forall T > 0.$$

Take L^2 limits uniformly in $t \le T$ in $(T)_\varepsilon$ to see that
(IV.3.24)

$$X_t(G_0\phi) = m_1 \times m_2(\phi) + N^0_t(\phi) - \int_0^t G_0\phi(x_1,x_2)C(ds,dx_1,dx_2) - \tilde{L}_t(\phi) \quad \forall t \ge 0 \text{ a.s.},$$

where each term is a.s. continuous in t, $N^0_t(\phi)$ is an L^2 martingale and the last two terms have square integrable total variation on compact time intervals.

To complete the proof we need to show there is a continuous increasing $M_F(\mathbb{R}^d)$-valued process $L_t(X)$ such that

$$(IV.3.25) \qquad L_t(X)(\phi) = \tilde{L}_t(\phi) \quad \forall t \ge 0 \text{ a.s. for all } \phi \in b\mathcal{B}(\mathbb{R}^d)_+.$$

Note that (IV.3.23) then identifies $L(X)$ as the collision local time of X as the notation suggests. Let D_0 be a countable dense set in

$$C_\ell(\mathbb{R}^d) = \{\phi \in C_b(\mathbb{R}^d) : \phi \text{ has a limit at } \infty\}$$

containing 1. Choose $\varepsilon_n \downarrow 0$ and ω outside a null set so that

$$(IV.3.26) \qquad \lim_{n\to\infty} \sup_{t\le n} |L_t^{\varepsilon_n}(\phi) - \tilde{L}_t(\phi)| = 0 \quad \text{for all } \phi \in D_0,$$

and (recall Corollary III.1.7)

$$(IV.3.27) \qquad \mathcal{R}_\delta \equiv \mathrm{cl}(\cup_{t\ge\delta} S(Z_t^1) \cup S(Z_t^2)) \text{ is compact for all } \delta > 0.$$

Let $\eta > 0$. The definition of L^ε shows that $K_\delta = \left\{\frac{x_1+x_2}{2} : x_i \in \mathcal{R}_\delta\right\}$ is a compact support for $L_\infty^\varepsilon(X) - L_\delta^\varepsilon(X)$. Our choice of ω implies $\tilde{L}.(1)$ is continuous and allows us to choose $\delta > 0$ so that $L_\delta^{\varepsilon_n}(1) < \eta$ for all n. Therefore

$$L_\infty^{\varepsilon_n}(X)(K_\delta^c) = L_\delta^{\varepsilon_n}(K_\delta^c) < \eta \quad \text{for all } n.$$

Therefore $\{L_t^{\varepsilon_n}(X) : n \in \mathbb{N},\ t \ge 0\}$ are tight and (IV.3.26) shows that for each $t \ge 0$, all limit points of $\{L_t^{\varepsilon_n}\}$ in the weak topology on $M_F(\mathbb{R}^d)$ coincide. Therefore there is an $M_F(\mathbb{R}^d)$-valued process $L_t(X)$ such that $\lim_{n\to\infty} L_t^{\varepsilon_n}(X) = L_t(X)$ for all $t \ge 0$ a.s., $L_t(X)$ is non-decreasing in t and satisfies

$$(IV.3.28) \qquad L_t(X)(\phi) = \tilde{L}_t(\phi) \text{ for all } t \ge 0 \text{ and } \phi \in D_0 \text{ a.s.}$$

In particular $L_t(X)(\phi)$ is continuous in $t \ge 0$ for all $\phi \in D_0$ a.s. and hence $L_t(X)$ is a.s. continuous in t as well. If $\psi_n \xrightarrow{bp} \psi$, then using Dominated Convergence in (IV.3.24) one can easily show there is a subsequence such that

$$\lim_{k\to\infty} \tilde{L}_t(\psi_{n_k}) = \tilde{L}_t(\psi)\ \forall t \ge 0 \text{ a.s.}$$

by showing this is the case for each of the other terms in (IV.3.24). (A subsequence is needed as one initially obtains L^2 convergence for the martingale terms.) It then follows from (IV.3.28) that (IV.3.25) holds and the proof is complete. ∎

Proof of Theorem IV.3.2. (a) As we may take $X = Z$ in Theorem IV.3.8, it remains only to show that $L_t(Z)$ is not identically 0. The L^2 convergence in Theorem IV.3.8 and a simple second moment calculation show that

$$\mathbb{E}(L_t(Z)(1)) = \lim_{\varepsilon\downarrow 0} \mathbb{E}(L_t^\varepsilon(Z)(1)) = \frac{1}{2} \iint \int_0^{2t} p_s(z_1 - z_2) ds\, m_1(dz_1) m_2(dz_2) \ne 0.$$

(b) We first give a careful argument if $d > 6$. Recall the definition of $G_\delta(X)$ given at the beginning of this Section and recall that $h(u) = (u \log^+(1/u))^{1/2}$. If $\delta_i(3,\omega)$ is as in Corollary III.1.5, then that result and the fixed time hitting estimate, Theorem III.5.11, show that for $x \in \mathbb{R}^d$ and $t > 0$,

$$\begin{aligned}
&\mathbb{P}(Z_s^i(B(x,\varepsilon)) > 0 \text{ for some } s \in [t, t+\varepsilon^2(\log^+(1/\varepsilon))^{-1}], \\
&\qquad\qquad \text{and } \delta_i(3,\omega) > \varepsilon^2(\log^+(1/\varepsilon))^{-1}) \\
&\le \mathbb{P}(Z_t^i(B(x, \varepsilon + 3h(\varepsilon^2(\log^+(1/\varepsilon))^{-1}))) > 0) \\
&\le C_d \gamma_i^{-1} t^{-d/2} m_i(1) \Big(\varepsilon + 3h(\varepsilon^2(\log^+(1/\varepsilon))^{-1})\Big)^{d-2} \\
(IV.3.29) \qquad &\le C_d' \gamma_i^{-1} t^{-d/2} m_i(1) \varepsilon^{d-2}.
\end{aligned}$$

Let $S_n = \{B(x_i^n, 2^{-n}) : 1 \le i \le c_d n^d 2^{nd}\}$ be an open cover of $[-n,n]^d$. If $\delta > 0$, $\mathcal{R}_\delta$ is as in (IV.3.27), and $\eta_n = 2^{-2n}(\log 2^n)^{-1}$, then

$$\begin{aligned}
&\mathbb{P}(G_\delta(Z^1) \cap G_\delta(Z^2) \neq \emptyset, \mathcal{R}_\delta \subset [-n,n]^d, Z_n^1 = Z_n^2 = 0, \delta_1(3) \wedge \delta_2(3) > \eta_n,) \\
&\le \sum_{0 \le j \le n\eta_n^{-1}} \sum_{1 \le i \le c_d n^d 2^{nd}} \mathbb{P}(Z_s^1(B(x_i^n, 2^{-n}))Z_s^2(B(x_i^n, 2^{-n})) > 0 \\
&\qquad\qquad \text{for some } s \in [\delta + j\eta_n, \delta + (j+1)\eta_n], \delta_1(3) \wedge \delta_2(3) > \eta_n) \\
&\le (n\eta_n^{-1} + 1)c_d n^d 2^{nd} (C_d')^2 (\gamma_1\gamma_2)^{-1} \delta^{-d} m_1(1) m_2(1) 2^{-n2(d-2)} \quad \text{by (IV.3.29)} \\
&\le c(d,\delta) m_1(1) m_2(1) n^{2+d} 2^{-n(d-6)} \to 0 \text{ as } n \to \infty.
\end{aligned}$$

As $n \to \infty$ the left-hand side of the above approaches $\mathbb{P}(G_\delta(Z^1) \cap G_\delta(Z^2) \neq \emptyset)$ by Corollary III.1.5 and (IV.3.27), and so the result follows by letting $\delta \downarrow 0$.

Finally we sketch the argument in the critical 6-dimensional case. First (IV.3.29) can be strengthened to

$$(IV.3.30) \quad \mathbb{P}\Big(\int_t^{t+\varepsilon^2} Z_s^i(B(x,\varepsilon))\, ds > 0\Big) \le c_d t^{-d/2} m_i(1) \varepsilon^{d-2} \quad \forall t \ge 4\varepsilon^2,\ d \ge 3.$$

This may shown using an appropriate nonlinear pde as in Section III.5. A short proof is given in Proposition 3.3 of Barlow, Evans and Perkins (1991). Now introduce a restricted Hausdorff measure $q^f(A)$ for $A \subset \mathbb{R}_+ \times R^d$ and $f : [0,\varepsilon) \to \mathbb{R}_+$ a non-decreasing function for which $f(0+) = 0$. It is given by

$$q^f(A) = \liminf_{\delta \downarrow 0} \Big\{ \sum_{i=1}^\infty f(r_i) : A \subset \cup_{i=1}^\infty [t_i, t_i + r_i^2] \times \prod_{j=1}^d [x_i^j, x_i^j + r_i],\ r_i < \delta \Big\}.$$

If $d > 4$ and $\psi_d(r) = r^4 \log\log(1/r)$ (as in Theorem III.3.9) then there are $0 < c_1(d) \le c_2(d) < \infty$ so that

$$(IV.3.31) \quad c_1 q^{\psi_d}(A \cap G(Z^i)) \le \int_0^\infty \int 1_A(s,x) Z_s^i(dx)\, ds \le c_2 q^{\psi_d}(A \cap G(Z^i)) \quad \forall A \in \mathcal{B}(\mathbb{R}_+ \times \mathbb{R}^d) \text{ a.s.}$$

This is Theorem 3.1 of Barlow, Evans, and Perkins (1991) and may be shown using the ideas presented in Section III.3. (It should be possible to prove $c_1 = c_2$ here.) If $q^d = q^{r^d}$, then a simple consequence of (IV.3.30) (cf. Corollary III.5.10) is

$$(IV.3.32) \quad q^{d-2}(A) = 0 \text{ implies } A \cap G(Z^1) = \emptyset \text{ a.s. for all } A \subset \mathbb{R}_+ \times \mathbb{R}^d, d \ge 3.$$

(IV.3.31) shows that $q^{d-2}(G(Z^2)) = 0$ if $d \ge 6$ and so (IV.3.32) with $A = G(Z^2)$ implies that $G(Z^1) \cap G(Z^2) = \emptyset$ a.s. ∎

Proof of Lemma IV.3.3. If $d = 2$ this is a simple consequence of Theorem III.3.4, so assume $d > 2$. We may assume that our space carries independent historical processes (H^1, H^2) associated with (Z^1, Z^2). Let h and $\delta_i(3, H^i)$ be as in the Historical Modulus of Continuity (Theorem III.1.3) and let $\bar{h}_d$ and $r_0(\delta, H^1)$ be as in Theorem III.3.4. Those results show that on

$$\{\omega : \delta_i(3, H^i) > 2^{-n},\ i = 1, 2, \text{ and } r_0(\delta, H^1) > h(2^{-n})\},$$

we have

$$\begin{aligned}
&\sup_{t\geq\delta, t\in[j2^{-n},(j+1)2^{-n}]} \iint 1(|z_1 - z_2| \leq h(2^{-n}))Z_t^1(dz_1)Z_t^2(dz_2)\\
&\leq \sup_{t\geq\delta, t\in[j2^{-n},(j+1)2^{-n}]} \iint 1(|y_1(j2^{-n}) - y_2(j2^{-n}| \leq 7h(2^{-n}),\\
&\qquad |y_1(t) - y_2(t)| \leq h(2^{-n}))H_t^1(dy_1)H_t^2(dy_2)\\
&\leq \sup_{t\in[j2^{-n},(j+1)2^{-n}]} \int \gamma_1 c(d)\bar{h}_d(h(2^{-n}))\\
&\qquad \times 1\Big(y_2(j2^{-n}) \in S(Z^1_{j2^{-n}})^{7h(2^{-n})}\Big)H_t^2(dy_2). \qquad (IV.3.33)
\end{aligned}$$

A weak form of Lemma III.1.6 (with s fixed) has also been used in the last line. If $H^{1,*} = \sup_t H_t^1(1)$, then (III.3.1) and the ensuing calculation show that for $n \geq N(H^1)$,

$$\begin{aligned}
&S(Z^1_{j2^{-n}})^{7h(2^{-n})}\\
&\subset \text{a union of } \gamma_1^{-1}(H^{1,*}+1)2^{n+2} \text{ balls of radius } 10h(2^{-n})\ \forall j\in\mathbb{N}. \qquad (IV.3.34)
\end{aligned}$$

Let $W_n(j) = Z^2_{j2^{-n}}(S(Z^1_{j2^{-n}})^{7h(2^{-n}})$. Condition on H^1 and assume that $n \geq N(H^1)$. Then (IV.3.34) implies

$$P^x(B_s \in S(Z^1_{j2^{-n}})^{7h(2^{-n}}) \leq c_d\gamma_1^{-1}(H^{1,*}+1)2^n h(2^{-n})^d s^{-d/2}. \qquad (IV.3.35)$$

Therefore

$$\begin{aligned}
&\int_0^\infty \sup_x P^x(B_s \in S(Z^1_{j2^{-n}})^{7h(2^{-n})})\,ds\\
&\leq (c_d\gamma_1^{-1}+1)(H^{1,*}+1)\int_0^\infty \min(2^n h(2^{-n})^d s^{-d/2}, 1)\,ds\\
&\leq c(d,\gamma_1)(H^{1,*}+1)2^{-n(1-2/d)}\log 2^n\\
&\equiv \gamma_2^{-1}\lambda_n^{-1}.
\end{aligned}$$

If $f_n(x) = \lambda_n 1(x \in S(Z^1_{j2^{-n}})^{7h(2^{-n})})$ and $G(f_n, t)$ is as in Lemma III.3.6, then $\gamma_2 G(f_n, j2^{-n}) \leq 1$ and so Lemma III.3.6 implies that on $\{n \geq N(H^1)\}$ and for $j2^{-n} \geq 1/n$,

$$\begin{aligned}
\mathbb{P}(W_n(j) \geq 17n\lambda_n^{-1}|H^1) &\leq e^{-17n}\mathbb{E}(e^{\lambda_n W_n(j)}|H^1)\\
&\leq e^{-17n}\exp\Big(m_2(1)2\lambda_n \sup_x P^x(B_{j2^{-n}} \in S(Z^1_{j2^{-n}})^{7h(2^{-n})})\Big)\\
&\leq e^{-17n}\exp(m_2(1)c'(\delta,\gamma_1,\gamma_2)) \quad (\text{by } (IV.3.35)). \qquad (IV.3.36)
\end{aligned}$$

The Markov property and (III.1.3) show that conditional on $\sigma(H^1)\vee\sigma(H^2_s, s\le j2^{-n})$, $t\mapsto H^2_{j2^{-n}+t}(\{y : y(j2^{-n})\in S(Z^1_{j2^{-n}})^{7h(2^{-n})}\})$ is equal in law to $\mathbb{P}_{W_n(j)\delta_0}(Z^2_\cdot(1)\in\cdot)$. Therefore if $\eta_n>0$ and $K_n = 17n\lambda_n^{-1}$, then

$$\begin{aligned}
&\mathbb{P}\Big(\sup_{t\in[j2^{-n},(j+1)2^{-n}]} H^2_t(\{y : y(j2^{-n})\in S(Z^1_{j2^{-n}})^{7h(2^{-n})}\}) > \eta_n,\ W_n(j)<K_n|H^1\Big)\\
&\le \mathbb{E}\Big(\mathbb{P}_{W_n(j)\delta_0}\Big(\sup_{t\le 2^{-n}}\exp(2^n\gamma_2^{-1}Z^2_t(1)) > \exp(2^n\eta_n\gamma_2^{-1})\Big)1(W_n(j)<K_n)|H^1\Big)\\
&\le \exp(-2^n\eta_n\gamma_2^{-1})\mathbb{E}\Big(\mathbb{E}_{K_n\delta_0}(\exp(2^n\gamma_2^{-1}Z^2_{2^{-n}}(1)))|H^1\Big) \quad \text{(weak } L^1 \text{ inequality)}\\
&\le \exp(-2^n\eta_n\gamma_2^{-1})\mathbb{E}(\exp(K_n 2^{n+1}/\gamma_2)|H^1) \qquad \text{(Lemma III.3.6)}.
\end{aligned}$$

Set $\eta_n = 35n/\lambda_n = c''(d,\gamma_1,\gamma_2)(H^{1,*}+1)2^{-n(1-2/d)}n^2$ and use (IV.3.36) in the above to conclude that on $\{n\ge N(H^1)\}$ and for $j2^{-n}\ge 1/n$,

$$\begin{aligned}
&\mathbb{P}\Big(\sup_{t\in[j2^{-n},(j+1)2^{-n}]} H^2_t(\{y : y(j2^{-n})\in S(Z^1_{j2^{-n}})^{7h(2^{-n})}\}) > \eta_n|H^1\Big)\\
&\le e^{-17n}\exp(m_2(1)c'(\delta,\gamma_1,\gamma_2)) + \exp\Big(-\frac{2^n\eta_n}{\gamma_2}+\frac{(34)2^n n}{\gamma_2\lambda_n}\Big)\\
&\le e^{-17n}\exp(m_2(1)c'(\delta,\gamma_1,\gamma_2)) + \exp\Big(-\frac{2^n n}{\gamma_2\lambda_n}\Big)\\
&\le e^{-17n}\exp(m_2(1)c'(\delta,\gamma_1,\gamma_2)) + \exp(-c(d,\gamma_1)2^{2n/d}(\log 2)n^2).
\end{aligned}$$

A conditional application of Borel-Cantelli now shows there is an $N(H)<\infty$ a.s. so that for $n\ge N(H)$,

$$\begin{aligned}
&\sup_{j2^{-n}\ge 1/n}\ \sup_{t\in[j2^{-n},(j+1)2^{-n}]} H^2_t(\{y : y(j2^{-n})\in S(Z^1_{j2^{-n}})^{7h(2^{-n})}\})\\
&\le c''(\delta,\gamma_1,\gamma_2)(H^{1,*}+1)2^{-n(1-2/d)}n^2.
\end{aligned}$$

Use this in (IV.3.33) to see that for a.a. ω if n is sufficiently large, then

$$\begin{aligned}
&\sup_{t\ge\delta}\iint 1(|z_1-z_2|\le h(2^{-n}))Z^1_t(dz_1)Z^2_t(dz_2)\\
&\le \gamma_1 c(d)c''(d,\gamma_1,\gamma_2)(H^{1,*}+1)\bar{h}_d(h(2^{-n}))2^{-n(1-2/d)}n^2.
\end{aligned}$$

An elementary calculation now completes the proof. ∎

4. A Singular Competing Species Model–Higher Dimensions.

In this Section we describe how to use collision local time to formulate and solve the competing species model introduced in Section IV.1 in higher dimensions. The actual proof of the main results (due to Evans and Perkins (1994,1998) and Mytnik (1999)) are too long to reproduce here and so this Section will be a survey of known results together with some intuitive explanations.

We use the notation of Section IV.1 with $E_i = \mathbb{R}^d$ and $A_i = \Delta/2$. In particular, $\Omega^2 = C(\mathbb{R}_+, M_F(\mathbb{R}^d))^2$ with its Borel σ-field $\mathcal{F}^2$ and canonical right-continuous filtration $\mathcal{F}_t^2$. In view of Remark IV.3.1(b), here is the natural extension of $(CS)_m^\lambda$ to higher dimensions.

Definition. Let $\lambda = (\lambda_1, \lambda_2) \in \mathbb{R}_+^2$ and $m = (m_1, m_2) \in M_F(\mathbb{R}^d)^2$. A probability $\mathbb{P}$ on $(\Omega^2, \mathcal{F}^2)$ satisfies $(GCS)_m^\lambda$ iff

$$X_t^i(\phi_i) = m_i(\phi_i) + \int_0^t X_s^i\Big(\frac{\Delta\phi_i}{2}\Big)ds + M_t^i(\phi_i) - \lambda_i L_t(X)(\phi_i),$$

where $M_t^i(\phi_i)$ is a continuous $\mathcal{F}_t^2$ – martingale under $\mathbb{P}$ such that

$$(GCS)_m^\lambda \quad M_0^i(\phi_i) = 0 \text{ and} \langle M^i(\phi_i), M^j(\phi_j)\rangle_t = \delta_{ij}\int_0^t X_s^i(\phi_i^2)ds$$

for $\phi_i \in \mathcal{D}(\Delta/2)$ and $i = 1, 2$.

The existence of $L_t(X)$ is implicit in $(GCS)_m^\lambda$. We will say $X = (X_1, X_2)$ satisfies $(GCS)_m^\lambda$ if X is a process whose law satisfies $(GCS)_m^\lambda$. Let $(GCS)_m^{\varepsilon,\lambda}$ denote the corresponding martingale problem in which $L_t(X)(\phi_i)$ is replaced by $L_t^{\varepsilon,i}(X)(\phi_i)$ for $i = 1, 2$ (recall Remark IV.3.1(c)).

Note first that the Domination Principle (Proposition IV.1.4) shows that if X satisfies $(CS)_m^\lambda$, we may assume there are a pair of independent super-Brownian motions (Z^1, Z^2) such that $X^i \le Z^i$ a.s. If $d \ge 6$, then Theorem IV.3.2(b) implies $G(X^1) \cap G(X^2) \subset G(Z^1) \cap G(Z^2) = \emptyset$ and so $L(X)$ must be 0 a.s. Corollary IV.1.3 (with $g_i^0 = 0$) now shows that (X^1, X^2) is a pair of independent super-Brownian motions. Conversely, using Theorem IV.3.2(b) it is easy to see that a pair of independent super-Brownian motions does satisfy $(GCS)_m^\lambda$ with $L(X) = 0$ if $\int g_{d-2}(|z_1 - z_2|)m_1(dz_1)m_2(dz_2) < \infty$. (The latter condition ensures that $\sup_{\varepsilon>0}\mathbb{E}(L_\delta^\varepsilon(X)(1))$ approaches 0 as $\delta \downarrow 0$, and Theorem IV.3.2(b) shows that $L_t^\varepsilon(X)(1) - L_\delta^\varepsilon(X)(1) \overset{a.s.}{\to} 0$ as $\varepsilon \downarrow 0$ for any $\delta > 0$.) Therefore we only consider the above martingale problem for $d \le 5$ when non-trivial solutions may exist.

Next we show that if $d = 1$, then $(GCS)_m^\lambda$ may be viewed as a generalization of $(CS)_m^\lambda$.

Proposition IV.4.1. Assume $d = 1$ and $m \in F$. The unique solution P_m^0 of $(CS)_m^\lambda$ also satisfies $(GCS)_m^\lambda$.

Proof. We need only show that

$$(IV.4.1) \qquad L_t(X)(dx) = \Big(\int_0^t u_1(s,x)u_2(s,x)ds\Big)dx \quad \mathbb{P}_m^0\text{a.s.}$$

Let $\phi \in C_b(\mathbb{R})$. Theorem IV.2.1 shows that $X_s^i(dx) = u_i(s,x)dx$ for all $s > 0$ $\mathbb{P}_m^0$-a.s. and Proposition IV.2.3 shows that $t \to u_i(t,\cdot)$ is a continuous map from $(0,\infty)$ to $C_K(\mathbb{R})$ $\mathbb{P}_m^0$-a.s. It is now easy to see that $\mathbb{P}_m^0$-a.s. for all $0 < \delta \le t$,

$$\begin{aligned}
&\lim_{\varepsilon\downarrow 0} L_t^\varepsilon(X)(\phi) - L_\delta^\varepsilon(X)(\phi) \\
(IV.4.2) \quad &= \lim_{\varepsilon\downarrow 0}\int_\delta^t \iint \phi\Big(\frac{x_1+x_2}{2}\Big)p_\varepsilon(x_1 - x_2)u_1(s,x_1)u_2(s,x_2)dx_1dx_2ds \\
&= \int_\delta^t \int \phi(x)u_1(s,x)u_2(s,x)dxds.
\end{aligned}$$

Note also by the Domination Principle,

$$\mathbb{P}^0_m\Big(L^\varepsilon_\delta(X)(1) + \int_0^\delta \int u_1(s,x)u_2(s,x)dxds\Big)$$

$$(IV.4.3) \qquad \le \iint \int_0^\delta p_{2s+\varepsilon}(x_1 - x_2) + p_{2s}(x_1 - x_2)dsdm_1(x_1)dm_2(x_2)$$
$$\le c\sqrt{\delta}m_1(1)m_2(1) \to 0 \text{ as } \delta \downarrow 0.$$

(IV.4.1) now follows from (IV.4.2) and (IV.4.3). ∎

Recall that $\mathbb{P}^\varepsilon_m$ is the unique solution of $(CS)^{\varepsilon,\lambda}_m$ which is equivalent to $(GCS)^{\varepsilon,\lambda}_m$. In view of Remark IV.3.1(c) we may expect $\mathbb{P}^\varepsilon_m$ to converge to a solution of $(GCS)^\lambda_m$ as $\varepsilon \downarrow 0$.

Notation. $M_{FS}(\mathbb{R}^d) = \{m \in M_F(\mathbb{R}^d) : \int_0^1 r^{1-d} \sup_x m(B(x,r))dr < \infty\}$.

If $m_1, m_2 \in M_{FS}$, then an integration by parts shows that

$$\sup_{z_2} \int g_{d-2}(|z_1 - z_2|)dm_1(z_1) < \infty$$

and so (m_1, m_2) satisfies the hypothesis (IC) of Theorem IV.3.8.

Theorem IV.4.2. (a) Assume $1 \le d \le 3$ and $m \in (M_{FS})^2$.
(i) Then $\mathbb{P}^\varepsilon_m \overset{w}{\Rightarrow} \mathbb{P}_m$ on Ω^2, where $((X_t)_{t\ge0}, (\mathbb{P}_\nu)_{\nu \in M_{FS}})$ is an $(M_{FS})^2$-valued Borel $(\mathcal{F}^2_t)$-strong Markov process and $\mathbb{P}_m$ satisfies $(GCS)^\lambda_m$.
(ii) If, in addition, $\lambda_1 = \lambda_2$. then $\mathbb{P}_m$ is the unique solution of $(GCS)^\lambda_m$.

(b) If $d = 4$ or 5, $m \in (M_F(\mathbb{R}^d) - \{0\})^2$ satisfies the hypothesis (IC) of Theorem IV.3.2, and $\lambda \ne (0,0)$, then there is no solution to $(GCS)^\lambda_m$.

Discussion.(b) Theorem IV.3.8 shows that the existence of a collision local time for for any potential solutions of $(GCS)^\lambda_m$ is to be expected if $d \le 5$ and Theorem IV.3.2 suggests it will be nontrivial for $d \le 5$. These results may lead one to believe that nontrivial solutions exist for $d \le 5$. It turns out, however, that it is not the existence of collisions between a pair of independent super-Brownian motion that is germane to the existence of the solutions to (GCS). Rather it is the existence of collisions between a single Brownian path, B, and an independent super-Brownian motion, Z. If $G(B) = \{(t, B_t) : t \ge 0\}$, then

$$(IV.4.4) \qquad \mathbb{P}(G(B) \cap G(Z) \ne \emptyset) > 0 \text{ iff } d < 4.$$

To see this for $d \ge 4$, recall from (IV.3.31) that $q^{\psi_d}(G(Z)) < \infty$ a.s. We had $d > 4$ there but the proof in Theorem 3.1 of Barlow, Evans and Perkins (1991) also goes through if $d = 4$. This shows that $q^d(G(Z)) = 0$ if $d \ge 4$ and so (IV.4.4) is true by Theorem 1 of Taylor and Watson (1985) (i.e., the analogue of (IV.3.32) for $G(B)$). For $d \le 3$ one approach is to use a Tanaka formula to construct a nontrivial inhomogeneous additive functional of B which only increases on the set of times when $B(t) \in S(Z_t)$ (see Theorem 2.6 of Evans-Perkins (1998)). The construction requires a mild energy condition on the initial distributions of B and Z but the required result then holds for general initial conditions by Theorem III.2.2. Alternatively, a short direct proof using Theorem III.3.4 is given in Proposition 1.3 of Barlow and Perkins (1994).

To understand the relevance of (IV.4.4), we demonstrate its use in a heuristic proof of (b). Assume X satisfies $(GCS)_m^\lambda$ for $d = 4$ or 5. Let $Z^i \geq X^i$ be a pair of dominating independent super-Brownian motions (from Proposition IV.1.4) and let H^i be the historical process associated with Z^i. The particle approximations in Example IV.1.1 suggest that X^1 is obtained from Z^1 by killing off some of the particles which collide with the X^2 population, and similarly for X^2. Use the notation of the Historical Cluster Representation (Theorem III.1.1) and let $\{y_1, \ldots, y_M\}$ be the finite support of $r_{t-\varepsilon}(H_t^1)$ for fixed $0 < \varepsilon < t$. These are the ancestors at time $t-\varepsilon$ of the entire Z^1 population at time t. Which of these ancestors are still alive in the X^1 population at time $t-\varepsilon$? By Theorem III.3.1, y_i has law $\mathbb{E}(H^i_{t-\varepsilon}(\cdot))/m_1(1)$ and so is a Brownian path stopped at time $t-\varepsilon$ and is independent of Z^2. (IV.4.4) shows that $G(y_i) \cap G(Z^2) = \emptyset$ a.s. Therefore each y_i will not have encountered the smaller X^2 population up to time $t-\varepsilon$ and so must still be alive in the X^1 population. Let $\varepsilon \downarrow 0$ to see that the entire family tree of the population of Z^1 at time t never encounters Z^2 and hence X^2. This means that no particles have been killed off and so $Z_t^1 = X_t^1$ a.s., and by symmetry, $Z_t^2 = X_t^2$ a.s. These identities hold uniformly in t a.s. by continuity. The fact that $\mathbb{P}(L(Z) \neq 0) > 0$ (Theorem IV.3.2) shows that Z does not satisfy $(GCS)_m^\lambda$ and so no solution can exist. In short, for $d = 4$ or 5, the only collisions contributing to $L_t(Z)$ are between particles whose family trees die out immediately and so killing off these particles has no impact on the proposed competing species model.

The above proof is not hard to make rigorous if there is a historical process associated with X^i so that we can rigorously interpret the particle heuristics. To avoid this assumption, the proof given in Section 5 of Evans and Perkins (1994) instead uses the ideas underlying the Tanaka formula in the previous Section. The proof outlined above would also appear to apply more generally to any killing operation based on collisions of the two populations. In (GCS) we would replace $\lambda_i L_t(X)(\phi_i)$ with $A_t^i(\phi_i)$, where $A^i_\cdot$ is an increasing continuous M_F-valued process such that $S(A^i(dt, dx)) \subset G(X^1) \cap G(X^2)$ a.s. The non-existence of solutions for $d = 4$ or 5 in this more general setting is true (unpublished notes of Barlow, Evans and Perkins) but the 4-dimensional case is rather delicate.

(a) Tightness of $\{\mathbb{P}_m^\varepsilon\}$ is a simple exercise using the Domination Principle and Theorem IV.3.2. To show each limit point satisfies $(GCS)_m^\lambda$, a refinement of Theorem IV.3.8 is needed for $d \leq 3$ (see Theorem 5.10 of Barlow, Evans and Perkins (1991)). This refinement states that in (IV.3.10) the rate of convergence to 0 in probability is uniform in X satisfying (H_2). In the proof of (IV.3.10), the only step for which this additional uniformity requires $d \leq 3$ (and which requires some serious effort) is (IV.3.14). To handle this term we use Theorem III.3.4 to first bound the integrals with respect to $X_s^i(dx_i)$ at least if $d \leq 3$. If $\mathbb{P}_m^{\varepsilon_n} \overset{w}{\Rightarrow}$, use Skorohod's theorem to obtain solutions X^{ε_n} of $(CS)_m^{\varepsilon_n,\lambda}$ which converge a.s. to X, say, as $n \to \infty$. We now may let $n \to \infty$ in $(CS)_m^{\varepsilon_n,\lambda}$ to derive $(GCS)_m^\lambda$ for X–the above uniformity and a simple comparison of $L^{\varepsilon_n}(X^{\varepsilon_n})$ with $L^{\varepsilon_n,i}(X^{\varepsilon_n})$ (see Lemma 3.4 of Evans and Perkins (1994)) show that $L^{\varepsilon_n,i}(X^{\varepsilon_n}) \to L(X)$ in probability as $n \to \infty$ and the other terms are easy to handle.

To proceed further seems to require considerable additional effort. The full convergence of the $\{P^{\varepsilon}_m\}$ to a nice strong Markov process is provided in Evans and Perkins (1998) (Theorems 1.6 and 8.2). Here we showed that each limit point has an associated pair of historical processes which satisfy a strong equation driven by a pair of independent historical Brownian motions whose supports carry a Poisson field of marks indicating potential killing locations. This strong equation has a solution which is unique, both pathwise and in law. (This general approach of using strong equations driven by historical processes will be used in another setting with greater attention to detail in the next Section.) This approach does show that the natural historical martingale problem associated with $(GCS)^{\lambda}_m$ is well-posed (Theorem 1.4 of Evans and Perkins (1998)). The uniqueness of solutions to $(GCS)^{\lambda}_m$ itself remains open in general as we do not know that any solution comes equipped with an associated historical process (from which we would be able to show it is the solution of the aforementioned strong equation). If $\lambda_1 = \lambda_2$, uniqueness of solutions to $(GCS)^{\lambda}_m$ was proved by Mytnik (1999) by a duality argument. Mytnik built a dual family of one-dimensional distributions (as opposed to a dual process) by means of an intricate and original Trotter product construction. One phase of the Trotter product requires solutions to a non-linear evolution equation with irregular initial data. As is often the case with duality arguments, it is non-robust and does not appear to handle the case where $\lambda_1 \neq \lambda_2$. It is somewhat disconcerting that after all of this effort the general question of uniqueness to our competing species model remains unresolved in general. I suspect the correct approach to these questions remains yet to be discovered and so was not tempted to provide a detailed description of the proofs here.

V. Spatial Interactions
1. A Strong Equation

We continue our study of measure-valued processes which behave locally like (A,γ,g)-DW superprocesses, i.e., where A, γ, and g may depend on the current state, X_t, of the process. In this Chapter we allow the generator governing the spatial motion, A, to depend on X_t. These results are taken from Perkins (1992), (1995). To simplify the exposition we set $\gamma=1$ and $g=0$, although as discussed below (in Section V.5) this restriction may be relaxed. Our approach may be used for a variety of dependencies of A_{X_t} on X_t but we focus on the case of state dependent diffusion processes. Let

$$\sigma : M_F(\mathbb{R}^d)\times\mathbb{R}^d\to\mathbb{R}^{d\times d},\quad b : M_F(\mathbb{R}^d)\times\mathbb{R}^d\to\mathbb{R}^d,\quad a=\sigma\sigma^*,$$

and set

$$A_\mu\phi(x)=\sum_i\sum_j a_{ij}(\mu,x)\phi_{ij}(x)+\sum_i b_i(\mu,x)\phi_i(x),\quad \text{for } \phi\in C_b^2(\mathbb{R}^d).$$

Here $a(\mu,x)$ and $b(\mu,x)$ are the diffusion matrix and drift of a particle at x in a population μ.

If $\mathrm{Lip}_1=\{\phi:\mathbb{R}^d\to\mathbb{R}:\|\phi\|_\infty\le 1,\ |\phi(x)-\phi(y)|\le\|x-y\|\ \forall x,y\in\mathbb{R}^d\}$ and μ, $\nu\in M_F(\mathbb{R}^d)$, the Vasershtein metric on $M_F(\mathbb{R}^d)$, introduced in Section II.7, is

$$d(\mu,\nu)=\sup\{|\mu(\phi)-\nu(\phi)|:\phi\in\mathrm{Lip}_1\}.$$

Recall that d is a complete metric on $M_F(\mathbb{R}^d)$ inducing the topology of weak convergence.

Our approach will be based on a fixed point argument and so we will need the following Lipschitz condition on b and σ:

Assume there is a non-decreasing function $L:\mathbb{R}_+\to\mathbb{R}_+$ such that

$$\begin{aligned}&(a)\ \|\sigma(\mu,x)-\sigma(\mu',x')\|+\|b(\mu,x)-b(\mu',x')\|\\ (\mathrm{Lip})\quad &\le L(\mu(1)\vee\mu'(1))[d(\mu,\mu')+\|x-x'\|]\quad \forall\mu,\mu'\in M_F(\mathbb{R}^d),\ x,x'\in\mathbb{R}^d.\\ &(b)\ \sup_x\|\sigma(0,x)\|+\|b(0,x)\|<\infty.\end{aligned}$$

Remark V.1.1. (a) (Lip) easily implies that for some non-decreasing $C:\mathbb{R}_+\to\mathbb{R}_+$,

$$(B)\qquad \|\sigma(\mu,x)\|+\|b(\mu,x)\|\le C(\mu(1))\quad\forall\mu\in M_F(\mathbb{R}^d),\ x\in\mathbb{R}^d.$$

(b) The results of Sections V.1-V.4 remain valid without (Lip)(b) (see Section 5 of Perkins (1992)).

Originally published in: *Ecole d'Eté de Probabilités de Saint-Flour XXIX – 1999*, Lecture Notes in Mathematics, Vol. **1781**, 281–317, DOI: 10.1007/3-540-47944-9_9,

Exercise V.1.1. Prove that (Lip) holds in the following cases.
(a) $\sigma(\mu,x) = f(\mu(\phi_1),\ldots,\mu(\phi_n),x)$, where ϕ_i are bounded Lipschitz functions on $\mathbb{R}^d$ and $f:\mathbb{R}^{n+d}\to\mathbb{R}^{d\times d}$ is Lipschitz continuous so that $\sup_x \|f(0,x)\| < \infty$.
(b) $b(\mu,x) = \sum_{k=1}^n \int b_k(x,x_1,\ldots,x_k)d\mu(x_1)\ldots d\mu(x_k)$
$\sigma(\mu,x) = \sum_{k=1}^n \int \sigma_k(x,x_1,\ldots,x_k)d\mu(x_1)\ldots d\mu(x_k)$,
where b_k and σ_k are bounded Lipschitz continuous functions taking values in $\mathbb{R}^d$ and $\mathbb{R}^{d\times d}$, respectively.

A special case of (b) would be $b(\mu,x) = \int b_1(x,x_1)d\mu(x_1)$ and $\sigma(\mu,x) = \int p_\varepsilon(x-x_1)d\mu(x_1)$. Here $b_1(x,x_1)\in\mathbb{R}^d$ models an attraction or repulsion between individuals at x and x_1, and particles diffuse at a greater rate if there are a number of other particles nearby.

To motivate our stochastic equation, consider the branching particle system in Section II.3 where $Y^\alpha \equiv B^\alpha$ are Brownian motions in $\mathbb{R}^d$, $X_0 = m \in M_F(\mathbb{R}^d)$, and $\nu^n = \nu = \frac{1}{2}\delta_0 + \frac{1}{2}\delta_2$. Recall from (II.8.3) that if $H_t^N = \frac{1}{N}\sum_{\alpha\sim t}\delta_{B^\alpha_{\cdot\wedge t}}$, then H^N converges weakly to a historical Brownian motion, H, with law $\mathbb{Q}_{0,m}$. Let $Z_0:\mathbb{R}^d\to\mathbb{R}^d$ be Borel. Now solve

$$(SE)_N(a)\ Z_t^\alpha = Z_0(B_0^\alpha) + \int_0^t \sigma(X_s^N, Z_s^\alpha)dB_s^\alpha + \int_0^t b(X_s^N, Z_s^\alpha)ds,\quad t < \frac{|\alpha|+1}{N}$$

$$(b)\ X_s^N = \frac{1}{N}\sum_{\beta\sim s}\delta_{Z_s^\beta}.$$

Such solutions are easy to construct in a pathwise unique manner on $[i/N,(i+1)/N)$ by induction on i. On $[i/N,(i+1)/N)$, we are solving a finite system of stochastic differential equations driven by $\{B_s^\alpha : \alpha\sim i/N,\ s\in[i/N,(i+1)/N)\}$ and with Lipschitz continuous coefficients. The latter uses

$$\begin{aligned} d\Big(\frac{1}{N}\sum_{\alpha\sim i/N}\delta_{x^\alpha}, \frac{1}{N}\sum_{\alpha\sim i/N}\delta_{\hat{x}^\alpha}\Big) &\le N^{-1}\sum_{\alpha\sim i/N}\|x^\alpha - \hat{x}^\alpha\| \\ &\le (H_{i/N}^N(1))^{-1/2}N^{-1/2}\|x-\hat{x}\|_2, \end{aligned}$$

where $\|x-\hat{x}\|_2 = \Big(\sum_{\alpha\sim i/N}\|x^\alpha-\hat{x}^\alpha\|^2\Big)^{1/2}$ and we have used Cauchy-Schwarz in the last inequality. This shows there is a pathwise solution to $(SE)_N$ on $[i/N,(i+1)/N)$. Now let the B^α's branch at $t=(i+1)/N$ and continue on $[(i+1)/N,(i+2)/N)$ with the new set of Brownian motions $\{B_s^\alpha : \alpha\sim(i+1)/N,\ s\in[(i+1)/N,(i+2)/N)\}$. These solutions are then pieced together to construct the $\{Z_t^\alpha : t < (|\alpha|+1)/N,\ \alpha\}$ in $(SE)_N$. If $N\to\infty$, we may expect $X^N \overset{w}{\Rightarrow} X$, where

$$(SE)\qquad (a)\ Z_t(\omega,y) = Z_0(y_0) + \int_0^t \sigma(X_s,Z_s)dy(s) + \int_0^t b(X_s,Z_s)ds$$

$$(b)\ X_t(\omega)(A) = \int 1(Z_t(\omega,y)\in A)H_t(\omega)(dy)\quad \forall A\in\mathcal{B}(\mathbb{R}^d)$$

The intuition here is that ω labels a tree of branching Brownian motions and y labels a branch on the tree. Then $Z_t(\omega,y)$ solves the sde along the branch y in the tree ω and $X_t(\omega)$ is the empirical distribution of these solutions. Our objective in this

Chapter is to give a careful interpretation of the stochastic integral in (SE)(a), prove that (SE) has a pathwise unique strong Markov solution and show that $X^N \overset{w}{\Rightarrow} X$.

2. Historical Brownian Motion

Throughout this Section we work in the setting of the historical process of Section II.8 where $(Y, P^x) \equiv (B, P^x)$ is d-dimensional Brownian motion. We adopt the notation given there with $E = \mathbb{R}^d$, but as B has continuous paths we replace $(D(E), \mathcal{D})$ with $(C, \mathcal{C})$, where $C = C(\mathbb{R}_+, \mathbb{R}^d)$ and $\mathcal{C}$ is its Borel σ-field. Let $\mathcal{C}_t = \sigma(y_s, s \le t)$ be its canonical filtration. If $Z : \mathbb{R}_+ \times C \to \mathbb{R}$, then

$$\begin{aligned} & Z \text{ is } (\mathcal{C}_t)\text{-predictable} \iff Z \text{ is } (\mathcal{C}_t)\text{-optional} \\ (V.2.1) \qquad & \iff Z \text{ is Borel measurable and } Z(t,y) = Z(t, y^t)\ \forall t \ge 0. \end{aligned}$$

This follows from Theorem IV.97 in Dellacherie and Meyer (1978) and the fact that the proofs given there remain valid if D is replaced by C. We will therefore identify Borel functions on $\hat{\mathbb{R}}^d = \{(t,y) \in \mathbb{R}_+ \times C : y = y^t\}$ with $(\mathcal{C}_t)$-predictable functions on $\mathbb{R}_+ \times C$. If $C^s = \{y \in C : y = y^s\}$ then this identification allows us to write the domain of the weak generator for the path-valued process W in (II.8.1) as

$$\begin{aligned} \mathcal{D}(\hat{A}) = \{\phi : \mathbb{R}_+ \times C \to \mathbb{R} : \ & \phi \text{ is bounded, continuous, and } (\mathcal{C}_t)\text{-predictable,} \\ & \text{and for some } \hat{A}_s\phi(y) \text{ with the same properties, } \phi(t,B) - \phi(s,B) \\ & - \int_s^t \hat{A}_r\phi(B)dr,\ t \ge s \text{ is a } (\mathcal{C}_t)\text{-martingale under } P_{s,y}\ \forall s \ge 0,\ y \in C^s\}. \end{aligned}$$

Recall here that $P_{s,y}$ is Wiener measure starting at time s with past history $y \in C^s$, and for $m \in M_F^s(C)$ (recall this means $y = y^s$ $m-$ a.s.) define $P_{s,m} = \int P_{s,y} m(dy)$.

For the rest of this Section assume $\tau \ge 0$ and $(K_t)_{t \ge \tau}$ satisfies $(HMP)_{\tau, K_\tau}$ (from Section II.8) on $(\Omega, \mathcal{F}, (\mathcal{F}_t)_{t \ge \tau}, \mathbb{P})$ with K_τ now possibly random with law ν, $\hat{\gamma} \equiv 1$, $\hat{g} \equiv 0$, and $\hat{A}$ equal to the generator of the path-valued Brownian motion described above. Assume that this probability space is complete, the filtration is right-continuous and $\mathcal{F}_\tau$ contains all the null sets. We also assume $\mathbb{E}(K_\tau(1)) < \infty$ so that $m(\cdot) = E(K_\tau(\cdot)) \in M_F^\tau(C)$ and we can still work with a martingale problem as opposed to a local martingale problem. Call such a process, K, an $(\mathcal{F}_t)$-historical Brownian motion starting at (τ, ν) (or (τ, m) if $\nu = \delta_m$). As in Theorem II.8.3, K is an $(\mathcal{F}_t)$-strong Markov process and has law $\mathbb{Q}_{\tau,\nu} = \int \mathbb{Q}_{\tau, K_0} d\nu(K_0)$. In this setting the superprocess property (II.8.5) becomes

$$(V.2.2) \qquad \mathbb{P}(K_t(\psi)) = P_{\tau,m}(\psi(B^t)) \quad \text{for } t \ge \tau,\ \psi \in b\mathcal{C}.$$

Note also that if $S \ge \tau$ is a finite valued $(\mathcal{C}_t)$-stopping time, then

$$(V.2.3) \qquad P_{\tau,m}(g|\mathcal{C}_S)(y) = P_{S,y^S}(g) \quad P_{\tau,m} - \text{a.a. } y\ \forall g \in b\mathcal{C}.$$

To see this write $g(y) = \tilde{g}(y^S, y(S+\cdot))$ and use the strong Markov property at time S.

Our main objective in this Section is the seemingly minor extension of $(HMP)_{\tau,K_\tau}$ presented in Proposition V.2.6 below, and the reader may want to skip ahead to this result and its Corollary V.2.7. The latter plays a key role in what follows. Note, however, that Proposition V.2.4 will also be used in our stochastic calculus on Brownian trees and the proof of Lemma V.2.2 illustrates a neat idea of Pat Fitzsimmons.

We first reduce the definition of $\mathcal{D}(\hat{A})$ to zero starting times.

Notation. $b\mathcal{P}$ is the space of bounded $\mathcal{C}_t$-predictable processes on $\mathbb{R}_+ \times C$.

Lemma V.2.1. $\phi \in \mathcal{D}(\hat{A})$ iff $\phi \in b\mathcal{P}$ is continuous and for some continuous $\hat{A}\phi \in b\mathcal{P}$,

$$n(t,y) = \phi(t,y) - \phi(0,y) - \int_0^t \hat{A}_r\phi(y)dr \text{ is a } (\mathcal{C}_t)\text{-martingale under } P^x \ \forall x \in R^d.$$

Proof. We need only show the above condition is sufficient for membership in $\mathcal{D}(\hat{A})$. Assume ϕ is as above and let $s > 0$. It suffices to show $n(t) - n(s)$, $t \geq s$ is a $(\mathcal{C}_t)_{t\geq s}$-martingale under $P_{s,y}$ for every $y \in C^s$. Let $t \geq r \geq s$ and ψ be a bounded continuous $\mathcal{C}_r$-measurable mapping on C. We must show that

$$P_{s,y}(n(t)\psi) = P_{s,y}(n(r)\psi) \quad \forall y \in C^s. \tag{V.2.4}$$

The left-hand side is

$$P^0(n(t, y/s/(y(s)+B))\psi(y/s/(y(s)+B))) \tag{V.2.5}$$

and so is continuous in y by Dominated Convergence. The same is true of the right-hand side. It therefore suffices to establish (V.2.4) on a dense set of y in C^s. Next we claim that

$$\text{the closed support of } P^{y_0}(B^s \in \cdot) \text{ is } \{y \in C^s : y(0) = y_0\}. \tag{V.2.6}$$

To see this first note that for every ε, $T > 0$, $P^{y_0}(\sup_{s\leq T}|B_s - y_0| < \varepsilon) > 0$ (e.g. by the explicit formula for the two-sided hitting time in Theorem 4.1.1 of Knight (1981)). Now use the classical Cameron-Martin-Girsanov formula to conclude that for any $\psi \in C(\mathbb{R}_+, \mathbb{R}^d)$, and ε, $T > 0$,

$$P^{y_0}(\sup_{s\leq t}|B_s - \int_0^s \psi(u)du - y_0| < \varepsilon) > 0.$$

The claim follows easily. It implies that (V.2.4) would follow from

$$P_{s,y^s}(n(t)\psi) = P_{s,y^s}(n(r)\psi) \quad P^{y_0} - \text{a.a. } y \text{ for all } y_0 \in \mathbb{R}^d.$$

By (V.2.3) this is equivalent to

$$P^{y_0}(n(t)\psi|\mathcal{C}_s)(y) = P^{y_0}(n(r)\psi|\mathcal{C}_s)(y) \quad P^{y_0} - \text{a.a. } y \text{ for all } y_0 \in \mathbb{R}^d.$$

This is immediate by first conditioning $n(t)\psi$ with respect to $\mathcal{C}_r$. ∎

Fitzsimmons (1988) showed how one can use Rost's theorem on balayage to establish sample path regularity of a general class of superprocesses. Although we have not needed this beautiful idea for our more restrictive setting, the next result illustrates its effectiveness.

Lemma V.2.2. Let ϕ, $\psi : \mathbb{R}_+ \times C \to \mathbb{R}$ be $(\mathcal{C}_t)$-predictable maps such that for some fixed $T \geq \tau$, $\phi(t,y) = \psi(t,y)\ \forall \tau \leq t \leq T\ P_{\tau,m}$ – a.s. Then

$$\phi(t,y) = \psi(t,y)\ K_t - \text{a.a. } y \quad \forall \tau \leq t \leq T \quad \mathbb{P} - \text{a.s.}$$

Proof. Return to the canonical setting of historical paths, $(\Omega, \mathcal{F}_H, \mathcal{F}^H[\tau, t+], \mathbb{Q}_{\tau,\nu})$ of Section II.8, with $\hat{g} = 0$, $\hat{\gamma} = 1$, and P^x=Wiener measure. Recall the $\hat{E}$-valued diffusion $W_t = (\tau + t, Y^{\tau+t})$ with laws $\hat{P}_{\tau,y}$ and the W-superprocess

(V.2.7) $$\hat{X}_t = \delta_{\tau+t} \times H_{\tau+t} \text{ with laws } \hat{\mathbb{P}}_{\tau,m}.$$

Note first that

(V.2.8)
if $g : [\tau, \infty) \times C \to \mathbb{R}_+$ is $(\mathcal{C}_t)_{t\geq\tau}$–predictable then $H_t(g_t)$ is $\mathcal{F}^H[\tau, t+]$–predictable.

To see this start with $g(t,y) = g_1(t)g_2(y^t)$, where g_1, g_2 are non-negative bounded continuous functions on $\mathbb{R}_+$ and C, respectively. Then (V.2.8) holds because $K_t(g_t)$ is a.s. continuous. A monotone class argument now proves (V.2.8) (recall (V.2.1)).

Let S be an $\mathcal{F}^H[\tau, t+])_{t\geq\tau}$-stopping time such that $\tau \leq S \leq T$ and let $\lambda > 0$. Then $\hat{S} = S - \tau$ is an $(\mathcal{F}_t^{\hat{X}})_{t\geq 0}$-stopping time. Define a finite measure μ on $\hat{E}$ by

$$\mu(g) = \mathbb{Q}_{\tau,\nu}(e^{-\lambda(S-\tau)} H_S(g_S)) = \hat{\mathbb{P}}_{\tau,\nu}(e^{-\lambda\hat{S}} \hat{X}_{\hat{S}}(g)),$$

where the second equality holds by Lemma II.8.1. Let $U_\lambda f$ be the λ-resolvent of W. If f is a non-negative function on $\hat{E}$, then the superprocess property ((II.8.5) and the display just before it) shows that

$$\begin{aligned}\langle \delta_\tau \times m, U_\lambda f\rangle &= \hat{\mathbb{P}}_{\tau,\nu}\Big(\int_0^\infty e^{-\lambda t} \hat{X}_t(f)\, dt\Big)\\ &\geq \hat{\mathbb{P}}_{\tau,\nu}\Big(e^{-\lambda\hat{S}} \int_0^\infty e^{-\lambda t} \hat{X}_{t+\hat{S}}(f) dt\Big)\\ &= \hat{\mathbb{P}}_{\tau,\nu}\Big(e^{-\lambda\hat{S}} \hat{\mathbb{P}}_{\hat{X}_{\hat{S}}}\Big(\int_0^\infty e^{-\lambda t} \hat{X}_t(f) dt\Big)\Big)\\ &= \hat{\mathbb{P}}_{\tau,\nu}(e^{-\lambda\hat{S}} \hat{X}_{\hat{S}}(U_\lambda f)) = \langle \mu, U_\lambda f\rangle.\end{aligned}$$

A theorem of Rost (1971) shows there is a randomized stopping time, V, on $C(\mathbb{R}_+, \hat{E}) \times [0,1]$ (i.e., V is jointly measurable and $\{y : V(y,u) \leq t\} \in \mathcal{C}_{t+}$ for all $u \in [0,1]$) such that for every non-negative Borel function g on $\hat{E}$,

(V.2.9) $$\begin{aligned}\mu(g) &= \int_0^1 \hat{P}_{\tau,m}(e^{-\lambda V(u)} g(W_{V(u)}))\, du\\ &\leq \int_0^1 P_{\tau,m}(g(\tau + V(u), Y^{\tau+V(u)}))\, du.\end{aligned}$$

If $g(t,y) = |\phi(t,y) - \psi(t,y)|$, then the right-hand side of (V.2.9) is zero by hypothesis and so

$$H_S(|\phi(S) - \psi(S)|) = 0 \quad \text{a.s.}$$

The Section Theorem (Theorem IV.84 of Dellacherie and Meyer (1978)) and (V.2.8) then show that

$$H_t(|\phi(t)-\psi(t)|)=0 \quad \forall t\in[\tau,T] \quad \mathbb{Q}_{\tau,\nu}-\text{a.s.}$$

As K has law $\mathbb{Q}_{\tau,\nu}$, the result follows. ∎

Lemma V.2.3. Let $n:[\tau,\infty)\times C\to\mathbb{R}$ be a $(\mathcal{C}_t)_{t\ge\tau}$-predictable L^2-martingale under $P_{\tau,m}$. Then

$$K_t(n_t)=K_\tau(n_\tau)+\int_\tau^t\int n(s,y)dM(s,y) \quad \forall t\ge\tau \quad \mathbb{P}-\text{a.s.}$$

and is a continuous square integrable $(\mathcal{F}_t)$-martingale.

Proof. Let $N>\tau$. Then (V.2.3) and the Section Theorem imply that

$$(V.2.10) \qquad n(t,y)=P_{t,y^t}(n(N)) \quad \forall\tau\le t\le N \quad P_{\tau,m}-\text{a.s.}$$

Now let

$$S=\{X:C\to\mathbb{R}:X\in L^2(P_{\tau,m}),\ n^X(t,y)\equiv P_{t,y^t}(X)\text{ satisfies}$$
$$K_t(n_t^X)=K_\tau(n_\tau^X)+\int_\tau^t\int n^X(s,y)dM(s,y)\ \forall t\ge\tau\ \mathbb{P}-\text{a.s.}\}.$$

Implicit in the above condition is that both sides are well-defined and finite. If $X\in C_b(C)$, then n^X is bounded and continuous on $\mathbb{R}_+\times C$ (recall (V.2.5)) and n^X is a continuous $(\mathcal{C}_t)$-martingale under P^x for all $x\in\mathbb{R}^d$ by (V.2.3). Lemma V.2.1 shows that $n^X\in\mathcal{D}(\hat{A})$ and $\hat{A}n^X=0$. $(HMP)_{\tau,K_\tau}$ therefore shows that $X\in S$.

Let $\{X_n\}\subset S$ and assume $X_n\xrightarrow{bp}X$. Then Dominated Convergence shows that $n^{X_n}\xrightarrow{bp}n^X$, $K_t(n_t^{X_n})\to K_t(n_t^X)$ $\forall t\ge\tau$, and (use (V.2.2))

$$\mathbb{P}\Big(\int_\tau^t\int(n^{X_n}(s,y)-n^X(s,y))^2K_s(dy)ds\Big)$$
$$=\int_\tau^t P_{\tau,m}((n^{X_n}(s,B^s)-n^X(s,B^s))^2)ds\to 0\text{ as }n\to\infty.$$

Therefore we may let $n\to\infty$ in the equation showing $X^n\in S$ to conclude that $X\in S$. This and $C_b(C)\subset S$ show that $b\mathcal{C}\subset S$.

Let X be a non-negative function in $L^2(P_{\tau,m})$ and set $X_n=X\wedge n\in S$. Monotone Convergence shows that $n^{X_n}\uparrow n^X\le\infty$ pointwise and $K_t(n_t^{X_n})\uparrow K_t(n_t^X)$ for all $t\ge\tau$. (V.2.2) shows that

$$\mathbb{P}\Big(\int_\tau^t\int(n^{X_n}(s,y)-n^X(s,y))^2K_s(dy)ds\Big)=\int_\tau^t P_{\tau,m}\Big((n^{X_n}(s,B^s)-n^X(s,B^s))^2\Big)ds$$
$$\le\int_\tau^t P_{\tau,m}((X_n-X)^2)ds \quad \text{(by (V.2.3))}$$
$$\to 0\text{ as }n\to\infty.$$

This allows us to conclude that $X \in S$, as above. It also shows that $\int_\tau^t \int n^X(s,y)dM(s,y)$ is a continuous L^2 martingale. In addition we have $\mathbb{P}(K_\tau(n_\tau^X)) = m(n_\tau^X) = P_{\tau,m}(X) < \infty$. All these results now extend to any X in $L^2(P_{\tau,m})$ by considering the positive and negative parts of X. Taking $X = n(N)$ we obtain the required result for $t \le N$, but with $\tilde{n}(t,y) = P_{t,y^t}(n(N))$ in place of $n(t,y)$. Now use (V.2.10) and Lemma V.2.2 to obtain the required result for $t \le N$ as none of the relevant quantities are changed for $t \le N$ off a $\mathbb{P}$-null set if we replace $\tilde{n}$ with n. Finally let $N \to \infty$. ∎

Definition. Let $(\hat{\Omega}, \hat{\mathcal{F}}, \hat{\mathcal{F}}_t) = (\Omega \times C, \mathcal{F} \times \mathcal{C}, \mathcal{F}_t \times \mathcal{C}_t)$ and let $\hat{\mathcal{F}}_t^*$ denote the universal completion of $\hat{\mathcal{F}}_t$. If T is a bounded $(\mathcal{F}_t)_{t\ge\tau}$-stopping time (write $T \in \mathcal{T}_b$ and note this means that $T \ge \tau$), the normalized Campbell measure associated with K_T is the probability $\hat{\mathbb{P}}_T$ on $(\hat{\Omega}, \hat{\mathcal{F}})$ given by

$$\hat{\mathbb{P}}_T(A \times B) = \mathbb{P}(1_A K_T(B))/m(1).$$

We denote sample points in $\hat{\Omega}$ by (ω, y). Therefore under $\hat{\mathbb{P}}_T$, ω has law $K_T(1)m(1)^{-1}d\mathbb{P}$ and given ω, y is then chosen according to $K_T(\cdot)/K_T(1)$. We will also consider $T \in \mathcal{T}_b$ as an $(\hat{\mathcal{F}}_t)$-stopping time and define $\hat{\mathcal{F}}_T$ accordingly.

Proposition V.2.4. (a) Assume $T \in \mathcal{T}_b$ and $\psi \in b\hat{\mathcal{F}}_T$, then

$$(V.2.11) \qquad K_t(\psi) = K_T(\psi) + \int_T^t \int \psi(y)dM(s,y) \quad \forall t \ge T \quad \mathbb{P}\text{-a.s.}$$

(b) Let $g : [\tau,\infty) \times \hat{\Omega}$ be $(\hat{\mathcal{F}}_t)$-predictable and bounded on $[\tau, N] \times \hat{\Omega}$ for all $N > \tau$. Then

$$(V.2.12) \quad \int\int_\tau^t g_s(\omega,y)ds K_t(dy) = \int_\tau^t \int \left[\int_\tau^s g_r(\omega,y)dr\right] dM(s,y) + \int_\tau^t K_s(g_s)ds$$
$$\forall t \ge \tau \quad \text{a.s.}$$

Proof (a) Assume first T is constant and $\psi(\omega,y) = \psi_1(\omega)\psi_2(y^T)$ for $\psi_1 \in b\mathcal{F}_T$ and $\psi_2 : C \to \mathbb{R}$ bounded and continuous. Then

$$\phi(s,y) = P_{s,y^s}(\psi_2(B^T))$$

is a bounded predictable $(\mathcal{C}_t)$-martingale under P^{y_0} for each $y_0 \in \mathbb{R}^d$ (use (V.2.2)) and is continuous on $\mathbb{R}_+ \times C$ (as in (V.2.5)). Lemma V.2.1 shows that $\phi \in \mathcal{D}(\hat{A})$ and $\hat{A}\phi = 0$. Therefore $(HMP)_{\tau,K_\tau}$ implies that for $t \ge T$,

$$K_t(\psi) = \psi_1 K_t(\phi_t) = \psi_1 K_T(\psi_2) + \psi_1 \int_T^t \int \phi(s,y)dM(s,y)$$

$$= K_T(\psi) + \int_T^t \int \psi(y)dM(s,y),$$

because $\psi_1\phi(s) = \psi$ for $s \geq T$.

The proof now proceeds by a standard bootstrapping. The result clearly holds for ψ as above and T finite-valued and then for general T by the usual approximation of T by a decreasing sequence of finite-valued stopping times (the continuity of ψ_2 helps here). A monotone class argument now gives the result for any $\psi(\omega, y) = \tilde{\psi}(\omega, y^T)$, where $\tilde{\psi} \in b(\mathcal{F}_T \times \mathcal{C})$. We claim that any $\psi \in b\hat{\mathcal{F}}_T$ is of this form. For any $\psi \in b\hat{\mathcal{F}}_T$ there is an $(\hat{\mathcal{F}}_t)$-predictable process X so that $\psi = X(T)$ (Dellacherie and Meyer (1978), Theorem IV.67). It suffices to show that $X(T, \omega, y) = X(T, \omega, y^T)$ because we then prove the claim with $\tilde{\psi} = X(T)$. For this, first consider $X(t, \omega, y) = 1_{(s,u]}(t)1_A(\omega)1_B(y)$ for $u > s \geq \tau$, $A \in \mathcal{F}_s$ and $B \in \mathcal{C}_s$. Then the above claim is true because $1_B(y) = 1_B(y^s)$ and so on $\{s < T(\omega) \leq u\}$, $1_B(y) = 1_B(y^T(\omega))$. The aforementioned standard bootstrapping now gives the claim for any $(\hat{\mathcal{F}}_t)$-predictable X and so completes the proof of (a).

(b) First consider $g(s, \omega, y) = \phi(\omega, y)1_{(u,v]}(s)$ where $\phi \in b\hat{\mathcal{F}}_u$, $\tau \leq u < v$. Then $\mathbb{P}$-a.s. for $t \geq u$,

$$\begin{aligned}
\int\int_\tau^t g_s(\omega, y) ds K_t(dy) &= K_t(\phi)(t \wedge v - t \wedge u) \\
&= \int_u^t\int \phi(\omega, y)(s \wedge v - s \wedge u) dM(s, y) \\
&\quad + \int_\tau^t 1(u < s \leq v) K_s(\phi) ds \quad \text{(by (a) and integration by parts)} \\
&= \int_\tau^t\int \left[\int_\tau^s g_r(\omega, y) dr\right] dM(s, y) + \int_\tau^t K_s(g_s) ds.
\end{aligned}$$

If $t < u$, the above equality holds because both sides are zero. The result therefore holds for linear combinations of the above functions, i.e., for $(\hat{\mathcal{F}}_t)$-simple g. Passing to the bounded pointwise closure we obtain the result for all $(\hat{\mathcal{F}}_t)$-predictable and bounded g. For g as in (b), we first get the result for $t \leq N$ by considering $g_{s\wedge N}$, and then for all t by letting $N \to \infty$. ∎

Remarks V.2.5. (a) If $g : [\tau, \infty) \times \hat{\Omega} \to \mathbb{R}$ is $(\hat{\mathcal{F}}_t^*)$-predictable and bounded, and μ is a σ-finite measure on $\hat{\Omega}$, then there are bounded $(\hat{\mathcal{F}}_t)$-predictable processes $g_1 \leq g \leq g_2$ such that $g_1(t) = g_2(t)$ $\forall t \geq \tau$ μ-a.e. This may be proved by starting with a simple g (i.e. $g(t, \omega, y) = \sum_{i=1}^n \phi_i(\omega, y)1_{(u_i, u_{i+1}]}(t) + \phi_0(\omega, y)1_{\{u_0 = t\}}$, where $\tau = u_0 < \ldots < u_{n+1} \leq \infty$, $\phi_i \in b\hat{\mathcal{F}}_{u_i}^*$) and using a monotone class theorem as on p. 134 of Dellacherie and Meyer (1978).

(b) If we take

$$\mu(A) = \mathbb{P}\left(\int_{\tau}^{\infty}\int 1_A(\omega, y) K_s(dy) ds\right)$$

in the above, then the right side of (V.2.12) is the same for g_1, g_2 and g. Here we have used the obvious extension of the stochastic integral with respect to M to $(\hat{\mathcal{F}}_t^*)$-predictable integrands. It follows from Proposition V.2.4 (b) that the left-hand side is the same for g_1 and g_2. By monotonicity it is the same for g and so (V.2.12) holds for $(\hat{\mathcal{F}}_t^*)$-predictable, bounded g. A straightforward truncation argument then gives it for $(\hat{\mathcal{F}}_t^*)$-predictable g satisfying

$$\text{(V.2.13)} \qquad \int_{\tau}^{t}\int \left[\left[\int_{\tau}^{s} |g_r| dr\right]^2 + |g_s|\right] K_s(dy) ds < \infty \quad \forall t > 0 \quad \mathbb{P}\text{-a.s.}$$

(c) In Proposition V.2.4 (a), if T is predictable, $\psi \in b\hat{\mathcal{F}}_{T-}^*$ (i.e. $(\hat{\mathcal{F}}^*)_{T-}$), $g(t, \omega, y) = \psi(\omega, y) 1_{[T,\infty)}(t)$, (so g is $(\hat{\mathcal{F}}_t^*)$-predictable) and we take

$$\mu(A) = \mathbb{P}\left(\int 1_A(\omega, y) K_T(dy) + \int_{T}^{\infty}\int 1_A(\omega, y) K_s(dy) ds\right)$$

in (a), then $\psi_i = g_i(T, \omega, y) \in b\hat{\mathcal{F}}_{T-}$ (g_i as in (a)) and the right side of (V.2.11) is unchanged if ψ is replaced by ψ_i. As above, the inequality $\psi_1 \le \psi \le \psi_2$ shows the same is true of the left side. Therefore (V.2.11) remains valid if T is predictable and $\psi \in b\hat{\mathcal{F}}_{T-}^*$.

(d) If $\phi : [\tau, \infty) \times \hat{\Omega} \to \mathbb{R}$ is bounded and $(\hat{\mathcal{F}}_t^*)$-predictable, then $K_t(\phi_t)$ is $(\mathcal{F}_t)$-predictable. To see this, first note that this is clear from (c) (with $T = u$) if $\phi(t, \omega, y) = \psi(\omega, y) 1_{(u,v]}(t)$ for some $\tau \le u < v$ and $\psi \in b\hat{\mathcal{F}}_{u-}^*$. A monotone class argument now completes the proof (see Theorem IV.67 in Dellacherie and Meyer (1978)).

Here is the extension of $(HMP)_{\tau, K_\tau}$ we mentioned earlier.

Proposition V.2.6. Assume $\phi : [\tau, \infty) \times C \to \mathbb{R}$ is a $(\mathcal{C}_t)_{t \ge \tau}$-predictable map for which there is a $(\mathcal{C}_t)_{t \ge \tau}$-predictable map, $\bar{A}_{\tau,m}\phi = \bar{A}\phi$, such that

(i) $P_{\tau,m}\left(\int_\tau^t \bar{A}\phi(s)^2\, ds\right) < \infty \quad \forall\, t > \tau$

(ii) $n(t, y) = \phi(t, y) - \int_\tau^t \bar{A}\phi(s, y) ds,\ t \ge \tau$ is an L^2 $(\mathcal{C}_t)_{t \ge \tau}$-martingale under $P_{\tau,m}$.

Then

$$K_t(\phi_t) = K_\tau(\phi_\tau) + \int_\tau^t \int \phi(s, y) dM(s, y) + \int_\tau^t K_s(\bar{A}\phi_s) ds \quad \forall t \ge \tau \text{ a.s.}$$

The stochastic integral is a continuous L^2 martingale, $K_\tau(\phi_\tau)$ is square integrable, and $\mathbb{P}\Big(\int_\tau^t K_s(|\bar{A}\phi_s|)ds\Big) < \infty$ for all $t \geq \tau$.

Proof. Note that (i), Cauchy-Schwarz and the superprocess property (V.2.2) show that (V.2.13) holds for $g = \bar{A}\phi$–in fact the expression there is integrable. Therefore we may use Remark V.2.5 (b) and Lemma V.2.3 to see that $\mathbb{P}$-a.s. for all $t \geq \tau$,

$$\begin{aligned} K_t(\phi_t) &= K_t(n_t) + K_t\Big(\int_\tau^t \bar{A}\phi(s)\,ds\Big) \\ &= K_\tau(n_\tau) + \int_\tau^t \int n(s,y)dM(s,y) + \int_\tau^t \int \Big[\int_\tau^s \bar{A}\phi_r(y)dr\Big] dM(s,y) \\ &\qquad + \int_\tau^t K_s(\bar{A}\phi_s)ds \\ &= K_\tau(\phi_\tau) + \int_\tau^t \int \phi(s,y)dM(s,y) + \int_\tau^t K_s(\bar{A}\phi_s)ds. \end{aligned}$$

Lemma V.2.3 shows that $K_\tau(\phi_\tau) = K_\tau(n_\tau)$ is square integrable and (V.2.2), (i) and (ii) show that

$$\mathbb{P}\Big(\int_\tau^t K_s(\phi_s^2)ds\Big) = \int_\tau^t P_{\tau,m}(\phi(s,B)^2)ds < \infty.$$

This shows the stochastic integral is an L^2 martingale and a similar argument shows that the drift term has integrable variation on bounded intervals. ∎

Notation. Let $\mathcal{D}(\bar{A}_{\tau,m})$ denote the class of ϕ considered in the above Proposition. Clearly $\mathcal{D}(\hat{A}) \subset \mathcal{D}(\bar{A}_{\tau,m})$ and $\bar{A}_{\tau,m}$ is an extension of $\hat{A}$ in that for any $\phi \in \mathcal{D}(\hat{A})$,

$$\int_\tau^t \bar{A}_{\tau,m}\phi(s,y)ds = \int_\tau^t \hat{A}\phi(s,y)ds \quad \forall t \geq \tau \quad P_{\tau,m} - \text{a.s.}$$

Exercise V.2.1. Assume $\tau = 0$ and K is an $(\mathcal{F}_t)$-historical Brownian motion starting at $(0,\nu)$–note we can treat K_0 as a finite measure on $\mathbb{R}^d$. Let $Z_0 : \mathbb{R}^d \to \mathbb{R}^d$ be a Borel map, $B \in \mathcal{B}(\mathbb{R}^d)$ and define $\tilde{Z}_0 : C \to C$ by

$$\tilde{Z}_0(y)(t) = y(t) - y(0) + Z_0(y(0)).$$

(a) If $\phi \in \mathcal{D}(\hat{A})$, show that $\tilde{\phi}(t,y) = \phi(t, \tilde{Z}_0(y))1_B(y(0)) \in \mathcal{D}(\bar{A}_{0,m})$ and

$$\bar{A}_{0,m}\tilde{\phi}(t,y) = \hat{A}\phi(t,\tilde{Z}_0(y))1_B(y(0)).$$

(b) Define $K_t'(F) = K_t(\{y : \tilde{Z}_0(y) \in F,\ y(0) \in B\})$ for $F \in \mathcal{C}$, and let ν' be the law of $K_0' = K_0(Z_0^{-1}(A) \cap B)$ for $A \in \mathcal{B}(\mathbb{R}^d)$. Show that K' is an $(\mathcal{F}_t)$-historical Brownian motion starting at $(0,\nu')$ and therefore has law $\mathbb{Q}_{0,\nu'}$.

Hint. Use (a) to show that K' satisfies $(HMP)_{0,K_\tau'}$.

Corollary V.2.7. Let $T \in \mathcal{T}_b$ and assume $n(t)$, $t \geq \tau$ is a $(\mathcal{C}_t)$-predictable square integrable martingale under $P_{\tau,m}$. Then $n(t \wedge T)$, $t \geq \tau$ is an $(\hat{\mathcal{F}}_t)$-martingale under $\hat{\mathbb{P}}_T$.

Proof. Let $s \geq \tau$, $A \in \mathcal{F}_s$ and $B \in \mathcal{C}_s$. Define

$$\phi(t,y) = (n(t,y) - n(s,y))1_B(y)1(t \geq s).$$

Then $\phi \in D(\bar{A}_{\tau,m})$ and $\bar{A}_{\tau,m}\phi = 0$. Therefore $K_t(\phi_t)$ is an $(\mathcal{F}_t)$-martingale by Proposition V.2.6 and so

$$\begin{aligned}\hat{\mathbb{P}}_T((n(T) - n(s\wedge T))1_A(\omega)1_B(y)) &= \mathbb{P}(K_T(\phi_T)1_{A\cap\{T>s\}})m(1)^{-1} \\ &= \mathbb{P}(K_s(\phi_s)1_{A\cap\{T>s\}})m(1)^{-1} = 0,\end{aligned}$$

the last because $\phi_s = 0$. ∎

Example V.2.8. Recall $C_K^\infty(\mathbb{R}^k)$ is the set of infinitely differentiable functions on $\mathbb{R}^k$ with compact support and define

$$\begin{aligned}D_{fd} = \{\phi : \mathbb{R}_+ \times C \to \mathbb{R} : \phi(t,y) = \psi(y_{t_1\wedge t}, \ldots, y_{t_n\wedge t}) \equiv \psi(\bar{y}_t),\\ 0 \leq t_1 \leq \ldots \leq t_n,\ \psi \in C_K^\infty(\mathbb{R}^{nd})\}.\end{aligned}$$

If ϕ is as above, let

$$\phi_i(t,y) = \sum_{k=1}^n 1(t < t_k)\psi_{(k-1)d+i}(\bar{y}_t) \quad 1 \leq i \leq d,$$

$$\phi_{ij}(t,y) = \sum_{k=1}^n \sum_{\ell=1}^n 1(t < t_k \wedge t_\ell)\psi_{(k-1)d+i,(\ell-1)+j}(\bar{y}_t) \quad 1 \leq i,j \leq d,$$

$$\nabla\phi(t,y) = (\phi_1(t,y), \ldots, \phi_d(t,y)), \text{ and } \bar{\Delta}\phi(t,y) = \sum_{i=1}^d \phi_{ii}(t,y).$$

Itô's Lemma shows that for any $m \in M_F(C)$, $D_{fd} \subset D(\bar{A}_{\tau,m})$ and $A_{\tau,m}\phi = \frac{\bar{\Delta}\phi}{2}$ for $\phi \in D_{fd}$. In fact this remains true if C_K^∞ is replaced with C_b^2 in the above. Note that D_{fd} is not contained in $\mathcal{D}(\hat{A})$ because $\bar{\Delta}\phi(t,y)$ may be discontinuous in t.

Theorem V.2.9. Let $m \in M_F^\tau(C)$. An $(\mathcal{F}_t)_{t\geq\tau}$-adapted process $(K_t)_{t\geq\tau}$ with sample paths in $\Omega_H[\tau,\infty)$ is an $(\mathcal{F}_t)$-historical Brownian motion starting at (τ,m) iff for every $\phi \in D_{fd}$,

$$M_t(\phi) = K_t(\phi_t) - m(\phi_\tau) - \int_\tau^t K_s\Big(\frac{\bar{\Delta}\phi_s}{2}\Big)ds$$

is a continuous $(\mathcal{F}_t)$-local martingale such that $M_\tau(\phi) = 0$ and

$$\langle M(\phi)\rangle_t = \int_\tau^t K_s(\phi_s^2)ds \text{ for all } t \geq \tau \text{ a.s.}$$

Proof. The previous Example and Proposition V.2.6 show that an $(\mathcal{F}_t)$-historical Brownian motion does satisfy the above martingale problem.

The proof of the converse uses a generalization of the stochastic calculus developed in the next section for historical Brownian motion. It is proved in Theorem 1.3 of Perkins (1995)–see also Section 12.3.3 of Dawson (1993) for a different approach to a slightly different result. We will not use the uniqueness here although it plays an central role in the historical martingale problem treated in Perkins (1995) and discussed in Section 5 below. ∎

3. Stochastic Integration on Brownian Trees

Consider the question of defining the stochastic integral appearing in (SE)(a). We first need a probability measure on $\hat{\Omega}$ under which y is a Brownian motion. An infinite family of such probabilities is given below. We continue to work with the $(\mathcal{F}_t)$-historical Brownian motion, K_t, on $(\Omega, \mathcal{F}, (\mathcal{F}_t)_{t\geq\tau}, \mathbb{P})$ starting at (τ, ν) but now set $\tau = 0$ for convenience and so may view K_0 and its mean measure, m, as finite measures on $\mathbb{R}^d$.

Definition. Let $\bar{\Omega}' = (\Omega', \mathcal{G}, \mathcal{G}_t, \mathbb{Q})$ be a filtered space and T be a $(\mathcal{G}_t)$-stopping time. An $\mathbb{R}^d$-valued $(\mathcal{G}_t)$-adapted process, B_t, on $\bar{\Omega}'$ is a $(\mathcal{G}_t)$-Brownian motion stopped at T iff for $1 \leq i,j \leq d$, $B_t^i - B_0^i$ and $B_t^i B_t^j - \delta_{ij}(t \wedge T)$ are continuous $(\mathcal{G}_t)$-martingales.

If T is a constant time, (V.2.2) shows that under $\hat{\mathbb{P}}_T$, y is a Brownian motion stopped at T. The next result extends this to stopping times.

Proposition V.3.1. If $T \in \mathcal{T}_b$, then under $\hat{\mathbb{P}}_T$, y is a $(\hat{\mathcal{F}}_t)$-Brownian motion stopped at T.

Proof. Apply Corollary V.2.7 with $n(t,y) = y_t^i - y_0^i$ and $n(t,y) = y_t^i y_t^j - y_0^i y_0^j - \delta_{ij} t$. This gives the result because $y_t^i = y_{t\wedge T}^i$ $\hat{\mathbb{P}}_T$-a.s. ∎

Notation. $f \in D(n,d)$ iff $f : \mathbb{R}_+ \times \hat{\Omega} \to \mathbb{R}^{n\times d}$ is $(\hat{\mathcal{F}}_t^*)$-predictable and

$$\int_0^t \|f(s,\omega,y)\|^2 ds < \infty \ K_t - \text{a.a. } y \quad \forall t \geq 0 \quad \mathbb{P} - \text{a.s.}$$

Definition. If $X, Y : \mathbb{R}_+ \times \hat{\Omega} \to E$, we say $X = Y$ K-a.e. iff $X(s,\omega,y) = Y(s,\omega,y)$ for all $s \leq t$ for K_t-a.a. y for all $t \geq 0$ $\mathbb{P}$-a.s.
If E is a metric space we say X is continuous K-a.e. iff $s \to X(s,\omega,y)$ is continuous on $[0,t]$ for K_t-a.a. y for all $t \geq 0$ $\mathbb{P}$-a.s.

If $T \in \mathcal{T}_b$ and $f \in D(n,d)$, then $\int_0^T \|f(s,\omega,y)\|^2 ds < \infty$ $\hat{\mathbb{P}}_T$-a.s. Therefore the classical stochastic integral $\int_0^t f(s,\omega,y)dy_s \equiv \hat{\mathbb{P}}_T - \int_0^t f(s,\omega,y)dy_s$ is uniquely defined up to $\hat{\mathbb{P}}_T$-null sets. The next result shows one can uniquely define a single process which represents these stochastic integrals for all T simultaneously.

Proposition V.3.2. (a) If $f \in D(n,d)$, there is an $\mathbb{R}^n$-valued $(\hat{\mathcal{F}}_t)$-predictable process $I(f,t,\omega,y)$ such that

$$(V.3.1) \quad I(f, t\wedge T, \omega, y) = \hat{\mathbb{P}}_T - \int_0^t f(s,\omega,y)dy(s) \quad \forall t \geq 0 \quad \hat{\mathbb{P}}_T - \text{a.s. for all } T \in \mathcal{T}_b.$$

(b) If $I'(f)$ is an$(\hat{\mathcal{F}}_t^*)$-predictable process satisfying (V.3.1), then

$$I(f,s,\omega,y) = I'(f,s,\omega,y) \quad K - \text{a.e.}$$

(c) $I(f)$ is continuous K-a.e.

(d) (Dominated Convergence) For any $N > 0$, if $f_k, f \in D(n,d)$ satisfy

$$\lim_{k\to\infty} \mathbb{P}\Big(K_N\Big(\int_0^N \|f_k(s) - f(s)\|^2 ds > \varepsilon\Big)\Big) = 0 \quad \forall\varepsilon > 0,$$

then

$$\lim_{k\to\infty} \mathbb{P}\left(\sup_{t\le N} K_t\left(\sup_{s\le t} \|I(f_k,s,\omega,y) - I(f,s,\omega,y)\| > \varepsilon\right)\right) = 0 \quad \forall\varepsilon > 0.$$

(e) For any $S \in \mathcal{T}_b$ if $f_k, f \in D(n,d)$ satisfy

$$\text{(V.3.2)} \qquad \lim_{k\to\infty} \mathbb{P}\Big(K_S\Big(\int_0^S \|f_k(s) - f(s)\|^2 ds\Big)\Big) = 0,$$

then

$$\sup_{t\le S} K_t(\sup_{s\le t} \|I(f_k,s) - I(f,s)\|^2) \xrightarrow{\mathbb{P}} 0 \text{ as } k \to \infty.$$

Proof. To avoid factors of $m(1)^{-1}$ we will assume $m(1) = 1$ throughout.

(b) Let

$$J(t,\omega) = \int \sup_{s\le t} \|I(f,s,\omega,y) - I'(f,s,\omega,y)\| \wedge 1\, K_t(dy).$$

Assume for the moment that J is $(\mathcal{F}_t)$-predictable, and let T be a bounded $(\mathcal{F}_t)$-predictable stopping time. Then

$$\mathbb{P}(J(T,\omega)) = \hat{\mathbb{P}}_T\left(\sup_{s\le T} \|I(f,s) - I'(f,s)\| \wedge 1\right) = 0,$$

because under $\hat{\mathbb{P}}_T$, $I(f, s\wedge T)$ and $I'(f, s\wedge T)$ are both versions of $\hat{\mathbb{P}}_T - \int_0^s f(s)dy(s)$. By the Section Theorem we see that $J(t,\omega) = 0$ $\forall t \ge 0$ $\mathbb{P}$-a.s., as required.

To prove J is $(\mathcal{F}_t)$-predictable, let $\phi(t,\omega,y)$ be the integrand in the definition of J. The projection of a $\mathcal{B}\times\hat{\mathcal{F}}_t^*$-measurable set onto Ω is $\hat{\mathcal{F}}_t^*$-measurable (Theorem III.13 of Dellacherie and Meyer (1978)) and so $\phi(t)$ is $(\hat{\mathcal{F}}_t^*)$-adapted. Therefore $\phi(t-)$ is $(\hat{\mathcal{F}}_t^*)$-predictable (being left-continuous) and hence so is

$$\phi(t) = \phi(t-) \vee (\|I(f,t,\omega,y) - I'(f,t,\omega,y)\| \wedge 1).$$

Remark V.2.5 (d) now shows that $J(t) = K_t(\phi(t))$ is $(\mathcal{F}_t)$-predictable.

(a), (c) For simplicity set $d = n = 1$ in the rest of this proof (this only affects a few constants in what follows). If $f(s,\omega,y) = \sum_{i=1}^n f_i(\omega,y)1_{(u_i,u_{i+1}]}(s)$, where $f_i \in b\hat{\mathcal{F}}_{u_i}$ and $0 = u_0 < \ldots < u_{n+1} \le \infty$, (call f $(\hat{\mathcal{F}}_t)$-simple), then define

$$I(f,t,\omega,y) = \sum_{i=0}^n f_i(\omega,y)\,(y(t\wedge u_{i+1}) - y(t\wedge u_i)).$$

This clearly satisfies the conclusions of (a) and (c).

Let $f \in D(1,1)$. As in the usual construction of the Itô integral we may choose a sequence of simple functions $\{f_k\}$ so that

$$\hat{\mathbb{P}}_k\left(\int_0^k |f(s)-f_k(s)|^2\,ds \ge \frac{1}{4}2^{-3k}\right) \le 2^{-k}. \tag{V.3.3}$$

Define

$$I(f,t,\omega,y) = \begin{cases} \lim_{k\to\infty} I(f_k,t,\omega,y) & \text{if it exists} \\ 0 & \text{otherwise} \end{cases}.$$

Clearly $I(f)$ is $(\hat{\mathcal{F}}_t)$-predictable.

Fix a bounded $(\mathcal{F}_t)$-stopping time T and let $m,\ell \ge n \ge T$. Use the fact that $I(f_m, t\wedge T, \omega, y)$ is a version of the $\hat{\mathbb{P}}_T$-Itô integral and standard properties of the latter to see that

$$\begin{aligned}
&\hat{\mathbb{P}}_T\left(\sup_{t\le T}|I(f_m,t,\omega,y)-I(f_\ell,t,\omega,y)|>2^{-n}\right)\\
&\le \hat{\mathbb{P}}_T\left(\int_0^T |f_m(s)-f_\ell(s)|^2 ds \ge 2^{-3n}\right)+2^{-n}\\
&= \mathbb{P}\left(K_T\left(\int_0^T |f_m(s)-f_\ell(s)|^2 ds > 2^{-3n}\right)\right)+2^{-n}\\
&= \mathbb{P}\left(K_n\left(\int_0^T |f_m(s)-f_\ell(s)|^2 ds > 2^{-3n}\right)\right)+2^{-n}\\
&\qquad\qquad\text{(by Proposition V.2.4(a))}\\
&\le \hat{\mathbb{P}}_n\left(\int_0^n |f_m(s)-f_\ell(s)|^2 ds > 2^{-3n}\right)+2^{-n} \le 3\cdot 2^{-n},
\end{aligned} \tag{V.3.4}$$

the last by (V.3.3) and an elementary inequality. This shows both that

$$\sup_{t\le T}|I(f_m,t,\omega,y)-I(f,t,\omega,y)| \to 0 \quad \text{as } m\to\infty \quad \hat{\mathbb{P}}_T\text{-a.s.} \tag{V.3.5}$$

and

$$\sup_{t\le T}\left|\left(\hat{P}_T-\int_0^t f_m(s,\omega,y)dy(s)\right)-\left(\hat{P}_T-\int_0^t f(s,\omega,y)dy(s)\right)\right| \to 0 \quad \text{as } m\to\infty \quad \hat{\mathbb{P}}_T\text{-a.s.}$$

It follows that

$$\hat{\mathbb{P}}_T - \int_0^t f(s,\omega,y)dy(s) = I(f,t\wedge T,\omega,y) \quad \forall t\ge 0 \quad \hat{\mathbb{P}}_T\text{-a.s.}$$

because the left side is constant on $t \geq T$ $\hat{\mathbb{P}}_T$-a.s. This gives (a).

Set $m = n$ and let $\ell \to \infty$ in (V.3.4), and use (V.3.5) to conclude that

$$\sup_{T \in \mathcal{T}_b, T \leq m} \hat{\mathbb{P}}_T\left(\sup_{s \leq T} |I(f_m, s, \omega, y) - I(f, s, \omega, y)| > 2^{-m}\right) \leq 3 \cdot 2^{-n},$$

that is

$$\sup_{T \in \mathcal{T}_b, T \leq m} \mathbb{P}\left(K_T\left(\sup_{s \leq T} |I(f_m, s, \omega, y) - I(f, s, \omega, y)| > 2^{-m}\right)\right) \leq 3 \cdot 2^{-m}.$$

An application of the Section Theorem ($\tilde{J}(t, \omega) = K_t(\sup_{s \leq t} |I(f_m, s) - I(f, s)| > 2^{-m})$ is $(\mathcal{F}_t)$-predictable, as in (b)) gives

$$\mathbb{P}\left(\sup_{t \leq m} K_t\left(\sup_{s \leq t} |I(f_m, s, \omega, y) - I(f, s, \omega, y)| > 2^{-m}\right)\right) \leq 3 \cdot 2^{-m}.$$

Two successive applications of Borel-Cantelli show that

$$\text{(V.3.6)} \qquad \lim_{m \to \infty} \sup_{s \leq t} |I(f_m, s, \omega, y) - I(f, s, \omega, y)| = 0 \quad K_t\text{-a.a. y} \quad \forall t \geq 0 \quad \text{a.s.}$$

This certainly implies (c).

(d) Fix $N > 0$ and assume $\{f_k\}$, f satisfy the hypotheses of (d). Argue exactly as in (V.3.4) with $\varepsilon > 0$ in place of 2^{-n}, but now use Remark V.2.5 (c) in place of Proposition V.2.4 (a) and take the sup over $(\mathcal{F}_t)$-predictable times to conclude

$$\sup_{T \leq N, T \text{ predictable}} \hat{\mathbb{P}}_T\left(\sup_{t \leq T} |I(f_k, t, \omega, y) - I(f, t, \omega, y)| > \varepsilon\right)$$

$$\leq \hat{\mathbb{P}}_N\left(\int_0^N |f_k - f(s, \omega, y)|^2 ds > \varepsilon^3\right) + \varepsilon.$$

The first term on the right-hand side approaches 0 as $k \to \infty$. As ε is arbitrary, the same is true for the left-hand side. As in (b),

$$(t, \omega) \to K_t\left(\sup_{s \leq t} |I(f_k, s, \omega, y) - I(f, s, \omega, y)| > \varepsilon\right)$$

is $(\mathcal{F}_t)$-predictable and the Section Theorem implies

$$\sup_{t \leq N} K_t\left(\sup_{s \leq t} |I(f_k, s, \omega, y) - I(f, s, \omega, y)| > \varepsilon\right) \xrightarrow{\mathbb{P}} 0 \quad \text{as} \quad k \to \infty \quad \forall \varepsilon > 0.$$

The random variables on the left are all bounded by $\sup_{t \leq N} K_t(1) \in L^1$ and so also converge in L^1 by Dominated Convergence for all $\varepsilon > 0$. This is the required conclusion in (d).

(e) Let $S, T \in \mathcal{T}_b$ with $T \le S$. Doob's maximal L^2 inequality shows that

$$\begin{aligned}\mathbb{P}\Big(K_T\Big(\sup_{s\le T}(I(f,s)-I(f_k,s))^2\Big)\Big) &\le c\hat{\mathbb{P}}_T\Big(\int_0^T (f(s)-f_k(s))^2\,ds\Big)\\ &= c\mathbb{P}\Big(K_S\Big(\int_0^T (f(s)-f_k(s))^2\,ds\Big)\Big)\\ &\le c\mathbb{P}\Big(K_S\Big(\int_0^S (f(s)-f_k(s))^2\,ds\Big)\Big).\end{aligned}$$

Remark V.2.5 (c) was used in the second line. Therefore (V.3.2) implies that

$$(V.3.7)\qquad \lim_{k\to\infty}\sup_{T\le S, T\in\mathcal{T}_b}\mathbb{P}\Big(K_T\Big(\sup_{s\le T}(I(f,s)-I(f_k,s))^2\Big)\Big)=0.$$

As in the proof of (b), $(t,\omega)\to K_t(\sup_{s\le t}(I(f,s)-I(f_k,s))^2)$ is $(\mathcal{F}_t)$-predictable. A simple application of the Section Theorem shows that

$$\begin{aligned}&\mathbb{P}\Big(\sup_{t\le S}K_t\Big(\sup_{s\le t}(I(f,s)-I(f_k,s))^2\Big)>\varepsilon\Big)\\ &=\sup_{T\le S,T\in\mathcal{T}_b}\mathbb{P}\Big(K_T\Big(\sup_{s\le T}(I(f,s)-I(f_k,s))^2\Big)>\varepsilon\Big),\end{aligned}$$

which approaches zero as $k\to\infty$ by (V.3.7). ∎

Lemma V.3.3. Let $g:\mathbb{R}_+\times\hat{\Omega}\to\mathbb{R}_+$ be $(\hat{\mathcal{F}}_t^*)$-predictable and $S\in\mathcal{T}_b$.
(a) $\mathbb{P}\Big(K_S\Big(\int_0^S g_s ds\Big)\Big)=\mathbb{P}\Big(\int_0^S K_s(g_s)ds\Big)$.
(b) $\mathbb{P}\Big(\sup_{t\le S}K_t\Big(\int_0^t g_s ds\Big)>\varepsilon\Big)\le\varepsilon^{-1}\mathbb{P}\Big(\int_0^S K_s(g_s)ds\Big)$ for all $\varepsilon>0$.

Proof. (a) By Monotone Convergence it suffices to consider g bounded. This case is then immediate from Remark V.2.5(b).
(b) From Remark V.2.5 (d) we see that $K_t\Big(\int_0^t g_s\,ds\Big)\le\infty$ is $(\mathcal{F}_t)$-predictable. By the Section Theorem,

$$\begin{aligned}\mathbb{P}\Big(\sup_{t\le S}K_t\Big(\int_0^t g_s ds\Big)>\varepsilon\Big) &= \sup_{T\le S,T\in\mathcal{T}_b}\mathbb{P}\Big(K_T\Big(\int_0^T g_s ds\Big)>\varepsilon\Big)\\ &\le \sup_{T\le S,T\in\mathcal{T}_b}\varepsilon^{-1}\mathbb{P}\Big(K_T\Big(\int_0^T g_s ds\Big)\Big)\\ &=\varepsilon^{-1}\mathbb{P}\Big(\int_0^S K_s(g_s)ds\Big)\quad\text{(by (a))}.\end{aligned}$$ ∎

Notation. If $X(t)=(X_1(t),\dots,X_n(t))$ is an $\mathbb{R}^n$-valued process on $(\hat{\Omega},\hat{\mathcal{F}})$ and $\mu\in M_F(\mathbb{R}^d)$, let $\mu(X_t)=(\mu(X_1(t)),\dots,\mu(X_n(t)))$ and

$$\int_0^t\int X(s)dM(s,y)=\Big(\int_0^t\int X_1(s)dM(s,y),\dots,\int_0^t\int X_n(s)dM(s,y)\Big),$$

whenever these integrals are defined. We also write $\int_0^t f(s,\omega,y)dy(s)$ for $I(f,t,\omega,y)$ when $f \in D(n,d)$ and point out that dependence on K is suppressed in either notation.

Proposition V.3.4. If $f \in D(n,d)$ is bounded, then

$$(V.3.8) \qquad K_t(I(f,t)) = \int_0^t \int I(f,s)dM(s,y) \quad \forall t \geq 0 \quad \mathbb{P}-\text{a.s.},$$

and the above is a continuous L^2 $(\mathcal{F}_t)$-martingale.

Proof. To simplify the notation take $n = d = 1$ and $m(1) = 1$. Assume first that

$$(V.3.9) \quad f(s,\omega,y) = \phi_1(\omega)\phi_2(y)1(u < s \leq v), \quad \phi_1 \in b\mathcal{F}_u,\ \phi_2 \in b\mathcal{C}_u,\ 0 \leq u < v.$$

If $n(t,y) = \phi_2(y)(y(t\wedge v) - y(t \wedge u))$, then $(n(t), \mathcal{C}_t)$ is an L^2-martingale under $P_{0,m}$ and $I(f,t) = \phi_1(\omega)n(t,y)$. Lemma V.2.3 shows that $\mathbb{P}$-a.s. for all $t \geq 0$,

$$\begin{aligned} K_t(I(f,t)) &= \phi_1(\omega)K_t(n_t) \\ &= \phi_1(\omega)\int_0^t \int n(s,y)dM(s,y) \\ &= \int_0^t \int I(f,s)dM(s,y). \end{aligned}$$

In the last line we can take ϕ_1 through the stochastic integral since $\phi_1 \in b\mathcal{F}_u$ and $n(s)$ vanishes for $s \leq u$.

Suppose (V.3.8) holds for a sequence of $(\hat{\mathcal{F}}_t)$-predictable processes f_k and f is an $(\hat{\mathcal{F}}_t^*)$-predictable process such that $\sup_k \|f_k\|_\infty \vee \|f\|_\infty < \infty$ and

$$(V.3.10) \qquad \lim_{k\to\infty} \mathbb{P}\Big(\int\int_0^N (f_k(s) - f(s))^2 ds K_N(dy)\Big) = 0 \quad \forall N \in \mathbb{N}.$$

We claim (V.3.8) also holds for f. If $N \in \mathbb{N}$, then use the fact that under $\hat{\mathbb{P}}_s$, $I(f)$ is an ordinary Itô integral to conclude

$$\begin{aligned} \mathbb{P}\Big(\int_0^N \int (I(f,s) - I(f_k,s))^2 K_s(dy)ds\Big) &= \int_0^N \hat{\mathbb{P}}_s((I(f,s) - I(f_k,s))^2)ds \\ &= \int_0^N \hat{\mathbb{P}}_s\Big(\int_0^s (f(r) - f_k(r))^2 dr\Big)ds \\ &= \int_0^N \mathbb{P}\Big(K_N\Big(\int_0^s (f(r) - f_k(r))^2 dr\Big)\Big)ds, \end{aligned}$$

where Remark V.2.5 (c) is used in the last line. This approaches zero as $k \to \infty$ by (V.3.10). Therefore $\int_0^t \int I(f,s)dM(s,y)$ is a continuous L^2 martingale and

$$(V.3.11) \qquad \lim_{k\to\infty} \Big\| \sup_{t\leq N} \Big| \int_0^t \int I(f_k,s) - I(f,s)dM(s,y) \Big| \Big\|_2 = 0 \quad \forall N \in \mathbb{N}.$$

Proposition V.3.2 (e) with $S = N$ and (V.3.10) imply

$$(V.3.12) \qquad \sup_{t\le N} K_t(|I(f_k,t) - I(f,t)|) \xrightarrow{\mathbb{P}} 0 \text{ as } k\to\infty \quad \forall N\in\mathbb{N}.$$

Now let $k \to \infty$ in (V.3.8) (for f_k) to see that it also holds for f.

We may now pass to the bounded pointwise closure of the linear span of the class of f satisfying (V.3.9) to see that the result holds for all bounded $(\hat{\mathcal{F}}_t)$-predictable f. If f is bounded and $(\hat{\mathcal{F}}^*_t)$-predictable, there is a bounded $(\hat{\mathcal{F}}_t)$-predictable $\tilde{f}$ so that

$$\mathbb{P}\Big(\int\int_0^N (f(s)-\tilde{f}(s))^2 ds K_N(dy)\Big) = 0 \quad \forall N\in\mathbb{N}$$

(see Remark V.2.5 (a)) and so the the result holds for f by taking $f_k = \tilde{f}$ in the above limiting argument. ∎

Theorem V.3.5 (Itô's Lemma). Let Z_0 be $\mathbb{R}^n$-valued and $\hat{\mathcal{F}}^*_0$-measurable, $f \in D(n,d)$, g be an $\mathbb{R}^n$-valued $(\hat{\mathcal{F}}^*_t)$-predictable process and $\psi \in C_b^{1,2}(\mathbb{R}_+\times\mathbb{R}^n)$. Assume

$$(V.3.13) \qquad \int_0^t K_s(\|f_s\|^2 + \|g_s\|)ds < \infty \quad \forall t>0 \text{ a.s.},$$

and let

$$(V.3.14) \qquad Z_t(\omega,y) = Z_0(\omega,y) + \int_0^t f(s,\omega,y)dy_s + \int_0^t g(s,\omega,y)ds.$$

If $\nabla\psi$ and ψ_{ij} denote the gradient and second order partials in the spatial variables, then

$$\int \psi(t,Z_t)K_t(dy) = \int \psi(0,Z_0)dK_0(y) + \int_0^t\int \psi(s,Z_s)dM(s,y)$$
$$(V.3.15) \qquad + \int_0^t K_s\Big(\frac{\partial\psi}{\partial s}(s,Z_s) + \nabla\psi(s,Z_s)\cdot g_s + \frac{1}{2}\sum_{i=1}^n\sum_{j=1}^n \psi_{ij}(s,Z_s)(ff^*)_{ij}(s)\Big)ds.$$

The second term on the right is an L^2 martingale and the last term on the right has paths with finite variation on compact intervals a.s. In particular $\int\psi(t,Z_t)K_t(dy)$ is a continuous $(\mathcal{F}_t)$-semimartingale.

Proof. Assume first that $\|f\|$ and $\|g\|$ are bounded. Let $T \in \mathcal{T}_b$ and $\mathbf{Z}_t = (t, Z_t)$. Itô's Lemma shows that $\hat{\mathbb{P}}_T$-a.s.

$$\psi(\mathbf{Z}_{t\wedge T}) = \psi(\mathbf{Z}_0) + \hat{\mathbb{P}}_T - \int_0^{t\wedge T} \nabla\psi(\mathbf{Z}_s)f(s)\cdot dy(s)$$
$$+ \int_0^{t\wedge T} \frac{\partial\psi}{\partial s}(\mathbf{Z}_s) + \nabla\psi(\mathbf{Z}_s)\cdot g(s) + \frac{1}{2}\sum_{i=1}^n\sum_{j=1}^n \psi_{ij}(\mathbf{Z}_s)(ff^*)_{ij}(s)ds \ \forall t\ge 0.$$

Let $\tilde{b}(s)$ denote the integrand in the last term. This shows that

$$\tilde{I}(t) = \psi(\mathbf{Z}_t) - \psi(\mathbf{Z}_0) - \int_0^t \tilde{b}(s)ds$$

is a $(\hat{\mathcal{F}}_t^*)$-predictable process satisfying (V.3.1) with $\nabla\psi(\mathbf{Z}_s)f(s)$ in place of $f(s)$. Proposition V.3.2 therefore implies that

$$\psi(\mathbf{Z}_t) = \psi(\mathbf{Z}_0) + I(\nabla\psi(\mathbf{Z})f, t) + \int_0^t \tilde{b}(s)ds \quad K-\text{a.e.} \tag{V.3.16}$$

Since $\|\tilde{b}\|$ and $\|\nabla\psi(\mathbf{Z})f\|$ are bounded and $(\hat{\mathcal{F}}_T^*)$-predictable we may apply Proposition V.3.4 and Remarks V.2.5 to see that $\mathbb{P}$-a.s. for all $t \geq 0$,

$$\begin{aligned}
&\int \psi(\mathbf{Z}_t)K_t(dy) \\
&= \int \psi(\mathbf{Z}_0)K_0(dy) + \int_0^t \int \psi(\mathbf{Z}_0)dM(s,y) + \int_0^t \int I(\nabla\psi(\mathbf{Z})f, s)dM(s,y) \\
&\quad + \int_0^t \int \Big[\int_0^s \tilde{b}(r)dr\Big]dM(s,y) + \int_0^t K_s(\tilde{b}(s))ds \\
&= \int \psi(\mathbf{Z}_0)K_0(dy) + \int_0^t \int \psi(\mathbf{Z}_s)dM(s,y) + \int_0^t K_s(\tilde{b}(s))ds.
\end{aligned}$$

In the last line we used (V.3.16) and the fact that this implies the stochastic integrals of both sides of (V.3.16) with respect to M coincide. This completes the proof in this case.

Assume now that f, g satisfy (V.3.13). By truncating we may choose bounded $(\mathcal{F}_t^*)$-predictable f^k, g^k such that $f^k \to f$ and $g^k \to g$ pointwise, $\|f^k\| \leq \|f\|$ and $\|g^k\| \leq \|g\|$ pointwise, and therefore

$$\lim_{k\to\infty} \int_0^t K_s(\|f_s^k - f_s\|^2 + \|g_s^k - g_s\|)ds = 0 \quad \forall t \geq 0 \text{ a.s.} \tag{V.3.17}$$

By (V.3.13) we may choose $S_n \in \mathcal{T}_b$ satisfying $S_n \uparrow \infty$ a.s. and

$$\int_0^{S_n} K_s(\|f_s\|^2 + \|g_s\|)ds \leq n \quad \text{a.s.}$$

Define Z^k as in (V.3.14) but with (f^k, g^k) in place of (f, g). Note that Lemma V.3.3(b) shows that $\sup_{t\leq S_n} K_t\Big(\int_0^t \|g_s\|ds\Big) < \infty$ for all n a.s. and so $Z_t(\omega, y)$ is well-defined K_t-a.a. y for all $t \geq 0$ a.s. The same result shows that

$$\sup_{t\leq S_n} K_t\Big(\int_0^t \|g_s^k - g_s\|ds\Big) \xrightarrow{\mathbb{P}} 0 \text{ as } k \to \infty \quad \forall n. \tag{V.3.18}$$

A similar application of Proposition V.3.2 (e) and Lemma V.3.3(a) and gives

$$\sup_{t\leq S_n} K_t(\sup_{s\leq t} \|I(f^k, s) - I(f, s)\|^2) \xrightarrow{\mathbb{P}} 0 \text{ as } k \to \infty \quad \forall n. \tag{V.3.19}$$

(V.3.18) and (V.3.19) imply

$$(V.3.20) \qquad \sup_{t\le T} K_t(\sup_{s\le t} \|Z^k(s) - Z(s)\|) \xrightarrow{\mathbb{P}} 0 \text{ as } k \to \infty \quad \forall T > 0.$$

We have already verified (V.3.15) with (Z^k, f^k, g^k) in place of (Z, f, g). The boundedness of ψ and its derivatives, together with (V.3.17) and (V.3.20), allow us to let $k \to \infty$ in this equation and use Dominated Convergence to derive the required result. (Clearly (V.3.13) implies the "drift" term has bounded variation on bounded time intervals a.s.) ∎

Corollary V.3.6. If f, g, and Z are as in Theorem V.3.5, then

$$X_t(A) = \int 1(Z_t(\omega, y) \in A) K_t(dy)$$

defines an a.s. continuous $(\mathcal{F}_t)$-predictable $M_F(\mathbb{R}^n)$-valued process.
Proof. Take $\psi(x) = e^{iu\cdot x}$, $u \in \mathbb{Q}^n$ to see that $\int e^{iu\cdot x} dX_t(x)$ is continuous in $t \in [0, \infty)$ for all $u \in \mathbb{Q}^n$ a.s. Lévy's Continuity Theorem completes the proof of continuity. Remark V.2.5 (d) shows that X is $(\mathcal{F}_t)$-predictable. ∎

4. Pathwise Existence and Uniqueness

As in the last Section we assume $(K_t, t \ge 0)$ is an $(\mathcal{F}_t)$-historical Brownian motion on $\bar{\Omega} = (\Omega, \mathcal{F}, \mathcal{F}_t, \mathbb{P})$ starting at $(0, \nu)$ with $\mathbb{E}(K_0(\cdot)) = m(\cdot) \in M_F(\mathbb{R}^d)$. Therefore K has law $\mathbb{Q}_{0,\nu}$ on $(\Omega_X, \mathcal{F}_X)$. Recall that σ, b satisfy (Lip) and $Z_0 : \mathbb{R}^d \to \mathbb{R}^d$ is a Borel map. If $\int_0^t \sigma(X_s, Z_s) dy(s)$ is the stochastic integral introduced in Proposition V.3.1, here is the precise interpretation of (SE):

$$(SE)_{Z_0,K}\ (a)\ Z(t,\omega,y) = Z_0(y_0) + \int_0^t \sigma(X_s, Z_s) dy(s) + \int_0^t b(X_s, Z_s) ds \quad K-\text{a.e.}$$

$$(b)\ X_t(\omega)(A) = \int 1(Z_t(\omega, y) \in A) K_t(\omega)(dy) \quad \forall A \in \mathcal{B}(\mathbb{R}^d)\ \forall t \ge 0 \text{ a.s.}$$

(X, Z) is a solution of $(SE)_{Z_0,K}$ iff Z is a $(\hat{\mathcal{F}}_t^*)$-predictable $\mathbb{R}^d$-valued process, and X is an $(\mathcal{F}_t)$-predictable $M_F(\mathbb{R}^d)$-valued process such that $(SE)_{Z_0,K}$ holds. Let $\bar{\mathcal{H}}_t$ denote the usual enlargement of $\mathcal{F}^H[0, t+]$ with $\mathbb{Q}_{0,\nu}$-null sets.

Theorem V.4.1.
(a) There is a pathwise unique solution (X, Z) to $(SE)_{Z_0,K}$. More precisely X is unique up to $\mathbb{P}$-evanescent sets and Z is unique K-a.e. Moreover $t \mapsto X_t$ is a.s. continuous in t.
(b) There are $(\bar{\mathcal{H}}_t)$-predictable and $(\hat{\bar{\mathcal{H}}}_t^*)$-predictable maps, $\tilde{X} : \mathbb{R}_+ \times \Omega_H \to M_F(\mathbb{R}^d)$ and $\tilde{Z} : \mathbb{R}_+ \times \hat{\Omega}_H \to \mathbb{R}^d$, respectively, which depend only on (Z_0, ν), and are such that that $(X(t,\omega), Z(t,\omega,y)) = (\tilde{X}(t, K(\omega)), \tilde{Z}(t, K(\omega), y))$ defines the unique solution of $(SE)_{Z_0,K}$.
(c) There is a continuous map $X_0 \to \mathbb{P}'_{X_0}$ from $M_F(\mathbb{R}^d)$ to $M_1(\Omega_X)$, such that if (X, Z) is a solution of $(SE)_{Z_0,K}$ on some filtered space $\bar{\Omega}$, then

$$(V.4.1) \qquad \mathbb{P}(X \in \cdot) = \int \mathbb{P}'_{X_0(\omega)} d\mathbb{P}(\omega).$$

(d) If T is an a.s. finite $(\mathcal{F}_t)$-stopping time, then

$$\mathbb{P}(X(T+\cdot)\in A|\mathcal{F}_T)(\omega)=\mathbb{P}'_{X_T(\omega)}(A)\quad \mathbb{P}-\text{a.s. for all } A\in\mathcal{F}_X.$$

For the uniqueness and continuity of $\mathbb{P}'_{X_0}$ we will need to prove the following stability result. Recall that d is the Vasershtein metric on $M_F(\mathbb{R}^d)$.

Theorem V.4.2. Let $K^1\le K^2\equiv K$ be $(\mathcal{F}_t)$-historical Brownian motions with $\mathbb{E}(K_0^i(\cdot))=m_i(\cdot)\in M_F(\mathbb{R}^d)$, and let $Z_0^i:\mathbb{R}^d\to\mathbb{R}^d$ be Borel maps, $i=1,2$. Set $T_N=\inf\{t:K_t(1)\ge N\}\wedge N$. There are universal constants $\{c_N,N\in\mathbb{N}\}$ so that if (X^i,Z^i) is the unique solution of $(SE)_{Z_0^i,K^i}$ then

$$\mathbb{P}\Big(\int_0^{T_N}\sup_{u\le s}d(X^1(u),X^2(u))^2ds\Big)\le c_N\Big[\int\|Z_0^1-Z_0^2\|^2\wedge 1dm_2+m_2(1)-m_1(1)\Big].$$

Here are the metric spaces we will use in our fixed point argument:

$$\begin{aligned}S_1=\{X:\mathbb{R}_+\times\Omega\to M_F(\mathbb{R}^d):X \text{ is } (\mathcal{F}_t)-\text{predictable and a.s. continuous },\\ X_t(1)\le N\ \forall t<T_N\ \forall N\in\mathbb{N}\},\end{aligned}$$
$$S_2=\{Z:\mathbb{R}_+\times\hat\Omega\to\mathbb{R}^d: Z \text{ is } (\hat{\mathcal{F}}_t^*)-\text{predictable and continuous } K\text{-a.e.}\},$$
$$S=S_1\times S_2.$$

Processes in S_1 are identified if they agree at all times except perhaps for a $\mathbb{P}$-null set and processes in S_2 are identified if they agree K-a.e. If $T\in\mathcal{T}_b$, $\theta>0$ and $Z_1,Z_2\in S_2$, let

$$(V.4.2)\quad d_{T,\theta}(Z_1,Z_2)=\hat{\mathbb{P}}_T\Big(\int_0^T(\sup_{u\le s}\|Z_1(u)-Z_2(u)\|^2\wedge 1)e^{-\theta s}ds\Big)^{1/2}\le\theta^{-1/2},$$

where, as usual, $\hat{\mathbb{P}}_T$ is the normalized Campbell measure associated with K_T. If $\bar\theta=(\theta_N,N\ge1)$ is a positive sequence satisfying

$$(V.4.3)\qquad \sum_1^\infty N\theta_N^{-1/2}<\infty,$$

define metrics $d_i=d_i^{\bar\theta}$ on S_i and $d_0=d_0^{\bar\theta}$ on S by

$$d_1(X_1,X_2)=\sum_{N=1}^\infty\mathbb{P}\Big(\int_0^{T_N}\sup_{u\le s}d(X_1(u),X_2(u))^2e^{-\theta_N s}ds\Big)^{1/2},$$
$$d_2(Z_1,Z_2)=\sum_{N=1}^\infty\sup_{T\le T_N,T\in\mathcal{T}_b}d_{T,\theta_N}(Z_1,Z_2)<\infty\quad(\text{by (V.4.2)}),$$
$$d_0((X_1,Z_1),(X_2,Z_2))=d_1(X_1,X_2)+d_2(Z_1,Z_2).$$

Note that if $u < T_N$, then $d(X_1(u), X_2(u)) \le X_1(u)(1) + X_2(u)(1) \le 2N$ and so by (V.4.3),

$$d_1(X_1, X_2) \le \sum_N 2N\theta_N^{-1/2} < \infty.$$

Lemma V.4.3. (S_i, d_i), $i = 1, 2$ and (S, d_0) are complete metric spaces.

Proof. This is straightforward. We will only show d_2 is complete. Suppose $\{Z_n\}$ is d_2 Cauchy. Let $N \in \mathbb{N}$. Then

$$\lim_{m,n\to\infty} \sup_{T \le T_N, T \in \mathcal{T}_b} \mathbb{E}\Big(\int \int_0^T \sup_{u \le s} \|Z_n(u) - Z_m(u)\|^2 \wedge 1 ds K_T(dy)\Big) = 0.$$

An application of the Section Theorem implies that

$$\sup_{t \le T_N} \int \int_0^t (\sup_{u \le s} \|Z_n(u) - Z_m(u)\|^2 \wedge 1) ds K_t(dy) \xrightarrow{\mathbb{P}} 0 \text{ as } m, n \to \infty.$$

Since $Z_n - Z_m$ is continuous K-a.e., this implies

$$\sup_{t \le T_N} \int \sup_{s \le t} \|Z_n(s) - Z_m(s)\|^2 \wedge 1 K_t(dy) \xrightarrow{\mathbb{P}} 0 \text{ as } m, n \to \infty.$$

A standard argument now shows that there is a subsequence $\{n_k\}$ so that

(V.4.4)

$Z_{n_k}(s, \omega, y)$ converges uniformly in $s \le t$ for $K_t -$ a.e. y for all $t \ge 0$ $\mathbb{P} -$ a.s.

Now define

$$Z(s, \omega, y) = \begin{cases} \lim_{k\to\infty} Z_{n_k}(s, \omega, y) & \text{if it exists} \\ 0 \in \mathbb{R}^d & \text{otherwise.} \end{cases}$$

Then Z is $(\hat{\mathcal{F}}_t^*)$-predictable and continuous K-a.e. (by (V.4.4)). Dominated Convergence easily shows that $d_2(Z_{n_k}, Z) \to 0$ as $k \to \infty$. ∎

Proof of Theorem 4.1(a) and Theorem 4.2.

For $i = 1, 2$ define $\Phi^i = (\Phi_1^i, \Phi_2^i) : S \to S$ by

$$\Phi_2^i(X, Z)(t) \equiv \tilde{Z}_t^i(\omega, y) = Z_0^i(y_0) + \int_0^t \sigma(X_s, Z_s) dy(s) + \int_0^t b(X_s, Z_s) ds \tag{V.4.5}$$

and

$$\Phi_1^i(X, Z)(t)(\cdot) \equiv \tilde{X}_t^i(\cdot) = \int 1(\tilde{Z}_t^i(\omega, y) \in \cdot) K_t^i(dy). \tag{V.4.6}$$

In (V.4.5) the stochastic integral is that of Section 3 relative to K. Note that this integral also defines a version of the K^1 stochastic integral. To see this, note it is trivial for simple integrands and then one can approximate as in the construction of the K-stochastic integral to see this equivalence persists for all integrands in $D(d, d)$ (defined with respect to K). Remark V.1.1(a) implies

$$\int_0^t K_s(\|\sigma(X_s, Z_s)\|^2 + \|b(X_s, Z_s)\|) ds \tag{V.4.7}$$

$$\leq \int_0^t K_s(1)(C(X_s(1))^2 + C(X_s(1)))ds < \infty \quad \forall t > 0 \text{ a.s.}$$

Corollary V.3.6 therefore may be used to show that $\tilde{X}$ is a.s. continuous. Note also that $\tilde{X}_t(1) = K_t(1) \leq N$ for $t < T_N$ and it follows that Φ^i takes values in S. Clearly a fixed point of Φ^i would be a solution of $(SE)_{Z_0^i,K^i}$.

To avoid writing factors of $m(1)^{-1}$, assume that $m(1) = 1$. Let $T \in \mathcal{T}_b$ satisfy $T \leq T_N$. Let $\tilde{Z}^i$ and $\tilde{X}^i$ be as in (V.4.5) and (V.4.6) for some $(X^i, Z^i) \in S$. Doob's strong L^2 inequality and Cauchy-Schwarz imply

$$\begin{aligned}
&\hat{\mathbb{P}}_T\Big(\sup_{s\leq t\wedge T} \|\tilde{Z}^1(s) - \tilde{Z}^2(s)\|^2 \wedge 1\Big) \\
&\leq c\hat{\mathbb{P}}_T\Big(\|Z_0^1 - Z_0^2\|^2 \wedge 1 + \int_0^{t\wedge T} \|\sigma(X_s^1, Z_s^1) - \sigma(X_s^2, Z_s^2)\|^2 \\
&\qquad + N\|b(X_s^1, Z_s^1) - b(X_s^2, Z_s^2)\|^2 ds\Big) \\
&\leq c\int \|Z_0^1 - Z_0^2\|^2 \wedge 1 dm_2 \qquad (V.4.8) \\
&\qquad + cL(N)^2 N\hat{\mathbb{P}}_T\Big(\int_0^{t\wedge T} (d(X_s^1, X_s^2) + \|Z_s^1 - Z_s^2\|)^2 \wedge C(N)^2 ds\Big),
\end{aligned}$$

where in the last line we used Proposition V.2.4 (a), (Lip) and Remark V.1.1 (a). Note that if $T > 0$ then $T_N > 0$ and so $K_T(1) \leq \sup_{t\leq T_N} K_t(1) \leq N$. If $T = 0$, $K_T(1)$ may be bigger than N but the integral in (V.4.8) is then zero. Therefore

$$\begin{aligned}
&\hat{\mathbb{P}}_T\Big(\sup_{s\leq t\wedge T} \|\tilde{Z}^1(s) - \tilde{Z}^2(s)\|^2 \wedge 1\Big) \\
&\leq c_N\Big[\int \|Z_0^1 - Z_0^2\|^2 \wedge 1 dm_2 + \mathbb{P}\Big(\int_0^{t\wedge T} d(X_s^1, X_s^2)^2 \wedge 1 ds\Big) \qquad (V.4.9) \\
&\qquad + \hat{\mathbb{P}}_T\Big(\int_0^{t\wedge T} \|Z_s^1 - Z_s^2\|^2 \wedge 1 ds\Big)\Big].
\end{aligned}$$

Multiply the above inequality by $e^{-\theta t}$ and integrate t over $\mathbb{R}_+$ to conclude that

$$\begin{aligned}
&\sup_{T\leq T_N, T\in\mathcal{T}_b} d_{T,\theta}(\tilde{Z}^1, \tilde{Z}^2)^2 \\
&\leq \frac{c_N}{\theta}\Big[\int \|Z_0^1 - Z_0^2\|^2 \wedge 1 dm_2 + \mathbb{P}\Big(\int_0^{T_N} (d(X_s^1, X_s^2)^2 \wedge 1)^2 e^{-\theta s} ds\Big) \\
&\qquad + \sup_{T\leq T_N, T\in\mathcal{T}_b} \hat{\mathbb{P}}_T\Big(\int_0^T (\|Z_s^1 - Z_s^2\|^2 \wedge 1) e^{-\theta s} ds\Big)\Big].
\end{aligned}$$

Therefore

$$\begin{aligned}
\sup_{T\leq T_N, T\in\mathcal{T}_b} d_{T,\theta}(\tilde{Z}^1, \tilde{Z}^2) &\leq \sqrt{\frac{c_N}{\theta}}\Big[\Big(\int \|Z_0^1 - Z_0^2\|^2 \wedge 1 dm_2\Big)^{1/2} \\
&\qquad + \mathbb{P}\Big(\int_0^{T_N} (d(X_s^1, X_s^2)^2 \wedge 1)^2 e^{-\theta s} ds\Big)^{1/2} \qquad (V.4.10)
\end{aligned}$$

$$+ \sup_{T \le T_N, T \in \mathcal{T}_b} d_{T,\theta}(Z^1, Z^2)\Big].$$

Take $\theta = \theta_N$ in (V.4.10), assume

$$\delta_0 = \sum_{N=1}^{\infty} \sqrt{\frac{c_N}{\theta_N}} < \infty, \tag{V.4.11}$$

and sum the resulting inequality over N to conclude that

$$d_2(\tilde{Z}^1, \tilde{Z}^2) \le \delta_0 \Big[\Big(\int \|Z_0^1 - Z_0^2\|^2 \wedge 1 dm\Big)^{1/2} + d_1(X^1, X^2) + d_2(Z^1, Z^2)\Big]. \tag{V.4.12}$$

Consider next a bound for $d_1(\tilde{X}^1, \tilde{X}^2)$. Let $\phi \in \mathrm{Lip}_1$. Then

$$|\tilde{X}_u^1(\phi) - \tilde{X}_u^2(\phi)|^2 \le 2\Big[\Big(\int \phi(\tilde{Z}_u^1) - \phi(\tilde{Z}_u^2) dK_u^1\Big)^2 + \Big(\int \phi(\tilde{Z}_u^2) d(K_u^1 - K_u^2)\Big)^2\Big]$$
$$\le 2\Big[\Big(\int \|\tilde{Z}_u^1 - \tilde{Z}_u^2\| \wedge 2 dK_u^1\Big)^2 + (K_u^2(1) - K_u^1(1))^2\Big].$$

Therefore if $N(u) = \int \sup_{s \le u} \|\tilde{Z}_s^1 - \tilde{Z}_s^2\| \wedge 1 K_u^1(dy)$, then

$$\sup_{u \le t \wedge T_N} d(\tilde{X}_u^1, \tilde{X}_u^2)^2 \le 8\Big[\sup_{u \le t \wedge T_N} N(u)^2 + \sup_{u \le t \wedge T_N} (K_u^2(1) - K_u^1(1))^2\Big]. \tag{V.4.13}$$

Claim N is an $(\mathcal{F}_t)$-submartingale. Let $U \le V$ be $(\mathcal{F}_t)$-stopping times. Then Remark V.2.5 (c) implies

$$\mathbb{E}(N(U)) = \mathbb{E}\Big(\int \sup_{s \le U} \|\tilde{Z}_s^1 - \tilde{Z}_s^2\| \wedge 1 \, dK_V\Big) \le \mathbb{E}(N(V)).$$

As $N(t) \le K_t(1) \in L^1$, the claim follows by a standard argument. On $\{T_N > 0\}$ we have $\sup_{t \le T_N} K_t(1) \le N$. Therefore (V.4.13), Cauchy-Schwarz and Doob's strong L^2 inequality imply

$$\begin{aligned}
&\mathbb{P}(1(T_N > 0) \sup_{u \le t \wedge T_N} d(\tilde{X}_u^1, \tilde{X}_u^2)^2) \\
&\le c\mathbb{P}([N(t \wedge T_N)^2 + (K_{t\wedge T_N}^2(1) - K_{t \wedge T_N}^1(1))^2] 1(T_N > 0)) \\
&\le c\mathbb{P}\Big(N \int \sup_{s \le t \wedge T_N} \|\tilde{Z}_s^1 - \tilde{Z}_s^2\|^2 \wedge 1 \, dK_{t \wedge T_N}^2 + N(K_{t\wedge T_N}^2(1) - K_{t\wedge T_N}^1(1))\Big) \\
&= cN\Big[\hat{\mathbb{P}}_{T_N}\Big(\sup_{s \le t \wedge T_N} \|\tilde{Z}_s^1 - \tilde{Z}_s^2\|^2 \wedge 1\Big) + m_2(1) - m_1(1)\Big] \quad \text{(by Remark V.2.5 (c))} \\
&\le c_N'\Big[\int \|Z_0^1 - Z_0^2\|^2 \wedge 1 \, dm_2 + m_2(1) - m_1(1) + \mathbb{P}\Big(\int_0^{t \wedge T_N} d(X_s^1, X_s^2)^2 \wedge 1 \, ds\Big) \\
&\qquad + \hat{\mathbb{P}}_{T_N}\Big(\int_0^{t \wedge T_N} \|Z_s^1 - Z_s^2\|^2 \wedge 1 \, ds\Big)\Big],
\end{aligned}$$

the last by (V.4.9). It follows that

$$\mathbb{P}\Big(\int_0^{T_N} \sup_{u\le t} d(\tilde X^1_u,\tilde X^2_u)^2 e^{-\theta_N t}dt\Big)$$
$$\le \int_0^\infty \mathbb{P}\Big(1(T_N>0)\sup_{u\le t\wedge T_N} d(\tilde X^1_u,\tilde X^2_u)^2\Big)e^{-\theta_N t}dt$$
$$\le \frac{c'_N}{\theta_N}\Big[\int \|Z^1_0-Z^2_0\|^2\wedge 1\,dm_2+m_2(1)-m_1(1)$$
$$+\mathbb{P}\Big(\int_0^{T_N}(d(X^1_s,X^2_s)^2\wedge 1)e^{-\theta_N s}ds\Big)+\hat{\mathbb{P}}_{T_N}\Big(\int_0^{T_N}(\|Z^1_s-Z^2_s\|^2\wedge 1)e^{-\theta_N s}ds\Big)\Big].$$

Therefore if

$$\delta'_0=\sum_1^\infty\sqrt{\frac{c'_N}{\theta_N}}<\infty, \tag{V.4.14}$$

then

$$d_1(\tilde X^1,\tilde X^2)\le\delta'_0\Big(\int \|Z^1_0-Z^2_0\|^2\wedge 1\,dm_2^{1/2}+(m_2(1)-m_1(1))^{1/2} + d_1(X^1,X^2)+d_2(Z^1,Z^2)\Big). \tag{V.4.15}$$

Now set $Z^1_0=Z^2_0=Z_0$ and $K^1=K^2=K$ and $\Phi^i=\Phi$. (V.4.12) and (V.4.15) imply

$$d_0(\Phi(X^1,Z^1),\Phi(X^2,Z^2))\le(\delta_0+\delta'_0)d_0((X^1,Z^1),(X^2,Z^2)).$$

Therefore if we choose $\{\theta_N\}$ so that (V.4.11) and (V.4.14) hold with $\delta_0,\delta'_0\le 1/4$, then Φ is a contraction on the complete metric space (S,d_0). It therefore has a unique fixed point (X,Z) which is a solution of $(SE)_{Z_0,K}$. Conversely if (X,Z) is a solution of $(SE)_{Z_0,K}$ then $X_s(1)=K_s(1)$ and, as in (V.4.7) and the ensuing argument, we see that $(X,Z)\in S$. Therefore (X,Z) is a fixed point of Φ and so is pathwise unique. This completes the proof of Theorem V.4.1 (a).

For Theorem 4.2, let (X^i,Z^i) satisfy $(SE)_{Z^i_0,K^i}$, $i=1,2$. Then $(X^i,Z^i)\in S$ by the above and so $\Phi^i(X^i,Z^i)=(X^i,Z^i)$. Therefore (V.4.12) and (V.4.15) imply

$$d_0((X^1,Z^1),(X^2,Z^2))\le(\delta_0+\delta'_0)\Big[d_0((X^1,Z^1),(X^2,Z^2)) + \Big(\int\|Z^1_0-Z^2_0\|^2\wedge 1\,dm_2\Big)^{1/2}+(m_2(1)-m_1(1))^{1/2}\Big].$$

As $\delta_0+\delta'_0\le 1/2$ by our choice of θ_N, this gives

$$d_1(X^1,X^2)\le d_0((X^1,Z^1),(X^2,Z^2)) \le\Big(\int\|Z^1_0-Z^2_0\|^2\wedge 1\,dm_2\Big)^{1/2}+(m_2(1)-m_1(1))^{1/2},$$

and hence

$$\mathbb{P}\Big(\int_0^{T_N} \sup_{u\le s} d(X^1(u),X^2(u))^2\,ds\Big)$$
$$\le e^{\theta_N N} d_1(X^1,X^2)^2 \le 2e^{\theta_N N}\Big(m_2(1)-m_1(1)+\int \|Z_0^1-Z_0^2\|^2\wedge 1\,dm_2\Big). \quad \blacksquare$$

Proof of Theorem 4.1(b). Let $(\tilde{X},\tilde{Z})$ be the unique solution of $(SE)_{Z_0,H}$, where H is the canonical process on $\bar{\Omega}_H = (\Omega_H, \mathcal{F}_H, \bar{\mathcal{H}}_t, \mathbb{Q}_{0,\nu})$. Clearly $(\tilde{X},\tilde{Z})$ depends only on Z_0 and ν, and satisfies the required predictability conditions. We must show that $(\tilde{X}(t,K(\omega)), \tilde{Z}(t,K(\omega),y))$ solves $(SE)_{Z_0,K}$ on $\bar{\Omega}$. Let $I_H(f,t,H,y)$, $f\in D_H$ denote the stochastic integral from Proposition V.3.2 on $\bar{\Omega}_H$ and $I(f,t,\omega,y)$, $f\in D$ continue to denote that with respect to K on $\bar{\Omega}$. We claim that if $f\in D_H$, then $f\circ K(t,\omega,y)\equiv f(t,K(\omega),y)\in D$ and

$$(V.4.16) \qquad I(f\circ K) = I_H(f)\circ K \quad K-\text{a.e.}$$

The first implication is clear because K has law $\mathbb{Q}_{0,\nu}$. (V.4.16) is immediate if f is a $(\widehat{\mathcal{H}}_t)$-predictable simple function. A simple approximation argument using Proposition V.3.2 (d) then gives (V.4.16) for all $f\in D_H$. It is now easy to replace H with $K(\omega)$ in $(SE)_{Z_0,H}$ and see that $(\tilde{X}\circ K, \tilde{Z}\circ K)$ solves $(SE)_{Z_0,K}$ on $\bar{\Omega}$. $\blacksquare$

For (c) we need:

Lemma V.4.4. Let $\mu_1,\mu_2\in M_F(\mathbb{R}^d)$ satisfy $\mu_1(\mathbb{R}^d)\le\mu_2(\mathbb{R}^d)$. There is a measure $m\in M_F(\mathbb{R}^d)$, $B\in\mathcal{B}(\mathbb{R}^d)$, and Borel maps $g_i:\mathbb{R}^d\to\mathbb{R}^d$ such that

(1) $\mu_1(\cdot)=m(g_1^{-1}(\cdot)\cap B)$, $\mu_2(\cdot)=m(g_2^{-1}(\cdot))$

(2) $\int \|g_1-g_2\|\wedge 1\,dm \le 2d(\mu_1,\mu_2)$

(3) $m(B^c)=\mu_2(1)-\mu_1(1)\le d(\mu_1,\mu_2)$.

Proof. (3) is immediate from (1).

If $\mu_1(\mathbb{R}^d)=\mu_2(\mathbb{R}^d)$, (1) and (2) with $B=\mathbb{R}^d$ and no factor of 2 in (2) is a standard "marriage lemma" (see, e.g., Szulga (1982)). Although the usual formulation has m defined on the Borel sets of $(\mathbb{R}^d)^2$ and g_i the projection maps from $(\mathbb{R}^d)^2$ onto $\mathbb{R}^d$, the above results follows as m and g_i can be carried over to $\mathbb{R}^d$ through a measure isomorphism between this space and $(\mathbb{R}^d)^2$.

If $\mu_1(\mathbb{R}^d)<\mu_2(\mathbb{R}^d)$, let $\mu_1'=\mu_1+(\mu_2(\mathbb{R}^d)-\mu_1(\mathbb{R}^d))\delta_{x_0}$, where x_0 is chosen so that $\mu_1(\{x_0\})=0$. By the above case there are m, g_i satisfying (1) and (2) with μ_1' in place of μ_1, $B=\mathbb{R}^d$, and no factor of 2 in (2). (1) follows easily with $B=g_1^{-1}(x_0)^c$. For (2) note that

$$\begin{aligned}\int \|g_1-g_2\|\wedge 1\,dm &\le d(\mu_1',\mu_2) \quad \text{(by the above case)}\\ &\le d(\mu_1',\mu_1)+d(\mu_1,\mu_2)\\ &= \mu_2(1)-\mu_1(1)+d(\mu_1,\mu_2)\le 2d(\mu_1,\mu_2). \quad \blacksquare\end{aligned}$$

Proof of Theorem 4.1(c). Assume first that $K_0 = m$ is deterministic and let (X, Z) solve $(SE)_{Z_0,K}$. The law of X on Ω_X, $\mathbb{P}_{m,Z_0}$, depends only on (m, Z_0) by (b). The next step is to show that it in fact only depends on $X_0 = m(Z_0^{-1}(\cdot))$. Define $\tilde{Z}_0(y)(t) = y(t) - y(0) + Z_0(y(0))$ and $K_t'(\phi) = K_t(\phi \circ \tilde{Z}_0)$. Then K' is an $(\mathcal{F}_t)$-historical Brownian motion starting at $(0, X_0)$ by Exercise V.2.1 with $B = \mathbb{R}^d$. Let (X', Z') be the unique solution of $(SE)_{Z_0',K'}$, where $Z_0'(y_0) = y_0$. Let I, I' denote the stochastic integrals on $D(d,d)$ and $D'(d,d)$, respectively with respect to K and K', respectively. If $f : \mathbb{R}_+ \times \hat{\Omega} \to \mathbb{R}^{d\times d}$, set $f \circ \tilde{Z}_0(t,\omega,y) = f(t,\omega,\tilde{Z}_0(y))$. We claim that if $f \in D'$, then $f \circ \tilde{Z}_0 \in D$ and

$$(V.4.17) \qquad I'(f) \circ \tilde{Z}_0 = I(f \circ \tilde{Z}_0) \quad K-\text{a.e.}$$

The first inclusion is trivial. To prove the equality note it is obvious if f is simple and then use Proposition V.3.2(d) to extend the equality to all f in D' by approximating by simple functions as in the construction of I. If $\hat{Z}_t = Z_t' \circ \tilde{Z}_0$, then $\hat{Z}$ is $(\hat{\mathcal{F}}_t^*)$-predictable, and

$$Z_t' = y_0 + I'(\sigma(X', Z'), t) + \int_0^t b(X_s', Z_s')ds \quad K' - \text{a.e.}$$

implies

$$\hat{Z}_t = \tilde{Z}_0(y)(0) + I'(\sigma(X', Z'), t) \circ \tilde{Z}_0 + \int_0^t b(X_s', Z_s') \circ \tilde{Z}_0 \, ds \quad K - \text{a.e.}$$

Now use (V.4.17) to get

$$\hat{Z}_t = Z_0(y_0) + I(\sigma(X', \hat{Z}), t) + \int_0^t b(X_s', \hat{Z}_s)ds \quad K - \text{a.e.,}$$

and also note that $\mathbb{P}$-a.s. for all $t \geq 0$,

$$X_t'(\cdot) = \int 1(Z_t' \in \cdot)K_t'(dy) = \int 1(\hat{Z}_t \in \cdot)K_t(dy).$$

Therefore $(X', \hat{Z})$ solves $(SE)_{Z_0,K}$ and so $X' = X$ $\mathbb{P}$-a.s. by (a). This implies they have the same law and so $\mathbb{P}_{m,Z_0} = \mathbb{P}_{X_0,\mathrm{id}} \equiv \mathbb{P}'_{X_0}$, thus proving the claim.

To show the continuity of $\mathbb{P}'_{X_0}$ in X_0 we will use the stability of solutions to (SE) with respect to the initial conditions (Theorem V.4.2). Let $X_0^i \in M_F(\mathbb{R}^d)$, $i = 1, 2$ and choose m, B, and $g_i \equiv Z_0^i$ as in Lemma V.4.4 with $\mu_i = X_0^i$. Let $K_t^2 = K_t$ be an $(\mathcal{F}_t)$-historical Brownian motion starting at $(0, m)$ and define

$$K_t^1(A) = K_t(A \cap \{y : y(0) \in B\}).$$

Then Exercise V.2.1 shows that $K^1 (\leq K^2)$ is an $(\mathcal{F}_t)$-historical Brownian motion starting at $(0, m(\cdot \cap B))$. If (X^i, Z^i) solves $(SE)_{Z_0^i,K^i}$, then $X^i(0) = X_0^i$, as the notation suggests, and so X_i has law $\mathbb{P}'_{X_0^i}$. Introduce the uniform metric $\rho_M(x^1, x^2) = \sup_t d(x_t^1, x_t^2) \wedge 1$ on Ω_X and let d_M denote the corresponding Vasershtein metric on $M_F(\Omega_X)$. This imposes a stronger (i.e., uniform) topology on Ω_X, and hence on $M_F(\Omega_X)$, but as our processes have compact support in time the strengthening is illusory. If T_N is as in Theorem V.4.2 and $\zeta = \inf\{t : K_t(1) = 0\}$, then $X_t^i = 0$ if $t \geq \zeta$ and so

$$\begin{aligned} d_M(\mathbb{P}'_{X_0^1}, \mathbb{P}'_{X_0^2}) &\leq \sup_{\phi \in \mathrm{Lip}_1} \int |\phi(X^1) - \phi(X^2)| \, d\mathbb{P} \\ &\leq \int \sup_t d(X_t^1, X_t^2) \wedge 1 \, d\mathbb{P} \\ &\leq \mathbb{P}(T_N \leq \zeta) + \mathbb{P}\Big(1(T_N > \zeta) \sup_{t \leq T_N} d(X_t^1, X_t^2) \wedge 1\Big) \\ &\leq \mathbb{P}(T_N \leq \zeta) + \mathbb{P}\Big(\int_{T_N}^{T_{N+1}} \sup_{t \leq u} d(X_u^1, X_u^2)^2 \wedge 1 \, du \, 1(T_N > \zeta)\Big)^{1/2}, \end{aligned}$$

the last because on $\{T_N > \zeta\}$, $T_N = N$ and $T_{N+1} = N + 1$. Theorem V.4.2 implies that

$$\begin{aligned} d_M(\mathbb{P}'_{X_0^1}, \mathbb{P}'_{X_0^2}) &\leq \mathbb{P}(T_N \leq \zeta) + \sqrt{c_{N+1}\Big(\int \|Z_0^1 - Z_0^2\|^2 \wedge 1 \, dm + m(1) - m_1(1)\Big)} \\ &\leq \mathbb{P}(T_N \leq \zeta) + \sqrt{c_{N+1} 3 d(X_0^1, X_0^2)}. \end{aligned}$$

The first term approaches zero as $N \to \infty$ and so the uniform continuity of $X_0 \to \mathbb{P}'_{X_0}$ with respect to the above metrics on $M_1(\Omega_X)$ and $M_F(\mathbb{R}^d)$ follows.

Returning now to the general setting of (a) in which K_0 and X_0 may be random, we claim that if $A \in \mathcal{F}_X$, then

$$\text{(V.4.18)} \qquad \mathbb{P}(X \in A | \mathcal{F}_0)(\omega) = \mathbb{P}'_{X_0(\omega)}(A) \quad \mathbb{P} - \text{a.s.}$$

Take expectations of both sides to complete the proof of (c). For (V.4.18), use (b) and the $(\mathcal{F}_t)$-Markov property of K to see that $\mathbb{P}$-a.s.,

$$\text{(V.4.19)} \qquad \mathbb{P}(X \in A | \mathcal{F}_0)(\omega) = \mathbb{P}(\tilde{X}(K) \in A | \mathcal{F}_0)(\omega) = \mathbb{Q}_{0,K_0(\omega)}(\tilde{X} \in A).$$

Recall that $(\tilde{X}, \tilde{Z})$ is the solution of $(SE)_{Z_0,H}$ on $(\Omega_H, \mathcal{F}_H, \bar{\mathcal{H}}_t, \mathbb{Q}_{0,\nu})$. Let $\bar{\mathcal{H}}^{K_0}$ be the augmentation of $\mathcal{F}^H[0, t+]$ with $\mathbb{Q}_{0,K_0}$-null sets. Claim that

$$\text{(V.4.20)} \qquad \text{For } \nu - \text{a.a. } K_0, \ (\tilde{X}, \tilde{Z}) \text{ satisfies } (SE)_{Z_0,H} \text{ on } (\Omega_H, \mathcal{F}_H, \bar{\mathcal{H}}_t, \mathbb{Q}_{0,K_0}).$$

The only issue is, as usual, the interpretation of $I(\sigma(\tilde{X}, \tilde{Z}), t)$ under these various measures. Let $I_{K_0}(f)$, $f \in D_{K_0}$ be this integral under $\mathbb{Q}_{0,K_0}$ and $I(f)$, $f \in D$ be the integral under $\mathbb{Q}_{0,\nu}$. Starting with simple functions and bootstrapping up as usual we can show for ν-a.a. K_0,

$f \in D$ implies $f \in D_{K_0}$ and in this case
$I(f, t, \omega, y) = I_{K_0}(f, t, \omega, y) \ \forall t \leq u$ for K_u – a.a. $y \ \forall u \geq 0 \ \mathbb{Q}_{0,K_0}$ – a.s.

It is now a simple matter to prove (V.4.20).

Since $X_0(\omega)(\cdot) = K_0(\omega)(Z_0^{-1}(\cdot))$ a.s. by $(SE)_{Z_0,K}$, (V.4.20) and the result for deterministic initial conditions established above, imply that

$$\mathbb{Q}_{0,K_0(\omega)}(\tilde{X} \in \cdot) = \mathbb{P}'_{X_0(\omega)}(\cdot) \quad \text{a.s.}$$

Use this in (V.4.19) to derive (V.4.18) and so complete the proof of (c).

To establish the strong Markov property (d) we need some notation and a preliminary result.

Notation. If $s \geq 0$ let $\hat{\theta}_s(y)(t) = y(t+s) - y(s)$ and $\mathcal{F}_t^{(s)} = \mathcal{F}_{s+t}$. If $Z_s(\omega, y)$ is $\hat{\mathcal{F}}_s^*$-measurable and $\tilde{Z}_s(\omega, y)(t) = Z_s(\omega, y) + \hat{\theta}_s(y)(t)$, for each $\phi : \mathbb{R}_+ \times \hat{\Omega} \to E$, let

$$\phi^{(s)}(t, \omega, y) = \phi(t - s, \omega, \tilde{Z}_s(\omega, y))1(t \geq s),$$

suppressing dependence on Z_s. If $B \in \mathcal{F}$ has positive $\mathbb{P}$ measure, let $\mathbb{P}_B(\cdot) = \mathbb{P}(\cdot|B)$.

Lemma V.4.5. Let $s \geq 0$, $B \in \mathcal{F}_s$ have positive $\mathbb{P}$ measure and assume Z_s is as above. Then

$$K_t^{(s)}(A) \equiv \int 1_A(\tilde{Z}_s(\omega, y))K_{t+s}(dy), \quad t \geq 0$$

is an $(\mathcal{F}_t^{(s)})$-historical Brownian motion under $\mathbb{P}_B$.
Moreover $\mathbb{P}_B(K_0^{(s)}(1)) \leq m(1)/\mathbb{P}(B) < \infty$.

The reader who has done Exercise V.2.1 should have no trouble believing this. The proof of (d) now proceeds along familiar lines. The continuity of $X_0 \to \mathbb{P}'_{X_0}$ allows us to assume $T \equiv s$ is constant by a standard approximation of T by finite-valued stopping times. We then must show that

(V.4.21) There is a process $\hat{Z}$ so that $(X_{s+\cdot}, \hat{Z})$ solves $(SE)_{\hat{Z}_0, K^{(s)}}$ on $(\Omega, \mathcal{F}, \mathcal{F}_t^{(s)}, \mathbb{P}_B)$, where $\hat{Z}_0(y) = y_0$ and $B \in \mathcal{F}_s$ satisfies $\mathbb{P}(B) > 0$.

If B is as above and $A \in \mathcal{F}_X$ then (V.4.21) and (V.4.1) show that

$$\mathbb{P}_B(X_{s+\cdot} \in A) = \int \mathbb{P}'_{X_s(\omega)}(A) d\mathbb{P}_B,$$

and this implies

$$\mathbb{P}(X_{s+\cdot} \in A|\mathcal{F}_s)(\omega) = \mathbb{P}'_{X_s(\omega)}(A) \text{ a.s.},$$

as required.

The proofs of Lemma V.4.5 and (V.4.21) are somewhat tedious and are presented in the Appendix at the end of the Chapter for completeness. It should be clear from our discussion of the martingale problem for X in Section 5 below that this is not the best way to proceed for this particular result.

Now return to the martingale problem for X. Recall that $a(\mu, x) = \sigma(\mu, x)\sigma(\mu, x)^*$ and

$$A_\mu \phi(x) = \sum_{i=1}^d \sum_{j=1}^d a_{ij}(\mu, x)\phi_{ij}(x) + \sum_{i=1}^d b_i(\mu, x)\phi_i(x), \quad \text{for } \phi \in C_b^2(\mathbb{R}^d).$$

The following exercise is a simple application of Itô's Lemma (Theorem V.3.5) and is highly recommended.

Exercise V.4.1. If (X, Z) satisfies $(SE)_{Z_0,K}$ show that

$$(MP)^{a,b}_{X_0} \quad \text{For all } \phi \in C^2_b(\mathbb{R}^d), \quad M^X_t(\phi) = X_t(\phi) - X_0(\phi) - \int_0^t X_s(A_{X_s}\phi)ds$$

$$\text{is a continuous } (\mathcal{F}_t)\text{-martingale such that } \langle M^X(\phi)\rangle_t = \int_0^t X_s(\phi^2)ds.$$

We will comment on the uniqueness of solutions to $(MP)^{a,b}_{X_0}$ in the next section. Uniqueness in (SE) alone is enough to show weak convergence of the branching particle systems from Section V.1 to the solution X of (SE). Of course (Lip) continues to be in force.

Theorem V.4.6. Let $m \in M_F(\mathbb{R}^d)$ and let X^N be the solution of $(SE)_N$ constructed in Section V.1. If $X_0 = m(Z_0 \in \cdot)$, then

$$\mathbb{P}(X^N \in \cdot) \overset{w}{\Rightarrow} \mathbb{P}'_{X_0} \text{ on } D(\mathbb{R}_+, M_F(\mathbb{R}^d)) \text{ as } N \to \infty.$$

Sketch of Proof. Tightness and the fact that all limit points are supported on the space of continuous paths may be proved as in Section II.4. One way to prove that all limit points coincide is to take limits in $(SE)_N$ and show that all limit points do arise as solutions of $(SE)_{Z_0,K}$ for some K. More general results are proved by Lopez (1996) using the historical martingale problem (HMP) discussed in the next section.

5. Martingale Problems and Interactions

Our goal in this Section is to survey some of the recent developments and ongoing research in the martingale problem formulation of measure-valued diffusions. In order to use the uniqueness of solutions to $(SE)_{Z_0,K}$ (both pathwise and in law) to show that the associated martingale problem $(MP)^{a,b}_{X_0}$ is well-posed, we would have to show that it is possible to realize any solution X of $(MP)^{a,b}_{X_0}$ as part of a solution (X, Z) to $(SE)_{Z_0,K}$ for some historical Brownian motion K. In general it is not possible to recover K from X (see Barlow and Perkins (1994) for this result for ordinary super-Brownian motion) and so this appears to require a non-trivial enlargement of our space.

Donnelly and Kurtz (1999) were able to resolve the analogous problem in the setting of their exchangeable particle representations through an elegant application of a general result of Kurtz (1998) on filtered martingale problems.

Theorem V.5.1 (Donnelly-Kurtz (1999), Kurtz (1998)). If $X_0 \in M_F(\mathbb{R}^d)$, (σ, b) satisfies (Lip), and $a = \sigma\sigma^*$, then $(MP)^{a,b}_{X_0}$ is well-posed.

Discussion. Although I had originally planned to present these ideas in detail (the treatment in Section 6.5 of Donnelly and Kurtz (1999) is a bit terse), I will have to settle for a few (even more terse) remarks here.

First, the ideas of Section V.4 are readily adapted to the exchangeable particle representation of Donnelly and Kurtz. Indeed it is somewhat simpler as (SE) is replaced by a countable system of (somewhat unappealing) stochastic differential equations with jumps. This pathwise uniqueness leads to the uniqueness of the martingale problem for the generator $\mathcal{A}$ of their exchangeable infinite particle system, $(X_k(t))$, and $X_t(1)$, the total population size. The underlying population is $X_t \in M_F(\mathbb{R}^d)$ where for each fixed t, $\frac{X_t(\cdot)}{X_t(1)}$ is the deFinetti measure of $(X_k(t))$ and so

$$(V.5.1) \qquad X_t(\phi) = X_t(1) \lim_{N\to\infty} (N)^{-1} \sum_{k=1}^{N} \phi(X_k(t)) \quad \text{a.s.}$$

Of course X will satisfy $(MP)^{a,b}_{X_0}$ and these particular solutions will be unique as the richer structure from which they are defined is unique. On the other hand, given an arbitrary solution X of $(MP)^{a,b}_{X_0}$, one can introduce

$$\nu_t(\cdot) = \mathbb{E}\Big(\prod_1^\infty \frac{X_t(\cdot)}{X_t(1)} 1(X_t(1) \in \cdot)\Big).$$

This would be the one-dimensional marginals of $((X_k(t)), X_t(1))$ if such an exchangeable system existed. Some stochastic calculus shows that ν_t satisfies the forward equation associated with $\mathcal{A}$:

$$\nu_t(\phi) = \nu_0(\phi) + \int_0^t \nu_s(\mathcal{A}\phi)\, ds, \quad \phi \in \mathcal{D}(\mathcal{A}).$$

The key step is then a result of Kurtz (1998) (Theorem 2.7), earlier versions of which go back at least to Echevaria (1982). It gives conditions on $\mathcal{A}$, satisfied in our setting, under which any solution of the above forward equation are the one-dimensional marginals of the solution to the martingale problem associated with $\mathcal{A}$. In our setting this result produces the required $((X_k(t)), X_t(1))$ from which X can be recovered by (V.5.1). Here one may notice one of many simplifications we have made along the way–to obtain (V.5.1) from the martingale problem for $\mathcal{A}$ we need to introduce some side conditions to guarantee the fixed time exchangeability of the particle system. Hence one needs to work with a "restricted" martingale problem and a "restricted" forward equation in the above. This shows that every solution to $(MP)^{a,b}_{X_0}$ arises from such an exchangeable particle system and in particular is unique in law by the first step described above.

The methods of the previous section also extend easily to the historical processes underlying the solutions obtained there. Let d_H be the Vasershtein metric on $M_F(C)$ associated with the metric $\sup_t \|y_t - y'_t\| \wedge 1$ on C. Let $\mathcal{F}^H_t = \mathcal{F}^H[0, t+]$ and assume

$\hat\sigma : \mathbb{R}_+ \times \hat\Omega_H \to \mathbb{R}^{d\times d}$, $\hat b : \mathbb{R}_+ \times \hat\Omega_H \to \mathbb{R}^d$ are $(\widehat{\mathcal{F}^H_t})$-predictable

and for some nondecreasing function L satisfy

(HLip) $\quad \|\hat\sigma(t, J, y)\| + \|\hat b(t, J, y)\| \le L(t \vee \sup_{s\le t} J_s(1))$ and

$$\|\hat\sigma(t, J, y) - \hat\sigma(t, J', y')\| + \|\hat b(t, J, y) - \hat b(t, J', y')\|$$
$$\le L(t \vee \sup_{s\le t} J_s(1) \vee \sup_{s\le t} J'_s(1)) \Big[\sup_{s\le t} d_H(J_s, J'_s) + \sup_{s\le t} \|y_s - y'_s\|\Big].$$

The historical version of (SE) is:

$$(HSE)_{Z_0,K}\,(a)\ Z_t(\omega,y)=Z_0(y)+\int_0^t\hat\sigma(s,J,y)dy(s)+\int_0^t\hat b(s,J,Z)ds\quad K-\text{a.e.}$$

$$(b)\ J_t(\omega)(\cdot)=\int 1(Z(\omega,y)^t\in\cdot)K_t(dy)\quad\forall t\geq 0\ \mathbb{P}-\text{a.s.}$$

As in Section V.4, a fixed point argument shows that solutions to $(HSE)_{Z_0,K}$ exist and are pathwise unique (Theorem 4.10 in Perkins (1995)). Recall the class D_{fd} of finite dimensional cylinder functions from Example V.2.8. If $\hat a_{ij}=\hat\sigma\hat\sigma^*_{ij}$, the corresponding generator is

$$A_J\phi(t,y)=\hat b(t,J,y)\cdot\nabla\phi(t,y)+\frac{1}{2}\sum_{i=1}^d\sum_{j=1}^d\hat a_{ij}(t,J,y)\phi_{ij}(t,y),\quad\phi\in D_{fd}.$$

Assume for simplicity that $K_0=m$ is deterministic and hence so is $J_0=m(Z_0\in\cdot)$. It is again a simple exercise (cf. Exercise V.4.1) to show that the solution J of $(HSE)_{Z_0,K}$ satisfies

$$\forall\phi\in D_{fd}\quad J_t(\phi)=J_0(\phi)+\int_0^t\int A_J\phi(s,y)J_s(dy)ds+M_t(\phi),$$

$(HMP)^{\hat a,\hat b}_{J_0}$ where $M_t(\phi)$ is a continuous $(\mathcal{F}_t)$-martingale such that

$$\langle M(\phi)\rangle_t=\int_0^t J_s(\phi_s^2)ds.$$

The situation in (HSE) is now symmetric in that a historical process J is constructed from a given historical Brownian motion K. If $\hat a(t,J,y)$ is positive definite it will be possible to reconstruct K from J so that (HSE) holds (the main steps are described below) and so we have a means of showing that any solution of (HMP) does satisfy (HSE) and hence can derive:

Theorem V.5.2 (Perkins (1995)). Assume (HLip), $J_0\in M_F^0(C)$, and $\hat a=\hat\sigma\hat\sigma^*$ satisfies

$$\langle\hat a(t,J,y)v,v\rangle>0\ \forall v\in\mathbb{R}^d-\{0\}\text{ for }J_t-\text{a.a. }y\ \forall t\geq 0.$$

Then $(HMP)^{\hat a,\hat b}_{J_0}$ is well-posed.

One can use the change of measure technique in Section IV.1 to easily obtain the same conclusion for $(HMP)^{\hat a,\hat b,\hat g}_{J_0}$ in which $A_J\phi(t,y)$ is replaced with $A_J\phi(t,y)+\hat g(t,y)\phi(t,y)$, where $\hat g:\mathbb{R}_+\times\hat\Omega_H\to\mathbb{R}$ is bounded and $(\hat{\mathcal{F}}^H_t)$-predictable. As J is intrinsically time-inhomogeneous, one should work with general starting times $\tau\geq 0$ and specify $J_{t\wedge\tau}=J^0\in\{H\in\Omega_H:H_{\cdot\wedge\tau}=H\}\equiv\Omega^\tau_H$. The resulting historical martingale problem $(HMP)^{\hat a,\hat b,\hat g}_{\tau,J^0}$ is again well-posed and if $(\hat a,\hat b,\hat g)(t,J,y)=(\tilde a,\tilde b,\tilde g)(t,J_t,y)$, the solution will be a time-inhomogeneous $(\mathcal{F}_t)$-strong Markov process.

To prove the above results one needs to start with a solution of $(HMP)^{\hat a,\hat b}_{J^0}$, say, and develop the stochastic integration results in Section V.3 with respect to J. This general construction is carried out in Section 2 of Perkins (1995) and more general stochastic integrals for "historical semimartingales" with jumps may be found in Evans and Perkins (1998), although a general construction has not been carried out to date. Under the above non-degeneracy condition on $\hat{a}$, this then allows one define a historical Brownian motion, K, from J so that $(HSE)_{Z_0,K}$ holds, just as one can define the Brownian motion, B, from the solution, X, of $dX = \sigma(X)dB + b(X)dt$.

Consider next the problem of interactive branching rates. If $\gamma : M_F(\mathbb{R}^d) \times \mathbb{R}^d \to \mathbb{R}_+$, then the extension of $(MP)^{a,b}_{X_0}$ which incorporates this state-dependent branching rate is

$$(MP)^{a,b,\gamma}_{X_0} \quad \forall \phi \in C^2_b(\mathbb{R}^d) \quad M^X_t(\phi) = X_t(\phi) - X_0(\phi) - \int_0^t X_s(A_{X_s}\phi)ds \text{ is a}$$

$$\text{continuous } (\mathcal{F}_t)\text{-martingale such that } \langle M^X(\phi)\rangle_t = \int_0^t X_s(\gamma(X_s)\phi^2)ds.$$

In general, uniqueness in law of X remains open. In the context of Fleming-Viot processes Dawson and March (1995) were able to use a dual process for the moments to resolve the analogous problem in which the sampling rates $\gamma(\mu, x, y)$ of types x and y may depend on the population μ in a smooth manner. Their result is a perturbation theorem analogous to that of Stroock and Varadhan (1979) for finite-dimensional diffusions but the rigidity of the norms does not allow one to carry out the localization step and so this very nice approach (so far) only establishes uniqueness for sampling rates which are close enough to a constant rate.

For our measure-valued branching setting, particular cases of state-dependent branching rates have been treated by special duality arguments (recent examples include Mytnik (1998) and Dawson, Etheridge, Fleischmann, Mytnik, Perkins, Xiong (2000a, 2000b)). If we replace $\mathbb{R}^d$ with the finite set $\{1, \dots d\}$ and the generator A_μ with the state dependent Q-matrix $(q_{ij}(x))$, the solutions to the above martingale problem will be solutions to the stochastic differential equation

$$\text{(V.5.2)} \qquad dX^j_t = \sqrt{2\gamma_j(X_t)X^j_t}dB_j(t) + \sum_{i=1}^d X^i_t q_{ij}(X_t)dt.$$

Some progress on the uniqueness of solutions to this degenerate sde has recently been made by Athreya, Barlow, Bass, and Perkins (2000).

If $\hat{\gamma} : \mathbb{R}_+ \times \hat{\Omega}_H \to (0,\infty)$ is $(\hat{\mathcal{F}}^H_t)$-predictable, then conditions on $\hat{\gamma}$ are given in Perkins (1995) under which $(HMP)^{\hat{\gamma},\hat{a},\hat{b},\hat{g}}_{\tau,J_0}$ is well-posed. In this martingale problem we have of course

$$\langle M(\phi)\rangle_t = \int_\tau^t \int \hat{\gamma}(s,J,y)\phi(s,y)^2 J_s(dy)ds, \quad \phi \in D_{fd}.$$

Although the precise condition is complicated (see p. 48-49 in Perkins (1995)), it basically implies that $\hat{\gamma}$ should be represented by a (possibly stochastic) time integral. It is satisfied in the following examples.

Example V.5.3. (a) $\hat{\gamma}(t,J,y) = \gamma(t,y(t))$ for $\gamma \in C^{1,2}_b(\mathbb{R}_+ \times \mathbb{R}^d)$, γ bounded away from zero. In this case there is no interaction but branching rates may depend on space-time location something which our strong equation approach does not directly allow.

(b) (Adler's branching goats). $\hat{\gamma}(t,J,y)=\exp\left\{-\int_0^t\int p_\varepsilon(y'_s-y_t)J_s(dy')e^{-\alpha(t-s)}ds\right\}$. The branching rate at y_t is reduced if our goat-like particles have grazed near y_t in the recent past.

(c) (General time averaging). $\hat{\gamma}(t,J,y)=\int_{t-\varepsilon}^t f_\varepsilon(t-s)\gamma(s^+,J,y)ds$, where $f_\varepsilon:\mathbb{R}_+\to\mathbb{R}_+$ is C^1, $\mathrm{supp}(f_\varepsilon)\subset[0,\varepsilon)$, $\int_0^\varepsilon f_\varepsilon(s)ds=1$, and γ satisfies its analogue of (HLip).

Given the difficulties already present in the finite-dimensional case (V.5.2), resolving the uniqueness of solutions to $(MP)_{X_0}^{\gamma,a,b}$ or $(HMP)_{\tau,J_0}^{\hat{\gamma},\hat{a},\hat{b}}$ would appear to be an interesting open problem (see Metivier (1987)).

V.6. Appendix: Proofs of Lemma V.4.5 and (V.4.21)

We start by proving (IV.4.21), assuming the validity of Lemma V.4.5, and then address the proof of the latter.

Assume Z_s, B, $\mathbb{P}_B$, and $K^{(s)}$ are as in Lemma V.4.5. Let $I_s(f)$, $f\in D_s(d,d)\equiv D_s$ refer to the stochastic integral with respect $K^{(s)}$ on $(\Omega,\mathcal{F},\mathcal{F}_t^{(s)},\mathbb{P})$. There is some possible confusion here because of the other probabilities $\mathbb{P}_B$. Note, however, that if $I_{s,B}(f)$, $f\in D_{s,B}$ denotes the integral under $\mathbb{P}_B$, then

$$\text{(V.6.1)}\qquad D_s\subset D_{s,B}\text{ and for }f\in D_s\text{ we may take }I_{s,B}(f)=I_s(f).$$

Let $I(f)$, $f\in D$ continue to denote the stochastic integral with respect to K. The expression "$K^{(s)}$-a.e." will always mean with respect to $\mathbb{P}$ (not $\mathbb{P}_B$). With these clarifications, and the notation $\phi^{(s)}$ introduced prior to Lemma V.4.5, we have:

Lemma V.6.1. (a) If $\psi:\mathbb{R}_+\times\hat{\Omega}\to\mathbb{R}^d$ is $\mathcal{B}\times\hat{\mathcal{F}}^*$-measurable, then

$$\psi^{(s)}=0\quad K-\text{a.e. iff }\psi=0\quad K^{(s)}-\text{a.e.}$$

(b) If $f\in D_s$, then $f^{(s)}\in D$ and $I_s^{(s)}(f)=I(f^{(s)})\quad K-\text{a.e.}$

Proof. (a) is a simple exercise in using the definitions.

(b) The same is true for the first implication in (b). To check the equality, first let

$$f(t,\omega,y)=\sum_1^n f_i(\omega,y)1_{(t_{i-1},t_i]}(t)+f_0(\omega,y)1_{\{0\}}(t),$$
$$\text{where } f_i\in b\hat{\mathcal{F}}^*_{t_i+s},\ 0=t_0<t_1\ldots<t_n\le\infty.$$

Then

$$\begin{aligned}\text{(V.6.2)}\qquad I_s^{(s)}(f,t,\omega,y)&=I_s(f,t-s,\omega,\tilde{Z}_s(\omega,y))1(t\ge s)\\&=\sum_1^n f_i(\omega,\tilde{Z}_s(\omega,y))\cdot[\tilde{Z}_s(\omega,y)((t-s)\wedge t_i)\\&\qquad-\tilde{Z}_s(\omega,y)((t-s)\wedge t_{i-1})]1(t\ge s)\\&=\sum_1^n f_i(\omega,\tilde{Z}_s(\omega,y))\cdot[y(t\wedge(t_i+s))-y(t\wedge(t_{i-1}+s))].\end{aligned}$$

We also have

$$f^{(s)}(t,\omega,y)=\sum_1^n f_i(\omega,\tilde Z_s(\omega,y))1_{(s+t_{i-1},s+t_i]}(t)+f_0(\omega,\tilde Z_s(\omega,y))1_{\{s\}}(t),$$

and so the required result follows for such a simple f from the definition of $I(f^{(s)})$ and (V.6.2).

If $f\in D_s$, then as in (V.3.3) there are $(\widehat{\mathcal{F}_t^{(s)}})$-simple functions $\{f_k\}$ such that

$$\mathbb{P}\Big(K_k^{(s)}\Big(\int_0^k\|f_k(t)-f(t)\|^2dt>2^{-k}\Big)\Big)<2^{-k}.$$

We know $f^{(s)},f_k^{(s)}\in D$ by the above and therefore for $k\ge s$,

$$\begin{aligned}&\mathbb{P}\Big(K_k\Big(\int_0^k\|f_k^{(s)}(t)-f^{(s)}(t)\|^2dt>2^{-k}\Big)\Big)\\&=\mathbb{P}\Big(K_{k-s}^{(s)}\Big(\int_0^{k-s}\|f_k(t)-f(t)\|^2dt>2^{-k}\Big)\Big)\\&\le\mathbb{P}\Big(K_k^{(s)}\Big(\int_0^k\|f_k(t)-f(t)\|^2dt>2^{-k}\Big)\Big)\quad\text{(use Remark V.2.5 (a))}\\&<2^{-k}.\end{aligned}$$

A double application of Proposition V.3.2 (d) now allows us to prove the required equality by letting $k\to\infty$ in the result for f_k. ∎

Proof of (V.4.21). Recall that (X,Z) is the solution of $(SE)_{Z_0,K}$. This gives us the Z_s which is used to define $K^{(s)}$ and $\phi^{(s)}$. Note that a solution to $(SE)_{\hat Z_0,K^{(s)}}$ with respect to $\mathbb{P}$ will also be a solution with respect to $\mathbb{P}_B$ (by (V.6.1)) and so we may assume that $B=\Omega$.

By Lemma V.4.5 and Theorem V.4.1 (a) there is a unique solution $(\hat X,\hat Z)$ to $(SE)_{\hat Z_0,K^{(s)}}$ on $(\Omega,\mathcal{F},\mathcal{F}_t^{(s)},\mathbb{P})$. Define

$$Z_t'(\omega,y)=\begin{cases}Z_t(\omega,y)&\text{if }t<s\\\hat Z_t^{(s)}(\omega,y)&\text{if }t\ge s,\end{cases}$$

and

$$X_t'(\omega)=\begin{cases}X_t(\omega)&\text{if }t<s\\\hat X_{t-s}(\omega)&\text{if }t\ge s\end{cases}\equiv\begin{cases}X_t(\omega)&\text{if }t<s\\\hat X_t^{(s)}(\omega)&\text{if }t\ge s\end{cases}$$

If $V(t)=\int_0^t b(\hat X_u,\hat Z_u)du$, then by $(SE)_{\hat Z_0,K^{(s)}}$,

$$\hat Z(t)=y_0+I_s(\sigma(\hat X,\hat Z),t)+V(t)\quad K^{(s)}-\text{a.e.},$$

and so Lemma V.6.1 implies that K-a.e.,

$$\begin{aligned}\hat{Z}^{(s)}(t) &= Z_s + I_s^{(s)}(\sigma(\hat{X},\hat{Z}),t) + V^{(s)}(t)\\ &= Z_s + I(\sigma(\hat{X},\hat{Z})^{(s)},t) + \int_0^{t-s} b(\hat{X}_u(\omega),\hat{Z}_u(\omega,\tilde{Z}_s))du\, I(t\geq s)\\ &= Z_s + I(\sigma(\hat{X}^{(s)},\hat{Z}^{(s)})1(\cdot\geq s),t) + \int_s^t b(\hat{X}_u^{(s)},\hat{Z}_u^{(s)})du\, 1(t\geq s).\end{aligned}$$

It follows that K-a.e. for $t \geq s$,

$$\begin{aligned}Z'(t) &= Z_s + I(\sigma(X'_\cdot,Z'_\cdot)1(\cdot\geq s),t) + \int_s^t b(X'_u,Z'_u)du\\ &= Z_s + \int_s^t \sigma(X'_u,Z'_u)dy(u) + \int_s^t b(X'_u,Z'_u)du.\end{aligned}$$

Therefore we see that K-a.e.,

$$Z'(t) = Z_0 + \int_0^t \sigma(X'_u,Z'_u)du + \int_0^t b(X'_u,Z'_u)du,$$

first for $t \geq s$ by the above, and then for all $t \geq 0$ by the fact that (X,Z) solves $(SE)_{Z_0,K}$. Also $\mathbb{P}$-a.s. for all $t \geq s$

$$\begin{aligned}X'_t(\cdot) &= \int 1(\hat{Z}_{t-s}\in\cdot)K_{t-s}^{(s)}(dy)\\ &= \int 1(\hat{Z}_{t-s}(\omega,\tilde{Z}_s(\omega,y))\in\cdot)K_t(dy)\\ &= \int 1(Z'_t(\omega,y)\in\cdot)K_t(dy),\end{aligned}$$

and the above equality is trivial for $t < s$. We have shown that (X',Z') solves $(SE)_{Z_0,K}$ and so $X' = X$ a.s. This means that $X_{t+s} = \hat{X}_t$ for all $t \geq 0$ a.s. and (V.4.21) is proved. ∎

Proof of Lemma V.4.5. We will show that

(V.6.3) $\qquad K_t^{(s)}$ satisfies $(HMP)_{0,K_0^{(s)}}$ on $(\Omega,\mathcal{F},\mathcal{F}_t^{(s)},\mathbb{P})$.

It follows immediately that the same is true with respect to $\mathbb{P}_B$ and so the first result follows. The last inequality then follows trivially.

Assume first that $Z_s(\omega,y) = Z_s(y)$, $Z_s \in b\mathcal{C}_s$. If $\phi \in \mathcal{D}(\hat{A})$ and $n(t,y) = \phi(t,y) - \int_0^t \hat{A}\phi(r,y)dr$, then for $t \geq s$,

$$\phi^{(s)}(t,y) = n(t-s,\tilde{Z}_s(y)) + \int_s^t (\hat{A}\phi)^{(s)}(r,y)dr.$$

If $y_0 \in C^s$, $s \le u \le t$, and $G = \{y : y^s \in F_1,\ \hat\theta_s(y)^{u-s} \in F_2\}$ for some $F_i \in \mathcal{C}$, then $G \in \mathcal{C}_u$ and

$$\begin{aligned}P_{s,y_0}(n(t-s,\tilde Z_s)1_G) &= 1_{F_1}(y_0)P_{s,y_0}(n(t-s,Z_s(y_0)+\hat\theta_s(\cdot))1_{F_2}(\hat\theta_s(\cdot)^{u-s}))\\ &= 1_{F_1}(y_0)\int n(t-s,Z_s(y_0)+y)1_{F_2}(y^{u-s})dP^0(y)\\ &= 1_{F_1}(y_0)\int n(u-s,Z_s(y_0)+y)1_{F_2}(y^{u-s})dP^0(y)\\ &= P_{s,y_0}(n(u-s,\tilde Z_s)1_G),\end{aligned}$$

reversing the above steps in the last line. This shows that for any $m_s \in M_F^s(C)$, $\phi^{(s)} \in \mathcal{D}(\bar A_{s,m_s})$ and $\bar A_{s,m_s}(\phi^{(s)}) = (\hat A\phi)^{(s)}$. Now apply Proposition V.2.6 to the historical Brownian motion $\{K_t : t \ge s\}$ to conclude

$$\begin{aligned}K_t^{(s)}(\phi_t) &= \int \phi_{s+t}^{(s)}K_{s+t}(dy)\\ &= K_s(\phi_s^{(s)}) + \int_s^{s+t}\int \phi^{(s)}(r,y)dM(r,y) + \int_s^{s+t} K_r((\hat A\phi)_r^{(s)})dr.\end{aligned} \tag{V.6.4}$$

That is,

$$\begin{aligned}\int \phi_t(\tilde Z_s(y))K_{t+s}(dy) &= \int \phi_0(\tilde Z_s(y))K_s(dy) + \int_s^{s+t}\int \phi(r-s,\tilde Z_s(y))dM(r,y)\\ &\quad + \int_s^{s+t}\int \hat A\phi(r-s,\tilde Z_s(y))K_r(dy)dr\ \forall t \ge 0 \text{ a.s.}\end{aligned} \tag{V.6.5}$$

If $Z_s(\omega,y) = \sum_{i=1}^n 1_{B_i}(\omega)Z^i(y)$ for $B_i \in \mathcal{F}_s$ and $Z^i \in b\mathcal{C}_s$, set $Z_s = Z^i$ in (V.6.5), multiply by $1_{B_i}(\omega)$ and sum over i to see that (V.6.5) remains valid if $Z_s(y)$ is replaced by $Z_s(\omega,y)$. Now pass to the pointwise closure of this class of $Z_s(\omega,y)$ and use the continuity of ϕ and $\hat A\phi$ to conclude that (V.6.5) remains valid if Z_s is $\hat{\mathcal{F}}_s^*$-measurable (Remark V.2.5 (c) allows us to pass from $\hat{\mathcal{F}}_s$ to $\hat{\mathcal{F}}_s^*$ on the right-hand side of (V.6.5) for all $t \ge 0$ simultaneously). Now reinterpret this general form of (V.6.5) as (V.6.4) and let $M^{(s)}(\phi)(t)$ denote the stochastic integral in (V.6.4). Then $M^{(s)}(\phi)(t)$ is an $(\mathcal{F}_t^{(s)})$-martingale and

$$\langle M^{(s)}(\phi)\rangle_t = \int_s^{s+t} K_r(\phi_r^{(s)2})dr = \int_0^t K_r^{(s)}(\phi_r^2)dr.$$

The last term on the right-hand side of (V.6.5) equals $\int_0^t K_r^{(s)}((\hat A\phi)_r)\,dr$ and the first term equals $K_0^{(s)}(\phi_0)$. This proves (V.6.3) and so completes the proof. ∎

References

R. Adler and R. Tribe (1998). Uniqueness for a historical SDE with a singular interaction, J. Theor. Prob. 11, 515-533.

D. Aldous (1993). Tree-based models for random distribution of mass. J. Stat. Phys. 73, 625-641.

S. Athreya, M. T. Barlow, R. Bass and E. Perkins (2000). Degenerate stochastic differential equations and super-Markov chains. Preprint.

P. Baras and M. Pierre (1984). Singularités éliminables pur des equations semilinéaires. Ann. Inst. Fourier 34, 185-206.

M.T. Barlow, S.N. Evans and E. Perkins (1991). Collision local times and measure-valued diffusions. Can. J. Math. 43, 897-938.

M.T. Barlow and E. Perkins (1994). On the filtration of historical Brownian motion. Ann. Prob. 22, 1273-1294.

M. Bramson, J.T. Cox, and J.F. Le Gall (2001). Super-Browian limits of voter model clusters. To appear, Ann. Prob.

M. Bramson, R. Durrett and G. Swindle (1989). Statistical mechanics of crabgrass. Ann. Prob. 17, 444-481.

D.L. Burkholder (1973). Distribution function inequalities for martingales. Ann. Prob. 1, 19-42.

Z. Ciesielski and S.J. Taylor (1962). First passage times and sojourn times for Brownian motion and exact Hausdorff measure of the sample path. Trans. Amer. Math. Soc. 103, 434-450.

J.T. Cox, R. Durrett, E. Perkins (1999). Rescaled particle systems converging to super-Brownian motion. In: Perplexing Problems in Probability–Festschrift in Honor of Harry Kesten, pp. 269-284, Birkhäuser, Boston.

J.T. Cox, R. Durrett, E. Perkins (2000). Rescaled voter models converge to super-Brownian motion. Ann. Prob. 28, 185-234.

D.A. Dawson (1978). Geostochastic calculus. Can. J. Statistics 6, 143-168.

D.A. Dawson (1992). Infinitely divisible random measures and superprocesses. In: Stochastic Analysis and Related Topics, pp. 1-130, Birkhäuser, Boston.

D. A. Dawson (1993). Measure-valued Markov Processes, Ecole d'Eté de Probabilités de Saint Flour 1991, Lect. Notes in Math. 1541, Springer, Berlin.

D. A. Dawson, A. Etheridge, K. Fleischmann, L. Mytnik. E. Perkins, and J. Xiong (2000a). Mutually catalytic branching in the plane. Preprint.

D. A. Dawson, K. Fleischmann, L. Mytnik. E. Perkins, and J. Xiong (2000b). Mutually catalytic branching in the plane: uniqueness. Preprint.

D.A. Dawson, I. Iscoe, E. Perkins (1989). Super-Brownian motion: path properties and hitting probabilities. Probab. Th. Rel. Fields 83, 135-205.

Originally published in: *Ecole d'Eté de Probabilités de Saint-Flour XXIX – 1999*, Lecture Notes in Mathematics, Vol. **1781**, 318–329, DOI: 10.1007/b93152,

D.A. Dawson and K.J. Hochberg (1979). The carrying dimension of a stochastic measure diffusion. Ann. Prob. 7, 693-703.

D.A. Dawson and P. March (1995). Resolvent estimates for Fleming-Viot operators and uniqueness of solutions to related martingale problems. J. Funct. Anal. 132, 417-472.

D.A. Dawson, E. Perkins (1991). Historical Processes. Mem. Amer. Math. Soc. 93 n. 454.

D.A. Dawson and V. Vinogradov (1994). Almost sure path properties of $(2, d, \beta)$-superprocesses. Stoch. Proc. Appl. 51, 221-258.

C. Dellacherie and P.A. Meyer (1978). Probabilities and Potential. North-Holland Math. Studies 29, North Holland, Amsterdam.

E. Derbez and G. Slade (1998). The scaling limit of lattice trees in high dimensions. Commun. Math. Phys. 193, 69-104.

J.S. Dhersin and J.F. Le Gall (1997). Wiener's test for super-Brownian motion and for the Brownian snake. Probab. Th. Rel. Fields 108, 103-129.

P. Donnelly and T.G. Kurtz (1999). Particle representations for measure-valued population models. Ann. Prob. 27, 166-205.

J. Dugundji (1966). Topology. Allyn and Bacon, Boston.

R. Durrett and E. Perkins (1999). Rescaled contact processes converge to super-Brownian motion for $d \geq 2$. Probab. Th. Rel. Fields 114, 309-399.

E.B. Dynkin (1991). A probabilistic approach to one class of nonlinear differential equations. Probab. Th. Rel. Fields 89, 89-115.

E.B. Dynkin (1993). Superprocesses and partial differential equations. Ann. Prob. 20, 1185-1262.

E.B. Dynkin (1994). An Introduction to Branching Measure-Valued Processes, CRM Monographs 6, Amer. Math. Soc., Providence.

E.B. Dynkin and S.E. Kuznetsov (1996). Superdiffusions and removable singulariities for quasilinear partial diferential equations. Comm. Pure. Appl. Math. 48,125-176.

E.B. Dynkin and S.E. Kuznetsov (1998). Trace on the boundary for solutions of nonlinear differential equations. Trans. Amer. Math. Soc. 350, 4499-4519.

P.E. Echevaria (1982). A criterion for invariant measures of Markov processes. Z. f. Wahrsch. verw. Gebiete 61, 1-16.

N. El Karoui and S. Roelly (1991). Propriétés de martingales, explosion et représentation de Lévy-Khintchine d'une classe de processus de branchement à valeurs mesures. Stoch. Process. Appl. 38, 239-266.

A. Etheridge (2000). An Introduction to Superprocesses, University Lecture Series vol.20, Amer. Math. Soc.

A. Etheridge and P. March (1991). A note on superprocesses. Probab. Th. Rel. Fields 89, 141-147.

S.N. Ethier and T.G. Kurtz (1986). Markov Processes: Characterization and Convergence. Wiley, N.Y.

S.N. Ethier, and T.G. Kurtz (1993). Fleming-Viot processes in population genetics. Siam. J. Cont. Opt. 31, 345-386.

S.N. Evans and E. Perkins (1991). Absolute continuity results for superprocesses. Trans. Amer. Math. Soc. 325, 661-681.

S.N. Evans and E. Perkins (1994). Measure-valued branching diffusions with singular interactions. Can. J. Math. 46, 120-168.

S.N. Evans and E. Perkins (1998). Collision local times, historical stochastic calculus and competing species. Elect. J. Prob. 3, paper 5.

W. Feller (1939). Die Grundlagen der Volterraschen Theorie des Kampfes ums Dasein in Wahrscheinlichkeitstheoretischer Behandlung. Acta Biotheoretica 5, 11-40.

W. Feller (1951). Diffusion processes in genetics. Proc. Second Berkeley Symp. Math. Statist. Prob., Univ. of California Press Berkeley, pp. 227-246.

P.J. Fitzsimmons (1988). Construction and regularity of measure-valued Markov processes. Israel J. Math. 64, 337-36.

P.J. Fitzsimmons (1992). On the martingale problem for measure-valued branching processes. In: Seminar on Stochastic Processes 1991, pp. 39-51, Birkhäuser, Boston.

I.I. Gihman and A.V. Skorohod (1975). The Theory of Stochastic Processes II. Springer, New York.

T.E. Harris (1963). The Theory of Branching Processes. Springer, Berlin.

R. van der Hofstad, G. Slade (2000). Convergence of critical oriented percolation to super-Brownian motion above 4+1 dimensions. Preprint.

D.N. Hoover and H.J. Keisler (1984). Adapted probability distributions. Trans. Amer. Math. Soc. 286, 159-201.

D.N. Hoover and E. Perkins (1983). Nonstandard construction of the stochastic integral and applications to stochastic differential equations I, II, Trans. Amer. Math. Soc. 275, 1-36, 37-58.

N. Ikeda and S. Watanabe (1981). Stochastic Differential Equations and Diffusion Processes. North Holland, Amsterdam.

I. Iscoe (1986). A weighted occupation time for a class of measure-valued branching processes. Probab. Th. Rel. Fields 71, 85-116.

I. Iscoe (1988). On the supports of measure-valued critical branching Brownian motion. Ann. Prob. 16, 200-221.

J. Jacod (1977). A general theorem of representation for martingales. In: Proceedings of Symposia in Pure Math. 31, pp. 37-54, Amer. Math. Soc, Providence.

J. Jacod and A.N. Shiryaev (1987). Limit theorems for Stochastic Processes. Springer-Verlag, Berlin.

A. Jakubowski (1986). On the Skorohod topology. Ann. Inst. H. Poincaré B22, 263-285.

S. Kakutani (1944). On Brownian motion in n-space. Proc. Imp. Acad. Tokyo 20, 648-652.

O. Kallenberg (1986). Random Measures. Academic Press, New York.

F. Knight (1981). Essentials of Brownian Motion and Diffusion. Amer. Math. Soc., Providence.

N. Konno and T. Shiga (1988). Stochastic differential equations for some measure-valued diffusions. Probab. Th. Rel. Fields 79, 201-225.

N. Krylov (1997). On a result of Mueller and Perkins. Probab. Th. Rel. Fields 108, 543-557.

N. Krylov (1997b). On SPDE's and superdiffusions. Ann. Prob. 25, 1789-1809.

T.G. Kurtz (1975). Semigroups of conditioned shifts and approximation of Markov processes. Ann. Prob. 3, 618-642.

T.G. Kurtz (1998). Martingale problems for conditional distributions of Markov processes. Elect. J. Prob. 3, paper 9.

J.F. Le Gall (1993). A class of path-valued Markov processes and its applications to superprocesses. Probab. Th. Rel. Fields 95, 25-46.

J.F. Le Gall (1993b). The uniform random tree in a Brownian excursion. Probab. Th. Rel. Fields 96, 369-383.

J. F. Le Gall (1994). A lemma on super-Brownian motion with some applications. In: The Dynkin Festschrift pp. 237-251. Birkhäuser, Boston.

J. F. Le Gall (1998). The Hausdorff measure of the range of super-Brownian motion. In: Perplexing Problems in Probability–Festschrift in Honor of Harry Kesten, pp. 285-314, Birkhäuser, Boston.

J. F. Le Gall (1999). Spatial Branching Processes, Random Snakes and Partial Differential Equations. Lectures in Mathematics ETH Zurich, Birkhäuser Verlag, Basel.

J. F. Le Gall and E. Perkins (1995). The Hausdorff measure of the support of two-dimensional super-Brownian motion. Ann. Prob. 232, 1719-1747.

J. F. Le Gall, E. Perkins and S.J. Taylor (1995). The packing measure of the support of super-Brownian motion. Stoch. Process. Appl. 59, 1-20.

Z. Li and T. Shiga (1995). Measure-valued branching diffusions: immigrations, excursions and limit theorems, J. Math. Kyoto Univ. 35, 233-274.

M. Lopez (1996). Path properties and convergence of interacting superprocesses. Ph.D. thesis, UBC.

M. Metivier (1987). Weak convergence of measure-valued processes using Sobolev-imbedding techniques. In: Proc. Trento 1985 SPDE and Applications, Lect. Notes. Math. 1236, 172-183, Springer, Berlin.

C. Mueller and E. Perkins (1992). The compact support property for solutions of the heat equation with noise. Probab. Th. Rel. Fields 93, 325-358.

C. Mueller and R. Tribe (1994). Stochastic pde's arising from the long range contact and long range voter processes. Probab. Th. Rel. Fields 102, 519-546.

L. Mytnik (1998). Weak uniqueness for the heat equation with noise. Ann. Prob. 26, 968-984.

L. Mytnik (1999). Uniqueness for a competing species model. Can. J. Math. 51, 372-448.

A. Pazy (1983). Semigroups of Linear Operators and Applications to Partial Differential Applications. Springer-Verlag, Berlin.

E. Perkins (1988). A space-time property of a class of measure-valued branching diffusions. Trans. Amer. Math. Soc. 305, 743-795.

E. Perkins (1989). The Hausdorff measure of the closed supoprt of super-Brownian motion. Ann. Inst. H. Poincaré Stat. 25, 205-224.

E. Perkins (1990). Polar sets and multiple points for super-Brownian motion. Ann. Prob. 18, 453-491.

E. Perkins (1991). On the continuity of measure-valued processes. In: Seminar on Stochastic Processes 1990, pp. 261-268, Birkhäuser, Boston.

E. Perkins (1992). Measure-valued branching diffusions with spatial interactions. Probab. Th. Rel. Fields 94, 189-245.

E. Perkins (1992b). Conditional Dawson-Watanabe processes and Fleming-Viot processes. In: Seminar on Stochastic Processes 1991, pp.142-155, Birkhäuser, Boston.

E. Perkins (1994). The Strong Markov Property of the Support of Super-Brownain Motion. In: The Dynkin Festschrift Markov Processes and their Applications, pp.307-326, Birkhäuser, Boston.

E. Perkins (1995). On the martingale problem for interactive measure-valued diffusions. Mem. Amer. Math. Soc. 115 n. 549.

E. Perkins (1995b). Measure-valued branching diffusions and interactions. Proceedings of the International Congress of Mathematicians, Zurich, 1994, pp.1036-1046, Birkhäuser Verlag, Basel.

E. Perkins and S.J. Taylor (1998). The multifractal spectrum of super-Brownian motion. Ann. Inst. H. Poincaré Stat. 34, 97-138.

S.C. Port and C. J. Stone (1978). Brownian Motion and Classical Potential Theory. Academic, New York.

M. Reimers (1989). One-dimensional stochastic partial differential equations and the branching measure-diffusion. Probab. Th. Rel. Fields 81, 319-340.

M. Reimers (1989b). Hyperfinite methods applied to the critical branching diffusion. Probab. Th. Rel. Fields 81, 11-27.

D. Revuz and M. Yor (1991). Continuous Martingales and Brownian Motion. Springer, Berlin.

S. Roelly-Coppoletta (1986). A criterion of convergence of measure-valued processes: application to measure branching processes. Stochastics 17,43-65.

L.C.G. Rogers and D. Williams (1987). Diffusions, Markov Processes and Martingales Vol. 2. Wiley, Chichester.

C. A. Rogers and S. J. Taylor (1961). Functions continuous and singular with respect to a Hausdorff measure. Mathematika 8, 1-31.

H. Rost (1971). The stopping distributions of a Markov process. Invent. Math. 14, 1-16.

W. Rudin (1974). Real and Complex Analysis. McGraw-Hill, New York.

A. Schied (1996). Sample path large deviations for super-Brownian motion. Probab. Th. Rel. Fields 104, 319-347.

M. Sharpe (1988). General Theory of Markov Processes. Academic, Boston.

T. Shiga (1990). A stochastic equation based on a Poisson system for a class of measure-valued diffusion processes. J. Math. Kyoto. Univ. 30, 245-279.

D.W. Stroock and S.R.S. Varadhan (1979). Multidimensional Diffusion Processes, Springer-Verlag, Berlin.

S. Sugitani (1987). Some properties for the measure-valued diffusion process. J. Math. Soc. Japan 41, 437-462.

A. Szulga (1982). On minimal metrics in the space of random variables. Teor. Veroyatn. Primen 27, 401-405.

S.J. Taylor (1961). On the connections between generalized capacities and Hausdorff measures. Proc. Cambridge Philos. Soc. 57, 524-531.

S.J. Taylor (1964). The exact Hausdorff measure of the sample path for planar Brownian motion. Proc. Cam. Phil. Soc. 60, 253-258.

S.J. Taylor and N.A. Watson (1985). A Hausdorff measure classification of polar sets for the heat equation. Math. Proc. Cam. Phil. Soc. 47, 325-344.

R. Tribe (1991). The connected components of the closed support of super-Brownian motion. Probab. Th. Rel. Fields 84, 75-87.

R. Tribe (1992). The behavior of superprocesses near extinction. Ann. Prob. 20, 286-311.

R. Tribe (1994). A representation for super Brownian motion, Stoch. Proc. Appl. 51, 207-219.

J. Walsh (1986). An Introduction to Stochastic Partial Differetial Equations, Ecole d'Eté de Probabilités de Saint Flour 1984, Lect. Notes. in Math. 1180, Springer, Berlin.

S. Watanabe (1968). A limit theorem of branching processes and continuous state branching processes, J. Math. Kyoto U. 8, 141-167.

T. Yamada and S. Watanabe (1971). On the uniqueness of solutions of stochastic differential equations. J. Math. Kyoto U. 11, 155-167.

Index